Electricity and Modern Physics

m.k.s. Version

Cover picture: The path of a beam of electrons through a cathode-ray deflection tube (p. 216).

Photo: Bromhead (Bristol) Ltd

Electricity and Modern Physics

m.k.s. Version

GEORGE A. G. BENNET

EDWARD ARNOLD (PUBLISHERS) LTD, LONDON

First published 1968
by Edward Arnold (Publishers) Ltd
25 Hill Street, London W1X 8LL

Reprinted 1968, 1969, 1970, 1971, 1972

ISBN 0 7131 2015 0

Printed in Great Britain by
Butler & Tanner Ltd, Frome and London

'And He is before all things, and by Him all things consist'
—Col. 1 : 17

Preface

A new approach to Physics

The last 20 years have seen sweeping changes in the organization and content of Physics courses at all levels. New teaching methods have been thought out, new apparatus developed, and a new attitude to the teaching of Physics has started to emerge in countries in many parts of the world. No longer is it regarded as sufficient just to impart a knowledge of the factual content of the subject; the purpose is rather to give the student also some insight into the excitement of the search to understand the structure of the Universe. In this book I have tried to give a comprehensive account of one branch of the subject, thoroughly in line with modern ideas. It is intended for students in the age range from 16 to 20 who have already done an introductory course in the subject. However, the formal presentation is complete in itself, and reference to textbooks of a more elementary standard should not be necessary.

The danger of superficiality

One consequence of the new approach has been an increase in the number of students doing Physics at an advanced level. Physics is now to be reckoned as a subject of major social significance—even perhaps as a 'popular' subject! There is a danger here of the teaching of the subject becoming over-simplified and superficial; its exacting discipline of precise thought and expression can be lost in the attempt to reach an ever-wider audience. This presents the author of a textbook with conflicting requirements. On the one hand the explanations of fundamental concepts must be straightforward and not clogged with discussions and mathematical derivations too difficult for the average student. On the other hand we know that future scientific developments, are going to be in the hands of a small minority of very able students; and their special needs must be kept in mind. For these the foundations of the subject must be handled in an intellectually satisfying manner, and the mere coverage of an examination syllabus is quite inadequate. I have sought to meet these requirements by making a very careful choice of the order of topics; in each chapter the advanced work is arranged to *follow* the more elementary sections in which fundamental concepts are presented. Any student should thus be able to read up to a certain point in a chapter and find everything within his grasp. All the hardest mathematical and theoretical work is safely confined to the later sections of the chapters, and can be left to a second reading. Occasionally the needs of logical presentation have forced me to include harder topics in an elementary section. This is indicated by the use of smaller type with a star (★) at the beginning. For the guidance of teacher and student I give here a table showing the point in each chapter up to which I suggest the text might be studied at a first reading.

Suggested work for a first reading

Chapter	*Up to the following sections (incl.)*
1	to end
2	2.8
3	to end
4	4.3
5	5.4
6	6.2
7	7.4
8	8.4
9	9.3
10	10.6
11	11.5
12	12.6
13	13.3
14	14.4
15	15.2
16	16.7
17	17.3

Having adopted this method of arranging the material, I have felt free to include in the later parts of some chapters discussions of important fundamental issues that are definitely too difficult for a first, or even a second, reading. This enables the course to be rounded off with introductory accounts of some of the great themes of Modern Physics. For example, the phenomenon of electromagnetic induction cannot be satisfactorily analysed without considering the significance of the *frames of reference* in which the calculations and measurements are being made. This leads on naturally to a discussion of some of the central concepts of the theory of Relativity (pp. 109 *et seq.*). Again, by including a section on transmission lines, it is possible to proceed to a description of plane electromagnetic waves *in vacuo* and to a simple calculation of their velocity in terms of the electric and magnetic space constants, ε_0 and μ_0 (pp. 262 *et seq.*).

In planning a physics course it is always a matter of some difficulty deciding how far to include details of the design of various technical devices. A textbook of Physics cannot be a handbook of electrical technology. I hope the selection I have made will commend itself to teachers and students. I have tried always to concentrate on the physical principles illustrated by suitable practical applications. In the case of electronic circuits I have aimed in each case to give enough detail to enable the student to set up the circuit and start experimenting with it himself. In transistor circuits I have limited myself to discussing the common-emitter arrangement, but have dealt with this in some detail. I have also included a brief account of switching and logic circuits, enough to enable a student to try out these also for himself.

The use of m.k.s. rationalized units

There are at least three distinct forms of m.k.s. treatment in vogue at the present time; each of these has powerful arguments in its favour. The reason for this proliferation is to be found in the unsatisfactory nature of the schemes originally proposed for teaching the m.k.s. system. These were thought out primarily from the point of view of the electrical engineer, for whom a treatment based on parallel fields was desirable. However, in an elementary course this approach seemed to drive the subject into a straitjacket that was both unnatural and also tended to conceal its proper experimental basis.

The difficulty came from the attempt to make *rationalizing* seem sensible at an elementary stage. There is no doubt about its usefulness once we get to more advanced work. I take the view that there is no fundamental significance in rationalizing. The concepts and laws of Physics cannot in any way be affected by the presence or absence of π in certain formulae. Rationalizing is *convenient*; that is all. Once this point of view is accepted we do not have to go out of our way to make rationalizing seem 'correct' by using only uniform fields in the elementary treatment of the subject. Magnetic fields can then be calculated, if we wish, using current elements and the Biot-Savart rule, and electric fields using point charges and the inverse-square law. In fact the uniform field in a parallel-plate capacitor is a simple enough concept at the most elementary level, and we may as well start the study of electric fields at this point. But it is very difficult to treat the field of a long solenoid as fundamental, and I have made no attempt to do this.

Following the usual modern teaching practice, the elementary theory of magnetic and electric fields is handled entirely using a single vector for each field—the flux density B for magnetic fields, and the electric intensity (or potential gradient) E for electric fields. The other vectors H and D are introduced for specialized purposes only in sections dealing with more advanced topics. Thus, magnetism is presented (in Chapter 5) as an effect of electric currents, point poles being eliminated entirely from the argument. In this chapter only the vector B is used (defined in terms of the force acting on a current). The properties of magnetic materials are discussed in the next chapter, at first in descriptive terms only; H is introduced halfway through the chapter, leading on to a discussion of Ampère's circuital rule for magnetic fields and the magnetic circuit. Likewise electric fields are described by means of the intensity E (defined as the force per unit charge, and immediately related to the potential gradient). The flux density D is introduced only for handling Gauss's theorem and the theoretical work associated with it. With most classes at this level teachers are probably content to leave out of the syllabus the *general* theory of electromagnetic fields inside materials. If it is sufficient to deal only with the limited theory of *uniform* fields (including symmetrical arrangements like toroidal solenoids), then the additional vectors H and D need not be introduced at all; and we can work (as in this book) with B and E only. But some students will be proceeding eventually to further courses in Engineering and Physics, and these should be provided in outline with the complete electromagnetic field theory in its traditional form employing four vectors. This more specialized work I have concentrated almost entirely in the latter half of Chapter 6.

It is just at this point in the presentation of the subject (when we introduce the theory of fields inside materials) that we have to make a choice between the Kennelly and Sommerfeld systems of definitions. Up to this point the differences between the two systems are not really significant. In this book I have chosen to adopt the Kennelly system, since this is more nearly in line with the traditional development of the subject, and—more important—this is the system generally used now by electrical engineers. It seems to me inexcusable to plan an m.k.s. physics course that would be out of step with the methods of the world of technology. The main argument in favour of an m.k.s. system is that it provides a single set of units and formulae for all purposes, to replace the multiplicity of systems that have come down to us from the past. At all costs we must avoid introducing a new diversity of systems.

Flexibility

A textbook needs to be adaptable for a wide variety of courses. Students come with differing amounts of background knowledge; their career needs vary; and individual teachers have their own interests and personal preferences as to order of treatment. The book should not tie down the class or teacher to a particular order. To give as much flexibility as possible I have planned this book as a set of more or less self-contained 'units'. There are numerous cross-references; and, within certain obvious limits, they can be taken in almost any order to suit the needs of any given class. These units are as follows:

I	Electric currents	(Chapters 1–4)
II	Magnetic fields	(Chapters 5–7)
III	Instruments and machines	(Chapters 8 and 9)
IV	Electric fields	(Chapters 10 and 11)
V	Alternating currents	(Chapter 12)
VI	Electronics and nuclear physics	(Chapters 13–17)

Questions

Examination questions and problems on the subject matter of each chapter have been gathered together under their respective chapter headings at the end of the book. These have been sorted into three grades of difficulty—**A, B** and **C.** '**A** questions' are straightforward introductory applications of new ideas; '**C** questions' are probably rather too hard for the average student. At the end of each group of questions cross-references are given to questions of similar type under other chapters.

Answers to the numerical problems are provided at the end of the book. Wherever reasonable I have quoted answers to the number of significant figures justified by the accuracy of the data. In recent years this has come to be regarded as the correct thing to do. However, few physicists (or examining boards) are entirely consistent in the way they present data. In doubtful cases I have usually given the answer to 2 significant figures; sometimes I have added a more exact answer in brackets to help a student to check his arithmetic. In some places I have written the final digit of a result as a suffix to help indicate accuracy. Thus '$2{\cdot}7_4$ volts' is intended to show that the answer is 2·74 volts with an uncertainty of perhaps 4 or 5 in the last place. An answer '2·74 volts' would mean that the uncertainty is not more than 1 in that place.

Slide-rules

An efficient physicist should be able to use a slide-rule with speed and accuracy. I have therefore attached some hints on this subject as an appendix to the text.

Acknowledgements

It would be impossible to list here all those works from which I have gleaned ideas and information in the course of writing this book. But those familiar with the subject matter will recognize that my presentation owes much to the report on 'The Teaching of Modern Physics' published by the Association for Science Education and the Nuffield Foundation; and I should like to acknowledge my indebtedness to those who worked on this report and to Mr J. L. Lewis of Malvern College in particular, whose dynamic approach has done so much to form a new outlook in the teaching of Physics in this country.

I should like to express my gratitude to a number of friends who have read parts of the manuscript and made valuable comments and suggestions: my colleagues at Clifton College, Messrs F. G. Mee, A. M. Joyce and G. P. Rendle; also Dr G. Lindsay Jones of Oundle School, Mr C. F. Tolman of Whitgift School and Mr D. C. Chaundy of Malvern College.

I have personally worked through all the problems, and the blame for any errors must be squarely laid at my door. However, this onerous task has been greatly shortened by the assistance of my former pupils, Mr P. F. Taylor of Queen's College, Cambridge, and Mr J. L. Hammond of St John's College, Cambridge; and also by the labours of numerous Cliftonians, on whom the problems have been tried, and who have often been quick to point out errors in my first solutions.

The encouragement and wise counsel of my Publishers have contributed in no small measure to this book and I am very grateful to them. I should also like to thank Mrs G. N. Burton and Mrs A. Brownlee, who have typed and re-typed these pages with unfailing speed and accuracy.

I should like to acknowledge with gratitude the help I have received from various busy people in finding suitable photographs of particle tracks and for permission to reproduce these: The Director of the Science Museum and Dr F. A. B. Ward, Keeper of the Department of Physics; Professor C. F. Powell, F.R.S., of the H. H. Wills Physics Laboratory, Bristol; and Professor A. B. Pippard, F.R.S., of the Cavendish Laboratory, Cambridge.

I am grateful to the various examining bodies listed below for permission to reprint questions from their past examination papers. The copyrights of these questions of course reside with the bodies concerned. Each question is individually acknowledged as indicated below.

Oxford and Cambridge Schools Examination Board (*O & C*)
Oxford Local Examinations Board (*O*)
Cambridge Local Examinations Syndicate:

Advanced and Scholarship level papers (*C*)
Papers set for H.M. Forces (*Cam. Forces*)
Cambridge Oversea H.S.C. (*Cam. Overseas*)

London School Examination Council (*L*)
Joint Matriculation Board (*N*)
Southern Universities Joint Board (*S*)
Welsh Joint Education Committee (*W*)
Associated Examining Board (*A.E.B.*)
Oxford Entrance Scholarships (*Ox. Schol.*)
Cambridge Entrance Scholarships (*Cam. Schol.*)

G. A. G. BENNET

Clifton,
August 1967

Contents

★ = advanced topics which may be omitted at a first reading

1: Atoms and Electric Currents

1.1 The atomic theory

To the present generation it is no doubt a commonplace that matter consists of *atoms*—very small, almost indestructible particles, whose combinations and interactions give rise to the properties of matter in bulk. However, it is only in this century that the atomic theory has entered the realm of well-accepted scientific truth; and it is important for the student to appreciate the kind of evidence on which our belief in atoms is based. None of the ordinary techniques used for making measurements on small pieces of matter in a school laboratory can possibly deal with single atoms. The diameter of an atom is far less than the resolving limit of any kind of microscope. No balance is sensitive enough to weigh a single atom. It is true that there are experiments in which we claim to be observing events involving single atoms. We may for instance 'observe' the disintegration of an atom in a cloud chamber (p. 302). But what we actually see are chains of water droplets intersecting at a point in the cloud chamber; each droplet must contain many millions of atoms. We say that each chain of droplets shows the track of a particle emerging from an atomic disintegration; but this is a human interpretation which we ourselves impose on the observations. Its validity depends on our correctly reading into the experiment a great deal more than is immediately obvious from this experiment alone.

Most of the evidence for the atomic structure of matter is even more indirect than this. We start by suggesting an imaginary model for the composition of matter. We then work out what behaviour to expect from matter constructed according to this model. In some respects the behaviour of actual matter may agree with this; in other respects it does not. The model is then varied and modified in ever greater and greater detail, until the agreement between the properties of the theoretical model and of real matter is exact. This is a process which still continues at the present day. No single piece of experimental evidence for the atomic theory is by itself conclusive; but the accumulation of evidence from many branches of Physics and Chemistry, always consistent with the same picture of the structure of matter, leaves very little room in the end for doubting the correctness of the theory.

An atomic theory of a kind first arose as a philosophical speculation amongst the ancient Greeks; but it was only in the early years of the nineteenth century that the first clear evidence for the theory was assembled, when Dalton pointed out that the laws of chemical combination were readily explained on an atomic basis. Subsequent developments showed that the atomic picture of chemical processes could be made wholly consistent. The chemical evidence for the atomic theory, compelling though it is, gives little indication of the properties of atoms; all it can do is to provide a table of the relative masses of the different kinds of atoms (*atomic weights*, they were called). Chemistry cannot provide any figure for the actual mass or size of an atom.

Little physical evidence in support of the atomic theory could be offered until about the middle of the nineteenth century. By then the work of Joule had established the true nature of heat as a form of energy. This paved the way for rapid developments in the atomic theory; this particular branch of it became known as the *kinetic theory of matter*. It was supposed that the atoms in a piece of matter are in a state of continuous random motion; in a solid this motion must be limited to oscillations about fixed mean positions if the material is to retain its shape permanently; but in a fluid the motion can be more general. The *heat energy* carried by the matter is supposed to be the *sum total* of the mechanical energy of its atoms. Its *temperature* is taken to be a measure of the *average energy* of the atoms. A simple confirmation of the theory is provided by observation of the movements of small particles suspended in a gas or liquid—the *Brownian movement*, as it is called. For instance, if a drop of water containing a little black water-colour pigment is placed on a slide and observed with a medium-power microscope, the black particles are found to be in a state of ceaseless agitation. It is easy to suppose that we are observing the particles sharing in the random motions of the atoms adjoining them.

Later developments of the kinetic theory showed that detailed observation of the Brownian movement in a gas enables the number of molecules per unit volume to be calculated; it comes to the enormous figure of $2 \cdot 7 \times 10^{25}$ molecules per cubic m at *standard temperature and pressure* (s.t.p.). By dividing this into the mass per cubic m (i.e. the density) of the gas, we obtain the mass of a single molecule. This gives the mass of a hydrogen molecule as $3 \cdot 4 \times 10^{-27}$ kg; if each molecule of hydrogen contains two atoms, the mass of a single hydrogen atom is then $1 \cdot 7 \times 10^{-27}$ kg.

The kinetic theory was developed first for gases, since the mathematical difficulties are in this case less than for solids or liquids. It is supposed that in a gas the average distance of a molecule from its nearest neighbours is very much greater than the diameter of the molecule. The interactions between molecules are then assumed to have negligible effects on their motions, except actually at the moments of collisions. The pressure of the gas is supposed to arise through the continual impacts of its molecules on the walls of the containing vessel. Using the known laws of mechanics, it is possible to predict how the pressure P exerted by such a molecular 'model' will depend on the volume V and absolute temperature T. The predictions are found to be in excellent agreement with the basic gas laws (*Boyle's law* and *Charles's law*). Furthermore, the theory provides a formula connecting the velocities of the molecules with the gas constant R (in the equation $PV = RT$). For oxygen at room temperature the average velocity is 0·5 km per second, for hydrogen about 2 km per second.

Other types of measurement enable estimates to be made of molecular diameters. The kinetic theory shows that certain properties of gases (e.g. their thermal conductivity, viscosity, etc.) should depend on the average distance travelled between collisions —a quantity known as the *mean free path*. This in turn depends on the diameter of the molecules; the larger they are, the more frequent must be the collisions between them. The measurements show that for air at s.t.p. the mean free path is 10^{-7} m, which gives the diameter of the molecules as about 3×10^{-10} m.

It is also possible to estimate the dimensions of atoms by a number of other methods. For instance, using the known values of the masses of atoms we can calculate how many atoms there must be in a piece of solid material of known mass and volume. It is reasonable to assume that in the solid state the atoms are closely packed together; and so we can use the figures to calculate the volume occupied by a single atom. Always the results obtained for atomic dimensions are of the same order of magnitude. There are, however, significant variations between the estimates made by different methods; and it is clear that the elementary concept of atoms as hard elastic spheres is inadequate. When an atom is fixed in a stable configuration, such as a chemical compound or the regularly spaced lattice of a crystal, the forces on it are in equilibrium; and the distance between the centres of two atoms is a precisely fixed quantity, which cannot be varied much by elastic stresses or changes of temperature. But when the molecules are free to move at random in a gas, they do not appear to behave as though they had sharply defined surfaces. If we must form mental pictures of atoms, it is better not to think of billiard balls but of puffs of cotton wool with hard centres! But such efforts of the imagination are of dubious scientific value. The point is that the molecular model only agrees with the observed properties of gases if we suppose that the closest distance of approach of two molecules depends on the violence of the collision.

There are many other lines of evidence that have been considered in the development of the theory, and these have led to endless refinements of our ideas about the properties of atoms. But enough has been said here to show the basis of the modern atomic theory; without ever observing individual atoms we have developed a single consistent picture of an atomic structure of matter, which suffices to give detailed explanations of a vast body of experimental facts. At the present time there certainly seems to be no possible ground for doubting the validity of the theory.

Even so it is hard enough to appreciate the minute scale and vast rapidity of atomic processes.

It may be useful therefore to consider some of the implications of the figures we have given. The volume of a molecule of diameter 3×10^{-10} m must be about $1{\cdot}5 \times 10^{-29}$ m^3. In 1 m^3 of a gas at s.t.p. the total volume occupied by the molecules themselves is thus about 4×10^{-4} m^3—about 1 part in 2500 of the actual volume of the gas. With a mean free path of 10^{-7} m each molecule travels about 300 times its own diameter between collisions. But since it moves at an average speed of 5×10^2 m s^{-1}, it must make some 5×10^9 collisions with other molecules every second! In spite of this it spends less than 1% of its time being affected in any way by molecular collisions. The mean free path is inversely proportional to the pressure of the gas. Thus at the lowest pressures readily attainable in the laboratory (about 10^{-8} mm of mercury) the mean free path is nearly 10 km. Some idea of the smallness of molecules can be gained by considering that even at this pressure there are over 10^{14} molecules per cubic m along the path. At the lowest densities believed to exist in parts of outer space there are still several molecules per cubic metre. But the mean free paths are then to be measured in light-years and collisions only occur at intervals of several centuries!

1.2 The structure of the atom

Our knowledge of the internal structure of atoms has been arrived at in much the same way as our belief in the atoms themselves. The ambition of physicists is to develop a theory of the atom that shall be able to explain all the observed properties of matter. For instance any satisfactory theory must be able to account for the varied behaviour of atoms in chemical processes. It must also explain the emission and absorption of light by matter.

In this book we are chiefly concerned with electric and magnetic effects, which, as we shall see, arise from the properties of the fundamental particles that go to make up the atom itself. In due course we shall consider some of the detailed evidence. But for the time being we must be content with a bare sketch of the atomic model that has been found to fit the facts.

Although the effective diameter of an atom is about 3×10^{-10} m, nearly all its mass appears to be concentrated in a very small central region, called the *nucleus*. This is rather less than 10^{-14} m in diameter, and itself contains fundamental particles of two kinds—*protons* and *neutrons*—packed closely together; these two particles have nearly the same mass. The rest of the space occupied by the atom contains particles of a third type, called *electrons*, which are in motion round the nucleus. The number of electrons in the outer parts of the atom is normally exactly the same as the number of protons in the nucleus, but each electron is only about 1/2000 as massive as the proton or the neutron.

The atom is held together by two different types of forces:

1 *Nuclear forces.* All the particles in the nucleus (known collectively as *nucleons*) are attracted to one another by very short-range forces; that is, the forces are negligible except when the particles are about 10^{-15} m apart; but at this separation the forces become very large, sufficient to hold the nucleons together in a closely packed structure.

2 *Electrical forces.* A different type of force is found to operate between electrons and protons. These forces are effective at much greater separations than the nuclear forces. The force acts as a repulsion between like particles (i.e. between electron and electron or between proton and proton), and as an attraction between unlike particles (i.e. between electron and proton). These forces are described by saying that the particles carry *electric charge*, the proton a positive charge and the electron an equal negative charge; the neutron is an uncharged particle. Thus, like charges repel, and unlike charges attract. Normally atoms are electrically neutral, since the number of electrons in them is equal to the number of protons and the total charge is thereby zero. In this condition the atom exerts little or no electrical force on charged particles in its neighbourhood.

Under certain circumstances an atom may lose electrons from the electron-cloud surrounding its nucleus, or it may gain surplus electrons from its surroundings. The whole atom then acquires a net positive or negative charge and behaves in many respects like a single charged particle. An atom in this charged condition is referred to as an *ion*. Hydrogen and metals have a tendency to lose electrons, forming positive ions; while many non-metals have a tendency to gain extra electrons, forming negative ions. Some simple chemical compounds are produced by the forces of attraction between such pairs of oppositely charged ions. In

The twelve lightest elements and their most important isotopes

Element	*Symbol*	*Atomic number*	*Atomic mass (a.m.u.)*	*Nucleus* No. of protons	*Nucleus* No. of neutrons	*No. of electrons*
Hydrogen	H	1	1·0078	1	0	1
Deuterium	D		2·0141	1	1	
Helium	He	2	3·0161	2	1	2
			4·0026	2	2	
Lithium	Li	3	6·015	3	3	3
			7·016	3	4	
Beryllium	Be	4	9·012	4	5	4
			10·012	4	6	
Boron	B	5	10·013	5	5	5
			11·009	5	6	
Carbon	C	6	12·000	6	6	
			13·003	6	7	6
			*14·003	6	8	
Nitrogen	N	7	14·003	7	7	7
			15·000	7	8	
Oxygen	O	8	15·995	8	8	
			16·999	8	9	8
			17·999	8	10	
Fluorine	F	9	18·998	9	10	9
Neon	Ne	10	19·992	10	10	
			20·994	10	11	10
			21·991	10	12	
Sodium	Na	11	22·990	11	12	11
Magnesium	Mg	12	23·985	12	12	
			24·986	12	13	12
			25·983	12	14	

* This isotope is radioactive. Short-lived radioactive isotopes have not been included in the table.

fact many of the processes of inorganic chemistry are concerned at the atomic level with transferring electrons from one atom or ion to another. In chemical terms, adding an electron to an atom or ion is described as *reduction*, and removing an electron as *oxidation*—a broadening of the original meanings of these terms. In a given atom only a small number of the outermost electrons are involved in any chemical effects. These are known as the *valence electrons.*

The chemical behaviour of an atom is decided by the number of electrons it contains when un-ionized. This in turn is fixed by the number of protons in the nucleus, which is called the *atomic number*. Classifying the atoms according to their atomic numbers reveals the sequence familiar in the periodic table of the chemical elements. The first few members of this sequence are shown in the accompanying table.

The mass of an atom is almost equal to that of the nucleons it contains; the electrons contribute only about 1/4000 of the total mass. The modern practice is to measure atomic masses relative to that of the commonest sort of *carbon* atom (whose nucleus contains 12 nucleons); this is taken arbitrarily as 12 *atomic mass units* (a.m.u.) exactly. The connection between this and the kilogram is found to be

$$1 \text{ a.m.u.} = 1{\cdot}66 \times 10^{-27} \text{ kg}$$

Expressed in a.m.u. any atomic mass is nearly equal to the number of nucleons in the atom, a figure known as the *atomic mass number*. (N.B. *Atomic mass number* has no units and is necessarily a whole number. *Atomic mass* must be expressed in appropriate units—kg or a.m.u.—and, when expressed in a.m.u., is *nearly* equal to the atomic mass number.)

The simplest possible atom is that of *hydrogen*, consisting of a single proton round which orbits a single electron. The nucleus of *deuterium* contains one proton and one neutron; and since one electron only is required to balance this electrically, it has the same chemical properties as ordinary hydrogen. It is sometimes called *heavy hydrogen*. It occurs as about 1 part in 4000 of natural hydrogen.

The atoms of a given chemical element must all contain the same number of protons. With the light elements in the table the number of neutrons is also approximately equal to this; but in most cases there are several possible alternatives. Groups of atoms having the same atomic number but differing atomic masses are known as *isotopes*. Samples of a given element derived from natural sources consist of mixtures of the possible isotopes of the element; the proportions are found to vary little from one sample to another. The measurement of atomic mass by chemical methods therefore gives a figure that is an average for the naturally occurring mixture of isotopes. Expressed in a.m.u. this figure is not necessarily a whole number, and is usually referred to by the older name of *atomic weight*. For instance natural *magnesium* is a mixture of isotopes with atomic mass numbers 24, 25 and 26, the lightest providing nearly 80% of the total. The mixture is in such proportions that the effective atomic weight is 24·32 a.m.u. It is this quantity that must be used in chemical calculations involving magnesium.

1.3 Electric currents

There are many substances in which there are charged atomic particles that can wander freely through the material. In such a substance electric charge can be transferred from one point to another by a general drift of the charged particles within it. Such a movement of electric charge is called an *electric current*. Materials through which an electric current will pass are known as *conductors*. Those substances in which there are no charged particles that are free to move are known as *non-conductors* or *insulators*.

The most important class of conductors is formed by the *metals*. In any solid the atoms themselves are necessarily fixed at permanent sites in the crystal. But the outermost (valence) electrons of each atom are partly shared with neighbouring atoms. It is this feature that gives a solid its cohesive strength. In most non-metals each valence electron is attached to a particular group of atoms; and no general movement of electrons through the material is possible. But with metals the bonds between atoms are such that some of the valence electrons are free to move from one atom to the next right through the crystal lattice.

Another important type of conductor is the class of liquids known as *electrolytes*. Many chemical compounds when they go into solution break up into positively and negatively charged ions, which then move independently in the liquid, and are therefore available to conduct an electric current through it.

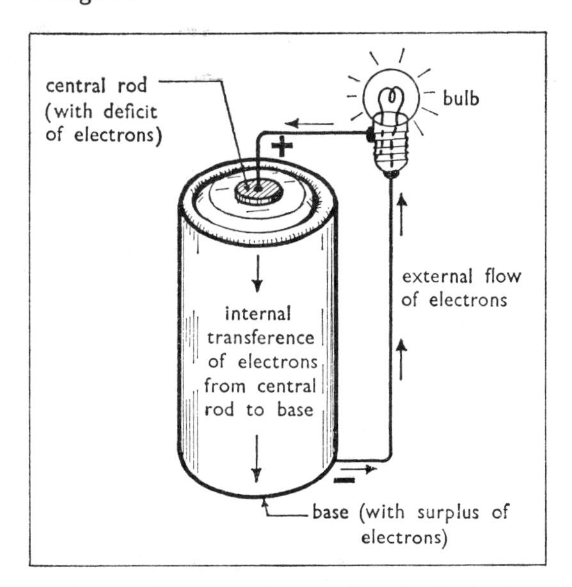

Fig. 1.1 The flow of electrons in a simple circuit

The electrical forces between charged particles are so great that it is impossible to unbalance the distribution of charge in a piece of matter to any appreciable extent. Any tendency for charge of one sign to pile up at some point generates forces of repulsion that soon bring the process to a halt. A continuous flow of electric current can thus only happen in a closed path or *circuit*, as it is called. The circuit must also include some device such as an electric battery or dynamo, whose function is to maintain the circulation of electric charge. This it does by an internal transference of electrons from one of its terminals to the other; in a battery this is brought about by chemical processes. One terminal thus gains a surplus of electrons and so carries a negative charge. At the other terminal there is a corresponding deficit of electrons and a net positive charge. When a complete conducting path is provided externally between the terminals, the electrical forces tend to even out the distribution of charge by a suitable flow of current; the battery or dynamo then acts to maintain the initial distribution of charge by a fresh transference of electrons between the terminals *inside* the device. A continuous circulation is thus created (Fig. 1.1).

Needless to say there is nothing obvious to indicate that a flow of *anything* is taking place in an electric circuit. The flow of charge is a theoretical explanation of the process that we believe in because it has proved to be perfectly consistent. The current must rather be detected by some definite effect to which it gives rise. Three types of effect may be distinguished:

(*i*) *A heating effect.* In all forms of electric conduction the movements of the charged particles are continually impeded by their interactions with stationary particles of matter in their paths. The energy lost in collisions increases the random motion of the particles of the conductor through which the current passes, and therefore raises its temperature. The heating effect of a current in the 'element' of an electric fire or the filament of a light bulb is too well known to require any demonstration. A full discussion of this process will be found in Chapter 2.

(*ii*) *A chemical effect.* The movement of electrons through a metal does not cause any chemical effect in it. But in an electrolyte the position is different. In this case the conduction takes place through the simultaneous movement of positive and negative ions in opposite directions. Where the current enters and leaves the solution through suitable conducting plates (called *electrodes*) some of the ions are discharged; and chemical changes are therefore observed as long as the current flows. The particular reactions that occur depend on the materials used. The effect may be demonstrated by joining two carbon rods to the terminals of a battery by means of copper wires; the circuit may then

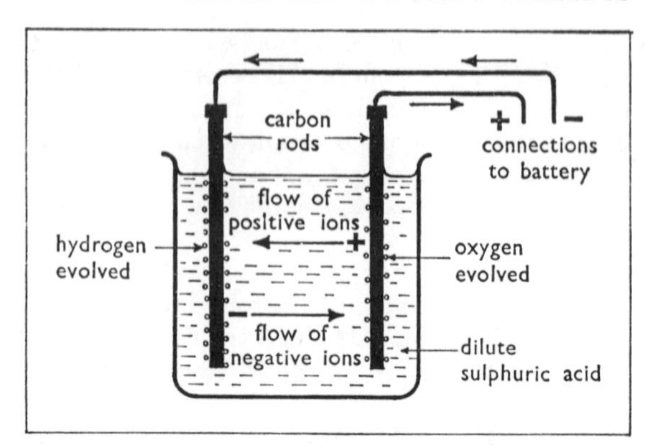

Fig. 1.2 The chemical effect of an electric current—carbon electrodes in dilute sulphuric acid

be completed by dipping the rods into a beaker of dilute sulphuric acid—which is an electrolyte (Fig. 1.2). It will then be noticed that bubbles of hydrogen are evolved at the surface of the negative rod and bubbles of oxygen on the positive rod. The action is clearly caused by the electric current, since it ceases when one of the copper wires is disconnected from the battery. If instead the rods are dipped into a solution of copper sulphate, oxygen is still evolved at the positive rod, but metallic copper is now found to be deposited on the negative one. This process is considered in detail in Chapter 3.

(*iii*) *A magnetic effect.* This may be demonstrated by causing an electric current to flow in the vicinity of a small pivoted magnet (i.e. a compass). In most positions close to the circuit the compass will be deflected slightly from its usual north-pointing direction. This effect is due to yet another type of force that acts between the fundamental particles of matter; it occurs between *charged particles in motion*, and acts in addition to the electrical forces already considered. Since charges in motion constitute an electric current, we may say that magnetic effects are due to forces acting between electric currents. This may be shown by passing a large current through two parallel flexible conductors (Fig. 1.3). If the currents are in the same direction, the magnetic force is seen to act so as to draw the conductors together. When the currents are in opposite directions, the wires are repelled from one another. Such effects are considered in more detail in Chapter 5.

It is not perhaps obvious how the properties of

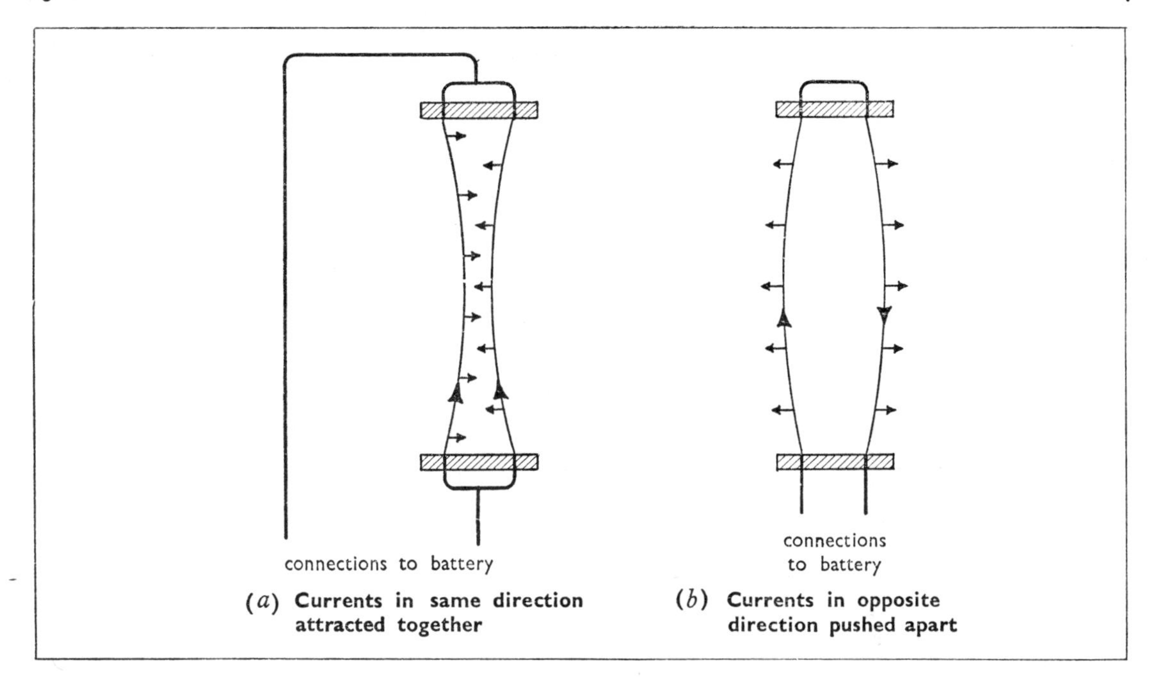

Fig. 1.3 The magnetic force between neighbouring electric currents

magnets can arise from the interactions of electric currents. But a little thought will show that on an atomic scale there are never-ending circulations of electric charge taking place. An electron in rapid motion round a nucleus constitutes an electric current. Indeed it is now understood that electron, proton and neutron must all be regarded as being permanently in a state of 'spin', which causes a magnetic effect even when they are not otherwise in motion. We should therefore expect all matter to show magnetic effects; and such is found to be the case, though with most materials the effects are very weak. It so happens that in iron and a few other substances the interactions between atoms are such as to magnify the effects considerably. This aspect of magnetism is dealt with more fully in Chapter 6.

The strength of an electric current could well be measured by the intensity of any one of the three effects to which it gives rise. However, the theory of currents is by now so well established that it is simplest to define an *electric current* in terms of the rate of passage of electric charge in a circuit.

> *An* **electric current** *is formed by electric charges in motion; it is equal to the quantity of electric charge passing a given point per unit time.*

The same electric current may be carried in various ways in different parts of the same circuit; at one point there may be a movement of positively charged particles in one direction; at another there may be an equal movement of negatively charged particles in the opposite direction—the net transference of charge could be the same in both cases; or there may be a simultaneous movement of positive and negative charges in opposite directions (as in electrolytic conduction). Any or all of these processes may occur in any one circuit.

By general agreement the *positive* direction of electric current is taken to be that in which a positive charge at that point of the circuit would flow. An arrow drawn on a circuit diagram indicates this positive direction; it implies that, if the current at that point is actually a flow of negatively charged particles, then the movement of the particles is opposite to that of the arrow. Fig. 1.4 shows how this is done on a conventional circuit diagram for a single-celled battery connected to a light bulb. Note that in the conventional symbol for a battery the *longer stroke* is taken to be its *positive* terminal.

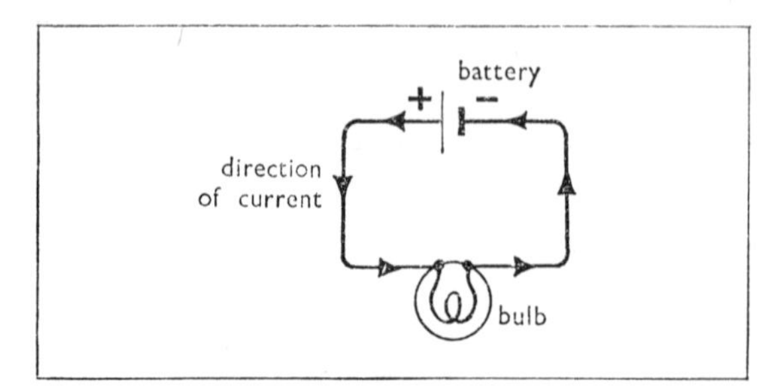

Fig. 1.4 The conventional direction of a current; positive ions move with the arrow, negative ions and electrons in the opposite direction

It may be possible one day literally to count the number of electrons passing some point of a circuit. We shall then be able to measure electric currents directly in terms of the definition given above. But at present this cannot be done. To measure a current and to fix a unit in which to measure it we must still rely on one of the three effects we have described. The most convenient of these is the magnetic effect. Indeed, as we shall see, most indicating meters used in electric circuits depend in some way on the magnetic effects of a current. The practical unit of electric current is called the *ampere* (abbreviated to *amp* or just A) after the French scientist who established the laws governing the magnetic forces between currents.

The **ampere** *is that steady current which, flowing in two infinitely long, straight, parallel conductors of negligible, circular cross-section placed 1 metre apart in a vacuum produces a force between them of* 2×10^{-7} *newtons per metre length of conductor.*

(The *newton* is the absolute unit of force in a system based on the metre, kilogram and second —the m.k.s. system.)
The prefixes used in the metric system are applied to electrical units also so as to provide convenient multiples and submultiples of the basic units. For example, the *milliamp* (mA) and *microamp* (μA) are derived units in regular use for recording small currents.

★ The definition of the ampere may in fact be cast in a number of different forms. The established laws of electromagnetism enable us to work out expressions for the magnetic forces acting between electric currents flowing in any imaginable configuration. These expressions each contain a numerical constant, whose value depends on our choice of units for measuring the electric currents and other quantities concerned. To fix the ampere it is only necessary to specify the value of this constant for some stated arrangement of conductors. The form of definition adopted in 1948 by the International Committee of Weights and Measures is given above. But it was particularly stated by the Committee that this definition was not to be regarded as having priority over any other demonstrably equivalent definition.

The current balance

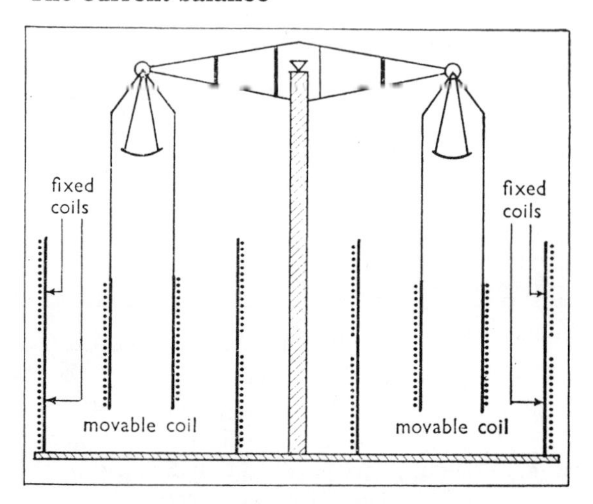

Fig. 1.5 The principle of a current balance

The apparatus used in the standardizing laboratories for fixing the value of the ampere is the *current balance.* The principle of this instrument is shown in Fig. 1.5. The beam, supported on its knife-edges at the centre, carries a single-layer, cylindrical coil at each end; these are joined in series, and the connection to them is by means of wires arranged to exert negligible torque on the beam. Concentric with each movable coil is a pair of fixed single-layer, cylindrical coils. All six coils are connected in series in such a way that the forces operating between them will combine in deflecting the beam in one direction. The beam is maintained in equilibrium by moving along it a rider of known mass. In this way the forces acting between the coils are measured in terms of a known weight. If the dimensions of the apparatus are also known, the way in which these forces depend on the current (in amperes) can be calculated from the laws of electromagnetism; and so the current in the coils is found. The accuracy attainable at present is about 5 parts in 10^6.

If an electric current is the rate of flow of electric charge, fixing the unit of current automatically establishes also a unit of charge. The practical unit of electric charge is known as the *coulomb* (abbreviated to C), after the French scientist who first investigated the forces acting between stationary electric charges.

> *The* **coulomb** *is the quantity of electric charge that passes a given point in a circuit when a current of* 1 *ampere flows for* 1 *second.*

Thus if a current of I amp flows in a circuit for t seconds, the quantity Q of electric charge that passes is given (in coulombs) by

$$Q = It$$

The charge on one electron turns out to be $-1{\cdot}60 \times 10^{-19}$ coulomb. Or we may say that one coulomb is numerically the charge on $6{\cdot}2 \times 10^{18}$ electrons.

(Sometimes the phrase 'quantity of electricity' is used where we have written 'quantity of electric charge'. This is an old-fashioned usage, and it will not be adopted in this book. We shall keep the word 'electricity' to refer to the subject under study. However, the student may well come across the older phrase in other books and in examination questions.)

The velocities of the current carriers

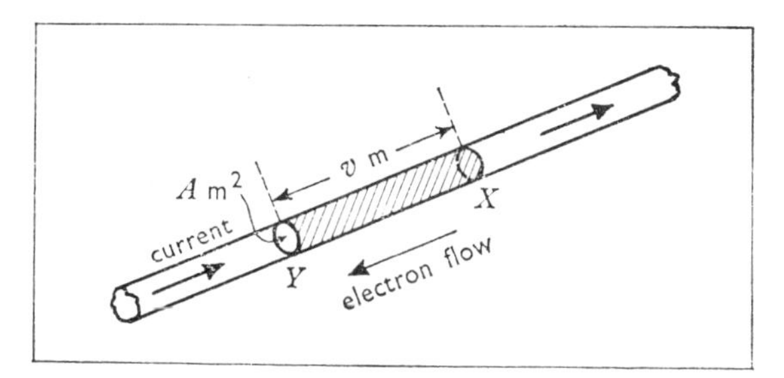

Fig. 1.6 Calculating the velocity of the charged particles in a conductor

Consider a length of conductor of cross-sectional area A m² containing n electrons per cubic metre (Fig. 1.6). The number of electrons in the conductor remains constant; when a current flows in it, electrons enter it at one end and the same number leave at the other. The motion of the individual electrons is doubtless somewhat irregular, but suppose their average velocity is v m s⁻¹. Then the number of electrons passing a point such as Y is the number contained in the part of the conductor between X and Y of length v m.

$$\text{volume between X and Y} = Av \text{ m}^3$$
$$\therefore \text{ number of electrons contained} = nAv$$

The electric current I amp flowing in the conductor is equal to the quantity of electric charge passing a point such as Y in one second. If the charge carried by one electron is $-e$ coulomb, then

$$I = nAve$$

$$\therefore\ v = \frac{I}{nAe}$$

The resistance to the motion of the electrons is roughly proportional to their average velocity v. A good conducting material is one for which n is large and v therefore small. Copper is one of the best conductors, having about 10^{29} free electrons per cubic metre.

In a complete circuit of conductors joined in series the current I is the same at all points. However, the velocity of the electrons and therefore the resistance to the flow may vary from one part of the circuit to another. In a copper wire of relatively large cross-section v is small and very little resistance is presented to the flow of current. But if the circuit includes a length of fine wire such as the filament of a light bulb, v is there much greater, and most of the available energy is dissipated in driving the current through this wire—which thus becomes white hot.

The actual values of the velocity v are amazingly small. Consider for instance a current of 1 amp flowing through a copper wire of cross-sectional area $2{\cdot}0 \times 10^{-7}$ m² (about ½ mm diameter).

$$v = \frac{I}{nAe} = \frac{1}{10^{29} \times 2 \times 10^{-7} \times 1{\cdot}6 \times 10^{-19}}$$
$$= 3{\cdot}1 \times 10^{-4} \text{ m s}^{-1}$$

Although the electrons move so slowly, the electric forces that set them in motion are propagated round a circuit with great rapidity—in fact very nearly with the speed of light (3×10^8 m s⁻¹). If a current of 1 amp flowed steadily along such a wire in a transatlantic cable (3000 km long), it would take about 300 years for a given electron to traverse the ocean. And yet a small displacement of the electrons at the London end sends a wave of electric forces along the cable, and a similar displacement of the electrons in New York occurs within $\frac{1}{100}$ second! Even in the circuit of a small electric torch it is doubtful whether a given electron could pass right round the circuit before the battery is run down.

The same analysis applies to conduction in electrolytes, except that the different types of ions may well move with different velocities. In the nature of things the cross-sectional areas of electrolyte are usually greater than those of metallic wires, and the velocities are correspondingly less (p. 38).

In due course we shall come across other ways by which electric charge can be transferred from one point to another. One way of great technical importance is by means of electron beams travelling in a vacuum. In a television picture tube such a beam is used to provide the moving spot of light that builds up the picture. In this case the electrons move along the tube unimpeded by any collisions, and travel at about a quarter the speed of light. The entire energy of the electrons is given up on impact with the front wall of the tube. Because the speed is so high the number of electrons in the beam is relatively small. Suppose there are N electrons per m length of the beam moving at a speed v m s^{-1}. The number of electrons passing a given point is then Nv per second; and the current I in the beam is given by

$$I = Nve$$

With electrons moving at, say, 8×10^7 m s^{-1} and a total beam current of 100 microamp, we have

$$N = \frac{I}{ve} = \frac{10^{-4}}{8 \times 10^7 \times 1{\cdot}6 \times 10^{-19}}$$

$$= 8 \times 10^6 \text{ electrons per m}$$

1.4 Electrical energy

The circulation of an electric current leads to the continual dissipation of energy throughout the circuit in the form of heat; it may also result in the production of other forms of energy—e.g. mechanical energy in an electric motor. All this energy must be continuously supplied by the battery, dynamo, etc., that maintains the flow of current. Such devices for generating electric current are therefore primarily means of converting energy to electrical form from some other form. Thus an electric battery is a means of producing electrical energy from chemical potential energy; a dynamo produces it from kinetic energy, a microphone from the vibrational energy of a sound wave, a photocell from light energy, and so on.

In the electric circuit this production of energy is made apparent by the existence of electrical forces acting on the charged particles in the circuit, giving rise to the current. This is described by saying that the battery, dynamo, etc. produces an *electromotive force,* usually abbreviated to *e.m.f.* The electromotive force is measured in terms of the energy imparted to the complete circuit per unit charge driven round the circuit by the source of e.m.f. (This includes the energy used in driving the current through the source of e.m.f. itself.)

The **electromotive force (e.m.f.)** *of a battery (or dynamo, etc.) is equal to the energy converted to electrical form per unit charge passing through the battery.*

Potential difference

A source of e.m.f. imparts energy to a circuit through its ability to set up an uneven distribution of electric charge, by transferring electrons internally from one of its terminals to the other. In this way electrical forces are brought into play which will act on any charged particles near the terminals. We describe this by saying that the source of e.m.f. produces an *electric field* between its terminals. A complete conducting path between the terminals allows the electric field to act inside the conductors forming the circuit, thus setting the free charges in them in motion. (The field also exists outside the conductors; but here it is not effective in causing a current—provided the insulation between the terminals does not break down!) Inside the conductors the field does work on the moving charged particles, causing the electrical energy provided by the source of e.m.f. to be converted into heat and perhaps other forms of energy.

A useful way of describing this field is to attach to each point of the circuit a figure, called its *potential,* in such a way that the flow of current (in the conventional sense) is from points at higher potential to points at lower potential. The difference of potential between two points can be regarded as the electrical 'pressure' acting to drive current from one point to the other. The work done by the electric field per unit charge passing between the points is taken as a measure of the *potential difference* (usually abbreviated to *p.d.*).

The **potential difference (p.d.)** *between two points acts so as to drive electric current* *from the point at higher potential to that at lower potential; it is equal to the energy converted from electrical to other forms per unit electric charge that passes.*

The unit in which p.d.'s are to be measured is fixed as soon as we have chosen a unit of energy. The absolute unit of energy in the m.k.s. system is the *joule*. This is the quantity of energy supplied (or work done) when a force of 1 newton moves its point of application a distance of 1 metre in the direction of the force. The joule is normally used at present for measurements of all types of energy—mechanical, electrical or heat. However, chemical energy and heat are also often measured in *calories*. The relationship between the joule and the calorie is

$$1 \text{ calorie} = 4{\cdot}1868 \text{ joules*}$$

To within 0·3% this can be written

$$1 \text{ calorie} = 4{\cdot}2 \text{ joules}$$

The unit of potential difference, based on the joule and the coulomb, is called the *volt* (abbreviated to V) after the Italian scientist Volta, who first investigated the action of electric batteries. The *millivolt* (mV) and *kilovolt* (kV) are derived units that are also regularly used.

The **volt** *is the potential difference between two points such that the energy converted from electrical to other forms is 1 joule per coulomb of electric charge that passes from one point to the other.*

Thus if a charge Q coulombs flows in a part of a circuit across which there is a p.d. of V volts, the electrical energy supplied is given (in joules) by

$$\text{energy} = QV$$

Since both p.d. and e.m.f. are defined in terms of the energy converted per unit electric charge, it is clear that the same unit is to be used to measure both quantities. Thus the volt is also the practical unit of e.m.f. If a charge Q coulombs flows in a circuit containing a source of e.m.f. E volts, the electrical energy supplied to the whole circuit is given (in joules) by

$$\text{energy} = QE$$

It is important to understand the distinction between *electromotive force* and *potential difference*, and to use the two concepts correctly. A source of e.m.f. is an energy converting device, the e.m.f. being the electrical manifestation of the conversion process. The e.m.f. acts *in* the circuit as a pressure tending to drive current round it. The action of the e.m.f. is to create an electric field between the terminals of the source of e.m.f.; we describe this by saying that the e.m.f. produces a potential difference between the terminals. The potential decreases as we pass round the circuit from the positive to the negative terminal. The potential difference *across* any part of the circuit acts as the electrical pressure tending to drive current through that part of the circuit and supplying electrical energy to it. But this derives from the source of e.m.f., and there could be no p.d. *across* any part of the circuit without an e.m.f. acting *in* the circuit. Because some of the energy generated by the source of e.m.f. is wasted as heat driving current through the source itself, the p.d. between its terminals is usually less than the e.m.f.

Although only *differences* of potential are of any significance in electric circuits, it is sometimes convenient to have a conventional *zero of potential* with reference to which we calculate the potentials of other points. Many pieces of electrical apparatus are operated with some part connected to the earth. It is then usual to regard the earth as being at zero potential. But the choice is quite arbitrary, and we are always free to fix the zero (if we need one at all) to suit our convenience.

Batteries

Electric *batteries* consist of several electric *cells* joined together so as to provide greater p.d.'s or greater currents than would be possible with a single cell. There are two ways of doing this:

(*i*) *Cells in series.* In this case the same current, and therefore the same charge, flows in each cell. Suppose a charge Q passes through the battery. The total energy imparted to the circuit is the sum of the quantities of energy provided by each cell. In Fig. 1.7*a* the total energy produced

$$= E_1Q + E_2Q + E_3Q$$

$$\therefore \text{ total e.m.f.} = \frac{\text{total energy}}{Q} = E_1 + E_2 + E_3$$

* This is the International Steam Table calorie, and is now taken by the ISO as the *definition* of the calorie. The old 15°C calorie is 4·1855 J, while the Thermochemical calorie is 4·1840 J.

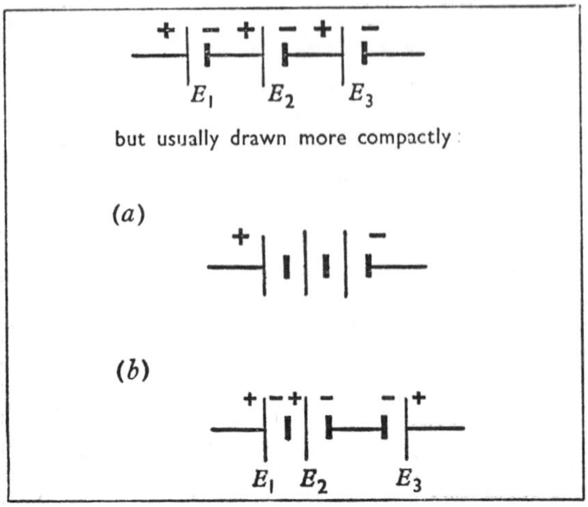

+ − + − + −

E_1 E_2 E_3

but usually drawn more compactly:

(a)

+ −

(b)

+ − + − − +

E_1 E_2 E_3

Fig. 1.7 Cells in series: (*a*) all acting in the same direction; (*b*) one cell reversed and acting in opposition to the others

Most electric cells have e.m.f.'s between 1 and 2 volts. By joining a number of them in series in this way batteries of large e.m.f. may be produced.

If one of the cells is reversed so that its electrical pressure acts in opposition to the others, its e.m.f. must be counted as negative in assessing the resultant e.m.f. of the combination. Thus in Fig. 1.7*b*

$$\text{resultant e.m.f.} = E_1 + E_2 - E_3$$

The effect of driving a current through a cell in opposition to its e.m.f. is to reverse the energy conversion that usually takes place in it; i.e. chemical energy is formed again from the electrical energy made available by the other cells.

(*ii*) *Cells in parallel.* Suppose we have *n identical* cells of e.m.f. E joined in parallel as in Fig. 1.8. In this case the current through the battery divides, equal parts of it passing through each cell. If a total charge Q passes through the battery, the charge passing through each cell is Q/n.

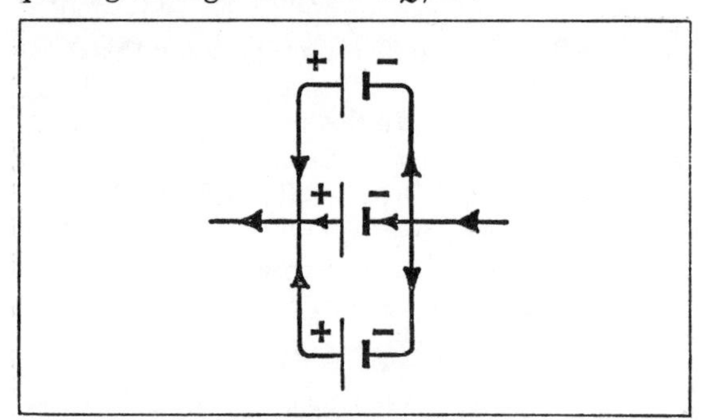

Fig. 1.8 Cells in parallel

$\therefore$ energy provided by each cell

$$= E \times \frac{Q}{n}$$

$$\therefore \text{ total energy provided} = E \times \frac{Q}{n} \times n$$

$$= EQ$$

$$\therefore \text{ total e.m.f.} = \frac{\text{total energy}}{Q} = E$$

The resultant e.m.f. is thus the same as that of one of the cells. The possible advantage of this arrangement is that it can provide a much larger current than could be obtained from a single cell.

When the cells are not identical, the position is far more complex. It then becomes difficult to predict how the current will be apportioned between them; and it is even possible for some of the cells to drive current backwards through the others. Such arrangements have no practical value.

Power

It is often necessary to know the *rate* at which energy is being converted from one form to another in some mechanical or electrical device. This is known as the *power* of the device.

> *The* **power** *of a device is the energy converted in it per unit time.*

Power can be expressed in terms of any suitable unit of energy—ft-lbf per second, calories per second, etc. The unit usually used, however, is the practical unit based on the joule. It is called the *watt* (abbreviated to W) after the British engineer who developed the steam engine.

> *The* **watt** *is a unit of power; it is a rate of conversion of energy of 1 joule per second.*

Thus a device of power W watts operating for t seconds converts a quantity of energy given (in joules) by

$$\text{energy} = Wt$$

Other units of power regularly used are the *kilowatt* (kW) and the *megawatt* (MW).

The traditional British system of units is one based on the foot, pound and second (the f.p.s. system). The unit of energy in this system is the *ft-lbf*; and the unit of power is the *horsepower* (h.p.), which is defined as a rate of conversion of energy of 550 ft-lbf s^{-1}. This unit is still to some extent used for specifying the powers of electric motors. The connection between the horse-power and the watt is

$$1 \text{ h.p.} = 746 \text{ watts}$$

With an error of only 0·5% this can written

$$1 \text{ h.p.} = 750 \text{ watts} = \tfrac{3}{4} \text{ kilowatt}$$

The electrical units introduced in this chapter can be regarded as being based on three fundamental quantities—the *coulomb*, the *joule* and the *second*. The other units are derived from them as follows:

1 amp = 1 coulomb per second
1 volt = 1 joule per coulomb
1 watt = 1 joule per second

The student will find it worthwhile to memorize these verbal equations. Many simple problems in electricity can be solved by re-writing the quantities in the problem in terms of the units of charge, energy and time.

Example. A 6-volt battery drives a current of 4 amps through a heating coil for 15 s. Calculate (*a*) the quantity of electric charge that flows in the circuit, (*b*) the total quantity of heat produced.

Re-writing the quantities in terms of coulombs, joules and seconds, the battery e.m.f. is 6 joules per coulomb and the current is 4 coulombs per second.

∴ the charge that passes = 4 × 15
= 60 coulombs
∴ the heat produced = 6 × 60
= 360 joules

2: Resistance

2.1 Ohm's law

An electric current is caused to flow through a conductor by applying a potential difference across it. Ohm was the first person to distinguish clearly between the two concepts, p.d. and current, and to investigate the relation between them.

The current in a conductor depends on a number of factors besides the p.d. across it; the most important of these is the temperature. But such things as the elastic strain to which the conductor is subjected, the illumination of its surface, or indeed almost any of its physical conditions, may also affect the current. However, Ohm found that if all these were kept constant, the current in a given conductor is directly proportional to the p.d. across it.

For electrolytic conduction the connection between p.d. and current is complicated by e.m.f.'s that arise at the electrodes, accompanying the conversion of electrical energy to chemical energy (p. 40). The same situation arises in other cases where the conductor is itself the site of an e.m.f. —for instance, the armature coil of a motor, or a coil in which a rapidly changing current induces an e.m.f. by self-induction (p. 103). For these reasons *Ohm's law* only applies strictly to steady currents in metallic conductors which are not themselves the site of any e.m.f.

> **Ohm's law.** *A steady current flowing through a metallic conductor, which is not itself the site of an e.m.f., is proportional to the potential difference between its ends provided the temperature and other physical conditions are constant.*

A consequence of Ohm's law is that the quantity

$$\frac{\text{potential difference}}{\text{current}}$$

is a constant for a given metallic conductor under steady physical conditions. It is known as its *resistance*. Even when we are dealing with a device (e.g. radio valve or discharge tube) for which Ohm's law does not apply, it is still often convenient to define this quantity as the resistance of the device; in such a case the resistance depends on the current.

> *The* **resistance** *of a conductor which is not itself the site of an e.m.f. is*
>
> $$\frac{\textit{the potential difference across it}}{\textit{the current through it}}$$
>
> *when a steady current is flowing; for a conductor that obeys Ohm's law this quantity is constant for all currents if the temperature and other physical conditions are steady.*

A knowledge of the resistances of the conductors making up a circuit enables us to calculate the distribution of currents and p.d.'s in it. The practical unit of resistance is called the *ohm* (abbreviated to 'Ω'—the Greek letter omega); it is defined so that the resistance will be in ohms when the p.d. is in volts and the current in amps.

> *The* **ohm** *is the resistance of the conductor through which a steady current of 1 amp passes when a potential difference of 1 volt exists across it, the conductor not itself being the site of an e.m.f.*

It equally follows from Ohm's law that the reciprocal quantity

$$\frac{\text{current}}{\text{potential difference}}$$

is a constant for a given conductor; this is called its *conductance*. The unit is the ohm^{-1} or *reciprocal ohm* (sometimes labelled the *mho*).

The experimental testing of Ohm's law is a matter of some difficulty. In principle the apparatus shown in Fig. 2.1 might be used.

The variable resistor or rheostat S can be used

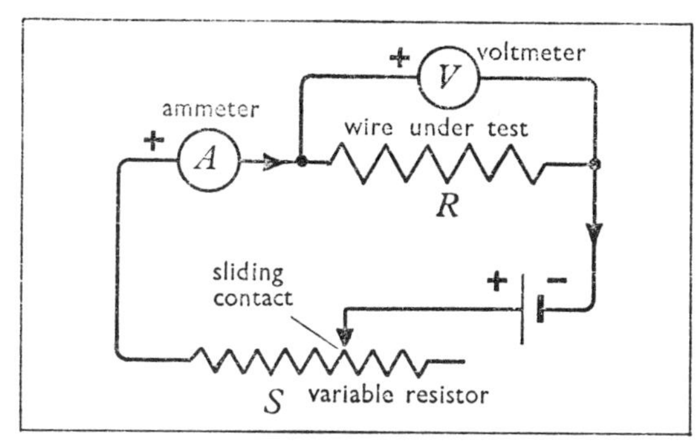

Fig. 2.1 The experimental test of Ohm's law

to adjust the current through the wire R under test, while the ammeter and voltmeter measure the values of current and p.d. for various settings of the rheostat. But it is not possible to use ordinary moving-coil ammeters and voltmeters for this purpose, since the scales of these instruments are constructed on the assumption that Ohm's law is correct; such an experiment would, therefore, do no more than confirm the correct calibration of the instruments used! We must instead find forms of ammeter and voltmeter whose action in no way assumes the truth of the law. For instance, the current may be measured by its magnetic effect, e.g. using a current balance (p. 8). If moving-coil ammeters and voltmeters are used, their scales must be calibrated by comparison with other instruments whose readings do not depend on Ohm's law.

It is instructive to repeat this experiment not only for coils of wire at steady temperatures for which Ohm's law holds, but also for a number of devices for which it does not hold. For example:

(*i*) *A fine wire*, such as an electric bulb filament. In this case the temperature changes considerably with the current, and Ohm's law does not apply.

(*ii*) *A selenium rectifier* (p. 329). Not being a metallic conductor, the laws controlling the current through such a device are different, and Ohm's law does not apply.

(*iii*) *An electric cell*—a Daniell cell is best (p. 41). Such a device is the site of an e.m.f. and again Ohm's law does not hold.

With each of the last two the experiment should be performed with the current flowing in both directions through the device. In all three cases graphs should be plotted of current I against p.d. V.

2.2 Internal resistance of cells

When a current is taken from a cell (or other source of e.m.f.), it is found that the p.d. across it falls. This is just what we should expect from considerations of energy. A numerical example will make this clear:

Suppose we have a cell of e.m.f. 1·5 volts. This means that for each coulomb of electric charge that passes through it 1·5 joules of energy are made available to the complete circuit in which it is connected. But some of this energy is needed to drive the current through the material of the cell itself; let us suppose that 0·3 joule per coulomb is so used. This means that 1·2 joules of energy are available in the circuit *outside* the cell for each coulomb of charge driven round it—in other words, that the p.d. across it is 1·2 volts.

The way in which the p.d. across a cell varies with the current through it may be investigated with the arrangement shown in Fig. 2.2.

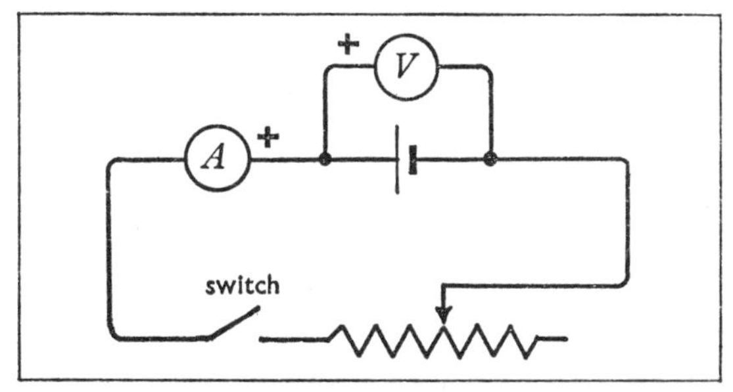

Fig. 2.2 Finding how the p.d. across a cell varies with the current through it

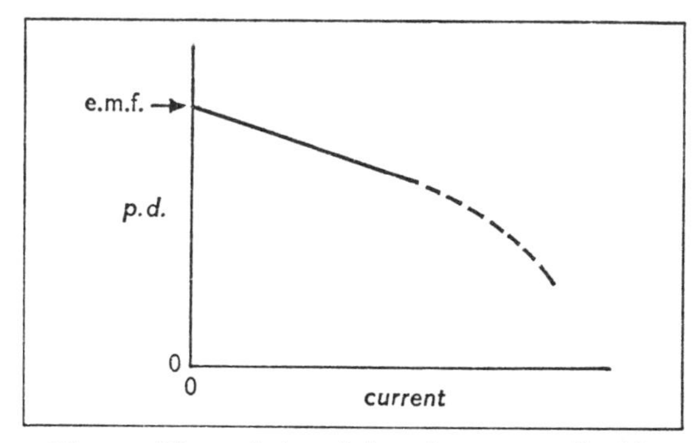

Fig. 2.3 The variation of the p.d. across a cell with the current through it

With most types of cell the results are not at all uniform; the chemical changes occurring in a cell when a current flows lead to changes in its e.m.f.—

an effect known as *polarization* (p. 40). The p.d. across the cell therefore depends not only on the current taken, but also on the length of time for which it has been flowing and on the previous treatment of the cell. However, a graph of p.d. against current is of the general form shown in Fig. 2.3. For zero current the p.d. is equal to the e.m.f.; the subsequent *drop in p.d.* is found to be approximately proportional to the current. This is a result similar to Ohm's law for metallic conductors; and in the same way we may go on to state that

$$\frac{\text{the drop in p.d.}}{\text{the current}}$$

is a constant (approximately) for the cell, called its *internal resistance.* The approximate character of this quantity should be noted; attempts to measure it with superb accuracy are futile—the results obtained often vary for a given cell by as much as 20% from one occasion to another. The difficulty arises because of the uncertainty in deciding exactly how much of the drop in p.d. is due to changes in e.m.f. and how much to the effects of internal resistance. The fact is that for most types of cell there is no way of predicting precisely what the p.d. will be except when the current is very small and the p.d. is nearly equal to the e.m.f. However, in spite of this there are many occasions when even an inexact knowledge of the internal resistance of a cell is very useful. For instance, when it is small compared with the total resistance of the circuit, the resulting uncertainty in calculating the current may be quite negligible. This is almost invariably the case with an accumulator, whose internal resistance is very small (about 0·005 ohm).

Internal resistance. *When a cell of e.m.f. E is supplying a current I, the p.d. V between its terminals is less than E by an amount proportional to I;*

$$\therefore (E - V) \propto I$$
$$\text{or} \quad (E - V) = Ir$$

where r is a constant for the cell known as its internal resistance.

From this definition we have—

$$E = V + Ir$$

If the cell is in a circuit connected to a resistance R,

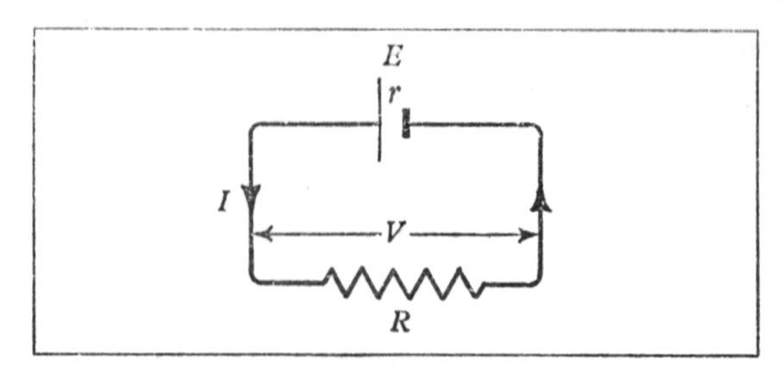

Fig. 2.4 The analysis of a complete circuit

as shown in Fig. 2.4, then we also have

$$V = IR$$
$$\therefore E = IR + Ir = I(R + r)$$

This is a relation of the same form as those obtained in defining resistance and internal resistance. Compare the three formulae obtained:

$$V = IR \quad \text{(outside the cell)}$$
$$(E - V) = Ir \quad \text{(for the cell itself)}$$
$$E = I(R + r) \quad \text{(the complete circuit)}$$

In each case we have the form—

electrical pressure = *current* × *resistance*

This single relation, correctly applied, is the only one that need be memorized in order to solve all ordinary circuit problems.

Example. A dry cell of e.m.f. 1·5 volts and internal resistance 2 ohms is connected to a torch bulb whose resistance at its operating temperature is 5·5 ohms. What is the p.d. across the cell?

Considering the complete circuit, we have

e.m.f. = current (I) × total resistance
$\therefore 1{\cdot}5 = I \times 7{\cdot}5$
$\therefore I = 0{\cdot}2$ amp

Now considering the cell only, we have

Drop in p.d. = current × internal resistance
$\therefore$ Drop in p.d. = 0·2 × 2 = 0·4 volt
$\therefore$ p.d. across cell = 1·1 volts

Alternatively we could reason:

p.d. across bulb = 0·2 × 5·5
= 1·1 volts

which is bound to be the same as the p.d. across the cell.

The following method for measuring the internal resistance of a cell has the merit that the occurrence of polarization in the cell can be detected, and some allowance may be made for it. The apparatus is connected as in Fig. 2.5. The voltmeter must

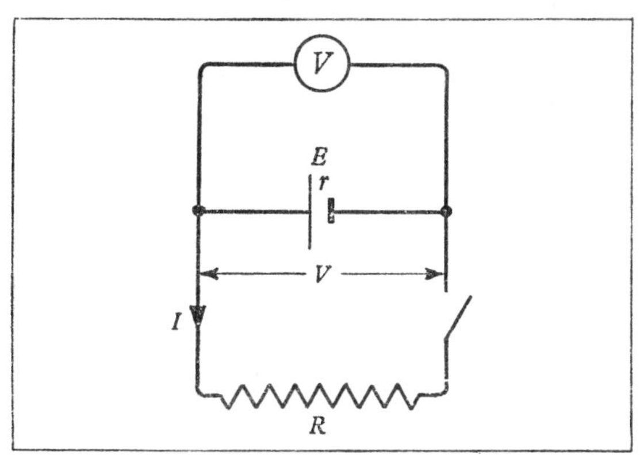

Fig. 2.5 Measuring the internal resistance of a cell

have a resistance much greater than the internal resistance r of the cell, so that a true reading of the e.m.f. E is obtained when the switch is open. The known resistance R must be of the same order of magnitude as the internal resistance of the cell, but must be large enough to avoid drawing an excessive current from it; a value between 2 and 5 ohms is probably suitable for a small dry cell. Having read the e.m.f., the switch is closed and the p.d. across the cell and resistor is read. We then have

for the whole circuit

$$E = I(R + r)$$

and for the external circuit

$$V = IR$$

Dividing,

$$\frac{E}{V} = \frac{R + r}{R} = 1 + \frac{r}{R}$$

$$\therefore r = R\left(\frac{E}{V} - 1\right)$$

The occurrence of polarization will be apparent from the steady fall of the reading of the voltmeter as the switch is left closed. If this is very marked, it may be allowed for by plotting a graph of p.d. against time from the moment of switching on (Fig. 2.6). The graph is produced back to the axis, where the intercept gives the value of the p.d. immediately after switching on, before any polarization has occurred.

The same technique may be used to measure the internal resistance when the cell is polarized. The readings of p.d. are continued with the switch closed until they are nearly constant; the switch is then opened, and the readings continued for a short time, while the cell recovers from its polarization. By producing the curve back the value of the e.m.f. before the cell had started to recover from its polarization may be deduced. From this and the corresponding value of V the internal resistance may be calculated as before. It will usually be found that polarization does not much affect the internal resistance of a cell.

2.3 Combinations of resistors

Many circuits contain networks of resistors and

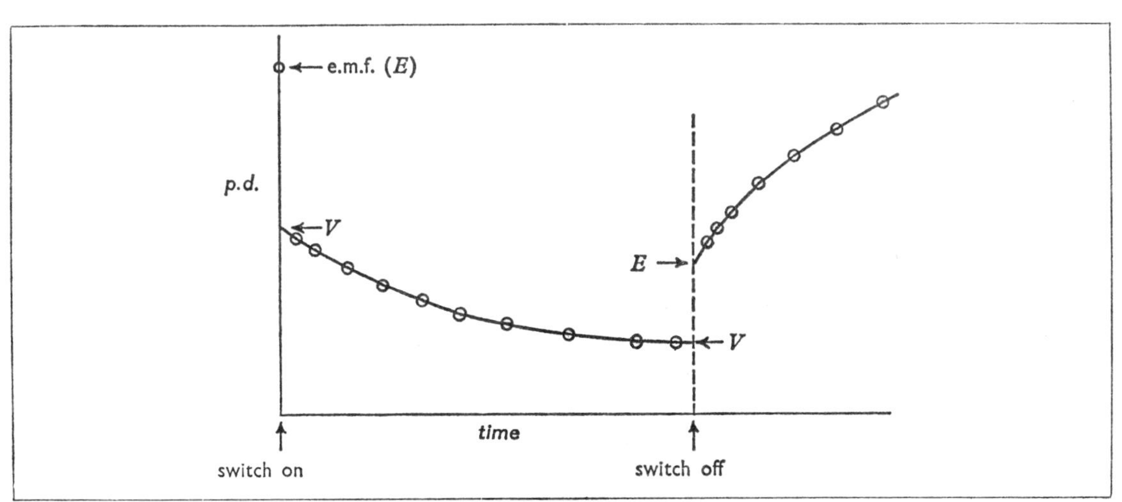

Fig. 2.6 Graph of p.d. against time, showing polarization effects

cells; and we must be able to calculate the combined effective resistance of various arrangements.

(*i*) *Resistors in series* (Fig. 2.7). In this arrangement the same current I flows through all the resistors, R_1, R_2 and R_3.

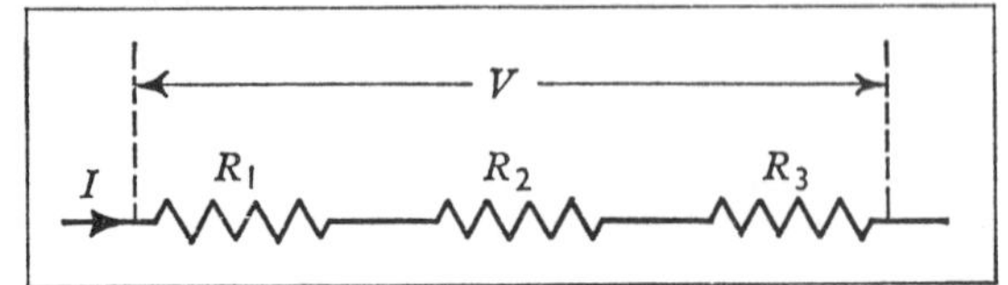

Fig. 2.7 Resistors in series

The combined effective resistance R is given by—

$$R = \frac{V}{I}$$

where V is the total p.d. across the combination. V is the sum of the separate p.d.'s across the individual resistors.

$$\therefore\ V = IR_1 + IR_2 + IR_3 = I(R_1 + R_2 + R_3)$$
$$\therefore\ R = R_1 + R_2 + R_3$$

(*ii*) *Resistors in parallel* (Fig. 2.8). In this arrangement the same p.d. V exists across all the resistors, R_1, R_2, R_3.

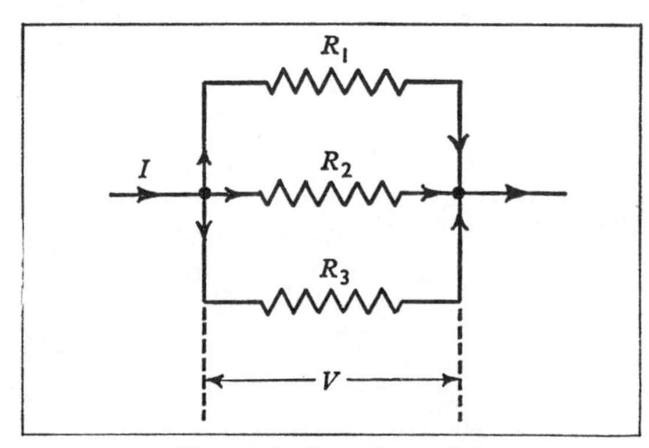

Fig. 2.8 Resistors in parallel

The combined effective resistance R is again given by—

$$R = \frac{V}{I}$$

where I is the total current through the combination. I is the sum of the separate currents through the individual resistors.

$$\therefore\ I = \frac{V}{R_1} + \frac{V}{R_2} + \frac{V}{R_3} = V\left(\frac{1}{R_1} + \frac{1}{R_2} + \frac{1}{R_3}\right)$$
$$\therefore\ \frac{I}{V} = \frac{1}{R} = \frac{1}{R_1} + \frac{1}{R_2} + \frac{1}{R_3}$$

Notice that this result may be expressed by saying that the combined *conductance* of conductors in parallel is equal to the sum of their separate *conductances*, since conductance is the reciprocal of resistance.

In two particular cases the last formula may be simplified, making for greater speed of calculation:

(*a*) The combined resistance of *n equal resistances* R in parallel is given by—

$$R = \frac{R}{n}$$

(*b*) For *two resistors* only in parallel, re-arrangement gives—

$$R = \frac{R_1R_2}{R_1 + R_2} = \frac{\text{product}}{\text{sum}}$$

The same rules may be used to calculate the combined effective internal resistance of an arrangement of cells. The rules for obtaining the combined e.m.f. of a complete battery have already been discussed (p. 12). Consider for instance a battery of three identical cells each of e.m.f. 1·5 volts and internal resistance 0·6 ohm.

Joined (*a*) in series, their combined e.m.f. is 4·5 volts and the internal resistance is 1·8 ohms.

Joined (*b*) in parallel, the e.m.f. is the same as that of one cell, 1·5 volts, and the internal resistance is equal to that of three 0·6-ohm resistances in parallel, namely 0·2 ohm.

★ With these rules it is possible to deduce the effective resistance of most arrangements of conductors, and to calculate the currents in many types of circuits. Cross-connected arrangements (such as unbalanced forms of the Wheatstone networks and potentiometer circuits discussed in Chapter 4) cannot generally be handled in this way, but must be treated by means of more advanced circuit theorems; see Kirchhoff's laws (p. 31).

2.4 Resistivity

The resistance of a conductor at a given temperature depends on its length and cross-sectional area and on the material of which it is made. It is easily shown by experiment that the resistance of a uniform conductor made of any given material is directly proportional to its length and inversely proportional to its cross-sectional area. On theoretical grounds this can be seen to follow from the results of the previous section. A conductor of length l can be regarded as a set of conductors in series each of unit length; suppose each such conductor has a resistance r. Then the total resistance is lr, which is proportional to the length. Likewise a conductor of cross-sectional area A can be regarded as a collection of A conductors in parallel each of unit cross-sectional area; suppose each conductor has a resistance r'. Then the total resistance is r'/A, which is inversely proportional to the cross-sectional area. Thus the resistance R of a conductor of length l and cross-sectional area A may be written in the form

$$R = \frac{\rho l}{A}$$

where ρ is a property of the material of the conductor. It is called its *resistivity* (or sometimes, *specific resistance*).

Resistivity. *The resistance R of a uniform conductor is directly proportional to its length l and inversely proportional to its cross-sectional area A;*

$$\therefore \; R \propto \frac{l}{A}$$

or $$R = \frac{\rho l}{A}$$

where ρ is a constant (for fixed temperature and other physical conditions) for the material of the conductor known as its resistivity.

Re-arranging this formula,

$$\rho = \frac{RA}{l}$$

This shows us that the dimensions of resistivity are those of [resistance × length]. When the length is measured in m and the cross-sectional area in m², the unit of resistivity is accordingly the *ohm m.* (Resistivities are also sometimes expressed in *ohm cm.* The dimensions of a conductor must then be expressed in *cm* to obtain its resistance in ohms from the formula $R = \rho l/A$.)

In order to measure the resistivity of a material we have first to select a wire of suitable resistance. Modern wire is manufactured to very high standards of uniformity of diameter, but we cannot rely on this in conducting the measurements. The diameter should be measured at several positions along the length of it—in case it tapers; and at each position in two directions at right angles—in case the cross-section is not exactly circular. If the figures differ very little, their mean is taken as the effective diameter. The accuracy of the value of resistivity obtained is limited chiefly by the measurement of the diameter. For example, suppose the wire used has a diameter of about 1 mm; using a micrometer screw-gauge, this should be found with a maximum error of 0·005 mm—i.e. ½% error. But this figure will be squared in the calculation of the area of cross-section. Therefore the resulting final error will be twice this, namely 1%. There is plainly no great difficulty in finding the other quantities needed to at least this accuracy: provided the resistance R is more than 1 ohm, a metre bridge method will suffice (p. 51), and, if the length l is more than 50 cm, an error of 2 mm in measuring this will be relatively unimportant. If the material under test is a pure metal, its temperature must be controlled and recorded; an uncertainty of 1°C could be significant. However, in the case of the special resistance alloys—constantan, eureka and manganin (p. 21)—this precaution is unnecessary, since their resistivities scarcely change at all with temperature.

Silver is the best conductor, closely followed by copper. The resistivity of a pure metal is considerably increased by even small traces of impurity; for this reason the copper used for electrical connecting wire is always electrolytically purified (p. 34). Aluminium has a resistivity about twice that of copper. But since its density is about a third as much, an aluminium conductor can be much lighter and cheaper than a copper one. The overhead power lines of the national grid system are made of aluminium; the lighter cables enable less expensive pylons to be used to support them. The cables carry a steel core to give them the tensile strength to enable very long loops to be suspended between pylons; the aluminium strands are twisted around this core.

The table shows the resistivities of a number of pure metals and alloys. The resistivities of non-metallic substances are vastly greater. Almost all of them are more than 10^4 ohm m; and those materials used for insulation of electrical circuits have resistivities between 10^{13} and 10^{16} ohm m. Intermediate between these extremes comes the important class of materials known as *semiconductors*—germanium, silicon, etc. In these substances only a very small proportion of the valence electrons are 'free' to move through the crystal lattice; the resistivity of germanium is 0·65 ohm m, and of silicon $2{\cdot}3 \times 10^3$ ohm m. However, these figures are only obtained with the purest specimens; and the addition of minute traces of impurities may increase the number of free charges in the crystal lattice considerably, and so *reduce* the resistivity (p. 324). By contrast, impurities in a metal do not much affect the number of conduction electrons, but have a marked effect on their freedom of movement and therefore *increase* the resistivity.

It is sometimes convenient to deal with the reciprocal of resistivity; this quantity is called the *conductivity* of a material. It is chiefly used in studying the behaviour of electrolytes. At low dilutions the conductivity is proportional to the concentration of ions in a solution; and the measurement of conductivity thus provides a valuable tool for studying the properties of ions.

Resistivities and temperature coefficients of resistivity of various common metals

Metal	Resistivity (ohm m)	Temp. coeff. of resistivity (per °C)
	Values at 20°C	
Pure metals		
Silver	$1{\cdot}63 \times 10^{-8}$	0·0040
Copper	1·69	0·0043
Aluminium	3·21	0·0038
Gold	2·42	0·0040
Platinum	11·0	0·0038
Iron (soft)	14·0	0·0062
Lead	20·8	0·0043
Mercury	94·1	0·0009
Alloys		
Nichrome	130×10^{-8}	0·00017
Eureka, Constantan	49	about 0·00002
Manganin	44	about 0·00002

The student should notice the use of suffixes in the words introduced in this chapter—

resist*or*	resist*ance*	resist*ivity*
conduct*or*	conduct*ance*	conduct*ivity*

This usage is spreading in scientific terminology, and its significance should be apparent from these examples.

'*-or*' signifies the *body* that possesses the property considered—in the above examples it is probably a metal wire.
'*-ance*' signifies the *property* that the body possesses.
'*-ivity*' signifies the corresponding *property of the material* of which the body is composed.

The physical basis of Ohm's law

We have already deduced an expression for the velocity of the charged particles in a conductor of cross-sectional area A carrying a current I.

$$v = \frac{I}{nAe} \qquad \text{(p. 9)}$$

where n is the number of charged particles per unit volume and e is their charge. It is reasonable to assume that the average velocity v is proportional to the *potential gradient* along the conductor (which is a measure of the electric field in it, p. 157). If the conductor is of length l and the p.d. across it is V, then

$$\text{potential gradient} = \frac{V}{l}$$

and we may therefore write

$$v = k.\frac{V}{l}$$

The constant k is the velocity for unit potential gradient (1 volt per m), sometimes called the *mobility* of the particles.

$$\therefore\ v = \frac{I}{nAe} = k.\frac{V}{l}$$

$$\therefore\ \text{the resistance} = \frac{V}{I} = \frac{l}{kneA}$$

Ohm's law therefore holds for a conductor provided both n and k are constant; i.e. we must suppose (i) that the number of particles in the conductor is unaffected by the current, (ii) that the average velocity v is indeed proportional to the potential gradient.

The resistivity ρ is then given by

$$\rho = \frac{1}{kne}$$

$$\therefore \text{ the conductivity} = kne$$

When the temperature of a metal is raised, the thermal vibrations of the atoms increase; this increases the interaction of the electrons with the crystal lattice, and therefore reduces k. On this account the resistance of a metal increases with temperature. On the other hand in electrolytes raising the temperature decreases the viscosity of the liquid, allowing the ions to move more freely. A temperature rise therefore increases the conductivity of an electrolyte (and so decreases its resistivity).

Semiconductors show a very considerable decrease of resistivity when the temperature rises. In these materials greater thermal vibration of the atoms 'frees' more electrons in the crystal lattice; and n is thus greatly increased. In some semiconductors the number of conduction electrons can also be increased by raising the p.d. These materials do not then obey Ohm's law; indeed the current in them may vary as the third or fourth power of the p.d. Substances with this property are used for making *surge suppressors.* These are devices for protecting equipment from sudden surges of current such as may occur at the moments of switching on and off. The surge suppressor is joined in parallel with the equipment to be protected; very little current flows through it at the normal p.d. But a sudden increase in p.d. causes an enormous rise in the suppressor current, and diverts the surge from the equipment.

2.5 Temperature coefficients

The resistivity of any pure metal is found to increase nearly uniformly with temperature. This variation is described with the aid of a temperature coefficient similar to the coefficients of expansion used in dealing with changes of length and volume.

> *The* **temperature coefficient of resistivity** *of a substance is the increase in its resistivity per deg. C rise of temperature divided by its resistivity at 0°C.*

If this temperature coefficient is α, and the resistivity at t°C is ρ_t, this definition may be expressed mathematically by—

$$\alpha = \frac{\rho_t - \rho_0}{\rho_0 t}$$

Re-arranging $\qquad \rho_t = \rho_0(1 + \alpha t)$

There is surprisingly little variation between the temperature coefficients of resistivity of many pure metals, as may be seen from the table on p. 20. Also the values of these coefficients occupy a range near the figure $3{\cdot}7 \times 10^{-3}$ per deg. C, which is the same as the coefficient of expansion of a gas (1/273 per deg. C). It follows that the resistivity of a pure metal is approximately proportional to its absolute temperature. This provides a very useful way of doing rough calculations of changes of resistance with temperature.

When measurements of resistivity are made over large temperature ranges it is found that the variation is not quite uniform. But for all pure metals a slight modification of the above expression can be made to agree very closely with the experimental values:

$$\rho_t = \rho_0(1 + \alpha t + \beta t^2)$$

where β is a coefficient that is very small in comparison with α.

For alloys the temperature coefficients of resistivity are generally less than for pure metals. For example, for nichrome (80% nickel, 20% chromium) it is about $0{\cdot}17 \times 10^{-3}$ per deg. C. This alloy is used for heating elements of electric fires, cookers, etc.; it has the property of forming a protective oxide layer when heated, which partly protects it against further oxidation at high temperatures and to some extent provides insulation between the coils of wire in the element. It is also commonly used in wire-wound resistors for electronic equipment, since its resistivity is unusually high. Eureka or constantan (60% copper, 40% nickel) and manganin (84% copper, 12% manganese, 4% nickel) are two alloys whose temperature coefficients of resistivity are very small; since they also have quite high resistivities, they are useful for constructing standard resistors. Semiconductors have large *negative* temperature coefficients of resistivity.

At temperatures near to the absolute zero some metals are found to lose the property of resistance altogether and become perfect conductors. They are then said to be *superconducting*. The onset of superconductivity occurs at a sharply defined transition temperature, which is lowered if the specimen is placed in a magnetic field. Electric currents have been maintained in rings of superconducting material for periods of several months, conclusively demonstrating the complete absence of resistance in metals in this condition. Above the transition temperature the same currents would have ceased in a few seconds in even the largest rings. Fig. 2.9 shows how the resistivity of such a metal varies from the absolute zero upwards.

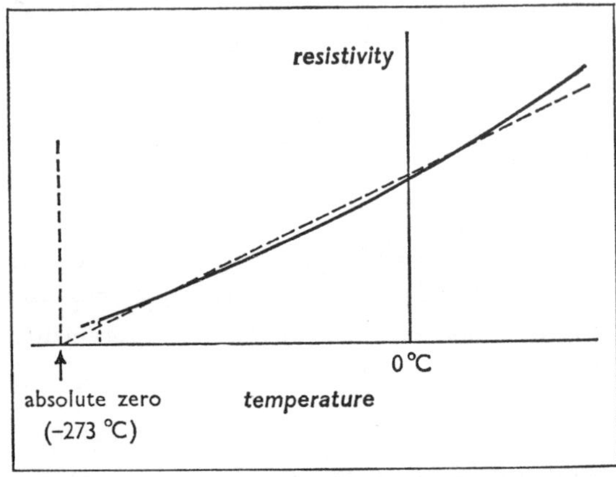

Fig. 2.9 The variation of resistivity with temperature

So far we have discussed only the temperature coefficient of *resistivity*. Most practical applications are concerned rather with the *temperature coefficient of resistance.* Since it is expressed as a fractional change, the latter coefficient also is a property of the material rather than of any particular specimen of it; i.e. the same value will be obtained for all specimens of a given material.

The **temperature coefficient of resistance** *of a substance is the increase in resistance of a specimen of it per deg. C rise of temperature divided by its resistance at 0°C.*

The difference between the two coefficients arises from the change in dimensions of a conductor with temperature. Since the thermal expansion of a metal is very small, the difference is not large and can be ignored in many applications; for copper it amounts to about ½%, which is not always negligible.

Any of the methods described in Chapter 4 for the measurement of resistance may be adapted to find the temperature coefficient of resistance of a specimen. The coil of wire is mounted inside a waterproof container, and immersed in a water bath. The resistance is then measured at a number of temperatures covering a suitable range. At each temperature time must be allowed for the wire to reach the temperature of the bath—i.e. the readings of resistance should be repeated until they reach a steady value. The temperature coefficient of resistance α is then found from the relation

$$\alpha = \frac{R_t - R_0}{R_0 t}$$

In solving problems using these temperature coefficients care must be taken to avoid unjustified approximations. The change of resistance with temperature is often rather large (about 40% between 0°C and 100°C); and it is important for the calculation to be based on the resistance at 0°C, as required by the definition of the temperature coefficient. The following example shows the method to be used.

Example. A copper coil has a resistance of 50·0 ohms at 20°C. What is its resistance at 80°C? (The temperature coefficient of resistance of copper $= 4{\cdot}28 \times 10^{-3}\ °C^{-1}$)

Using symbols as above—

$$R_{20} = R_0(1 + 20\alpha)$$

Also $$R_{80} = R_0(1 + 8\rho\alpha)$$

Dividing,

$$\frac{R_{80}}{R_{20}} = \frac{1 + 80\alpha}{1 + 20\alpha}$$

$$\therefore\ R_{80} = \frac{50(1 + 80 \times 4{\cdot}28 \times 10^{-3})}{1 + 20 \times 4{\cdot}28 \times 10^{-3}}$$

$$= \frac{50 \times 1{\cdot}342}{1{\cdot}086} = 61{\cdot}9 \text{ ohms}$$

An approximate method of calculation, such as is often used in problems on thermal expansion (where the coefficients are much smaller), would give—

increase in resistance

$$= \alpha \times (\text{initial resistance}) \times (\text{temp. rise})$$
$$= 4{\cdot}28 \times 10^{-3} \times 50 \times 60$$
$$= 12{\cdot}9 \text{ ohms}$$
$$\therefore\ R_{80} = 62{\cdot}9 \text{ ohms}$$

The approximation clearly introduces a significant error. But to an accuracy of 2% this method is quicker and generally adequate.

2.6 Power calculations

The circulation of an electric current is a process of conversion of energy (p. 10). In the source of e.m.f. electrical energy is produced continually from some other form of energy; while throughout the circuit it is converted again to other forms. In particular, in any of the resistors of the circuit the flow of current results in the production of heat energy. We shall now calculate expressions for the power produced in electric circuits.

Consider a resistance of R ohms through which a current of I amps is flowing and across which there is a p.d. of V volts (Fig. 2.10).

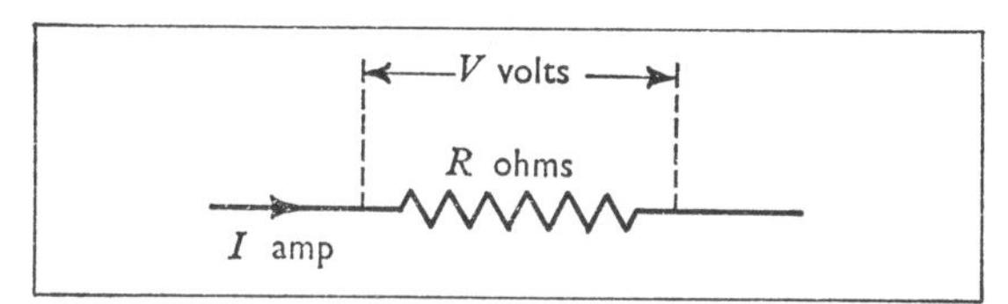

Fig. 2.10 The power dissipated in a resistor

Since there is a current of I amps flowing through the conductor, this means that I coulombs of electric charge are flowing through it per second. Since there is a p.d. of V volts across it, this means that V joules of energy (in this case heat) are produced in the conductor per coulomb of charge flowing through it. Hence, VI joules of energy are produced per second. Therefore, the power production W in the conductor is given by

$$W = VI \text{ watts}$$

(since W watts $= W$ joules/second)

Now from the definition of resistance we have

$$R = \frac{V}{I}$$

substituting for V,

$$W = I^2R$$

This expresses the result first discovered experimentally by Joule:

Joule's law. *The rate of production of heat by an electric current in a given conductor at a given temperature is proportional to the square of the current flowing in it.* ($W \propto I^2$)

Substituting also for I,

$$W = \frac{V^2}{R}$$

Thus we have three alternative expressions for the power production W:

$$W = VI = I^2R = \frac{V^2}{R}$$

The last two of these formulae can only apply to the production of *heat* in resistors. But the first formula, not involving the resistance R, may be used to give the rate of production of energy of all forms in any circuit-component in which a current I flows at a potential difference V. For example, if a current of 0·5 amp is flowing through the armature of a small motor at a p.d. of 20 volts, the total power supplied is 10 watts; this includes heat and mechanical energy. If the armature resistance is 12 ohms, then

the rate of production of *heat* $= I^2R = 3$ watts

$\therefore$ the rate of production of *mechanical energy* $= 10 - 3 = 7$ watts

Commercial calculations of electrical energy

The calculation of quantities of electrical energy for commercial purposes is most conveniently carried out by means of a special unit devised for the purpose, called the *kilowatt-hour* (kW h). In Great Britain it is defined by Act of Parliament as *the Board of Trade Unit* of electrical energy—often colloquially called 'the unit', for short.

The **kilowatt-hour** *is the quantity of energy used in 1 hour by a device of power 1 kilowatt.*

As the name of the unit implies, to calculate quantities of energy in kW h it is only necessary to multiply the power consumption in kilowatts (i.e. the rate at which the energy is being used) by the time in hours for which it is used:

Energy (in kW h) = power (in kW) × time (in hr)

The connection between the kilowatt-hour and the joule is found as follows: a device of power 1 kW uses energy at a rate of 1000 joules per second.

$\therefore$ energy used in 1 hr $= 1000 \times 3600$ joules

$\therefore$ 1 kW h $= 3{\cdot}6 \times 10^6$ joules.

★ The maximum power theorem

A source of e.m.f. E delivering a current I generates a total power W given by

$$W = EI$$

Some of this power is wasted producing heat in the source of e.m.f. and the remainder is delivered for use in the rest of the circuit (often called the *load*). It is of importance to know under what conditions the power w delivered to the load will be a maximum.

Suppose the source of e.m.f. has internal resistance r and the load is a resistance R. Then

$$I = \frac{E}{R + r}$$

$$\therefore \quad w = I^2R = \frac{E^2R}{(R + r)^2}$$

To find the value of R that makes this expression a maximum we need to differentiate it with respect to R and equate to zero.

$$\therefore \quad \frac{dw}{dR} = \frac{E^2(r - R)}{(R + r)^3}$$

$$= 0, \quad \text{when } R = r$$

It is easily verified that this stationary value is indeed a maximum.

A source of e.m.f. delivers maximum power when the resistance of the load is equal to the internal resistance of the source.

The condition for maximum *efficiency* is rather different. Thus

$$\text{efficiency} = \frac{w}{W} = \frac{R}{R + r}$$

To make this a maximum we need R to be large compared with r. In the maximum *power* condition the efficiency is 50%, i.e. half the power generated is wasted as heat in the source of e.m.f. When large power outputs are concerned, this situation is clearly unacceptable, and we shall aim rather at maximum efficiency—chiefly by making the internal resistance of the source small. However, when we need to extract the maximum possible value from a small source of power, it may be desirable to aim at the maximum power condition. An example of this situation is an amplifier delivering power to a loudspeaker (p. 245).

2.7 Types of resistor

Many different methods have been developed for constructing fixed and variable resistors suitable for the numerous purposes for which they are required in radio engineering and science. The most important types are described briefly below.

(*i*) *Carbon composition resistors.* By mixing carbon, which is a conductor, with non-conducting materials it is possible to produce a mixture which can have almost any value of resistivity. The usual mixture is of carbon black, resin binder and a refractory powder (of the kind used in making fire bricks); this is milled and sifted, and then pressed into short cylindrical blocks. Heating in a kiln solidifies the blocks. The ends are sprayed with metal, to which tinned copper connecting wires are soldered (Fig. 2.11).

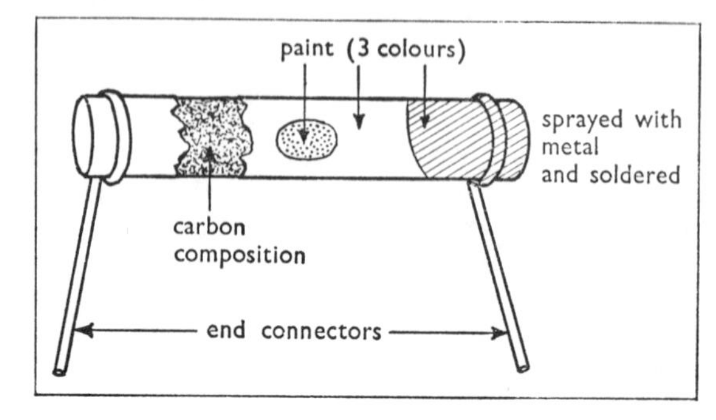

Fig. 2.11 A carbon composition resistor

The resistors are then measured and sorted into resistance 'groups'. They are sold as having 5%, 10% or 20% tolerance. Their stability is very poor; the values are likely to drift by about 5% in the course of a year, and any stress (such as excessive current or heating in other ways) may cause a permanent change in resistance of 30% or more. A further defect is that the resistance may vary with the p.d. applied, which is another way of saying that Ohm's law does not hold for this composition. However, these defects are quite unimportant for many radio purposes, and their cheapness and small size lead to their being widely used.

★ The values of low accuracy resistors are usually marked on them with the Standard Colour Code:

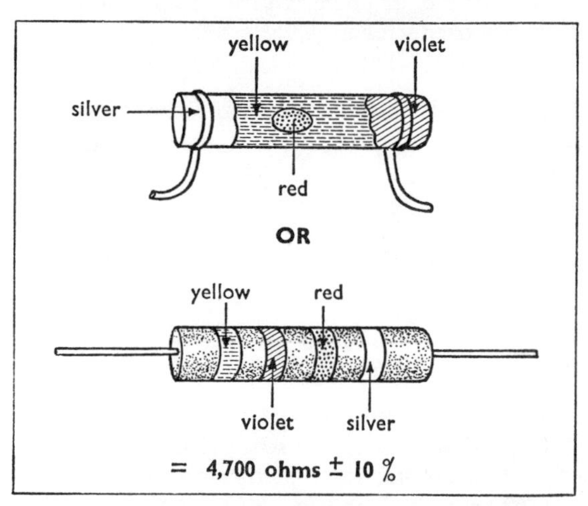

Fig. 2.12 The use of the Standard Colour Code

The table shows the digits denoted by each colour. (The sequence of colours is approximately that of the spectrum.)

Colour	*Digit*
Black	0
Brown	1
Red	2
Orange	3
Yellow	4
Green	5
Blue	6
Violet	7
Grey	8
White	9

The colour of the body of the resistor or of the first ring or dot from the end indicates the first digit; the colour of the end or of the second ring or dot gives the second digit; and a single spot in the middle or the third ring or dot shows the number of ensuing noughts (Fig. 2.12). A fourth colour is used to show the tolerance; gold 5%, silver 10%, and no colour 20%.

(*ii*) *Cracked-carbon resistors.* By heating small ceramic rods to a temperature of about 1000°C in methane vapour, decomposition or 'cracking' of the vapour occurs, and a uniform layer of carbon is deposited. The resistance between the ends at this stage is between 10 and 1000 ohms, depending on the thickness of the layer. This is then increased by anything up to 10,000 times by cutting a helical groove in the film, thereby altering the shape of the actual conducting path. In this way, the resistance may be adjusted to the required value to within 2% or better. Connection to the ends is made by metal caps fitted over them (Fig. 2.13). These resistors maintain their values to within about 2% under most ordinary stresses.

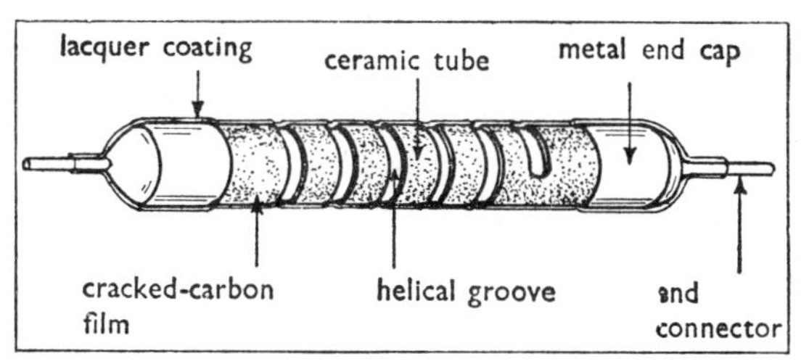

Fig. 2.13 A cracked-carbon resistor

(*iii*) *Wire-wound resistors.* For precision purposes this is the only form of resistor ever used. Two common designs are shown in Fig. 2.14. Enamelled nichrome wire is used in the ordinary commercial type, since it has a very high resistivity and may be used in fine gauges without much danger of damage by corrosion; the overall accuracy and stability of such resistors is about 1%. Manganin is used for high-precision work, since it has a very low temperature coefficient of resistance, and also produces very low thermal e.m.f.'s (p. 45) when connected in a circuit with copper wires. The accuracy and stability can be better than 0·01% in ordinary use, and rather better than this by careful control of the temperature. The value of a *standard resistor* is sometimes adjusted to the exact value required by incorporating a large resistance in parallel with it; large variations may be made in the length of wire in the latter resistor to effect only small changes in the total resistance. (See No. 3 in the questions on the work of this chapter.)

Adjustable standard resistors are made by mounting a number of precision wire-wound resistors in a box with suitable plugs or switches to enable the

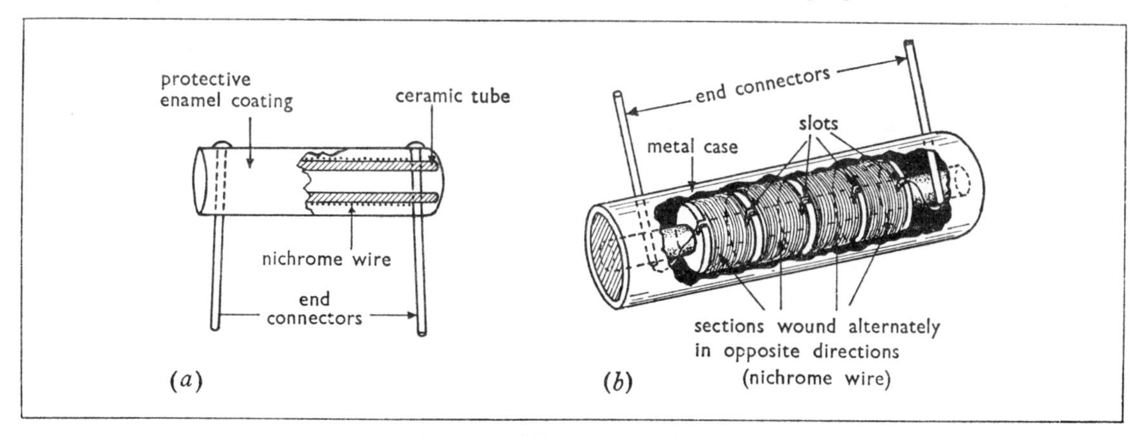

Fig. 2.14 Wire-wound resistors

required combination of resistors to be brought into the circuit. The student should examine the different designs in use in an electrical laboratory.

With any type of resistor, and particularly with standards, great care should be taken not to exceed the maximum power dissipation for which it is designed. Excessive heating invariably changes the resistance value and may damage the insulation or even fuse the wire. Even if the power dissipation is kept within the rated amount, adequate ventilation must be allowed to avoid undue rise of temperature.

Variable resistors

The simple ones used in radio circuits consist of a strip of carbon composition in the form of a broken circle to which a sliding connection is made by means of a rotating arm; fixed connections are made to the two ends of the strip. A similar wire-wound design has the wire wound on thin card, which is bent into an arc of a circle; the rotating arm makes contact with the turns along the edge of the card.

For dealing with larger currents a straight ceramic tube is wound with the resistance wire, and the sliding contact is carried on a metal bar parallel with the tube.

A variable resistor can be used in two ways in circuits:

(*a*) *As a rheostat*—for controlling the current in a low-resistance device; only one of the end connections is used in this case (Fig. 2.15*a*). To provide a large range of adjustment of current the rheostat must have a maximum resistance much greater than that of the rest of the circuit in series with it.

(*b*) *As a potential divider*—for controlling the p.d. across some device. For instance, in Fig. 2.15*b* is an arrangement that might be used for comparing the readings of two voltmeters. The battery is joined across the whole length of resistance wire. Any required fraction of the total p.d. can then be tapped off between the sliding contact and one end of the wire. This arrangement enables the p.d. to be varied continuously right down to zero.

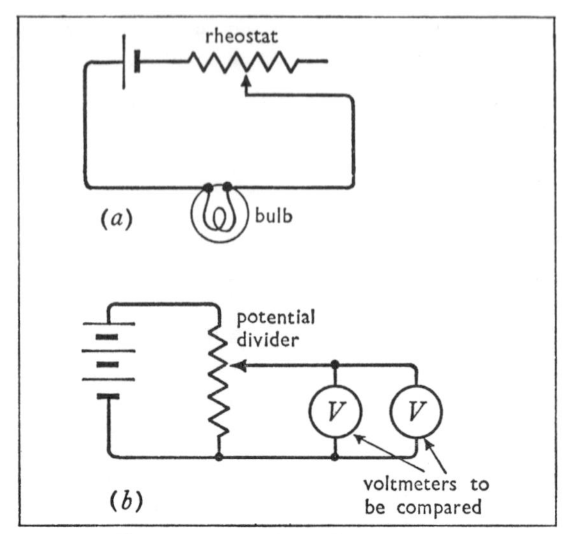

Fig. 2.15 Uses of a variable resistor:
(*a*) as a rheostat to control a current;
(*b*) as a potential divider to control p.d.

★ Wire-wound resistors for use in circuits with varying currents have to be wound *non-inductively* (p. 104); this means that the magnetic field produced by the coil must be a minimum. With high-frequency alternating currents it is also necessary to reduce the effective capacitance between the ends of the coil (p. 176); this means that the electric field between the ends must be a minimum, or in other words that the ends of the coil (between which there may be a high p.d.) must be kept well separated from each other.

The simplest form of non-inductive winding is made by doubling the wire to be used, and then winding the doubled piece on a suitable core (Fig. 2.16). The two halves of the coil then produce equal and opposite magnetic effects. But this arrangement is not suitable for high frequencies, since the capacitance between the ends is large. Fig. 2.14*b* shows a non-inductive winding in sections; the sections are wound alternately in opposite directions.

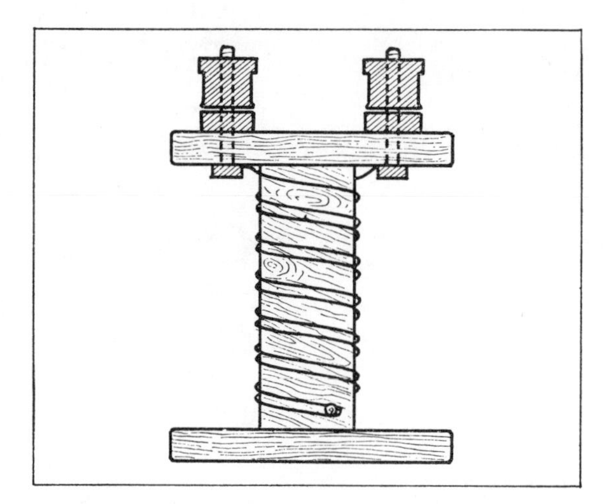

Fig. 2.16 A simple non-inductive winding for a resistance coil

2.8 Alternating currents

We have dealt so far only with cases in which the currents are steady and in one direction only: this is called *direct current* (d.c.). For large-scale power distribution there are however many advantages in using *alternating current* (a.c.). In this case the direction of the e.m.f. acting in the circuit reverses many times a second and so therefore does the current.

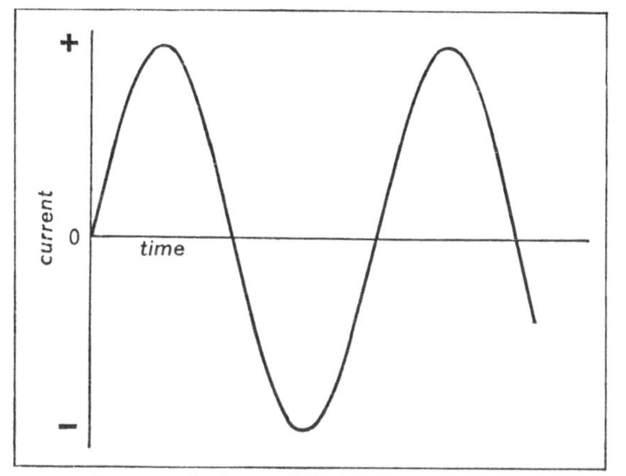

Fig. 2.17 A sinusoidal alternating current

The variation of current with time (the waveform) may follow almost any repetitive pattern; but the most important for practical purposes is the sinusoidal waveform shown in Fig. 2.17. This is the waveform to which the public electricity supply approximates closely. The sinusoidal waveform is so called because the variation of current i with time t is according to the *sine function*:

$$i = i_0 \sin \omega t$$

where i_0 and ω are constants. Since the maximum and minimum value of $\sin \omega t$ are $+1$ and -1, i_0 in this equation is the maximum value (or *peak value*) of the current in either direction. Also, a sine function goes through a complete cycle of variation when the angle advances by 2π radians; in the above equation ωt advances by 2π when t advances by $2\pi/\omega$. This is called the *period* of one oscillation. Its reciprocal, i.e. the number of oscillations per second, is called the *frequency* f of the oscillation.

Thus $$f = \frac{\omega}{2\pi}$$

$$\therefore \omega = 2\pi f$$

Frequencies are often expressed in *cycles per second*, abbreviated to *c/s*. But the internationally agreed name for the unit is now the *hertz* (abbreviated to Hz), and in this book we shall normally call it that. However, the student will often come across the older name in examination questions and elsewhere.

> *The* **hertz** *is the frequency of an oscillating quantity that performs one complete oscillation per second.*

The frequency of the public supply in Great Britain (and in most of Europe) is 50 Hz (or 50 c/s); in the U.S.A. a frequency of 60 Hz is used.

For many applications it is quite immaterial whether alternating or direct current is used. Electric heating and lighting for instance are unaffected by the direction of flow of the current. The power dissipation in a resistor fluctuates (100 times per second with the 50 Hz mains); but this is too rapid to allow much oscillation of the temperature of an electric fire element or of the filament of a bulb, though the variation of temperature is quite detectable. A slight hum can usually be heard from an electric fire at 100 oscillations per second (a note close to A flat on the musical scale); this is due to the slight alternate contractions and expansions of the element. Likewise the small variations in light intensity from an electric bulb can be detected by moving some object rapidly from side to side in the light; instead of seeing a continuous 'blur', a series of separate impressions is made on the eye, these being 1/100 s apart in time. The effect is even more marked with fluorescent lighting; the discharge through the mercury vapour actually stops when the applied p.d. falls below a certain value, and the only light then left is the weak continuing glow from the fluorescent coating of the tube. This causes much greater flicker than a filament lamp. But these effects are rarely of sufficient importance to make any practical difference. For some purposes direct current is essential; e.g. electro-plating, and other electrolytic processes, and electronic equipment—radio, television, etc. But in such cases simple and efficient devices have been developed for *rectifying* the alternating current to give direct current at any desired p.d. Alternating current motors are less efficient than d.c. ones; but this is not important with small and medium-sized motors. Only with the largest motors, such as those used in electric trains, has the d.c. motor any marked advantage.

But the generation and distribution of alternating current is so much cheaper and more efficient than direct current that its use for large-scale power purposes is now universal.

There are several ways in which we can express the *magnitude* of an alternating current:

(*i*) *The peak value* of the current is its maximum value (in either direction); in the equation

$$i = i_0 \sin \omega t$$

i_0 is the peak value.

(*ii*) *The mean value* or *half-cycle average* of the current is its average value taken over *half a cycle* while the current flows in one direction only (the average current over the whole cycle is of course zero).

(*iii*) *The root-mean-square (r.m.s.) value* of an alternating current is the square-root of the mean value of the square of the current taken over a whole cycle. At first sight this may seem an unnecessarily complicated way of specifying an alternating current, but it is in fact much the most useful way. The reason for this may be seen as follows. The power dissipated in a resistance is proportional to the square of the current

$$w = i^2 R$$

With alternating current the value of i and so of w varies from moment to moment.

$\therefore$ The mean value of w

$$= (\text{mean value of } i^2) \times R$$
$$= I^2 R$$

where

$$I = \sqrt{(\text{mean value of } i^2)} = \text{r.m.s. current}$$

If direct current I were passing through the resistance, the same power I^2R would be dissipated in it. From this point of view the r.m.s. value is the most convenient way of specifying an alternating current.

Alternating potential differences may be analysed in the same way: the *peak value*, the *mean value* and the *r.m.s. value* of an alternating p.d. are all employed on occasions, but the r.m.s. value is generally the most useful for the same reason as before. The power w dissipated in a resistance R when the p.d. across it is v is given by

$$w = \frac{v^2}{R}$$

$\therefore$ the mean value of w

$$= \frac{(\text{the mean value of } v^2)}{R}$$
$$= \frac{V^2}{R}$$

if $\quad V = \sqrt{(\text{mean value of } v^2)} = \text{r.m.s. voltage}$

Again, if a steady p.d. V were placed across this resistance, the same power V^2/R would be dissipated in it.

The practical result of all this is then very simple: provided we use r.m.s. values of current and p.d., the analysis of the currents flowing in networks of resistances and the calculations of power dissipated can be performed in exactly the same way as with direct currents. Likewise light bulbs and heating elements designed for a given d.c. voltage will work correctly on alternating supplies of the same r.m.s. voltage. Only when a circuit contains devices (e.g. transformers) whose functioning depends on the use of alternating currents, must different methods of calculation be employed. This is considered in detail in Chapter 12.

The relation between the peak and r.m.s. values of a current or p.d. depends on the waveform involved. For the case of a sinusoidal waveform it may be found as follows. The current i is given by

$$i = i_0 \sin \omega t$$

By definition, the r.m.s. current I is given by

$$I = \sqrt{(\text{mean value of } i^2)}$$
$$= i_0\sqrt{(\text{mean value of } \sin^2 \omega t)}$$

Using a well-known trigonometrical identity we have

$$\sin^2 \omega t = \tfrac{1}{2} - \tfrac{1}{2} \cos 2\omega t$$

The mean value of the second term on the right-hand side is zero, since the positive and negative parts of a sine or cosine curve are equal.

$\therefore$ the mean value of $\sin^2 \omega t = \tfrac{1}{2}$

$$\therefore I = i_0\sqrt{\tfrac{1}{2}} = \frac{i_0}{\sqrt{2}} = 0{\cdot}707 \times i_0$$

Fig. 2.18 shows the relation graphically. The full line is the graph of $\sin \omega t$; it oscillates both sides of zero between $+1$ and -1. The graph of $\sin^2 \omega t$ is shown dotted; it is always positive and oscillates between 0 and 1 at *twice* the frequency of $\sin \omega t$. Its mean value is $\tfrac{1}{2}$.

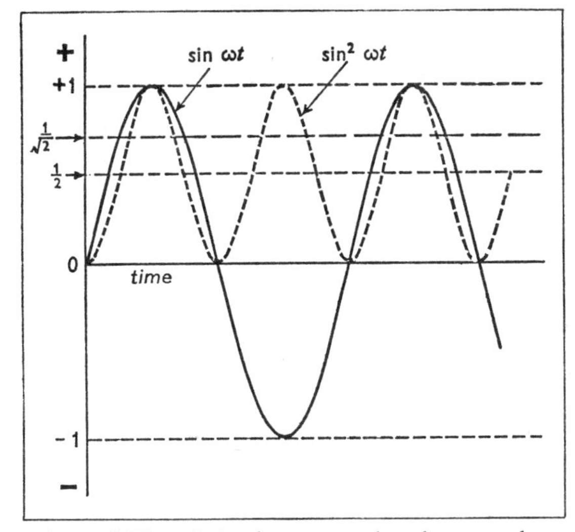

Fig. 2.18 The relation between peak and r.m.s. values

Similarly, the r.m.s. value V and the peak value v_0 of the p.d. are connected by

$$V = \frac{v_0}{\sqrt{2}} = 0{\cdot}707 \times v_0$$

★ The connection between the peak and mean values of current or p.d. again depends on the waveform concerned. For the particular case of a sinusoidal current with a period of oscillation T, we have

$$\text{half-cycle average} = \frac{\int_0^{\frac{1}{2}T} i\,dt}{\frac{1}{2}T}$$

$$= \frac{i_0\int_0^{\pi/\omega} \sin \omega t\,dt}{\pi/\omega}$$

$$= \frac{\omega i_0}{\pi}\left[-\frac{1}{\omega}\cos \omega t\right]_0^{\pi/\omega}$$

$$= \frac{2}{\pi} i_0$$

$$\therefore \text{mean current} = \frac{2}{\pi} i_0 = 0{\cdot}637 \times i_0$$

$$\text{Similarly,}\quad \text{mean p.d.} = \frac{2}{\pi} v_0 = 0{\cdot}637 \times v_0$$

Combining the two results we have

$$\frac{\text{r.m.s. current } I}{\text{mean current}} = \frac{0{\cdot}707 i_0}{0{\cdot}637 i_0} = 1{\cdot}11$$

$$= \frac{\text{r.m.s. voltage } V}{\text{mean voltage}}$$

Summary

The r.m.s. value of a current I is equal to the steady current that would dissipate power in a given resistance at the same average rate.

The r.m.s. voltage V is equal to the steady voltage which, joined across a given resistance, would lead to the same average dissipation of power in it. Except with special a.c. equipment, for a.c. circuits as for d.c., we have

$$\frac{V}{I} = R$$

and

$$W = VI = I^2R = \frac{V^2}{R}$$

In the particular case of *sinusoidal currents and voltages only*, whose peak values are i_0 and v_0, respectively

$$I = \frac{i_0}{\sqrt{2}} = 0{\cdot}707 i_0$$

$$V = \frac{v_0}{\sqrt{2}} = 0{\cdot}707 v_0$$

$$\text{mean current} = \frac{2}{\pi} i_0 = 0{\cdot}637 i_0$$

and

$$\text{mean voltage} = \frac{2}{\pi} v_0 = 0{\cdot}637 v_0$$

$$I = 1{\cdot}11 \times \text{mean current}$$

$$V = 1{\cdot}11 \times \text{mean voltage}$$

2.9 Fuses and filaments

When an electric current flows in a wire, heat is evolved in it and its temperature rises until the rate of loss of heat from its surface is equal to the rate at which the heat is supplied. The temperature then remains steady. The tungsten filament of an electric lamp is so fine that this final temperature is well in excess of 2000°C. On the other hand the wires used for joining up an electric circuit must be thick enough so that the temperature they reach is normally very little above that of the surroundings. The function of a fuse is to protect the circuit against the effects of excessive currents such as can arise from short circuits. It consists of a short length of relatively fine wire contained in a porcelain holder or a glass tube; its diameter is selected so that it quickly reaches its melting point and breaks the circuit if the current exceeds the maximum safe value.

The filament of a small electric lamp is of such low thermal capacity that it reaches its final steady temperature in a fraction of a second. It is important in a lighting circuit that the fuse should heat up much more slowly than the lamp filaments; otherwise the fuse might blow every time the lights are switched on! When a lamp is cold, the filament has less than a tenth of its normal resistance at its working temperature. Therefore for an instant after switching on there is likely to be a surge of current that may well be in excess of the normal maximum for the circuit. Fortunately the wire of the fuse takes longer to heat up than the much finer filaments and usually survives the initial surge.

The behaviour of fuse wires and filaments may be discussed mathematically as follows. Consider a wire of radius r and length l through which a current I is flowing. If its resistivity is ρ, its resistance R is given by

$$R = \frac{\rho l}{\pi r^2}$$

and the rate of production of heat W in it is given by

$$W = \frac{I^2 \rho l}{\pi r^2}$$

When the wire has reached a steady temperature, W is also the rate of loss of heat from its surface by conduction, radiation, etc. Assuming this is proportional to the surface area we can write

$$W = 2\pi r l . h$$

where h is the rate of loss of heat per unit area of surface; it is a function of the temperature and is the same for all wires of the same material losing heat in similar circumstances.

$$\therefore \quad \frac{I^2 \rho l}{\pi r^2} = 2\pi r l . h$$

$$\therefore \quad I^2 = 2\pi^2 r^3 \times \frac{h}{\rho}$$

Without assuming anything yet about the variation of h and ρ with temperature, this expression leads to two important conclusions:

(i) The temperature reached by a given wire depends only on the current through it, and for a given current is not dependent on the length l. This last statement needs some qualification. The ends of a wire can lose heat by conduction through its supports, and are usually cooler than the centre. The temperature of the central part is only independent of the length provided the wire is long enough for end effects not to matter.

(ii) For wires of a given kind heated to the same temperature

$$I^2 \propto r^3$$

$$\text{or} \qquad r \propto I^{2/3}$$

Thus if we know the current at which a given fuse wire melts, this result may be used to determine what diameter wire of the same metal should be used as a fuse at any other current. The same expression can also be used to compare the radii of the filaments of lamps to be run at the same temperature but with differing currents.

To obtain maximum luminous efficiency the temperature of a lamp filament should be as high as possible. Tungsten is therefore used, since this has a melting point of 3400°C. However, even tungsten cannot be used in a vacuum above 2100°C if serious evaporation of the filament is to be avoided. Most lamps are nowadays filled with an inert gas. This reduces evaporation and enables temperatures of 2300°C or more to be reached. Also the tungsten that does evaporate is now carried away by convection and deposited on the top of the bulb instead of producing a uniform blackening of the glass, as in the vacuum type. However, the heat loss from the filament is so increased by the presence of the gas that a straight wire would need to be excessively fine and fragile to reach the required temperature. To get over this difficulty a coiled filament is used. This has the effect of reducing the exposed surface area, and so enables the same temperature to be reached with a more robust wire. The same design also makes for a more compact source of light—a matter of some importance in projector and headlamp bulbs. Such bulbs are for preference low-voltage, high-current devices so that the filament can be as short and thick as possible.

★ To calculate the actual temperature reached by a wire it is necessary to make assumptions about the variation of the resistivity ρ and the heat loss per unit area h. For a pure metal we can assume that ρ is approximately proportional to the absolute temperature T. Thus

$$\rho = kT$$

where k is a constant. In problems of this kind it is

often assumed that the heat loss is entirely by radiation. In fact this is hardly ever the case even approximately, and certainly not with gas-filled bulbs. In vacuum bulbs the heat loss through the supports is usually considerable. However, with a certain type of vacuum bulb with a long filament a high proportion of the heat is lost by radiation, and we can reasonably ignore the rest.

According to Stefan's law the rate of radiation of heat H per unit area from the surface of a *black body* is given by

$$H = \sigma T^4$$

where σ is Stefan's constant. A tungsten filament is by no means a black body; but taking its emissivity as ε (it is usually less than 0·5) we may write approximately

$$h = \varepsilon\sigma T^4$$

This assumes also that the heat radiated back to the filament from its surroundings is negligible—a small error in comparison with the other approximations! Substituting these expressions for ρ and h we have

$$I^2 = 2\pi^2 r^3 \times \frac{\varepsilon\sigma T^3}{k}$$

$$\therefore\ T^3 = \frac{kI^2}{2\pi^2 r^3 \varepsilon\sigma}$$

$$\therefore\ T \propto I^{2/3}$$

The temperature may instead be expressed in terms of the p.d. V across the wire by substituting

$$I = \frac{V}{\rho l/\pi r^2}$$

Writing $\rho = kT$, the expression then simplifies to

$$T^5 = \frac{V^2 r}{2\varepsilon\sigma k l^2}$$

$$\therefore\ T \propto V^{2/5}$$

Since the power W is assumed to be proportional to T^4, the variation of W with I and V is then given by

$$W \propto (I^{2/3})^4 = I^{8/3} \simeq I^{2\cdot 7}$$

and

$$W \propto (V^{2/5})^4 = V^{8/5} = V^{1\cdot 6}$$

Eliminating W between the two expressions we find

$$I^5 \propto' V^3$$

or

$$I = A.V^{3/5}$$

where A is a constant. This relation may be tested experimentally by measuring pairs of values of V and I for a suitable bulb. Taking logs

$$\log I = \log A + 0{\cdot}6 \log V$$

A graph of $\log I$ against $\log V$ should then be a straight line of gradient 0·6.

Alternatively we may use the measurements to test the basic assumption that the power W $(= VI)$ is proportional to T^4. If the resistance R $(= V/I)$ is proportional to T, we may use it as a measure of the absolute temperature T of the filament. We can then test the relation

$$W \propto R^n$$

by plotting a graph of $\log W$ against $\log R$. With most bulbs the gradient n is very much less than 4.

2.10 Kirchhoff's laws

Some circuits cannot be broken down into sets of series and parallel combinations of conductors. It is then necessary to make use of generalized forms of the circuit laws already considered. These are known as *Kirchhoff's laws.*

Kirchhoff's laws. *When steady currents are flowing in a network of conductors:*
1. *The algebraic sum of the currents entering any point in the network is zero.*
2. *The algebraic sum of the products of current and resistance taken round any closed path in the network is equal to the algebraic sum of the e.m.f.'s acting in that path.*

The application of these laws to the solution of problems is best explained by considering an actual example.

Example. A dynamo of e.m.f. 30·0 volts and internal resistance 1·00 ohm is joined in parallel with a battery of accumulators of e.m.f. 24·0 volts and internal resistance 0·50 ohm; the two are connected to a resistance of 13·0 ohms (Fig. 2.19). Calculate the current in each part of the circuit.

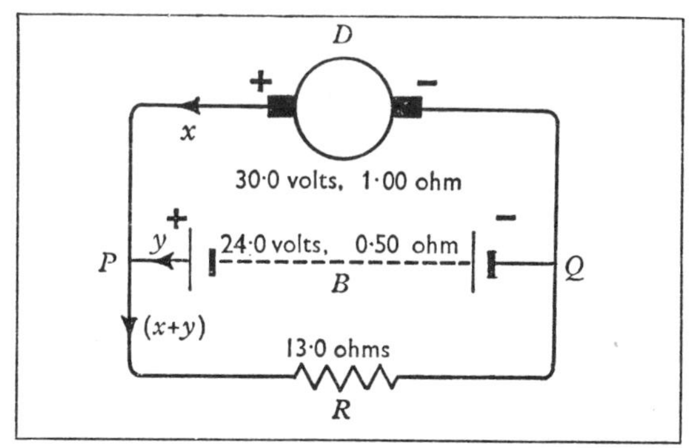

Fig. 2.19 A problem illustrating the use of Kirchhoff's laws

Let the currents through the dynamo and battery be x and y, respectively, in the directions shown in the diagram. Kirchhoff's first law applied to the point P shows that the current leaving P must be equal to that entering it; i.e. the current through the resistor must be $(x + y)$.

Applying the second law—

(i) to the closed path $\overrightarrow{\text{DPRQD}}$—

$$x + 13(x + y) = 30$$
$$\therefore\ 14x + 13y = 30$$

(ii) to the closed path $\overrightarrow{\text{BPRQB}}$—

$$0{\cdot}5y + 13(x + y) = 24$$
$$\therefore\ 13x + 13{\cdot}5y = 24$$

The solution of these simultaneous equations is

$x = 4{\cdot}6$ amps (through the dynamo)
$y = -2{\cdot}7$ amps (through the battery)

$\therefore$ current in the resistance $= x + y = 1{\cdot}9$ amps

The negative value obtained for y indicates that the current is in the *opposite* direction to that shown in the diagram—i.e. the battery is *charging*.

The basis of Kirchhoff's laws

(*i*) *The first law.* We have defined an electric current as a flow of electric charge, which we have assumed to be indestructible. If then the first law did not apply at some point of a circuit, the rates at which charge entered and left the point would be unequal, and there would be a steady accumulation or loss of charge there. This apparently does not happen. Closer examination shows that there is here a fundamental issue to be settled by experiment. The absolute measurement of a current is conducted in terms of the magnetic field it produces—e.g. by using a current balance; it is assumed that the current is *proportional to the magnetic field.* The validity of our concept of a current as a flow of charge therefore rests on demonstrating that Kirchhoff's first law applies to currents *measured magnetically.*

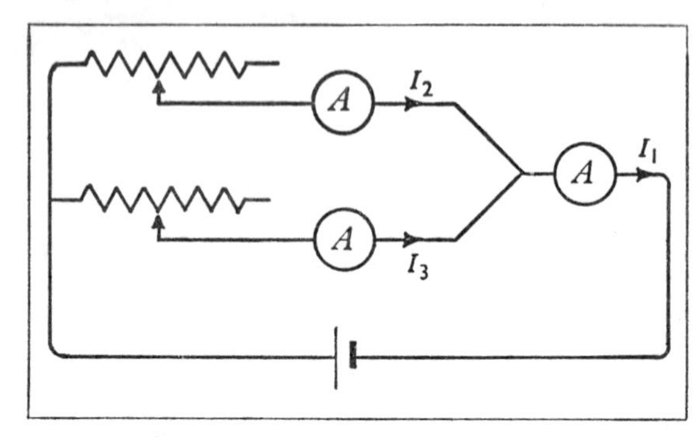

Fig. 2.20 Testing Kirchhoff's first law

This fundamental experiment may be performed with the apparatus in Fig. 2.20. We expect to find for all settings of the rheostats

$$I_1 = I_2 + I_3$$

(*ii*) *The second law.* Suppose one branch of some network consists of a source of e.m.f. E of internal resistance r.

If the current through it is I, then the p.d. V between the ends of it is given by

$$V = E - Ir$$

The algebraic sum of such p.d.'s taken round a closed path back to the same point again must be zero.

$$\therefore\ \Sigma E - \Sigma Ir = 0$$
$$\therefore\ \Sigma Ir = \Sigma E$$

which is the symbolic statement of the second law.

3: Electrolysis and Thermoelectricity

3.1 Conduction in liquids

Except for the case of molten metals (such as mercury at room temperature), the passage of an electric current through a liquid causes chemical changes; the process is called *electrolysis*. Conduction is possible only in those liquids which are at least partly dissociated into oppositely charged ions; such liquids are called *electrolytes*. Solutions of many inorganic chemical compounds (common salt, sulphuric acid, etc.) are examples of this type of liquid. There are however many substances (sugar, for instance) which dissolve without splitting up into ions. Solutions of these do not conduct electric currents, and are called *non-electrolytes*.

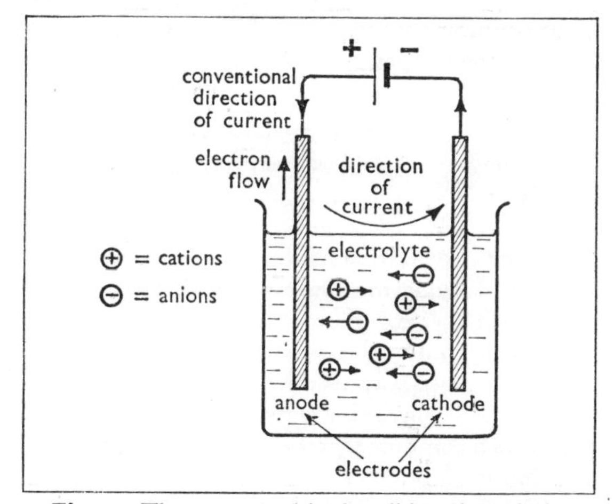

Fig. 3.1 The terms used in describing electrolysis

The plates through which the current enters and leaves an electrolyte are known as *electrodes*. That towards which the positive ions travel is called the *cathode*; and the other, towards which the negative ions travel, is called the *anode* (Fig. 3.1). The flow of current through an electrolyte takes place by the movement of all the ions in it—the positive ions in one direction and the negative in the other. The positive ions are referred to as *cations*; these are mostly formed from metals or hydrogen. The negative ions are referred to as *anions*. Thus in a circuit diagram (Fig. 3.1) the arrows which show the direction of the current (in the conventional sense) actually indicate the direction of movement of the cations; the anions move in the opposite direction.

An electrode must normally be made of metal, or perhaps some form of carbon (such as graphite); and in such a material conduction of current can take place only by the movement of the free electrons it contains (p. 5); in the electrolyte, however, there are no free electrons, and conduction takes place only by the movement of the ions. At the surface of the electrodes, where the current passes between electrode and electrolyte, the mechanism of conduction must change. This can happen in one of two ways; either some of the ions are discharged, electrons being transferred between ions and electrode; or else fresh ions are formed from the material of the electrode, and these pass into solution. The discharging of an ion usually causes the substance concerned to come out of solution and be liberated either as a deposit on the electrode or in bubbles of gas, as the case may be. The chemical changes that occur in electrolysis are thus seen to arise at the surface of the electrodes. Two examples will show the way in which these processes operate.

(*i*) *The electrolysis of copper sulphate solution with inert electrodes* (*carbon or platinum*). The copper sulphate in solution consists of the ions $Cu(H_2O)_4^{2+}$ and SO_4^{2-}. Also the water itself provides small quantities of the ions $(H_3O)^+$ and OH^- (Fig. 3.2). Both sorts of positive ions travel towards the cathode, but at the surface of the cathode the $Cu(H_2O)_4^{2+}$ ions are discharged in preference to the $(H_3O)^+$ ions (although for very large currents some hydrogen may be liberated). A layer of metallic copper is thus deposited on the cathode.

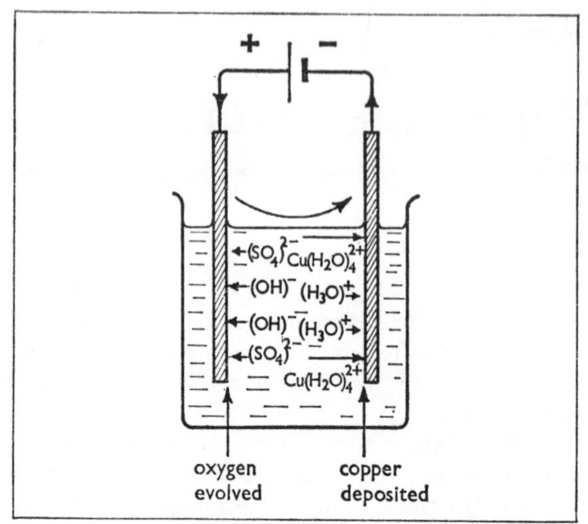

Fig. 3.2 The electrolysis of copper sulphate solution with inert electrodes (carbon or platinum)

Of the two negative ions in solution it is found that OH^- is discharged in preference to SO_4^{2-}; this leads to the evolution of oxygen at the anode, since the hydroxyl groups (OH) react together in pairs according to the reaction—

$$2(OH^-) \rightarrow H_2O + O + 2e^-$$

(e^- here stands for an electron)

The net result in this case is that the copper is steadily removed from the solution, which therefore becomes paler. At the same time it becomes increasingly acid, since the SO_4^{2-} ions remain in solution, and further dissociation of the water provides the balancing $(H_3O)^+$ ions.

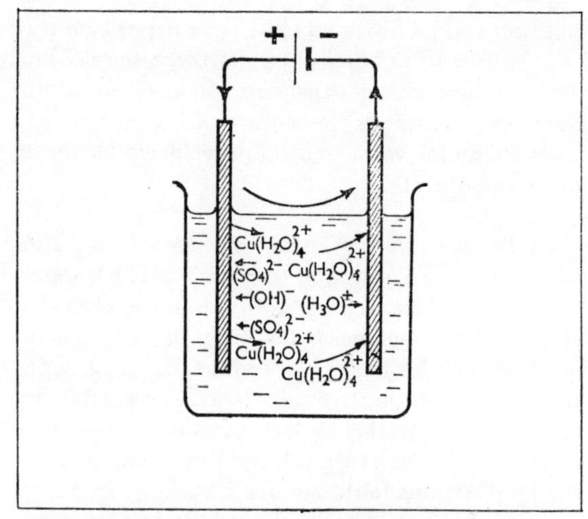

Fig. 3.3 The electrolysis of copper sulphate solution with copper electrodes

(*ii*) *The electrolysis of copper sulphate solution with copper electrodes* (Fig. 3.3). At the cathode, copper is deposited as in the previous case. But at the anode, instead of OH^- ions being discharged, fresh $Cu(H_2O)_4^{2+}$ ions are formed from the copper of the anode itself. The formation of each ion leaves two surplus electrons behind on the anode; the same charge (carried by two electrons) is imparted to each of the $Cu(H_2O)_4^{2+}$ ions discharged at the cathode. Since the current is the same at the surfaces of both electrodes, the concentration of the electrolyte remains constant; and the net result is just the transfer of copper from the anode to the cathode. The chief practical importance of this process lies in its ability to purify the copper. The anode is made of a block of impure copper and the cathode is a small sheet of very pure copper; then, as the current passes, pure copper only is taken from the anode and deposited on the cathode. There is a large demand for electrolytically purified copper, since this degree of purity is needed for electrical cables and connectors. Even the smallest trace of impurity can markedly increase the resistivity of the copper.

In modern chemical terminology reactions in which electrons are *removed* from the ions concerned are classified as *oxidations*, while those in which electrons are *added* to ions are classified as *reductions*. Consider for instance the reaction produced by adding metallic zinc to dilute sulphuric acid, with the evolution of hydrogen. Each zinc atom passes into solution as a zinc ion (Zn^{2+}); it imparts the two surplus electrons to two hydrogen ions $(H_3O)^+$, converting them into atoms of hydrogen, which combine together in pairs and come out of solution as bubbles of hydrogen. The chemical equation is:

$$Zn + \underbrace{2(H_3O)^+ + SO_4^{2-}}_{\text{sulphuric acid}} \rightarrow Zn^{2+} + H_2 + SO_4^{2-} + 2H_2O$$

By the rule given above this constitutes an oxidation of the zinc and a reduction of the hydrogen. This is of course a broadening of the original meanings of the words 'oxidation' and 'reduction'. The student should consult a textbook of chemistry for full explanations of the modern meanings of these terms. It will be seen that in chemical reactions an oxidation is accompanied by a corresponding reduction, since one component of the reaction receives an electron at the expense of another.

In electrolysis the reactions at the electrodes can likewise be classified in this way. At the anode, electrons are removed from the ions and it follows that all anodic reactions are oxidations. Similarly at the cathode, electrons are added to the ions, and therefore all cathodic reactions are reductions. From the chemical point of view the significance of electrolysis is very largely in this—that it enables an oxidation and its corresponding reduction to be conducted at different points in the same solution, so that the products of the actions can be kept apart. In this way many chemical reactions can be brought about which without electrolysis would be impossible. Examples of industrial electrolytic processes may be found in textbooks of chemistry.

3.2 Faraday's laws

The factors affecting the quantities of matter liberated or dissolved in electrolysis were investigated by Faraday. They may be summarized in the following laws:

Faraday's laws of electrolysis. *When an electric current causes electrolysis:*
1. *The mass of any substance liberated or dissolved is proportional to the total electric charge that passes.*
2. *The masses of different substances liberated or dissolved by the passage of the same electric charge are in the ratio of their chemical equivalents.*

These may be put in mathematical terms as follows:

Let M = mass liberated or dissolved;
Q = total charge passing.

The first law then is

$$M \propto Q$$

or

$$M = zQ$$

where z is a constant for any given substance known as its *electrochemical equivalent* (usually abbreviated to *e.c.e.*).

The **electrochemical equivalent (e.c.e.)** *of a substance is the mass of it liberated or dissolved in electrolysis per unit electric charge that passes.*

The charge Q that passes will normally be calculated from the current I and the time t for which it flows. Thus $Q = It$, where Q is in coulombs if I is in amps and t in seconds. So we may write

$$M = zIt$$

The second law states that for constant charge Q

$$M \propto E$$

where E = the chemical equivalent of the substance. Another way of expressing this is

$$z \propto E$$

In other words, if z_1 and z_2 are the e.c.s.'s of two substances whose chemical equivalents are E_1 and E_2 respectively, then

$$\frac{z_1}{z_2} = \frac{E_1}{E_2}$$

The chemical equivalent of an element is related to its atomic weight A and valency V by

$$E = \frac{A}{V}$$

Testing the first law

The full testing of the laws of electrolysis involves a lengthy experimental programme, which would not normally be carried out in a school laboratory. Usually the student will be expected to make a single determination of the e.c.e. of copper, for example. The constancy of the values obtained by different members of a class can then be taken to affirm the truth of the first law. The electrolysis of copper sulphate solution using copper electrodes is usually used for this purpose. To obtain reasonable results great care has to be taken in cleaning the cathode, scouring it with pumice or sand and washing it in dilute sulphuric acid; if this is not done, the deposit is not stable and is liable to flake off. The cathode is dried and weighed at the start and again at the end of the experiment. The current is maintained as steady as possible and is read on an ammeter in series at regular intervals. To obtain even 1 g of deposited copper a charge of about 3000 coulombs must be passed—i.e. 1 amp flowing for 50 minutes; and this is a minimum for satisfactory measurements. The current used cannot be very large, since if the current density is more than about 200 amps per m^2 of cathode surface, the deposit is powdery and unstable.

If the truth of the first law is assumed, and the e.c.e. of copper is known, we can use such an experiment as a means of measuring quantities of electric charge. An electrolytic tank used in this way is referred to as a *voltameter* (after the Italian scientist Volta); this word is not to be confused with the very similar word *voltmeter*.

Testing the second law

The method used by Faraday was to connect in series a number of different voltameters, using a wide variety of substances. Being in series, the same charge must pass through all the voltameters even if the current varies during the experiment. At each electrode the mass of material liberated or dissolved is found; and these quantities are shown to be proportional to the chemical equivalents of the substances concerned—i.e. the masses divided by the chemical equivalents should be constant.

An alternative way of regarding the second law of electrolysis makes use of the idea of measuring quantities of matter in *kilogram-equivalents*.

A **kilogram-equivalent** *of any material is the quantity of it whose mass, measured in kilograms, is numerically equal to its chemical equivalent.*

E.g. 1 kg-equiv. of hydrogen = 1·008 kg
1 kg-equiv. of oxygen = 8·000 kg
1 kg-equiv. of silver = 107·9 kg

If we now consider 1 kg-equivalent of a number of substances that can be liberated or dissolved in electrolysis, by definition the masses of these quantities are in the ratio of their chemical equivalents. According to the second law of electrolysis these masses will be liberated or dissolved by the passage of *the same quantity of electric charge.* This quantity is known as *Faraday's constant* (denoted by F).

Faraday's constant *is the quantity of electric charge that liberates or dissolves 1 kilogram-equivalent of any substance in electrolysis.*

Electrolytic measurements give Faraday's constant as

$$F = 9{\cdot}65 \times 10^7 \text{ coulombs per kg-equiv.}$$

A knowledge of this figure and the chemical equivalent of any substance is sufficient for all electrolytic calculations.

Example. How long will it take to deposit 1 kg of copper in electrolysis at a current of 10·00 amps? The atomic weight of copper is 63·6 and its valency 2.

The chemical equivalent of copper

$$= \frac{63{\cdot}6}{2} = 31{\cdot}8$$

i.e. $9{\cdot}65 \times 10^7$ coulombs deposit 31·8 kg of copper.

∴ 1 kg of copper is deposited by

$$\frac{9{\cdot}65 \times 10^7}{31{\cdot}8} \text{ coulombs}$$

Charge passes at a rate of 10·00 coulombs per second.

$$\therefore \text{ time required} = \frac{9{\cdot}65 \times 10^7}{10 \times 31{\cdot}8} \text{ s} = 84{\cdot}3 \text{ hours}$$

Thus the e.c.e. z of a substance of chemical equivalent E is given by

$$z = \frac{E}{F}$$

The previous expressions for the mass M deposited in electrolysis may therefore be written

$$M = zQ = zIt$$
$$= \frac{E}{F} \times Q = \frac{E}{F} \times It$$

The significance of measuring quantities in kg-equivalents is that 1 kg-equivalent of any substance in ionized form carries the *same number* of electronic charges. This number, known as *Avogadro's constant*, is found to be $6{\cdot}02 \times 10^{26}$ per kg-equiv.; we shall denote it by N.* Thus

$$F = Ne$$

* It is necessary to beware at this point of the confusion that may arise from using mixed systems of units. In many branches of Science it is common to employ a *c.g.s. system* (based on the *centimetre, gram* and *second*). In particular this is usual in Chemistry. Quantities of material may then be expressed in *gram*-equivalents, instead of kilogram-equivalents, and Faraday's constant (F) and Avogadro's constant (N) have (numerically) 1/1000 of their values in the m.k.s. system.

$F = 9{\cdot}65 \times 10^4$ coulombs per *gram*-equivalent
$N = 6{\cdot}02 \times 10^{23}$ per *gram*-equivalent

No difficulty need arise as long as the student remembers that these quantities are not dimensionless numbers, but should always be expressed with the correct units in the same way as any other physical quantity.

where e is the charge on an electron. Since N may be found from independent measurements, this relationship provides one of the best ways of obtaining the electronic charge e ($1{\cdot}60 \times 10^{-19}$ coulomb).

It is instructive to compare the concept of a *kg-equivalent* with the related ideas of a *kg-molecule* (or *kg-mole*, as it is usually called) and a *kg-atom.*

A *kg-mole* of any substance is its molecular weight in kilograms. It contains N *molecules.*

A *kg-atom* of any element is its atomic weight in kilograms. It contains N *atoms.*

Thus Avogadro's constant N has the three equivalent definitions:

$N =$ the number of molecules per kg-mole of any substance
= the number of atoms per kg-atom of any element
= the number of electronic charges carried in electrolysis per kg-equivalent of any substance

The deduction of Faraday's laws of electrolysis on the ionic theory

Consider an element liberated or dissolved in electrolysis whose atomic weight is A and valency V. Hence its chemical equivalent E is given by

$$E = \frac{A}{V}$$

Let the mass of a single atom of carbon-12 $= 12m$; i.e. we are taking m as the atomic mass unit, $1{\cdot}66 \times 10^{-27}$ kg (p. 4).

We may therefore take the mass of an atom of hydrogen as m (nearly); or of a helium atom as $4m$; or of a silver atom as $108m$; etc. The differences between the masses of atoms and their ions is negligible, so that we may say that the mass of an ion of a substance of atomic weight A is Am.

In the ionic theory the charge of an ion is assumed to be proportional to its valency: e for a univalent ion, $2e$ for a divalent ion, and Ve for an ion of valency V.

Now if a charge Q is passed,

$$\text{the number of ions discharged} = \frac{Q}{Ve}$$

$$\therefore \text{ the mass } M \text{ deposited} = \frac{Q}{Ve} \times Am$$

Re-arranging, $$M = \frac{m}{e} \times E \times Q$$

In this equation m/e is a universal constant. Thus the ionic hypothesis requires that the mass liberated should be proportional to Q and to E, in agreement with the laws of electrolysis.

The equation we have obtained also shows the significance of F from the atomic point of view. We had previously obtained the equation

$$M = \frac{E}{F} \times Q$$

Comparing the two expressions we have

$$F = \frac{e}{m}$$

$$\therefore \text{ the charge per unit mass of an ion} = \frac{Ve}{Am} = \frac{F}{E}$$

In particular the charge per unit mass of the proton (hydrogen ion)

$$= \frac{9{\cdot}65 \times 10^7}{1{\cdot}008} = 9{\cdot}58 \times 10^7 \text{ coulombs per kg}$$

Since $e = 1{\cdot}60 \times 10^{-19}$ coulomb

this gives the mass of the proton as

$$\frac{1{\cdot}60 \times 10^{-19}}{9{\cdot}58 \times 10^7} = 1{\cdot}67 \times 10^{-27} \text{ kg}$$

in excellent agreement with the deductions of the kinetic theory of gases (p. 2).

3·3 The ionic theory

The ionic hypothesis was first introduced by the Swedish chemist Berzelius in 1812. But it took nearly a century of development before it could be presented in a form that met with general acceptance in the scientific world. Following closely on the introduction of the atomic theory by Dalton at the beginning of the nineteenth century Berzelius suggested that the forces holding atoms together in chemical compounds were electrostatic in nature. He supposed that the atoms of metallic elements formed positive particles and non-metallic elements negative ones; and that they were then held together by the electrostatic forces between them. The development of the theory ran into difficulties since, as we now know, other types of force also operate in holding together the atoms in chemical compounds.

In 1833 Faraday investigated the phenomenon of electrolysis. The laws he discovered make an ionic explanation of the process almost inevitable, just as in chemistry the laws of Constant Composition and Multiple Proportions demand an atomic explanation. But the exact nature of the ions and how they came into being could not at that time be clearly explained; and most scientists still remained sceptical. It was Faraday who coined the word *ion* to describe the charged particles that were supposed to be responsible for carrying the current through an electrolyte.

In the 1870's Kohlrausch made extended measurements of the conductivity of electrolytes at varying concentrations. These measurements enabled him to calculate the proportion of molecules ionized in the solutions. It appeared that electrolytes fell into two classes, which Kohlrausch called *strong* and *weak*. The strong electrolytes were completely dissociated into ions under all conditions; such were the solutions of strong acids and alkalis and the salts formed from them. Other electrolytes were only slightly dissociated at large concentrations; but at sufficient dilution even these were completely dissociated. Kohlrausch also used his measurements to give the ionic velocities in the solutions (p. 9). For a potential gradient in the electrolyte of 1 volt per m the velocities obtained were of the order of 10^{-7} m s^{-1}.

It was not until 1887 that a Danish student, Arrhenius, propounded a theory of ionic dissociation that was adequate to explain the phenomena of electrolysis, at least for weak electrolytes. Previous versions of the ionic theory had supposed that the production of ions was initiated by the application of a p.d. between the electrodes, and that the ions did not exist before the flow of current started. But this idea is contradicted by the observation that in many cases of electrolysis current flows no matter how small the applied p.d.; if the p.d. had to supply energy to produce ionic dissociation, we would expect some minimum p.d. to be needed before current could start to flow.

Arrhenius suggested that dissociation into ions starts as soon as an electrolyte goes into solution, and is not dependent on the application of a p.d. Each molecule in solution has a certain chance of dissociating. Since we are concerned in any practical case with enormous numbers of molecules, this means that there will be a steady production of ions at a rate proportional to the number of unionized molecules present. At the same time random movements of the ions in the liquid will bring oppositely charged ions together, when they have a chance of recombining to form uncharged molecules again. The rate of recombination can be taken to be proportional to the numbers of each sort of ion present; the greater the degree of ionization of the substance dissolved, the greater the rate of recombination and the smaller the rate of dissociation (an application of the Chemical Law of Mass Action). A state must soon be reached in which ionization and recombination are proceeding at the same rate. The proportion of molecules ionized then remains constant. Owing to the rapidity of atomic processes this apparent equilibrium is normally reached almost at once.

If some of the ions in the solution are discharged and liberated by electrolysis (or precipitated by some chemical action), for a moment recombination proceeds less rapidly, since there are fewer ions to recombine. The continually proceeding dissociation at once tends to restore the proportion of molecules ionized to its original value. There is thus always a steady supply of ions available to conduct the current, even if as in some cases the proportion of molecules ionized is very small.

Kohlrausch's results indicated that at infinite dilution all electrolytes would be completely dissociated. This is what we should expect on the theory of Arrhenius, since at very low concentrations collisions of ions will be rare events, and the rate of recombination will be negligible.

With strong electrolytes the process is complicated by the tendency of the ions to group together in clusters when they are in strong concentrations; thus the particles in motion become larger with increasing concentration with a consequent decrease in effective mobility. An exact theory was eventually provided on this basis by Debye and Hückel, though this was not until 1923.

Further evidence for the ionic theory is provided by other properties of electrolytes. Many of their distinctive properties depend only on the presence of a particular ion, and are not affected much by changing the other ions accompanying it. For example, the general properties of acids arise from the $(H_3O)^+$ ions they form in solution; it is these ions that give the sour taste of an acid. Likewise the general properties of solutions of alkalis arise from the hydroxyl ions (OH^-) they contain. For instance in contact with the skin these ions combine with the fats present, producing the thin film of soap that causes the characteristic slippery feeling of alkalis. Other examples are provided by the characteristic colours of solutions containing particular metallic ions; e.g. the blue colour of solutions of the $Cu(H_2O)_4^{2+}$ ion; this colour is not much affected by changing the anion accompanying it (unless this also produces colouring).

Another independent line of evidence for the ionic theory is provided by the phenomenon of *osmosis* and the changes of boiling and freezing points related with it. Theory shows that the magnitude of the osmotic pressure, the depression of the freezing point of the solution, and the elevation of the boiling point should all depend on the *number of particles* present in the solution. Measurements on electrolytes show that the numbers of particles are much greater than the numbers of molecules of solute; there must therefore be some dissociation of the molecules. The student should consult a textbook of physical chemistry for details of such measurements and of the theory involved.

3·4 The simple cell

If any two different metals are dipped into a tank of any electrolyte, and an electrical connection is made between them, it is found that an e.m.f. acts so as to drive current round the circuit so formed. The resulting electrolysis is accompanied by the usual chemical changes. Such an arrangement is called an *electric cell.* It is in fact a device for producing electrical energy from chemical potential energy; the magnitude of the e.m.f. depends on the energy changes involved at the two electrode surfaces. Many combinations of materials have been used to make electric cells; one arrangement, traditionally known as the *simple cell,* is shown in Fig. 3·4.

It consists of electrodes of zinc and copper in an electrolyte of dilute sulphuric acid. Since zinc reacts to some extent with cold sulphuric acid, it is usual to rub mercury into the zinc electrode forming an amalgam of zinc on the surface. This almost completely prevents the *local action* of the acid on the zinc. When the circuit is completed and a current passes, hydrogen is evolved at the surface of the copper electrode and zinc goes into solution from the other. The current flows (in the conventional sense) from the copper to the zinc outside the cell; and from zinc to copper inside. The copper electrode is thus the *positive pole* or terminal of the cell and the zinc is the *negative pole* or terminal. The total chemical action of the cell is the same as the familiar action of dilute sulphuric acid on zinc —an action that releases a certain amount of energy, some of which, in the operation of the simple cell, is changed into electrical energy.

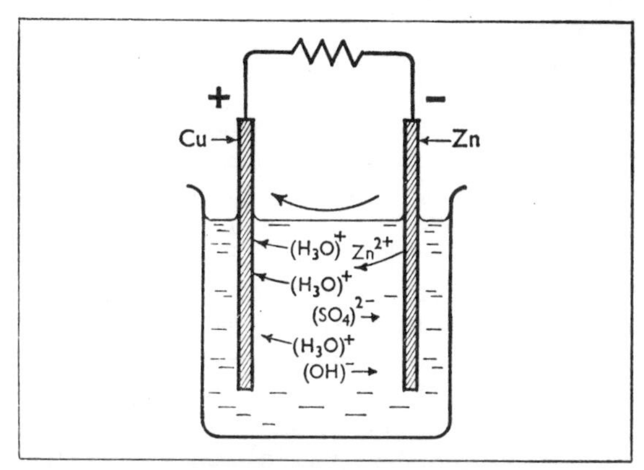

Fig. 3·4 The simple cell

In terms of the motions of the ions the explanation of the action of the cell is as follows. At the zinc electrode Zn atoms are ionized and pass into solution as Zn^{2+} ions leaving two electrons each on the electrode. For each Zn^{2+} ion formed two $(H_3O)^+$ ions are discharged at the copper electrode, taking two electrons from it. The action continues just as long as electrons can pass from the zinc to the copper through some piece of metal (containing free electrons). If the circuit is broken and the electron flow stopped, charge accumulates on the electrodes, and the electrostatic forces prevent any further action.

Note. In a cell the positive electrode is the cathode, since the positive ions (metals and hydrogen) are driven to it. This is the opposite to the examples of electrolysis considered earlier in this chapter. The student should not make the mistake of learning that the cathode is the negative electrode, since this is incorrect for the electrolysis taking place in cells. Rather—*the cathode is the electrode at which the positive ions arrive.* When a current is driven through an electrolyte by some external source of e.m.f. the positive ions are attracted to the negative electrode, which is then the cathode; but in a cell the cathode becomes positive because the positive ions are driven to it by the chemical action.

At the zinc electrode of a simple cell energy is imparted to the circuit, while at the copper electrode some of this energy is used in liberating the hydrogen. The rest of the energy is available to drive current round the circuit. Some of it is used in driving the current through the electrolyte of the cell (raising its temperature), and some of it in driving the current round the rest of the circuit. The energy per unit charge (i.e. the electric potential) at various points is shown in the graph in Fig. 3.5. Since we are concerned only with differ-

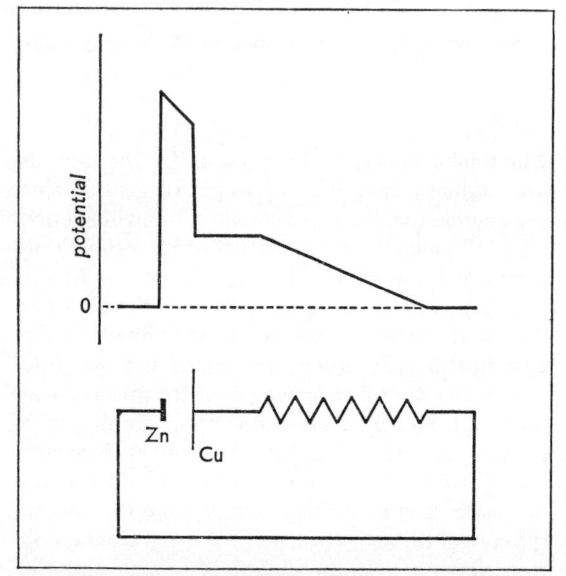

Fig. 3.5 The variation of potential round a circuit containing a simple cell

ences of potential, the zinc electrode is here arbitrarily taken as the zero of potential. As with all sources of e.m.f. the p.d. between the terminals of the cell falls when a current is flowing on account of the internal resistance; only when the current is zero is the p.d. equal to the e.m.f.

Quite apart from this drop in p.d. because of the internal resistance there is also a decrease in the actual e.m.f. of the cell on account of the effect known as *polarization*. As soon as a current flows, the electrolysis changes the composition of the cell; the copper electrode is covered with a layer of hydrogen and the concentrations of the ions are altered in the neighbourhood of both electrodes. This gives rise to a *back e.m.f.* acting in opposition to the e.m.f. of the cell and partly cancelling it out. The law known as *Le Chatelier's principle* shows that a decrease in e.m.f. is to be expected when a current flows. (See textbooks of chemistry for an explanation of this principle.) Unpolarized, the e.m.f. of the simple cell is about 1·1 volts, but this may drop to about 0·4 volt after a current has been passing for a few seconds. To some extent this defect in the cell as a practical device can be overcome by adding a suitable oxidizing agent to the electrolyte—potassium dichromate is often used. Even then the electrolyte needs constant stirring for the potassium dichromate to remain effective; and it cannot be left assembled for any length of time, since the oxidizing agent slowly attacks the electrodes.

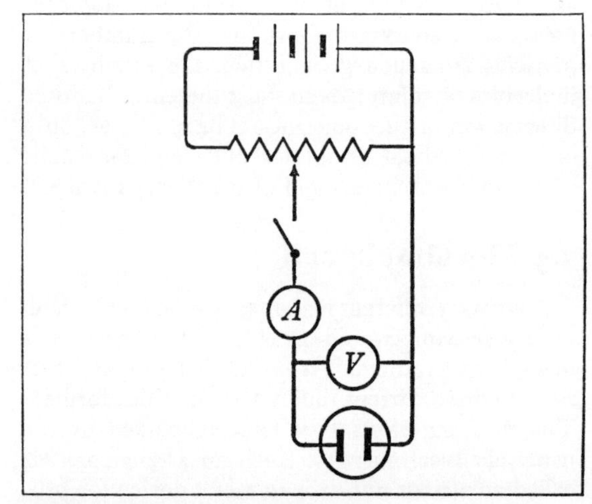

Fig. 3.6 Demonstrating the back e.m.f. produced by polarization

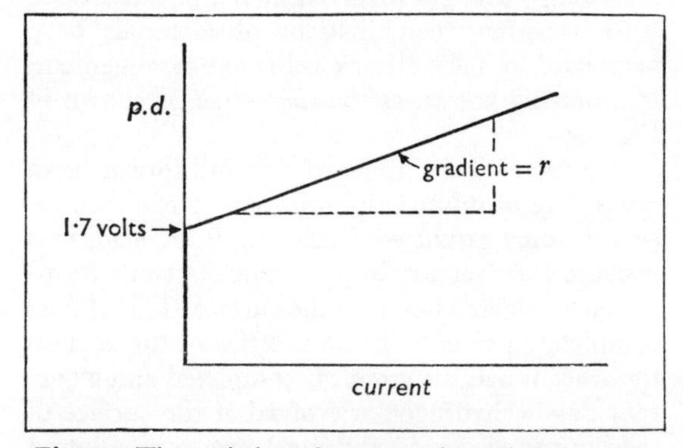

Fig. 3.7 The variation of current with p.d. for a water voltameter

The existence of a back e.m.f. due to polarization may be demonstrated in any electrolysis in which the products change the effective composition of the cell. Fig. 3.6 shows a circuit that may be used to show the effect for the electrolysis of dilute sulphuric acid with platinum electrodes. The potential divider joined across the 6-volt battery enables the p.d. applied to the voltameter to be varied from 0 to 6 volts. With the switch closed the p.d. is slowly increased from zero. At first only a very small current flows and no evolution of hydrogen and oxygen can be observed at the electrodes. Only when the p.d. has been raised above about 1·7 volts will a larger current start to flow and gas be evolved. Increasing the p.d. above this point causes a steady increase of current, as shown in the graph of Fig. 3.7. We suppose this behaviour to be due to the existence of a back e.m.f. of maximum value 1·7 volts arising from the polarization of the voltameter. If now the switch is opened, this e.m.f. gives a reading on the voltmeter, which at first is 1·7 volts but rapidly falls off as the electrodes and electrolyte are restored to their original condition. From energy considerations we should expect a minimum p.d. to be required for this electrolysis to proceed. The net result is the decomposition of water; to do this a quantity of energy at least equal to the heat of reaction of hydrogen and oxygen must be supplied in electrical form. Measurements show that the combination of 1 kg-equivalent each of hydrogen and oxygen releases $1{\cdot}47 \times 10^8$ joules of heat. To produce this quantity of hydrogen and oxygen in electrolysis requires the passage of $9{\cdot}65 \times 10^7$ coulombs of electric charge. The p.d. required (i.e. the energy per unit charge) is thus

$$\frac{1{\cdot}47 \times 10^8}{9{\cdot}65 \times 10^7} \text{ joules per coulomb}$$
$$= 1{\cdot}5 \text{ volts}$$

The actual p.d. varies slightly with the type of electrode used; for platinum it is 1·7 volts.

Only when the products of electrolysis do not affect the composition of the cell is there no back e.m.f. Such for instance is the electrolysis of copper sulphate with copper electrodes; in this case electrolysis takes place continuously even for very small applied p.d.'s; and it is found that Ohm's law applies to this voltameter fairly closely. In other cases allowance must be made for the back e.m.f. E. Fig. 3.7 shows that the current I is proportional to the amount by which the p.d. V is in excess of E. Thus

$$\frac{V - E}{I} = r$$

a constant which we may call the resistance of the voltameter. This is very similar to the concept of the internal resistance of a cell (p. 16). The difference is that the current is normally being driven through the voltameter *in opposition* to the e.m.f. E, so that V is greater than E. This should be regarded as a special application of the general result discussed on p. 16:

electrical pressure = current × resistance

In the present case the electrical pressure effective in driving current through the voltameter is the *difference* between the applied p.d. and the back e.m.f.

3·5 Types of cell

(*i*) *The Daniell cell.* This consists of electrodes of zinc and copper each immersed in an acid solution of its own ions; the two solutions are kept from mixing by a porous partition which, however, allows ions to pass slowly through it so that a flow of current between the solutions is possible. The amalgamated zinc rod (−) stands in an acid solution of zinc sulphate inside the porous pot (Fig. 3.8). This in turn stands in a glass jar containing concentrated copper sulphate solution. The cathode (+) is a cylindrical plate of copper surrounding the porous pot.

When the cell is connected in a closed circuit, Zn^{2+} ions go into solution at the anode and

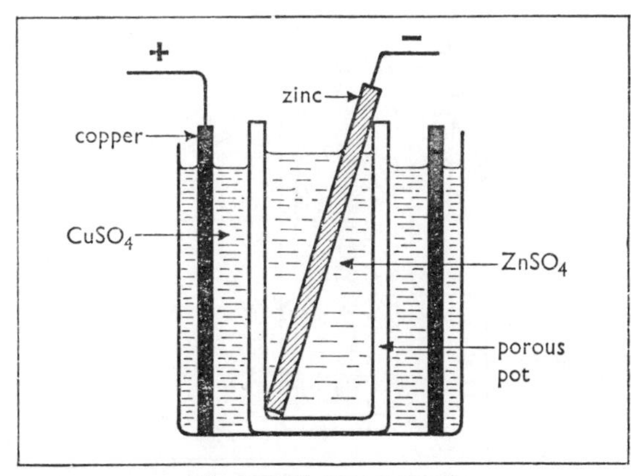

Fig. 3.8 The Daniell cell

$Cu(H_2O)_4^{2+}$ ions are discharged and copper is deposited at the cathode. The concentration of the copper sulphate solution can be maintained by keeping a few crystals of copper sulphate in the outer compartment. Since the action of the cell does not in any way change its composition, no polarization occurs and the e.m.f. of the cell remains constant in use, though it is subject to some variation with temperature.

The cell cannot be left assembled for any length of time, since it is gradually 'poisoned' by the diffusion of the copper ions into the inner compartment, where they are at once deposited on the zinc rod. The diffusion of the zinc ions into the outer space is of less consequence; it is therefore important to keep the level of the liquid in the inner space higher (if anything) than that outside so as not to increase the inward movement of copper ions.

The e.m.f. of the Daniell cell is about 1·08 volts; it is sometimes used as a subsidiary standard in simple laboratory experiments. But for work of the highest precision the Weston cadmium cell (see below) is now universally used as a standard.

(*ii*) *The Leclanché cell.* This has electrodes of zinc and carbon in an electrolyte of ammonium chloride solution. The carbon electrode is fixed in a porous pot which is packed tightly with a mixture of powdered carbon and an oxidizing agent, manganese dioxide (Fig. 3.9). The powdered carbon, being porous, extends the effective surface area of the electrode. The porous pot stands in a weak solution of ammonium chloride in a glass jar, in one side of which is immersed the zinc rod. The porous pot in this case is not designed to keep the solution out of the inner compartment, but merely to keep the powder tightly packed round the carbon rod. The e.m.f. is about 1·5 volts.

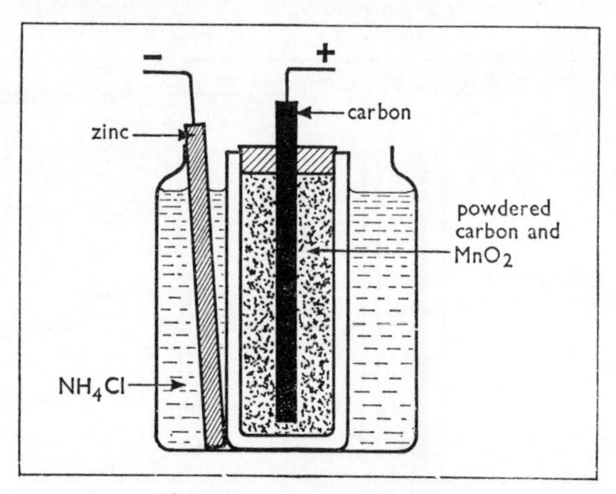

Fig. 3.9 The Leclanché cell

When a current passes, Zn^{2+} ions go into solution at the anode, and $(H_3O)^+$ ions are discharged at the cathode; this causes some polarization, which is gradually removed by the reduction of the manganese dioxide—

$$2MnO_2 + 2H \rightarrow Mn_2O_3 + H_2O$$

Subsequently the manganese dioxide is re-formed from the oxygen of the atmosphere dissolved in the solution—

$$2Mn_2O_3 + O_2 \rightarrow 4MnO_2$$

The Zn^{2+} ions eventually form with the ammonium chloride an insoluble complex, which is precipitated harmlessly to the bottom of the jar. To maintain the cell, the zinc rod must occasionally be replaced; evaporation must be made up (tap water can be used for this), and some ammonium chloride must be thrown in. The life of the cell is then almost indefinite.

It can deliver a current of about 1 amp for a minute or so before polarization becomes excessive, and it quickly recovers afterwards. It is therefore well suited for intermittent use, e.g. for door-bells, etc. In the laboratory it is useful for applications in which constancy of e.m.f. is of no importance, e.g. Wheatstone bridge circuits (4.2); it is cheaper than an accumulator and is less likely to cause damage, since its internal resistance (about 1 ohm) is sufficient to limit the current that can be taken from it to about 1·5 amps.

(*iii*) *The Leclanché 'dry' cell.* Chemically this cell is identical with the Leclanché cell already considered; but it is adapted to be portable. The case itself is the zinc electrode (—). The carbon and manganese dioxide is held packed round the central carbon rod by a bag of fabric (Fig. 3.10). The electrolyte is absorbed either in a jelly or in a porous mixture of plaster of Paris and flour; glycerine is often added to it, since this is hygroscopic and helps to maintain the electrolyte in a moist condition. The case is sealed by layers of sawdust and pitch. The cell ceases to function when one of three things happens, (*a*) the electrolyte actually goes dry, (*b*) the zinc perforates so that the moisture runs out and evaporates, (*c*) the manganese dioxide is entirely used up and permanent polarization occurs.

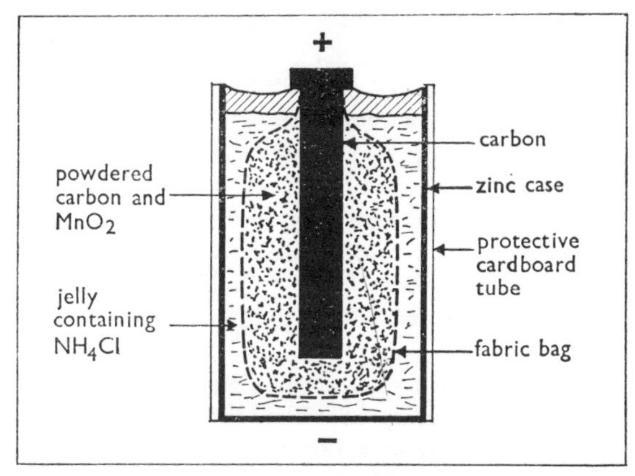

Fig. 3.10 The Leclanché 'dry' cell

(*iv*) *The Weston cadmium cell.* This is designed specially for standards purposes. As in the Daniell cell polarization is eliminated (for very small currents) by having each electrode in a solution of its own ions. The electrodes are of mercury (+) and cadmium–mercury amalgam (—), and the electrolyte is a saturated solution of cadmium sulphate. A paste of mercurous sulphate and cadmium sulphate solution is placed above the mercury electrode (Fig. 3.11). The e.m.f. is 1·0186 volts at 20°C, and varies little with temperature. Its internal resistance is deliberately designed to be high (about 500 ohms), and it is usually used in series with a high protective resistance ($\frac{1}{2}$ megohm). Its design protects it against polarization provided the current is no more than about 10 microamps; but if more

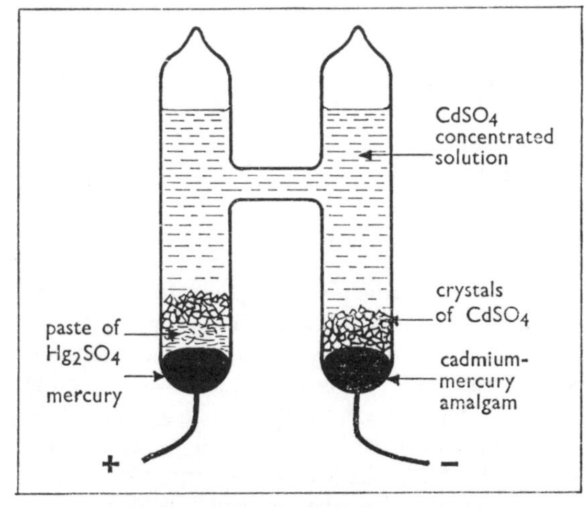

Fig. 3.11 The Weston cadmium cell

than this is passed through it, permanent changes of its e.m.f. can occur.

The Weston cell provides the internationally recognized secondary standard of p.d. (the primary standard is simply the definition). Carefully constructed and used its e.m.f. will remain constant to within 1 part in 10^6 for several years, and to 1 part in 10^5 for very long periods (p. 61).

3.6 Storage cells

In certain types of cell the action is truly reversible. A current driven through it in opposition to its e.m.f. re-forms the original materials of the electrodes, and the cell can therefore be used again and again. It is in fact a device for temporary storage of electrical energy. We shall consider two of the common types of storage cell. In all such cells the chemical actions are of very complicated kinds, and we shall deal with them only in outline.

(*i*) *The lead-acid accumulator.* The negative electrode is of lead and the positive one of lead dioxide (PbO_2). To increase their effective surface area, in both electrodes the active material is in the form of a spongy paste which is pressed into a grid made of a lead–antimony alloy. The electrolyte is sulphuric acid, initially of specific gravity 1·21. The initial e.m.f. of the cell is about 2·1 volts; this falls rapidly to 2·0 volts, at which figure it remains nearly constant until the cell is almost 'discharged'; re-charging should always be begun if the e.m.f. falls as low as 1·8 volts.

During discharge the lead in the negative plate and the PbO_2 in the positive plate both change into insoluble lead sulphate ($PbSO_4$). In this process the SO_4^{2-} ions are removed from the solution and the specific gravity falls. To re-charge the cell current must be driven through it in opposition to its e.m.f.; the chemical actions are then reversed and the sp. gr. of the solution rises again.

It will be seen that the e.m.f. of the cell provides little indication of its state of charge since it remains close to 2·0 volts for 90% of the discharge period; so the specific gravity of the acid is normally used to indicate this. The cell should be re-charged when the sp. gr. falls below 1·18.

When re-charging, the e.m.f. of the cell rises quickly to about 2·2 volts, and this increases again to nearly 2·7 volts when re-charging is complete. At this point the electrolysis consists of the evolution of hydrogen and oxygen, and this 'gassing' at the electrodes is often taken as an indication that charging is finished. Distilled water is added occasionally to make up the loss caused by the decomposition of the water.

A lead–acid accumulator is easily damaged by misuse. The strong acid solution makes the internal resistance of the cell very small—less than 0·01 ohm. This is one of its most useful properties, since it means that the p.d. will be almost constant for all normal currents. But it also means that a short-circuit will give a most damaging current of several hundred amps. In extreme cases this causes irregular heating of the plates, buckling and an internal short-circuit; even if this does not happen it weakens the structure of the plates and makes the material flake off and collect as a sludge at the bottom of the tank. The other common source of damage is more insidious. There is always a tendency for the lead sulphate to change into a white allotropic form, which is non-porous and clogs the plates rendering them useless; this happens most rapidly when the sp. gr. of the acid is low. It is known colloquially as 'sulphating'. To minimize this effect (*a*) discharging should never be allowed to proceed too far, (*b*) the cell should never be left idle for long periods, particularly in a low state of charge, (*c*) the plates should never be allowed to remain uncovered by electrolyte.

(*ii*) *The nickel–iron (NiFe) accumulator.* The active material in the positive plates is nickel oxide (Ni_2O_3) and in the negative plates finely divided iron. In both plates the material is held in pockets of finely perforated steel. This gives much greater mechanical strength than in the lead–acid cell. The electrolyte is a solution of potassium hydroxide (sp. gr. 1·2). When discharging, the nickel oxide (Ni_2O_3) is changed into a lower oxide (NiO), and the iron is changed into an iron oxide (FeO); these actions are reversed when the cell is re-charged. The electrolyte appears to take no direct part in the chemical action, and so only a minimum quantity of it need be used.

The e.m.f. of the cell is about 1·2 volts falling to 1·0 volt when fully discharged. On re-charging the back e.m.f. is about 1·5 volts, rising finally to 1·85 volts. This means that it cannot so readily be used for voltage stabilization as a lead–acid cell. For instance in a car battery/dynamo circuit a lead–acid battery maintains the p.d. close to 12 volts whatever the e.m.f. of the dynamo, which changes with the engine speed. The equivalent alkaline accumulator (of 10 cells) would allow up to 50% increase in p.d. at high speed when the dynamo would be driving a charging current through the battery. Also the cost of the alkaline cell is greater than the lead–acid type. However, its greater mechanical strength and its freedom from troubles of chemical pollution (such as 'sulphating' in the lead–acid type) are leading to its increasing use for many applications.

An alkaline accumulator using cadmium instead of iron is also often used. This has the same e.m.f. (1·2 volts) as the nickel–iron but does not require such a large charging p.d., and is therefore comparable in its properties to the lead–acid type. The cost of these cells is however much higher.

From the point of view of the energy used, the efficiency of the charge–discharge process is about 75% for a lead–acid or a nickel–cadmium accumulator and about 60% for a nickel–iron one. However, to store 1 kilowatt-hour of electrical energy requires about 125 lb of lead–acid accumulator and about 70 lb of the nickel–iron or nickel–cadmium types.

The *capacity* of any accumulator is specified by the total quantity of charge that passes through it during the effective discharge process. It is usually measured in *ampere-hours*, this being the quantity of charge that passes when a current of 1 amp flows for 1 hour (i.e. 3600 coulombs). Knowledge of this quantity enables the time required to charge or discharge an accumulator to be approximately worked out.

Example. How long will it take to charge a 10-amp-hr accumulator at a 2-amp rate?

$$\begin{matrix}\text{Current} \\ \text{(in amps)}\end{matrix} \times \begin{matrix}\text{time} \\ \text{(in hours)}\end{matrix} = \begin{matrix}\text{capacity} \\ \text{(in amp-hr)}\end{matrix}$$

$$\therefore \text{ time required} = \frac{10}{2} = 5 \text{ hours}$$

Although in principle the capacity of an accumulator should be a constant property of it, it is found in practice that it falls drastically if the rated current of the cell is exceeded. If a cell is quoted as having a capacity of 10 amp-hr at a 1-amp rate, this means that its capacity will be this figure provided the current used is not more than 1 amp. At a 2-amp rate the capacity might fall to half this.

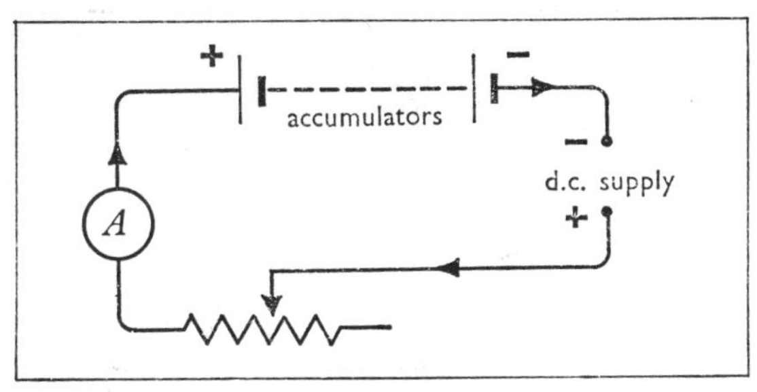

Fig. 3.12 Charging circuit for a battery of accumulators

A typical charging circuit for a set of accumulators to be charged from a d.c. supply is shown in Fig. 3.12. Owing to the low internal resistance of the cells a series resistance is always included to limit the current to the required value. Suppose the accumulators in the figure have a total e.m.f. of 36 volts and that a 50-volt d.c. supply is used; then to obtain a charging current of 5 amps we should calculate as follows:

The resultant e.m.f. in the circuit

$$= 50 - 36 = 14 \text{ volts}$$

$$\therefore \text{ the series resistance required} = \frac{14}{5} = 2{\cdot}8 \text{ ohms}$$

The total power supplied in this example

$$= (\text{the p.d. of the supply}) \times (\text{current})$$
$$= 50 \times 5 = 250 \text{ watts}$$

The power being stored in the batteries

$$= (\text{the battery e.m.f.}) \times (\text{current})$$
$$= 36 \times 5 = 180 \text{ watts}$$

The rest of the power (70 watts) is lost as heat in the resistance.

3·7 Thermoelectricity

There is a group of phenomena associated with temperature differences between the parts of an electric circuit. We shall consider *two* of these effects.

(i) The Seebeck effect

If a circuit consists of two different metals A and B (Fig. 3.13), an e.m.f. acts in it when the junctions of the metals are maintained at different temperatures. This is known as the *Seebeck effect.* A pair of junctions of this kind is known as a *thermocouple.*

In atomic terms the effect may be explained as follows: we suppose that the number of free electrons per unit volume and their average velocity varies from one metal to another. Thus, at the junction of two metals, there may be a tendency for electrons to migrate in one direction across the junction. Such migration cannot continue for long since it quickly sets up an opposing electric field that prevents any further movement across the boundary. The result is that a fixed potential difference arises between the two metals—known as the *contact p.d.* The velocities of the electrons in a metal depend to some extent on the temperature; we should therefore expect the contact p.d. between any two given metals to vary with the temperature. In Fig. 3.13 any difference between the two contact p.d.'s, v' and v, will act to drive current round the circuit, the resultant e.m.f. e being given by

$$e = v' - v$$

Although the contact p.d. between two metals may have a value up to about 0·5 volt, the thermoelectric e.m.f.'s are only of the order of a few millivolts for a temperature difference between the junctions of 100 deg. C.

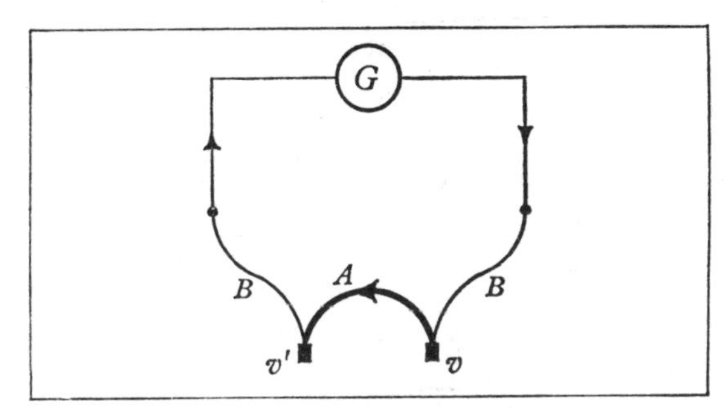

Fig. 3.13 A thermocouple

Practical arrangements using the Seebeck effect often need to employ more than two metals to make up the circuit. It is important to know what the effect of the additional junctions will be. Consider a circuit made up of three metals, A, B and C (Fig. 3.14). If the circuit is all at the same temperature, there is no possible source of energy that could drive current round it; therefore the algebraic sum of the three contact p.d.'s (taken in the same sense round the circuit) must be zero.

$$v_1 + v_2 + v_3 = 0$$

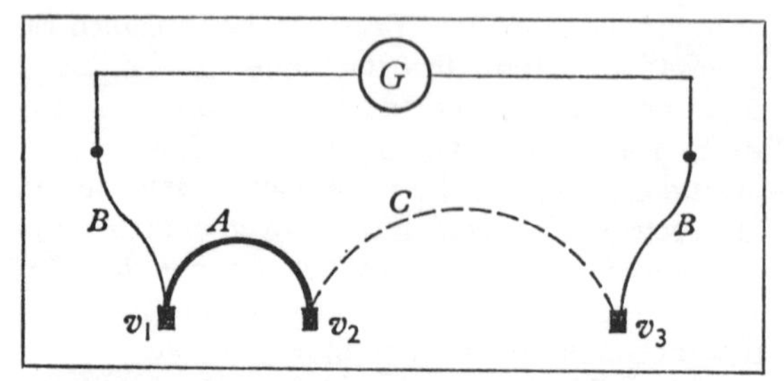

Fig. 3.14 The effect of introducing a third metal

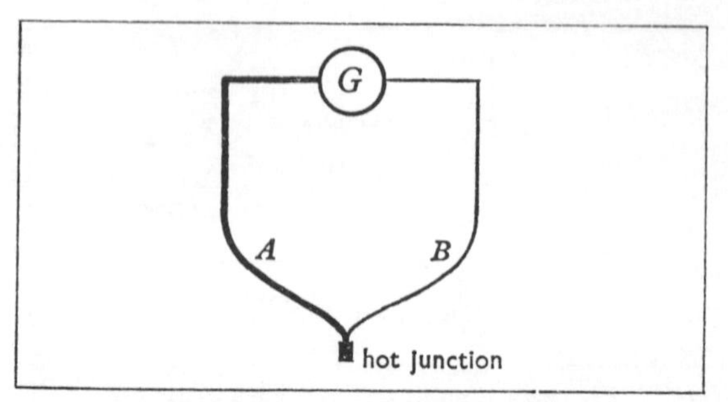

Fig. 3.15 A thermocouple with the cold junction in the galvanometer (at room temperature)

When the temperature of the junction A–B is changed, the contact p.d. at this point changes to a new value v_1', but v_2 and v_3 remain the same. The thermoelectric e.m.f. e acting in the circuit is then given by

$$e = v_1' + v_2 + v_3$$

but from above we have

$$v_2 + v_3 = -v_1$$
$$\therefore e = v_1' - v_1$$

This is the same e.m.f. as we should get if the metal C were eliminated from the circuit by bringing its junctions with A and B into contact. Thus, breaking a circuit at some point and inserting a piece of any metal, whose temperature is the same as that of the wires at the break, does not alter the total e.m.f. in the circuit. This is very useful in practical measurements. It means, for instance, that the existence of a film of solder between the metals A and B in Fig. 3.13 does not alter the total e.m.f. It also means that this circuit could just as well be connected as in Fig. 3.15 without affecting the e.m.f.; in this case the galvanometer takes the place of the metal C in Fig. 3.14, and it makes no difference what metals enter into its construction. It is essential for the whole conducting path inside the galvanometer to be at the same temperature; but provided this is so, we can treat the arrangement as though it formed a simple thermocouple of two metals only, taking the temperature of the second junction of A and B as that of the galvanometer.

The values of the e.m.f. of a given thermocouple over adjacent temperature ranges are additive: if the e.m.f. is ${}_ae_b$ when the junctions are at temperatures a and b, and ${}_be_c$ with the junctions at temperatures b and c, then the e.m.f. ${}_ae_c$ with the junctions at temperatures a and c is given by

$${}_ae_c = {}_ae_b + {}_be_c$$

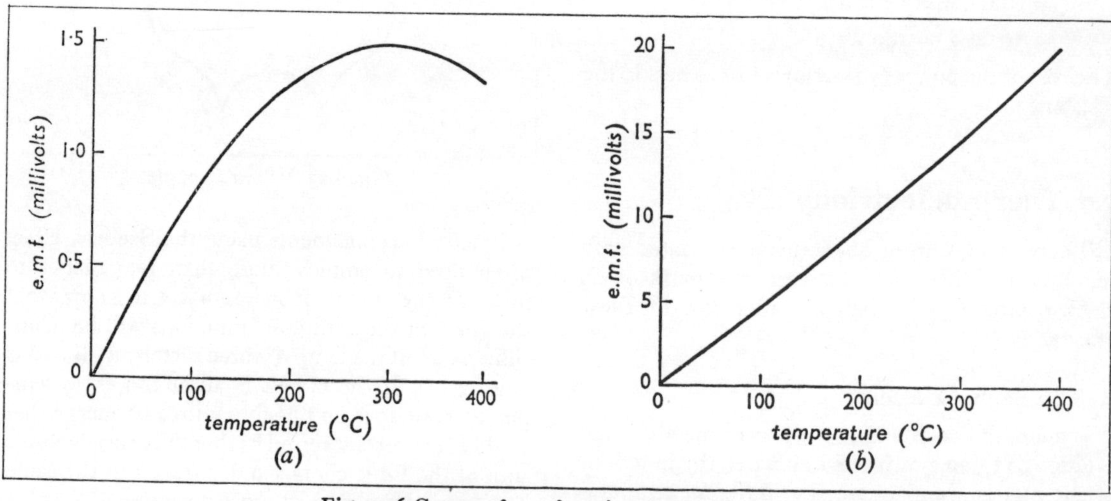

Fig. 3.16 Curves of e.m.f. against temperature:
(*a*) For a copper–iron thermocouple;
(*b*) For a copper–constantan thermocouple

This may be shown by considering the circuit formed by joining the two thermocouples with e.m.f.'s ${}_ae_b$ and ${}_be_c$ in series; the reasoning is left as an exercise for the student.

Thermoelectric thermometers

When one junction of a thermocouple is kept at a fixed temperature (0°C say) the variation of the e.m.f. e with the temperature t of the other junction usually follows a *parabolic* law:

$$e = at + bt^2$$

Fig. 3.16 shows the curves for copper–iron and copper–constantan thermocouples. The point at which the curve becomes horizontal is known as the *neutral temperature*; near this point the arrangement is clearly of no use for temperature measurement. But at other temperatures a thermocouple is a useful thermometer for many applications. Such an instrument must always be calibrated at a third fixed point (besides 0°C and 100°C) in order to find the constants a and b. Copper–constantan produces about 10 times the e.m.f. of copper–iron and is the more satisfactory combination for temperature measurement; also its e.m.f./temperature curve is almost linear over a large range. Another commonly used thermocouple is that employing the nickel–chromium alloys, *alumel* and *chromel*. A combination of platinum and a platinum–rhodium alloy is used for standard temperature measurements between 630°C and 1063°C.

The thermoelectric series

For a given temperature range the direction of the thermoelectric e.m.f. for different combinations of metals may be predicted by arranging the metals in a series; between the temperatures 0°C and 100°C the thermoelectric series for a selection of metals is:

> antimony, nichrome, iron, zinc, copper, gold, silver, lead, aluminium, mercury, platinum–rhodium, platinum, nickel, constantan (eureka), bismuth.

At the cold junction, the e.m.f. acts so as to drive current from the metal earlier in the list towards that later in the list. The series also indicates in a general way the magnitude of the e.m.f.; the further apart the two metals lie, the greater is the e.m.f. they produce. The largest e.m.f. (with metals) is obtained with an antimony–bismuth thermocouple.

(ii) The Peltier effect

When an electric current is driven round a circuit consisting of two different metals, heat is evolved at one junction and absorbed at the other. This is known as the *Peltier effect*. It is complementary to the Seebeck effect. Thus the energy for the e.m.f. of a thermocouple comes from the heat made available by the difference of temperature between the junctions. When this e.m.f. drives a current round the circuit, the Peltier effect leads to the evolution of heat at the cold junction and the absorption of heat at the hot junction—thus tending to equalize the temperatures. Indeed, by the *second law of thermodynamics* the net transfer of heat must be from the point at higher temperature to the point at lower temperature.

The rate of evolution or absorption of heat is proportional to the current; and the Peltier effect is therefore *reversible*. If the current is driven round the circuit *in opposition* to a Seebeck e.m.f., heat is evolved at the hot junction and absorbed at the cold one—tending to magnify the temperature difference. Starting with the junctions of a thermocouple at the same temperature, the Peltier effect will produce a temperature difference such that the resulting Seebeck e.m.f. acts *in opposition* to the current. (This can be regarded as an example of Le Chatelier's principle.)

The difficulty in demonstrating the small *Peltier effect* lies in separating it from the much larger *Joule effect* (p. 23), whereby heat is evolved in all parts of a circuit in which a current flows. The usual result is that, instead of a rise in temperature at one junction and a fall at the other, we get the temperature rising at *both* junctions—but slightly faster at one than the other. To satisfy ourselves that this is not because of differences of resistance at the junctions, we need to make use of the reversibility of the Peltier effect. Reversing the current in the circuit leaves the Joule effect unaltered ($W \propto I^2$), but reverses the Peltier effect; the temperature difference should then come the other way round.

This may be done with the circuit in Fig. 3.17. First, with the thermocouple at room temperature current is driven through it by the accumulator. After a short time the two-way switch S_1 is moved

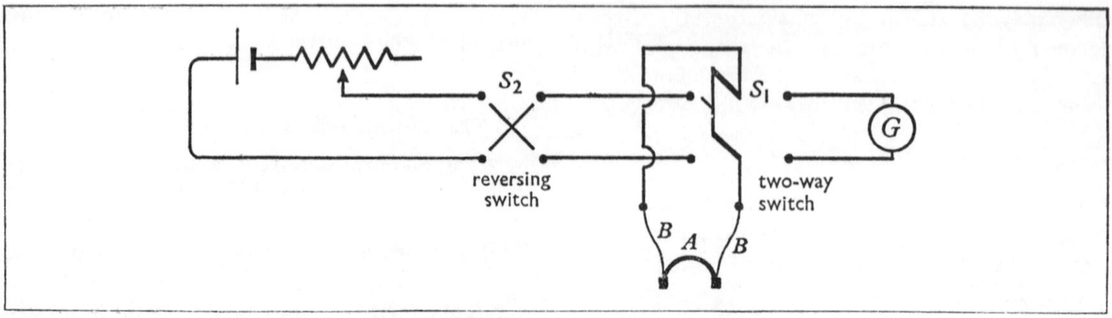

Fig. 3.17 Demonstrating the Peltier effect

over so that the thermocouple is connected to the galvanometer. The deflection observed is a measure of the temperature difference produced between the junctions. When the junctions have once more regained room temperature, the reversing switch S_2 is changed over, and the experiment is repeated with the current in the opposite direction. Now, when the thermocouple is joined to the galvanometer, the deflection is in the opposite direction, indicating that the temperature difference has reversed.

The Peltier effect is normally far too small to produce any significant cooling effect. However, some *semiconductor* materials (p. 20) show a vastly greater Peltier effect than any metal. With a suitably chosen pair of semiconductors a considerable cooling effect may be produced when the current flows the right way through the junction. At present the cost of these materials is high; but no doubt future developments will lead to an efficient refrigerator based on this principle.

4: Circuit Measurements

4.1 Resistance measurement

The direct measurement of a resistance consists in finding values for a steady current I through it and the corresponding p.d. V across it. The resistance R is then given by

$$R = \frac{V}{I}$$

Using an ammeter and voltmeter, either of the arrangements in Fig. 4.1 may be used. In (a) the resistance obtained initially is actually that of the ammeter and unknown coil in series. A subsidiary experiment must be performed with the voltmeter joined across the ammeter (as shown dotted) in order to find the resistance of the latter; the value of R is then the difference of the two results. Likewise in (b) the resistance obtained initially is that of the voltmeter in parallel with R; again a subsidiary experiment must be performed to find the resistance of the voltmeter so that the necessary correction can be applied. In each case several sets of readings are taken with different settings of the rheostat. Only by doing this can we be sure that the currents used are not causing excessive heating of the coil R with consequent change of its resistance.

The accuracy of the method is limited to that of the instruments used. The ordinary ammeter or voltmeter in regular use in a laboratory should not be relied on to an accuracy better than 2%. If it is an instrument of really good construction, it may be calibrated for the occasion (pp. 55, 58), and then its readings might be correct to within 0·5%. But the student should notice that, while the absolute accuracy is probably the same over most of the scale of a dial-reading instrument, the percentage accuracy increases with the deflection of the pointer. For instance a typical ammeter reads from 0 to 1·2 amps; and the smallest scale division represents 0·02 amps. If we can estimate to one tenth of the smallest scale division, the absolute accuracy is 0·002 amps. Now, at a scale reading of 0·2 amps, this gives a percentage accuracy of *reading* the instrument of 1%; but at a scale reading of 1·0 amp this becomes 0·2%. Thus for accurate work a dial-reading instrument should never be used at much less than half of full-scale deflection. Special standard ammeters and voltmeters are also obtainable which can be calibrated reliably to give results accurate to 0·1% at the best parts of their scales. These instruments have very long pointers (and therefore long scales) and a Vernier device to facilitate accurate readings.

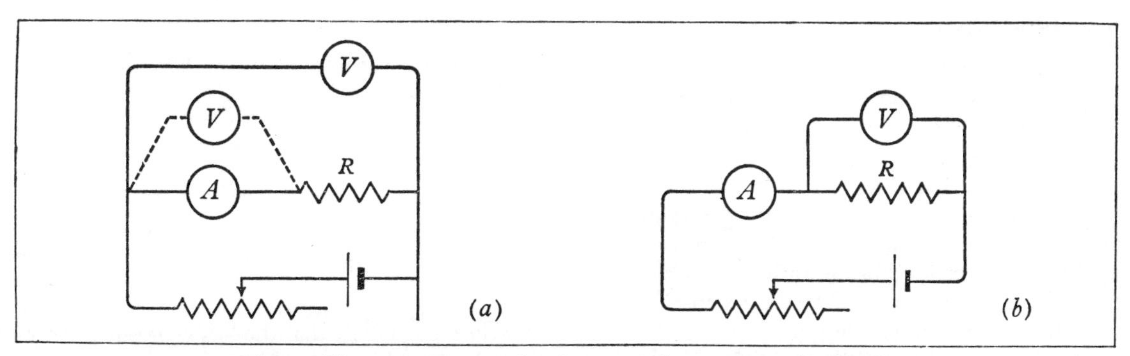

Fig. 4.1 The measurement of resistance with ammeter and voltmeter

Substitution method

As with many scientific measurements, the direct measurement of resistance is not easy to perform accurately. There are, however, a number of satisfactory methods for *comparing* two resistances; one of the simplest of these uses a substitution technique (Fig. 4.2). The reading of the ammeter

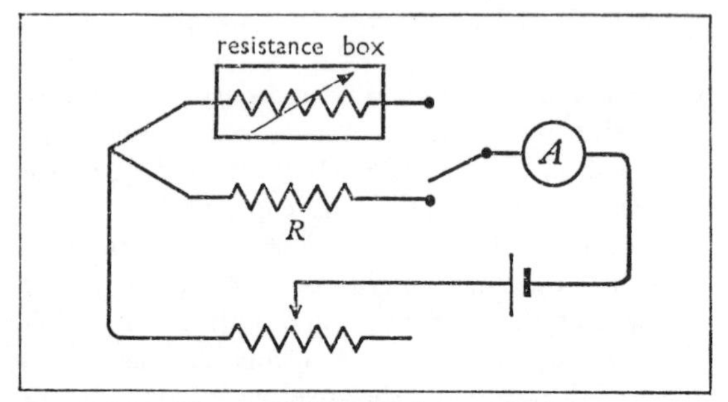

Fig. 4.2 A substitution method

is first adjusted to a suitable value; then the unknown coil of resistance R is disconnected and a standard resistance box joined in its place. The resistance of the box is adjusted until the ammeter reading is brought back to its former value; the total box resistance is then equal to R. The result does not depend on the correct calibration of the meter, but it does depend on its readings being precisely repeatable. With a good meter used at full-scale deflection an accuracy of 0·2% should be obtained. The method has the special merit that almost any circuit can be adapted for the purpose, and the value of a resistance may therefore be found under the conditions of the circuit in which it is to be used.

4.2 The Wheatstone bridge

For the quick and accurate comparison of resistances the Wheatstone network or 'bridge' is widely used. The arrangement is shown in Fig. 4.3. The four resistances P, Q, R, S are joined in a quadrilateral ABCD. A source of e.m.f. is connected across one diagonal AC, and a galvanometer and tapping key across the other BD. One or more of the resistances is adjusted until no deflection of the galvanometer can be detected when the tapping key is pressed. The bridge is then said to be *balanced.* When this condition holds, it may be

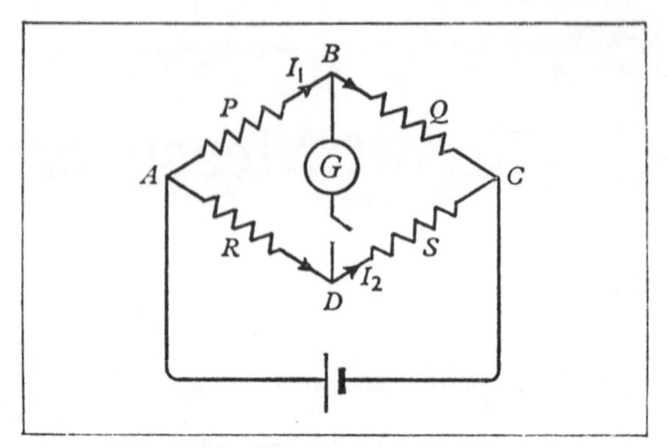

Fig. 4.3 The Wheatstone bridge network

shown that the relation connecting the four resistances is

$$\frac{P}{Q} = \frac{R}{S}$$

Thus if Q is a known resistance, and the ratio of R to S is known, P may be found. The proof of this result is as follows:

When the bridge is balanced, there is no current through the galvanometer; therefore the same current flows through P and Q; let this be I_1. Likewise the same current flows through R and S; let this be I_2. Also with no current through the galvanometer B and D must be at the same potential.

$\therefore$ the p.d. across AB = the p.d. across AD

$$\therefore I_1P = I_2R$$

Also, the p.d. across BC = the p.d. across DC

$$\therefore I_1Q = I_2S$$

Dividing $$\frac{P}{Q} = \frac{R}{S}$$

The condition for balance is unaffected by interchanging the positions of the cell and galvanometer in the circuit. Thus, if the cell is joined across BD and the galvanometer across AC, similar reasoning gives the balance condition as

$$\frac{P}{R} = \frac{Q}{S}$$

and this may be re-arranged to give as before

$$\frac{P}{Q} = \frac{R}{S}$$

It does not follow that the bridge is equally sensitive in the two arrangements; but with modern

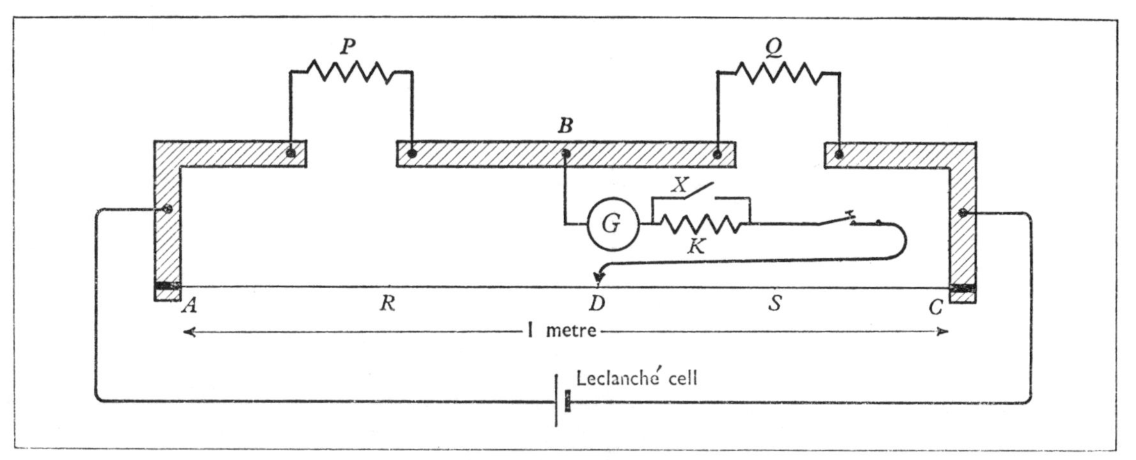

Fig. 4.4 The metre bridge

galvanometers there is usually sufficient sensitivity in both arrangements. The accuracy of the method is limited by that of the known resistances of the bridge, and should be better than 0·2%.

Two practical versions of the Wheatstone bridge are in common use:

(*i*) *The Metre bridge.* In this arrangement the resistors R and S of the Wheatstone network consist of the two parts AD and DC of a length of uniform resistance wire stretched over a metre scale (Fig. 4.4); the knife-edge contact maker D can be placed at any point on the wire so that the ratio R/S (= AD/DC) can be adjusted to balance the bridge. In the other arm of the bridge, P is the unknown resistance and Q a standard resistance coil. The connections between the parts of the bridge are made by means of heavy copper strips of low resistance. The wire ADC is heavily soldered to these strips; a connection resistance at these points of as little as 0·005 ohm might be equivalent to adding an extra millimetre to the wire, and could introduce a significant error. Errors on this account are in fact almost unavoidable; but it is possible to reduce them considerably by repeating the measurements with the standard resistance and the unknown interchanged in the gaps of the bridge. It can be shown that the average of the two results obtained is free from this sort of error, as long as the balance point is close to the centre of the wire; if the balance point is more than 20 cm from the centre, the errors increase rapidly, and the method is not satisfactory. It is therefore important to use only standards that are nearly equal to the unknown. It should then be possible to keep the errors below 0·5%. K is a protective resistor included in series with the galvanometer to prevent damage to it when preliminary attempts to find the balance point are being made. It can be by-passed by closing the switch X when an approximate balance point has been found; the maximum sensitivity of the galvanometer is then available for making the final adjustment.

(*ii*) *The Post Office box.* In this arrangement three of the resistances of the Wheatstone network, P, Q and R, consist of plug or switch type resistance boxes already connected together inside the one case (Fig. 4.5). The unknown resistance, the galvanometer (with its protective resistor) and the cell are connected externally to the appropriate terminals of the box. The resistors P and Q can have only the three possible values, 10, 100 and 1000 ohms; R can be adjusted in steps of 1 ohm from 0 to 10,000 ohms. Thus the ratio P/Q can have any of the values: 100, 10, 1, 0·1, 0·01, depending on the order of magnitude of the unknown resistance S. A balance is then obtained by varying R.

In testing the balance of the bridge, it is important to press the tapping key in series with the cell before that in series with the galvanometer; owing to the effects of self-induction (p. 103) there is a short interval after connecting the cell during which the currents in the circuit are growing to their final steady values. If the galvanometer is connected during this interval, a spurious deflection

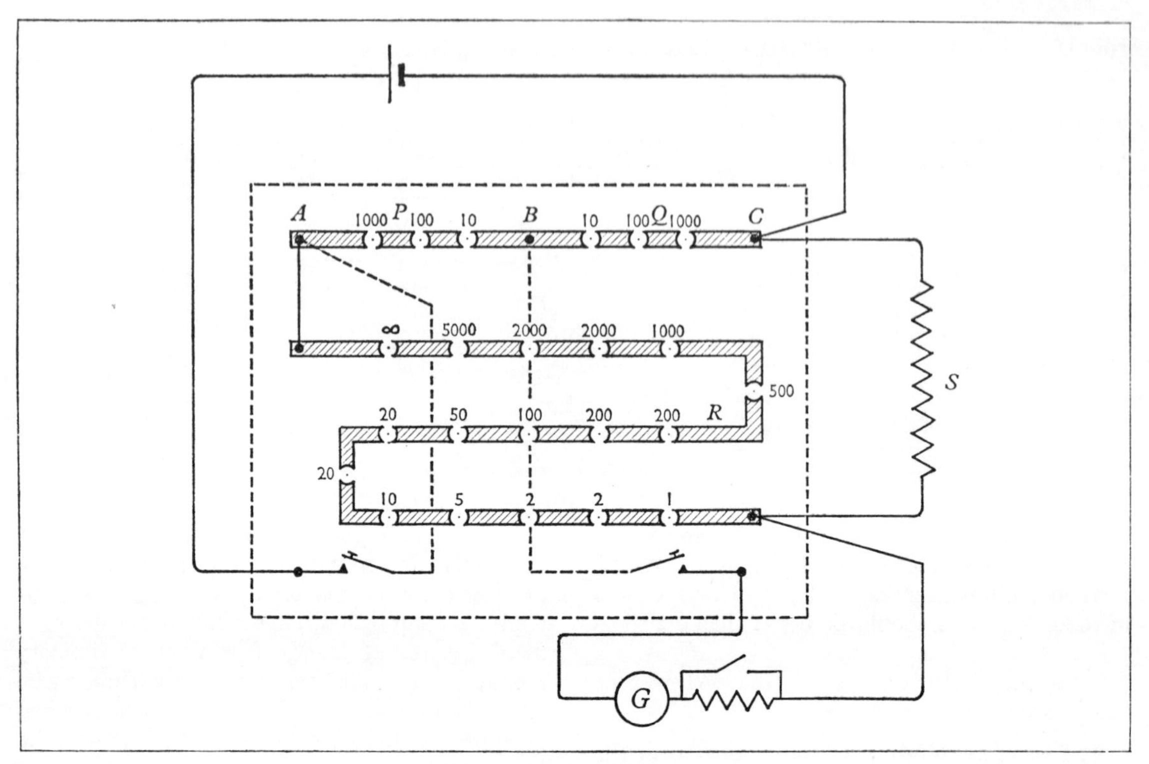

Fig. 4.5 The Post Office box

may be observed due to a momentary fluctuation of current, even though the bridge is actually balanced for steady currents.

A factor limiting the accuracy of the Wheatstone bridge is the presence of contact resistances at the points where the resistors are joined in the circuit. It is difficult to ensure that a screwed-down or soldered connection has a resistance of less than 0·002 ohm—and the circuit inevitably includes several such resistances. If all the contacts are clean and firm and all the resistances are greater than 10 ohms, it should be possible to obtain an accuracy of 0·1%—always assuming that the standards are known to this accuracy. When the resistance to be found is less than 1 ohm, the errors due to connection resistances of this sort are likely to be relatively large and the bridge method should be avoided. The potentiometer method given below is the most suitable one for measuring very small resistances.

★ Measurements on electrolytes

The condition for the balance of the Wheatstone bridge is not dependent on the steadiness of the e.m.f. of the cell. Indeed, provided the resistances used are free from inductive and capacitative effects, an alternating e.m.f. may be used; the galvanometer is then replaced by a pair of headphones or some other a.c. detecting device. The best operating frequency for headphones usually seems to be between 500 and 1000 cycles per second.

An a.c. bridge of this kind can be used to find the resistivity (or the conductivity) of an electrolyte. The use of direct current in this case would lead to polarization (see p. 41) and the introduction of unwanted e.m.f.'s. The use of alternating current prevents this. The conductivities of two electrolytes may be *compared* by measuring the resistances between two electrodes fixed in any suitable vessel, filled first with one liquid and then with the other. A U-tube is often used for this (Fig. 4.6*a*). For absolute measurements of conductivity the section of electrolyte used must be of known length and cross-sectional area, as in the Kohlrausch cell (Fig. 4.6*b*). The electrodes are two flat circular plates fitting closely in a cylindrical glass vessel. The leads making connection to the plates are insulated by glass rods so that only the region *between* the plates is used for conduction. The conductivities of electrolytes increase rapidly with temperature. It is therefore important to control the temperature in these experiments with great care, and small currents should be used.

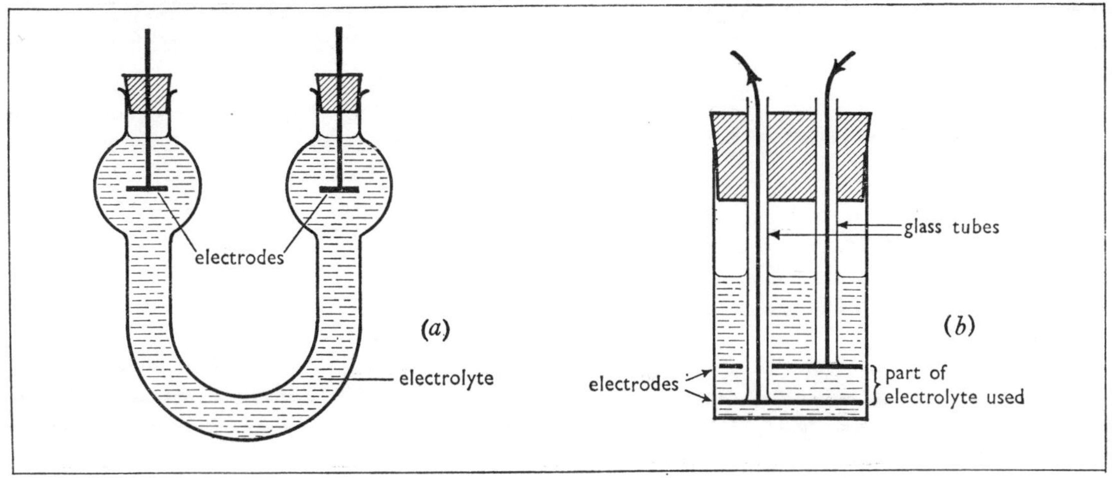

Fig. 4.6 (*a*) U-tube with fixed electrodes for comparing conductivities of electrolytes
(*b*) The Kohlrausch cell for measuring the conductivity of an electrolyte

4·3 The potentiometer

The potentiometer is a circuit arrangement used primarily for comparing potential differences; it can be adapted to measure currents and resistances. Under the best conditions it can do these measurements with an accuracy of 1 part in 10^6, and is therefore the only method now used for very precise d.c. measurements.

In its simplest form it consists of a uniform resistance wire AB stretched over a scale, through which a steady current is passed from one or more accumulators (Fig. 4.7). The p.d. V to be found is connected at X and Y. Its positive side (at X) is joined to the positive end A of the wire; and the negative side of the p.d. is joined through a galvanometer and tapping key to the contact maker D. In the loop of the circuit AXYD the p.d. between the points A and D on the wire acts in opposition to the p.d. V under test; if the two p.d.'s are equal, no current flows through the galvanometer. The position of the contact maker D is adjusted to bring about this condition. There is then no deflection of the galvanometer when the tapping key is pressed; and the instrument is said to be *balanced*. The length of wire l between the points A and D is measured.

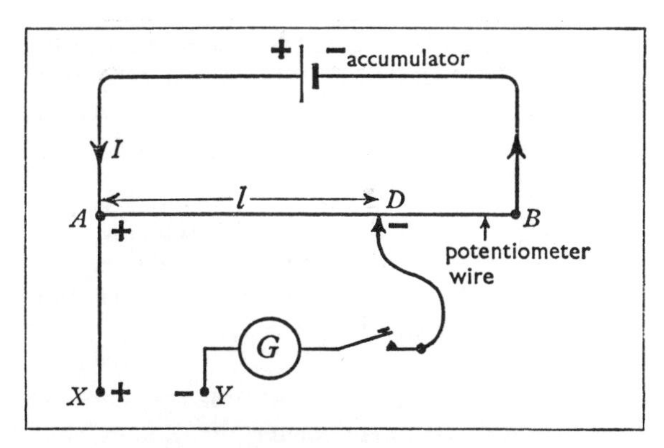

Fig. 4.7 The potentiometer circuit

Let the current through the potentiometer wire be I and let its resistance per unit length be r; then

$$\text{p.d. } V \text{ under test} = \text{p.d. between A and D} = Irl$$

If the wire is uniform and the current through it is steady, I and r are constants, and we may write

$$V \propto l$$

Two p.d.'s V_1 and V_2 may be connected in turn to the wire in this way and the corresponding balance lengths l_1 and l_2 found; then

$$\frac{V_1}{V_2} = \frac{l_1}{l_2}$$

We have assumed that the resistance between the terminal at A and the beginning of the wire at the zero point of the scale is negligible. This may be checked by repeating the measurements with a small resistance in series with the potentiometer wire, which will reduce the total p.d. across it.

Both the balance lengths will be changed, but their ratio should be unaltered. This procedure can also be regarded as a check on the uniformity of the potentiometer wire.

If a cell is joined to the points X and Y, no current will flow through it when the potentiometer is balanced, and the p.d. between its terminals is therefore precisely equal to its e.m.f. Thus by using a standard cell—normally the Weston cadmium cell (p. 43)—other p.d.'s may be found by direct comparison with the secondary electrical standard.

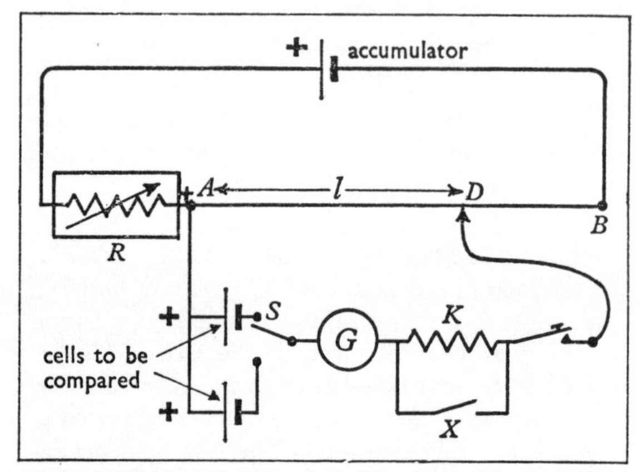

Fig. 4.8 The comparison of e.m.f.'s with a potentiometer

Fig. 4.8 shows the circuit used to compare the e.m.f.'s of two cells. The switch S enables either cell to be connected into the circuit. K is the usual protective resistance in series with the galvanometer to protect it from damage while preliminary attempts are being made to find the balance point for the contact maker D. It can then be shorted out by closing the switch X so as to give the maximum galvanometer sensitivity for the final adjustment. R is a plug resistance box, which is used as described above to obtain different pairs of values for the balance lengths l_1 and l_2, thus giving several independent estimates of the ratio l_1/l_2. Using a potentiometer wire 1 metre long and ordinary laboratory apparatus, the balance lengths can be found to the nearest half-millimetre; and the errors should then be less than 0·1%. Refined methods can greatly improve on this.

The potentiometer circuit can be regarded as a sort of voltmeter. But even in its simplest forms it is vastly more accurate than the best dial-reading instruments. It may be useful to list here the respects in which the potentiometer is superior to the ordinary type of voltmeter:

1. The scale of the potentiometer, stretched along the wire, may in principle be made as long as we choose, so that its sensitivity may be increased almost indefinitely. The scale of the voltmeter is limited by the maximum range of movement of the tip of the pointer.

2. The adjustment of a potentiometer is a 'null' one, and does not in any way depend on the calibration of the galvanometer. The galvanometer is used only to detect the current, not to measure it. The accuracy of an ordinary voltmeter is limited by its calibration.

3. When the potentiometer is balanced, it takes no current from any other circuit to which it is connected, and does not therefore disturb the p.d.'s it is measuring; if it is connected to a cell, no drop of p.d. occurs on account of internal resistance. A voltmeter takes some current, and will often materially affect the p.d. it is required to measure.

4. Since no current flows in the galvanometer circuit at balance, it follows that there is no potential difference between A and X or between D and Y (Fig. 4.7), whatever may be the resistance of the galvanometer circuit. Thus, fine connecting wires and sharp knife-edges may be used at A, D, X, and Y without in any way affecting the result. The only effect of a high resistance in the galvanometer circuit is somewhat to reduce its sensitivity for detecting a lack of balance. With a voltmeter a connection resistance comparable with that of the voltmeter itself would have a considerable effect on the readings.

5. The measurements of a potentiometer may be made by direct reference to the secondary electrical standard, namely the Weston cadmium cell. A voltmeter may only be calibrated against this standard by using a potentiometer.

On the other side it may be said that the voltmeter is direct reading, convenient in size, and always ready for use. The potentiometer is relatively cumbersome and slow to use.

In common with other electrical apparatus of any complexity a potentiometer circuit is capable of developing faults that can be quite baffling to those not used to finding them. A simple checking routine that enables many types of faults to be spotted at once is the following: with a large protective resistance in series with the galvanometer place the contact maker first at one end of the wire and then at the other; the deflections should be *in opposite directions.*

If this result is not obtained, it should be possible to locate the fault approximately from the behaviour of the galvanometer. The student is invited, as an exercise, to try and analyse the nature of the fault in each of the following cases before looking at the 'solutions' given below.

1. Both deflections are in the same direction, the one with the contact maker at the zero of the scale being the smaller of the two.

2. Both deflections are in the same direction, the one with the contact maker at the zero of the scale being the larger of the two.

3. The same deflection is obtained for all positions of the contact maker.

4. No deflection is obtained for any position of the contact maker.

5. The deflections of the galvanometer are opposite for the two positions of the contact maker, but the pointer flickers unsteadily.

Solutions:

1. The p.d. under test is connected the wrong way round; reverse its connections.

2. The accumulator that drives the current through the potentiometer produces too small a p.d. across the wire; either this accumulator needs re-charging or, if there is a resistance in series with the wire, this must be reduced.

3. There is a broken connection somewhere in the part of the circuit in series with the potentiometer wire and accumulator.

4. There is a broken connection somewhere in the galvanometer circuit (AXYD).

5. There is a loose connection somewhere in the circuit; shaking each of the wires forming the circuit while observing the effect on the galvanometer reading will probably quickly reveal the position of the fault.

Calibrating a voltmeter

Fig. 4.9 shows the arrangement used to calibrate a voltmeter for readings in the range 0·2 to 2·0 volts. It is really two distinct circuits, which are joined together at the points X and Y. The top circuit is the usual potentiometer arrangement. The lower circuit is designed to provide an adjustable p.d. that will be measured both by the voltmeter and the potentiometer; the variable resistor is connected as a potential divider so that the p.d. can be varied continuously from zero upwards.

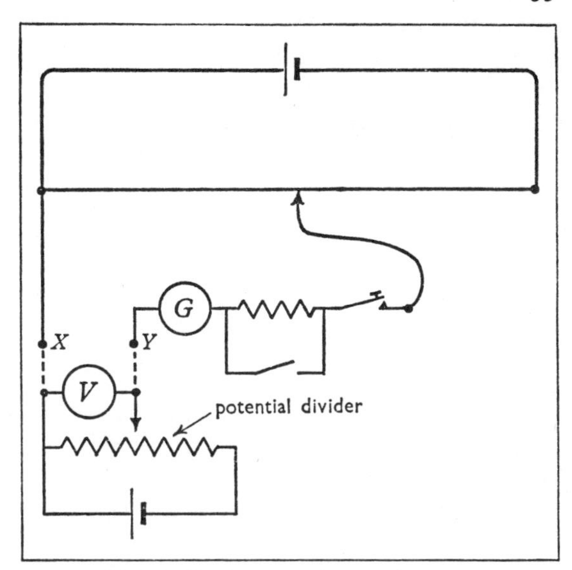

Fig. 4.9 The calibration of the voltmeter

The potentiometer is calibrated by finding the balance length for a standard cell connected to X and Y in place of the voltmeter and its associated circuit. This is done at intervals during the experiment so that a check is kept on the steadiness of the current through the potentiometer wire. From the balance length and known e.m.f. of the cell the potential drop per cm of potentiometer wire is found; the p.d. across any other balance length on the wire may then be calculated. Since the potentiometer is capable of much higher accuracy than any voltmeter, we may use it to obtain the error of the voltmeter at a number of points of its scale.

When the range of a voltmeter exceeds the p.d. that can reasonably be joined across the potentiometer wire, the circuit must be adapted as in Fig. 4.10. The resistors R_1 and R_2 form a *fixed* potential divider by means of which a known fraction of the p.d. across the voltmeter may be tapped off and measured by the potentiometer. Suppose for instance the voltmeter reads up to 12 volts and a 2-volt potentiometer is available. By choosing

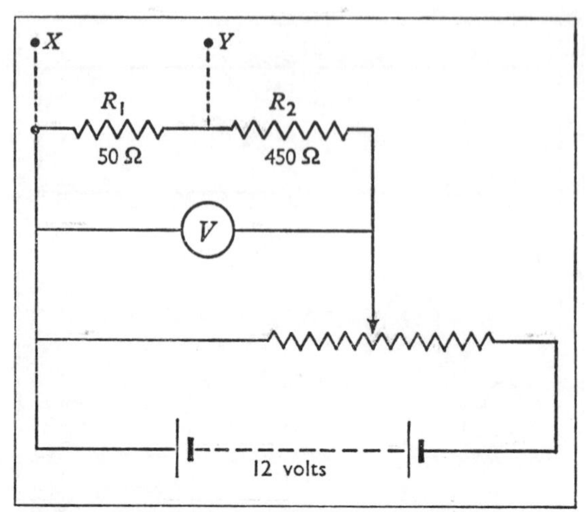

Fig. 4.10 The use of a potential divider to tap off a known fraction of a p.d. to be measured

R_1 50 ohms and R_2 450 ohms the p.d. between X and Y will be $\frac{1}{10}$ of that across the voltmeter, and will always lie in the range 0 to 1·2 volts.

The arrangement in Fig. 4.9 is suitable for p.d.'s down to about 0·2 volt; below this the balance length on a potentiometer wire 1 or 2 metres long is too small to be measured with sufficient accuracy. Thus with a single 2-volt accumulator joined across a 1-metre potentiometer wire the balance length for a p.d. of 0·1 volt would be 5 cm; an uncertainty of $\frac{1}{2}$ mm in this represents an error of 1%. It is then necessary to join a resistance in series with the wire so as to reduce the total p.d. across it. The balance length for a small p.d. can thus be made large enough to be found accurately. The technique is similar to that employed for finding the e.m.f. of a thermocouple discussed below.

Measurements on a thermocouple

The e.m.f.'s of thermocouples are generally in the range from 0 to 20 mV. To measure temperatures to the nearest 0·1°C changes in e.m.f. of about 4 μV must be detected. For this purpose a large resistance must be joined in series with the potentiometer wire (Fig. 4.11). This is really equivalent to extending the length of the potentiometer wire, thereby increasing the sensitivity of the instrument to the required point. For example, suppose we have a potentiometer wire 2 metres long of resistance 2 ohms per metre. A resistance of 1000 ohms joined in series is equivalent to using an extra 500 metres of potentiometer wire. If

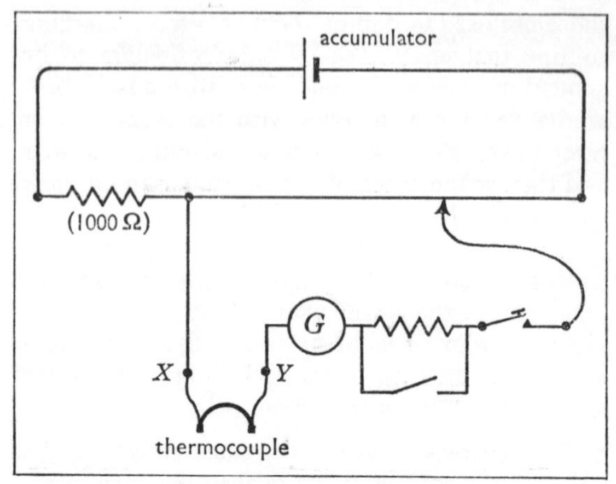

Fig. 4.11 The comparison of thermocouple e.m.f.'s with a potentiometer

the current is supplied by a 2-volt accumulator, the p.d. across the original 2 metres of wire is now about 8 mV, and 1 mm represents about 4 μV change in e.m.f. To make use of this increased sensitivity it is necessary to have a galvanometer that will respond to the very small p.d.'s concerned. The arrangement in Fig. 4.11 enables thermal e.m.f.'s to be *compared* one with another. It can therefore be used for *temperature measurements* with a thermocouple. If the p.d. across the accumulator is measured with an ordinary voltmeter, the

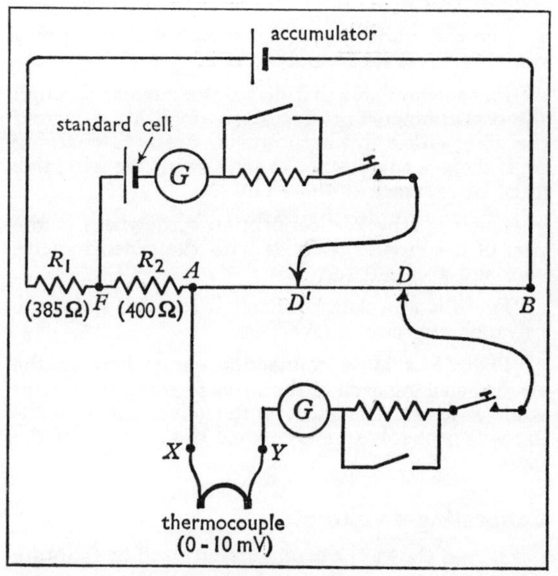

Fig. 4.12 The measurement of a thermocouple e.m.f. against the e.m.f. of a standard cell

actual values of the e.m.f.'s can be found with an accuracy equal to that of the voltmeter.

Further modification of the circuit is needed if the values are to be found by direct comparison with the e.m.f. of a standard cell. The resistance in series with the wire must then be divided into two nearly equal parts R_1 and R_2 (Fig. 4.12). These are chosen so that the p.d. across R_2 is nearly 1·0186 volts (the e.m.f. of a Weston cell). The standard cell and its associated circuit are joined between F and the contact maker D′, and a balance position can then be found with D′ somewhere on the wire. The thermocouple is joined between A and the contact maker D. For example, if the thermocouple e.m.f. ranges from 0 to 10 mV and the potentiometer wire is 2 metres long and of resistance 2 ohms per metre, the p.d. across the wire needs to be about 1/200 of the e.m.f. of the accumulator (2 volts). The total resistance in series therefore needs to be about 200 times the resistance of the wire—i.e. about 800 ohms. To give a balance point for the standard cell somewhere on the wire, suitable values of R_1 and R_2 would be 385 ohms and 400 ohms, respectively. But in practice such calculations are used to give an idea of the orders of magnitude required, and then the exact values of the two resistances are chosen by trial and error. First one of them is adjusted until the balance length for the e.m.f. to be measured is sufficiently large; then both the resistances are varied, keeping $(R_1 + R_2)$ constant, until the standard cell can be balanced at some point on the wire. The method of calculating the e.m.f. from the readings is shown in the following example.

Example. In the circuit of Fig. 4.12, for a certain temperature difference between the junctions of the thermocouple, the balance length on the wire is 159·2 cm. The standard cell is balanced when joined across R_2 (400 ohms) and 60 cm of the wire. Calculate the e.m.f. of the thermocouple, given that the e.m.f. of the standard cell is 1·0186 volts.

Since the resistance of the potentiometer wire is 2 ohms per metre, R_2 has the same resistance as 200 metres of the same type of wire. The effective balance length for the standard cell is therefore 20,060 cm. The two e.m.f.'s are in the ratio of their balance lengths:

$$\therefore \frac{E}{1{\cdot}0186} = \frac{159{\cdot}2}{20{,}060}$$

where E is the e.m.f. of the thermocouple.

$$\therefore E = \frac{1{\cdot}0186 \times 159{\cdot}2}{20{,}060} \text{ volt}$$

$$= 8{\cdot}083 \text{ mV}$$

In the precise measurement of p.d.'s, and particularly if these are very small, the greatest care has to be taken to eliminate stray thermal e.m.f.'s in series with the p.d. under test. For instance, if the potentiometer wire and knife-edge contact maker are made of different materials, the junction will act as part of a thermocouple if its temperature is raised above that of the rest of the circuit by the proximity of the observer's hand. The same effect may arise at any junction of dissimilar metals in the galvanometer circuit, unless care is taken to preserve uniform temperatures. In the construction of high-precision apparatus it is sometimes the practice to use silver-plated conductors throughout the circuit. However, in using ordinary equipment the student must be aware of this source of errors and conduct the experiment accordingly.

4.4 Measuring currents with a potentiometer

The current to be found is arranged to pass through a known resistance R (Fig. 4.13). The p.d. V across this is measured with a potentiometer in the usual way by comparison with the e.m.f. of a standard cell. The current I is then given by

$$I = \frac{V}{R}$$

The standard resistance R should normally be chosen so that the p.d. V across it is between 1 and 2 volts, since this is the range in which a simple potentiometer circuit is most effectively used.

For precise work resistors to be used with a potentiometer in this way are made with *four* terminals (Fig. 4.14). Two of these, known as the *current terminals*, are at the ends of the resistance wire, and are used to pass the current into and out of it. The other two, known as the *voltage terminals*, are joined by fine wires to two points A and B on the resistance wire; and it is between these points that the specified resistance exists. The p.d. between the voltage terminals is measured with the

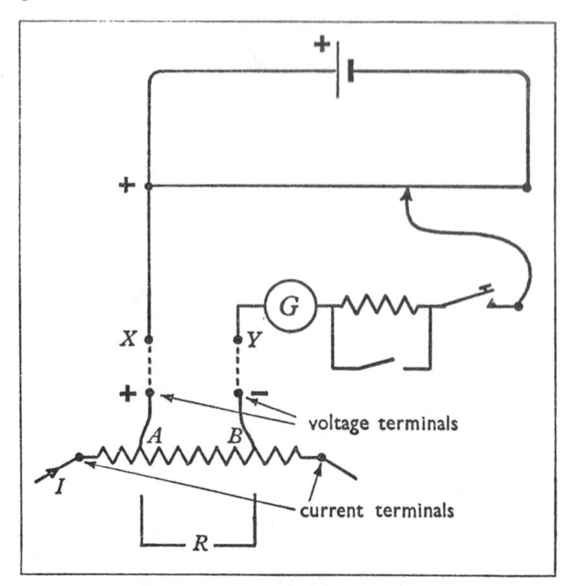

Fig. 4.13 The measurement of current with a potentiometer

potentiometer and standard cell; this is equal to the p.d. between A and B, since at balance no current passes through the fine wires and voltage terminals. With a two-terminal resistor the actual value of the resistance always includes the somewhat uncertain resistance of the contacts through which the current is passing. Even with heavy brass connectors the contact resistances could amount to 0·005 ohm. Only with resistances over 5 ohms could the error on this account be certainly less than 0·1%.

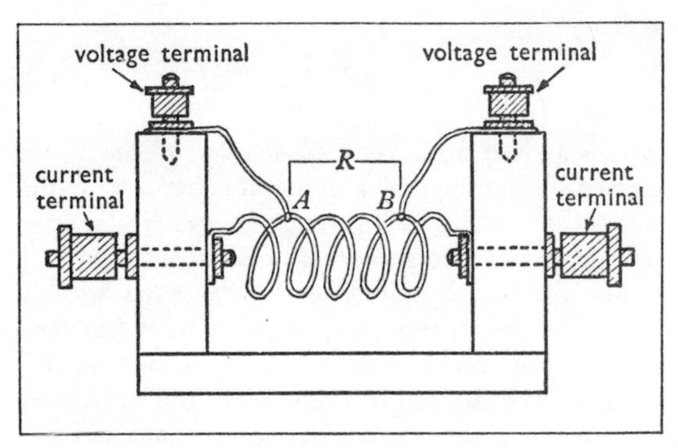

Fig. 4.14 The construction of a '4-terminal' resistor

Calibrating an ammeter

The potentiometer is used to measure the p.d. V across a suitable standard resistance R in series

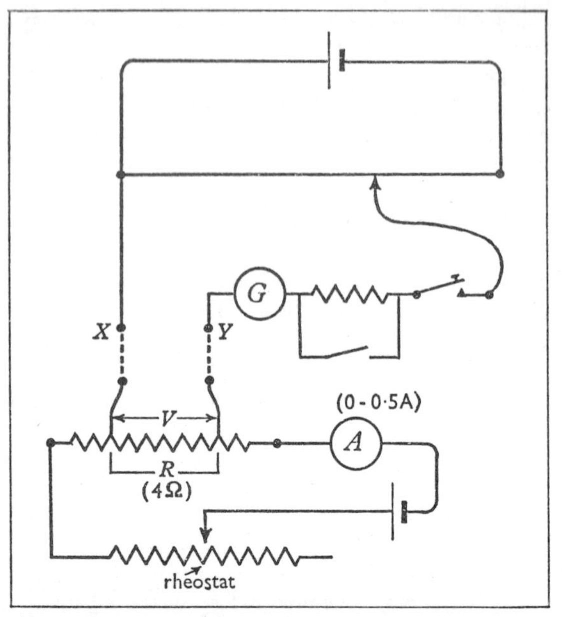

Fig. 4.15 The calibration of an ammeter

with the ammeter. The current I is then given by

$$I = \frac{V}{R}$$

this can be compared with the ammeter reading, so giving its error at any chosen point of its scale.

Fig. 4.15 shows the arrangement that would be used for calibrating an ammeter reading up to 0·5 amp. It should be regarded as two distinct circuits joined together at the points X and Y, the usual potentiometer circuit above, and the circuit providing the current for the ammeter and standard resistance below. The adjustment of the current is made with the variable resistance, used in this case as a rheostat, since the ammeter is a low-resistance device (p. 26). The resistance R is chosen so that the balance point will be near the end of the wire for the maximum current to be measured; i.e. V should be about 2 volts, when I is 0·5 amps

$$\therefore R = \frac{V}{I} = \frac{2}{0{\cdot}5} = 4 \text{ ohms}$$

The potentiometer is calibrated, as usual, by finding the balance length on the wire for a standard cell joined to X and Y in place of the ammeter circuit. From this the potential drop per cm of the wire is worked out, and the p.d. for any given balance length can then be calculated.

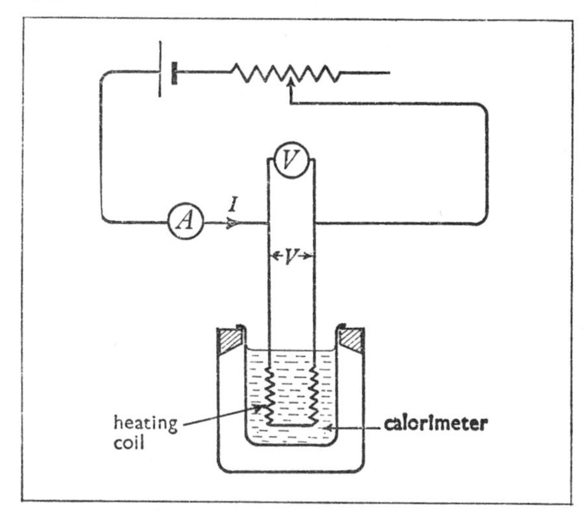

Fig. 4.16 The measurement of the heat supplied to a calorimeter with ammeter and voltmeter

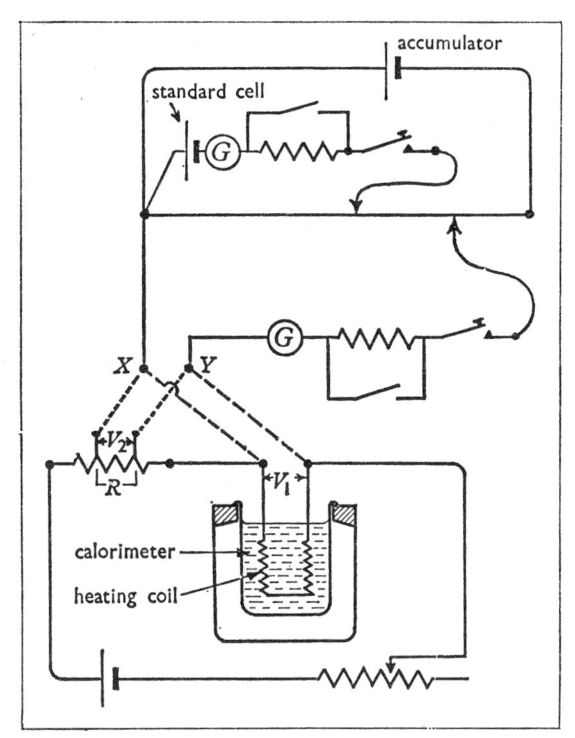

Fig. 4.17 The measurement of the heat supplied to a calorimeter with a potentiometer

Electrical calorimetry

The most satisfactory means of supplying a measured quantity of heat to a calorimeter and its contents is by an electric current. When the accuracy required is no better than 2%, the electrical measurements can be carried out using an ammeter to give the current I through the heating coil and a voltmeter the p.d. V across it. If the current is switched on for a length of time t, then

$$\text{heat supplied} = VIt$$

The heat is given in *joules* when V, I and t are measured in *volts*, *amps* and *seconds*, respectively.

Fig. 4.16 shows the circuit used. In this circuit a small part of the current recorded by the ammeter passes through the voltmeter instead of the heating coil. If the resistance of the voltmeter is large compared with that of the heating coil, this can be ignored; otherwise a correction must be applied. The current i through the voltmeter is given by

$$i = \frac{V}{\text{resistance of voltmeter}}$$

and this must be subtracted from the ammeter reading to give the actual current I through the heating coil.

When the accuracy required is higher than that of the meters available, the electrical measurements must be made with a potentiometer (Fig. 4.17). A suitable standard resistance R is connected in series with the heating coil; and the p.d.'s across R and the heating coil are both measured with the potentiometer by comparison with the e.m.f. of a standard cell. If these p.d.'s are V_1 and V_2 as shown, then

$$I = \frac{V_2}{R}$$

and the heat supplied in t seconds is given by

$$\text{heat supplied} = V_1It = \frac{V_1V_2t}{R}$$

The measurements are thus carried out by direct reference to the secondary electrical standards—standard cell and standard resistor. Accuracies approaching 1 part in 10^5 can thus be reached, which exceeds the accuracy of the rest of the measurements in any calorimeter experiment.

The measurement of temperature in such experiments is also often performed by electrical methods. In precise work ***resistance thermometers*** are usually used (p. 60). (The design of calorimeters and the methods used for eliminating heat losses are matters outside the scope of this book; and the student should refer to textbooks of *Heat* for details about them.)

4.5 Measuring resistance with a potentiometer

The potentiometer provides the most accurate known method for the comparison of two resistances. The arrangement used is shown in Fig. 4.18. For precision work both resistors should be of the four-terminal type described above (Fig. 4.14). A suitable current I is passed through the

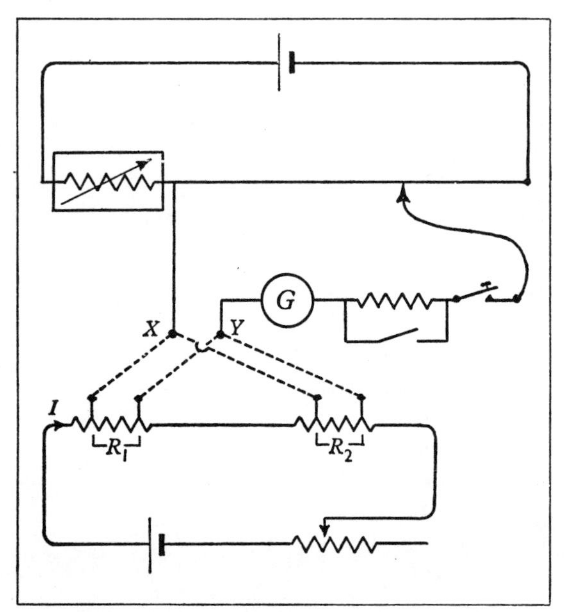

Fig. 4.18 The comparison of resistances with a potentiometer

two resistors, R_1 and R_2, connected in series. The potentiometer terminals X and Y are joined in turn across each resistor, and the corresponding balance lengths l_1 and l_2, are found. Now,

$$\frac{l_1}{l_2} = \frac{\text{p.d. across } R_1}{\text{p.d. across } R_2} = \frac{R_1}{R_2}$$

provided the current I does not change between the two measurements. To check this it is advisable to repeat both measurements several times. Only the ratio of the two p.d.'s need be known, so that the potentiometer does not need to be calibrated with a standard cell as with measurements of p.d. and current. It is simplest to choose the current I so that the two p.d.'s fall in the range from 1 to 2 volts, in which a simple potentiometer circuit is most effectively used. However, it may well be that the currents must be much smaller than this in order to avoid overheating the coils. In this case the potentiometer circuit must be adapted to the required range of p.d.'s by joining suitable resistances in series with the wire. In any case the measurements should always be repeated with various values of the current I through the resistors. Only by doing this can we find out for certain whether R_1 and R_2 are independent of current in the range of currents used. The potentiometer method provides the most satisfactory means of measuring very small resistance (less than 1 ohm); the use of the four-terminal technique makes it possible to avoid the errors due to the uncertain connection resistances at the terminals.

Resistance thermometers

Over much of their range resistance thermometers are appreciably more sensitive than any other thermometer even including the standard gas thermometers; they are also very convenient to use. Gas thermometers must still be used to measure the subsidiary fixed points required for calibrating other thermometers, but precise laboratory measurements of temperature are now largely performed with some form of resistance thermometer—chiefly the ***platinum resistance thermometer***. A large part of the International Temperature Scale is now defined with reference to this instrument.

The platinum resistance thermometer

The fine platinum wire whose resistance X is to be measured is contained in a hard glass U-tube inside the outer protective sheath. The ends of the fine wire are welded to *thick* platinum wires, which in turn are welded to silver leads connecting to the terminals at the top of the thermometer (Fig. 4.19). The fine wire must be of the highest purity; very small traces of impurity can appreciably affect its temperature coefficient of resistance. For this reason the silver leads cannot be joined direct to the fine wire, since the diffusion of silver into its ends would significantly affect its properties. Since the resistance X is to be measured with a potentiometer, four leads are used—two current leads and

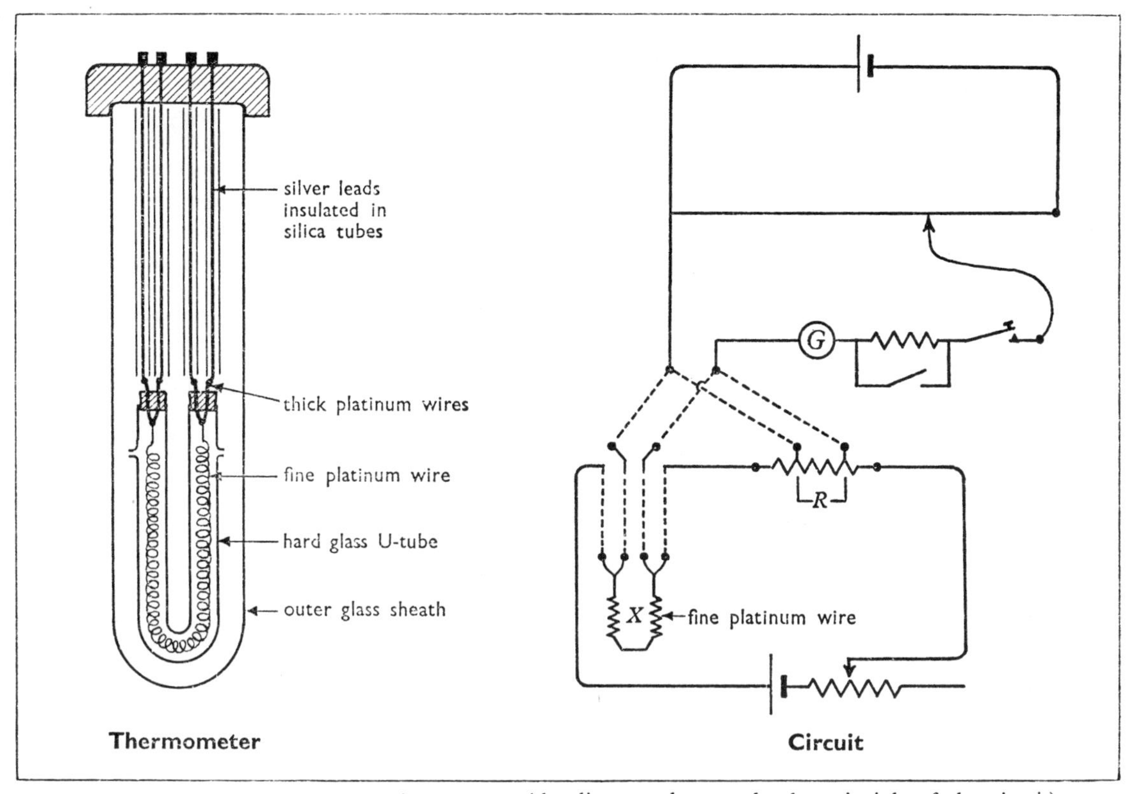

Fig. 4.19 A platinum resistance thermometer (the diagram shows only the principle of the circuit)

two voltage leads—so that the resistance actually measured is that of the fine platinum wire only. The current leads are used to join the wire in series with a standard resistance R; and the resistances are compared with a potentiometer in the usual way.

$$\therefore \frac{X}{R} = \frac{\text{balance length for } X}{\text{balance length for } R}$$

In order to measure a temperature, the instrument is first calibrated by finding the resistance of the platinum wire at the *ice-point* (X_0) and at the *steam-point* (X_{100}). Then, if X_t is its resistance at the unknown temperature t, the value of t on the Centigrade scale of the platinum resistance thermometer is given by

$$\frac{t}{100} = \frac{X_t - X_0}{X_{100} - X_0}$$

The variation of the resistance of a platinum wire with temperature is not quite linear, so that the value of t does not coincide with the Standard Scale temperature except in the immediate neighbourhood of the two fixed points. The deviation of the Platinum Scale temperature from the Standard Scale is calculated from measurements made at subsidiary fixed points, such as the *sulphur-point* and *oxygen-point*. Details of this technique will be found in textbooks of *Heat*.

4.6 Electrical standards

The accuracy of d.c. measurements in a laboratory depends ultimately on our knowledge of the secondary electrical standards, namely, the e.m.f. of a standard cell and the resistance of a standard coil. The measurement of these quantities by direct reference to the definitions of the volt and the ohm is a matter of considerable difficulty, and is rarely attempted outside the standardizing laboratories of the world. The present accuracy obtainable is about 1 part in 10^5; but the accuracy with which standard cells and coils can be *compared* amongst themselves is about 1 part in 10^6. The present practice is therefore to state the values of the e.m.f.'s of standard cells and the resistances of standard coils to 1 part in 10^6, while recognizing that the values given may be in error in terms of the definition of the volt and the ohm by up to 1 part in 10^5. International uniformity is preserved by frequent inter-comparisons of the standards produced by the various standardizing laboratories. Thus, at the present time the values of the volt and the ohm are *fixed by agreement* to within 1 part in 10^6; but future improvements in standardizing techniques may force us in due course to revise these values by anything up to 1 part in 10^5. Indeed there is no certainty that the agreed international values are not drifting by up to this amount from year to year—except that it is unlikely that all the standard cells and coils in the standardizing laboratories of the world should drift suddenly by the same amount.

At one time an accuracy of even 1 part in 10^5 was unattainable in the absolute measurements, and it was then the practice to choose arbitrary standards of the basic electrical quantities. The *international ohm* was stated to be the resistance of a specified column of mercury, while the *international ampere* was stated in terms of the deposition of silver in a specified silver voltameter. The *international volt* and the *international watt* were then fixed by reference to these. These units were defined so as to agree with the *absolute ohm, amp, volt, etc.* as far as these could be established at the time (1908). However, improvements in standardizing techniques gradually revealed serious divergences between the international and absolute units, the differences amounting to about 1 part in 2000 in some cases. The use of this system therefore became increasingly inconvenient; and in 1948 the old international units were abolished, and the system outlined above was adopted instead.

Each of the standardizing laboratories preserves its own group of standard cells and standard coils. No single cell or coil could be relied on to remain constant over a period of years. Therefore a group of standards is used, which are regularly compared with each other to check their relative constancy. At the National Physical Laboratory in Great Britain a group of about 20 standard cells is used; these are selected from an even larger batch of cells, the group selected being that which shows the least relative variation over the previous few years. The mean e.m.f. of the group is then assumed to have remained constant since the last international assessment. The composition of the working group is always changing, since it is found that cells develop instability after about 15 years; also it takes 2 or 3 years for a new cell to settle down completely. ('Instability' in this connection means variations of e.m.f. of a few parts in a million!) Likewise, a group of about five resistors is selected from a larger batch, and their mean resistance is assumed to remain constant from year to year.

Periodically, each standardizing laboratory measures the values of its standards by absolute measurements in terms of the definitions, using the best techniques available at the time. The national standards from all over the world are collected together and compared, using cells and coils selected from the national standard groups. Corrections are then applied to the national standards to bring them all into conformity. By this means the standards in use in different parts of the world are kept in agreement to the highest accuracy possible, while their absolute accuracy continually improves with new advances in standardizing techniques.

In the National Physical Laboratory three methods are used to fix the values of the standards:

(i) *The Lorenz rotating disc apparatus* (p. 107) is used to measure the *resistance* of the standard coils.

(ii) *A standard mutual inductance* of known dimensions (p. 113) is used for the direct measurement of inductance and capacitance; with a suitable a.c. bridge it can also be used to measure the *resistance* of the standard coils, as a check on the results of the Lorenz disc experiment.

(iii) *The current balance* (p. 8) can be used for the direct measurement of *current*; it can therefore be used to find the p.d. across an already standardized resistance coil through which this current flows. Using a potentiometric technique this p.d. can be compared with the *e.m.f. of a standard cell.*

5: Magnetic Effects of a Current

5.1 Lines of Force

Magnetic forces arise between charged particles in motion (p. 6). All atoms are therefore inherently magnetic because of the circulating electrons they contain. To a large extent the orbits of the electrons form equal and opposite pairs cancelling out their magnetic effects, but with many atoms there is some residual magnetic effect, caused chiefly by the 'spin' of the electrons. With most substances the effects observed are very small. But with *iron, nickel, cobalt* and a few other substances interactions between atoms produce intense magnetic properties. These are called *ferromagnetic* materials, and it is from these that magnets may be made.

The properties of ferromagnetic materials are extremely complex and we shall defer the study of them until the next chapter. However, they all arise from the behaviour of the circulating systems of charge within the atoms; and we therefore need in this chapter to investigate the nature of the magnetic behaviour of charged particles in motion, i.e. of electric currents. We give first an elementary account of the properties of magnets, since a small pivoted magnet (or *compass*) provides a convenient tool for studying magnetic effects.

In a magnet the magnetic properties appear to be concentrated in certain regions only. These regions are called the *poles* of the magnet. Thus if a bar magnet is plunged into a bowl of iron tacks, scarcely any adhere to the central parts of the bar, while many tacks cluster round its ends. A magnet seems to influence the space around it so that forces act on any other magnet placed nearby. Any region of space in which a magnet would be acted on by forces is called a *magnetic field.*

Magnetic poles are of two kinds. It is easily shown that the forces acting on the two kinds of poles placed in turn at the same point in a magnetic field are in opposite directions. The forces between two poles are such that *like poles repel each other, while unlike poles are attracted together.* In

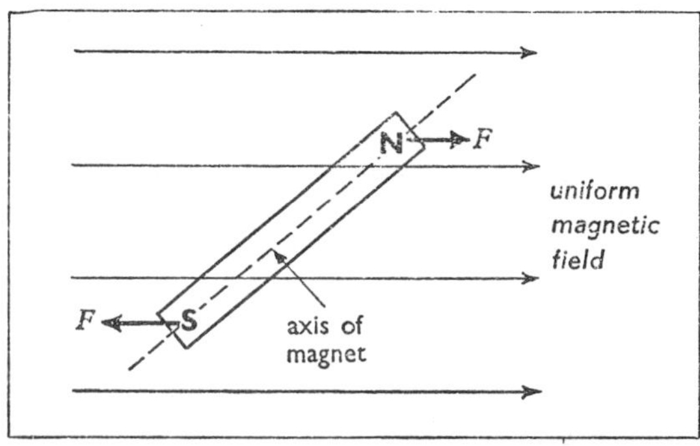

Fig. 5.1 A magnet in a uniform field experiences a couple —a pair of equal, opposite and parallel forces

any magnet it is found that poles always occur *in equal and opposite pairs.* This may be shown by placing the magnet in a uniform magnetic field; the forces on it are then found to constitute a *couple*—i.e. a pair of equal, opposite and parallel forces (Fig. 5.1). This couple will tend to turn the magnet until its *axis* is parallel to the forces. A couple can only cause rotation and cannot produce movement in any direction of the magnet as a whole. It is only in a non-uniform field that the forces on the poles can be unequal so that the magnet may then be moved in the direction of the resultant force. For instance, the field near another magnet is extremely non-uniform; it varies considerably from point to point round the magnet. Another magnet placed in this field therefore experiences a resultant force of attraction or repulsion, as well as a couple tending to turn it in line with the field.

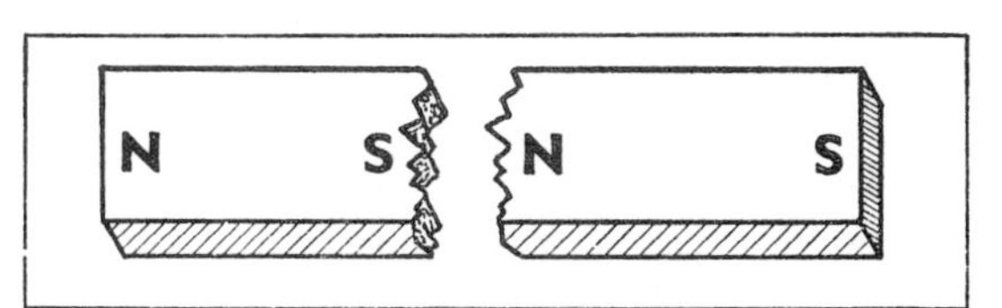

Fig. 5.2 There is no way of producing an isolated magnetic pole

When a piece of magnetized material is broken, fresh poles appear on either side of the break (Fig. 5.2). We can imagine this process continued to the limit until the specimen is broken down into its constituent atoms. We should then find that the individual atom is magnetized and behaves *as though* there were equal north and south poles located within it. In fact the poles have no physical reality; we can regard them as a way of analysing the behaviour of the circulating system of charge in the atom. A circular coil of wire carrying a current behaves the same way. Placed in a magnetic field forces act on it that tend to turn it so that its axis is parallel to the field. The nature of the couple can be shown to be exactly the same as that which would be experienced by a hypothetical magnetic *dipole*, consisting of two equal and opposite poles a short distance apart along its axis (p. 82). Likewise the magnetic field of a circular coil of wire is very similar to that of the equivalent dipole. Indeed it may be shown that the magnetic fields and forces produced by any magnet are identical with those that would arise from electric currents flowing in suitable configurations. It is this observation that justifies our belief that all magnetic effects, even those in magnetic materials, are caused by electric currents.

The earth itself has a weak magnetic field. This is believed to be caused by electric currents circulating within its core. The currents are probably generated by convection in the liquid core maintained by radioactive heating of the earth's interior. By analogy with other geographical terms we call the points in which the magnetic axis of the earth meets the surface the *north* and *south magnetic poles*. In the same way we talk about the *magnetic equator*—the great circle in a plane at right angles to the magnetic axis; and the *magnetic meridian* at a point—the vertical plane in which the earth's magnetic field acts at the point. A freely suspended magnet (i.e. a compass) therefore tends to align itself so that its magnetic axis is in the magnetic meridian. This provides us with a simple way of labelling the two kinds of pole: the *north pole of a magnet* is defined as the one that is drawn towards the earth's north magnetic pole; and the other end of the magnet is called its *south pole*. (Notice that this implies that in the magnetized core of the earth the northern end behaves like a south-type pole.) The magnetic axis of the earth is slowly changing its position. At present (1967) the earth's north magnetic pole is located in northern Canada.

The *direction* of a magnetic field is defined as the direction of the force with which the field acts on a *north* magnetic pole. Magnetic fields may be described pictorially by means of lines of force. A *line of force* is an imaginary line drawn in a field whose direction at every point coincides with the direction of the field there.

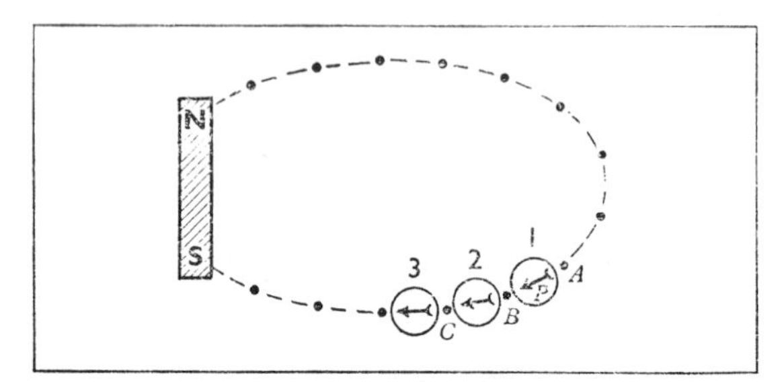

Fig. 5.3 Plotting a line of force using a small compass

The plotting of lines of force in the laboratory is usually done with the aid of a very small compass. Fig. 5.3 shows the method. To draw the line of force passing through the point P the compass B put down at this point and pencil dots A and is are made on the paper opposite the ends of it. The compass is then moved along to the second position with its south pole pointing towards B. A third dot C is marked in opposite the north pole. In this way a chain of dots is obtained, which may be joined up to give the line of force. The method breaks down at positions where the lines of force are sharply curved; this happens particularly near *neutral points*, i.e. points where the total field is *zero*. These occur near a magnet at places where the field of the magnet is exactly equal and opposite to the earth's field. The arrow on a line of force shows the direction of the field. Since a north pole would be repelled from the north pole of a magnet and attracted towards its south pole, the arrows are drawn emerging from the north pole and entering the south pole. (In the earth's field the lines of force are directed *towards* the north magnetic pole.)

Near the magnetic equator the earth's field is approximately parallel to the surface—i.e. horizontal. At the magnetic poles the field is vertical—vertically *downwards* at the north magnetic pole. In Great Britain the field is inclined downwards at about 68° to the horizontal. The angle that the field makes with the horizontal at a point is called the *inclination* or the *angle of dip*. The angle between the geographical and magnetic meridians at a point is called the *declination*. The knowledge of these angles at any point on the earth's surface is important for navigational purposes.

The production of a magnetic field by an electric current was discovered by Oersted in 1819. A descriptive picture of the fields due to circuits of various shapes may be obtained by plotting lines of force in the usual way with a small compass. The diagrams for three cases are shown in Fig. 5.4. Notice in particular that the lines of force near a straight wire are circles concentric with the wire; the field has no component parallel to the wire. The direction of the lines of force is according to a **right-hand screw rule:**

> *The direction of the lines of force round a current is that in which a right-handed screw would be turned in order to advance it in the direction of the current.*

The same rule may be applied to give the direction of the field of a *coil*. The imaginary right-handed screw may be placed, as shown, at the position of any section of the wire forming the coil; this gives the direction of the field near this section, and enables us to decide which way the lines of force pass through and round the coil.

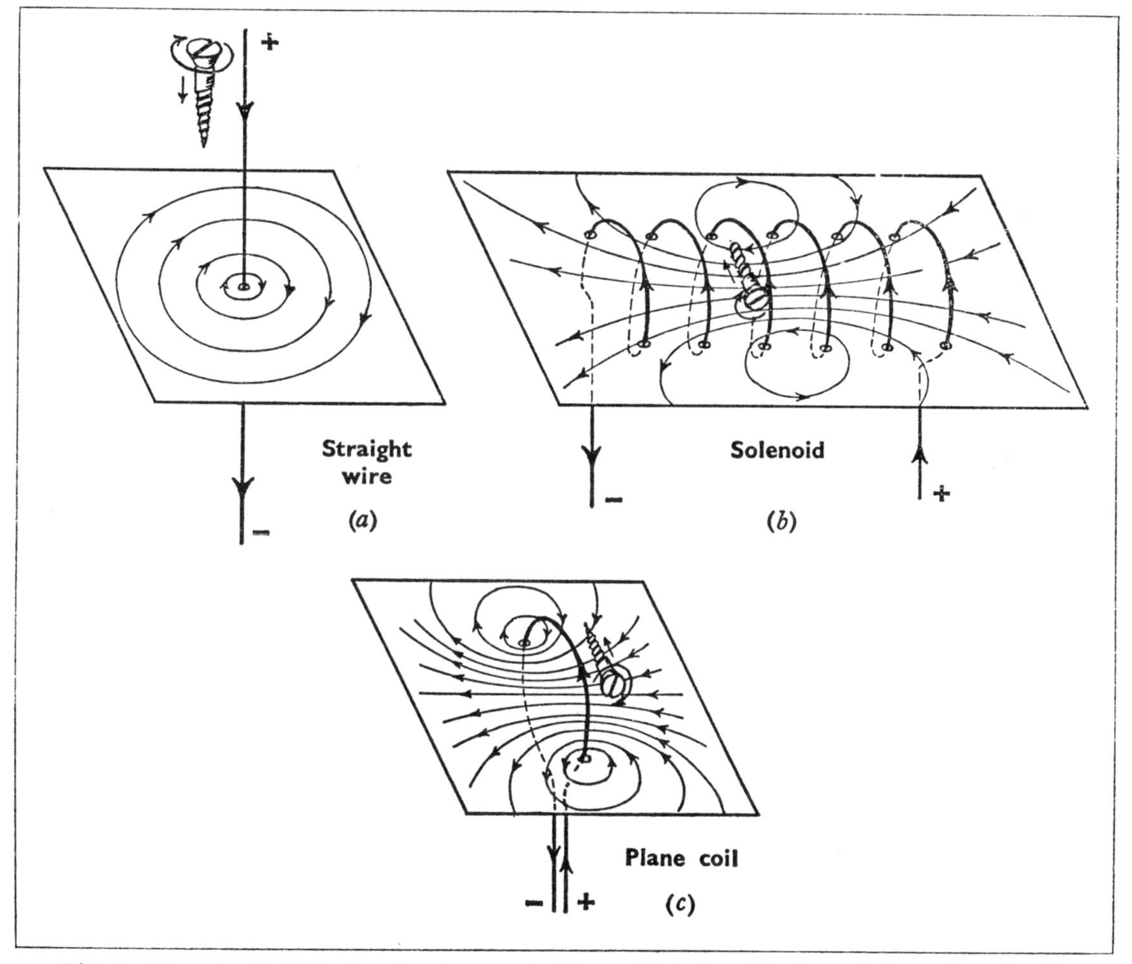

Fig. 5.4 The magnetic field of an electric current: (*a*) a straight wire; (*b*) a long tubular coil or solenoid; (*c*) a plane coil

The direction of the field is given in each case by the right-hand screw rule

In all cases the lines of force in the field of a current form closed loops. The lines emerge from the left-hand end of the tubular coil of wire (or *solenoid*) in Fig. 5.4*b* and enter its right-hand end. The left-hand end of the solenoid therefore behaves like the north pole of a bar magnet, and the right-hand end like the south pole. Likewise the plane coil in Fig. 5.4*c* produces a field resembling that of a thin disc-shaped magnet with a north pole on one face and a south pole on the other.

5.2 The motor effect

When a length of wire carrying a current is placed at right-angles to a magnetic field, it is found that a force acts on it whose line of action is at right-angles to both the field and the current. The direction in which the force acts is traditionally described by *Fleming's left-hand rule.*

Fleming's left-hand rule. *If the thumb and first two fingers of the left hand are put mutually at right-angles, and the First finger is pointed in the direction of the Field while the seCond finger is in the direction of the Current, then the thumb gives the direction of the force.* (Fig. 5.5)

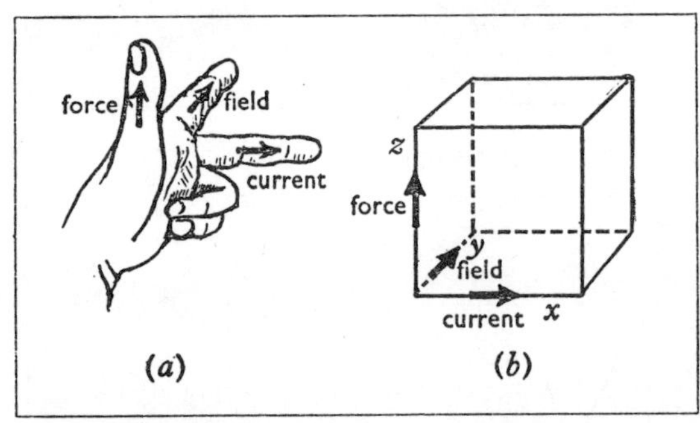

Fig. 5.5 Fleming's left-hand rule for the force acting on an electric current in a magnetic field

This basic motor effect may be demonstrated as shown in Fig. 5.6. A length of wire is stretched loosely between the poles of a horse-shoe magnet. When a current is passed through it, the wire moves up or down according to the directions of current and field, in agreement with Fleming's rule.

The force F is found to be proportional to the current I flowing in the wire and to the length l of wire in the field; it also of course depends on the

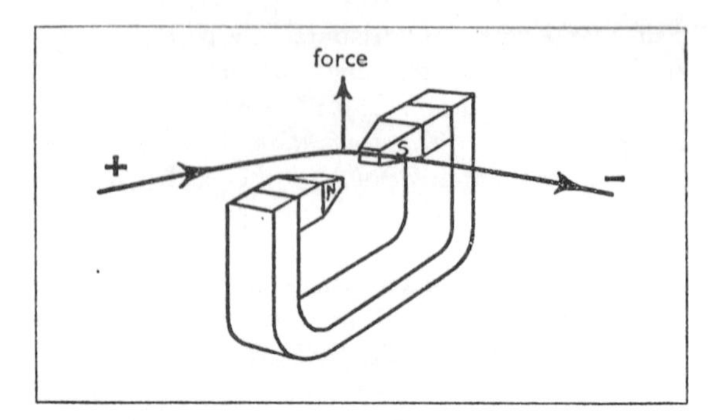

Fig. 5.6 Demonstrating the motor effect

magnetic field, and may be used as a means of measuring, and so defining, the magnitude of the field. The quantity so obtained is called the *flux density* of the field, and is denoted by the symbol B. (The reason for using this name will become apparent later in this chapter, p. 73.) Sometimes B is referred to as the *magnetic induction* of the field.

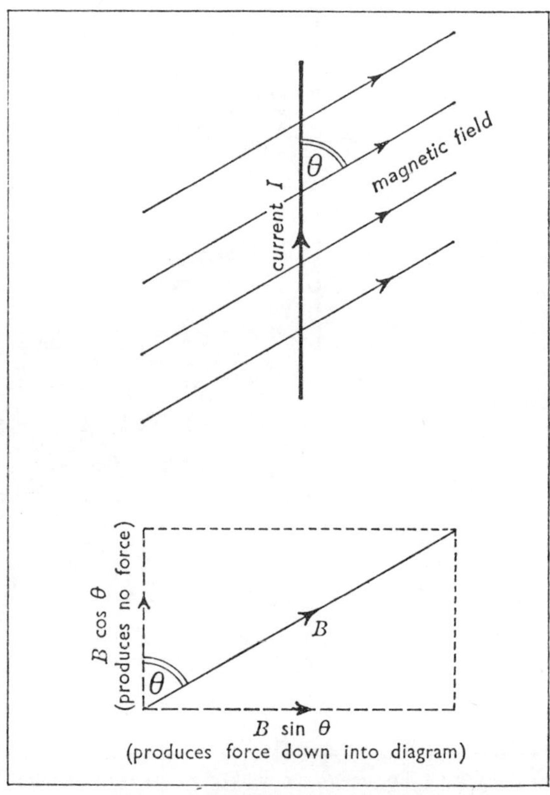

Fig. 5.7 Only the component of the field perpendicular to the current produces any force

The **flux density** *B of a magnetic field at a point is the force per unit length that acts on a wire carrying unit current lying at right-angles to the magnetic field.*

Thus the force F acting on the wire when placed at right-angles to the field is given by

$$F = BIl$$

It is found that the force drops to zero if the wire is turned parallel to the field. In the general case the force can be regarded as due to the *component* of the field *perpendicular* to the current ($B.\sin\theta$ in Fig. 5.7); the component parallel to the current produces no force. Thus in Fig. 5.7 the force F is given by

$$F = BIl.\sin\theta$$

In the m.k.s. system of units the force F must be measured in *newtons* and the length l in *metres.* The corresponding unit of flux density B can then be expressed dimensionally as the *newton amp^{-1} metre^{-1}*. A special name has now been selected for this unit, and it is called the *tesla* (T) —after an American scientist. However, in Great Britain it is still common to express it in terms of another unit, to be introduced later (p. 106), and it is referred to as the *weber per square metre* (Wb m^{-2}).

The couple acting on a coil in a magnetic field

We shall consider only the case of a rectangular coil. It is clear that opposite sides of the coil will experience equal and opposite forces; thus a system of couples acts on it tending to turn it to a position in which all the forces are in the plane of the coil. Since the magnetic force on a wire is always at right-angles to the field this position must be with the plane of the coil at right-angles to the field.

Consider a rectangular coil (a metres $\times$ b metres) of n turns carrying a current I free to rotate about an axis in its own plane, this axis being at right-angles to a field of flux density B (Fig. 5.8).

The forces acting on the top and bottom of the coil (through which the axis passes) are parallel to the axis and produce no turning moment. The force F acting on each of the other two sides is given by

$$F = BnIb$$

These two sides remain at right-angles to the field as the coil rotates so that $\sin\theta = 1$ in all positions. Let the angle between the axis *of the coil* and the field be ϕ. The perpendicular distance between the two forces is $a.\sin\phi$.

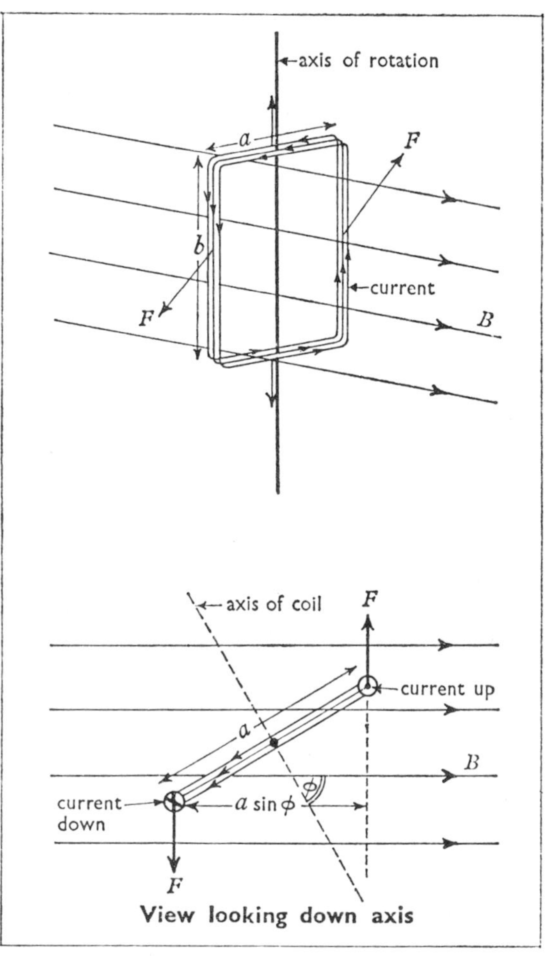

Fig. 5.8 The couple acting on a coil in a magnetic field

$$\therefore \text{ the moment of the couple } C = F\,a.\sin\phi$$
$$= BnIb\,a.\sin\phi$$
$$\therefore C = BAnI\sin\phi$$

where A is the area of the coil ($= ab$). By an extension of this reasoning the expression may be proved quite generally for a coil of area A and of any shape.

Magnetic forces on charged particles

Since an electric current consists of moving charged particles, it is reasonable to suppose that the forces acting on a wire carrying a current are the summed effects of the forces operating on the

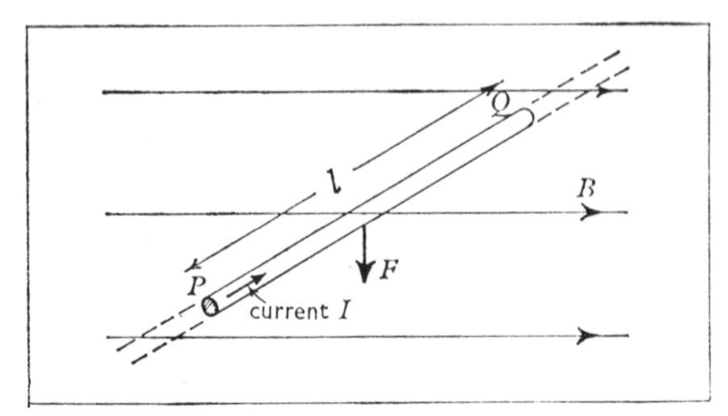

Fig. 5.9 The force on a conductor is the summed effect of the forces acting on the charged particles in it

individual particles in motion. On this assumption we can work out an expression for the force acting on a single charged particle in a magnetic field.

Consider a conductor PQ of length l and cross-sectional area A containing n free electrons per unit volume (Fig. 5.9). The number of electrons in the conductor remains constant while current flows through it. The current I is the passage of electric charge per unit time past any given point. If the charge on one electron is e, we have shown (p. 9) that the current I is given by

$$I = nAve$$

where v = the average drift velocity of the electrons.

Now the force F on the conductor PQ is given by

$$F = BIl \sin\theta = BnAvel \sin\theta$$

$$\therefore \text{ force acting on each particle} = \frac{BnAvel \sin\theta}{nAl}$$

$$= Bev \sin\theta$$

The force is at right-angles to the field and to the path of the particle in a direction given by Fleming's left-hand rule (not forgetting that the current in the conventional sense is opposite to the motion of an electron).

The effect of this force can be observed directly in the deflection of beams of charged particles in magnetic fields, e.g. electron beams in cathode-ray tubes (p. 216). Since the magnetic force acting on a charged particle is at right-angles to its path, it causes no change of speed but only a change of direction. If the motion is exactly at right-angles to a uniform field, the path is turned into a *circle*. In the special case of motion along a line of force, the force reduces to zero ($\theta = 0$). In general, with the motion inclined to the field, the path is a *helix* round the lines of force. Thus the effect of a magnetic field on the motions of all charged particles is to 'trap' them in the field, in the sense that their paths are always curved so that they do not move far in directions at right-angles to the field.

An important natural effect of the action of magnetic fields on charged particles occurs in the outer parts of the earth's atmosphere. Charged particles in these regions are caused to spiral back and forth round the lines of force of the earth's field, and so are effectively trapped high in the atmosphere for considerable times. These regions thus contain dense streams of high-speed particles; they are called the *Van Allen radiation belts* after their discoverer. One of the most striking results of modern research has been to show that in controlling the large-scale behaviour of matter in the Universe magnetic fields are quite as important as gravitational ones. While gravitational forces predominate in the interactions of large units of matter, such as stars and planets, the movements of charged particles are chiefly controlled by magnetic fields, fields which themselves are produced by currents of the charged particles. It appears that much of the matter in the Universe is in a condition referred to as a *plasma*; this means a gaseous state in which the particles are very largely ionized. A familiar example is an electric arc passing through the air between two conductors. The interiors of the sun and other stars and the gaseous nebulae of interstellar space are all in this condition; and the movements of matter within them are considerably affected by the magnetic fields they contain.

5.3 The magnetic field of a current

In order to calculate the field at some point due to a complete circuit it is necessary first to have an expression for the field of a *short section of wire* carrying a current; it is then possible (in principle) to divide up any circuit into a sufficient number of short sections and find the total field at a given point by a process of summation or integration. Consider a short section of wire at O of length l; let the current in it be I, and let its distance from the points P or P′ at which the field is required be x (Fig. 5.10); let θ be the angle that the section of wire makes with the line OP. It is found that the

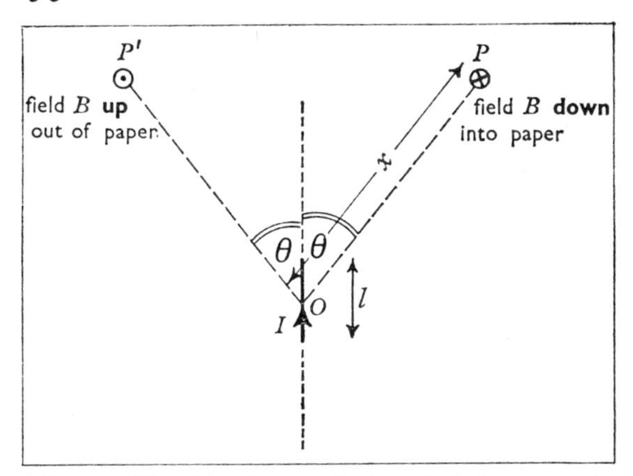

Fig. 5.10 The Biot–Savart rule for calculating magnetic fields due to currents

field B at P may be calculated with a formula known as the *Biot–Savart rule*:

$$B \propto \frac{Il \sin \theta}{x^2}$$

The line of action of B is at right-angles to the plane containing the section of wire and OP, and is in a direction given by the right-hand screw rule. Note the conventional symbols used in the diagram to show the direction of lines of force passing *through* the plane of the paper. (In the symbol for the line of force coming up out of the paper the dot may be thought of as the point of an arrow approaching the reader; the cross in the other symbol can be regarded as the tail feathers of an arrow receding from the reader.)

★ The full experimental test of the Biot–Savart rule is a matter of no little complexity. Since currents are normally measured by the magnetic fields they produce, we may assume that B is proportional to the current I. (This question has already been discussed on p. 32 in connection with the testing of Kirchhoff's laws.) Also, the Biot–Savart formula is applied by adding together the fields of successive elements of a circuit; this implies that the field of a given small section of a circuit is proportional to its length l. In any case the matter may be tested by measuring the fields of a set of coils of varying numbers of turns but of identical dimensions and carrying the same current; the field can be shown to be proportional to the number of turns, which in this case is a measure of the length of wire in a coil.

To investigate the dependence of B on the distance x we need to construct a set of flat circular coils of varying radius r, but all containing *the same length of wire l*. The fields at the centre of the coils may then be measured with a suitable instrument. All points of any one coil are at the same distance r from its centre; and the line joining each part of the wire to the centre makes the same angle with the wire (namely, a right-angle). If then the same current is passed through each coil in turn, the field at the centre is found to be inversely proportional to the square of the radius—in agreement with the Biot–Savart rule.

It remains to determine how B depends on the angle θ that the section of wire makes with the line OP (Fig. 5.10). However, we cannot in this case design a complete circuit so that θ is the same for all parts of it, while x and l are kept constant (except for the case of $\theta = 90°$, as on the axis of a flat circular coil); nor of course can we study the field due to a small part of a circuit only. We have rather to work back to possible expressions involving θ using measurements made on the fields of complete circuits in which θ varies from one point to another. The problem is a very intractable one; and in the early development of the subject several solutions were obtained, each of which appeared in all cases to give correctly the field of a complete circuit. It was eventually shown that there is no unique solution, but rather a range of possible formulae, all of which give the same result for the field of a *complete* circuit. It becomes then a matter of convenience which of these expressions we use; and in common with everyone else since Maxwell's day we have chosen the simplest one, namely

$$B \propto \sin \theta$$

We have still to choose the constant of proportionality in the Biot–Savart formula. This constant may be regarded as a property of the medium in which the conductor is embedded, whose value of course depends on what units are used for the quantities in the formula. It is called the *permeability* of the medium and is usually denoted by the Greek letter μ. The permeability of a vacuum is denoted by μ_0 (pronounced 'mu-nought'); it is also called the *magnetic space constant*. Magnetic forces are very little affected by the medium through which they are acting except where it consists of iron or some other *ferromagnetic material*. Therefore the permeabilities of air and other media (except ferromagnetics) have very nearly the same value μ_0 as for a vacuum.

In an m.k.s. system of units the distances l and x are to be measured in metres, the current I in amps, and the flux density B in teslas (or webers per square metre). In this book we are also adopting what is called a 'rationalized' system of units and formulae. By this we mean that the values of constants of the system, such as μ_0, are so adjusted that π disappears from certain commonly used formulae and re-appears instead in others. This is highly convenient in advanced applications. Thus π does not appear in the much-used formula for the *uniform* field inside a solenoid, quoted below; but it does appear in the expression for the field of a straight wire—a situation having axial symmetry in which we might well expect to find formulae involving π. Rationalizing is brought about by inserting a 4π in the denominator of the Biot–Savart formula. This of course has the effect of making μ_0 come 4π times as great as in an 'unrationalized' m.k.s. system of units, since the actual value of the flux density B cannot be affected by our arbitrary decision to rationalize the system. We therefore write the Biot–Savart formula for the field of a current element as

$$B = \frac{\mu_0 I l \sin\theta}{4\pi x^2}$$

The value of μ_0 for the m.k.s. rationalized system of units is then decided by the definition of the ampere, as shown below. It has the dimensions of Bx^2/Il, and may therefore be expressed in newton amp^{-2}.

Any other dimensionally equivalent combination of units may, of course, be used; for instance, if we express B in weber metre^{-2} we can see that μ_0 can be given in units of weber amp^{-1} metre^{-1}. Probably the most commonly used unit nowadays is one based on the unit of inductance, the *henry* (p. 112); in these terms μ_0 is expressed in henry metre^{-1}.

The fields of various current configurations

(*i*) *At the centre of a plane circular coil.* Consider a plane circular coil of n turns each of radius r metres carrying a current of I amps. Using the Biot–Savart rule, in this case we have

$$\theta = 90° \quad \text{and} \quad \sin\theta = 1$$

for all parts of the coil.

Also $x = r$; and the total length $l = 2\pi nr$.

$$\therefore B = \frac{\mu_0 I . 2\pi nr}{4\pi r^2}$$

$$\therefore B = \frac{\mu_0 nI}{2r}$$

Thus for plane circular coils of the same number of turns and carrying the same current the field at the centre is inversely proportional to the radius r.

In general the calculation of fields is a great deal more complicated than this, requiring the use of calculus. Two other important results will be quoted at this point, since they are simple to use in numerical problems; but their proofs will be deferred until later (p. 73 et seq.).

(*ii*) At a distance r metres from *a long straight wire* carrying a current of I amps:

$$B = \frac{\mu_0 I}{2\pi r}$$

(*iii*) *Inside a long solenoid* of n turns and of length l metres, carrying a current of I amps:

$$B = \frac{\mu_0 nI}{l}$$

$$= \mu_0 I \times (\text{no. of turns/metre})$$

The magnetic space constant μ_0

The modern definition of the ampere (p. 8) envisages an arrangement of two long, straight, parallel conductors. Let us then derive an expression for the forces acting between such a pair of conductors a distance r metres apart, each carrying a current of I amps.

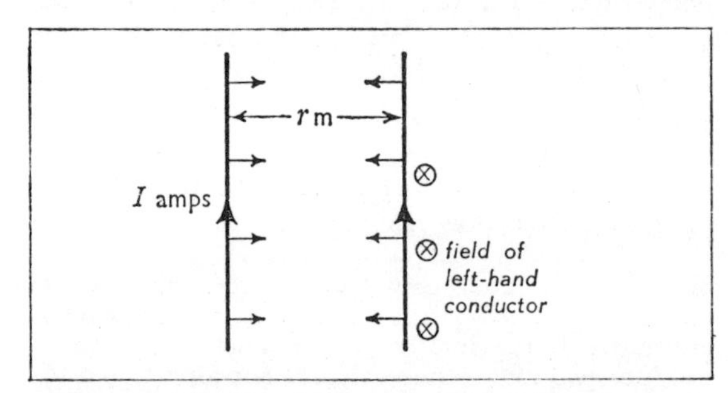

Fig. 5.11 The force between two parallel conductors

In Fig. 5.11 the field produced by the left-hand conductor at the position of the right-hand one is down into the paper. The flux density B of this field is given by

$$B = \frac{\mu_0 I}{2\pi r}$$

Applying Fleming's left-hand rule, the force on the right-hand conductor is found to be to the left—i.e. there is an *attraction* between the two conductors. (If the currents are in opposite directions, there is a repulsion between them.) The force F (in newtons) acting on a length l metres of one conductor is given by

$$F = BIl = \frac{\mu_0 I^2 l}{2\pi r}$$

In the definition of the ampere the following figures are specified:

$$l = r \ (= 1 \text{ metre})$$
$$F = 2 \times 10^{-7} \text{ newton}$$
$$I = 1 \text{ amp}$$

Substituting, we obtain the value of μ_0 that is implied in the definition of the ampere:

$$\mu_0 = 4\pi \times 10^{-7} \text{ weber amp}^{-1} \text{ metre}^{-1}$$
$$\text{(or henry metre}^{-1}\text{)}$$

Flux density as a vector

The flux density B of a magnetic field is only completely specified by stating its direction as well as its magnitude. In other words it must be treated as a vector. Indeed, in defining it we have already assumed that the force per unit length acting on a wire in a magnetic field is a measure of the *component* of the flux density at right-angles to the wire; and thus we have assumed its vector character. All the techniques used in handling other vectors (forces, velocities, etc.) may be applied in making calculations with magnetic fields.

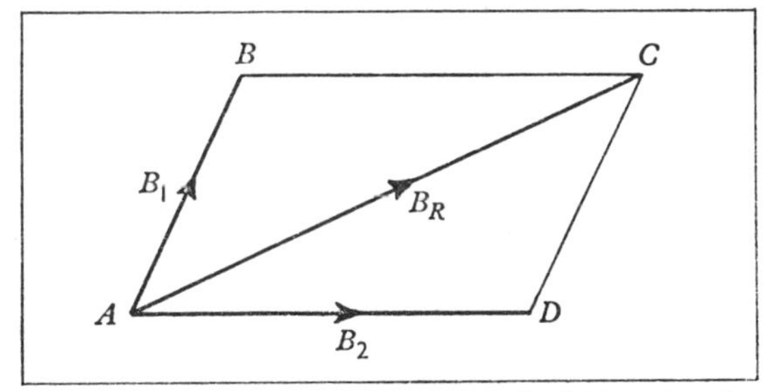

Fig. 5.12 Compounding two magnetic fields using the parallelogram rule for vectors

The resultant of two magnetic fields

This may be found by applying the parallelogram rule for vector addition. Thus in Fig. 5.12 the sides AB and AD of the parallelogram are drawn parallel to the two fields B_1 and B_2 to be combined; and their lengths are made proportional to the flux densities of the fields. The diagonal AC then represents in magnitude and direction the resultant B_R of the two fields.

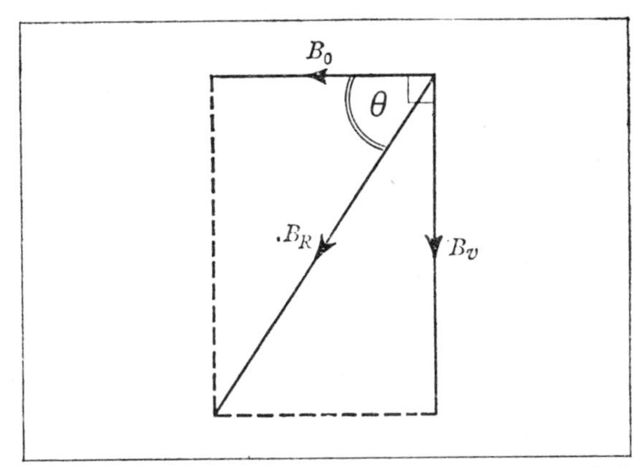

Fig. 5.13 The relation between the horizontal and vertical components of the earth's magnetic field

The components of a magnetic field in two directions

As with other vectors a magnetic field may be resolved into two components, whose combined effect would be the same as the original field alone. For instance it is often convenient to resolve the magnetic field of the earth into its horizontal and vertical components. If the angle of dip is θ, then the relation between the total field B_R at a point and its horizontal and vertical components B_0 and B_v is as shown in the parallelogram of vectors (a rectangle in this case) of Fig. 5.13. The two components are given by

$$B_0 = B_R \cos\theta$$

and

$$B_v = B_R \sin\theta$$

Also

$$\tan\theta = \frac{B_v}{B_0}$$

In the absence of other forces a small freely pivoted magnet will eventually settle down with its axis parallel to the resultant magnetic field. However, a compass is usually only free to rotate in a horizontal plane, and therefore it will settle down in line with the *horizontal* component of the field. For instance, suppose a horizontal magnetic field B is produced at the position of a small compass by means of a near-by coil or magnet: suppose this field is at right-angles to the magnetic meridian. With no additional field the compass points in the direction of the horizontal component B_0 of the earth's field. With the two fields acting together the compass will come to rest after rotating through

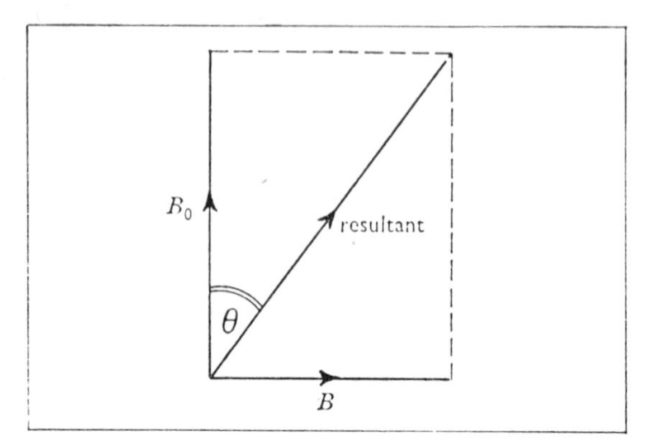

Fig. 5.14 The vector diagram for a deflection magnetometer

the angle θ shown (Fig. 5.14), which brings its axis parallel to the resultant of B_0 and B.

$$\therefore \tan\theta = \frac{B}{B_0}$$

Arrangements of this sort enable an unknown magnetic field to be compared with the field of the earth.

Alternatively, if the field B is that at the centre of (say) a plane circular coil, its value may be calculated in terms of the current in the coil. The deflection θ of the compass may then be used to give the horizontal component B_0 of the earth's field. Compasses designed to be used in this manner are sometimes called *deflection magnetometers.*

Another way of using a freely pivoted magnet as a magnetometer is to measure its period of torsional oscillation T in a horizontal plane. The stronger the field, the more rapidly the magnet oscillates. It may be shown (p. 85) that the horizontal component of the flux density B is inversely proportional to T^2. Thus if the periods of oscillation of a given magnet are T_1 and T_2 in two fields of flux densities B_1 and B_2 respectively, we have

$$\frac{B_1}{B_2} = \left(\frac{T_2}{T_1}\right)^2$$

A magnet used in this way is called a *vibration magnetometer* (or *oscillation magnetometer*).

5.4 Magnetic flux

The nature of the flux density B suggests a new way of describing the forces acting on a circuit. An examination of a few simple cases shows that the action of magnetic forces on an electric circuit is such as to make *the number of lines of B linked with it* a maximum; this includes both the lines of B due to the circuit itself and those arising from other sources. Consider, for instance, a short coil suspended in a magnetic field. The number of lines of B linked with it will be a maximum when the field along the axis of the coil due to its own current is parallel to the applied field; and this is the position into which the coil tends to turn. If the field is non-uniform, the forces act to move the coil towards the region of maximum flux density. There are also magnetic forces acting on the coil tending to make it expand sideways and contract along its length. All these forces act to increase the number of lines of B linked with the coil.

So far, the phrase *number of lines of flux density B* is not one to which a precise mathematical meaning has been given; its meaning would depend on how close together we chose to draw the lines. We need to look for some quantity that can be calculated from the flux density B and carries the same significance as the phrase 'number of lines of B' in the sense in which we have used it above. This quantity is called the *flux* of the field.

*The **flux** of a magnetic field through a small plane surface is the product of the area of the surface and the component of the flux density B normal to it.*

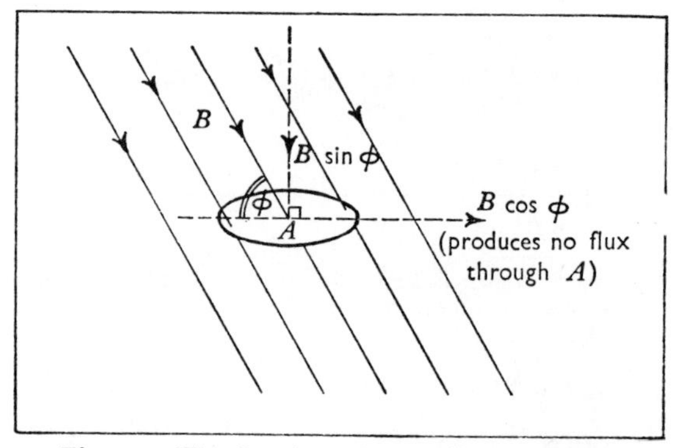

Fig. 5.15 The flux through a small plane area A

When B is parallel to the surface, no lines of B pass through it, and the flux through it is zero. In the general case in which the flux density B is inclined to the surface at an angle ϕ (Fig. 5.15), only the component of B normal to the surface

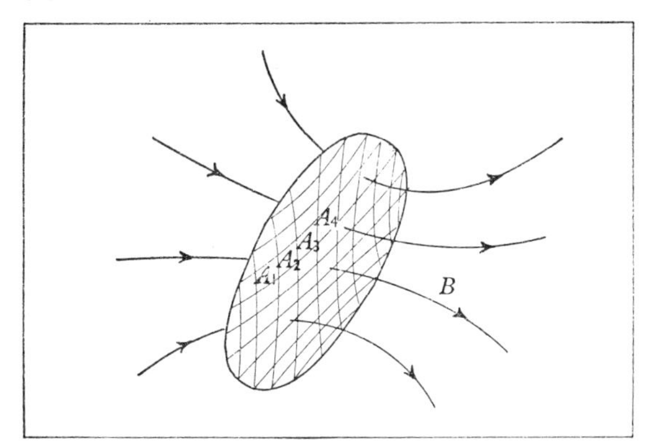

Fig. 5.16 Calculating the flux through an extended area, such as the cross-section of a coil

($B \sin \phi$) contributes to the flux. If the area of the small surface is A, then the flux N through it is given by

$$N = BA \sin \phi$$

To calculate the flux through a large area, like the cross-section of a coil, we imagine the space divided up by a network of lines (Fig. 5.16) into areas A_1, A_2, A_3, etc., each of which is sufficiently small to be treated as plane. The total flux through the cross-section is then the summed or integrated total of the amounts through the individual elements of area. Needless to say, with non-uniform fields such calculations can be of extreme difficulty; but the idea behind it is not a difficult one, and this is all that concerns us at the moment.

Thus, the rule is that *the magnetic forces acting on a circuit tend to move it so as to make the total flux linked with it a maximum*; this includes the flux due to the current in the circuit itself.

The m.k.s. unit of flux is called the *weber*. It is the flux of a uniform magnetic field of flux density 1 tesla through a plane surface of area 1 square metre placed normal to the flux density. This can be taken as a provisional definition of the weber. However, in Chapter 7 it will be possible to give a more satisfactory definition which is directly related to the method usually employed for measuring magnetic flux.

The reason for using the name *flux density* for the quantity B should now be apparent. Thus

flux density = flux per unit area

and 1 tesla = 1 weber metre^{-2}

Flux linkage

Suppose there is a flux N through the cross-section of a coil of n turns. This threads through each of the n turns, and the total flux linking the coil is nN. This quantity is called the *flux linkage* of the field with the coil.

flux linkage = (flux through cross-section of coil) × (number of turns)

To avoid confusing the *flux linkage with a coil* and the *flux through the cross-section of the coil* it is helpful to state flux linkages in *weber-turns*, thereby showing that the number of turns in the coil has been taken into account.

It is conventional to use verbs of motion in talking about flux; we refer to the flux 'entering', 'emerging from', 'threading through' this and that. But it must be understood that this is only figurative language. The position is a stationary one; and nothing actually flows along the lines of B!

5·5 Formulae for magnetic fields

We shall now make use of the Biot–Savart rule to derive expressions for the magnetic fields of various configurations of currents. Put in differential form the rule states that the flux density δB at a point P due to a current element of length δl carrying a current I amps is given by

$$\delta B = \frac{\mu_0 I \, \delta l \sin \theta}{4\pi x^2}$$

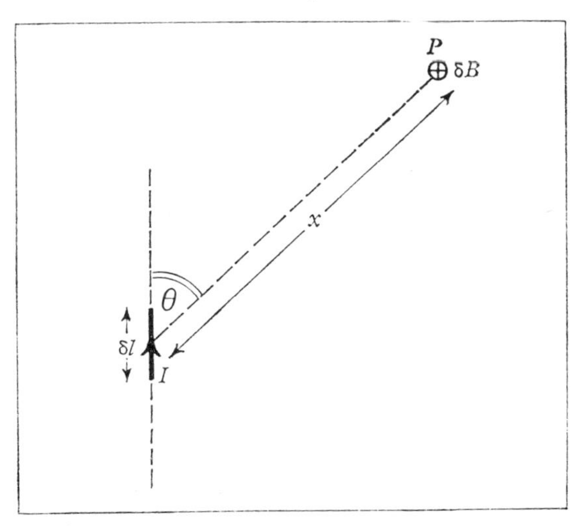

Fig. 5.17 The Biot–Savart rule in differential form

where x and θ are as specified in Fig. 5.17; the direction of the field component δB is given, as usual, by the right-hand screw rule.

(*i*) *A straight wire.* Consider a point P at a distance r from the wire (Fig. 5.18). The total field B at P is given by

$$B = \int dB = \frac{\mu_0 I}{4\pi}\int \frac{\sin\theta . dl}{x^2}$$

the integration being taken over the whole length of wire. Now, from the figure we have

$$l = l_0 - r.\cot\theta$$

$$\therefore dl = r.\text{cosec}^2\theta\, d\theta$$

Also $\quad x = r.\text{cosec}\,\theta$

$$\therefore B = \frac{\mu_0 I}{4\pi}\int_{\phi_1}^{\beta} \frac{\sin\theta . r.\text{cosec}^2\theta . d\theta}{r^2.\text{cosec}^2\theta}$$

$$= \frac{\mu_0 I}{4\pi r}\int_{\phi}^{\beta} \sin\theta . d\theta$$

$$= \frac{\mu_0 I}{4\pi r}\Big[-\cos\theta\Big]_{\phi}^{\beta}$$

$$= \frac{\mu_0 I}{4\pi r}\{-\cos\beta + \cos\phi_1\}$$

Writing this in terms of ϕ_2 we have

$$\phi_2 = 180° - \beta$$

$$\therefore \cos\phi_2 = -\cos\beta$$

$$\therefore B = \frac{\mu_0 I}{4\pi r}(\cos\phi_1 + \cos\phi_2)$$

The field of a complete circuit made up of sections of straight wire may then be summed using this expression. For example, the field of a square coil may be found by this means.

(*ii*) *A very long straight wire.* Using the above result, for a very long wire in the limit both ϕ_1 and ϕ_2 tend to zero. We then have

$$B = \frac{\mu_0 I}{2\pi r}$$

It is useful to know under what conditions a wire may be considered long enough to justify the use of this expression.

Suppose $\phi_1 = \phi_2 = \phi$, and that the angles are small. In this case we may write

$$\cos\phi = 1 - \frac{\phi^2}{2} \quad (\phi \text{ in radians})$$

$$\therefore B = \frac{\mu_0 I}{2\pi r}\left(1 - \frac{\phi^2}{2}\right)$$

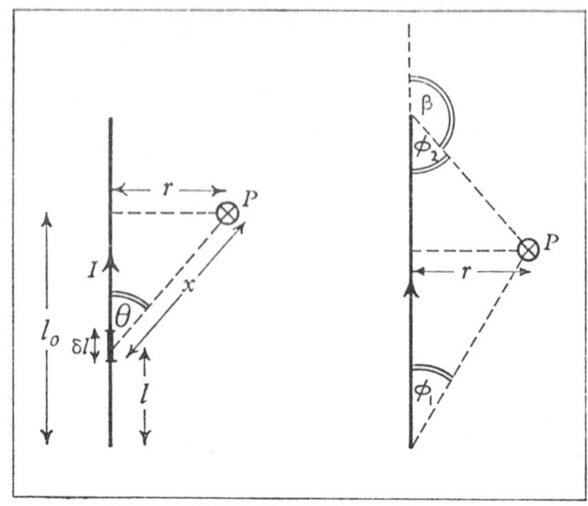

Fig. 5.18 The field of a straight wire

The fractional error in taking $B = \mu_0 I/2\pi r$ is $\phi^2/2$.

Now $\quad \phi \simeq \tan\phi = \dfrac{r}{l/2} = \dfrac{2r}{l}$

where l = total length of wire. If the maximum error allowable is (say) 1%, then we must have

$$\frac{\phi^2}{2} = \frac{2r^2}{l^2} < \frac{1}{100}$$

$$\therefore \quad < \frac{1}{14}\, l \text{ (approx.)}$$

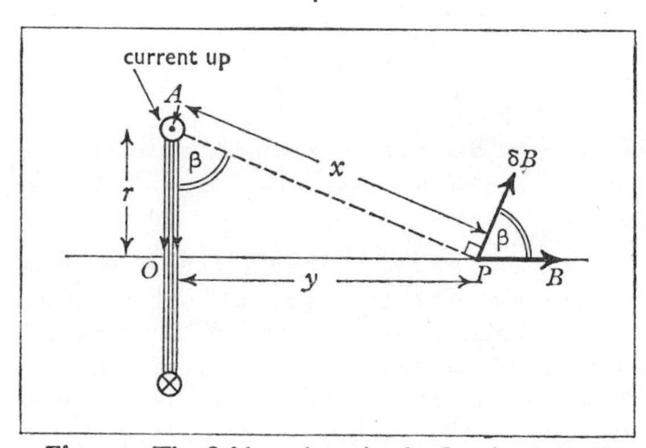

Fig. 5.19 The field on the axis of a flat circular coil

(*iii*) *On the axis of a flat circular coil.* Consider a point P on the axis of the coil at a distance y from its centre O (Fig. 5.19). The field δB due to an element of the coil at A of length δl is in the direction shown, and is given by

$$\delta B = \frac{\mu_0 I\, \delta l}{4\pi x^2}$$

The component of δB along the axis

$$= \delta B.\cos\beta$$

$$= \frac{\mu_0 I\,\delta l}{4\pi x^2}.\frac{r}{x}$$

$$= \frac{\mu_0 Ir.\delta l}{4\pi(y^2+r^2)^{\frac{3}{2}}}$$

By symmetry the total field B at P is along the axis, and is therefore given by

$$B = \frac{\mu_0 Ir}{4\pi(y^2+r^2)^{\frac{3}{2}}}\int \mathrm{d}l = \frac{\mu_0 Ir}{4\pi(y^2+r^2)^{\frac{3}{2}}}.2\pi nr$$

where n = the number of turns in the coil.

$$\therefore B = \frac{\mu_0 r^2 nI}{2(y^2+r^2)^{\frac{3}{2}}}$$

The variation of B with y is shown in Fig. 5.20. The positions of the points of inflexion of this curve are of some interest. They occur where

$$\frac{\mathrm{d}^2 B}{\mathrm{d}y^2} = 0$$

which is easily verified to be where $y = \pm r/2$.

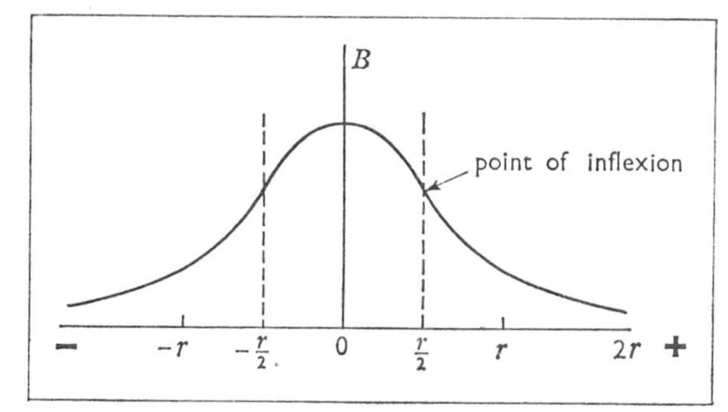

Fig. 5.20 The variation of B along the axis, showing the position of the point of inflexion at which the field changes uniformly

Near these points the curve approximates closely to a straight line; i.e. the field decreases uniformly with increasing y. This result is made use of in the *Helmholtz system of coils* (Fig. 5.21) in order to produce a magnetic field of exceptional uniformity. Two identical flat circular coils are mounted on a common axis a distance apart equal to their radius. Near the point on the axis midway between them the field B is then very nearly constant over an appreciable region, and is given by

$$B = 2 \times \frac{\mu_0 r^2 nI}{2(r^2/4+r^2)^{\frac{3}{2}}} = \frac{8\mu_0 nI}{5\sqrt{5}.r}$$

where n = the number of turns in *one* coil.

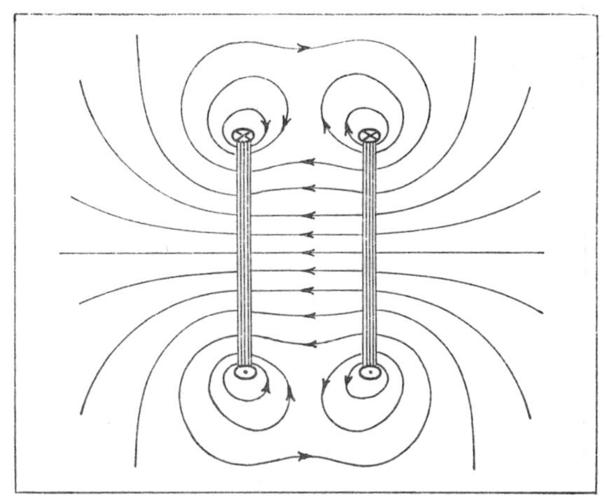

Fig. 5.21 The Helmholtz system of coils

(iv) On the axis of a solenoid. Consider a solenoid of length l wound uniformly with n turns each of radius r (Fig. 5.22). We may imagine the solenoid divided up into elements of length δy, as shown; each element is in effect a flat circular coil containing $n.\delta y/l$ turns. Using the previous result, the field δB at P due to an element at a distance y from it is given by

$$\delta B = \frac{\mu_0 r^2 n\,\delta y I}{2l(y^2+r^2)^{\frac{3}{2}}}$$

Now $\qquad y = r\cot\theta$

where θ is the angle shown in the diagram.

$$\therefore \mathrm{d}y = -r.\mathrm{cosec}^2\theta.\mathrm{d}\theta$$

Also $\quad y^2 + r^2 = x^2 = r^2.\mathrm{cosec}^2\theta$

$$\therefore (y^2+r^2)^{\frac{3}{2}} = r^3.\mathrm{cosec}^3\theta$$

$$\therefore \mathrm{d}B = \frac{-\mu_0 r^3 nI\,\mathrm{cosec}^2\theta}{2lr^3\,\mathrm{cosec}^3\theta}.\mathrm{d}\theta$$

$$= \frac{-\mu_0 nI}{2l}.\sin\theta.\mathrm{d}\theta$$

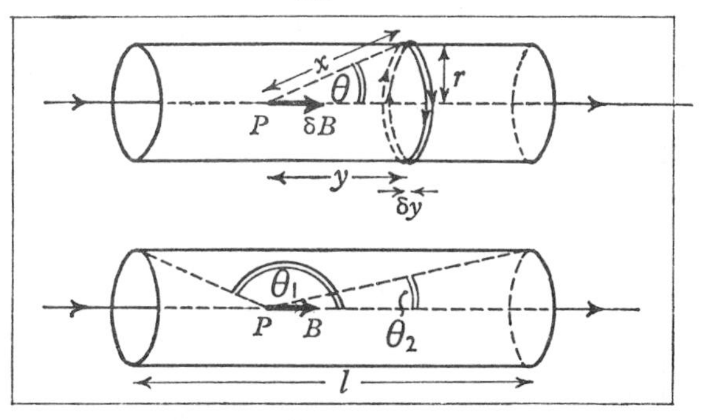

Fig. 5.22 The field of a solenoid

$$\therefore B = \frac{\mu_0 n I}{2l} \int - \sin\theta . d\theta$$

$$= \frac{\mu_0 n I}{2l} \Big[\cos\theta \Big]_{\theta_1}^{\theta_2}$$

$$\therefore B = \frac{\mu_0 n I}{2l} (\cos\theta_2 - \cos\theta_1)$$

(*v*) *On the axis of a very long solenoid.* In the limit, for a very long solenoid, we have

$$\theta_2 = 0 \qquad \cos\theta_2 = 1$$

and $$\theta_1 = 180° \qquad \cos\theta_1 = -1$$

Then $$B = \frac{\mu_0 n I}{l}$$

It may be shown that the field is in fact *uniform* over the cross-section of a very long solenoid (p. 88).

6: Magnetic Materials

6.1 Ferromagnetism

When a piece of unmagnetized iron is placed in a magnetic field, it becomes at least temporarily magnetized. One way of doing this is to place the iron near the poles of a magnet. For example, the north pole of the magnet in Fig. 6.1*a* induces a south pole on the end nearest to it of the bar of iron (and a north pole on the other end). The induced south pole of the bar is then attracted to the north pole of the magnet. It is by this process that a magnet picks up an unmagnetized piece of iron, such as an iron tack.

The induced magnetism of the iron bar produces a magnetic field that is superimposed on that of the magnet, with the result shown in Fig. 6.1*a*. The lines of force are concentrated together at the ends of the bar, where the field is greatly increased. But the space along the sides is relatively denuded of lines of force, and the resultant field here is very weak. In this way a block of iron can act as an effective shield for magnetic fields. If some object is completely enclosed in iron sheet, it is to some extent screened from the action of external magnetic fields.

Another way of magnetizing a bar of iron is to place it inside a solenoid. When a current passes, it produces a magnetic field along the axis of the coil, and the bar is magnetized accordingly (Fig. 6.1*b*). The magnetic field of the bar then considerably augments that of the solenoid. An iron core with a magnetizing coil wound around it is called an *electromagnet*; its strength can be adjusted by controlling the current, and it can be used to exert large mechanical forces (Fig. 6.2).

Magnetic materials differ very much in the ease with which they can be magnetized and demagnetized. In the early investigations of magnetism the extremes of magnetic properties were represented by *soft iron* and *hard steel.* Soft iron is very easily magnetized and demagnetized. Hard steels require large magnetic fields to magnetize them, and then retain their magnetization well. Soft iron is therefore a suitable material for the core of an electromagnet, since it requires only a small current to magnetize it. Hard steels are used for making permanent magnets. Modern magnetic materials show much more extreme properties than soft iron and hard steel, but the words 'soft' and 'hard' are still used (in a magnetic sense only) to describe the properties of these new materials. Among the modern 'soft' materials is the alloy known as *mumetal* (74% nickel, 20% iron, 5% copper, 1% manganese); *physically* it is a hard substance, but *magnetically* it is very soft. Even the earth's weak magnetic field is sufficient to reverse the magnetiza-

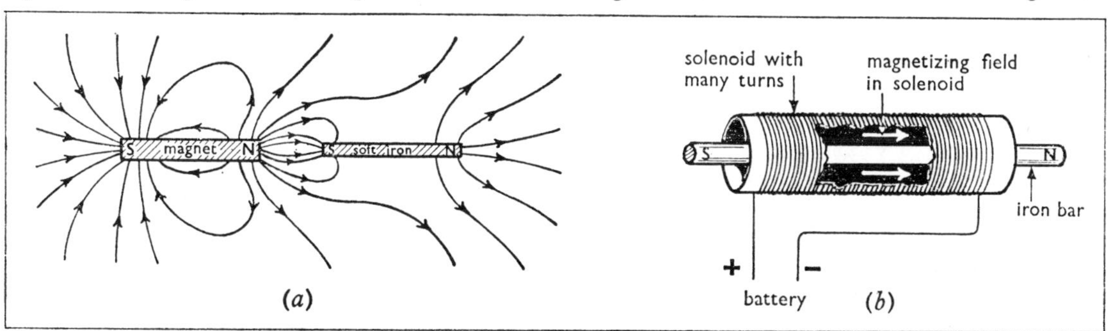

Fig. 6.1 Magnetizing an iron bar: (*a*) in the field of a magnet; (*b*) by the field inside a solenoid

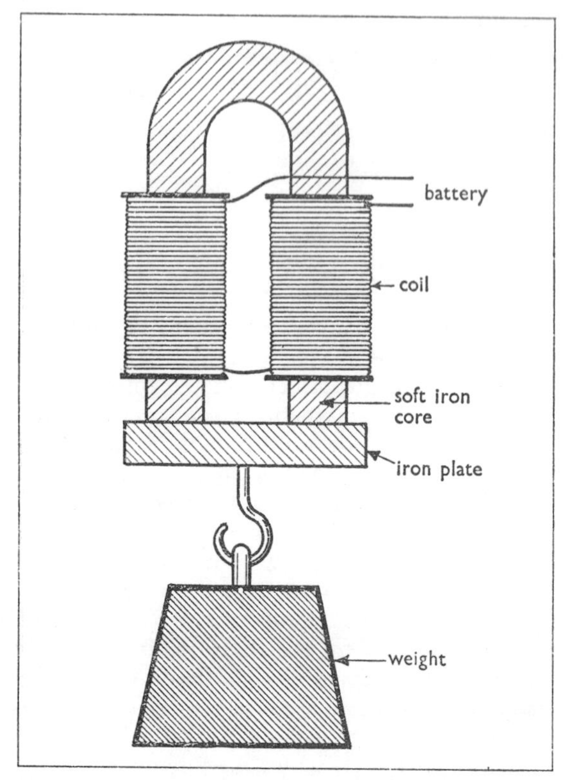

Fig. 6.2 An electromagnet

tion of a rod of mumetal when it is turned over in the magnetic meridian. Mumetal is one of the best materials to use for making screens to shield pieces of apparatus from stray magnetic fields. Typical of the modern 'hard' magnetic materials are the alloys in the *alnico* range. These are magnetized in manufacture and are then very hard to demagnetize. Compact and powerful magnets can be made with them.

Although the atoms of many elements behave individually as magnetic dipoles, with most materials in bulk only a very small magnetic effect is observed. It appears that the thermal agitation of the atoms is generally sufficient to cause the magnetic axes of the atoms to be arranged almost entirely at random even when strong magnetic fields are applied in an attempt to align them in one direction (p. 94). However, in iron, nickel, cobalt, a few of the rare earths, and some other substances another effect operates. These are called *ferromagnetic* materials. Their characteristic property is that the interactions between neighbouring atoms are such as to align them with their magnetic axes parallel to one another; and this is so even when no external magnetic fields are applied. Within any one crystal of the substance there is thus a tendency for all the atoms to be lined up in one direction. The actual result is not as simple as this, however, because in a *large* uniformly magnetized crystal strong additional forces would come into play tending to break down the alignment of the atoms. The nature of these forces may be demonstrated if an attempt is made to magnetize a bundle of parallel iron rods—steel knitting needles, for instance. This may be done by bringing the bundle up to one pole of a powerful magnet (Fig. 6.3). The needles are all magnetized in the same sense; and, when they are released, the forces of repulsion between their lower ends cause them to splay out so that their axes are no longer parallel. The same forces operate within a crystal of a ferromagnetic material opposing the natural tendency of the atoms to arrange themselves with their axes parallel. The equilibrium structure of the crystal is one in which it divides up into a series of regions called *domains*; in any one domain all the atoms are aligned with their magnetic axes in one direction, but the direction varies from one domain to another in such a way that the domain axes form a series of closed loops (Fig. 6.4*a*). The material

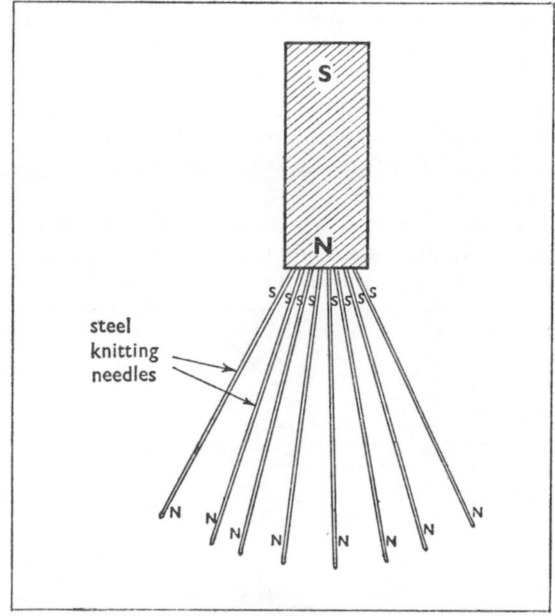

Fig. 6.3 When the needles are magnetized, their lower ends repel one another, and the needles splay out

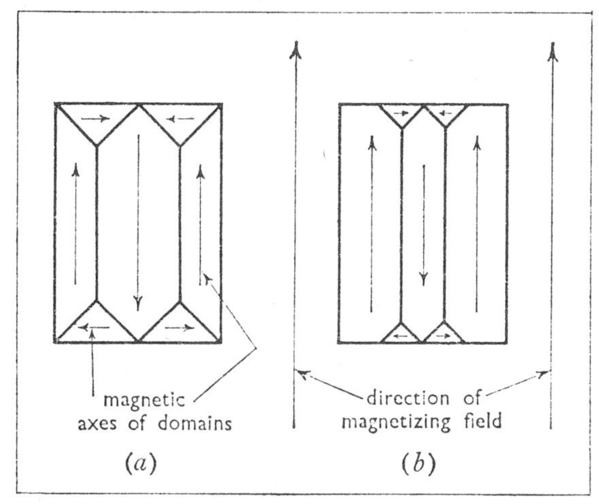

Fig. 6.4 (*a*) An 'unmagnetized' crystal divides into a set of domains whose axes form a pattern of closed loops
(*b*) In a magnetizing field the domain boundaries move so that more atoms are in line with the field

is then strained at the boundaries between domains, but this configuration is more stable than a uniformly magnetized crystal would be.

Thermal agitation always tends to disrupt the ordered domain structure. At high enough temperatures the local alignment of the atoms breaks down, and the material ceases to exhibit its characteristic magnetic properties. The temperature at which this happens is known as the *Curie point*. For iron this is 770° C. But for some ferromagnetic alloys it is below 100° C.

In a freshly formed crystal the magnetic fields of the domains cancel out and virtually no external field can be detected; in this condition we say that the specimen is *unmagnetized,* although on the atomic level there is a high degree of local magnetization. If now it is placed in a magnetic field, couples act on the atoms tending to align their magnetic axes with the field. For small fields the result is a general shifting of the domain boundaries in such a way that the proportion of atoms lined up with the field increases (Fig. 6.4*b*). The specimen now produces a magnetic field adding to the field that magnetized it. To a large extent this shifting of the domain boundaries is reversible; when the magnetizing field is removed, the boundaries return to their original positions and the magnetization disappears. But when the applied field is further increased, the axes of whole domains rotate abruptly one after another and the magnetization increases rapidly. These latter changes are mostly irreversible, so that the material retains much of its magnetization when the field is removed.

It is possible to obtain single large crystals of iron and other ferromagnetic materials. But generally these substances solidify in polycrystalline form—i.e. myriads of microscopic crystals orientated at random throughout the specimen. Each crystal divides separately into domains, and responds more or less independently to an applied field. The observed properties of the specimen are thus the summed effects of the random arrangement of crystals.

The magnetic properties of iron (or any other ferromagnetic material) may be represented graphically as in Fig. 6.5. Starting with an unmagnetized specimen a magnetic field is applied and increased slowly from zero. At first a small amount of magnetization is produced by the shifting of domain boundaries (O to A on the curve). A larger field causes permanent alterations of domain axes, and the magnetization then increases more rapidly (A to B on the curve). But beyond a certain strength of magnetic field virtually all the domains are aligned, and no further increase of magnetization is possible; the magnetization is then said to be *saturated* (C). When the magnetizing field is reduced again, the magnetization does not return to zero, but remains not far below its saturation

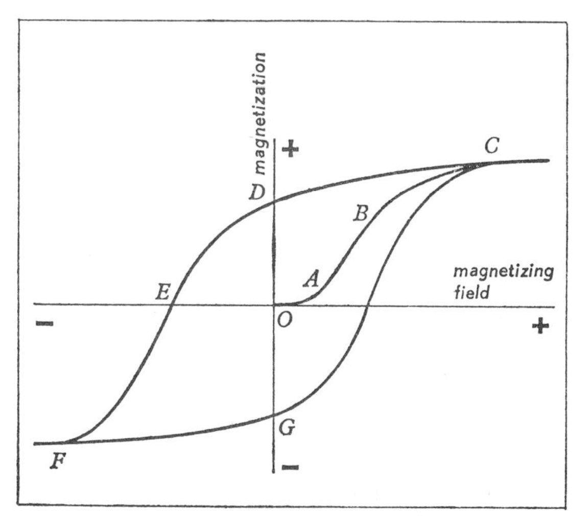

Fig. 6.5 A magnetization curve for a ferromagnetic material

value (D); an appreciable reverse field has to be applied before it is much reduced again (E). A large enough reverse field changes over the alignment of the domains, and a rapid change of magnetization then occurs until the specimen is saturated in the reverse direction (F). A similar sequence is followed when the field is once more reversed to produce saturation in the original direction (FGC). Thus the magnetization of a specimen does not depend in a simple way on the magnetizing field, but appears to lag behind the changes in the field. This phenomenon is known as *magnetic hysteresis*; and a magnetization curve, like that in Fig. 6.5 obtained by varying the magnetizing field cyclically, is called a *hysteresis loop*. Taking a piece of material through a cycle of magnetization like this requires the expenditure of a certain amount of energy, which appears in the form of heat in the specimen. For a typical soft magnetic material this amounts to about 2×10^{-2} joule per kilogram of material; but for a hard magnetic substance the *hysteresis loss* may be thousands of times greater than this.

The magnetization curves obtained for various ferromagnetic materials are all qualitatively similar to Fig. 6.5. Iron alloys do not vary much in their saturation magnetizations. But the strength of the magnetic field required to reverse the magnetization may vary over a range of 10^4 or more between the softest and hardest materials.

A bar magnet formed from a perfectly hard magnetic material could have uniform magnetization from one end to the other (Fig. 6.6*a*). But in practice uniform magnetization cannot be achieved because of the demagnetizing action of the forces operating within the magnet itself. The result is that the magnetization is 'splayed out' at the ends of the bar, and the effective *magnetic length* of the magnet is appreciably less than its physical length (Fig. 6.6*b*). In a bar of soft magnetic material these internal forces are sufficient to break down the alignment of the domains almost completely, and very little magnetization remains once the magnetizing field is removed.

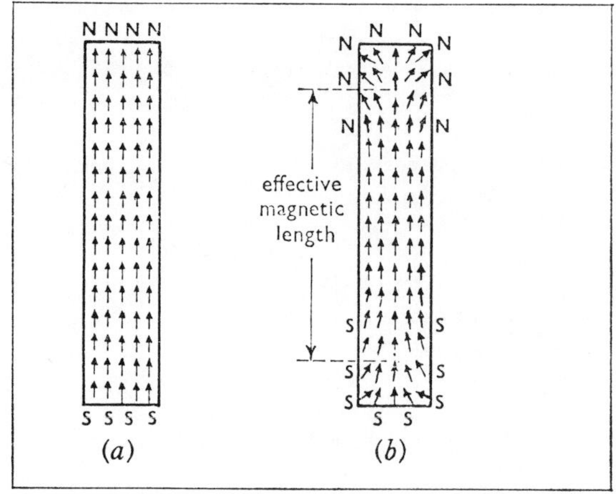

Fig. 6.6 (*a*) A perfectly hard magnetic material could be made into a uniformly magnetized bar
(*b*) In practice the magnetization splays out at the ends

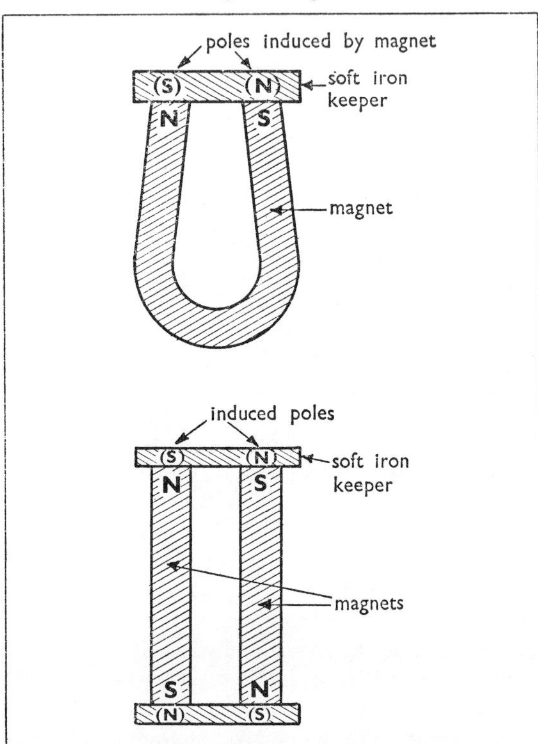

Fig. 6.7 The use of soft iron keepers to reduce the demagnetizing effect in magnets for storage purposes

The demagnetizing effect in a magnet is sufficient to produce a gradual loss of magnetization even with modern hard magnetic materials; the process is assisted by any mechanical shocks that the magnet receives. In order to avoid demagnetization when not in use permanent magnets are stored with soft iron *keepers* across their ends (Fig. 6.7). By arranging the magnetic material in a closed loop the demagnetizing action is reduced to a minimum. Even a soft magnetic material retains

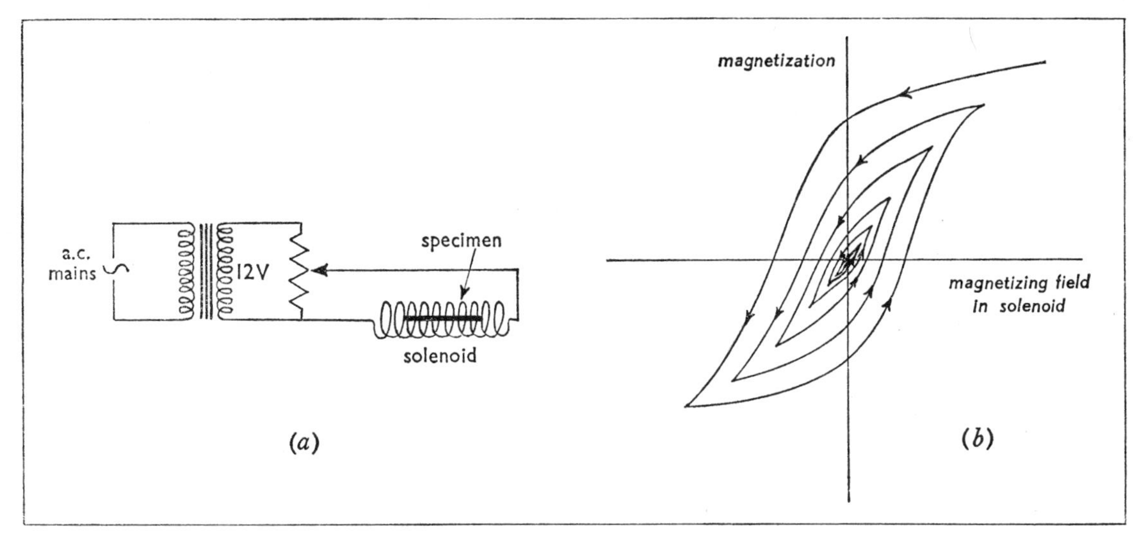

Fig. 6.8 (*a*) Arrangement for demagnetizing an iron specimen
(*b*) As the a.c. field is reduced, the hysteresis loops collapse towards the point where the specimen is completely demagnetized

much of its magnetization if it is arranged in a closed magnetic loop. This may be demonstrated with an electromagnet like that in Fig. 6.2. If the iron plate fits really closely against the poles, the magnetization of the core is little reduced when the current is switched off; and the plate remains firmly attached to it. But a very small gap in the magnetic loop is sufficient to produce a large demagnetizing action; and as soon as the plate is prized away a small distance the attraction between core and plate vanishes almost completely. It is clearly important when we are investigating the magnetization of ferromagnetic materials in varying fields to experiment with the materials arranged in the form of closed magnetic loops. Otherwise it is necessary to apply corrections for the demagnetizing action of the internal forces within the specimen.

In spite of this it is often difficult to remove the last traces of magnetization in a piece of iron, when we require to have it in the unmagnetized condition. We then adopt the method shown in Fig. 6.8. The bar to be demagnetized is placed inside a solenoid connected to an a.c. supply, and the current in it is gradually reduced to zero. The bar is thus taken round a series of minor hysteresis loops of steadily decreasing amplitude; and the final result is complete demagnetization.

6.2 Relative permeability

If the space in and around a coil is filled up with a uniform medium other than a vacuum the magnetization of the medium augments the field due to the coil at every point in it. The factor by which the flux density is increased is called the relative permeability of the medium and is denoted by the symbol μ_r. Thus in a uniform magnetic medium the Biot–Savart formula (p. 70) for the field due to a current element becomes

$$B = \frac{\mu_r \mu_0 Il \sin\theta}{4\pi x^2}$$

Thus the permeability μ of the medium is given by

$$\mu = \mu_r \mu_0$$

To distinguish this quantity from the *relative* permeability μ_r we call μ the *absolute* permeability of the medium. Notice that the relative permeability μ_r is a dimensionless quantity; for a vacuum it is equal to 1, by definition. But the value of the absolute permeability μ depends on the units used, and for a vacuum is equal to $4\pi \times 10^{-7}$ weber amp^{-1} metre^{-1} (or henry metre^{-1}) (p. 71).

The above definition of permeability is adequate as long as we are dealing with a uniform medium in which μ has a constant value at all points. However, with a ferromagnetic medium hysteresis occurs, and the magnetization depends on the past

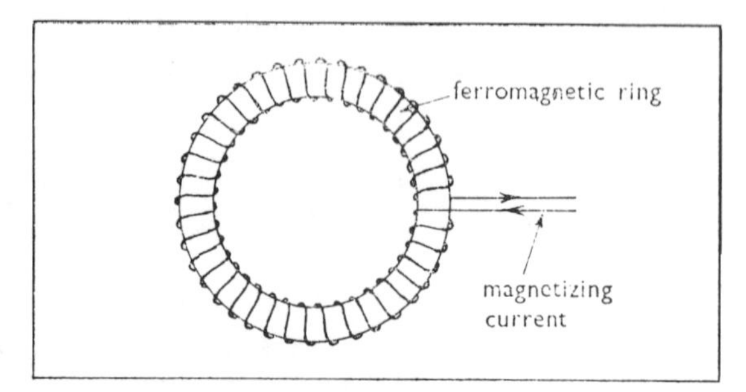

Fig. 6.9 An arrangement for producing uniform magnetization of a ferromagnetic specimen

history of the material at each point. The permeability may then vary from point to point in the medium. Any experiment on such materials must therefore be designed in such a way as to ensure that the magnetization of the material is uniform. This is best done by making the specimen in the form of a circular ring of constant cross-section. The ring may be magnetized by passing current through a solenoid wound uniformly round its circumference (Fig. 6.9).

If the ring is sufficiently large, the field is constant over the cross-section of the solenoid and is given by the same expression as for a long straight solenoid.

$$B = \frac{\mu n I}{l} \qquad \text{(p. 70)}$$

where $n =$ the total number of turns in the solenoid and $I =$ the current; the length l of the coil is taken as the mean circumference $2\pi R$. In the absence of the magnetic material in the coil the flux density B_0 would be given by

$$B_0 = \frac{\mu_0 n I}{l}$$

Thus B_0 is proportional to the current I and may be calculated from it. Then

$$\mu_r = \frac{B_r}{B_0}$$

and the measurement of the relative permeability under different conditions of magnetization depends on finding the flux density B in the ring for different values of the magnetizing current. A method of doing this is considered in the next chapter (p. 118).

When we have to deal with incomplete loops of ferromagnetic material (as in the magnetic systems of galvanometers and electric motors), the flux density depends on the widths of the gaps as well as on μ_r; and in general the calculation of flux densities in magnetic materials is a matter of some difficulty. Later in this chapter we shall consider the principles of such calculations (p. 88), and we shall also discuss a more general definition of permeability suitable also for situations of non-uniform magnetization (p. 85).

6.3 Magnetic moment

We have stated hitherto in this book that the magnetic properties of a magnet could be accounted for in terms of the circulating systems of currents in the atoms of which it is made up. We must now examine the experimental basis for this assertion. It must be demonstrated that the forces acting on, say, a bar magnet and the fields produced by it are identical with those that occur with a solenoid of the same shape and size and carrying a suitable current. When these points are established we shall have reasonable grounds for believing that moving electric charges (i.e. electric currents) are the *only* sources of *all* magnetic fields.

We have shown that the couple C acting on a coil of n turns of area A carrying a current I in a magnetic field of flux density B is given by

$$C = BAnI \sin\phi$$

where ϕ is the angle between the axis of the coil and the field. It makes no difference for this purpose whether the turns of the coil are wound closely together or are spread out along a tube as in a solenoid. As far as the coil is concerned the couple depends on the product AnI.

We therefore need to show that the couple exerted by a magnetic field on a magnet is proportional to $\sin\phi$ and to B. There are a number of ways by which this can be done. We shall leave it as an exercise to the student to work out the details of possible methods. For instance, the magnet may be attached to a beam pivoted inside a flat coil or a solenoid; the couple acting on it can then be measured by the moment of a balancing weight suspended from the beam. Alternatively, the magnet may be hung from a fine wire; the couple acting on it is then measured by the twist of the wire. Whatever method is adopted, it is important to keep the magnetic field relatively small. In large fields the magnetization of the magnet might be affected, and the above theory would then break down.

The results show that the couple acting on a magnet is indeed identical in all positions with that acting on a solenoid as long as they are both in air (or a vacuum); and we can deduce the value of the product AnI for the equivalent solenoid. But there is one respect in which the coil and magnet are not equivalent, and this becomes apparent when we imagine the space around them filled up by a uniform magnetic medium of relative permeability μ_r. Such a medium could well fill the space inside the coil; but the space occupied by the magnet could never be filled up by anything else. The magnet is therefore really equivalent to a solenoid which we prevent from being filled up by the material in the space around it. If we possessed a magnetic liquid, no doubt it would be possible to investigate experimentally the properties of such an 'empty' solenoid. As it is, the result must be left to calculation; this shows that the effect of emptying the space inside it is to reduce all the forces acting on the solenoid by the factor μ_r. The couple acting on a magnet is not then strictly proportional to B but rather to the quotient B/μ_r.

We usually find it convenient to keep together the combination $\mu_r\mu_0$ $(= \mu)$ in our fundamental formulae; so we shall write the expression for the couple C acting on a magnet as

$$C = M.\frac{B}{\mu}.\sin\phi$$

where M is a property of the magnet that decides the moment of the couple that acts on it in a given field. It is called the *magnetic moment* of the magnet. We have defined it in such a way as to be independent of the permeability of the material with which it is surrounded.

The couple acting on the equivalent 'empty' solenoid is given by

$$C = \frac{B}{\mu_r} AnI \sin\phi$$

Comparing the two expressions we see that for the solenoid and magnet to be equivalent their properties must be related by

$$M = \mu_0 AnI$$

This quantity can be called the *magnetic moment* of the coil.*

★ *The field of a magnet*

We must also show that the field produced by a magnet is identical with that of the equivalent coil. In most magnets the magnetization is by no means uniform; and it is therefore difficult to calculate the field close to the magnet. However, there are two simple results that may be tested by experiment:

(*i*) The total flux N emerging near the ends of a very long magnet should be equal to that passing through the centre of the equivalent long solenoid. If B is the flux density at the centre of the solenoid, we have

$$N = BA = \frac{\mu_0 AnI}{l} = \frac{M}{l}$$

It is shown in Chapter 7 how the flux through the centre of a magnet may be measured; and this provides a means of testing this prediction. The quantity M/l is sometimes called the *pole strength* of the magnet. It is equal to the flux emerging from the poles of the magnet only if the magnet is *very long*; in general the emerging flux is less than the pole strength.

(*ii*) At distances great compared with the length of a magnet we expect to find that its field is the same as that of the equivalent small coil. We have shown that the flux density B at a distance y along the axis of a short coil of n turns of radius r is given by

$$B = \frac{\mu_0 r^2 nI}{2(y^2 + r^2)^{\frac{3}{2}}} \qquad \text{(p. 75)}$$

If the coil is in a medium other than a vacuum, this expression is to be multiplied by the factor μ_r. However, 'emptying' the medium out of the space inside the coil has the effect of dividing the field again by the same factor, so that in these circumstances the flux density is unaffected by varying the medium,

* Some authors prefer to define M so that μ does not appear in the expression for the couple C. They can then write more simply

$$C = MB \sin\phi$$

and the magnetic moment of a coil becomes AnI. However, if this system is used, it must be borne in mind that M should be divided by the factor μ_r where this differs from unity. Admittedly this situation will arise only rarely in practice. But the present author prefers to define M so as to be independent of μ_r. This is incidentally in accord with the traditional development of the subject. If the couple C is expressed in terms of the magnetizing force H of the field, discussed in the next section, the formula obtained is just as simple as in the alternative system; and there is nothing to choose between the two systems on grounds of simplicity, unless one is determined to proceed without using the magnetizing force H. In an elementary treatment this can be done; but it becomes artificial and unsatisfactory in advanced work.

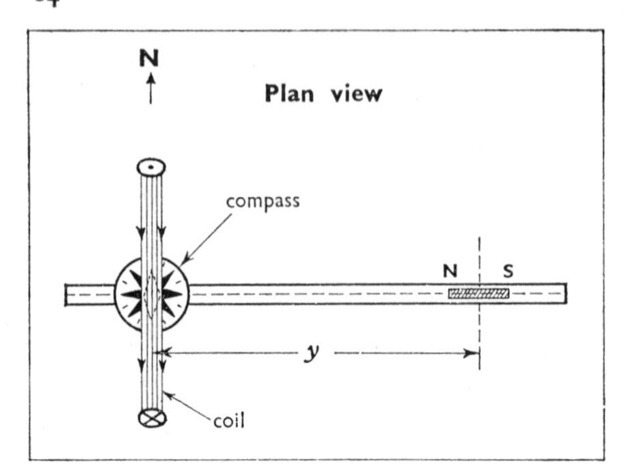

Fig. 6.10 Measuring the field along the axis of a magnet

and is still given by the above expression. If $y \gg r$, we can write

$$B = \frac{\mu_0 r^2 n I}{2y^3}$$

$$= \frac{\mu_0 A n I}{2\pi y^3}$$

where A = the cross-sectional area of the coil (πr^2). Writing $M = \mu_0 A n I$, we expect to find the field of a *short* magnet of moment M to be given by

$$B = \frac{M}{2\pi y^3}$$

This result may be tested with the apparatus shown in Fig. 6.10. The coil is mounted with its plane approximately in the magnetic meridian, and the magnet is laid along a scale on the axis of the coil. A small compass is placed at the centre. When the magnet is placed in position, the compass deflects out of the magnetic meridian. The current in the coil is then adjusted until the compass returns to its original position. The field produced by the magnet at the centre of the coil must now be equal and opposite to that of the coil, and can therefore be calculated.

6.4 The magnetizing force *H*

Up to this point we have measured magnetic fields by means of the forces that act on electric currents. The quantity so obtained we have called the *flux density B*. But we may also use the couple acting on a magnet of known moment M as a measure of the strength of the magnetic field in which it is placed. This couple C is given by

$$C = M.\frac{B}{\mu}.\sin\phi \qquad \text{(p. 83)}$$

where μ is the absolute permeability of the medium and ϕ is the angle between the axis of the magnet and the field. A measurement of the couple C therefore yields a value of the quantity

$$\frac{B}{\mu}$$

Now the individual atom behaves as a small permanent magnet; or, we may equally think of it as a small circulating current—as long as we realize that the space inside it is necessarily a vacuum. The couple that acts on it in a magnetic field therefore depends on the above quantity, and we regard it as a measure of the magnetizing effect of the field. For this reason it is called the *magnetizing force*; it is usually specified by the separate symbol H. (Sometimes H is referred to as the *intensity* of the magnetic field.) Thus

$$H = \frac{B}{\mu} = \frac{B}{\mu_r \mu_0}$$

and the couple C acting on a magnet is given by

$$C = MH \sin\phi$$

It is important to preserve the distinction between the two measures of a magnetic field, B and H. In a vacuum there is a constant ratio between them, and we have

$$B = \mu_0 H$$

H is then defined in terms of B without any further measurement being required. But in a ferromagnetic medium H cannot be calculated from B, since the value of μ_r depends on the complete past magnetic history of the medium. However, we may in principle measure H at any point by finding the couple that acts on a small magnet of known moment. The absolute permeability μ is then given by the ratio B/H.*

We thus have the following logical sequence of definitions:

In a vacuum, the **magnetizing force** *H at a point is the flux density B divided by the magnetic space constant* μ_0.

The **magnetic moment** *of a magnet is measured by the couple that would act on it when*

* The suggestion of measuring B and H *inside* a material raises certain logical difficulties, since strictly we must specify the nature of the cavities in which the measurements are to be made. These points are discussed further on p. 90.

it is placed in a vacuum with its axis at right-angles to a uniform magnetic field of unit magnetizing force.

The **magnetizing force** H *at a point in a medium is measured by the couple that would act on a small magnet of unit moment placed at that point with its axis at right-angles to the field.*

The **absolute permeability** μ *at a point in a medium is the ratio of the flux density* B *to the magnetizing force* H *at that point.*

In all these definitions the magnets could of course be replaced by the equivalent (empty) coils.

A number of results may be expressed more conveniently in terms of H rather than B. For instance, the magnetizing force H produced by an electric current does not depend on the permeability of the medium in which the coil is embedded (provided the medium has uniform magnetic properties throughout). The formulae for various configurations of currents are given by the following formulae; these are derived from the results given on p. 70, using the symbols with the same meanings as before.

(*i*) *The Biot–Savart rule* for the field of a single *current element*

$$H = \frac{Il \sin \theta}{4\pi x^2}$$

(*ii*) *At the centre of a plane circular coil*

$$H = \frac{nI}{2r}$$

(*iii*) *Near a long straight wire*

$$H = \frac{I}{2\pi r}$$

(*iv*) *Inside a long solenoid*

$$H = \frac{nI}{l}$$

$$= I \times (\text{no. of turns/metre})$$

These expressions show that the dimensions of H are those of

$$\left[\frac{\text{current}}{\text{length}}\right]$$

The m.k.s. unit of magnetizing force is therefore called the amp metre^{-1} (pronounced 'amp per metre'); and it is defined in terms of the *uniform* field that exists inside a long solenoid.

The **ampere per metre** *is the magnetizing force inside a long solenoid uniformly wound with n turns of wire per metre of its length carrying a current* I *such that the product* nI *is* 1 *ampere-turn per metre.*

The above expressions correctly give the contribution to the magnetizing force H from the current flowing in a given configuration. In general there is also a contribution to H from the magnetization of the medium. However, if the medium is of uniform permeability throughout, the latter contribution to the field vanishes, and the value of H depends only on the current and the geometry of the coil. Such for instance is effectively the case when the flux is confined to a closed magnetic ring (p. 82).

★ *The vibration magnetometer*

The couple acting on a magnet in a magnetic field tends to align it with its axis parallel to the field. If it is displaced from this position and is free to rotate, it executes angular oscillations about its equilibrium position. The period of oscillation depends on the magnetizing force H and on the magnetic moment M and may be used as a means of measuring one of these quantities in terms of the other. The magnet may be suspended on a fine nylon or silk thread. The couple due to the twisting of the thread must be much less than that arising from the magnetic field, otherwise the magnet cannot be regarded as 'free'. The magnet must be small enough so that the field under test can be taken as uniform over its length. It is usual to hang the magnet inside an empty beaker or some other container to protect it as far as possible from draughts.

If the magnet is displaced from its equilibrium position through an angle ϕ, the couple C acting on it is given by

$$C = -MH \sin \phi$$

A minus sign is used, since the couple acts in the direction of *decreasing* ϕ. The angular acceleration $\ddot{\phi}$ is given by

$$C = I\ddot{\phi}$$

$$\therefore \ddot{\phi} = -\frac{MH}{I} \sin \phi$$

$$\simeq -\frac{MH}{I} \phi, \text{ if } \phi \text{ is small.}$$

This is of the form

$$\ddot{\phi} = -\omega^2 \phi$$

where $$\omega^2 = \frac{MH}{I}$$

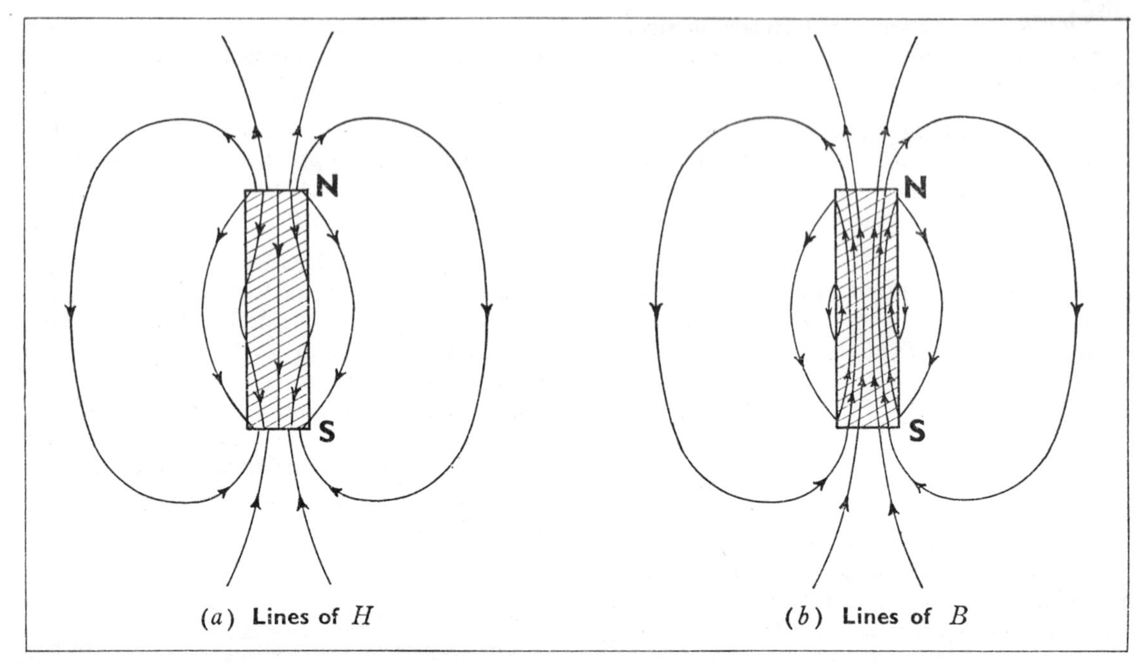

Fig. 6.11 The field of a uniformly magnetized bar: (*a*) lines of H; (*b*) lines of B. Outside the magnet the two descriptions are the same

The motion is therefore of simple harmonic type and the period T is given by

$$T = \frac{2\pi}{\omega}$$

$$= 2\pi\sqrt{\frac{I}{MH}}$$

The moment of inertia I may be calculated from the mass and dimensions of the magnet, and the measurement of T therefore enables M to be found if the magnetizing force H is known, or vice versa.

When the magnetometer is used to *compare* two fields I and M need not be known; the only requirement is that the magnetization must remain constant when the field is varied. Re-arranging the formula we then have

$$T^2 = \frac{4\pi^2 I}{MH}$$

$$\therefore\ H = \frac{4\pi^2 I}{M} \cdot \frac{1}{T^2} = kn^2$$

where $n(= 1/T)$ is the *frequency* of vibration and k is a constant of the apparatus. Thus

$$H \propto n^2$$

If the frequencies of oscillation are n_1 and n_2 in two fields of intensity H_1 and H_2 respectively, we have

$$\frac{H_1}{H_2} = \left(\frac{n_1}{n_2}\right)^2$$

The magnetizing force H is plainly a vector and may be represented by a line of force diagram in much the same way as B. It is instructive to compare the two sorts of diagram for the field in and near a uniformly magnetized bar (Fig. 6.11). Outside the bar the two sorts of diagram are identical, since B and H are in the same direction at every point and bear a constant ratio to one another. But inside, the lines of H run in the opposite direction to those of B. The lines of B follow the same pattern as for the equivalent solenoid of the same outline as the magnet. They therefore always form *continuous closed loops*—from the north pole of the magnet to the south pole outside it, and back again from south pole to north pole inside. This feature of the lines of B is sometimes referred to as the *continuity of magnetic flux*. However, the magnetizing force H inside the material is a measure of the couple that acts on a magnetic dipole inside the bar; and, as we have seen, the magnetic forces here act so as to tend to demagnetize the bar. The lines of H therefore run counter to the lines of B, i.e. from north pole to south pole inside the magnet as well as outside it; and they do not form closed loops. Only when the magnetic material is formed into a ring do the lines of both B and H run in a similar pattern of closed loops.

6.5 The magnetic circuit

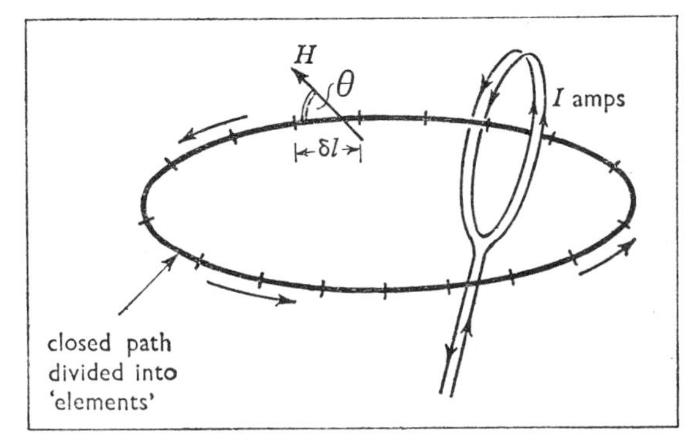

Fig. 6.12 Ampère's rule

Up to this point we have used the Biot–Savart rule as our starting point for the calculation of magnetic fields. A different way of analysing the magnetic fields of electric currents was devised by Ampère. It is of value for calculating the fields in certain cases, and provides further insights into the nature of the electromagnetic field. Ampère made use of a quantity known as the *line integral of the magnetizing force* round a closed path in the field. To find this quantity for a closed path such as that shown in Fig. 6.12 we imagine the path divided into short elements such as δl. For each element we then form the product of the length δl of the element and the component of H parallel to it. The line integral round the path is the sum or integral of these products taken right round the path. Thus, if the angle between H and the element δl is θ, we can write

$$\text{line integral} = \int H.\cos\theta.\mathrm{d}l$$

the integral being taken right round the path. This is usually expressed briefly by the notation

$$\oint H.\mathrm{d}l$$

Ampère's rule. *The line integral of the magnetizing force round any closed path in a magnetic field is equal to the total current linked with the path.*

If the path in question is linked with a coil of n turns carrying a current I, the rule states

$$\oint H.\mathrm{d}l = nI$$

This result may be proved quite generally from the Biot–Savart rule; indeed the two rules can be shown to be alternative and equivalent statements of the law giving the magnetic field of an electric current. However, the mathematics involved is difficult and beyond the scope of this book. We shall content ourselves with the demonstration that Ampère's rule gives the same result as the Biot-Savart rule for two important cases we have already considered.

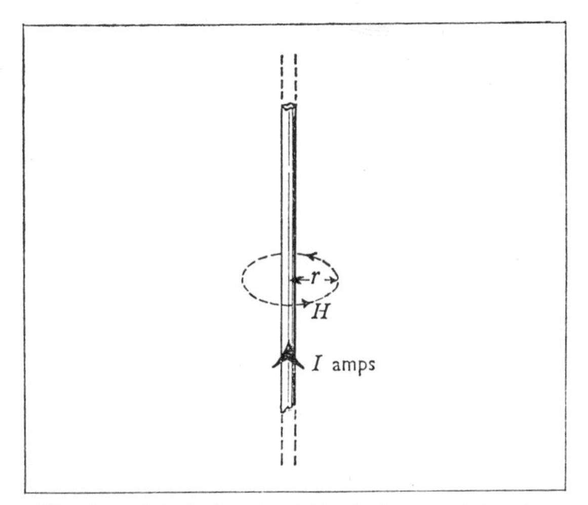

Fig. 6.13 Calculating the field of a long straight wire

(*i*) *The field near a long straight wire.* Consider a closed circular path of radius r concentric with the wire, which we suppose to carry a current of I amps (Fig. 6.13). By symmetry the magnetizing force may be assumed to be of constant magnitude H everywhere along this path. Ampère's rule then gives

$$\oint H.\mathrm{d}l = H.2\pi r = I$$

$$\therefore H = \frac{I}{2\pi r} \qquad \text{(cf. p. 85)}$$

(*ii*) *The field inside a long solenoid.* Consider a coil of n turns uniformly spaced round the circumference of a torus (Fig. 6.14). By symmetry the magnetizing force is of constant magnitude H at all points on the closed circular path shown, which is of total length l. In this case the current I in the coil links n times with the path. Ampère's rule therefore gives

$$\oint H.\mathrm{d}l = Hl = nI$$

$$\therefore H = \frac{nI}{l} \qquad \text{(cf. p. 85)}$$

In this case the field H varies over the cross-section of the coil, since the length l of a suitable circular path depends on the part of the cross-section considered.

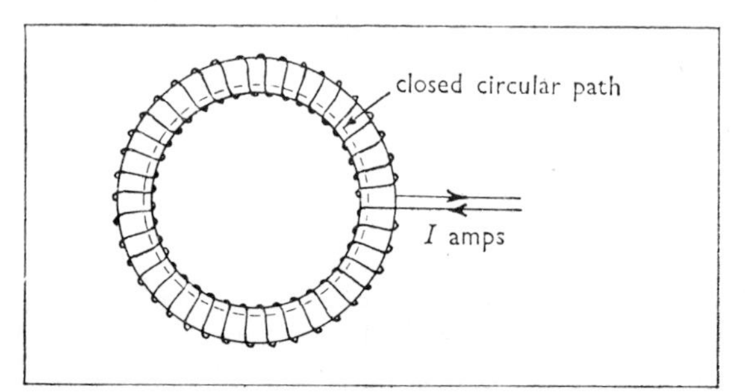

Fig. 6.14 Calculating the field of a long solenoid

We may pass from this to the case of a long *straight* solenoid by supposing the radius of the torus to be increased indefinitely, while keeping the cross-section of the coil and the number of turns per unit length (n/l) constant. The expression for the field does not change, but in the limit the length l of the path becomes constant for all parts of the cross-section. This shows that the field is constant over the whole cross-section of a long straight solenoid.

Magnetomotive force

Ampère's rule also provides a means of calculating the strengths of fields in a number of important cases where the magnetic flux is effectively confined by magnetic materials into one or more closed loops; such loops are called *magnetic circuits.* Examples are provided by the magnetic systems of transformers and chokes, electric motors and dynamos, electromagnets, moving-coil galvanometers, etc. In some respects the properties of these magnetic circuits are analogous to those of electric circuits.

In an *electric* circuit an *electro*motive force is required in some part of the circuit in order to produce an electric current. The effect of the e.m.f. is to produce an electric field, which in turn acts on the charged particles in the circuit to produce the current. The e.m.f. is equal to the work done by the forces of the electric field per unit charge driven round the circuit. It can be shown that this is equal to the line integral of the electric field round the circuit (e.g. see p. 266). By analogy in a *magnetic* circuit we say that a current flowing in a coil wrapped round some part of the magnetic circuit produces a *magnetomotive force* (*m.m.f.*) measured by the line integral of H round the circuit. Ampère's rule shows that the m.m.f. acting in a magnetic circuit linked n times with a current I is given by

$$\text{m.m.f.} = nI$$

The two types of circuit obviously differ in that charged particles can move round the electric circuit, whereas nothing actually flows round a magnetic circuit. However, it turns out that the equations connecting the *flux* in a magnetic circuit with the *m.m.f.* acting in it are strikingly similar to those describing the relation between *current* and *e.m.f.* in an electric circuit; and it is this that makes the analogy a useful one.

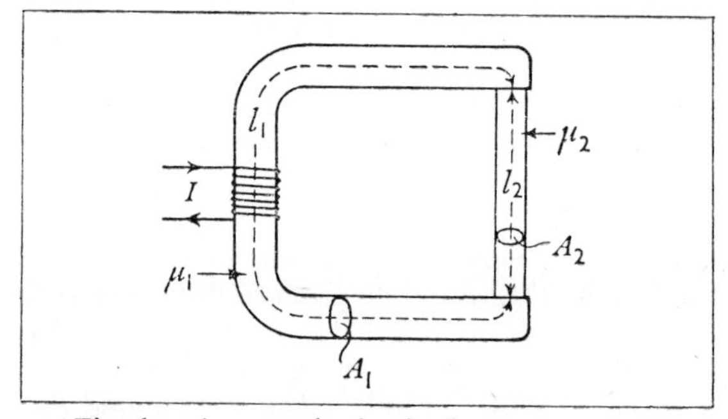

Fig. 6.15 A magnetic circuit of two components

Consider the magnetic circuit shown in Fig. 6.15. It consists of two sections of magnetic material, of lengths l_1 and l_2, cross-sectional areas A_1 and A_2, and absolute permeabilities μ_1 and μ_2 as shown. A coil of n turns carrying a current I is wrapped round some part of the circuit. We shall assume that the flux N through the coil remains entirely confined to the magnetic circuit. The flux densities in the two parts are therefore given by

$$B_1 = \frac{N}{A_1} \quad \text{and} \quad B_2 = \frac{N}{A_2}$$

and the fields intensities are

$$H_1 = \frac{B_1}{\mu_1} = \frac{N}{\mu_1 A_1}$$

and
$$H_2 = \frac{B_2}{\mu_2} = \frac{N}{\mu_2 A_2}$$

In this case the line integral of H round the circuit is given by

$$\oint H.dl = H_1 l_1 + H_2 l_2$$

$$\therefore \text{m.m.f.} = nI = N\left(\frac{l_1}{\mu_1 A_1} + \frac{l_2}{\mu_2 A_2}\right)$$

This may be compared with the expression for an electric circuit

e.m.f. = current × total resistance

where the resistance R of a conductor of length and cross-sectional area A is given by

$$R = \frac{\rho l}{A} \qquad \text{(p. 19)}$$

The equivalent magnetic quantity $l/\mu A$ is called the *reluctance* of the piece of magnetic material. (Notice that μ has a function analogous to $1/\rho$, which is the conductivity of the material. The permeability is thus seen to be a measure of the ease with which flux can pass through the material; $1/\mu$ can be called the *reluctivity* of the material.) The relation obtained above can thus be generalized to read

m.m.f. = flux × total reluctance

The usefulness of this result depends on whether the flux N is really constant round the circuit. In practice there is often appreciable flux leakage. A magnetic circuit is really analogous to an uninsulated electric circuit immersed in a weakly conducting fluid! It must also be borne in mind that in a ferromagnetic material the permeability μ is not a constant, but varies from one point to another of the hysteresis loop. However, the concept of the magnetic circuit provides a means of calculating magnetic fields that is sufficiently accurate for many practical applications. The following example will show how the method is used.

Example. An iron ring of mean perimeter 40 cm and of cross-sectional area 4 cm² is wound with 1500 turns of wire. The relative permeability of the iron may be assumed to remain constant at 2000. Calculate the current required to produce a flux density of 0·3 tesla in the iron.

How is the result affected if a saw cut 1 mm wide is made in the iron ring?

The length l round the ring = 0·4 m
The cross-sectional area = 4×10^{-4} m²
∴ the total flux in the ring = $0{\cdot}3 \times 4 \times 10^{-4}$
$= 1{\cdot}2 \times 10^{-4}$ webers

The reluctance of the ring

$$= \frac{0{\cdot}4}{2000 \times 4\pi \times 10^{-7} \times 4 \times 10^{-4}}$$

$$= \frac{10^7}{8\pi} = 4{\cdot}0 \times 10^5 \text{ amp weber}^{-1}$$

If the current is I amps, then the m.m.f.

$$= 1500I = 1{\cdot}2 \times 10^{-4} \times \frac{10^7}{8\pi}$$

$$\therefore I = \frac{1}{10\pi} = 0{\cdot}032 \text{ amp.}$$

When the saw cut is made,
the reluctance of the air gap

$$= \frac{10^{-3}}{4\pi \times 10^{-7} \times 4 \times 10^{-4}}$$

$$= \frac{10^7}{1{\cdot}6\pi}$$

The reluctance of the iron ring itself is not significantly reduced.

$$\therefore \text{total reluctance} = \frac{10^7}{8\pi} + \frac{10^7}{1{\cdot}6\pi} = \frac{10^7}{8\pi}(1 + 5)$$

$$= \frac{10^7}{8\pi} \times 6 \text{ amp weber}^{-1}$$

The presence of the saw cut thus increases the reluctance by a factor of 6. To produce the same flux in the iron (and in the gap) therefore requires 6 times the current

∴ current = 0·032 × 6 = 0·19 amp

This example illustrates well the considerable effect of introducing even a very small air gap into a magnetic circuit.

Permanent magnet calculations

Consider the case of a magnetic circuit that includes a permanent magnet (e.g. the magnetic system of a moving-coil galvanometer). In the absence of electric currents the m.m.f. in the circuit is zero. The total reluctance must therefore also

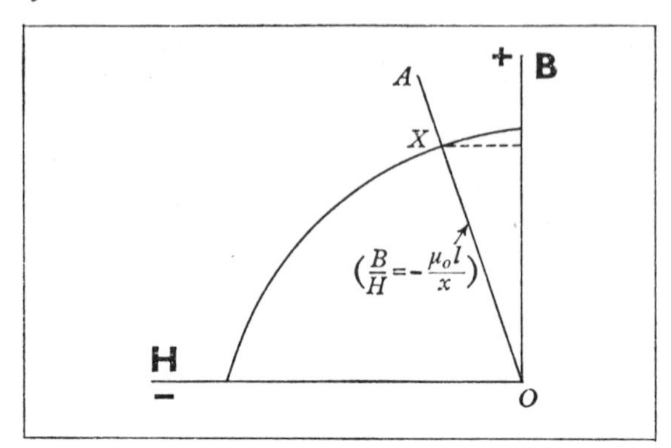

Fig. 6.16 Calculating the flux density in a permanent magnet

be zero. This is possible because the reluctance of a permanent magnet is *negative*; B and H in the magnet are in opposite directions (p. 86), and $\mu(= B/H)$ is therefore negative.

Suppose the air gap in such a circuit is of total width x, and that the magnet is of length l. We shall suppose that the reluctance of any soft iron pole pieces is negligible. Then

$$\text{total reluctance} = \frac{l}{\mu A} + \frac{x}{\mu_0 A} = 0$$

$$\therefore \mu = -\frac{\mu_0 l}{x}$$

To calculate the flux density in the magnet and air gap we then have to make use of the B–H curve for the magnet in question. Fig. 6.16 shows part of the hysteresis curve for a typical hard magnetic material. A line OA is drawn through the origin of gradient $-\mu_0 l/x$. This meets the curve at the point X, which shows the conditions under which the magnet will operate in the given magnetic circuit, since at this point μ will have the required value. The flux density in the magnet can then be read off from the axis.

6.6 The relation between B and H

We have assumed above that the definitions of B and H have some meaning *inside* a piece of magnetic material. It is by no means obvious that this is so. How, for instance, could we measure the forces acting on a coil of wire or a magnet inside a solid piece of iron? The difficulty is removed by considering the nature of the field inside very narrow slots imagined cut in the material.

Consider a slot cut at right-angles to the magnetization (Fig. 6.17*a*). If the reluctance of the slot is negligible it will not affect the field in the material significantly. The flux is continuous right through the region of the slot, and therefore the value of B in the slot (which can in principle be measured) is equal to that in the surrounding material. Likewise, if a sufficiently narrow slot is cut with its plane parallel to the magnetization (Fig. 6.17*b*), there will be negligible disturbance of the field in the neighbourhood. The line integral of the magnetizing force H must then be the same along a path passing through the slot as along a neighbouring parallel path in the material. The value of H in this slot is therefore equal to that in the surrounding material. In general with a slot cut in any direction, measurement of the flux density *perpendicular to the plane* of the slot gives the component of B in that direction; while the magnetizing force *in the plane* of the slot may be measured to give the component of H in that plane. In principle then, by cutting a suitable slot and placing in it a coil or magnet we can determine the component of B or H in any required direction. The slot must be narrow enough to produce negligible disturbance of the field in the material around it.

It must be admitted that this is a strangely abstract procedure, far removed from the pos-

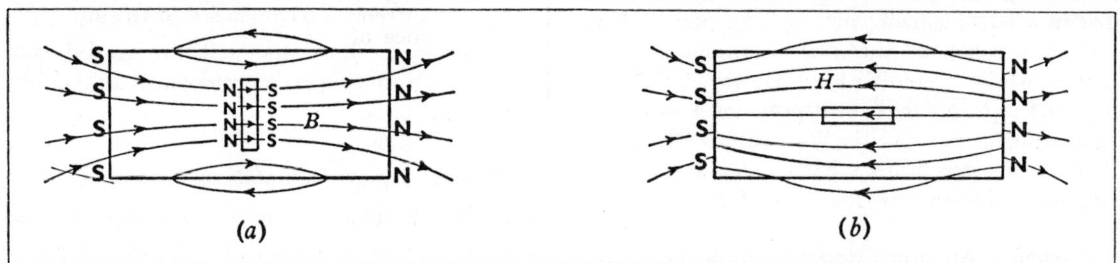

Fig. 6.17 B and H in magnetic material:
(*a*) Perpendicular to the slot, B is the same as in the surrounding material
(*b*) In the plane of the slot, H is the same as in the surrounding material

sibilities of genuine laboratory techniques—particularly as calculation shows that 'negligible disturbance' in a piece of iron would require slots less than 10^{-8} m thick! But the purpose of this reasoning has been to demonstrate that the definitions we have framed can be made consistent and meaningful under all circumstances. Having satisfied ourselves on this point, we are then free to look around for practical methods of measuring B and H in different cases. We might have chosen to define a measure of the magnetic field inside a material in any number of ways; small cavities of *any* shape could well have been specified. The reason for choosing these particular ways of defining the magnitude of the field is that they give quantities (B and H) whose properties are such that their values can be found from measurements made *outside* the specimen. Thus B is a quantity whose flux is continuous under all circumstances; no other measure of the field would have this property. We may therefore measure the flux emerging from different parts of a bar of iron and so obtain the value of the flux passing through the *inside* of the specimen; hence the *average* flux density in the bar may be measured. Likewise H is a quantity whose line integral round any closed path depends only on the electric currents linked with the path; no other measure of the field would have this property. By measuring the value of H *outside* a bar of iron and the current flowing round it we can derive the line integral of H *inside* the bar; hence the *average* magnetizing force in the bar may be measured.

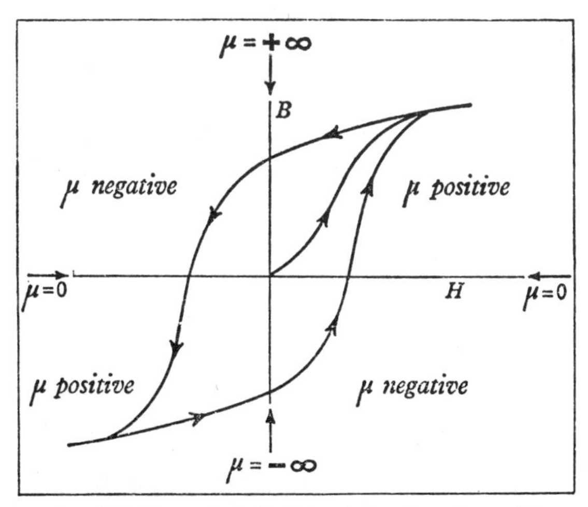

Fig. 6.18 In one hysteresis cycle μ takes all possible values between + and − infinity

The reason for introducing two alternative ways of measuring a magnetic field is to be found in the irregular behaviour of ferromagnetic materials. The magnetic properties of such a material depend on its past magnetic history and on the point in the hysteresis cycle to which it has been taken. Thus, a given value of B in the material may be accompanied by a large range of possible values of H; in a hysteresis loop showing the variation of B with H the value of μ moves over the entire range between + and − infinity (Fig. 6.18); it is certainly not to be regarded as a constant of the material! A knowledge of H in a ferromagnetic substance does not necessarily enable us to work out the value of B; instead, a separate measurement must be made to obtain B, using the forces, couples or e.m.f.'s that arise in an electric circuit placed in the field. We can then work out the value of μ if we wish, since $\mu = B/H$; but the value obtained applies only for the particular conditions in which the experiments were held.

Generally, in dealing with ferromagnetic materials, μ is an indeterminate quantity which we must avoid using as far as possible. When it is necessary to know the relation between B and H, reference must be made to the hysteresis curves showing the variation of B with H for suitable magnetization cycles. If we had no magnetic hysteresis to contend with, μ would be a constant which could be stated with certainty for any given material, and then the use of two alternative ways of measuring magnetic fields would be a scarcely justifiable luxury! B or H could be used for any calculation, according to choice. As it is, the use of both B and H is forced on us by the facts of ferromagnetism, and great care must be taken to use each of the vectors in the right place.

Fortunately the majority of practical applications are concerned with situations in which the flux is confined to particular magnetic loops, and in these circumstances it is possible to adopt a clear-cut subdivision of the functions of B and H:

(*i*) The line integral of the magnetizing force H depends solely on the current flowing in the magnetizing coils, so that the current can be regarded as the cause of H.

(*ii*) The magnetization of the material is caused

by the magnetizing force H; the combined action of the magnetized material and the current is expressed by the flux density B. Thus H can be regarded as the cause of B.

(*iii*) Finally the forces and couples acting on current-carrying conductors (as well as the e.m.f.'s induced in them by electro-magnetic induction) depend on the value of B, and we can regard B as the cause of them.

Thus to the electrical engineer the sequence of cause and effect is as follows:

currents $\longrightarrow H \longrightarrow$ magnetization $\longrightarrow B \longrightarrow$
forces on currents (and induced e.m.f.'s)

This way of reasoning helps to ensure that we use B and H in the right places, but the sequence is largely a man-made convention, and is sometimes a considerable over-simplification. For instance, when we are dealing with permanent magnets, there is no magnetizing current (except the atomic currents, represented by the magnetization of the material); but the magnetizing force H is certainly not zero. Likewise the flux produced by a magnet does not depend only on its magnetization, but also on the distribution of other sources of magnetic field around it; it is affected for instance by alterations to any soft-iron pole pieces attached to it.

Indeed there have always been those who have asserted that B and H are both figments of our imaginations—Ampère was one of these. The only observed effects, they say, are the interactions of electric currents or of charged particles in motion, and we have no need to invent magnetic fields to occupy the space they move in, let alone suppose the fields to have physical properties apart from the interacting particles. This is a philosophical viewpoint that can be consistently maintained, though it leads to considerable mathematical intricacies. However, the traditional use of the vectors B and H to describe the phenomena of magnetism is also completely consistent; and perhaps this is all that is meant by saying that B and H have physical 'reality'.

A classical illustration

In developing the mathematical theory of electric and magnetic fields Maxwell was continually making comparisons with the theory of the flow of fluids; the formal mathematics has much in common in the two cases. He found this a helpful aid to the imagination in unravelling the abstract ideas of electromagnetism. Indeed many of the words used in describing electric and magnetic fields are directly derived from this source (e.g. 'flux'). We shall use the obvious features of the behaviour of fluids to show that we should expect to find two distinct ways of measuring a magnetic field—or, indeed, any other sort of 'vector' field.

Let us consider the flow of water in a river. There are two ways by which the movement of the water at any point might be specified; each is appropriate to a particular sort of calculation.

(*i*) We could draw *lines of flow*, and state the *velocity* of the water at each point. If we wished to calculate the time taken by a given particle in the water to get from one point to another, it is this approach we should need.

(*ii*) We could state the *mass of water* crossing unit area (placed normal to the flow) per second. If we wished to calculate the total flux of water through a given cross-section of the river, this is the way we should want to measure the flow.

The two methods of measurement are not independent; they are connected by a property of the 'medium'—namely the density of the water ρ. Thus, the mass crossing unit area per sec $= \rho \times$ velocity. This connection is so simple that we should not normally bother to make use of both methods of measurement; but if ρ were not a simple property of the medium, we might have to do so. Similar considerations apply to the analysis of other sorts of vector field; and in this respect magnetic (and electric) fields resemble the 'velocity field' of a fluid. There will always be two ways of measuring such a field, each appropriate to its own type of calculation, and each related to its own class of phenomena:

(*i*) *A line effect*—velocity in a fluid, H in magnetism.

(*ii*) *An area effect*—mass crossing unit area per second in a fluid, B in magnetism.

H is the vector that obeys *circuit* laws, like Ampère's rule. B is the vector that obeys *surface* laws, like the law of continuity of magnetic flux.

When the medium has simple properties, the knowledge of a single constant (ρ in fluids, μ in magnetism) enables us to deduce one vector from the other. But when this is not the case, as in ferromagnetism, then both effects must be considered in every situation. Indeed hysteresis effects may well make it impossible to deduce B from H. If both quantities need to be known, they must be found independently by separate methods.

6.7 Magnetization and susceptibility

Consider a uniformly magnetized bar of length l, cross-sectional area A and magnetic moment M. We can imagine the bar cut up into a number of segments, which could then be re-arranged in any manner whatever. For instance we could form a magnet of length $l/2$ and cross-sectional area $2A$. Provided the magnetic axes of the segments are always kept parallel and their magnetization remains unaltered, the total magnetic moment M will stay constant. This can readily be seen by considering the equivalent solenoid of the magnet. The couple acting on each turn of the solenoid in a given field depends on its area and the current flowing in it. Provided these remain constant and the axes of the turns are kept parallel the separate turns could be re-positioned in any way we like without affecting the total couple acting on the solenoid, and therefore without affecting its magnetic moment.

This reasoning shows that the magnetic moment per unit volume of a specimen is independent of its shape and can be used as a measure of its *intensity of magnetization*, which we shall denote by J.

The **intensity of magnetization** *of a specimen is its magnetic moment per unit volume.*

Thus for the bar just considered,

$$J = \frac{M}{\text{volume}} = \frac{M}{Al} = \frac{\text{pole strength } (M/l)}{\text{cross-sectional area}}$$

We have seen that for a bar magnet that is sufficiently long compared with its width the flux emerging from the end of it is equal to the pole strength M/l (p. 83). For a long, thin, uniformly magnetized specimen we can therefore write

$$\begin{aligned} J &= \text{magnetic moment per unit volume} \\ &= \text{pole strength per unit area of cross-section} \\ &= \text{flux per unit area of cross-section.} \end{aligned}$$

In the case of a long, thin magnet J is thus equal to the flux density B in the body of the magnet. This shows that J is to be measured in the same units as B, namely webers metre^{-2}.

Usually the intensity of magnetization of a bar is far from uniform because of the demagnetizing effect of its own field. But in long, thin specimens the demagnetizing effect is small except at the very ends. The intensity of magnetization J is then nearly uniform and is obtained by dividing the magnetic moment by the volume. However, this is not the best way to approach the measurement of J. It is better to work with a specimen formed into a closed loop. The resultant magnetic moment of such a loop is in fact zero; but it is a straightforward matter to measure the flux density B in the loop, and J may be calculated from it.

In the general case the relation between B, H and J may be derived as follows. Imagine a small slot of infinitesimal thickness cut in the material perpendicular to its magnetization (as in Fig. 6.17*a*, p. 90). The flux density in the slot arises from two sources:

(*i*) From the magnetized material on either side of the slot. The flux emerging per unit area of the faces is equal to the intensity of magnetization J.

(*ii*) From the more distant regions of the magnetized material and from electric currents and other magnets in the neighbourhood. This is the only part of the field that would exist in a slot cut with its faces parallel to the magnetization; and the contribution to the flux density on this account is $\mu_0 H$. We therefore have

$$B = \mu_0 H + J \qquad \text{(I)}$$

In the absence of other currents and magnets, the magnetizing force H is simply the demagnetizing field produced by the magnet itself, and is therefore in the opposite direction to the intensity of magnetization J. The flux per unit area of cross-section is thus always rather less than J, since part of the potentially available flux is cancelled out by the magnet's own demagnetizing field. It is sometimes stated that a magnet can be regarded as a constant source of magnetic flux. But this is true only in so far as the internal field H does not vary appreciably. With an ideally hard magnetic material J would be constant, but the flux produced would still vary to some extent with H, and might therefore be affected by alterations to any pole pieces attached to the magnet—quite apart from the effect of other sources of magnetic field in the neighbourhood.

In the special but important case of a closed loop of magnetic material these difficulties do not arise. In this case H is caused only by the currents flowing in the coils wound on the loop and is readily calculated if the currents are known. A measurement of B then gives also the value of J. If the currents are switched off and the material is left magnetized, we have

$$H = 0$$

and

$$B = J$$

In soft magnetic materials the hysteresis loop is so narrow that over a certain range of the variables it approximates to a straight line passing through the origin. The intensity of magnetization J can then be regarded as approximately proportional to the magnetizing force H inside the specimen. Thus

$$J \propto H$$

or

$$J = \kappa\mu_0 H$$

where κ is a property of the medium called its *susceptibility*. By introducing μ_0 into the formula in this way we make κ a numerical constant, since J and $\mu_0 H$ have the same dimensions.

Even when the magnetization curve in no way approximates to a straight line, it is convenient to define the ratio $J/\mu_0 H$ as the susceptibility of the material. In such a case the variation of κ with H may be found from the magnetization curves.

Equation I above enables us to deduce the relation between the relative permeability μ_r and susceptibility κ of a medium. We have

$$J = \kappa\mu_0 H$$

Substituting in equation I,

$$\therefore B = \mu_0 H(1 + \kappa)$$

But

$$B = \mu_r \mu_0 H$$

$$\therefore \mu_r = 1 + \kappa$$

In ferromagnetic materials $\kappa \gg 1$, and we can then write

$$\mu_r \simeq \kappa$$

In other materials κ is very small, and μ_r is then always close to unity. It is in this connection that the concept of susceptibility is chiefly of use. For example in air $\mu_r = 1{\cdot}0000004$, and we prefer to write $\kappa = 4 \times 10^{-7}$.

Paramagnetism and diamagnetism

All materials exhibit magnetic properties to some extent, although the effects are generally very much smaller than with ferromagnetics. Those substances in which the individual atoms or ions have a net magnetic moment are called *paramagnetics*. The magnetic axes of the atoms or ions tend to align themselves parallel to a magnetic field; but thermal agitation opposes this tendency so that the magnetic susceptibilities of these materials are very small—mostly between 10^{-3} and 10^{-4} at 0°C. The susceptibilities vary inversely with the absolute temperature. At very low temperatures the thermal agitation is weak enough for some paramagnetic materials to exhibit saturation in large fields, just like iron at ordinary temperatures. The only difference between ferromagnetics and paramagnetics is that the latter have no tendency to form a domain structure. At temperatures above their Curie points (p. 79), when the domain structure breaks down, ferromagnetics become paramagnetic, and have susceptibilities of the same order of magnitude as other paramagnetics.

In a magnetic field the forces acting on a paramagnetic specimen are similar to those acting on a piece of soft iron. Thus it is attracted towards the stronger parts of a non-uniform field. This attraction is easily demonstrated with a paramagnetic liquid, such as a concentrated solution of manganous sulphate, contained in a U-tube. If one meniscus is placed between the poles of a powerful electromagnet, as shown in Fig. 6.19, a difference of level of about 1·5 cm is produced by a field of 0·5 tesla.

Those materials in which the atoms or ions have

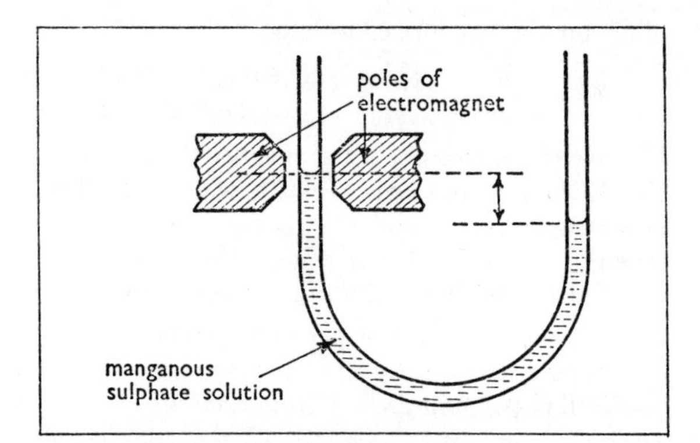

Fig. 6.19 The movement of a paramagnetic liquid in a magnetic field

no inherent magnetic moment show an opposite magnetic effect. They are known as *diamagnetics*. In these substances the magnetic moments of the electron orbits are exactly balanced under normal conditions. But when a magnetic field is applied, the motions of the electrons are modified slightly in such a way as to *reduce* the field in the material. This is in fact an example of the process of *electromagnetic induction* taking place on an atomic scale (see Chapter 7). The susceptibilities of diamagnetic materials are therefore negative, mostly about -10^{-5}, numerically much less than typical paramagnetic susceptibilities, and generally independent of temperature. Presumably the diamagnetic effect takes place in all materials, including paramagnetics, but is normally swamped by the much greater paramagnetic effect.

A diamagnetic material is *repelled* from the stronger parts of a non-uniform field; but the effect is difficult to demonstrate. If the experiment in Fig. 6.19 is repeated with a diamagnetic liquid, such as water, there is a just perceptible fall in the left-hand meniscus when a field of 0·5 telsa is switched on.

7: Electromagnetic Induction

7.1 Induced e.m.f.'s

When an electric current flows in a magnetic field, forces act on the circuit which are proportional to the flux density B of the field. The net effect of these forces is always to tend to make the *number of lines B* linking the circuit a maximum (p. 72); this quantity we have called the *flux linkage* of the magnetic field with the circuit (p. 73). In the most general terms we may say that the action of magnetic forces on an electric circuit is to change some of the available electrical energy in the circuit into mechanical forms of energy. This fact led Faraday to anticipate that the reverse process might also occur—namely, the production of electrical energy from mechanical. He eventually found the process he was looking for in a sequence of experiments similar to those described below. The production of an e.m.f. by magnetic means he called *electromagnetic induction.*

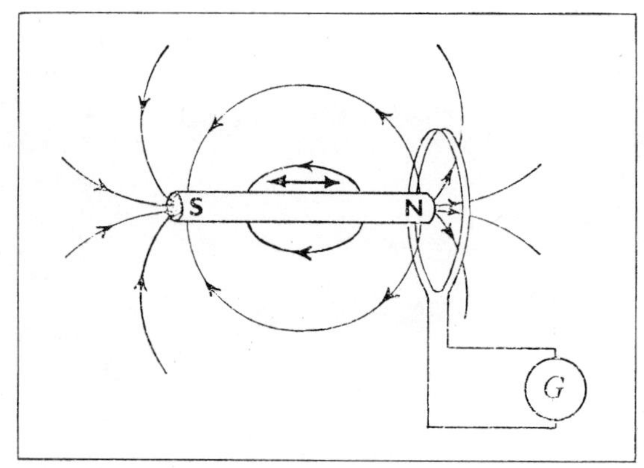

Fig. 7.1 An e.m.f. is induced in the coil when the magnet moves so that the flux linkage with the coil changes

A coil of many turns is connected to a galvanometer with no other source of e.m.f. in the circuit, and a magnet is moved about in its neighbourhood (Fig. 7.1). While the magnet is stationary, no e.m.f. acts in the circuit; but as soon as we *move* the magnet so that the flux linkage with the coil changes a current flows, which continues as long as the change in flux is occurring. The more rapidly a certain change of flux linkage is effected, the larger is the pulse of e.m.f. acting in the circuit. By detailed experiments Faraday showed that the e.m.f. induced is directly proportional to the *rate of change* of the flux linkage; this is known as *Faraday's law* of electromagnetic induction.

The magnetic field for this experiment may alternatively be provided by a second coil carrying a current (Fig. 7.2). Movement or rotation of the two coils relative to one another produces the same effect as the movement of the magnet in the last experiment. We may also alter the flux linkage of the field with the galvanometer circuit without any mechanical movement by simply switching the current on or off in the other coil. This effect may be increased by placing a soft iron bar inside the coils; and the effect is much greater still when both coils are wound on a closed ring of soft iron (Fig. 7.3). The effect of the soft iron (whose permeability is large) is to increase the flux through the core for a given current; and thus quite a large

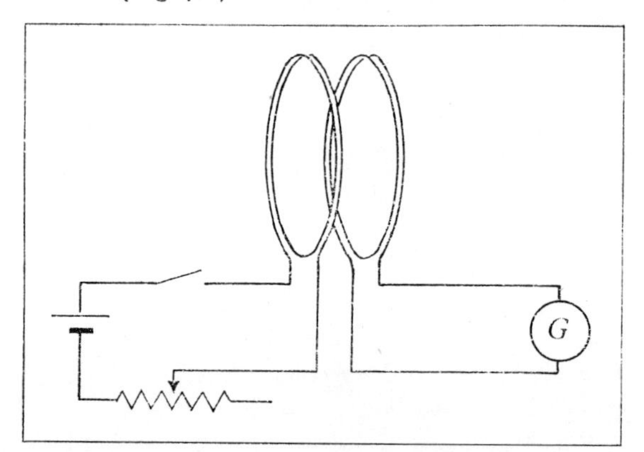

Fig. 7.2 An e.m.f. is induced in the right-hand coil when the current in the left-hand coil is switched on or off

pulse of e.m.f. is produced in one coil at the moments when the current in the other is switched on or reversed.

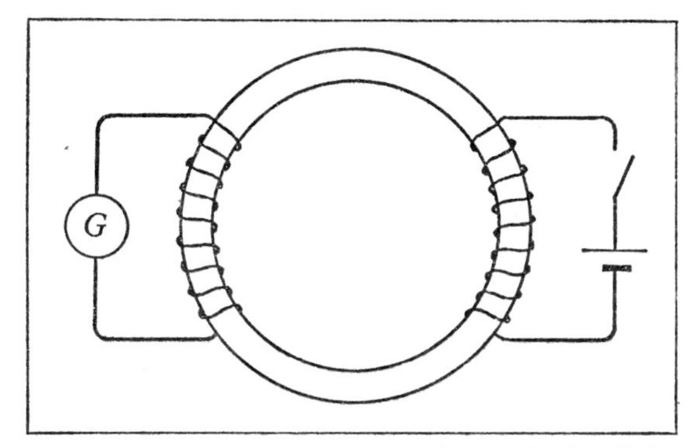

Fig. 7.3 Faraday's anchor ring experiment; the flux linkage of the coils is greatly increased by the iron core

It is found that the *direction* of the induced e.m.f. is such that the resulting current *opposes* the change of flux. This result is known as *Lenz's law*. It may be tested in any of the experiments described above by establishing the directions of the magnetic fields and currents in the coils. In each case the opposition to the change of flux linkage will be found to take two forms:

(*a*) *Magnetic:* the magnetic field produced by the current tends to maintain the flux linkage unaltered.

(*b*) *Mechanical:* the flow of current gives rise to forces which also act in such a way as to tend to prevent the changes in the flux linkage. It is in fact by this means that the available mechanical energy is transformed into electrical energy; i.e. the mechanical energy expended in overcoming these opposing forces is equal to the electrical energy generated in the circuit.

Thus in Fig. 7.1 if the magnet is moved into the coil north pole first, this increases the flux threading through the coil from the near side. By Lenz's law the induced e.m.f. should drive a current round the circuit in such a direction that the magnetic field produced will be in opposition to that of the magnet—thereby tending to delay the increase of flux. At the same time the field produced by the current acts on the north pole in such a way as to *repel* it from the coil, tending to prevent the change of flux by mechanical means. There is also a force acting on the *coil* due to the action of the magnet on the current, and this also tends to repel the coil from the magnet. These forces are normally far too small to produce any detectable effect; but we may use the direction rules for magnetic forces to show that the current is in such a direction as to produce the forces described above. When the north pole is withdrawn again from the coil, the current flows in the opposite direction, thereby tending (i) to prevent the flux linkage decreasing, (ii) to attract the north pole back to the coil, (iii) to attract the coil towards the magnet.

In Fig. 7.2 when the current is switched on in the left-hand coil, current flows round the other coil in the opposite sense, so as to tend to prevent the increase of flux. It is easily verified that the mechanical effect of this system of currents is such as to repel the coils away from one another. At the moment of switching off the induced current flows in the reverse direction, thus acting *magnetically* so as to delay the decrease in flux, and *mechanically* so as to draw the coils closer together.

In any practical case we can deduce the direction of an induced e.m.f. by considering either the magnetic or the mechanical action, whichever is the more convenient. Notice that the opposition to the change of flux occurs only if a current actually flows. In the above examples the galvanometers could be disconnected, and the terminals of the coil joined instead to the deflector plates of a cathode-ray oscilloscope. Pulses of e.m.f. would still arise in the coil as before, and the spot of light would be momentarily deflected accordingly; but practically no current would now flow and there would certainly be no detectable effect opposing the changes of flux. However, even with incomplete circuits like this we may still work out the directions of the e.m.f.'s by supposing the circuits to be completed with a conducting path for the currents.

Faraday's and Lenz's laws suffice to describe the facts of electromagnetic induction.

The laws of electromagnetic induction. *When the flux linkage of a magnetic field with a circuit is changing,*

1. *An e.m.f. is induced which is proportional to the rate of change of the flux linkage (Faraday's law);*
2. *The direction of the e.m.f. is such that the effects of any current it produces tend to oppose the change of flux (Lenz's law).*

Lenz's law is in fact a necessary consequence of

the law of conservation of energy. If the e.m.f.'s acted in any other way the result would be the production of electrical and mechanical energy from nothing. Thus, in Fig. 7.1 if the current flowed in the opposite sense to that which actually occurs when the north pole is moved towards the coil, the forces would act so as to draw the pole and coil together. If the magnet were free to move, it would then acquire kinetic energy as it moved ever faster and faster towards the coil, and at the same time electrical energy would be generated for use in the circuit connected to the coil! In fact the forces must be in the opposite sense so that work has to be done in moving the magnet towards the coil; and this work re-appears as electrical energy in the coil. However, the dependence of Lenz's law on the law of conservation of energy does not free us from the scientific responsibility of testing it by experiment. The correct scientific approach is rather: if experiment gave e.m.f.'s acting in the other direction, so much the worse for the law of conservation of energy! Even so fundamental a principle as this could be refuted by a single adverse observation.

So far our examples of electromagnetic induction have been of *complete* circuits in which e.m.f.'s were generated by the changing flux. But any conductor moving in a magnetic field may be considered in imagination to be part of a complete circuit; and so we must suppose that an e.m.f. is induced in such a conductor. For instance, the wing of an aeroplane flying horizontally is continually cutting the flux of the vertical component of the earth's field. Perilous though the experiment might be, we can at least *imagine* a momentary contact of the wing-tips with a pair of stationary wires through which a circuit could be completed. A current would then flow through the wing in such a direction that the magnetic force on it would tend to *retard* the plane (by Lenz's law). In the northern hemisphere this requires a current towards the port (left) wing-tip. From the point of view of the 'rest of the circuit' the port wing-tip therefore behaves as the positive terminal of the source of e.m.f.

The rate of change of flux in such a circuit is equal to the rate at which the conductor *cuts* the flux. We might in fact state *Faraday's law* in the alternative form:

Faraday's law (*alternative form*).
When a conductor is cutting the flux of a magnetic field, an e.m.f. is induced in it which is proportional to the rate of cutting flux.

The e.m.f. in a complete circuit could then be found by summing the effects of all the parts of the circuit calculated according to this 'flux-cutting' rule. But it is probably best to state the law, as we have done, in terms of the effect in a complete circuit, since this is the only thing that can readily be tested by experiment. However, this alternative way of thinking of the process of electromagnetic induction is often easier to use in calculations of e.m.f.'s.

In terms of the flux-cutting form of the law, the direction rule for electromagnetic induction may then be expressed in a *right-hand rule* (for dynamos) similar to Fleming's left-hand rule for the motor effect (p. 66):

Fleming's right-hand rule.
If the thumb and first two fingers of the right hand are put mutually at right-angles, and the first finger is pointed in the direction of the field while the thumb is in the direction of motion, then the second finger gives the direction of the current (*i.e. the direction positive charges move in it*).

The e.m.f. induced in a conductor moving in a magnetic field is readily explained on the atomic level in terms of the forces acting on the charged particles in it. We have seen (p. 67) that a charged particle moving across a magnetic field experiences

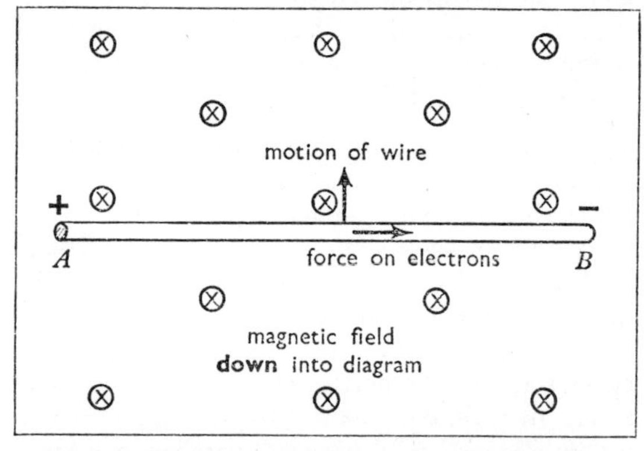

Fig. 7.4 The e.m.f. in a conductor arises from the forces acting on the charged particles in it

a force at right-angles to its path and to the field. A moving conductor contains equal numbers of protons and electrons on all of which the magnetic field acts. However, only the electrons are free to move inside the conductor.

In Fig. 7.4 the use of Fleming's left-hand rule shows that the force acting on a negative particle moving with the wire will be from A to B; and the end B therefore becomes negative, leaving A positive. If the wire is part of a complete circuit, the electrons continue in motion from A to B and round the circuit; and a steady current flows. It can be verified that this direction is in agreement both with Lenz's law and with the right-hand rule (for dynamos) given above.

The experimental test of Faraday's law

Simple qualitative experiments are sufficient to test Lenz's law of electromagnetic induction, since it is necessary only to verify the directions of the quantities involved. But to test Faraday's law we must find ways of measuring both the e.m.f.'s produced and the rates at which the changes causing them occur. We shall discuss the principles of three different methods.

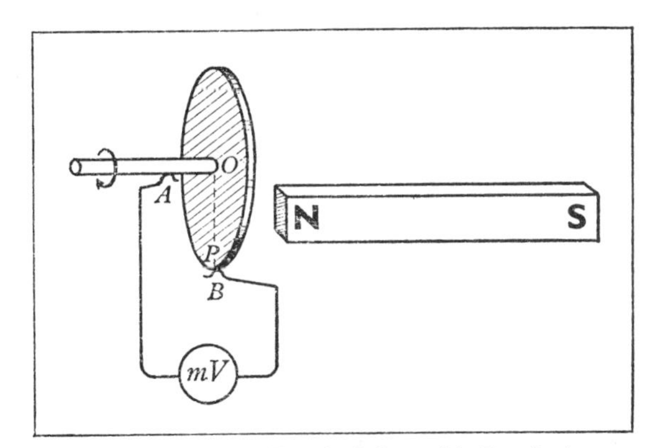

Fig. 7.5 Testing Faraday's law with Faraday's rotating disc

(*i*) *Using Faraday's rotating disc.* A metal disc is rotated at a steady speed in a magnetic field. Any given radius of the disc (such as OP in Fig. 7.5) continually cuts the flux at a steady rate; therefore a steady e.m.f. is generated in the circuit formed by the millivoltmeter and the radius momentarily between the sliding contacts A and B. The rate of change of the flux linkage with this circuit is proportional to the speed of rotation of the disc. It is

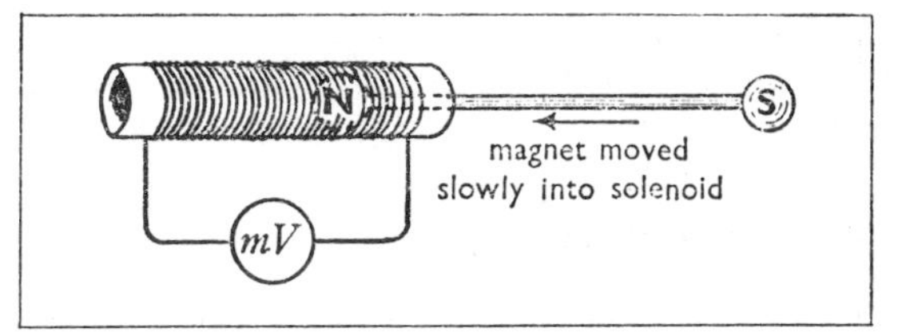

Fig. 7.6 Testing Faraday's law with a long magnet and solenoid

therefore necessary to show that the e.m.f. is proportional to the speed.

(*ii*) *Using a long magnet and a solenoid.* A long thin magnet is mounted so that one pole of it can be moved at a steady slow rate along the axis of a solenoid (Fig. 7.6). This can be done by mounting the magnet on a carriage that is advanced by the rotation of a screwed rod, driven at a steady speed. As the pole advances, the number of turns of the solenoid through which the flux from the pole passes steadily increases; that is, there is a steady rate of change of the flux linkage with the solenoid as a whole, and so a steady e.m.f. is produced in it. The rate of change of flux linkage is proportional to the speed of the magnet; and it is necessary to show that the e.m.f. is proportional to this.

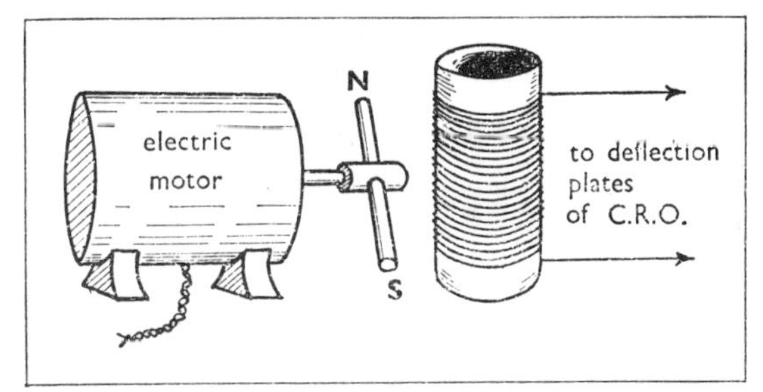

Fig. 7.7 Testing Faraday's law with a coil and rotating magnet

(*iii*) *Using a rotating magnet and cathode-ray oscilloscope.* A bar magnet is attached to the shaft of a variable-speed electric motor. A coil with a large number of turns is clamped near the magnet in such a way that the flux through it reverses every half-revolution of the magnet (Fig. 7.7). An alternating e.m.f. is thus generated in the coil, whose peak value is measured by the length of the trace on the cathode-ray tube screen. According to Faraday's law this peak e.m.f. should be proportional to the maximum rate of change of flux through the coil. This in turn is proportional to the speed of the motor, which can be measured by the frequency of the induced e.m.f. We have thus to investigate whether the length of the trace on the screen is proportional to the frequency.

7.2 Eddy currents

When a conductor moves in a magnetic field, e.m.f.'s are induced in all parts of it that cut the magnetic flux. Previously we have been studying the summed effects of these e.m.f.'s acting in circuits consisting of loops of wire, etc. But if the conductor is a large lump of metal, significant e.m.f.'s can act round closed paths inside the lump of metal itself. Although the e.m.f.'s are not usually very large, the resistance of the current paths is so low that large currents may flow. These induced currents circulating inside a piece of metal are known as *eddy currents*. By Lenz's law they must flow in such directions as to *oppose* the motion. They can indeed act as a very effective brake on the motion of a body, the mechanical energy being turned into the electrical energy of the eddy currents, which in turn is converted into heat inside the metal.

The effect of eddy currents may be demonstrated by swinging a pendulum with a thick copper bob between the poles of an electromagnet (Fig. 7.8). When the electromagnet is switched on, the flux through the bob varies rapidly along its path, and considerable eddy currents are generated in it, which produce a very marked braking effect. The eddy currents can be prevented in this case by using a bob with a series of slots cut in it so that the currents are now only free to circulate inside the relatively narrow teeth left between the slots. Very little braking effect can then be detected.

A similar demonstration may be performed with a copper cylinder hung from a cotton thread (Fig. 7.9) and set spinning between the poles of the magnet. The eddy current system acts as a very efficient brake in this case also. If the cylinder is made up of a stack of discs insulated from one another, the eddy currents are greatly reduced, since they can now circulate only inside the in-

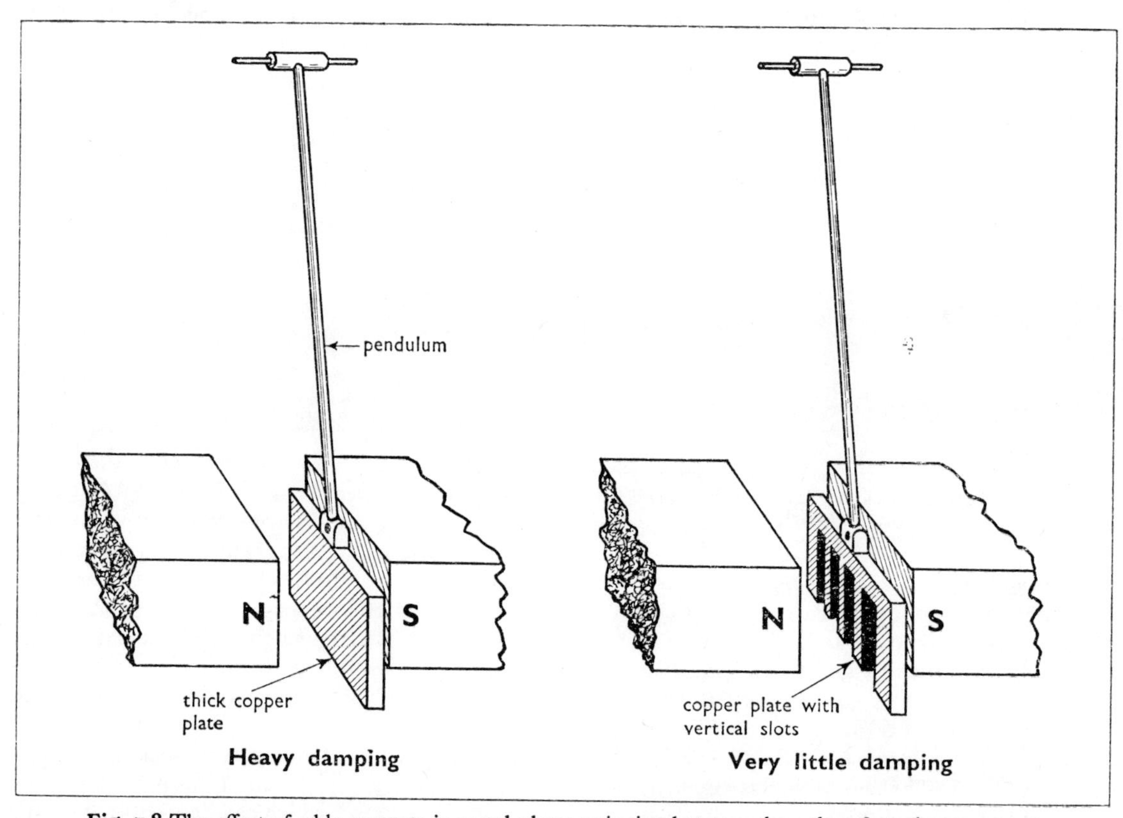

Fig. 7.8 The effect of eddy currents in metal plates swinging between the poles of an electromagnet

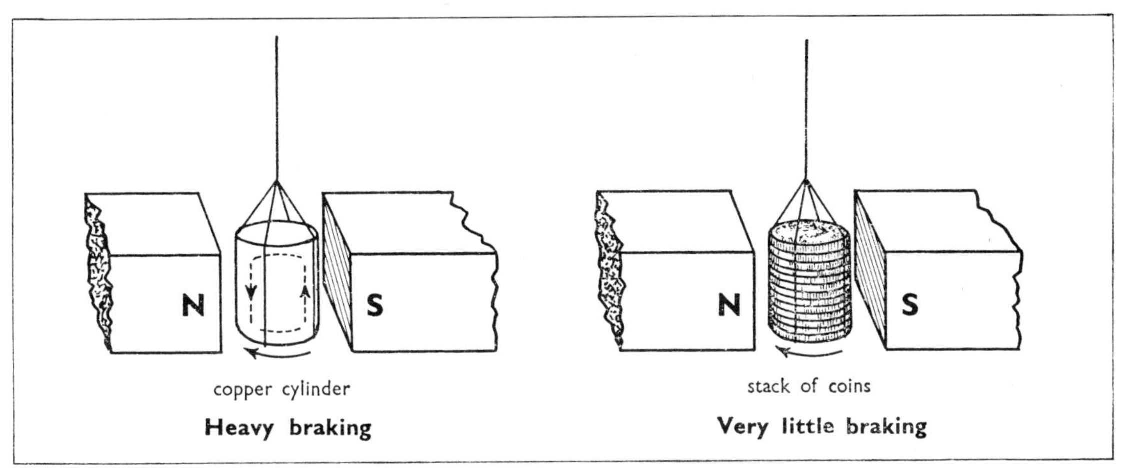

Fig. 7.9 The effect of eddy currents in a rotating metal cylinder

dividual discs and not up and down the whole length of the cylinder. A pile of coins may be used for this purpose; the dirt on the coins acts as sufficient insulation between them, and there is certainly very little braking effect.

Eddy currents arise also in a stationary piece of metal if the magnetic flux through the parts of it is changing. Thus, if a thick brass plate is held just below a magnet performing torsional oscillations in the earth's field, the movement is seen to be heavily damped on account of the interactions between the magnet and the eddy currents it induces in the plate.

The effect of eddy currents must particularly be allowed for in the design of iron-cored apparatus—motors, dynamos, transformers, etc.—if considerable losses of energy from this cause are to be avoided. It is necessary in these cases to build up the iron parts of the apparatus out of a stack of iron sheets, called *laminations*; these are insulated from one another by a thin layer of paper stuck on one side of each lamination. The loss of energy through eddy currents is approximately proportional to the square of the lamination thickness, so that by suitable design the loss from this cause can be reduced to any desired figure. For low-frequency power apparatus it is usual to choose laminations about 0·5 mm thick.

The eddy current loss is also approximately proportional to the square of the frequency of oscillation of the magnetic flux. The higher the frequency of an alternating current, the more is the loss from eddy currents in the core of any coil used, and the more the core must be subdivided to keep the eddy current loss down. If thin laminations are not sufficient, a bundle of iron wires may be used or even a core packed with iron dust. But at the frequencies used for radio work even this is not very satisfactory.

Another way of reducing eddy currents is to find magnetic substances that have very large resistivities; such for instance are the iron compounds known as *ferrites*. Many of these substances have such high resistivities that they may be classed as insulators, while their magnetic properties are similar to those of iron. They may be used to make high permeability cores for coils in radio apparatus and introduce very little loss even at high frequencies.

Sometimes eddy current losses are deliberately used as a means of supplying heat to an otherwise inaccessible metal object. For instance in the manufacture of radio valves to achieve a permanent vacuum it is necessary to raise the temperature of all parts of the valve as much as possible during evacuation to drive off the 'occluded' gases that might otherwise remain adsorbed on the surfaces. For this purpose the metal parts need to be heated to a much higher temperature than the glass envelope could stand. This is done by means of eddy currents. A coil carrying a high-frequency current is placed round the valve, thereby generating large eddy currents in the metal parts of it.

According to Lenz's law the effect of eddy

currents must be to tend to prevent an alternating magnetic flux from penetrating into a piece of metal. At high frequencies this process is so efficient that we may use a *non-magnetic* metal sheet as a screen for magnetic fields. It is the usual practice in radio apparatus to place aluminium cans round coils and other components that carry high-frequency currents. Eddy currents induced in the cans are sufficient to prevent an alternating magnetic field passing through the can and inducing unwanted e.m.f.'s in other parts of the circuit. (The same cans are also effective in screening the components from stray *electric* fields—p. 164.) However, this magnetic screening would not be effective at low frequencies. A mains transformer working at a frequency of 50 Hz produces an appreciable alternating field in its neighbourhood at this frequency, and this sometimes causes unwanted interference in other parts of a circuit. But nothing would be achieved by surrounding the transformer with a metal screen (of any reasonable thickness), since at this frequency the eddy currents would not be sufficient to give any significant reduction of the field penetrating the screen. The only effective screen in this case would be one made from a soft magnetic material, such as *mumetal* (p. 78).

A similar effect controls the flow of alternating current through any conductor at high frequencies. The lines of force of the magnetic field of a conductor carrying a current are circles concentric with the wire; and inside the conductor the field is of the same form. If the current is a high-frequency alternating one, e.m.f.'s are induced locally in the wire that make the distribution of current over its cross-section non-uniform. The distribution of current must in fact be such as to reduce the alternating flux in the body of the material as much as possible. Thus at high frequencies the current flows mostly in a layer very near the surface. This phenomenon is known as the *skin effect*.

At a frequency of 1 MHz (10^6 Hz) the thickness of the layer through which the current flows is only about 0·05 mm. Theory shows that the thickness of the 'skin' is inversely proportional to the square root of the frequency, so that at the frequencies used for television and radar (10^8 to 10^{10} Hz) it is comparable with the wavelength of visible light! One consequence of the skin effect is that at high frequencies the resistance of a conductor increases very considerably over its d.c. value, since its effective cross-sectional area is then much reduced.

7·3 Mutual induction

Two coils, electrically unconnected, can be 'linked' together by magnetic flux, any changes of the current in either coil causing e.m.f.'s to act in the other. This particular form of electromagnetic induction is called *mutual induction*. In d.c. circuits effects due to mutual induction can only arise at the moments when the current in one circuit is switched on or off; pulses of e.m.f. then occur in the other. But if an alternating current flows in one coil, it produces a continually varying flux through the other, causing an alternating e.m.f. in it of the same frequency. By this means electrical energy can be transferred from one circuit to another through the linking action of the magnetic flux.

At high frequencies this process is very efficient even with air-cored coils. But at low frequencies

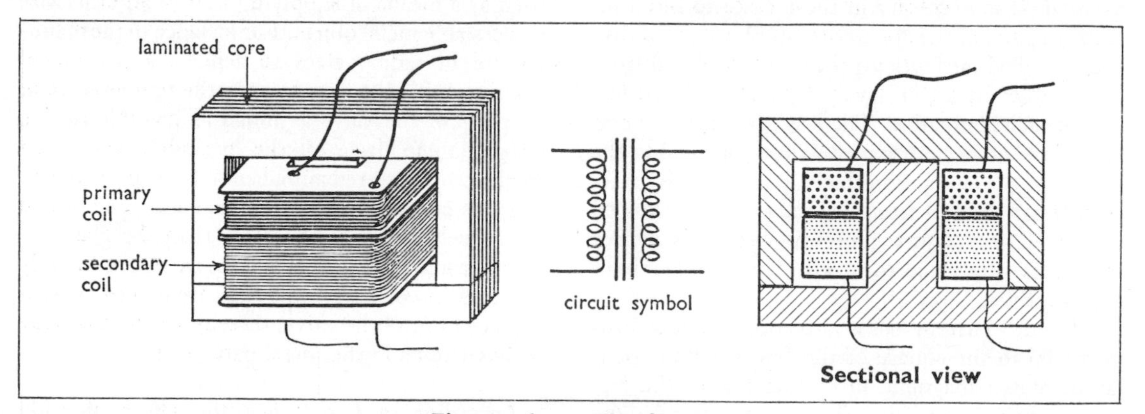

Fig. 7.10 An a.c. transformer

the transfer of energy from one coil to the other can only be really effective if the coils are wound on a closed loop of soft iron, which must be laminated to reduce eddy currents. This is the basis of the design of the a.c. transformer (Fig. 7.10). One coil, called the ***primary***, is connected to an a.c. supply, such as the 50 Hz mains. This coil has a large number of turns, and only a small current is needed in it to produce a large alternating flux in the core. An e.m.f. is accordingly induced in the other coil, called the ***secondary***; the magnitude of this e.m.f. is proportional to the number of turns of the secondary linked with the flux in the core. It can be shown (p. 202) that, provided the transformer is not overloaded,

$$\frac{\text{secondary p.d.}}{\text{primary p.d.}} = \frac{\text{number of turns in secondary}}{\text{number of turns in primary}}$$

Thus by suitable choice of this *turns ratio* we can produce any desired alternating p.d. at the terminals of the secondary. The a.c. transformer is an exceedingly efficient device. The large ones used in power supply systems handling hundreds of kilowatts are usually more than 99% efficient; and even the smaller ones used in the power supplies of radio sets have very small losses. Thus we can write, approximately

$$\text{power input to primary} = \text{power output from secondary}$$

$$\therefore \text{primary p.d.} \times \text{primary current} \simeq \text{secondary p.d.} \times \text{secondary current}$$

$$\frac{\text{secondary current}}{\text{primary current}} = \frac{\text{number of turns in primary}}{\text{number of turns in secondary}}$$

This is the *reciprocal* of the turns ratio. If the p.d. is transformed *up* in a certain ratio, the current is transformed *down* in the same ratio, and vice versa. The nature of the magnetic interaction between the coils of a transformer is discussed further in Chapter 12 (p. 202).

Transformers are often wound with several secondary coils, electrically insulated from one another. The mains transformer of a radio set is usually made this way. A typical example is shown in diagrammatic form in Fig. 7.11. In addition to the main high-tension winding there are also two low-voltage windings. These provide power for the heaters of the valves and the indicator lamps. The primary winding has several tapping points near one end. By correct choice of the number of turns in the primary coil the secondary e.m.f.'s can be

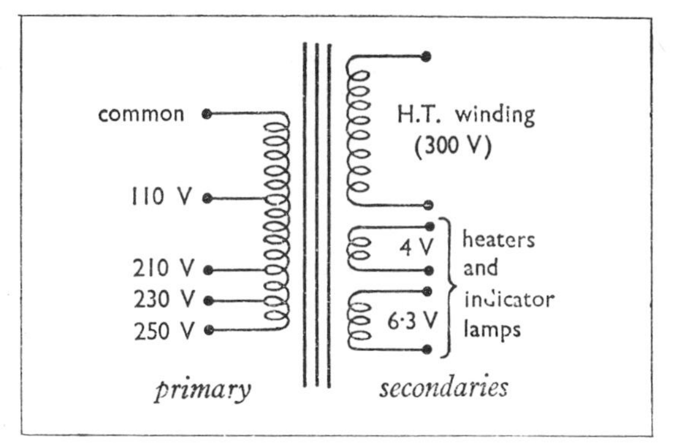

Fig. 7.11 A typical transformer for a radio set; several secondaries are provided, and the primary is tapped for use with different supply p.d.'s

kept at nearly constant values for a range of possible values of supply voltage. The heat loss in a coil of a transformer is proportional to its resistance and to the square of the current ($W = I^2R$). To minimize such losses the high-current, low-voltage windings are made of thick wire; while the low-current, high-voltage coils can well be of relatively fine wire.

7.4 Self-induction

In the process of ***mutual induction*** e.m.f.'s are induced in a circuit by changes in currents flowing in other circuits. But e.m.f.'s are also induced in a circuit by changes in its own current, since this also produces magnetic flux threading through the circuit. This process is known as *self-induction.* From Lenz's law we should expect the e.m.f.'s of self-induction to *oppose* the changes of current that caused them.

When a steady current is flowing in a coil no e.m.f. can arise from self-induction, and the value of the current depends only on the resistance and applied p.d. Thus in d.c. circuits the effects of self-induction are only apparent at switching on or off—while the current is rising to its maximum value or falling again to zero.

At switching on the induced e.m.f. acts in opposition to the applied p.d. so that it delays the growth of current. At first the back e.m.f. due to self-induction is nearly equal to the applied p.d. (Fig. 7.12); then as the current approaches its final value, the rate of increase of current becomes less, and the back e.m.f. falls to zero. With a very large

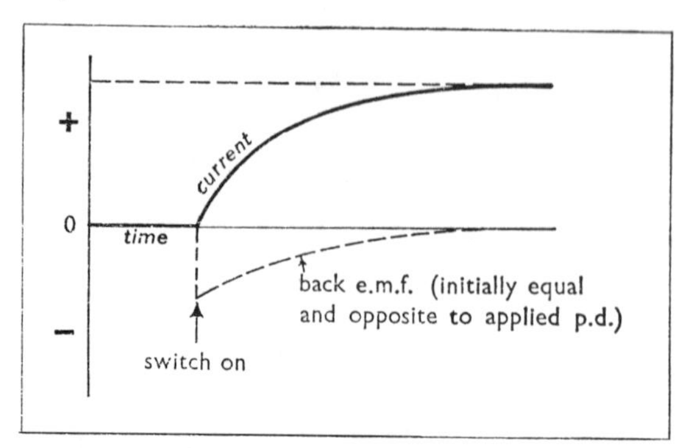

Fig. 7.12 The growth of current in an inductance

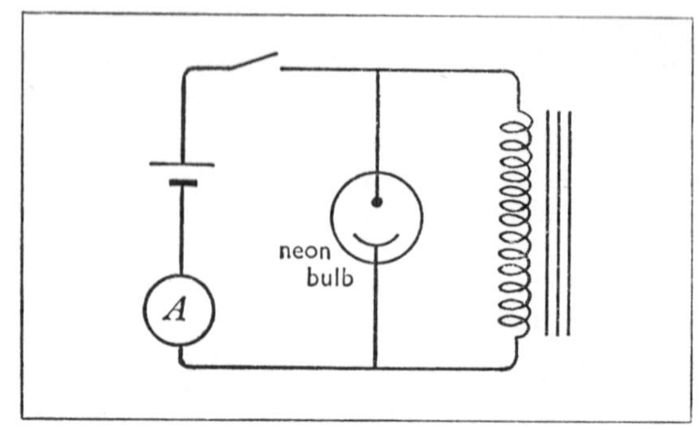

Fig. 7.13 A large e.m.f. is produced on switching off an inductive circuit

iron-cored coil the growth of current may well take several seconds. This may be demonstrated with the circuit of Fig. 7.13.

At switching off the behaviour of an inductive circuit like this is more dramatic. When the switch is opened, the current is obliged to drop almost instantly to zero, and the flux linked with the coil changes extremely rapidly. A very large pulse of e.m.f. is thus induced in it, which acts so as to tend to keep the current going. The peak value of this e.m.f. is often many hundreds of times greater than the original applied p.d., sufficient in Fig. 7.13 to light the neon lamp. If two fingers are placed on the switch terminals, an appreciable shock can be felt even when only a 2-volt cell is used in the circuit. (Those with weak hearts do well to take this last result on trust!)

This effect poses problems for the designer of switches. If the contacts are separated too slowly at switching off, the e.m.f. due to self-induction may be sufficient to cause an arc between them. The danger is greatest for circuits carrying large currents and containing highly inductive components. In extreme cases the arc can fuse the switch arms altogether. In the domestic light switch it is sufficient to use a spring-loaded switch arm, which results in rapid separation of the contacts. Even in this case trouble sometimes arises when fluorescent lighting is used, since inductive control coils have to be used in such circuits. In some large switch systems jets of compressed air are made to blow out the arc that forms as the contacts separate. With the largest circuit breakers the contacts are mounted in a tank of non-inflammable oil under pressure, which rapidly quenches the arc. Even so, the decomposition and vaporization of the oil produced by the arc leads to the sudden evolution of large volumes of gas, constituting a sizable explosion; and the frame of the tank must be very strong. In the National Grid network the cost of the circuit breakers is a large part of the total capital outlay.

The behaviour of an inductive coil (or *inductor*) with alternating current is quite complex, and is discussed in detail in Chapter 12. But in general terms we can see that the e.m.f.'s due to self-induction must act continually in such a way as to oppose the applied p.d. at every instant. The actual value of the current in such a case depends then not only on the resistance of the circuit, but also on the extent of the self-inductive effect; the larger this is, the smaller the current becomes. This may be demonstrated with a coil of many turns with a removable (laminated) iron core. The coil is joined in series with an electric bulb, and the arrangement is connected to the a.c. mains (Fig. 7.14). With the core removed the self-inductive effect at 50 Hz is fairly small, and the bulb will

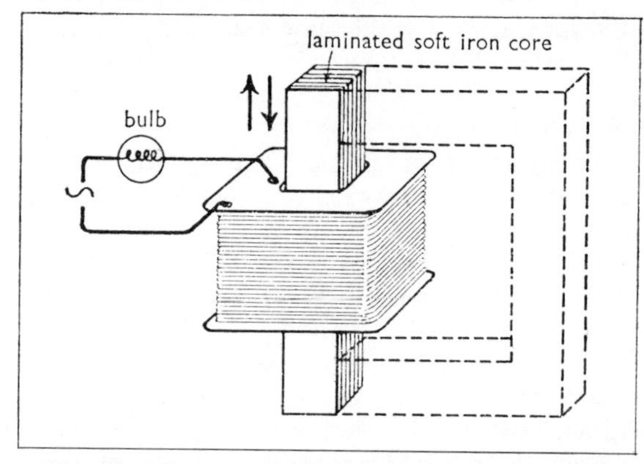

Fig. 7.14 The control of an alternating current with an inductive coil

just be slightly dimmer than usual, due chiefly to the extra resistance of the coil in the circuit. But when the iron core is introduced, the current at once falls to a small value. If it can be arranged so that a closed loop of laminated soft iron is formed through the coil (as suggested by the dotted lines in Fig. 7.14) the effect is even more marked. An inductor used in this way is called a *choke*; an alternating current can be controlled by this means without a great deal of heat loss, such as would occur if a resistor were used instead.

At the moments of switching on and off the behaviour of an a.c. circuit is more complicated than a d.c. one; but the general effects are similar —a slow growth in the magnitude of the alternating current at switching on, and a large pulse of e.m.f. at switching off, with consequent danger of arcing. However, the effects are less violent than in d.c. circuits, and this enables a.c. switches and thermostats to be of lighter construction than is required for direct current.

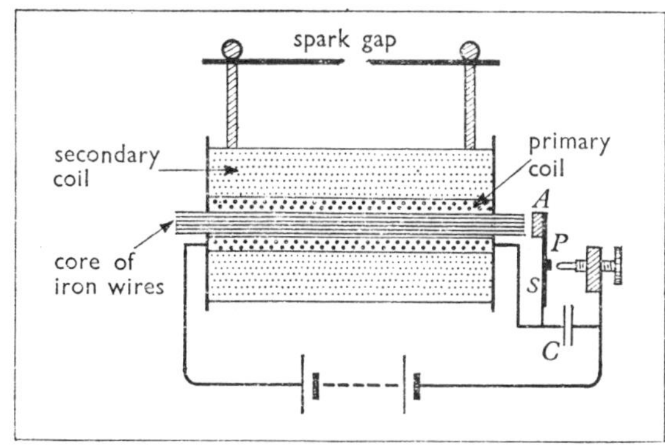

Fig. 7.15 An induction coil

The induction coil

In the *induction coil* an automatic make-and-break arrangement is included in series with an iron-cored coil so that a rapid succession of large pulses of e.m.f. is produced. Its construction is shown in Fig. 7.15. The core is made up of a bundle of soft iron wires; two coils well insulated from each other are wound on top of this. The *primary* coil is of thick-gauge wire and is designed to carry a large current supplied by a battery; the *secondary* coil has a very large number of turns of fine wire. The turns ratio between primary and secondary may be 100 or more.

When the primary circuit is completed, the current in it starts to grow, magnetizing the iron core; the soft iron armature block A is then attracted to the core. This breaks the circuit at the contact P, and the core demagnetizes again. The spring strip S now draws the armature back so as to close the contact once more; and so the cycle is repeated indefinitely. Each time the circuit is broken the current falls very rapidly to zero, producing a large pulse of e.m.f. in the primary. But the flux that links with the primary coil links many more times with the secondary, producing in it an even larger pulse of e.m.f. Peak p.d.'s of 50 to 100 kV can be produced by this means.

The contacts of the make-and-break arrangement move apart such a short distance in the induction coil that a special method must be adopted to control the arcing that would otherwise occur between them. For this purpose a capacitor C is joined across them. The property of a capacitor is to store charge on its plates—positive on one plate and negative on the other—when a p.d. is applied between them (p. 153). Thus, for a short time after opening the contacts current continues to flow charging up the capacitor. This gives time for the contacts to move far enough apart for arcing not to occur.

7.5 Calculating induced e.m.f.'s

We can now derive the mathematical statement of the laws of electromagnetic induction. If the flux linkage of a magnetic field with a circuit is N weber-turns (p. 73), we may use calculus notation to express Faraday's law as

$$e \propto \frac{dN}{dt}$$

where e = the induced e.m.f. It is usual in this connection to take the *positive* direction of e.m.f. as that which would give rise to flux through the circuit in the *same sense as* N. Lenz's law requires any current due to an induced e.m.f. to produce an *opposing* change of flux, which on this system of signs will be expressed by making e and dN/dt of opposite sign.

$$\therefore e = -k\frac{dN}{dt}$$

It only remains to determine the positive constant k. Its value depends on the units used, but we shall show that in any self-consistent system of units

—such as the m.k.s. system—we have the simple result

$$k = 1$$

Experiments show that the principle of conservation of energy holds for processes of electromagnetic induction of all kinds. To deduce the value of k it is only necessary to consider one imaginary case in which the e.m.f. may be calculated from the energy used; since k is a universal constant in any one system of units, the value deduced from the special case can then be used in all other cases also.

In Fig. 7.16 AB is a wire which is being propelled at speed v by a force F in contact with the smooth metal rails PQ and RS. A magnetic field of flux density B acts downwards perpendicular to the plane of the system. An e.m.f. therefore acts in the circuit. Lenz's law shows that current will flow in the direction shown, since this direction gives a magnetic force F' in opposition to the motion. The current i will also act so as to reduce the magnetic flux linking the circuit. This means that the actual e.m.f. induced is somewhat less than if the current were zero. To simplify the analysis we shall suppose that the reduction in the e.m.f. on this account is negligible; in other words we shall assume that the resistance of the circuit is made sufficiently high so that the magnetic field produced by the current i is negligible compared with the applied magnetic field B.

Now $$F' = Bil$$

where l is the distance between the rails (p. 67).

Also $$i = \frac{e}{r}$$

where r is the total resistance of the circuit. If the wire is moving at steady speed, then

$$F = F' = Bil = \frac{Bel}{r}$$

The rate of doing work $= F \times$ (distance moved per second)

$$= Fv = \frac{Belv}{r}$$

The work done re-appears as electrical energy in the circuit, and is finally changed into heat.

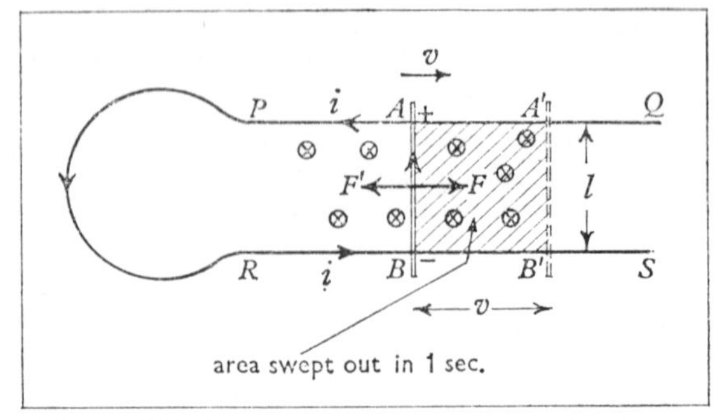

Fig. 7.16 Deducing the constant of proportionality in Neumann's law

The total rate of production of heat $= \dfrac{e^2}{r}$

$$\therefore \frac{e^2}{r} = \frac{Belv}{r}$$

$$\therefore e = Blv$$

Now the rate of change of flux linking the circuit is equal to the flux through the area ABB'A' swept out by the wire in one second; this area is lv.

$$\therefore \frac{\mathrm{d}N}{\mathrm{d}t} = Blv$$

Thus, e and $\mathrm{d}N/\mathrm{d}t$ are numerically equal; but, as explained above, we must regard them as of opposite sign, since the induced current produces a field in opposition to the change of flux. Thus $k = 1$, and we have

$$e = -\frac{\mathrm{d}N}{\mathrm{d}t}$$

This mathematical statement of the laws of electromagnetic induction is sometimes known as *Neumann's law*; it is a direct deduction from Faraday's and Lenz's laws.

Up to this point we have made use of a provisional definition of the m.k.s. unit of magnetic flux, the *weber*, in terms of the unit of flux density (the *tesla* or *weber per square metre*—p. 73). However, the most direct way of measuring the flux linking a circuit is to find the e.m.f. induced when the flux linkage is caused to change; the average flux density can then be deduced from this kind of measurement by dividing the flux by the cross-sectional area through which it passes. It is therefore best to treat the weber as the more fundamental unit and to define it in terms of induced e.m.f.'s. The unit of flux density can then be regarded as derived from the weber.

If the magnetic flux linking a single-turn coil is 1 **weber,** *an e.m.f. of* 1 *volt will be induced in the coil when the flux linkage is reduced to zero at a uniform rate in* 1 *second.*

The use of Neumann's law is best understood by considering some practical examples.

Example (*i*). Calculate the e.m.f. induced between the wing-tips of an aeroplane of wing span 30 metres flying horizontally at 250 m s^{-1}, if the flux density of the vertical component of the earth's magnetic field is 4×10^{-5} T.

The area swept out by the wing per second

$$= 30 \times 250 \text{ m}^2$$

∴ The flux cut by the wing per second

$$= \frac{dN}{dt} = 4 \times 10^{-5} \times 30 \times 250 \text{ Wb s}^{-1}$$

$$= 0{\cdot}3 \text{ Wb s}^{-1}$$

∴ e.m.f. = 0·3 volt

The student can check that in the northern hemisphere the port (left) wing-tip would be *positive.*

Example (*ii*). A horizontal gramophone turntable (made of brass) of diameter 30 cm rotates at $33\frac{1}{3}$ revolutions per minute in a uniform vertical magnetic field of flux density 0·010 T. Calculate the e.m.f. induced between the centre and rim of the turntable.

A given radius sweeps out an area

$$(\pi \times 15^2) \text{ cm}^2 \text{ per revolution}$$
$$= 225\pi \times 10^{-4} \text{ m}^2 \text{ per rev.}$$

The number of revolutions per second $= \dfrac{100}{3 \times 60}$

∴ The rate of cutting flux by a given radius

$$= \frac{dN}{dt} = 0{\cdot}01 \times \frac{100}{3 \times 60} \times 225\pi \times 10^{-4} \text{ Wb s}^{-1}$$

$$= 3{\cdot}9 \times 10^{-4} \text{ Wb s}^{-1}$$

∴ e.m.f. = $3{\cdot}9 \times 10^{-4}$ volt = 0·39 mV

If the magnetic field acts vertically downwards, with the usual direction of rotation of gramophones the rim will be positive.

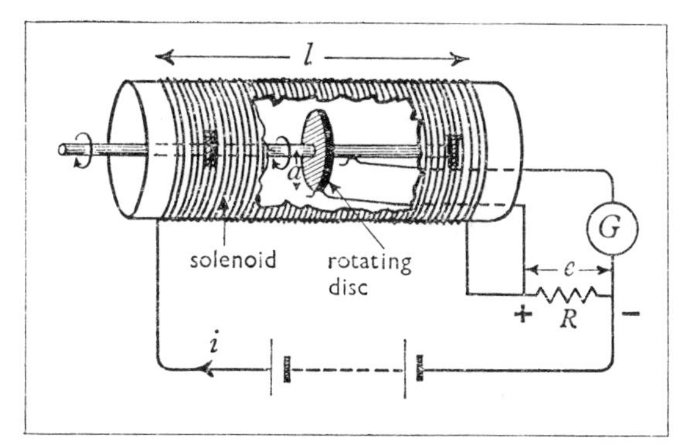

Fig. 7.17 The Lorenz disc method for the absolute measurement of resistance

The absolute measurement of resistance

It was shown by Lorenz that electromagnetic induction in a disc rotating in a magnetic field could be made the basis of an absolute determination of resistance. The apparatus that may be used is shown in Fig. 7.17. The axle is arranged to coincide with the axis of a long solenoid; and the apparatus is turned so that the plane of the disc is exactly in the magnetic meridian. The e.m.f.'s induced in it are then due to the field of the solenoid only. Sliding contacts are made to bear on the rim and axle, as shown; if the distances of these contacts from the central axis are a and b, and the disc performs r revolutions per second, then the area swept out per second by the part of a radius of the disc between the two contacts is given by

$$\text{area swept out} = \pi(a^2 - b^2) \times r$$

The flux density B in the solenoid is given by

$$B = \frac{\mu_0 ni}{l} \qquad \text{(p. 70)}$$

where n is the total number of turns in the coil, l its length and i the current flowing in it.

The e.m.f. e induced is equal to the rate of cutting flux by the part of the radius concerned. This e.m.f. is 'balanced' against the p.d. across the small resistance R in series with the solenoid, through which the same current i therefore flows; the adjustment is effected by varying the speed of rotation until the galvanometer gives no deflection. We then have

$$e = iR = \frac{\mu_0 n\pi r(a^2 - b^2)i}{l}$$

$$\therefore R = \frac{\mu_0 n\pi r(a^2 - b^2)}{l}$$

The resistance R is thus found in terms of the dimensions of the apparatus and the speed of rotation r.

To make the measurement fully satisfactory the simple apparatus in Fig. 7.17 needs many refinements. Friction at the sliding contacts produces heat which leads to thermoelectric e.m.f.'s (p. 45); these are by no means negligible compared with the small induced e.m.f. e. In practice only a small part of the rather large current i can be allowed to pass through the resistance R; a resistance network is therefore used in the circuit instead of the single resistor R. Allowance must also be made for the slight non-uniformity of the magnetic field inside the solenoid. However, in a suitably refined form this experiment has proved one of the most satisfactory yet devised for measuring the resistances of the coils maintained as standards at the national standardizing laboratories.

The e.m.f. induced in a rotating coil

Consider a flat coil of n turns each of area A rotating about an axis in its own plane at right-angles to a uniform magnetic field of flux density B. The flux linkage N of the magnetic field with the coil is given by

$$N = BAn \cos \phi \qquad \text{(p. 73)}$$

where ϕ is the angle between the *axis* of the coil and the field. If the angular speed of rotation is ω radians per second, then

$$\phi = \omega t$$

(measuring the time t from the position where $\phi = 0$).

$$\therefore N = BAn \cos \omega t$$

Differentiating, we obtain an expression for the induced e.m.f. e.

$$\therefore e = -\frac{dN}{dt} = BAn\omega \sin \omega t = -BAn\omega \sin \phi$$

The e.m.f. is thus an alternating quantity, whose peak value e_0 occurs at positions where $\phi = 90°$ (i.e. with the plane of the coil parallel to the field); it is given by

$$e_0 = BAn\omega$$

The r.m.s. value E of the e.m.f. is therefore given by

$$E = \frac{BAn\omega}{\sqrt{2}} \qquad \text{(p. 29)}$$

The frequency f is equal to the number of revolutions per second; and since the coil turns through 2π radians per revolution, we have

$$\omega = 2\pi f \text{ radians per second}$$

Example. Calculate the peak value of the e.m.f. induced in a circular coil of 1000 turns of radius 5·0 cm rotating at 3000 r.p.m. about an axis in its own plane at right-angles to a magnetic field of flux density 0·02 T.

$$\text{The number of rev per second} = \frac{3000}{60} = 50$$

$\therefore$ Angular speed $\omega = 2\pi \times 50 = 100\pi$ rad s^{-1}

The peak e.m.f. e_0 is given by

$$\begin{aligned} e_0 &= BAn\omega \\ &= 0{\cdot}02 \times \pi \times 5^2 \times 10^{-4} \times 1000 \times 100\pi \text{ volts} \\ &= 5\pi^2 = 49 \text{ volts} \end{aligned}$$

When such a rotating coil is connected in a complete circuit so that a current flows, then according to Lenz's law we should expect to find a couple acting on the coil opposing its rotation. In fact we can show that the couple fluctuates in just such a way that at any instant the rate at which mechanical energy must be supplied to maintain the rotation is equal to the rate of production of electrical energy. Let the current flowing in the coil at any instant be i; the electromagnetic torque C opposing the rotation is therefore given by

$$C = BAni \sin \phi \qquad \text{(p. 67)}$$

In order to maintain the rotation, the rate of supply of mechanical energy $= C\omega$

$$= BAni\omega \sin \phi$$

But the rate of production of electrical energy is given by exactly the same expression, since

$$e = BAn\omega \sin \phi$$

and the rate of production of electrical energy

$$= ei = BAni\omega \sin \phi$$

Induced e.m.f.'s and the forces on charged particles

We have already seen how the e.m.f. induced in a wire that is cutting magnetic flux may be

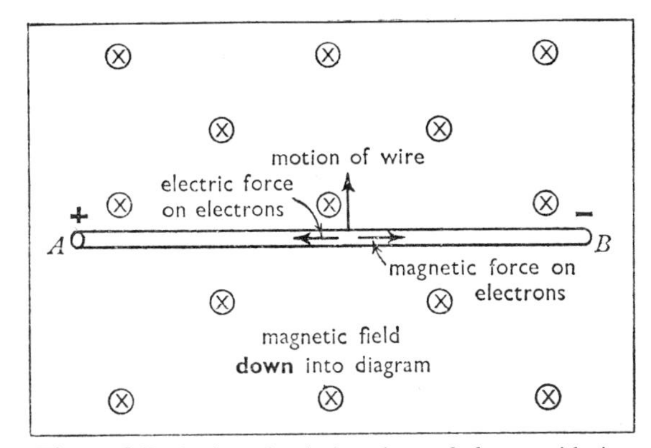

Fig. 7.18 Deducing the induced e.m.f. by considering the balanced forces that must act on the electrons in the conductor

explained in terms of the forces acting on the charged particles in it (p. 98). This explanation we can now make fully quantitative. When the wire is first set in motion we suppose that the magnetic forces cause a migration of electrons in one direction along it, thereby setting up a p.d. between its ends. This migration will continue until the magnetic forces are exactly balanced by the electric forces tending to restore the electrons to their original uniform distribution (Fig. 7.18). Suppose the wire is moving with velocity v in a plane at right-angles to a magnetic field of flux density B, as shown. The electrons share the motion of the wire, and the magnetic force F_m acting on one of them is given by

$$F_m = Bev \qquad \text{(p. 68)}$$

where e = the electronic charge. The electrical force F_e on an electron is given by

$$F_e = e \times \text{(potential gradient)}$$

$$= e \times \frac{V}{l} \qquad \text{(p. 157)}$$

where V is the p.d. induced between the ends of the wire (a distance l apart). The electrons reach equilibrium when these forces are balanced. Then

$$F_e = F_m$$

$$\therefore \frac{eV}{l} = Bev$$

$$\therefore V = Bvl$$

Now vl is the area swept out by the wire per second, and therefore the p.d. V is equal to the rate at which the flux is cut by the wire—in agreement with Faraday's law of electromagnetic induction (p. 98). A single physical principle thus proves sufficient to explain both the motor effect and the phenomenon of electromagnetic induction.

The moving wire in Fig. 7.18 must be regarded as an *open-circuited* source of e.m.f., since there is no closed path for any current to follow; in this case therefore the p.d. V between the ends A and B is equal to the e.m.f. generated. It is worth considering how this e.m.f. might possibly be detected. Suppose we decide to connect a voltmeter between A and B. If the voltmeter is moving along with the conductor AB, its connecting leads are cutting the flux at the same rate as AB, and the same e.m.f. V is induced in them. The system then consists of two equal sources of e.m.f. in parallel and no current will flow in the loop; the voltmeter does not therefore register anything. To detect the e.m.f. it is necessary rather for the voltmeter to be kept stationary while the wire is moving (or vice versa); and the two parts of the circuit must be joined together by flexible or sliding connectors.

Frames of reference

Whenever we talk about an e.m.f. induced in a conductor it is important to state what motion the measuring instruments have. In other words we must state in what *frame of reference* the calculations and measurements are being conducted. Consider for instance an aeroplane flying in the earth's magnetic field. It is meaningless to talk of the e.m.f. induced in the wing of the plane or the p.d. between its wing-tips—*unless* we specify also the frame of reference for which the statements are true. For an observer travelling in the plane there is no e.m.f. (provided the field is a uniform one), and there is no p.d. between the wing-tips. Indeed in this frame of reference no evidence can be found that the plane is *moving* through the field. After all, we cannot attach a label to a line of force and watch it drifting past! Only if the field is non-uniform would there be any e.m.f. detectable in a circuit loop fixed in the plane. But another observer making measurements in a frame of reference fixed in the ground analyses the situation rather differently. His observations show that the magnetic field is stationary in *his* frame of reference. He calculates that an e.m.f. must be induced in the wing of the aeroplane as it cuts the flux. To settle the point he contrives to make momentary connection between the wing-tips and his measuring instruments on the ground; and he duly detects the predicted p.d. The observer in the plane agrees that an e.m.f. should be detected in this experiment, but accounts for it in a different way. To him the plane is stationary with respect to the magnetic field, and the voltmeter on the ground with its connecting leads is receding rapidly through the flux; the e.m.f. is therefore induced in the ground-based part of the circuit. Both observers agree, as they must, about the effect—namely the voltmeter reading; but they disagree about the site of the e.m.f. in the circuit. Each believes that the e.m.f. arises in the part of the circuit that is moving with respect to his frame of reference. And a third observer in a train moving with respect to both the others would regard the e.m.f. as arising partly in one section of the circuit and partly in the other.

★ *Relativity*

The above situation is in fact a very simple application of the principle of Relativity. This principle states that the laws of physics must be the same for all observers who are moving at constant velocities with respect to one another. There is no 'preferred observer' whom we can assert to be 'at rest' and whose observations have any sort of priority over the observations of others. Experiments may be conducted with equal validity in any non-accelerating frame of reference, and the same laws will be discovered as in any other similar frame of reference. Two observers may analyse a given situation in terms of forces, velocities, masses, fields, etc., in different ways; but the connections between these quantities, expressed by the laws of physics, must be the same for both. One very important set of laws is that which shows how to relate the values of physical quantities for one observer with those measured by another. This set of laws makes up the Special Theory of Relativity propounded by Einstein in 1905. It is beyond the scope of this book to consider this theory in detail, though some of its results are discussed in other parts of the book (pp. 232 and 311). Generally speaking the analysis of events in the Special Theory of Relativity differs little from the older Newtonian analysis used in elementary mechanics, *except* when the speeds of objects approach that of light. When this happens, almost all the simple laws of mechanics break down. Forces and masses appear to vary as between one frame of reference and another, velocities can no longer be added vectorially, and many of the laws of physics have to be expressed in quite different ways in order that they may take exactly the same form for all non-accelerating observers—as required by Einstein's principle. All these results have been extensively tested by experiment, at any rate for the motions of the elementary particles of matter with which we are concerned at the moment; and the Special Theory of Relativity is generally regarded as one of the corner-stones of modern physics. The older 'classical' forms of physical laws are found in many cases to be approximations to the laws of Relativity appropriate to slow speeds.

As it happens, the laws of electromagnetism are amongst the few that remain unaltered in form in the theory of Relativity. However, the theory does lead us to expect that the analysis of the electromagnetic field into its electric and magnetic components may differ as between one observer and another. Indeed, as we have seen above, this situation arises at quite slow speeds. At such speeds two observers will agree about the resultant force acting on any particle or the resultant e.m.f. in a circuit. But their analyses of the force into its electric and magnetic components or their ideas on the sites at which e.m.f.'s arise in a circuit will in general differ. (At speeds comparable with that of light even the resultant force on a particle will be assessed differently by the two observers.) It is therefore important always to ensure that there is no possible doubt about the frame of reference with respect to which any electromagnetic calculation or measurement is being conducted. What in one frame of reference is simply a magnetic field in another may be a combination of an electric and a magnetic field, and so on.

To see how this works out it is necessary to consider what we mean by the two sorts of field. All electromagnetic effects may be analysed ultimately in terms of the forces acting on charged particles. If the charged particle is at rest (in a given frame of reference) it experiences only the electric force F given by

$$F_e = eE$$

where e is the charge on the particle and E is the potential gradient of the electric field (p. 156). If the particle is moving with velocity v (in the given frame of reference) it experiences in addition the magnetic force F_m given by

$$F_m = Bev$$

where B is the component of the flux density at right-angles to the velocity. The resultant force on the particle is thus the *vector sum* of these two components, which we can express symbolically by

$$\text{electromagnetic force} = F_e + F_m = e(E + v \times B)$$

What we are doing in effect is *defining* the electric force as that which acts on a stationary charge (in our chosen frame of reference) and the magnetic force as the additional component of force that arises when the charge is in motion.

It should now be clear why two observers will disagree about the division of the electromagnetic field into its electric and magnetic components. If a charged particle is at rest in one frame of reference it is moving in the other, and vice versa. As long as the relative speed of the two observers is small compared with that of light, they will agree about the resultant electromagnetic force, but will apportion it differently between its two components.

Consider for instance the simple situation of a rod cutting the flux of a magnetic field, as in Fig. 7.18. In a frame of reference fixed in the laboratory there is a magnetic flux density B as shown, *and no electric field*. In this frame of reference the charged particles moving as part of the rod are acted on only by the magnetic force F_m, protons towards A and electrons towards B. If the rod were made of insulating material, the atoms in it would be slightly strained (i.e. *polarized*, p. 161); but there would be no general movement of charge along it, and there would still be virtually no electric field. The situation is different if the rod is of conducting material. A general migration of charge will then take place along it, and this will create an electric field such that the force F_e exactly balances the magnetic force F_m. This electric field then exists also in the region around the wire. For instance if the ends A and B were joined to the plates of a capacitor (p. 153) moving with the rod there would be an electric field in the gap between its plates.

The description in the above paragraph has been for a frame of reference fixed in the laboratory.

Now let us analyse the same situation in a frame of reference fixed in the rod AB and moving with it through the laboratory. If the rod is of insulating material, the *stationary* charged particles in it are observed to be acted on by forces acting parallel to the rod, protons towards A and electrons towards B, and we should find that the material is slightly strained electrically (polarized). We should therefore conclude that we are in a region of electric field created by some outside agency, since there is manifestly no source of electric field in the rod itself. We could also do experiments to establish the existence of a magnetic field by studying the forces on charged particles *moving* in our frame of reference—e.g. we could measure the couple acting on a coil carrying a current. (At slow speeds the flux density B would prove to be negligibly different in the two frames of reference.) If now the rod is of conducting material, the electric field in our frame of reference will cause a migration of charge along it; this will set up an opposing electric field, such that the resultant electric field in the rod is exactly zero. If a capacitor were joined between A and B, there would be no resultant electric field in the gap between its plates.

Both the above descriptions of the electromagnetic field around the rod AB are equally valid, and both are in accordance with the universal laws of electromagnetism, as required by the principle of Relativity.

In this example a pure magnetic field in one frame of reference is interpreted as an electric field (in addition to the magnetic field) in the other. Similarly it can be shown that a pure electric field in the first frame of reference would appear as a magnetic field (in addition to the electric field) in the second. Indeed the very existence of magnetic forces can be explained entirely in this manner. A stationary charged particle near a current-carrying conductor experiences no resultant force, as long as the numbers of electrons and protons in the conductor remain equal. But, if the charged particle is moving with respect to a frame of reference fixed in the conductor, the electric forces on the particle cease to be balanced. According to the theory of Relativity, the force between the moving charged particle and the stationary protons in the conductor is slightly increased as assessed in the given frame of reference; so also is the force between the particle and the electrons in the conductor, *but by a different amount*, since the electrons also are in motion. The apparent difference between the two sets of electric forces is what we interpret as a magnetic force. The detailed analysis of this result is beyond the scope of this book; but enough has been said to show how the principle of Relativity has taken a place as one of the great unifying ideas of modern physics. In fact, by applying this principle the entire theory of electromagnetism may be derived from a single experimental result, namely the law of force for electric charges (p. 161).

7.6 Inductance

The way in which we measure the self-inductive effect in a coil follows from Neumann's law of electromagnetic induction:

$$e = -\frac{dN}{dt}$$

We shall deal first with the case of coils without cores of iron or other ferromagnetic material. The flux linkage N of the magnetic field with the coil is then always proportional to the current i flowing in it.

$$\therefore N = Li$$

where L is a constant depending on the geometry of the coil.

∴ The induced e.m.f. e is given by

$$e = -\frac{dN}{dt} = -L\frac{di}{dt}$$

Thus the induced e.m.f. is proportional to the rate of change of current in the coil. The constant L is called the *self-inductance* of the coil—or simply its *inductance.*

Inductance. *The e.m.f. e induced in a coil by self-induction is proportional to the rate of change of the current i in the coil;*

$$\therefore e \propto -\frac{di}{dt}$$

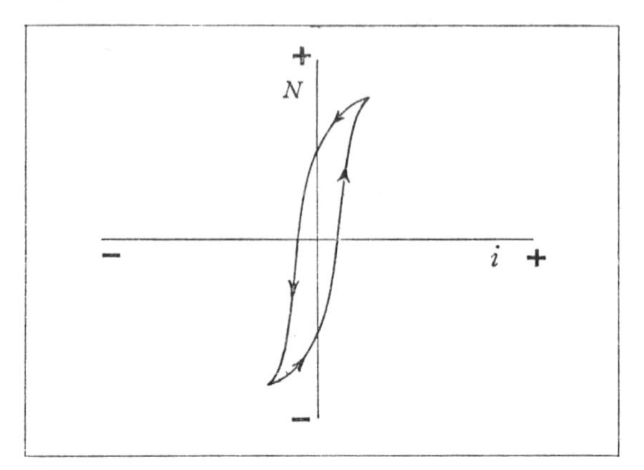

Fig. 7.19 The variation of flux N with current i for a coil with a ferromagnetic core. The inductance L is the gradient of this curve, and varies from point to point

or
$$e = -L\frac{di}{dt}$$

where L is a constant for the coil known as its self-inductance.

The m.k.s. unit of inductance is called the *henry* (after an American scientist). Thus L is given in henrys when e is in volts, i in amps and t in seconds.

The **henry** *is the self-inductance of a conductor in which an e.m.f. of 1 volt is induced when the current in it changes at the rate of 1 amp per second.*

The effect of a ferromagnetic core

In this case the result is much more complicated, since the flux linkage N is controlled by hysteresis in the core, and is not now simply proportional to the current i. Consideration of the nature of the hysteresis loop of a soft magnetic material shows that we should expect the relationship between N and i to be of the form shown in Fig. 7.19. The induced e.m.f. e is again given by

$$e = -\frac{dN}{dt} = -\frac{dN}{di}\frac{di}{dt}$$
$$= -L\frac{di}{dt} \text{ (by definition)}$$
$$\therefore L = \frac{dN}{di}$$

This is the gradient of the curve in Fig. 7.19, and varies from point to point of the cycle through which the core is taken. The relationship between current and applied p.d. is thus a very complicated one. If the coil is connected to a sinusoidal p.d. such as the a.c. mains, the current flowing through it is by no means of this waveform. Also the r.m.s. current is not proportional to the r.m.s. p.d. In some applications of chokes this non-linearity would be very unsatisfactory. However, it can be reduced to reasonable proportions by providing a narrow air gap in the ferromagnetic core, as in Fig. 7.20*a*. The demagnetizing effect that arises on this account reduces the flux through the core for a given current, and therefore decreases the inductance. But it also has the effect of tilting the curve of N against i as in Fig. 7.20*b*. Provided the core is not taken to saturation, the curve now approximates to a straight line; the inductance is therefore nearly constant, and only a little wave-form distortion is introduced by the choke.

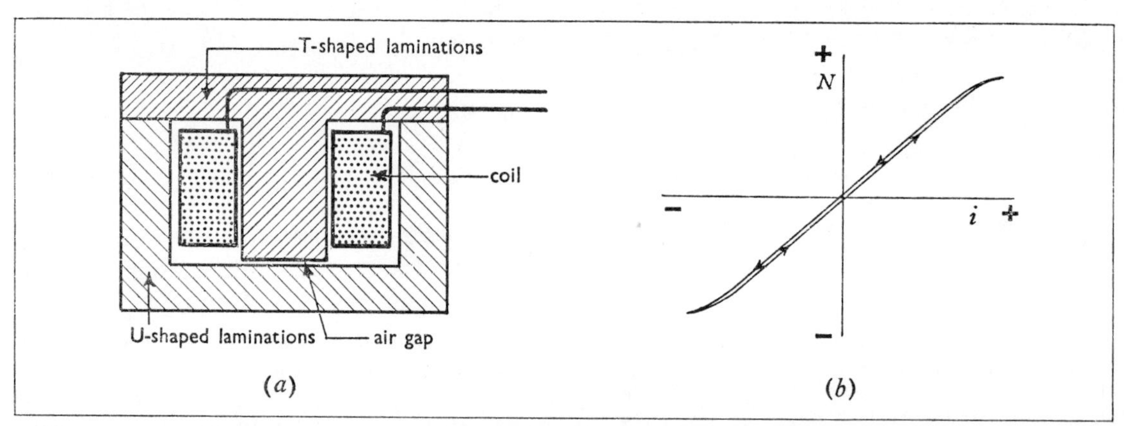

Fig. 7.20 By providing an air gap in the iron core of a choke the relation between N and i becomes nearly linear

Mutual inductance

A similar analysis may be used for the mutual induction between two coils. Thus in the absence of ferromagnetic material the flux N_2 linking the second coil due to a current i_1 in the first is given by

$$N_2 = {}_1M_2 . i_1$$

where ${}_1M_2$ is a constant depending on the geometry of the coils. Extending the notation in the same way, we have for the other coil

$$N_1 = {}_2M_1 . i_2$$

It may be shown (though the proof is beyond the scope of this book) that the two constants are equal; i.e.

$${}_1M_2 = {}_2M_1$$

so that we may combine the two equations into

$$N = Mi$$

where N is the flux linking either coil due to a current i in the other.

Proceeding as before, the e.m.f. e induced by mutual induction between the two coils is given by

$$e = -\frac{dN}{dt} = -M\frac{di}{dt}$$

and the constant M is known as the *coefficient of mutual inductance* between the coils—or more briefly as their *mutual inductance.*

The similarity of the defining equations shows that mutual inductance is to be measured in the same units as self-inductance. Indeed the definition of the henry can just as well be framed in terms of the mutual inductance of two coils.

Mutual inductance. *The e.m.f. e induced in a coil by the magnetic field due to the current in another coil is proportional to the rate of change of current* (di/dt) *in the latter coil*;

$$\therefore e \propto -\frac{di}{dt}$$

or

$$e = -M\frac{di}{dt}$$

where M is a constant for the two coils known as their mutual inductance.

When two coils have a common ferromagnetic core, we find as before that their mutual inductance is not a constant property but varies with the point on the hysteresis loop to which the core is taken. When it is essential to avoid waveform distortion over a wide range of currents, the core must be provided with an air gap. This is done, for instance, in the output transformer of a high-quality amplifier.

★ *The forces between coils*

Electromagnetic forces always act so that they produce maximum flux linkage with any coil carrying a current (p. 73). When two coils act on one another, it can be shown that the force F in a particular direction (call it the x-direction) on the second coil is given by

$$F = i_2\frac{dN_2}{dx} = i_1i_2\frac{dM}{dx}$$

and this is equal and opposite to the force acting on the first coil. Similarly the couple C tending to produce rotation about a given axis is given by

$$C = i_1i_2\frac{dM}{d\theta}$$

where θ is the angle of rotation about the axis.

These results show that the equality of the two coefficients of mutual induction depends basically on Newton's third law of motion. The symmetry of the above expressions requires

$${}_1M_2 = {}_2M_1$$

if the forces and couples with which the coils act on one another are to form equal and opposite pairs.

Calculations of inductance

In most cases these present considerable mathematical difficulties. We shall illustrate the technique by dealing with two important examples. The method is to find the total flux N linking the coil for a given current i, and then to use the expression

$$N = Li$$

This is equivalent to using the defining formula

$$e = -L\frac{di}{dt}$$

except with iron-cored coils, where in any case L is not an exactly defined quantity on account of hysteresis.

(*i*) *A long solenoid*

Consider a long solenoid of cross-sectional A, containing a total of n turns and of total length l. If we ignore the reduction of field that takes place near the ends, we can write

$$\text{flux density } B \text{ in the core} = \frac{\mu_0 ni}{l} \qquad \text{(p. 70)}$$

$\therefore$ the flux through the core $= \dfrac{\mu_0 Ani}{l}$

$\therefore$ the flux linkage N with the coil $= \dfrac{\mu_0 An^2 i}{l}$

$$\therefore L = \frac{N}{i} = \frac{\mu_0 An^2}{l}$$

We may extend the calculation to give the mutual inductance M between a long solenoid and a short secondary coil of n' turns wound round its centre. The flux linkage N' with the second coil is given by

$$N' = \frac{\mu_0 Ann'i}{l}$$

$$\therefore M = \frac{N'}{i} = \frac{\mu_0 Ann'}{l}$$

In this case the approximation made in ignoring the end effects is a much better one, since we are now concerned only with the flux through the *centre* of the long solenoid.

(ii) A pair of long, parallel wires

This is an important practical case, since the transmission lines used for many purposes are of this kind. To simplify the analysis we shall suppose that the current flowing in the wires is of sufficiently high frequency for the current to be effectively confined to the surface layers of the wires (p. 102); there is then no magnetic field *inside* the wires. We shall also assume that the separation d of the wires is much greater than their radius a; this means that the current in one wire will not significantly affect the uniform distribution of current over the surface of the other. (When the wires are close together, an effect similar to the skin effect produces a non-uniform distribution of current over their surfaces—the so-called *proximity effect.*)

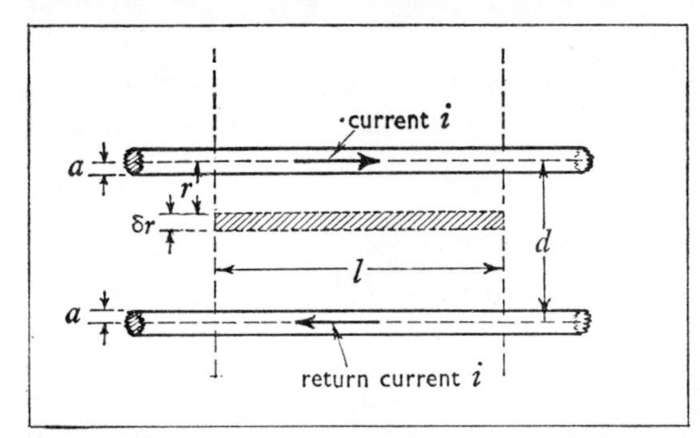

Fig. 7.21 Deducing the inductance of two parallel wires

If the current in the wires is i, the flux density B produced by *one* wire at a distance r from it is given by

$$B = \frac{\mu_0 i}{2\pi r} \qquad \text{(p. 70)}$$

The flux δN passing through a rectangular element of width δr and length l in the plane of the wires (Fig. 7.21) due to the current in *one* wire is given by

$$\delta N = \frac{\mu_0 i}{2\pi r} \times l\,\delta r$$

Integrating this between the limits a and $(\mathrm{d} - a)$ we obtain the flux passing through the gap between the wires in the length l. Each wire produces the same contribution to the flux, so that the total flux N linking the length l of the transmission line is given by

$$N = 2 \times \frac{\mu_0 il}{2\pi} \int_a^{\mathrm{d}-a} \frac{\mathrm{d}r}{r}$$

$$= \frac{\mu_0 il}{\pi} \log_e \left(\frac{\mathrm{d} - a}{a}\right)$$

$$\simeq \frac{\mu_0 il}{\pi} \log_e \frac{\mathrm{d}}{a} \quad (\text{since } a \ll \mathrm{d})$$

The inductance L of this length of cable is therefore given by

$$L = \frac{N}{i} = \frac{\mu_0 l}{\pi} \log_e \frac{\mathrm{d}}{a}$$

and the inductance per unit length is

$$\frac{L}{l} = \frac{\mu_0}{\pi} \log_e \frac{\mathrm{d}}{a}$$

In almost all practical cases the factor $\log_e (\mathrm{d}/a)$ lies between 3 and 6. The inductance of such a transmission line therefore usually falls between 1·2 and 2·4 μH per metre.

The units of μ_0

The above expressions for inductance show that μ_0 has the dimensions of

$$\left[\frac{\text{inductance}}{\text{length}}\right]$$

and the usual practice nowadays is to express it in henry metre^{-1}. Thus

$$\mu_0 = 4\pi \times 10^{-7}\ \mathrm{H\,m^{-1}}$$
$$= 1{\cdot}256 \times 10^{-6}\ \mathrm{H\,m^{-1}}$$

The energy stored in an inductance

Because of the e.m.f. of self-induction that acts when the current in a coil changes, electrical energy must be supplied in setting up the current against this e.m.f.

The rate of supplying electrical energy $= -e.$

$$= L\frac{\mathrm{d}i}{\mathrm{d}t}.i$$

$$\therefore \text{ the energy supplied in time } \delta t = L\frac{\mathrm{d}i}{\mathrm{d}t}.i\delta t$$

$$= Li\,\delta i$$

where $\delta i =$ the corresponding increase of current.

$$\therefore \text{ total energy supplied} = \int_0^i Li\,\mathrm{d}i = \tfrac{1}{2}Li^2$$

If L is in *henrys* and i in *amps*, the energy will be given in *joules*.

The growth of current in an inductance

Consider a coil of inductance L and resistance R; this may be treated for purposes of analysis as two separate circuit components, as shown in

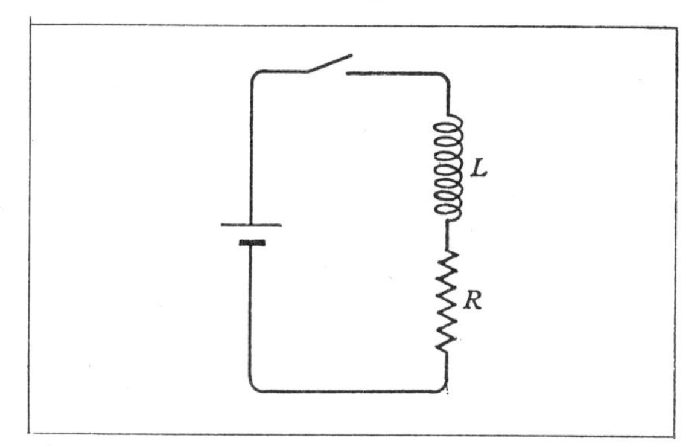

Fig. 7.22 A coil with inductance and resistance may be analysed as two 'pure' circuit elements, L and R, in series

Fig. 7.22. The p.d. applied to maintain the current i in the inductance must be equal and opposite to the back e.m.f. due to self-induction $(-L(\mathrm{d}i/\mathrm{d}t))$, while that across the resistance must be iR. The total applied p.d. is therefore given by

$$L\frac{\mathrm{d}i}{\mathrm{d}t} + iR$$

We suppose that $i = 0$ at the moment $t = 0$, and that at this moment a steady p.d. V is joined across the coil. By direct differentiation it can be checked

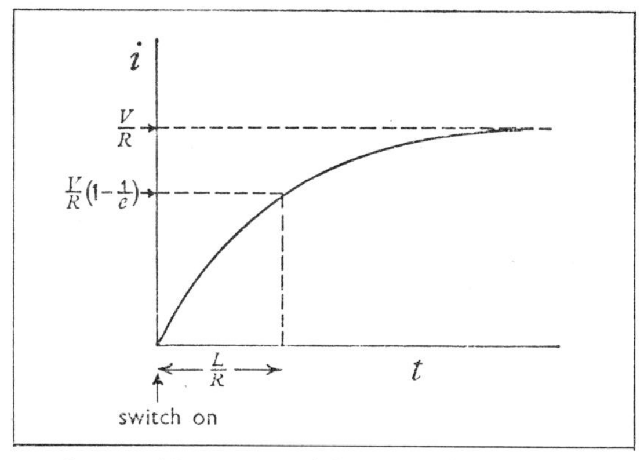

Fig. 7.23 The exponential growth of current in an inductive coil. The time constant is L/R

that the required solution of this differential equation is

$$i = \frac{V}{R}\left(1 - \mathrm{e}^{-\frac{t}{L/R}}\right)$$

(e here is the base of natural logs, 2.718 . . .).

The variation of i with t is shown in Fig. 7.23. The current tends asymptotically to the final value V/R. The quantity L/R has the dimensions of time, and is known as the *time constant* of the circuit. If L is in *henrys* and R in *ohms*, then the time constant is in *seconds*. It is a measure of the time taken by the current to grow. To be exact, the current grows to within $1/\mathrm{e}(= 0{\cdot}37)$ of its final value in this time interval; i.e. when $t = L/R$ the current is

$$\frac{V}{R}(1 - 0{\cdot}37) = 0{\cdot}63\frac{V}{R}$$

7.7 The measurement of flux

When the flux linkage N of a magnetic field with a closed circuit changes, an e.m.f. is induced in it and a current flows. It may be shown that the *total charge* that is driven round the circuit by the induced e.m.f. depends only on the final change of flux and on the resistance of the circuit.

The *magnitude* of the e.m.f. e is given by

$$e = \frac{\mathrm{d}N}{\mathrm{d}t}$$

If the resistance of the complete circuit is r, the current i is given by

$$i = \frac{e}{r}$$

The current is equal to the rate of flow of charge q;

$$\therefore i = \frac{dq}{dt}$$

$$\therefore \frac{dq}{dt} = \frac{1}{r}\frac{dN}{dt}$$

Integrating over the time interval during which the change of flux occurs, the left-hand side of the equation gives the total charge q that flows;

$$\therefore q = \frac{1}{r}\int dN$$

$$\therefore q = \frac{N_2 - N_1}{r} = \frac{\text{change of flux linkage}}{r}$$

If the change of flux is brought about fairly rapidly the charge q may be measured with a galvanometer.

When a pulse of current passes through the galvanometer coil an impulse is given to it that sets it swinging. It may be shown that the amplitude of the first oscillation is directly proportional to the total electric charge that passed during the pulse, provided the duration of this was small compared with the period of oscillation (p. 132). Used in this way the instrument is called a *ballistic galvanometer*. Measurements of this type are only possible if the damping of the galvanometer movement is very low. This usually means that only mirror galvanometers are suitable for the purpose. It is also necessary to ensure that the resistance r of the circuit is sufficiently high. If r is too small, the currents induced in the coil when it rotates may cause heavy damping.

Thus if the amplitude of the first 'swing' of the ballistic galvanometer is θ, we have

$$q \propto \theta$$

or

$$q = b\theta$$

where b is a constant that must be found by a separate calibration experiment. The instrument may then be used to find the change in the flux linked with a search coil joined in series with it; this gives the average flux density B at the position of the coil.

For instance, the search coil may be placed between the poles of an electromagnet. The flux N_1 linked with it is then given by

$$N_1 = BAn$$

where n is the number of turns in the coil and A is its cross-sectional area. If the coil is now suddenly removed to a position where the field is negligible ($N_2 = 0$), the charge q measured by the galvanometer is given by

$$q = \frac{BAn}{r}$$

$$\therefore B = \frac{rq}{An}$$

The cross-section of the search coil in such a measurement must be small enough so that the field can be regarded as uniform over the area A. The number of turns n is chosen so that the charge q is of the right order of magnitude to be measured by the ballistic galvanometer (probably about 1 microcoulomb). The search coil is joined up through long, twisted, flexible leads, and is connected in series with a high enough resistance to reduce the damping to a low value (Fig. 7.24). To save time between observations a tapping key and a suitable low resistance are joined across the galvanometer, as shown. The galvanometer coil can then be brought quickly to rest by pressing the key.

An alternative technique is to turn the search coil over in the field; the change of flux linkage with it is then $2BAn$. This needs two people to manage it properly—one to turn the coil, and the other to take the reading.

Obviously we are obliged to do this in measurements on the earth's field. For this purpose a coil of large cross-section and many turns is needed. It is best to pivot it on an axle supported in a frame that can be turned to bring the axle horizontal or vertical as required. An arrangement of this sort is called an *earth inductor*. The deflection θ produced by turning the coil through 180° is proportional to the component of the flux density normal to the initial position of the plane of the coil;

$$B \propto \theta$$

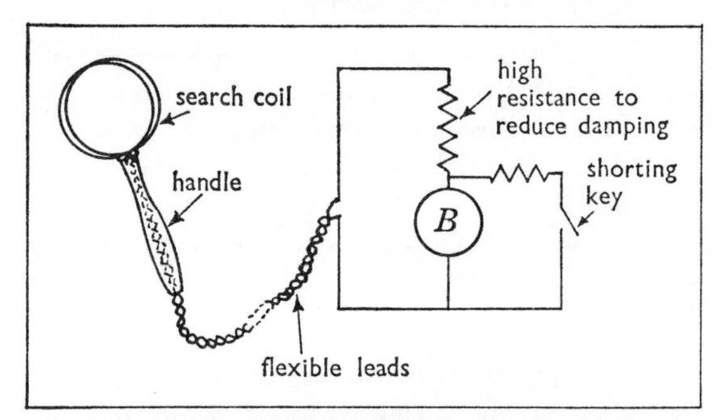

Fig. 7.24 The use of a search coil and ballistic galvanometer to measure flux

The angle of dip may be found even without first establishing the constant of proportionality. Suppose the deflections obtained for the horizontal and vertical components of the earth's field are θ_0 and θ_v, respectively, corresponding to the components of the flux density B_0 and B_v, then the angle of dip ϕ is given by

$$\tan\phi = \frac{B_v}{B_0} = \frac{\theta_v}{\theta_0}$$

If a search coil is made that will fit closely over the cross-section of a magnet, it may be used to investigate the flux in the body of the magnet at different positions. The coil, connected to a ballistic galvanometer, is placed over the magnet in the position required, and is then quickly slipped off and removed to a sufficient distance for the final flux through it to be treated as negligible. The throw of the galvanometer is proportional to the average flux through the coil at the point tested. A maximum reading is, of course, obtained with the coil round the centre of the magnet. By repeating the measurements at a number of positions along the magnet and beyond the ends of it, the 'leakage' of flux from different parts of it may be investigated. With a bar magnet less than half the total flux will be found to emerge from the end faces.

Calibrating a ballistic galvanometer

The full theory of the ballistic galvanometer is discussed in the next chapter (p. 132). It is shown there how to calculate the ballistic sensitivity from measurements of various constants of the instrument. However, it is usually simpler and quicker to use one of the following direct methods.

(*i*) *Using a standard capacitor.* When a capacitor of capacitance C is charged to a p.d. V, the charge q stored on the plates is given by

$$q = CV \qquad \text{(p. 175)}$$

q is in *coulombs*, if C is in *farads* and V in *volts*.

The quantity of charge stored in a standard capacitor charged to a measured p.d. can thus be calculated. When the terminals of the capacitor are joined to the ballistic galvanometer, the deflection θ produced by the charge q may be observed. The details of this technique are discussed in another chapter (p. 154). The method has the merit of simplicity. But the disadvantage is that the ballistic sensitivity of the instrument is in this case found for a circuit of infinite resistance. If it is then to be used in another circuit of sufficiently low resistance to affect the damping significantly, the sensitivity is altered, and allowance must be made for this.

Example. A flat circular coil of 200 turns of diameter 25 cm is laid on a horizontal table-top and connected by flexible wires to a mirror galvanometer, the complete circuit having a resistance of 800 ohms. When the coil is quickly turned over the spot of light swings out to a maximum reading of 30 divisions. A standard 0·1-μF capacitor is charged to a p.d. of exactly 6 volts and discharged through the same galvanometer; a maximum reading of 20 divisions is obtained. Assuming that the damping of the galvanometer is negligible in both cases, calculate the vertical component of the flux density of the earth's magnetic field.

If the vertical component of the flux density of the earth's field is B teslas, the flux linking the coil

$$= B \times \pi \times (0{\cdot}125)^2 \times 200 \text{ Wb}$$

$\therefore$ The charge q passing through the galvanometer is given by

$$q = \frac{2 \times B \times \pi \times (0{\cdot}125)^2 \times 200}{800}$$

$$= \frac{B \times \pi \times (0{\cdot}125)^2}{2} \text{ coulomb}$$

The charge on the capacitor $= 6 \times 10^{-7}$ coulombs.

$$\therefore q = \frac{30}{20} \times 6 \times 10^{-7} = 9 \times 10^{-7} \text{ coulombs}$$

$$\therefore B = \frac{18 \times 10^{-7}}{\pi \times (0{\cdot}125)^2} = 3{\cdot}7 \times 10^{-5} \text{ T}$$

(*ii*) *Using a standard mutual inductance.* It was shown on p. 114 that the mutual inductance M between a long solenoid and a short secondary coil wound round its centre is given by

$$M = \frac{\mu_0 A n n'}{l}$$

where the symbols are used with the same meanings as on p. 114. If the dimensions of the coil are measured in m, this gives the mutual inductance in henrys. It is not difficult to measure the quantities in this formula with at least the accuracy justified for a ballistic galvanometer experiment (about $\frac{1}{2}$%). A measured current i is then passed through the primary coil; when this is switched off the change

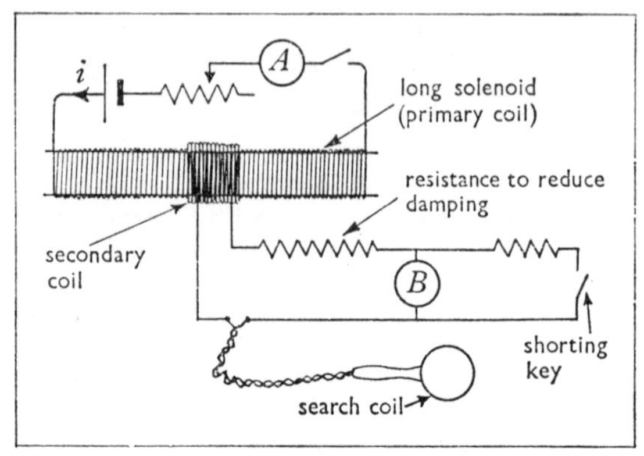

Fig. 7.25 The calibration of a ballistic galvanometer and search coil using a standard mutual inductance

of flux linkage N with the secondary coil is given by

$$N = Mi$$

The secondary is joined in series with both the ballistic galvanometer and whatever search coil is to be used subsequently with it (Fig. 7.25); a high resistance is usually included also to keep the damping of the galvanometer low and to act as a means of adjusting its ballistic sensitivity.

The deflection θ is noted when the known current in the primary coil is *switched off*, producing a known change of flux linkage N. Then

$$N = \lambda\theta$$

where λ is a constant which remains unaltered as long as the resistance of the galvanometer circuit is kept constant. The same constant of proportionality may then be used to calculate the changes of the flux linkage with the search coil from the galvanometer deflections.

This method therefore has the advantage that the instrument is calibrated directly for flux measurement in the very circuit in which it is to be used. The resistance of the circuit does not need to be known.

7.8 Plotting hysteresis curves

A search coil and ballistic galvanometer can be used for studying the hysteresis loops of ferromagnetic materials. In order to avoid demagnetizing effects and flux leakage it is necessary to make the specimen in the form of a closed ring. The search coil wrapped round the ring is now of course a fixture. In this case we use the coil to measure the changes of flux when the magnetization of the ring is suddenly altered. The magnetizing coil consists of a large number of turns of wire *uniformly* wound round the ring. If the average circumference of the ring is l, and the current in the coil is i, the magnetizing force is given by

$$H = \frac{ni}{l} \quad \text{(p. 85)}$$

where n = the total number of turns in the coil.

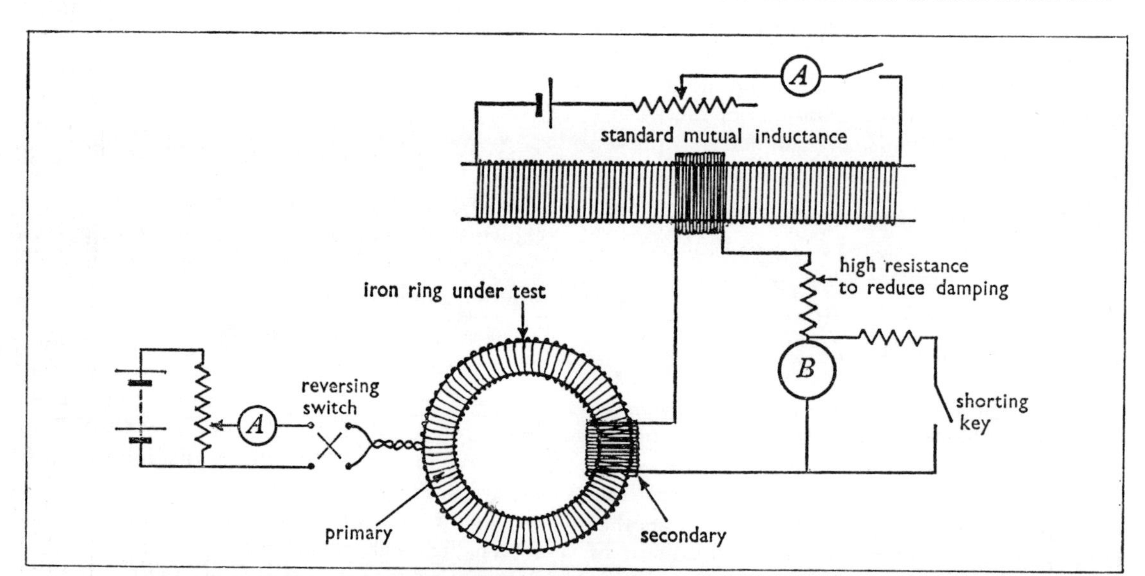

Fig. 7.26 Circuit for investigating the B–H curves of an iron ring; the mutual inductance is used to calibrate the ballistic galvanometer

The secondary coil (of n' turns) need only be wound round part of the ring. It is connected in series with the ballistic galvanometer and a suitable high resistance. The galvanometer may conveniently be calibrated using a standard mutual inductance, as described above; the secondary of the mutual inductance is then also joined in series with the instrument (Fig. 7.26).

When the flux density in the ring changes from B_1 to B_2, the charge q flowing through the galvanometer is given by

$$q = \frac{(B_2 - B_1)An'}{r}$$

where A = the area of cross-section of the ring, and r = the resistance of the whole galvanometer circuit. The charge q is given by the throw of the galvanometer, and so the change in flux density $(B_2 - B_1)$ is found.

The procedure is as follows. The current in the primary is first adjusted to such a value that the ferromagnetic ring is certainly saturated; call this current I_m. Suppose the flux density at this current

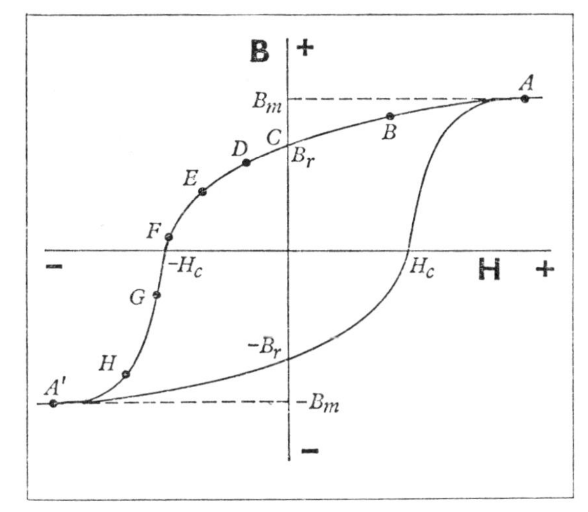

Fig. 7.27 Plotting a B–H hysteresis loop

is B_m. The current is then quickly reversed, so that the magnetization becomes saturated in the opposite direction. The change of flux density given by the throw of the galvanometer is therefore $2B_m$. So B_m is found, and the points A and A′ on the B–H curve can be marked in (Fig. 7.27). The current is then brought back to the value I_m before each subsequent reading. The points B, C of the curve are plotted by reducing the current in succession from I_m to each of the other values chosen. Each time the throw of the galvanometer gives the difference between B_m and the new value of B. In this way the curve between A and C can be plotted. The flux density B_r that remains (at the point C) when the magnetizing field has been reduced to zero is known as the *remanence* of the material.

The part of the curve between C and A′ is plotted by first bringing the material to the point C before each reading; i.e. the current is adjusted to the value I_m and is then switched off. The points D, E, F, G, H are then plotted in turn by increasing the current in the reverse direction from zero up to the values required. The throw of the galvanometer now gives in each case the difference between B_r and the new value of B. The reverse magnetizing force H_c required to reduce the flux density to zero is known as the *coercivity* of the material. The curve from A′ back to A can then be drawn in, if required, symmetrically the other side of the origin.

The table gives the values of saturation flux density B_m, remanence B_r and coercivity H_c for a selection of common magnetic materials. The values of B_m vary over a range of about 4 to 1 for different iron alloys; and in most materials

$$B_r \simeq \tfrac{2}{3}B_m$$

The big differences between the materials lie in their coercivities, which vary over a range of about 20,000 to 1.

Properties of ferromagnetic materials
(The values obtained depend on the manner of preparation of the specimen, and are therefore only approximate)

Material	B_m ($Wb\ m^{-2}$)	B_r ($Wb\ m^{-2}$)	H_c ($A\ m^{-1}$)
Iron (99·95% pure)	2·2	1·3	80
Cast iron	0·5	0·3	500
Mild steel	1·4	0·9	200
Silicon iron	1·6	0·8	60
Mumetal	0·9	0·6	3
Carbon steel	—	1·0	5,000
Cobalt steel	—	1·0	20,000
Alnico	—	1·0	60,000
Nickel	0·7	0·3	400
Cobalt	1·5	0·4	800

8: Measuring Instruments

8.1 The moving-coil galvanometer

An instrument for detecting or measuring an electric current is called a *galvanometer*. When such an instrument is to be used merely to *detect* currents (as in Wheatstone bridge and potentiometer circuits), it need only be provided with an arbitrary scale marked off in (say) millimetre divisions. Used as a measuring instrument, it may be calibrated to read directly the current through it in amps or the p.d. across it in volts, in which case it would be called an *ammeter* or a *voltmeter*. In fact, as we shall see, most ordinary ammeters and voltmeters are simply galvanometers modified by joining suitable resistances in parallel or in series with them. Whether a given instrument is called a galvanometer, an ammeter or a voltmeter depends on the use to which it is to be put and therefore on the manner in which it is connected up in a circuit. The same basic instrument is used in each case. The most common design for modern d.c. instruments is that employing the 'moving-coil' principle.

The pivoted type of instrument

The moving-coil galvanometer consists of a coil wound on a light frame, which is pivoted about an axis in the plane of the coil so that it is free to rotate between the poles of a permanent magnet (Figs. 8.1 and 8.2). The coil is pivoted in jewelled

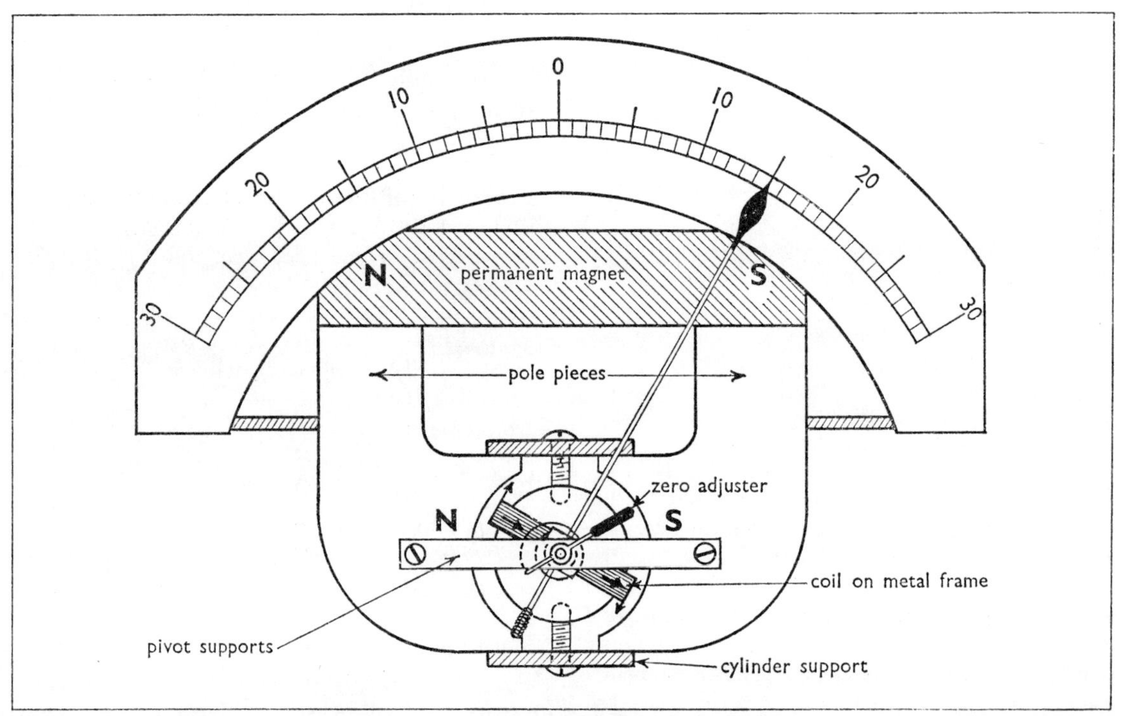

Fig. 8.1 A moving-coil galvanometer (pivoted type)

bearings, and its motion is restrained by two hair-springs, one at each end; the hair-springs are also used to conduct the current into and out of the coil. The two springs spiral in opposite directions so that any tendency for them to coil or uncoil with change of temperature does not lead to a shift of the zero position of the coil. When a current passes through the coil a couple acts on it tending to turn it to the position in which its plane would be at right-angles to the field (p. 67). The coil therefore rotates until the restoring couple due to the hair-springs is equal and opposite to the deflecting couple due to the current. Attached to the coil is a light, balanced pointer, which moves over a suitable scale; this may have its zero at the centre or at one end, depending on the use to which the instrument is to be put; with the centre-zero type of instrument current may be passed through it in either direction, but with an end zero care must be taken to join the instrument in a circuit the right way round.

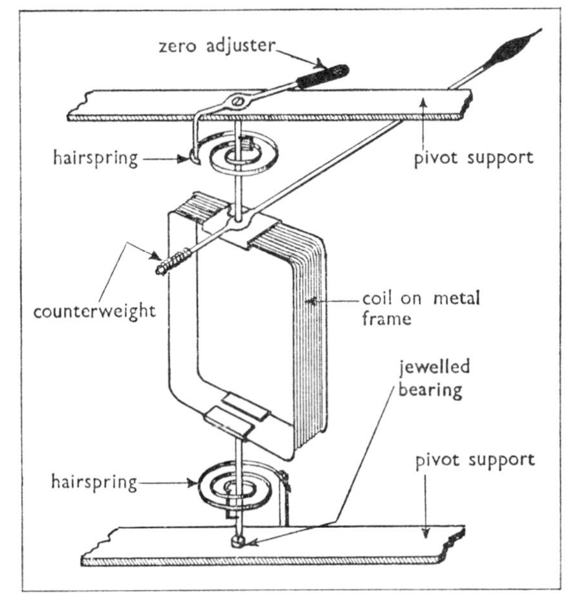

Fig. 8.2 The coil, pivot and hair-spring assembly of a moving-coil galvanometer. The connections to the coil are made through the hair-springs and pivot supports

The adjustment of the zero is made by rotating a small arm to which the outer end of one of the hair-springs is attached. The rest position of the coil is that in which the tensions of the two hair-springs are equal and opposite.

The magnet is fitted with concave soft iron pole pieces, whose faces are concentric with the axis of the coil. Also, a soft iron cylinder is fixed centrally between the pole pieces so that it partly fills the space inside the coil. This forms a narrow cylindrical gap in which the sides of the coil can move. The effect of this is to produce what is called a *radial* field; that is, the lines of force in the gaps are along radii to the central axis. Thus, as the coil rotates its plane is always parallel to the field in the gap. In this way the direction of the forces acting on the sides of the coil is in all positions such that the couple is a maximum. Also, by making the width of the gap uniform, the magnetic field in it becomes almost constant in magnitude. The result is that the deflecting couple acting on the coil does not depend on its position in the gap, but only on the current flowing in it. Now, the restoring couple due to the hair-springs is proportional to the deflection. Hence, the use of a radial field makes the deflection proportional to the current, and the instrument then has a uniform scale. This makes it particularly straightforward to calibrate and to read, since the divisions of the scale are evenly spaced, and the fixing of one point on it and the zero decides also all the other markings.

To bring the coil quickly to rest when a current is passed through it some sort of damping of the movement has to be provided. If the only damping were the small amount due to the air and the friction of the jewelled bearings, the coil would oscillate for an excessive time and measurements with the instrument would be very tedious to make. The most common arrangement is known as *electromagnetic damping*. This is provided by winding the coil on a light metal frame; it can be regarded as an additional coil of one 'turn' of very low resistance. When the coil swings, eddy currents are induced in the metal frame (p. 100), and this quickly brings it to rest. Excessive damping would make the coil unnecessarily slow in reaching its final deflection. The ideal design of frame is that which gives *critical damping*—i.e. the minimum amount that just prevents oscillation. The movement of the instrument is then said to be 'dead-beat'.

The same principle is often applied to provide some protection for the galvanometer against mechanical damage. When the instrument is not in use, a copper strip is connected between its terminals, so that the coil and strip form a closed circuit. Any sudden rotation of the coil, such as might result from an accidental blow, then causes currents to be induced in it; this to some extent damps the movement so that it is less likely to lead to damage.

The suspended coil instrument

In very sensitive instruments the slight friction that would occur even with good jewelled bearings would make the movement irregular. It is therefore necessary to dispense altogether with bearings; instead the coil is suspended from a narrow phosphor-bronze strip (Fig. 8.3). Connection to the coil is made through this strip and through a loose spiral of fine wire attached to the bottom. The arrangement of the coil and magnet is the same as that used in the pivoted type of instrument, though in some designs the central soft iron cylinder is dispensed with. Very careful levelling of these instruments is needed to ensure that the coil can move quite freely in the gap. Needless to say, the coil is very easily disturbed by vibrations; and any sudden jolt may well break the suspension strip. A special clamp must therefore be provided to grip the coil when the galvanometer is to be moved.

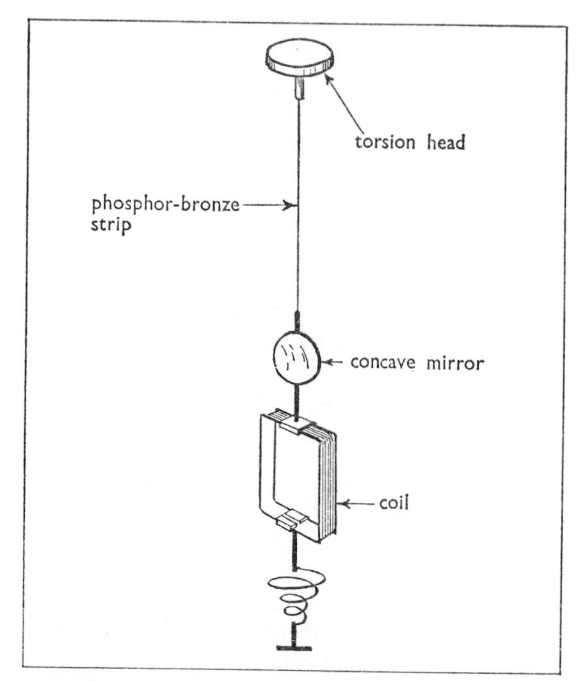

Fig. 8.3 The coil and mirror assembly of a suspended coil instrument

In recent years a more robust design has been developed in which the coil is suspended between *two* taut phosphor-bronze strips. These are maintained in sufficient tension to keep the coil centred between the pole pieces even without special levelling. This form of galvanometer is usually designed so that short-circuiting of the coil provides exceptionally heavy electromagnetic damping. No other clamping is then necessary to protect the coil from damage when the instrument is moved.

Some suspended coil instruments are made with a light metal pointer moving over a scale. But in the most sensitive designs instead of the pointer a small mirror is fixed to the coil. This is used to cast an image of an illuminated cross-wire on to a scale. Fig. 8.4 shows a common optical arrangement. M is a concave mirror with a radius of curvature equal to the distance of the coil from the scale. Thus the image should be sharply in focus with both the cross-wire and scale at this distance from the mirror. The lens L is used as a condenser, whose function is to concentrate the light from the bulb B in the direction of the mirror. The vertical cross-wire is stretched over the lens face. The image cast by the mirror on the scale is thus a bright disc (the image of the lens face) with a dark line marking its centre. The torsion head carrying the suspension strip may be rotated to bring the undeflected position of the spot of light to the required point on the scale.

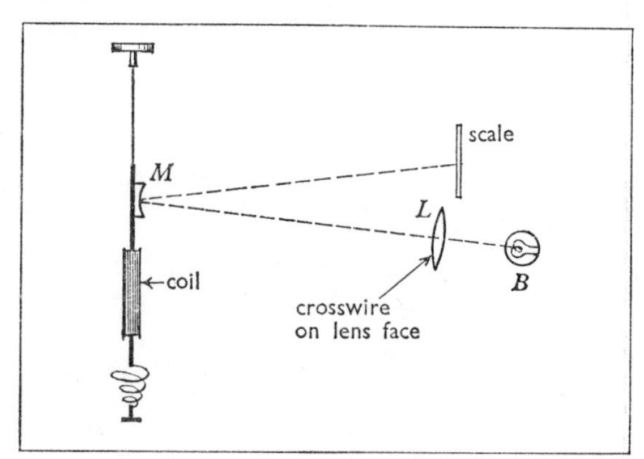

Fig. 8.4 The arrangement of lamp and scale with a mirror galvanometer

It is not usual in the suspended type of instrument to incorporate any device for producing extra damping; the coil is wound on a frame of light plastic or stiffened fabric. However, the circuit itself provides a certain amount of electromagnetic damping. When the coil swings, currents are induced in it which damp the movement. If we want the damping to be critical, the effective resistance of the circuit must be adjusted to bring this about. With the coil connected direct to a very low-resistance device, such as a thermocouple, the damping may well be excessive and the movement very sluggish; this can be avoided by connecting a suitable resistance in series with the galvanometer, although this leads to some loss of sensitivity. The value of this resistance is best found by trial and error. On the other hand, in a high-resistance circuit the coil might oscillate excessively; in this case a suitable resistance can be joined in parallel with the instrument to provide a path for the induced currents. Again there is some loss of sensitivity. (See Fig. 8.19, p. 133.)

8.2 Sensitivity

Let us consider the general case of a coil whose axis is inclined at an angle ϕ to a magnetic field of flux density B teslas, i.e. we are not limiting ourselves yet to the special case of a coil in a radial field. Suppose the coil has n turns each of area A m^2, carrying a current of i amps. Then the deflecting couple C (in newton-metres) is given by

$$C = BAni \sin \phi \qquad \text{(p. 67)}$$

The restoring couple C' due to the springs is proportional to the deflection θ of the coil, since Hooke's law is closely obeyed for this type of spring.

$$\therefore \quad C' = k\theta$$

where k is called the *suspension constant*; it is the couple acting per unit deflection. The coil comes to rest in such a position that $C = C'$.

$$\therefore \quad k\theta = BAni \sin \phi$$

$$\therefore \quad \theta = \frac{BAn \sin \phi}{k} i$$

In this expression, A, n and k are necessarily constant. To achieve a uniform scale ($\theta \propto i$) we need to ensure that $B \sin \phi$ is constant for all positions of the coil. As we have already shown, this is achieved by the use of a radial field; for in this case $\phi = 90°$, so that $\sin \phi = 1$. Also B = a constant for the uniform gap.

Thus, with a radial field

$$\frac{\theta}{i} = \text{a constant}$$

which is called the *current sensitivity* of the instrument. It is, however, quite common to express the *quality* of the instrument by the reciprocal of the sensitivity i/θ; this would probably be stated in *microamps per division*, and affords a direct indication of the smallest current that could be detected. For a good pivoted type galvanometer this figure might be about 10 μA per division, so that used with a Wheatstone bridge or potentiometer it could probably detect a current of rather less than 1 μA.

Thus, for an instrument with a radial field

$$\text{current sensitivity} = \frac{\theta}{i} = \frac{BAn}{k}$$

To design a galvanometer of maximum sensitivity:

(*i*) B must be as great as possible in the gaps. Apart from choosing the most suitable material from which to make the permanent magnet, a large value of B is obtained by making the gaps as narrow as possible.

(*ii*) The area of the coil A must be as large as possible. In practice, however, beyond a certain point a large coil is not an advantage, since it also has a large moment of inertia; this makes the period of oscillation of the coil inconveniently long. Except for special sub-standard instruments, it is the modern practice to make the coil very small.

(*iii*) The number of turns n of the coil should be as large as possible. Two factors, however, set a limit to n: first, increasing it beyond a certain point would require a wider gap for the coil to move in and so would reduce the value of B; second, increasing n also increases the resistance of the coil, and in some applications this reduces the usefulness of the instrument.

(*iv*) The suspension constant k should be as small as possible. However, this also increases the period of oscillation, and too weak a spring system would make the instrument unduly slow acting.

Sometimes it is more useful to express the sensitivity of the galvanometer in terms of the p.d. v across it. The *voltage sensitivity* is defined as the quantity θ/v. Now, if the resistance of the galvanometer is r, we have

$$i = \frac{v}{r}$$

$$\therefore \quad \theta = \frac{BAn}{k} \frac{v}{r}$$

$$\therefore \text{ voltage sensitivity} = \frac{\theta}{v} = \frac{BAn}{kr}$$

Whether we analyse the behaviour of a galvanometer in terms of its current or voltage sensitivity depends on the application considered. In general we need to take into account the resistance not only of the instrument but of the whole circuit in which it is to be used. The following example shows this:

Example. Two moving-coil galvanometers are identical in all respects except that one has a coil of 100 turns of resistance 20 ohms, and the other a coil of 200 turns of resistance 50 ohms. Compare (*a*) their current sensitivities, (*b*) their voltage sensitivities, (*c*) their deflections when joined to a source of constant small e.m.f. of internal resistance 25 ohms.

(*a*) Now the current sensitivity is proportional to n, since in this case the other factors in the equation are the same for both instruments.

If the current sensitivities are s_1 and s_2, respectively, we have

$$\frac{s_1}{s_2} = \frac{100}{200} = \frac{1}{2}$$

(*b*) The voltage sensitivities are proportional to n/r; if these are s_1' and s_2' respectively, we have

$$\frac{s_1'}{s_2'} = \frac{100}{200} \times \frac{50}{20} = \frac{5}{4}$$

(*c*) In this part of the problem we have:

$$\text{deflection} \propto ni$$

$$\text{also} \quad i \propto \frac{1}{\text{total resistance}}$$

$$\therefore \text{ deflection} \propto \frac{n}{\text{total resistance}}$$

If the deflections are θ_1 and θ_2 respectively, then

$$\frac{\theta_1}{\theta_2} = \frac{100\,(25 + 50)}{200\,(25 + 20)} = \frac{5}{6}$$

This shows the general principle to be followed. When the circuit is of low resistance compared with that of the galvanometer, high *voltage* sensitivity is necessary; this usually applies in potentiometer and Wheatstone bridge circuits. But in a circuit of high resistance, the *current* sensitivity must be high.

★ In analysing the behaviour of the most sensitive types of galvanometer it is useful to express the deflection in terms of the *power* that must be dissipated in the coil to produce a given deflection. The reason for this may be seen if we consider how the resistance R of the coil will depend on the number of turns n. For a coil wound on a frame of given dimensions, doubling n is only possible by halving the cross-sectional area a of the wire.

$$\text{Now,} \qquad R \propto \frac{l}{a}$$

where l = length of wire in the coil.

For a given frame we have

$$l \propto n \quad \text{and} \quad a \propto \frac{1}{n}$$

$$\therefore R \propto n^2$$

or put the other way round

$$n \propto \sqrt{R}$$

For a set of galvanometers identical except for the numbers of turns in their coils, the deflection θ is given by

$$\theta \propto ni$$

$$\therefore \theta \propto \sqrt{R}.i$$

$$\text{or} \qquad \theta \propto \sqrt{i^2R} \quad \text{i.e. } \theta \propto \sqrt{W}$$

where W is the power dissipated in the coil.

In the measurement of small currents and p.d.'s we must therefore design the circuit so that *maximum power* is delivered to the galvanometer. It may be shown that this is achieved by making the resistance of the galvanometer equal to that of the rest of the circuit in which it is connected (p. 23).

The most sensitive modern galvanometers give unit deflection for a power of about 10^{-19} watts. For a current measurement maximum sensitivity is obtained with a high-resistance coil (power $= i^2R$); the practicable limit is about 1000 ohms. The current for unit deflection in this case is about 10^{-11} amp. In a measurement of p.d. maximum sensitivity requires a low-resistance coil (power $= v^2/R$). Using a resistance of 10 ohms the p.d. for unit deflection is then about 10^{-9} volt. In principle it might be possible to design instruments more sensitive than this; but the Brownian movement of the coil (due to random collisions of air molecules with it) causes irregular vibrations which completely mask the deflection due to the current. No method has yet been devised for measuring p.d.'s less than about 10^{-9} volt; but it is possible to measure currents down to about 10^{-15} amp by electrostatic means (p. 186).

Measurements on a galvanometer

The measurement of the current or voltage sensitivity of a galvanometer may be carried out with the circuit shown in Fig. 8.5. The resistance R in series with the galvanometer is made large enough so that (*a*) the amount of damping is reasonable, (*b*) the combined resistance $(R + G)$ is very large compared with the small resistance r. We can then

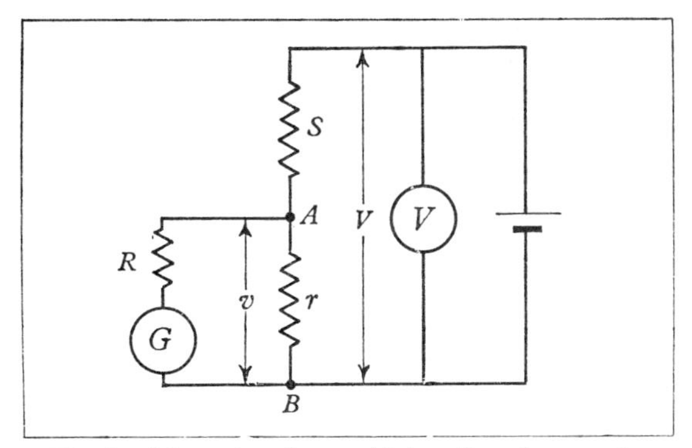

Fig. 8.5 Measuring the sensitivity of a galvanometer

assume that the effective resistance between A and B is r, with negligible error. S and r form a potential divider so that the p.d. v between A and B is a calculable fraction of the p.d. V across the accumulator.

$$\therefore \frac{v}{V} = \frac{r}{S+r}$$

The current i through the galvanometer is then given by

$$i = \frac{v}{R+G}$$

Measurement of the deflection θ produced by this current gives the current sensitivity; and this quantity divided by the resistance G of the galvanometer is the voltage sensitivity. The measurement should be repeated for several points of the scale.

In a practical example we might have:

$G = 100$ ohms $\quad R = 900$ ohms
$r = 2$ ohms $\quad S = 100{,}000$ ohms
$V = 2$ volts

$$\therefore \frac{v}{2} = \frac{2}{100{,}000} \quad \text{(since 2 is small compared with 100,000)}$$

$$\therefore v = 4 \times 10^{-5} \text{ volt}$$

$$\therefore i = \frac{4 \times 10^{-5}}{1000} = 4 \times 10^{-8} \text{ amp}$$

If the deflection θ is 20 divisions,

$$\text{the current sensitivity} = \frac{\theta}{i} = 5 \times 10^{8} \text{ div per amp}$$

$$= 500 \text{ div per } \mu\text{A}$$

$$\text{the voltage sensitivity} = 5 \text{ div per } \mu V$$

To measure the sensitivity by this method we need first to know the galvanometer resistance G. Provided the coil can be clamped, this may readily be found with a Wheatstone bridge arrangement, in which the galvanometer under test forms one arm of the bridge; a second galvanometer is used to obtain the balance. Even with the coil clamped great care must be taken to avoid passing excessive currents; very fine wire is sometimes used in a galvanometer coil, and even a small current may overheat it.

When the coil cannot be clamped, the following method due to Lord Kelvin may be used. A Wheatstone type network is connected up as shown in Fig. 8.6, with the galvanometer as one of the four resistances; a Post Office box would be suitable for this purpose. The resistance in series with the cell must be large enough to give a safe deflection of the galvanometer near the end of its scale. Instead of joining a galvanometer across the middle of the bridge, the points B and D are simply connected together through a tapping key. Now, if the usual bridge balance condition holds,

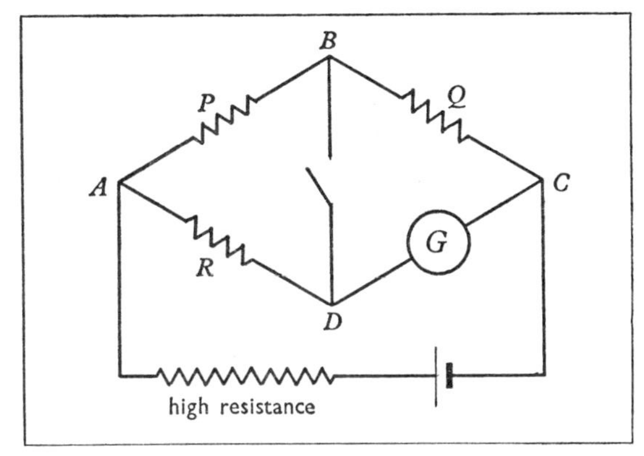

Fig. 8.6 Measuring the resistance of a galvanometer

$$\frac{P}{Q} = \frac{R}{G}$$

no current flows between B and D even when the tapping key is pressed. Therefore pressing the key can have no effect on the currents in other parts of the network; thus, when the bridge is balanced no *change* of the galvanometer deflection occurs when the key is pressed. The other resistances in the bridge are adjusted to bring this about, and so G is found.

8.3 D.C. meters

In most modern instruments the basis of the construction is an ordinary moving-coil galvanometer of the pivoted type. This is made as sensitive as possible consistent with reasonable robustness. The required range in amps or volts is then achieved by connecting resistances in parallel or in series with it as described below. A good ammeter or voltmeter is one that disturbs as little as possible the current or p.d. it is required to measure; it must also take as little power as possible. For this reason an ammeter is normally a low-resistance device so that the p.d. across it is small. Likewise, a voltmeter is normally a high-resistance device so that the current taken by it is small.

Ammeters

A low-resistance shunt S is joined in parallel with a suitable galvanometer (Fig. 8.7). If the resistance of the galvanometer is G, and the current through it is i, then the currents through S and G are inversely proportional to their resistances

$$\therefore \frac{I-i}{i} = \frac{G}{S}$$

where I is the total current to be measured. Rearranging,

$$\frac{I}{i} = \frac{S+G}{S} = N, \text{ say}$$

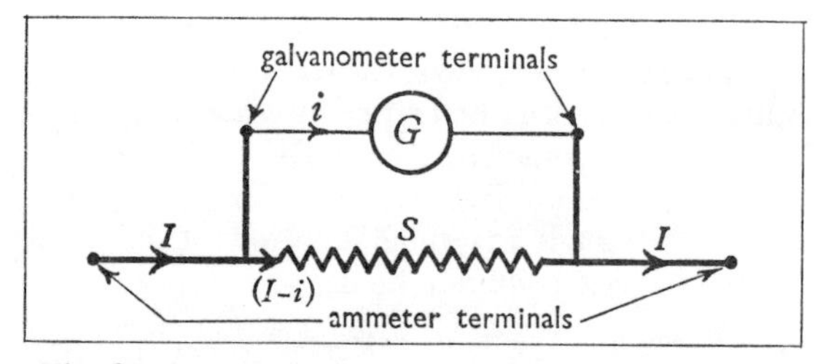

Fig. 8.7 The connection of a shunt to a galvanometer to make an ammeter

The total current I is thus always N times the current i registered by the galvanometer. The value of S may in this way be chosen so as to make the range of the instrument any desired multiple of that of the unshunted galvanometer. The best type of instrument to make into an ammeter is one that has *high-voltage sensitivity*, since this will require only a small p.d. across it to drive it to full-scale deflection, and the resulting ammeter is then of low total resistance.

If the same instrument is to be used on a number of different ranges, it is possible to provide it with several separate shunts; one of these is attached to the galvanometer terminals for each range. But a better method is a universal shunt of the type shown in Fig. 8.8. The resistance S is first adjusted

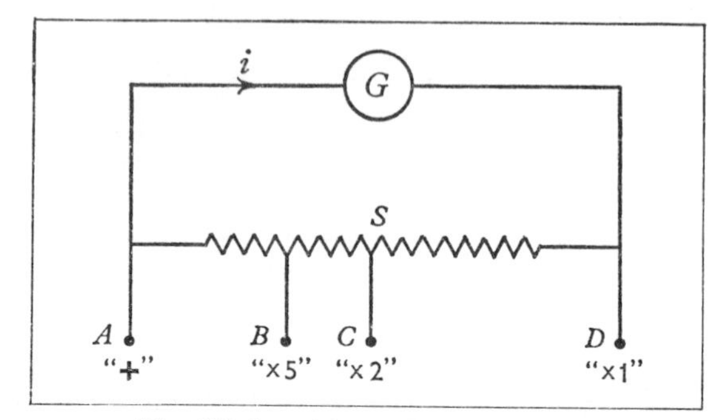

Fig. 8.8 A multi-range ammeter shunt

to provide the lowest required range of the ammeter; connection for this range is made to the terminals A and D. Further tappings are then made at points along S, as shown, to give the higher ranges. For instance, if the resistance of the part of the shunt between A and B is $S/5$, it may be shown that the ammeter range obtained by making connection to A and B is 5 times the basic range. Similarly, if the point C is halfway along AD, the range obtained by using the terminals A and C is 2 times the basic range. Such a multi-range instrument would have the terminals A, B, C, D marked '+, ×5, ×2, ×1', in that order. The ammeter is always connected so that the current enters at the '+' terminal (this is then the positive side of the meter); the other terminal is selected according to the range required.

Voltmeters

A high resistance R is joined in series with a suitable galvanometer (Fig. 8.9). If the resistance of the galvanometer is G, and the current through it i, then the p.d. V across the combination is given by

$$V = (R+G)i$$

Thus the p.d. V is proportional to the current i registered by the galvanometer, whose scale may therefore be calibrated to read p.d.'s directly. The series resistance R is chosen to make the

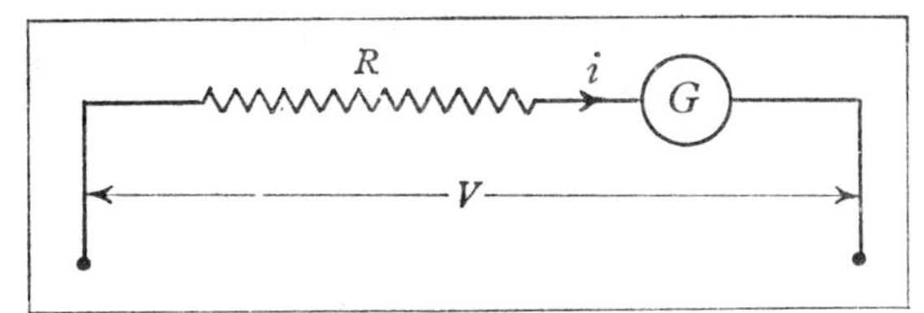

Fig. 8.9 The use of a series resistance to make a galvanometer into a voltmeter

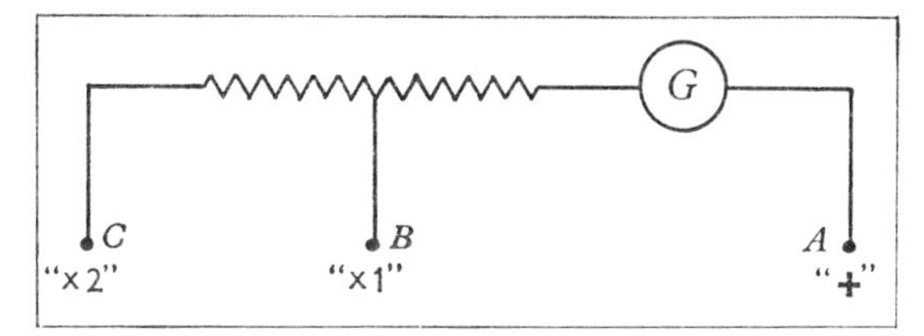

Fig. 8.10 The arrangement of resistors to make a multi-range voltmeter

instrument operate over the desired range. The best type of galvanometer to make into a voltmeter is one that requires only a small current to drive it to full-scale deflection—i.e. one having *high current sensitivity*.

Voltmeters are often specified by their resistance per volt at full-scale deflection. Thus, if i_0 is the current and V_0 the p.d. that gives full-scale deflection, then

$$\text{resistance/volt} = \frac{R+G}{V_0} = \frac{1}{i_0}$$

this quantity is therefore a constant for a given meter, the same no matter what series resistance is used with it. A good voltmeter should have a resistance of at least 500 ohms per volt. A knowledge of this quantity simplifies many voltmeter calculations.

Example. A voltmeter whose range is 0–200 V has a resistance of 1500 ohms per volt (full-scale). What resistance should be joined in series with it to give a range of 0–2000 V?

The resistance of the instrument on the 200 V range is 200 × 1500

$$= 3 \times 10^5 \text{ ohms}$$

The total resistance on the 2000 V range should be 2000 × 1500

$$= 3 \times 10^6 \text{ ohms}$$

∴ The extra resistance required in series

$$= 3 \times 10^6 - 3 \times 10^5 \text{ ohms}$$
$$= 2{\cdot}7 \times 10^6 \text{ ohms}$$

The arrangement of resistors for a multi-range voltmeter is shown in Fig. 8.10. The lowest range of the instrument is obtained by making connection to terminal A (the '+' terminal) and terminal B (marked '×1'). By connecting to A and C (marked '×2') the p.d. to be found is given by twice the scale reading; to achieve this the resistance between A and C must be twice that between A and B.

★ *The ohmmeter*

In a multi-range combined ammeter and voltmeter it is usual to make provision also for the instrument to measure resistance directly. The circuit used is shown in Fig. 8.11*a*; the dry battery is incorporated inside the case of the meter. The variable resistance S is of such a value that the combined resistance of the galvanometer and S is small compared with the series resistance r. The ohmmeter is used as follows. First, the external terminals of the meter,

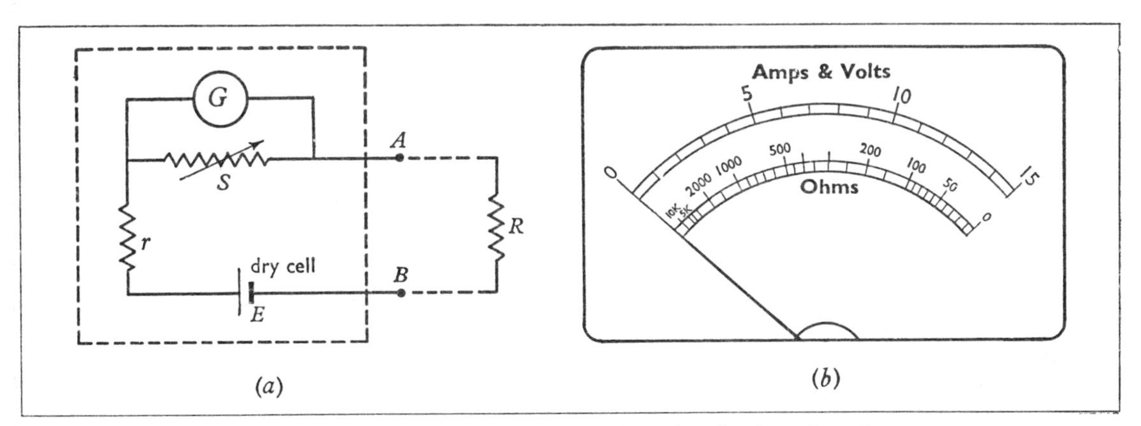

Fig. 8.11 (*a*) The ohmmeter circuit; (*b*) the 'ohms' scale of a universal meter

A and B, are short-circuited, and the resistance S is varied to produce full-scale deflection. Then the resistance R to be measured is joined to A and B and the current falls. As shown below, the use of this procedure makes the latter current almost independent of the e.m.f. E of the battery. Thus the scale may be calibrated to read the resistance R directly. This type of scale runs 'backwards'; i.e. the resistance is zero for maximum current and infinite for zero current (Fig. 8.11*b*).

Let the full-scale current flowing through the meter and its shunt be I_0, and let this fall to I when the resistance R is in circuit. Then,

$$E = I_0 r_0 = I(r_0 + R)$$

where r_0 is the total internal resistance of the meter circuit. Re-arranging,

$$R = r_0\left(\frac{I_0}{I} - 1\right)$$

Since the resistance of the galvanometer and shunt is small compared with r, r_0 is almost unaffected by the small changes in the value of S that occur when the meter is adjusted; and it may therefore be treated as a constant. Knowing r_0 and the currents, the values of R may be calculated in terms of the current readings, and the corresponding points of the scale marked accordingly. Notice that when

$$I = \tfrac{1}{2}I_0 \qquad R = r_0$$

i.e. the reading is r_0 for half-scale deflection.

In addition to the moving-coil instruments considered above, nearly all the instruments designed for a.c. work may also be used for d.c. measurements. Moving-iron meters in particular are much cheaper and more robust than moving-coil instruments, and are used to some extent with direct currents when no great sensitivity or precision is demanded.

8.4 A.C. meters

We have seen in an earlier chapter that there are three ways used for expressing the magnitude of an alternating current or p.d. (p. 28). Thus, we may measure:

(*a*) the peak value;
(*b*) the mean (or half-cycle average) value;
or (*c*) the r.m.s. value.

It is important to know which aspect of an alternating quantity a given meter measures. As long as we are dealing only with sinusoidal waveforms, it makes no difference whether an instrument has its scale marked to give peak, mean or r.m.s. values, since in this case the three quantities bear known ratios to one another (p. 29):

$$\frac{\text{r.m.s. value}}{\text{peak value}} = \frac{1}{\sqrt{2}} = 0{\cdot}707$$

$$\frac{\text{mean value}}{\text{peak value}} = \frac{2}{\pi} = 0{\cdot}637$$

$$\frac{\text{r.m.s. value}}{\text{mean value}} = \frac{\pi}{2\sqrt{2}} = 1{\cdot}11$$

This last ratio is known as the *form factor* of the alternating current. Its value can be used to give a rough indication of the purity of a generated waveform that is supposed to be sinusoidal; a value different from 1·11 indicates a deviation from the correct sinusoidal pattern.

In fact a.c. instruments are normally calibrated to read r.m.s. values, on the assumption that the form factor will be 1·11. However, when using an instrument whose deflection depends on mean values, we must bear in mind that with a non-sinusoidal waveform the reading is not truly the r.m.s. value, but rather a quantity 1·11 times the mean value.

The deflecting force of an a.c. instrument is a rapidly pulsating quantity; the pointer reaches a steady deflection because its inertia prevents it following the oscillations. The reading therefore depends on the mean value of the deflecting force. For an instrument to indicate r.m.s. values the deflecting force must depend on the *square* of the current through it (or the square of the p.d. across it in the case of a voltmeter). With such an instrument the same deflection will be obtained with a steady current of I amp or with an alternating current of I amp r.m.s. It may be calibrated using direct current; and then the readings obtained with alternating current will be r.m.s. values.

(*i*) *Moving-iron instruments.* The deflection is produced by the forces acting on a small piece of soft iron, which is magnetized by the current flowing in a coil. Two types in common use are shown in Fig. 8.12. In the *attraction type* (*a*) a small soft iron plate P is pivoted on the spindle through O that carries the pointer. The current to be measured passes through the coil C; this partly magnetizes the plate P causing it to be attracted into the coil where the magnetic field is a maximum. This is the same for both directions of current in C. In the *repulsion type* (*b*) two soft iron rods A and B are

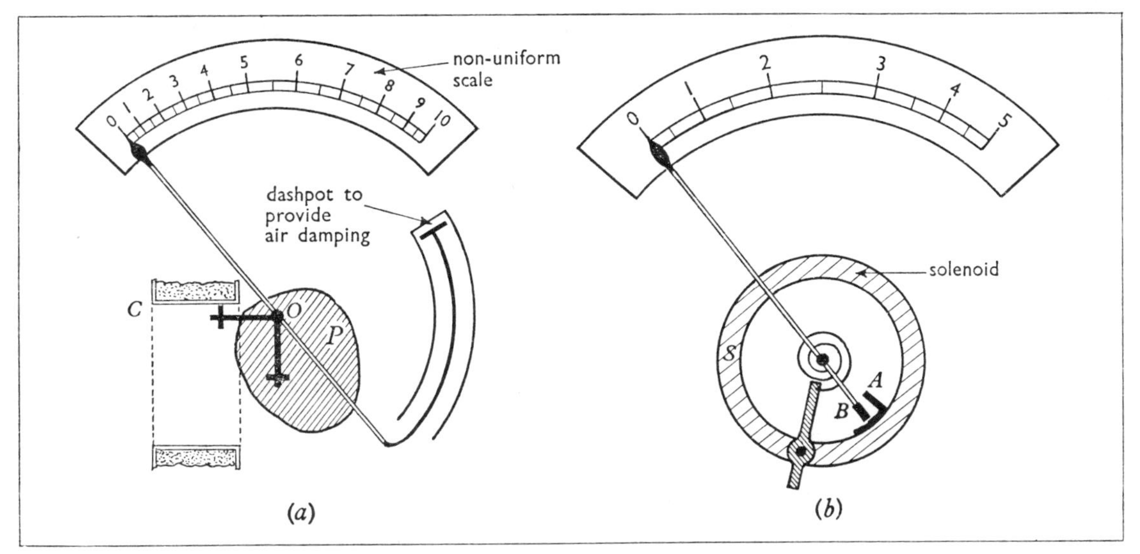

Fig. 8.12 Moving-iron meters: (*a*) attraction type; (*b*) repulsion type

mounted parallel to the axis of the solenoid S. A is fixed to the framework of the instrument and B is attached to the pointer. Whichever way the current flows in the solenoid, A and B are magnetized in the same sense and therefore repel one another.

The restoring couple is sometimes provided by hair-springs, as in (*b*), in the same way as in the moving-coil instrument. But a cheaper arrangement is to use *gravity control*, as in (*a*). In this case the moving parts are not balanced about their axis of rotation, and their weight is used to provide the restoring couple. When a current flows in the solenoid, the pointer comes to rest in the position in which the moment of the restoring force is equal and opposite to the moment of the mean deflecting force. With gravity control the zero is adjusted by levelling. Neither method of control gives a uniform scale, since the deflecting force is not proportional to the current. However, in some applications this is even an advantage, since we may arrange the scale so that the part of it to be used is highly expanded, making the instrument more sensitive in the required range.

These instruments are usually provided with some form of pneumatic damping, as shown in Fig. 8.12*a*. Attached to the moving parts is a light aluminium piston or vane, which moves inside a curved 'cylinder', known as a *dashpot.*

The readings of moving-iron meters ideally indicate r.m.s. values, since in both types of instrument the deflecting force depends on the *square* of the current through the solenoid. This may be seen as follows:

(*a*) In the attracted type the force acting on the moving piece of iron depends on the product of the magnetization induced in it and the strength of the field which acts on it. With an 'ideal' magnetic material the magnetization would be proportional to the magnetizing field. Thus the deflecting force in a given position is proportional to the square of the field strength, and therefore depends on the square of the magnetizing current.

(*b*) In the repulsion type the deflecting force is proportional to the product of the magnetizations of the two soft iron rods. The magnetization of each of these is 'ideally' proportional to the magnetizing field, and so also to the current in the solenoid. Again, the deflecting force depends on the square of the current. In practice these simple results do not hold exactly on account of hysteresis effects in the iron, which make the magnetization not truly proportional to the current.

(*ii*) *Hot-wire instruments.* In these instruments the deflection depends on the heating effect of an electric current. This is proportional to the *square* of the current ($W = I^2R$), and the readings therefore indicate r.m.s. values. The hot-wire ammeter is shown in Fig. 8.13. The current to be measured passes through the fine platinum-alloy wire AB,

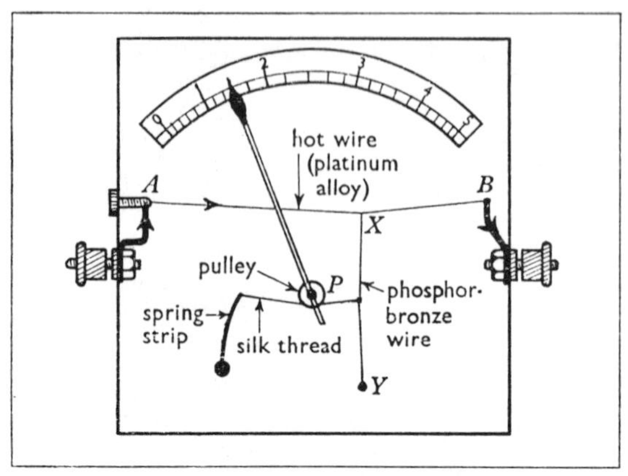

Fig. 8.13 The hot-wire ammeter

raising its temperature and causing it to expand and sag. The phosphor-bronze wire XY therefore sags also, allowing the spring S to pull in the silk thread; this is wrapped round the pulley P, which therefore turns, moving the pointer. The silk thread cannot be attached direct to the wire AB, since at maximum current this would reach a high enough temperature to scorch the thread. The instrument is subject to continual wandering of the zero due to changes in the surrounding temperature, and for this reason is not really satisfactory except for qualitative measurements. In any case the expansion of a wire is by no means a sensitive way of measuring its temperature!

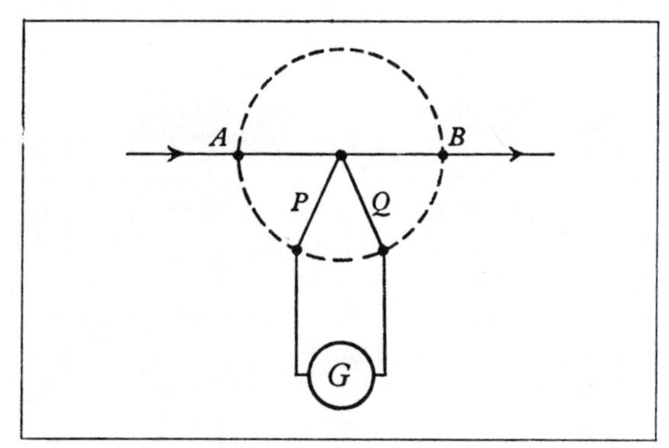

Fig. 8.14 The hot wire (AB) and junction of dissimilar metals (P and Q) in a thermocouple instrument

A better principle is used in the *thermocouple instrument* (Fig. 8.14). The current to be measured passes through the fine wire AB, raising its temperature. Attached to its centre are the ends of two other fine wires P and Q of dissimilar metals; this forms one junction of a thermocouple, the other junction being at the temperature of the surroundings (p. 46). Thus, by its heating effect the alternating current in AB is made to produce a direct current that can be measured with a sensitive moving-coil instrument. The hot junction is enclosed in a small evacuated bulb, and so is not affected by convection currents. It is particularly suitable for high-frequency measurements, since the short wire AB has very low impedance.

(*iii*) *Rectifier instruments.* A rectifier is a device which allows current to pass through it in one direction only; in the reverse direction it has a very high resistance. One example is the selenium rectifier (p. 329). Connected to an alternating supply a selenium rectifier passes a pulsating but uni-directional current which will give a steady deflection on a moving-coil instrument joined in series with it (Fig. 8.15). Current flows on the

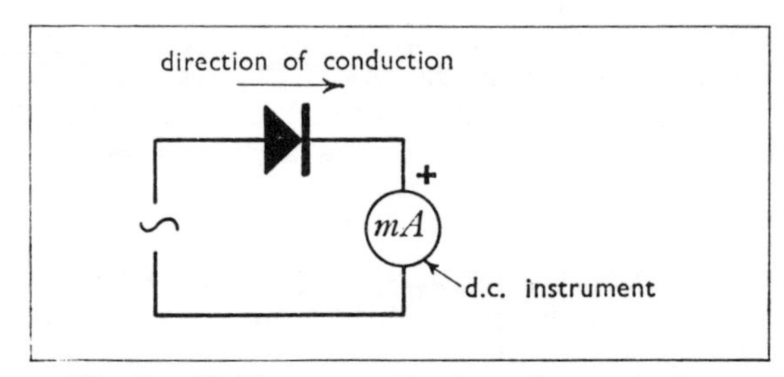

Fig. 8.15 Half-wave rectification using a selenium rectifier

positive half-cycles only, and the deflection is proportional to the *half-cycle average* or *mean value* of the current. A more satisfactory arrangement is to use four rectifiers joined in a bridge circuit (Fig. 8.16). This gives *full-wave* rectification.

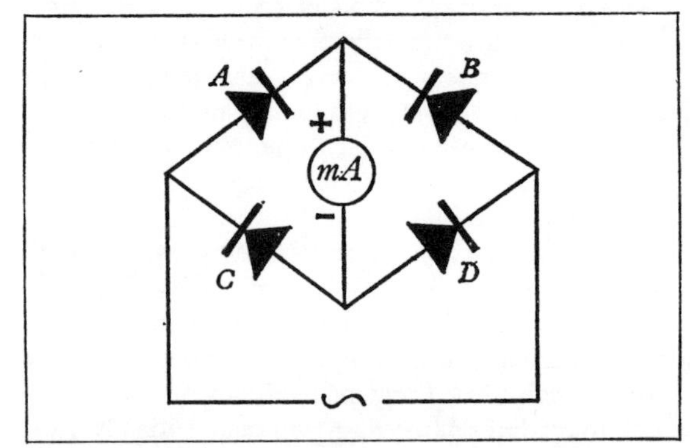

Fig. 8.16 Full-wave rectification using a bridge circuit

During one half-cycle current passes through A and D, and during the next half-cycle through B and C; in both half-cycles the current flows through the meter in the same direction, and the reading is again proportional to the half-cycle average or mean current.

A rectifier conducts very little for *small* p.d.'s in the forward direction; and it is therefore usual to make use of a step-up transformer to ensure that the applied p.d. is large enough to give efficient rectification. This in turn means that a given instrument can be calibrated correctly for only a limited range of frequencies.

Being built round a moving-coil galvanometer a rectifier instrument is more sensitive than other kinds of a.c. meter. This arrangement is used in universal multi-range instruments, that incorporate a.c. as well as d.c. ranges. All the shunts, series resistances, rectifiers and transformers are then included in the one case with the meter, and the connections for the various ranges are provided by switches. The a.c. ranges are calibrated to read r.m.s. values, but this calibration applies only to *sinusoidal* waveforms, and then only in a limited frequency range.

(*iv*) *Dynamometer instruments.* The dynamometer consists of a coil pivoted about an axis in its own plane and free to rotate in the field of a pair of fixed coils (Fig. 8.17). The coil is pivoted in

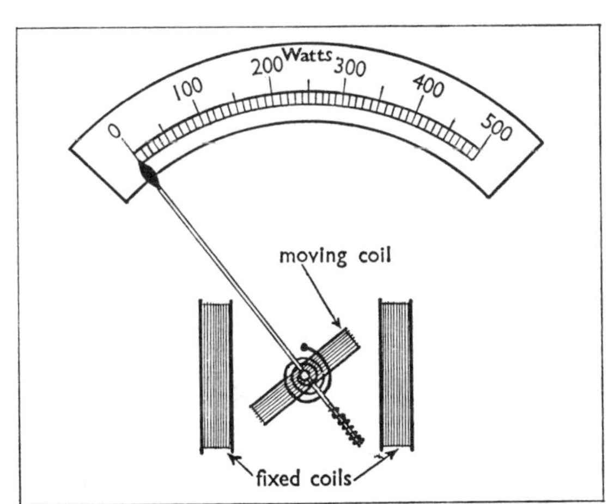

Fig. 8.17 The arrangement of fixed and moving coils in a dynamometer

jewelled bearings and its motion is controlled by two hair-springs, just as in the moving-coil d.c. instrument. No iron or other magnetic material is used, however, and the field produced by the fixed coils is therefore strictly proportional to the current in them. The absence of magnetic material also makes the field much smaller than in the d.c. instrument; and the dynamometer is therefore rather insensitive. It is also liable to disturbance by stray magnetic fields from other apparatus, though this source of error can be avoided by enclosing it in a mumetal shield (p. 78).

Used as an *ammeter* the fixed and moving coils are joined in series, both being of low resistance. The deflecting couple is obviously in the same direction whichever way the current flows. It is in fact proportional to the *current* in the moving coil and to the *field* of the fixed coils; both of these are proportional to the current i to be measured.

$$\therefore \text{ deflecting couple} \propto i^2$$

and the instrument measures r.m.s. values. The scale obeys a *square law*, expanding towards its upper end.

It is worth noting that the *current balance* (p. 8) is merely a form of dynamometer, in which the geometry is such that the forces between the coils can be calculated from first principles. It is normally used for measuring steady currents. But it can equally be employed to make an absolute measurement of the r.m.s. value of an alternating current.

The commonest use of a dynamometer is as a *wattmeter*. For this purpose the moving coil is joined in series with a high resistance *across* the supply, and the fixed coils are joined in series with the load (Fig. 8.18). Thus the fixed coils carry the

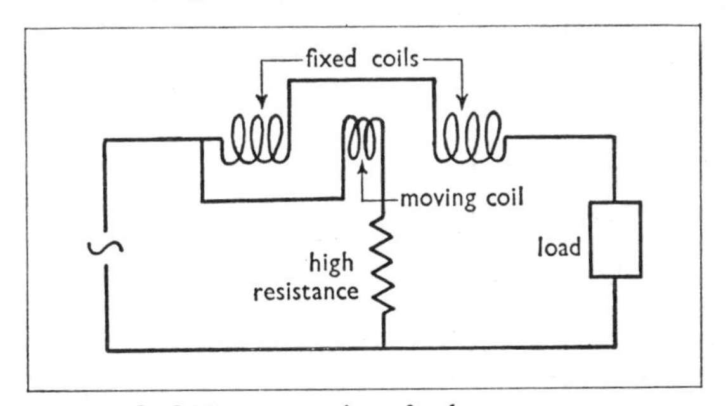

Fig. 8.18 The connection of a dynamometer as a wattmeter

same current i as the load, while the moving coil carries a current proportional to the p.d. v of the supply. The deflecting couple is proportional to the product of these two currents.

$$\therefore \text{ deflecting couple} \propto vi$$

Now, vi = instantaneous power supplied

$\therefore$ mean deflecting couple $\propto$ mean power supplied

This result applies even if v and i are not in phase, as happens in an inductive or capacitative circuit (p. 200). In this case the product of the r.m.s. values of p.d. and current does *not* give the mean power; but the wattmeter reading still does. Since the deflecting couple is proportional to the power, the instrument has an approximately uniform scale when used as a wattmeter.

Other types of a.c. measuring instrument are described in later chapters. They are briefly listed here for the sake of completeness.

(*v*) *The electrostatic voltmeter* (p. 149). This may be used as an a.c. instrument indicating r.m.s. values. It is only really suitable for rather large p.d.'s; but it has the merit of taking negligible current. The same applies to the *gold-leaf electroscope* (p. 149), which can be regarded as a form of electrostatic voltmeter.

(*vi*) *The diode valve voltmeter* (p. 242). This kind of instrument records *peak values* of the applied p.d., whatever the waveform may be. It can be adapted for use at almost any frequency.

(*vii*) *The cathode-ray oscilloscope* (p. 236). With a good modern tube the deflection of the spot of light is very nearly proportional to the p.d. between the deflector plates. With an alternating p.d. joined to the Y-plates the height of the trace is therefore proportional to the *peak p.d.* This instrument has the obvious merit that it enables the waveform to be studied at the same time as its magnitude is being measured.

8.5 The theory of the ballistic galvanometer

When a current passes for a very short time through the coil of a galvanometer, it gives it an impulse that sets it swinging. Provided the duration of the pulse of current is short compared with the period of oscillation of the coil, the amplitude of oscillation does not depend on the manner in which the current varies during the pulse; it is simply proportional to the total charge q that passed.

Using the symbols adopted on p. 67, the couple C acting on the coil when a current i flows in it is given by

$$C = BAni$$

$$\therefore \text{ the angular impulse} = \int C\,dt = BAn\int i\,dt$$

$$= BAnq$$

This is equal to the angular momentum imparted to the coil.

$$\therefore Banq = I\omega$$

where I = the moment of inertia of the suspended system; and ω = the angular velocity produced.

It is assumed here that the deflection is still negligible by the time the pulse of current has finished, so that the action of the restoring couple can be ignored during the impulse.

$$\therefore \text{ the kinetic energy gained} = \tfrac{1}{2}I\omega^2 = \frac{(BAnq)^2}{2I}$$

When the coil eventually comes to rest momentarily, this kinetic energy has been changed into elastic potential energy in twisting the suspension fibre. The couple C' due to the fibre is given by

$$C' = k\theta$$

where k is the suspension constant (p. 123).

The work done in twisting the fibre

$$= (\text{average couple}) \times \theta = \tfrac{1}{2}k\theta^2$$

$$\therefore \tfrac{1}{2}k\theta^2 = \frac{(BAnq)^2}{2I}$$

$$\therefore \theta = \frac{BAn}{\sqrt{kI}}\cdot q \qquad (1)$$

i.e. $\theta \propto q$

The constants in this expression are not usually known. However, they may be expressed in terms of other readily measurable properties of the instrument.

The current sensitivity s of the instrument is given by

$$s = \frac{BAn}{k} \qquad \text{(p. 123)}$$

Also, it is shown below that the period of oscillation T of the coil is given by

$$T = 2\pi\sqrt{\frac{I}{k}}$$

Equation (1) above can then be written

$$\theta = \frac{BAn}{k}\sqrt{\frac{k}{I}}.q = s.\frac{2\pi}{T}.q$$

$$\therefore\ \theta = \frac{2\pi s}{T}.q \qquad (2)$$

The period T may be found by timing a number of complete oscillations with a stop-watch, and the sensitivity s can be measured by the method given on p. 125. However, this analysis has ignored one important point: the damping of the movement of the coil is never really negligible, with the result that the actual deflection θ_1 obtained at the end of the first quarter-swing is always rather less than the theoretical value θ that we have calculated. We need therefore to consider in more detail the equation of motion of the coil.

Once the impulse due to the pulse of current has finished, there are two couples acting on the coil:

(*i*) the restoring couple C', proportional to θ.

(*ii*) the damping couple C'', which is due to air resistance and to the effects of any currents induced in the coil by its movement in the magnetic field; this couple is proportional to the angular speed $\mathrm{d}\theta/\mathrm{d}t$.

In order to get the signs right we must then write:

$$C' = -k\theta;$$

and

$$C'' = -\lambda\frac{\mathrm{d}\theta}{\mathrm{d}t}$$

where λ is the *damping constant.*

The equation of motion then is

$$-k\theta - \lambda\frac{\mathrm{d}\theta}{\mathrm{d}t} = I\frac{\mathrm{d}^2\theta}{\mathrm{d}t^2}$$

$$\therefore\ I\frac{\mathrm{d}^2\theta}{\mathrm{d}t^2} + \lambda\frac{\mathrm{d}\theta}{\mathrm{d}t} + k\theta = 0$$

It may be verified by direct differentiation that a solution of this equation is

$$\theta = a\,\mathrm{e}^{-\lambda t/2I}\sin\left\{\sqrt{\left(p^2 - \left(\frac{\lambda}{2I}\right)^2\right)}\,t\right\}$$

where $p^2 = k/I$.

In the present case we deliberately keep the damping very small, so that we can ignore the

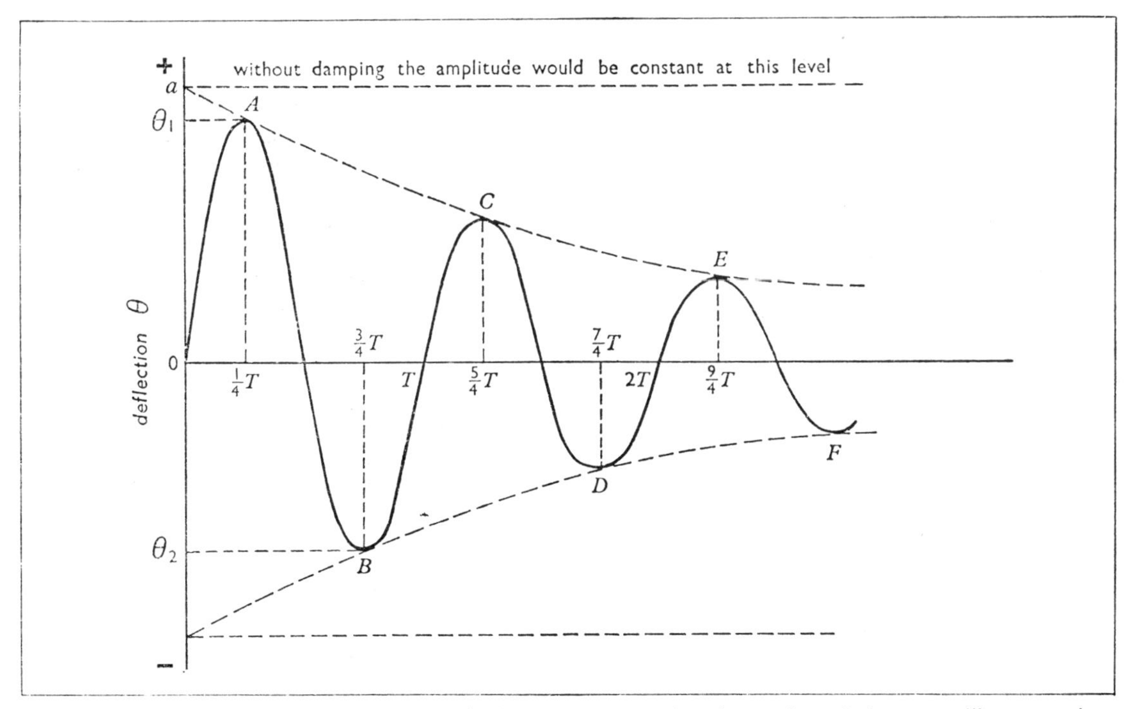

Fig. 8.19 The motion of a galvanometer coil after the passage of a charge through it—an oscillatory motion with exponentially decreasing amplitude

second term under the square-root sign in comparison with the first.

$$\therefore\ \theta = a\,\mathrm{e}^{-\lambda t/2I}\sin pt$$

The variation of θ with t is shown in Fig. 8.19. It is an oscillatory motion with exponentially decreasing amplitude. From the oscillatory factor ($\sin pt$) we get the period of oscillation T since

$$p = \frac{2\pi}{T}$$

$$\therefore\ T = \frac{2\pi}{p} = 2\pi\sqrt{\frac{I}{k}}$$

If there were no damping ($\lambda = 0$), the amplitude of oscillation would be a, as shown by the dotted line. This is the quantity that is really given in equation (2) above:

$$\therefore\ a = \frac{2\pi s}{T}.q$$

To allow for the exponential damping factor, we proceed as follows. Let the numerical values of the deflections at successive turning points (A, B, C, . . . in the graph) be θ_1, θ_2, θ_3, . . . , respectively. The values of t at these points are

$$\tfrac{1}{4}T,\ \tfrac{3}{4}T,\ \tfrac{5}{4}T,\ \ldots \quad \text{etc.}$$

$$\therefore\ \theta_1 = a\,\mathrm{e}^{-\lambda T/8I}$$

$$\theta_2 = a\,\mathrm{e}^{-3\lambda T/8I}$$

$$\theta_3 = a\,\mathrm{e}^{-5\lambda T/8I} \quad \text{etc.}$$

This series is a geometrical progression, the ratio of successive terms being constant.

$$\therefore\ \frac{\theta_1}{\theta_2} = \frac{\theta_2}{\theta_3} = \ldots = \mathrm{e}^{\lambda T/4I}$$

This common ratio is known as the *decrement* of the motion. The logarithm of this to the base e is called the *logarithmic decrement*, which we shall denote by δ.

$$\therefore\ \delta = \frac{\lambda T}{4I}$$

The deflection θ_1 at the first maximum, which we observe in order to measure the charge q, is therefore given by

$$\theta_1 = a\,\mathrm{e}^{-\delta/2} = \frac{2\pi s}{T}\mathrm{e}^{-\delta/2}.q$$

$$\therefore\ q = \frac{T}{2\pi s}\mathrm{e}^{\delta/2}.\theta_1$$

Note that, provided the damping is constant we still have

$$q \propto \theta_1$$

To find the logarithmic decrement δ it is best to observe the deflection at two well-separated turning points. For instance, we might record the 1st and 11th, θ_1 and θ_{11}.

$$\therefore\ \frac{\theta_1}{\theta_{11}} = \left(\frac{\theta_1}{\theta_2}\right)^{10} = (\mathrm{e}^{\delta})^{10} = \mathrm{e}^{10\delta}$$

$$\therefore\ \delta = \frac{1}{10}\log_e(\theta_1/\theta_{11})$$

$$= \frac{2{\cdot}303}{10}\log_{10}(\theta_1/\theta_{11})$$

If δ is sufficiently small we may write

$$\mathrm{e}^{\delta/2} = 1 + \frac{\delta}{2} \qquad \text{(Appendix p. 392)}$$

giving the simpler expression

$$q = \frac{T}{2\pi s}\left(1 + \frac{\delta}{2}\right).\theta_1$$

It must be remembered that the value of δ depends on the resistance of the galvanometer circuit; and it should therefore be measured for the actual circuit in which the instrument is to be used.

The quantity θ_1/q is called the *ballistic sensitivity* of the galvanometer. The above analysis shows how it may be found from measurement of the three constants of the instrument—s, T and δ. However, this process of calibration is rather a tedious one; and, having satisfied ourselves on the theoretical point that the charge passed is indeed proportional to the first 'throw', it is usually simpler to use one of the direct methods of calibration discussed on p. 117.

9: Generators and Motors

9.1 A.C. generators

Consider a flat coil rotating about an axis in its own plane at right-angles to a magnetic field (Fig. 9.1). The flux linkage of the magnetic field with the coil alternates in the manner shown in the graph. In positions (i) and (iii) the plane of the coil is normal to the field, and the flux linkage reaches its maximum value, but is momentarily unchanging; thus the rate of change of flux is zero. At these positions the e.m.f. is therefore zero. In positions (ii) and (iv) the flux linkage is momentarily zero, but it can be seen from the graph that it is changing at the maximum rate—decreasing in position (ii) and increasing in position (iv). At these positions (with the coil parallel to the field) the e.m.f. therefore reaches its maximum magnitude. By Lenz's law the e.m.f. acts so that the current driven by it opposes the changes of flux; it is therefore positive at (ii) and negative at (iv). Thus an alternating e.m.f, is generated in the coil, whose frequency is equal to the number of revolutions per second. (An expression for the actual value of the e.m.f. was deduced on p. 108.)

Fig. 9.2 shows the principle of a possible form of simple a.c. generator or *alternator*. The coil rotates between the poles of an electromagnet, the current for which comes from a separate d.c. supply. The connections to the coil must be made through sliding contacts, and the figure shows diagrammatically the system that is used in such a case. The ends of the coil are joined to two insulated copper rings (known as *slip-rings*) on the rotating shaft. The connection is then made through fixed graphite blocks (called *brushes*), which are pressed onto the slip-rings by springs.

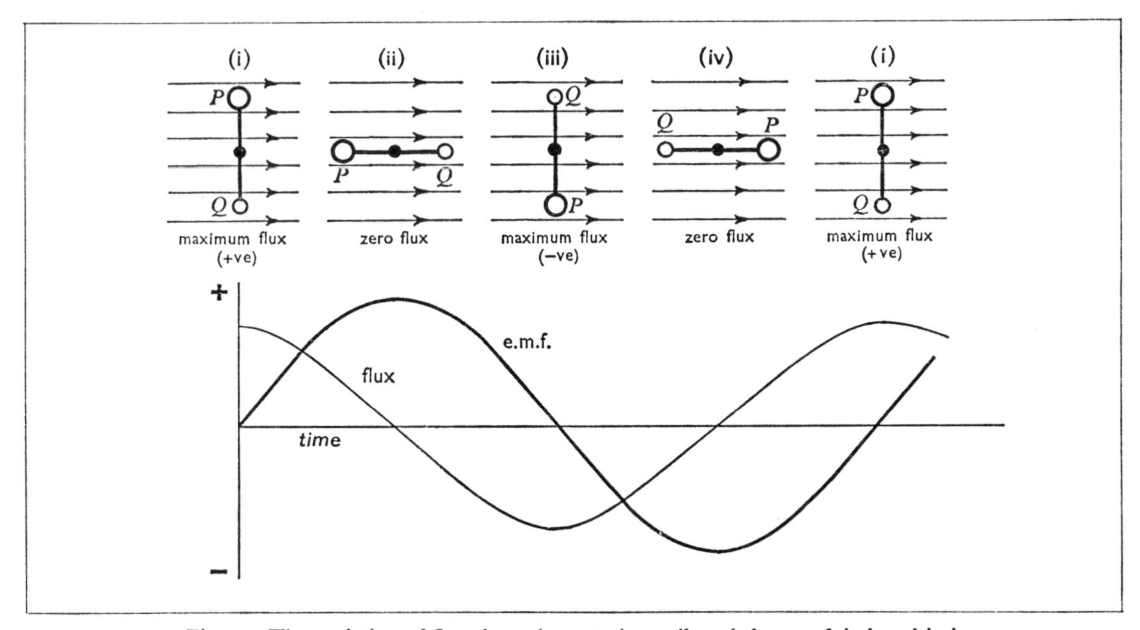

Fig. 9.1 The variation of flux through a rotating coil, and the e.m.f. induced in it

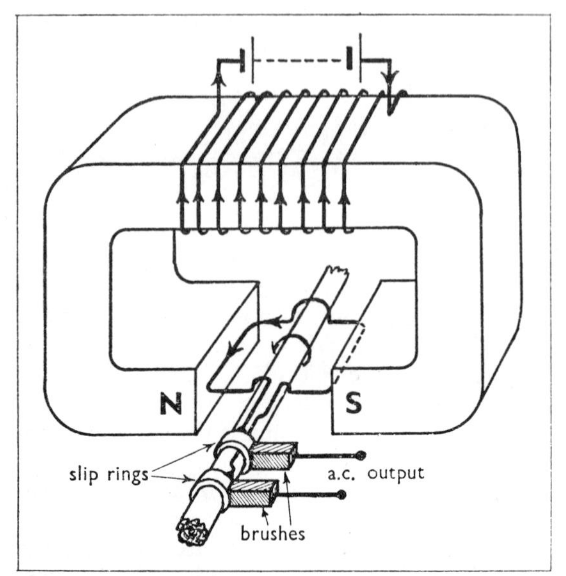

Fig. 9.2 Generating an alternating current by rotating a coil in a magnetic field

In practice, however, it is more satisfactory to keep the coil fixed and to rotate the magnetic field in relation to it. Fig. 9.3 shows the arrangement used in diagrammatic form; the coil is wound in two parts in series, between which the electromagnet rotates. The advantage is that the slip-rings and brushes are now required to carry only the relatively small current needed to magnetize the rotating electromagnet. In a large alternator the current generated in the fixed coils may run into thousands of amps; this can now flow through fixed connections, whereas it would be almost impossible to design brushes to carry currents of this magnitude.

To increase the flux through the coils they are wound on soft iron cores, which are laminated to reduce eddy currents. The rotating electromagnet is called the *rotor*, while the fixed system of coils with their core is called the *stator*.

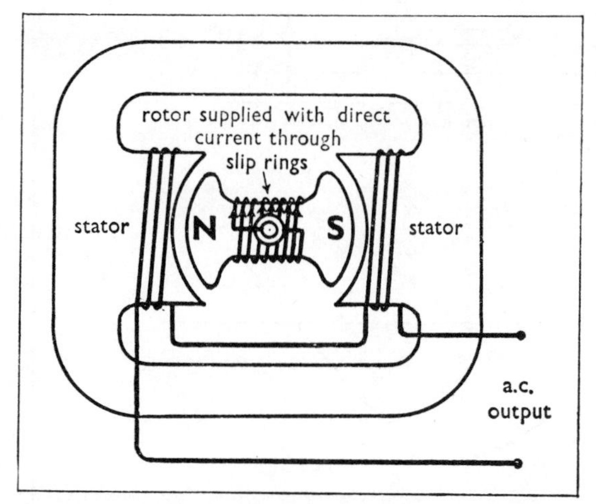

Fig. 9.3 A simplified a.c. generator or alternator (2-pole, single-phase type)

However, if only a single pair of stator coils is used, the torque required to keep the rotor turning would be very uneven, and vibrations would be set up. By Lenz's law the current flowing in the stator coils gives rise to forces *opposing* the rotation. The torque required to maintain the rotation therefore pulsates with the current. There are two ways of reducing this effect.

(i) The stator coil can be split up into 4, 6 or 8 parts spaced on suitable pole pieces round the inside of the stator. The rotor is then arranged to have the same number of poles, north and south poles being arranged alternately round it. Fig. 9.4 shows the arrangement of a four-pole alternator. The fluctuations of torque still occur, but are less serious; also the speed of rotation only needs to be half as much in order to produce the same frequency of alternating current.

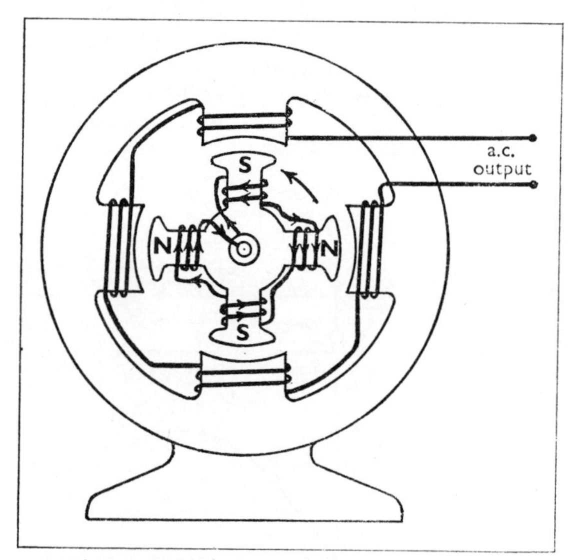

Fig. 9.4 The arrangement of coils in a 4-pole, single-phase alternator

(ii) The same rotor can be used to generate alternating current in *three separate sets* of stator coils spaced on pole pieces round the inside of the stator. The reaction between any one set of stator coils and the rotor is still a pulsating one, but their combined reactions are constant, since the peaks of current occur at different moments in the three sets of coils. Each set of coils is connected to a separate distribution line; but the three lines share a common return wire. This is known as a *three-phase system.* Fig. 9.5 shows the arrangement of a two-pole, three-phase alternator. The simpler system employing a single distribution circuit connected to a single set of stator coils is called a *single-phase system.*

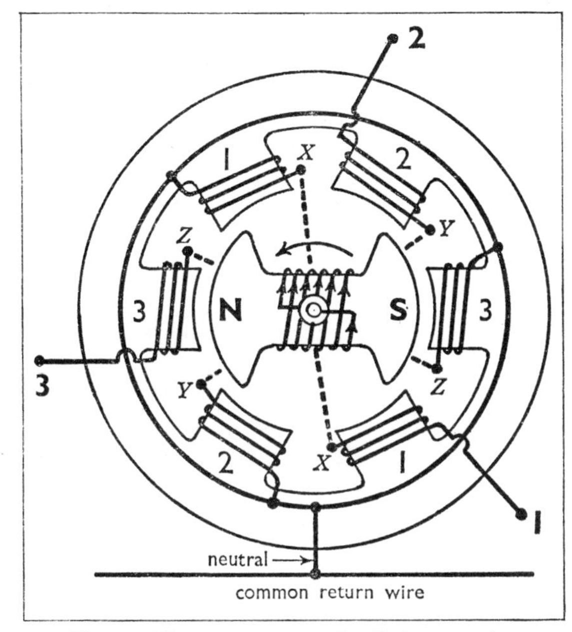

Fig. 9.5 The arrangement of coils in a 2-pole, 3-phase alternator

The public electricity supply employs a three-phase system, since this makes for greater efficiency not only in generation but also in distribution; this is discussed further in Chapter 12 (p. 205).

9.2 D.C. generators

When a coil rotates in a magnetic field, an alternating e.m.f. is generated in it. To produce a current flowing in one direction only, connection to the coil must be made through some form of rotary switch, which reverses the connections twice in every revolution.

Fig. 9.6*a* shows in diagrammatic form the method that might be used for a single coil. On the same shaft as the coil are mounted two insulated semi-cylindrical plates, one connected to each end of the coil. Contact with these is made by two carbon brushes, which bear on the plates on opposite sides of the shaft; the arrangement is called a *commutator*. It will be noticed that the positions of the brushes are such that the gaps of the commutator come under them at the moments when the plane

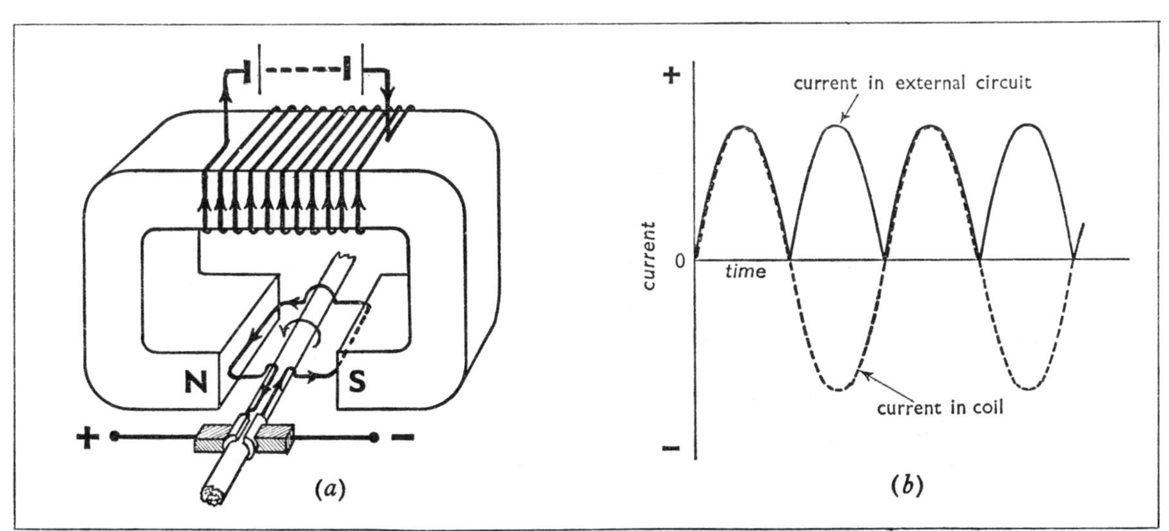

Fig. 9.6 Generating direct current with a rotating coil and commutator

of the coil is at right-angles to the field; in this position the e.m.f. is momentarily zero, but is just about to reverse. In Fig. 9.6*b* the dotted curve shows the alternating current in the coil, while the full line gives the current in the external circuit. This current is unidirectional, but pulsating.

To increase the flux the coils are wound on a soft iron core, which is laminated to reduce eddy currents (p. 101). The rotating coils and their core are referred to as the *armature*. The magnetic field is normally produced by an electromagnet called the *field* magnet. This is provided with concave pole pieces; also the armature coils are sunk in slots in the core, thereby reducing the air gap and providing the maximum possible flux through the core. The current for the field coils is usually taken from the dynamo itself. Fig. 9.7 shows the construction of a simplified d.c. dynamo employing a single-coil armature with the field coil joined in parallel with it (or *in shunt*, as it is called).

When a current is taken from the dynamo, by Lenz's law a torque arises in the armature *opposing* the rotation. The mechanical energy supplied to maintain the rotation against the action of this torque re-appears as the electrical energy generated in the dynamo circuit.

The e.m.f. produced by a single coil would be by no means steady, as we have seen. In practice the armature of a d.c. dynamo contains many coils, which are sunk in slots uniformly spaced round the armature core. They are connected to a commutator of many segments in such a way that at any given instant about two-thirds of the coil system is contributing to the e.m.f. The design of d.c. machines is a specialized topic of no little complexity; the student who wants to pursue it further should consult textbooks of electrical engineering.

★ *Different ways of connecting the field coils*

The current for the field coils of a dynamo may be provided in several different ways; each gives a dynamo with its own special characteristics.

(*i*) *Separately excited.* The field current in this

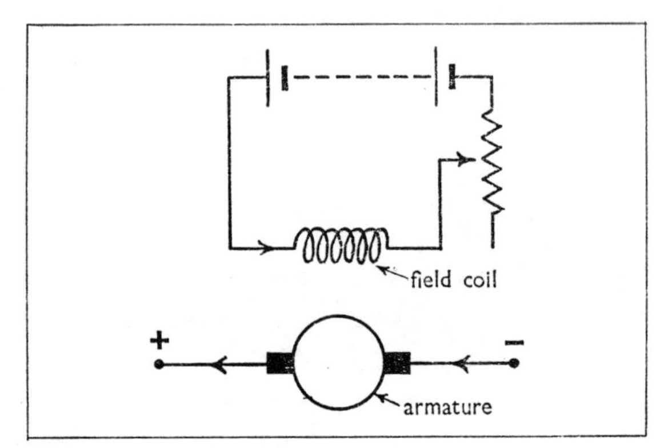

Fig. 9.8 A separately excited dynamo

case is provided by a separate source of e.m.f. The e.m.f. generated in the machine is proportional to the flux density produced by the field coil. As with

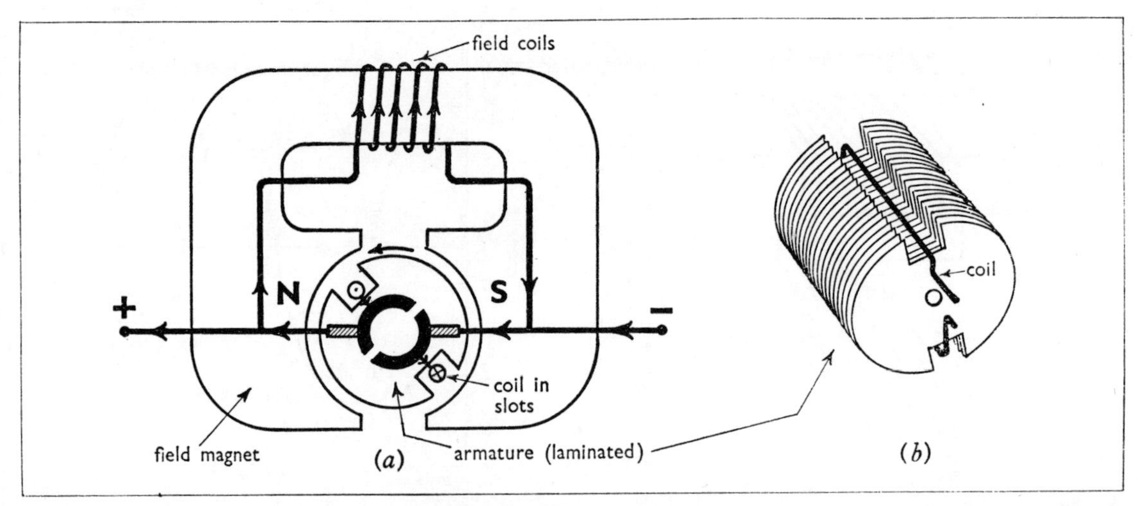

Fig. 9.7 A simplified d.c. dynamo (with shunt-wound field coils). The armature core consists of a stack of insulated iron discs (or laminations)

other sources of e.m.f., the p.d. falls when a current is taken on account of the internal resistance of the armature.

(*ii*) *Series-wound.* In this case the entire current produced by the machine passes through the field coil, which therefore needs only a small number of turns to give the required flux; heavy gauge wire is used to keep down the evolution of heat in these coils to a small value (Fig. 9.9). When no current

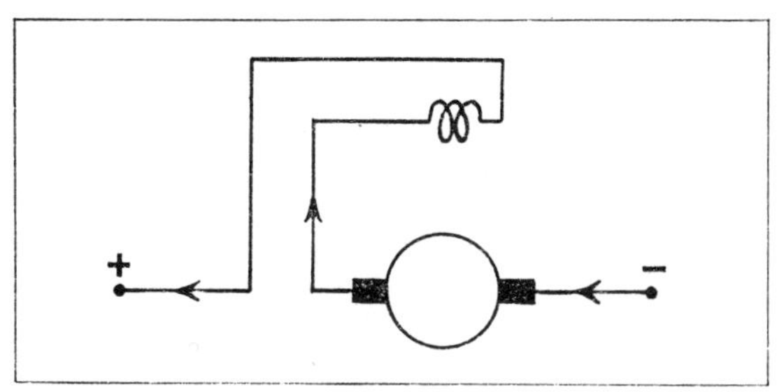

Fig. 9.9 A series-wound dynamo

is taken, the flux is only that due to the residual magnetism of the core, and the e.m.f. is small. As the current increases, so also does the flux produced by the field coils and with it the e.m.f. For a given speed of rotation the e.m.f. is thus approximately proportional to the current taken, until the point is reached at which the magnetic core is saturated. Owing to the unsteadiness of its p.d. the series-wound dynamo does not find many practical applications.

(*iii*) *Shunt-wound.* In this case the field coil is joined in *parallel* with the armature, usually with a rheostat in series to control the field current (Fig. 9.10). To avoid using a large current in the

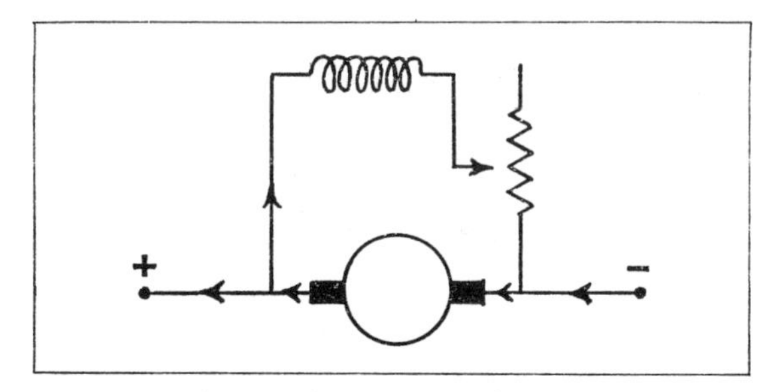

Fig. 9.10 A shunt-wound dynamo

field coil, this is made up of a large number of turns of relatively fine wire. For a given speed and setting of the rheostat, the field current (and with it the flux density) is proportional to the p.d. across the armature. The maximum e.m.f. is therefore obtained when no current is being taken by the external circuit. When the current taken increases, the p.d. drops on account of the internal resistance of the armature, and also because of the resulting drop in field current. The fall of p.d. with current is therefore greater than with a separately excited dynamo, but is still small enough to make this the most useful sort of dynamo for many purposes. The e.m.f. can readily be controlled with the field rheostat; in some designs this adjustment is made automatically so as to keep the p.d. nearly constant.

(*iv*) *Compound.* In this case the field magnet carries two windings: a series winding of a few turns of heavy gauge wire, and a shunt winding of many turns of fine wire (Fig. 9.11). When no current is

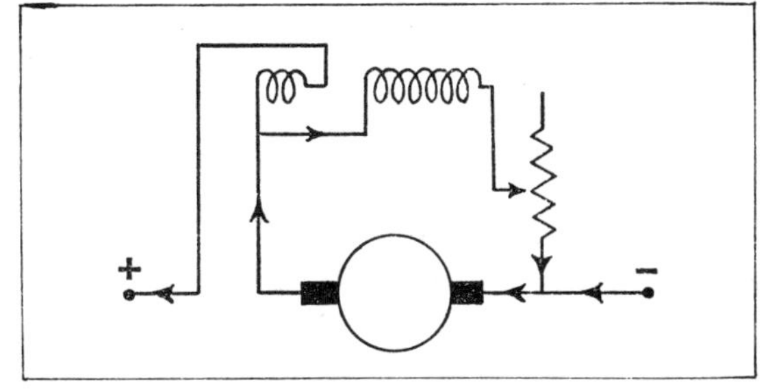

Fig. 9.11 A compound dynamo; the field magnet is equipped with both series and shunt windings

being taken, the flux is provided entirely by the shunt windings. However, as the current increases, the series windings cause an increase of flux, and with it of the e.m.f. By careful choice of the numbers of turns in the two windings it can be arranged for this rise in e.m.f. exactly to balance the internal drop of p.d., so that the p.d. at the output terminals remains nearly constant for all currents. The machine is then said to be *level compounded.*

9·3 D.C. motors

A coil carrying a current in a magnetic field experiences a couple that tends to turn it with its plane at right-angles to the field—i.e. so that the flux through it is a maximum. To produce continuous rotation a commutator must be used to reverse the connections to the coil whenever it passes through the equilibrium position. Fig. 9.12 shows the action of a commutator for a single-coil motor. The ends of the coil are connected to two insulated semi-cylindrical plates on the motor shaft; contact with these is made through carbon *brushes* in the same way as for a d.c. generator. Fig. 9.12 gives a sectional view of the coil and commutator in four different positions. In position A the couple acts on the coil so as to turn it to position B. In this position the gaps in the commutator come opposite the brushes. The momentum

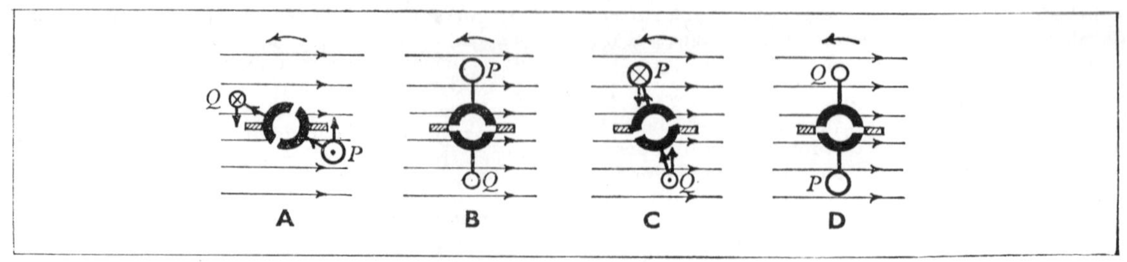

Fig. 9.12 The action of a commutator in producing continuous rotation

of the coil must be sufficient to carry it through to position C, when the current in it is reversed; the couple then acts to produce a further half revolution to position D. To give the maximum torque over as much as possible of the rotation the flux is increased by sinking the coil in slots in a soft iron cylindrical core (laminated to reduce eddy currents); and concave pole pieces are fitted to the electromagnet that produces the field. Fig. 9.13

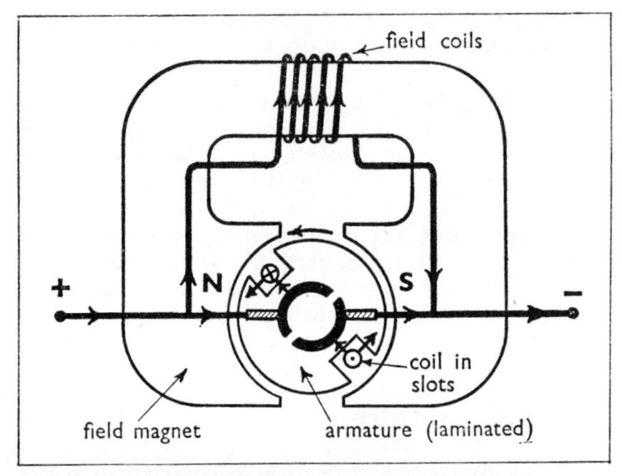

Fig. 9.13 A simplified d.c. motor (with shunt-wound field coils)

shows the construction of a simplified d.c. motor with a single-coil armature. In practice the armature is made with many coils sunk in slots spaced round the core and connected to a commutator of many segments. The coils are connected so that at any given moment at least two-thirds of them contribute to the torque. In fact in all respects the construction of a d.c. motor is identical with that of the equivalent d.c. dynamo (compare Fig. 9.13 with Fig. 9.7). Any d.c. machine may be used equally as a dynamo or a motor. The difference between the two is only a question of function. When mechanical power is being supplied to the machine, and is being changed into electrical power, we call it a *dynamo*; and when electrical power is being converted into mechanical power, we call it a *motor*.

Closer thought shows that the d.c. machine must always behave *simultaneously* both as a dynamo and as a motor. Any current flowing in the armature gives rise to a torque. Thus, when current is taken from a dynamo, a torque arises, as in a motor; by Lenz's law this must *oppose* the rotation. Likewise, whenever the armature of a motor rotates, an e.m.f. is induced in it, just as in a dynamo; by Lenz's law the e.m.f. must act *in opposition* to the applied p.d. The fact is that a rotating coil carrying a current in a magnetic field both generates an e.m.f. and experiences a couple. The difference between its functions as a dynamo and a motor is in the *direction of the current.*

AS A DYNAMO: the current is in the direction of the e.m.f.—which is of course responsible for driving the current through the armature and round the circuit. The product of e.m.f. and current is equal to the electrical power generated, which in turn is equal to the mechanical power *input* needed to maintain the rotation against the opposing electromagnetic torque.

AS A MOTOR: the current is driven through the armature by the applied p.d. in opposition to the induced e.m.f. The product of induced e.m.f. and current is now equal to the mechanical power *output.*

This can be compared with the similar behaviour of a reversible electric cell (such as a lead–acid accumulator). Connected to an electric circuit containing no other source of e.m.f. it drives current round the circuit, and so converts chemical energy into electrical energy. But if another source of e.m.f. greater than that of the accumulator is connected so as to drive current through it in the reverse direction, then the result is the conversion of electrical energy into chemical; and the accumulator is restored to its 'charged' condition (p. 43). An accumulator is a device for the mutual conversion of electrical and *chemical* energy; whereas the d.c. machine is for the mutual conversion of electrical and *mechanical* energy. But the action of the two devices in an electric circuit is essentially the same. A practical example may help to make this clear.

Consider an electric delivery van worked from a 100-volt battery climbing over the brow of a hill. As it ascends the hill, it travels fairly slowly; the back e.m.f. generated in its motor is proportional to the speed, and in this case is less than the battery e.m.f.; suppose it is 95 V. The difference between the two e.m.f.'s, namely 5 V, drives a large current through the low-resistance armature, and provides the torque to keep the van moving up the hill. The net result is the conversion of the chemical energy of the battery into potential energy of the van. As the van approaches the brow of the hill, the torque required decreases, and the vehicle accelerates. The increased speed raises the back e.m.f. generated in the armature to a figure very near the battery e.m.f.; it might be 99 V. The difference between the two, now 1 V only, drives a current one-fifth as great through the motor; but this is sufficient to maintain the speed on the smaller gradient and to supply the small amount of power needed to overcome friction. As the van starts to descend the other side of the hill, it speeds up further, until the induced e.m.f. is equal to the battery e.m.f. At this point the loss of potential energy is just sufficiently rapid to make up for the frictional loss; no current flows, and the motor is 'idling'. When the speed increases still further, the e.m.f. becomes more than 100 V, and the d.c. machine, now behaving as a dynamo, drives current in the reverse direction through the battery. Much of the potential energy acquired by the van as it ascended the hill is now being changed back into chemical energy. The current flowing in the machine now produces a torque which opposes the motion and therefore acts as a brake.

If the e.m.f. of a dynamo is E, the current through the armature I, and the p.d. across it V, then

$$E - V = Ir$$

where r = the internal resistance of the armature. Multiplying this equation by I and re-arranging, we have

$$EI = VI + I^2r$$

this can be interpreted as:

power generated = electrical power output
+ rate of heating armature

Likewise with a motor, using the same symbols

$$V - E = Ir$$

since in this case $V > E$, and the current flows in opposition to the back e.m.f. E. Multiplying by I and re-arranging,

$$VI = EI + I^2r$$

and this means:

power supplied = mechanical power output
+ rate of heating armature

To achieve high efficiency in a d.c. machine the armature resistance r clearly needs to be as low as possible.

Starting a motor

At the moment of starting the e.m.f. E is zero, and it only increases slowly to its final value as the motor gathers speed. If no other resistances were included in the circuit, the starting current I' would therefore be given by

$$V = I'r$$

In a large motor with low armature resistance I' could reach a large and damaging value, unless steps were taken to limit it. For instance, a motor operating on a 200 V d.c. supply might have an armature resistance of 0·05 ohm. Then

$$I' = \frac{200}{0{\cdot}05} = 4000 \text{ amps!}$$

When starting it is therefore necessary to join a suitable resistance in series with the armature. Suppose in the present example we must limit the current to 20 amps. The total resistance required must then be

$$\frac{200}{20} = 10 \text{ ohms}$$

As the motor speed increases the back e.m.f. rises from zero and the starting resistance is progressively switched out of the circuit; near full speed the back e.m.f. alone is sufficient to limit the current to a safe value, and the starting resistance is switched out of the circuit altogether. Fig. 9.14 shows the arrangement used for this purpose. With small d.c. motors (less than $\frac{1}{4}$ h.p.) the resistance of the armature is usually sufficient without the aid of any starting resistance.

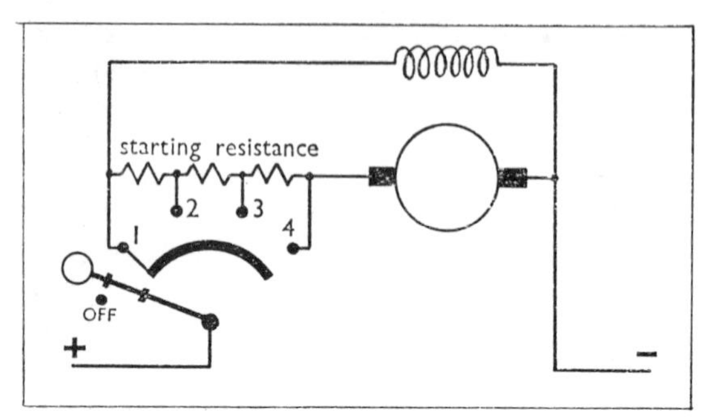

Fig. 9.14 The starting resistance and switches used with a large electric motor

Reversing an electric motor

The direction of rotation of a d.c. motor is not altered by reversing the polarity of the supply; this would merely reverse both the armature and field currents, leaving the torque still in the same direction. To reverse the rotation we need rather to reverse the connections to one only of the coils —say, the field coil.

However, very small motors, such as those used in toy electric trains, are often made with *permanent magnets*. Reversing may then be effected by simply changing the connections at the d.c. supply.

9.4 The properties of d.c. motors

The field coils of d.c. motors may be connected in the same variety of ways found in dynamos (Figs. 9.8 to 9.11).

Series-wound, shunt-wound and compound motors each have their own characteristic properties, making them suitable for particular applications. We shall consider these properties under the following headings:

(*i*) *The steadiness of the speed under changing loads.* The speed of a d.c. motor settles down at such a value that the back e.m.f. is a little less than the applied p.d. If the load increases, the armature current must rise to provide the extra torque. This is brought about by a drop in speed, which in turn causes a drop in the back e.m.f., thereby allowing the current to rise. However, the drop in e.m.f. required is usually very small. For example, in a motor working on a 100-volt supply the back e.m.f. might be 99 V. This has only to drop to 98 V to double the armature current.

Now for a given motor the back e.m.f. E is proportional to the magnetic flux in the core N and to the speed ω.

$$E \propto N\omega$$

With a *shunt-wound motor* the flux N is constant, since the field current does not vary. (The same applies with a *permanent magnet* motor.)

$$\therefore E \propto \omega \quad \text{in this case}$$

The small drop in E that must accompany an increase of load is brought about by a proportionately small drop in ω. Thus, provided it is not overloaded, the speed of a shunt-wound motor varies little under changing loads.

However, with a *series-wound motor* the field current is the same as the armature current. Since the changes in E are small we can regard it as approximately constant.

$$\therefore N\omega = \text{constant (approximately)}$$

Provided the field magnet is not saturated, we also have approximately

$$N \propto I$$

where I is the current.

$$\therefore \omega \propto \frac{1}{I}$$

Thus, if the current needs to double to cope with an increase of load, the speed drops to nearly half. It is clear that the speed of a series-wound motor is very unsteady. Furthermore, if the load is suddenly disconnected, as might happen by the accidental breaking of a coupling belt, the current required would drop to a low value and the motor might race dangerously. It is therefore necessary to provide protective devices to guard against this sort of mishap.

(*ii*) *The ease with which the speed can be controlled.* In the *shunt-wound motor* the speed is very easily controlled by altering the field rheostat. A decrease in the field current causes a fall in the flux N; this leads to an increase in speed in order to return the back e.m.f. E to its former value, just less than the applied p.d. V. Since the field current is not large, the controlling rheostat can be of quite light construction.

In the *series-wound motor* speed control is not so

simple. It is possible to use a variable resistance in *parallel* with the field coil, so that some of the total current is by-passed. However, such a resistance must be designed to carry a substantial fraction of the total current.

Both types of motor may be controlled by means of rheostats in series with the armature. But this is usually only practicable with low-power motors taking small currents.

(*iii*) *The magnitude of the starting torque.* With all electric motors the starting current can be allowed to be greater than that used under steady running conditions. Since the torque is proportional to the current I, this means that all d.c. motors produce their best torque under starting conditions. This is a very useful property. But the torque C is also proportional to the magnetic flux N.

$$\therefore C \propto NI$$

With a *shunt-wound motor*, the flux N is constant for a given setting of the field rheostat.

$$\therefore C \propto I \quad \text{in this case}$$

With the *series-wound motor*, the flux N depends on the total current, and provided the magnet is not saturated, we have approximately

$$N \propto I$$
$$\therefore C \propto I^2 \text{ in this case}$$

Suppose the current allowable at starting is k times the normal current used at full speed. The starting torque of a *shunt-wound motor* would then also be *k times* the steady-running torque—which is good. However, the starting torque of an equivalent *series-wound motor* using the same current would be *k^2 times* its steady-running torque—which is much better than with the shunt-wound model.

As might be expected, the compound motor combines the properties of the shunt-wound and series-wound types; its exact behaviour depends on the proportion of the field provided by each of the windings. Thus, by making it predominantly series-wound it can have the good starting characteristics of this type of motor, while the small shunt winding suffices to prevent it from racing dangerously if the load is removed. This type of motor is regularly used for traction purposes, e.g. for electric trains and trolley-buses. Series-wound motors are suitable for heavy-duty work where initial torque is all important—e.g. for cranes and winches—or for high-speed, direct-coupled motors, such as fans and grind wheels. Shunt-wound motors are used wherever control of speed is essential; e.g. for lathes and record players. (However, many record players employ synchronous a.c. motors in which the speed is locked to the frequency of the supply.)

It is interesting to compare the properties of an electric motor with those of an internal combustion engine. The electric motor runs satisfactorily at all speeds; it gives its best torque at starting, and is highly efficient at top speed. The engine of an ordinary car is only reasonably efficient within a limited range of engine speeds, and this efficiency is much less than that of the electric motor. Also at low speeds the torque falls to zero and the engine 'stalls'. A diesel engine is even more limited in its performance, though it is more efficient at its best running speed. Thus cars and buses require gear-boxes to match a limited range of engine speeds to a much larger range of possible travelling speeds. No gear-boxes would be needed if electric motors could be used. However, the means of storing electrical energy (e.g. accumulators) are extremely bulky. At present, electrically operated vehicles have only proved satisfactory as delivery vans and milk floats—except where a permanent supply network can be installed, as with trolley-buses and electric trains. If an efficient means could be found of converting the energy of the primary fuels, such as oil, directly into electrical energy, the electric motor might really come into its own for driving vehicles. This is attempted in the diesel-electric engines, which are in use to some extent on the railways. Here a diesel motor running at its optimum speed drives a d.c. dynamo, which then supplies power for one or more electric motors that drive the train. But it is questionable whether this multiple system is any more efficient than other modern engines, such as the gas turbine.

9.5 A.C. motors

Since a.c. power supplies are now widely used, the vast majority of motors encountered in industrial and domestic use are a.c. kinds. The subject is a vast one, and we can do no more here than indicate the principles behind some of the more important types.

(*i*) *The series-wound, commutator motor.* If the direction of the current through both the armature and the field windings of an ordinary d.c. motor is reversed, the torque remains in the same direction. Thus, in principle, a d.c. motor may be used with alternating current. However, the shunt-wound type cannot be used in this way, because self-inductive effects in the field coils would make the current in them lag behind the p.d. (p. 199); the result would be that the peaks of magnetic field would occur at different moments from the peaks of armature current, and the torque developed would be very small. But in the series-wound type the magnetic field and the armature current are necessarily in phase. Provided the core of the field magnet is laminated, the same motor may be adapted to run on alternating or direct current. Like the d.c. version, its a.c. counterpart is essentially a high-speed motor, whose speed is rather variable under changing loads.

(*ii*) *The three-phase synchronous motor.* This type of motor is in principle a three-phase alternator operated as a motor instead of as a generator (p. 137). The three-phase supply is connected to the three sets of stator coils (Fig. 9.5, p. 137). It may be shown that if the magnitudes of the currents in the three coils are equal, then the resultant magnetic field they produce is of constant magnitude and rotates at a steady speed. With the stator in Fig. 9.5, the field would rotate once for each cycle of alternation of the current—i.e. 50 times a second with the usual a.c. supply. A direct current from an auxiliary supply is passed through the rotor coils, as usual. If now the rotor can be brought to rotate at exactly the same speed as the field, then the magnetic forces will hold it 'locked in' to this speed, provided the load is not too great. The motor then necessarily remains in synchronism with the supply. The problem is to get its speed up to that of the field in the first place. This is sometimes done with an auxiliary starter motor, and sometimes the one rotor is made to contain also windings of the type used in an induction motor (see below). These provide the starting torque and cease to function when the speed approaches synchronism.

(*iii*) *The three-phase induction motor.* In this type of motor a three-phase stator is used just as in the synchronous motor or in the alternator. Thus, the magnetic field in and round the rotor is a rotating one of constant magnitude. The construction of the rotor is shown in Fig. 9.15. It consists of a number of copper bars welded at both ends onto copper rings that encircle the ends of the rotor. The arrangement is referred to as a *squirrel-cage.* The bars are embedded as usual in a laminated iron core. When the rotating field sweeps past the bars, large currents are induced that circulate internally round the system. They could be described as a system of controlled eddy currents.

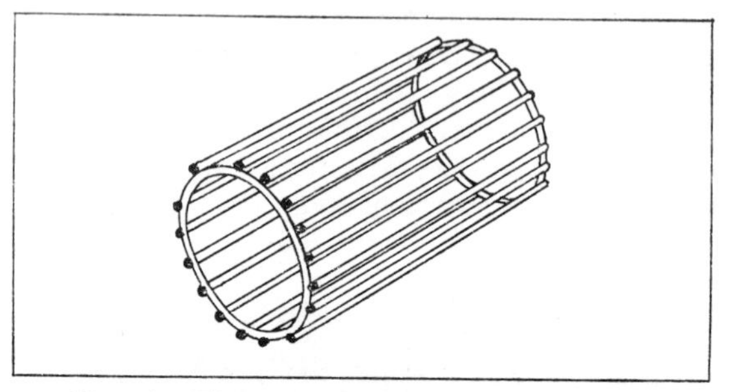

Fig. 9.15 The rotor of a squirrel-cage induction motor

The action of the magnetic field on these currents must be to speed up the rotor towards the point at which there is no relative motion of rotor and field. The motor can never in fact reach synchronism, since if there were no relative motion no currents would be induced in the rotor, and there would be no torque. The rotor therefore 'slips' behind the field; the greater the load, the more the rotor speed falls below that of the field.

(*iv*) *The single-phase induction motor.* This type of motor uses a squirrel-cage rotor of similar construction to that of the three-phase machine. A single-phase stator is used that gives an approximation to a rotating field. First, it may be noted that a simple alternating (non-rotating) field can be regarded as the resultant of two fields of equal amplitude rotating at the same speed in opposite directions. Fig. 9.16 shows successive positions of the two component fields; their resultant is clearly an alternating quantity with always the same line of action. With the rotor at rest, there can be no resultant torque. But if the rotor can be set moving in one direction, then it can be shown that the torque on it due to the 'component' of the field that rotates that way grows at the expense of the torque due to the other component. The starting torque is provided by using an auxiliary set of

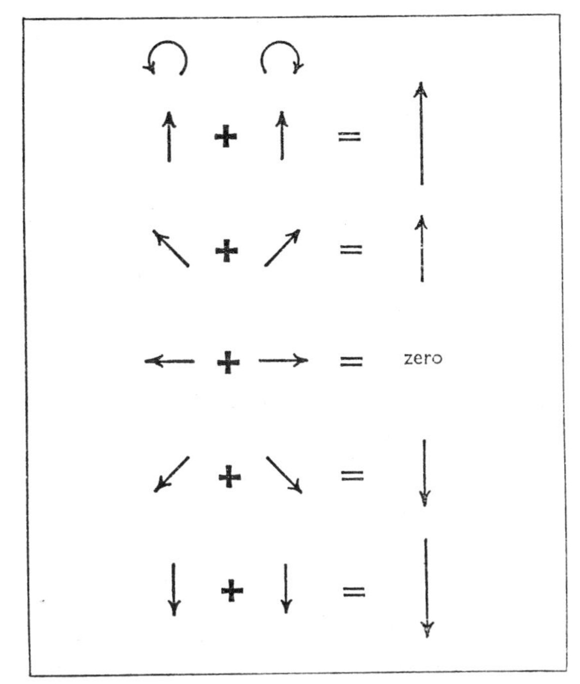

Fig. 9.16 An alternating field can be regarded as the sum of two fields of equal amplitude rotating at the same speed in opposite directions

coils arranged to give a field at right angles to that of the main stator coils (Fig. 9.17). By including a

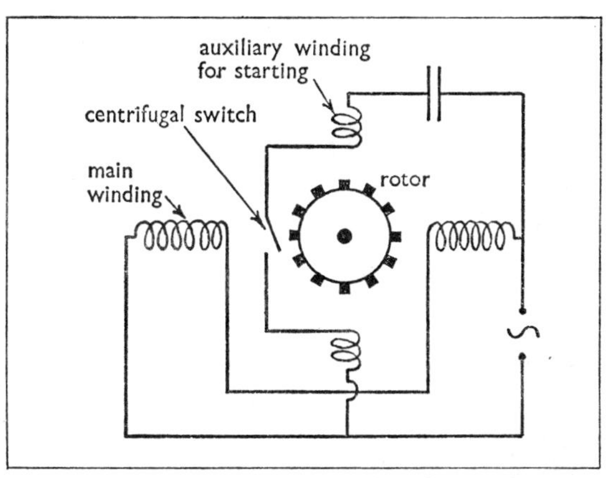

Fig. 9.17 The use of 'phase splitting' to produce a rotating field to start a single-phase induction motor

capacitor in series with the auxiliary windings the current in these is made to lead approximately 90° on the current in the main windings (p. 197). This system is known as *phase splitting*. We have now in effect a two-phase stator, which produces a rotating field—though not of constant magnitude as in the three-phase version. As soon as the rotor has reached sufficient speed to maintain its motion in the right direction, a centrifugal switch operates to disconnect the auxiliary coil. The efficiency of this motor is not as high as that of most other types, but it is adequate for low-power applications.

10: Electrostatics

10.1 Electrical forces and fields

Electrostatics is the study of the behaviour of stationary electric charges and the electric fields associated with them. An *electric field* is a region of space surrounding a system of electric charges; in such a region electric forces will act on any electric charge placed there. Leaving aside the special nuclear forces, there are *three* types of force (including the electric force) that can act between two charged particles—e.g. a proton and an electron:

(1) An *electrostatic* force;
(2) a *magnetic* force, if the particles are in motion (p. 6);
(3) a *gravitational* force.

All three forces are of the type that varies with distance according to an inverse square law—i.e. doubling the distance apart of the two particles reduces the force between them to a quarter as much, etc. But the three types of force are of very different orders of magnitude. The electrostatic force is incomparably greater than the gravitational force—the ratio is about 2×10^{39} for a proton and an electron. The magnetic force depends on the speeds of the particles, but is always less than the electric force—much less, provided the speeds do not approach that of light. (In the limiting case of two charged particles travelling in parallel paths at the speed of light the magnetic and electric forces would be equal.)

The magnitude of the electrostatic force is such that it is impossible to disturb the balance of protons and electrons in a piece of matter to any appreciable extent at all. For instance if we take a piece of copper wire 1 mm in diameter and remove from it 1 in 10^{13} of the electrons it contains, the electric field produced by the surplus positive charge left on it will be sufficient to disrupt and ionize the molecules of air in contact with the wire, thereby discharging it again. With a thicker wire the proportion that could be removed would be even less.

For this reason in experimental electrostatics we handle very small electric charges; a few microcoulombs will be considered large in this context, and we shall often experiment with charges of the order of 10^{-12} coulomb! On the other hand because of the large forces involved the potential differences between such charges and their surroundings will often run into tens of thousands of volts. The act of pulling off a nylon vest may separate only 10^{-10} coulomb of charge, but can produce a p.d. between the vest and its wearer of many thousands of volts, leading to a breakdown of the insulation of the intervening air and the familiar crackling sound! Even in a lightning discharge the quantity of charge concerned is only about 20 coulombs; this much flows through a 250-volt, 1-kW electric fire in 5 seconds.

There is thus a very big difference between the orders of magnitude of the charges and p.d.'s that we handle in electrostatics and in current electricity. This makes experimental work in the two departments of the subject appear very different, and the student must be prepared for some surprises. For instance many substances that we regard as excellent insulators at the low p.d.'s of current electricity will be treated as excellent conductors in electrostatics; a wooden ruler may have a resistance of 10^8 ohms; but this is sufficiently low to conduct away a charge of 10^{-10} coulomb at a p.d. of 10^4 volts in about a microsecond. In fact most ordinary insulators are quite inadequate for work in electrostatics on account of the film of moisture they carry on their surfaces except under especially dry conditions; glass, wood, ebonite, paper, the human body—are usually 'excellent' conductors. Without special drying the only adequate insulators are water-repellent substances, such as paraffin wax, nylon and many modern plastics. In most experiments we can assume that the bench, the walls, the experimenter and the earth are in good electrical contact. The only electric fields that exist will be between these objects and others that have been specially insulated and charged.

Frictional electricity

Historically, electric charges were first produced by friction. When good insulating bodies are rubbed together they are found to become 'charged', that is forces of attraction and repulsion are observed to act between them. It is at once apparent from simple experiments that there are two kinds of charge, and that the two kinds are opposite in the sense that they cancel one another out. This justifies the usual description of the two sorts of charge as being positive and negative. The choice of sign is entirely a conventional matter, and frictional effects were originally used to provide the definition:

> *A* **positive electric charge** *is one of the kind that is produced on the surface of a glass rod that has been rubbed with silk.*

As we now know, choosing our signs in this way leads to the electron having a negative charge and the proton a positive one. We suppose that when glass is rubbed with silk, electrons are transferred from the glass surface to the silk, thereby leaving a deficit of electrons on the glass—that is, a positive charge. It is found that an ebonite rod rubbed with cat's fur acquires a negative charge. Traditionally experiments in electrostatics have been conducted using these two insulating materials—glass and ebonite. However, the experimental difficulties are enormous; the materials have to be carefully dried in a heated cupboard before use, and even then the normal humidity of a temperate climate deposits a conducting layer of moisture on them within a matter of minutes! It is better to use modern insulating materials with water-repellent properties; many of these will retain a surface charge for a considerable period. Polythene, for instance, rubbed with a woollen duster acquires a *negative* charge. Perspex and cellulose acetate (the substance used for making the base of a photographic film) acquire a *positive* charge; and there are many other suitable materials. It is still necessary to establish the sign on one of these insulating materials by comparison with the positive charge that appears (fleetingly) on a glass rod rubbed with silk; but the rest of our experiments can then be conducted using modern plastics.

Normally frictional charges are observed only on good insulators, since with a conductor any charge at once leaks away to earth. But, if a conductor is carefully insulated from the earth and rubbed with an insulator, charges are separated just as with a pair of insulators. Frictional processes in the exhaust of an aeroplane can lead to the separation of

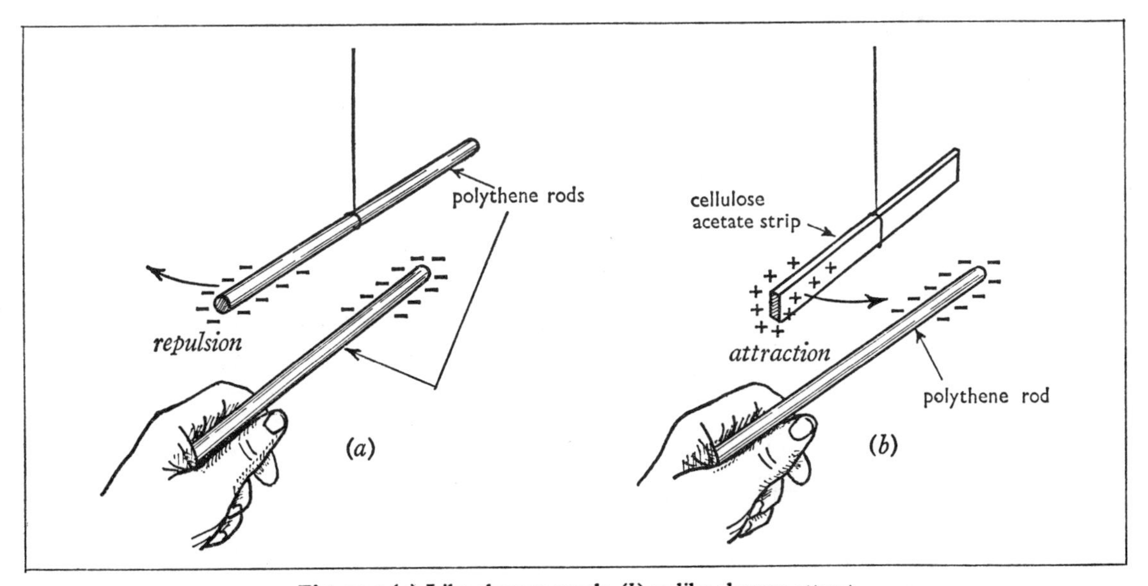

Fig. 10.1 (*a*) **Like charges repel;** (*b*) **unlike charges attract**

large charges; accidents have been caused by the sudden discharge that occurs as the plane touches down. This is now prevented by adding substances to the rubber of the tyres to make them slightly conducting; the charge on the plane then leaks away gently during the first moments of contact with the ground.

The forces acting between different sorts of charge may be investigated as shown in Fig. 10.1. A polythene rod and a strip of cellulose acetate are suspended by threads; they are then charged by rubbing with a duster. The nature of the electrostatic forces may be demonstrated by bringing near them other rods that have been similarly charged. Such experiments will establish that charges are indeed of two kinds and that the forces between them are such that *like charges repel* one another, while *unlike charges attract* one another.

If an insulated *conducting* body is brought near one of the rods, as in Fig. 10.2*a*, the electric forces acting on the free charges in the conductor will cause them to separate as indicated; as we should expect, the rod and the conductor are then attracted together. The same will happen if the conductor is earthed, as in Fig. 10.2*b*; this is really the same as Fig. 10.2*a*, except that the earth itself must now be regarded as part of the conductor. The equal and opposite charges produced on a conductor in this way by the presence of a charged body are called *induced* charges. This is the process that occurs when small specks of paper are picked up by a fountain pen rubbed on the sleeve. The specks of paper are conductors since they are probably moist and are in effective electrical contact with the table-top and so with the earth. Charges are induced on them by the charge on the pen (equal and opposite charges being repelled further away on the earth's surface), and the forces of attraction are sufficient to pick them up.

10.2 Electroscopes and electrometers

The forces that act between stationary charges are used in a number of instruments for measuring p.d.'s by electrostatic means. Such devices are called *electroscopes* or *electrometers*—or, if they are calibrated to read directly in volts, *electrostatic voltmeters.*

The instrument with which most of the early investigations in electrostatics were made is the gold-leaf electroscope (Fig. 10.3). It consists of a rectangular piece of fine gold-leaf attached at its top edge to the flattened side of a brass rod. This is fixed in an insulating plug in the top of a conducting box with glass windows. A small metal disc, called the *cap* of the electroscope, is usually mounted at the top of the brass rod. When a p.d. is produced between the cap and the case, a small charge flows on to the gold-leaf and an opposite charge is induced on the inside of the case. The electrostatic forces draw the gold-leaf out at an angle to the vertical, as shown. The instrument is mostly used

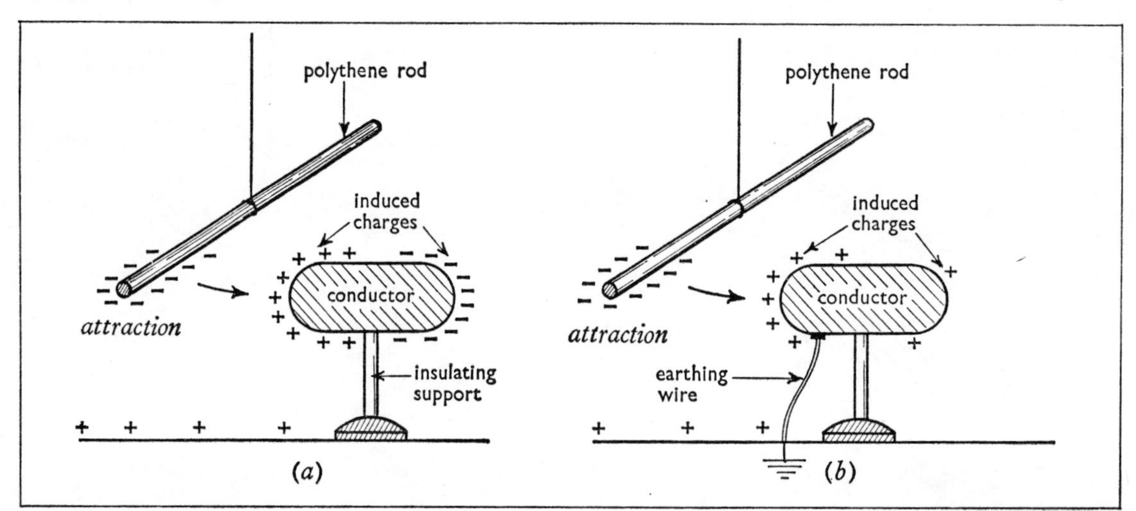

Fig. 10.2 The attraction between induced and inducing charges, with the conductor (*a*) insulated, (*b*) earthed

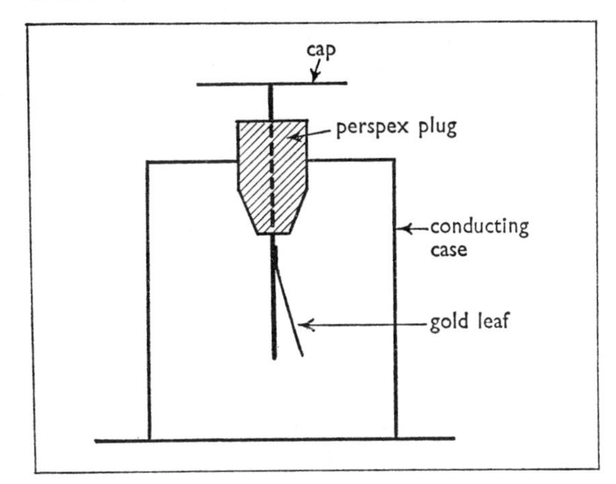

Fig. 10.3 The gold-leaf electroscope

for qualitative observations; but, if necessary, it can be calibrated against another voltmeter so that the angle of deflection indicates the p.d. An electroscope with a scale for reading the deflections should then more properly be called an *electrometer*.

The gold-leaf electroscope suffers from the defect of being rather fragile. More robust is the *Braun electroscope* (or *electrometer*, if it carries a scale). This consists of a light but rigid vane A pivoted at its centre (Fig. 10.4). When uncharged this comes to rest in a vertical position close to the fixed vane B. When a p.d. is applied between the vanes and the metal rim of the instrument, the like charges on the vanes repel one another and the pivoted vane A is deflected as shown.

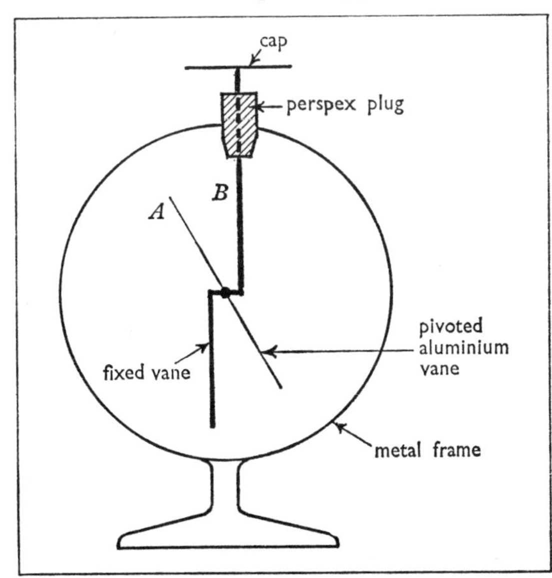

Fig. 10.4 The Braun electroscope

A rather more substantial instrument is the *electrostatic voltmeter* shown in Fig. 10.5. This consists of two sets of parallel plates (or vanes) interleaved with one another. One set of vanes is fixed to the frame of the instrument; the other (sometimes only a single vane) is pivoted on an axle between jewelled bearings, and carries a light pointer.

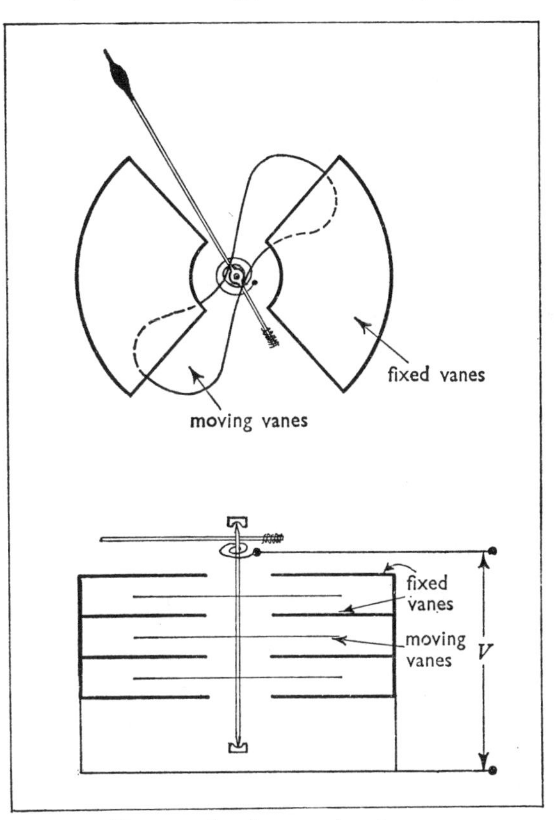

Fig. 10.5 An electrostatic voltmeter

The motion is controlled by two hair-springs; these fix the zero position with the vanes fully opened out. When a p.d. is applied, the electrostatic forces draw together the movable and fixed vanes. The pointer comes to rest at the position in which the deflecting couple due to the electrostatic forces is equal and opposite to the restoring couple due to the hair-springs. These instruments have the great merit of drawing no current from a circuit in which they are connected (apart from the momentary charging current). However, like other electrometers they are essentially high-voltage instruments; it is not easy to design an electrometer of any kind that will deflect at all for p.d.'s below about 50 volts.

In principle it is possible to derive theoretical expressions for the deflecting forces in the various types of electrometer. But in practice the mathematical difficulties are too great, and the instruments must be calibrated against the scales of other voltmeters. However, in the *attracted disc electrometer*, described later in this chapter (p. 169), the geometry is sufficiently simple for the calculations to be done; and this instrument can be used to make absolute measurements of p.d.

The deflecting force of an electrometer acts in the same direction whichever way round the p.d. is connected. It is therefore suitable for measuring alternating p.d.'s as well as steady ones.

Although an electrometer is used primarily for measuring p.d.'s, it may also be used to measure quantities of electric charge. Provided there are no other charged objects producing electrostatic forces in the neighbourhood, the p.d. across the instrument is directly proportional to the charge on its electrodes. Its scale might in principle be calibrated directly to read quantities of charge. It follows that an electrometer can also be used to measure small currents that are flowing onto or off the electrodes; the rate of change of the deflection gives the rate of flow of charge—which is equal to the current. A special development of the instrument for this purpose is the *pulse electrometer* described in the next chapter (p. 186).

To achieve maximum sensitivity in the measurement of charge and current it is necessary to reduce the scale of the apparatus as much as possible and to make the moving part very light. This is the idea behind the design of the *fibre electrometer* (Fig. 10.6). A very fine metal-plated quartz fibre forms the moving part; this is attached at both ends to a wire support which is held by an insulating plug in the metal tube that forms the case of the instrument. The fibre is deflected away from its support when a p.d. is applied between it and the case. The restoring force is provided by the elasticity of the fibre, rather than by gravity as with the gold-leaf electroscope. The tube incorporates a low power microscope with a graduated scale in the focal plane of the eyepiece to enable the position of the image of the fibre to be registered. The

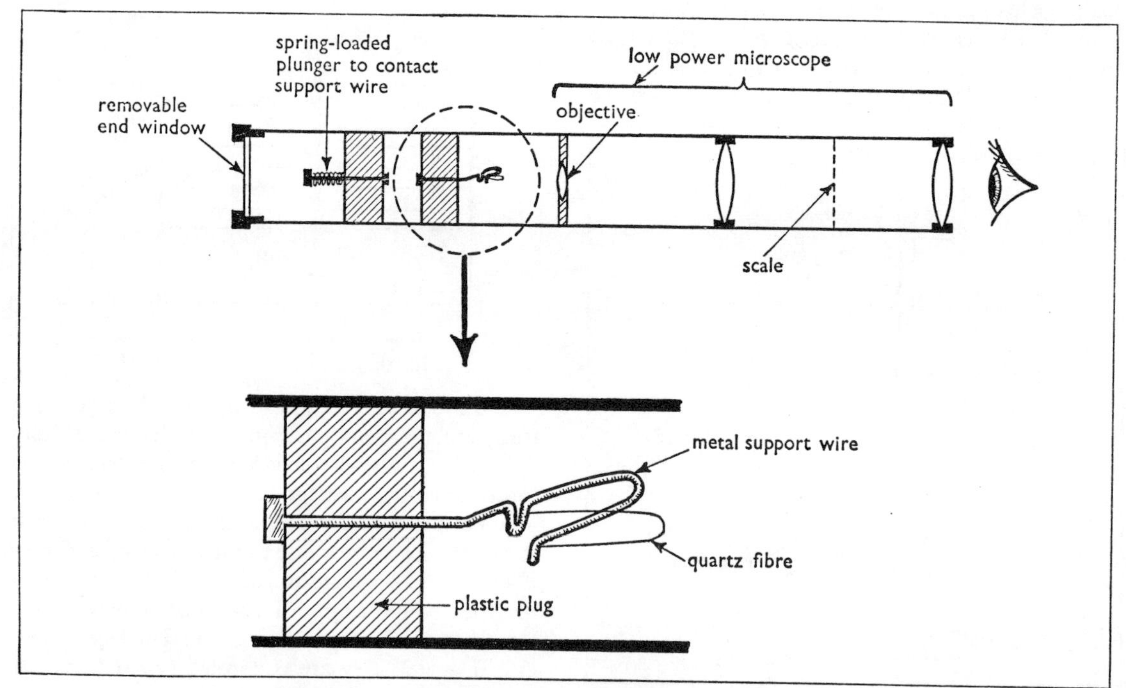

Fig. 10.6 A quartz fibre electrometer or radiation dosimeter

scale usually covers a range of p.d.'s between 100 V and 200 V. But a flow of charge of 10^{-10} coulomb onto or off the electrodes may be sufficient to take the fibre right across the scale. A particular use of this type of electrometer is as a *radiation dosimeter*, for monitoring the total dose of dangerous radiation received by people working with radioactive materials (p. 283).

Some preliminary experiments with a gold-leaf electroscope

(*i*) *The effect of charged objects held nearby.* The electroscope is stood on the bench top so that the case is effectively earthed; and the cap is touched with the finger to ensure that it is discharged. A positively charged rod is now brought near the cap, and the leaf deflects (Fig. 10.7). To a student

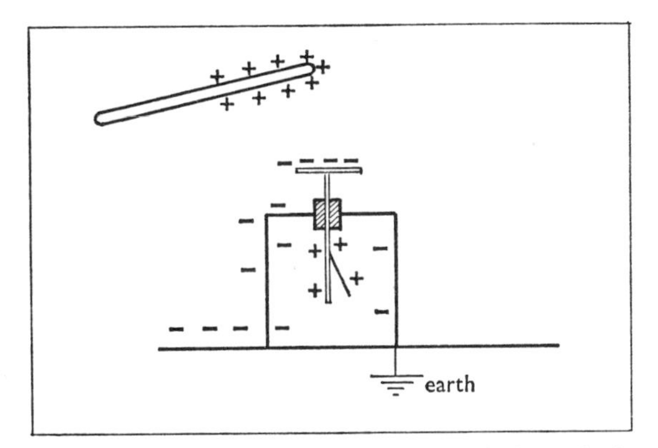

Fig. 10.7 The potential of the insulated electrode is raised by the presence of a positively charged object

brought up on current electricity this behaviour may seem surprising! It must be remembered that the positive charge produces an electric field that tends to drive the charge towards the negative charges on other objects nearby; if a conducting path were provided between the two, the charge would flow along it, and differences of potential would be manifest between the parts of it. However, the electric field is still present even in the absence of free charges for it to act on; and we may describe this by saying that the positively charged rod raises the potential of the space round about it and so of all insulated objects close to it. The potential of the cap and leaf of the electroscope therefore rises (relative to the earth), the field separates the charges on it as shown, and the leaf deflects. When the charged rod is removed, the leaf collapses again; the total charge on the electrode remains zero throughout.

If the electroscope already carries a net charge, bringing up a charged object augments or decreases the deflection according to the signs of the charges. For instance, some of the negative charge on a polythene rod may be given to the electroscope by scraping the rod against the edge of the cap—watch that the leaf is not torn off in the process! When the polythene rod is removed, the leaf remains at a negative potential and duly deflects. (Alternatively, a high-voltage battery could be joined momentarily between cap and base.) If now the positively charged rod is brought up, the leaf starts to collapse, since the positive charge on it reduces the negative potential of the electrode. In a certain position the potential may become zero, and the leaf be undeflected. If the rod is brought any closer than this, the leaf deflects again; but its potential is now positive—in spite of the net negative charge it carries. Evidently the potential of an insulated electrode depends not only on the charge it carries, but also on the distribution of other charged objects round about it.

The behaviour described above demonstrates incidentally that the two kinds of electric charge are indeed *opposites*—in the sense that one cancels out the effect of the other. They may therefore rightly be described as *positive* and *negative.*

Notice the effect of bringing an earthed object (e.g. the observer's hand) near a charged electroscope. Whichever sign of charge the electroscope carries, the deflection *falls*. This is because the charge induced on the earthed object is always of opposite sign to that on the electroscope. When we need to prove that a given body carries, say, a negative charge, it is not enough to bring it near a positively charged electroscope and show that the leaf collapses; an earthed object near the cap might cause the same effect, and the body under test might merely be a poor insulator earthed by the observer's hand. Rather, we must use a negatively charged electroscope and demonstrate an *increase* of deflection when the given body is brought near.

(*ii*) *Charging by induction.* We have already seen that a charged object induces charges of opposite sign on *earthed* objects near it. This effect may be used to induce a charge on an electroscope; it is called *charging by induction.* We proceed as follows (Fig. 10.8).

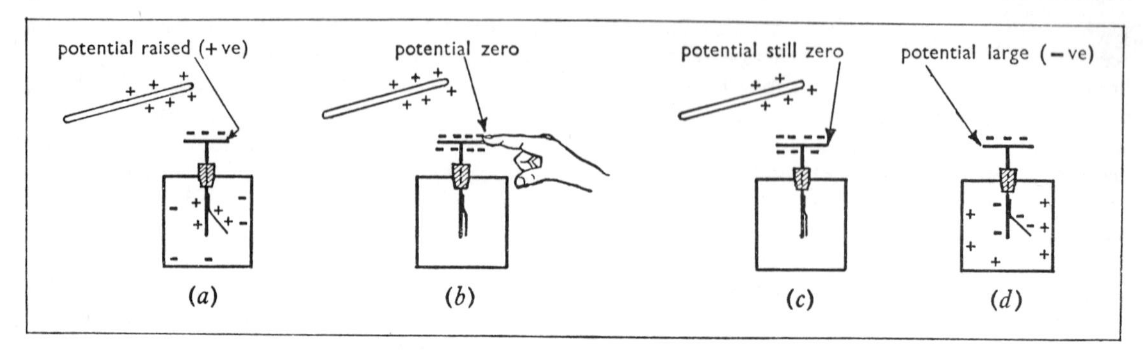

Fig. 10.8 Charging an electroscope by induction

(*a*) Bring up a positively charged rod (say) near the cap; its potential is raised, and a deflection occurs.

(*b*) Earth the cap by touching it with the finger; there is now no p.d. between cap and case, and the deflection falls to zero; i.e. negative charge has been induced on the cap (flowing on to it from the earth), and the combined effect of this charge and the positive charge nearby is to make the potential of the cap zero (earth potential).

(*c*) Remove the finger, so that the induced negative charge remains insulated on the cap; its potential is still zero.

(*d*) Remove the positively charged rod; the negative charge on the electroscope cap now takes it to a large negative potential, and the instrument deflects again. This potential is in fact nearly equal (but of opposite sign) to that initially produced by bringing up the positively charged rod.

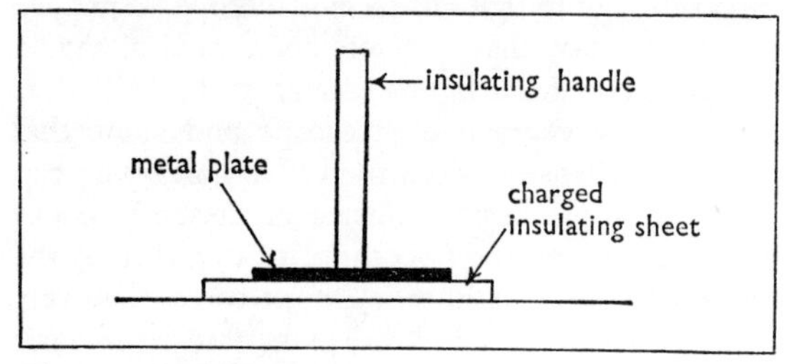

Fig. 10.9 The electrophorus; a large charge is induced on the metal plate when it is earthed

The *electrophorus* is a device for producing charge by this process. The maximum charge that can be produced by induction in one step is an amount equal to the inducing charge, and this the electrophorus very nearly does. It consists of a circular metal plate attached to an insulating handle (Fig. 10.9). This is placed on a plane insulating sheet on the surface of which frictional charges have been produced by rubbing with a suitable cloth. The plate is then earthed with the finger. The frictional charge induces an equal and opposite charge on its surroundings, and owing to its close proximity almost the entire amount appears on the electrophorus. It might be thought that the plate would be in such close contact with the insulating sheet that the charge on the latter would leak away through it. But it seems that even with apparently plane surfaces the contact is only at a few points and a negligible amount of charge is transferred direct. When the electrophorus is removed from the sheet, the large charge it now carries raises its potential to a high value, sufficient perhaps to give $\frac{1}{2}$-inch sparks when it is brought near another conductor. It will be noticed that the same charge on the insulating sheet may be used again and again to induce further quantities of charge on the electrophorus; and this may be transferred repeatedly to some other conductor. This does not contravene the principle of conservation of energy, since work is done on each occasion in withdrawing the electrophorus in opposition to the electrostatic forces that are attracting it to the charged sheet.

Numerous ingenious devices have been invented for mechanically repeating processes similar to the above so that large quantities of charge are eventually produced with correspondingly high potentials. The *Wimshurst machine*, found in most school laboratories, is an example of such a device.

(*iii*) *Testing the quality of an insulator.* The electroscope must first be given a suitable charge. Then the cap is touched with the object under test, the other end of it being held in the hand (earthed). Only if the deflection remains constant for some time can we take the object as an insulator for the purposes of electrostatic experiments. Many substances that we commonly class as insulators—paper, wood, rubber, etc.—will under these conditions appear to be quite good conductors. The quality of many 'insulators' depends on how dry they are. Generally air is an excellent insulator; but if any cause of ionization is present, it enables an appreciable current to pass. This may be demonstrated by bringing a speck of radioactive matter near the electroscope or by 'illuminating' the air around it with a beam of X-rays, when the leaf quickly collapses. The same happens if a match flame is held near the cap; the temperature is high enough in the flame for large numbers of ions to be produced.

10.3 Capacitors

An arrangement of two insulated conductors, such as a pair of parallel metal plates, is called a *capacitor* (the old-fashioned name, *condenser*, is still sometimes used instead). A capacitor can be used in an electric circuit as a temporary store of electric charge. When a p.d. is applied between the plates, positive charge flows onto one plate and negative charge onto the other. If the plates are then insulated, these charges will remain until connection is made between them, thereby discharging the capacitor. In the arrangement shown in Fig. 10.10 the charge stored is very small; at a p.d. of 10^4 volts it might not be more than 10^{-8} coulomb. But this can be increased by bringing the plates closer together and by increasing their area; this may be demonstrated qualitatively with an electroscope.

Two large rectangular plates on insulating stands are arranged to form a parallel-plate capacitor. One of these is earthed, and the other is connected to the cap of the electroscope (the case being earthed). A suitable charge is given to the insulated plate; this charge will remain constant, provided the insulation is good enough. The plates are now moved closer together. The deflection of the electroscope decreases, showing that the p.d. across the capacitor has fallen. Thus, to maintain the p.d. at its original value a *larger* charge would have to be placed on the capacitor plates. The plates

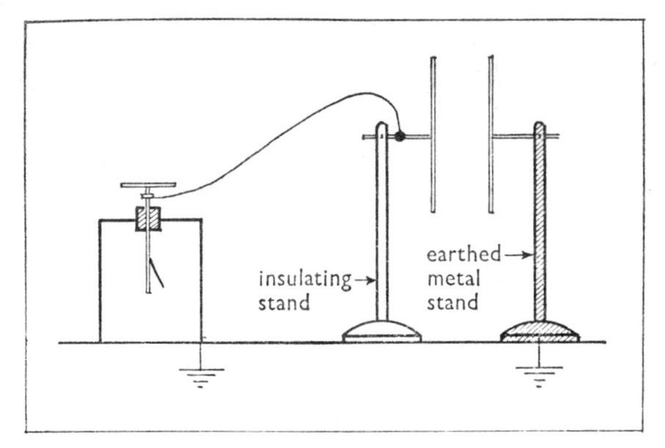

Fig. 10.10 Investigating the properties of a parallel-plate capacitor

are now kept a constant distance apart and moved so that the area overlapping is varied. When this is increased, the deflection of the electroscope again decreases. The insulator between the plates also affects the charge stored at a given p.d. Thus, if a slab of insulator is introduced between the plates, the deflection of the electroscope falls; to maintain the original p.d. the charge stored would therefore need to be increased. Most insulators increase the charge by a factor of between 2 and 10, if they fill the space between the plates, though a few materials produce a much greater effect than this.

The detailed theory of capacitors is discussed in the next chapter. The design of practical capacitors is described in section 11.2 (p. 178).

When a capacitor is joined to a steady source of e.m.f., there is a momentary flow of current onto one plate and off the other. This may be demonstrated with two ballistic galvanometers (p. 132) arranged as in Fig. 10.11. First the switch S is moved to contact A and current flows through the large resistance R; this serves to establish which way the galvanometers deflect when the current flows through them as shown. The switch is now moved to the 'neutral' position between the contacts, and the galvanometers are brought to rest by short-circuiting their coils with the keys K_1 and K_2. When the switch S is moved to contact B, there is a momentary pulse of current through the galvanometers, and the initial deflection is observed to be in the same direction as occurred with the switch at contact A. The deflections show that charge flowed onto the left-hand plate and that an equal quantity of charge flowed off the right-hand

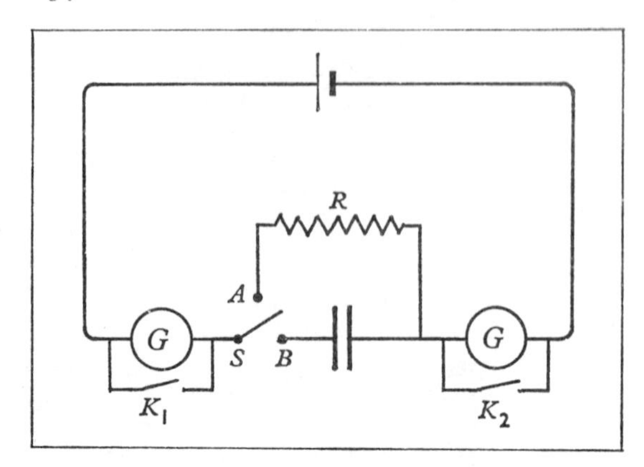

Fig. 10.11 Demonstrating the momentary flow of current onto the plates of a capacitor

one. If now the battery is disconnected and the wires leading to it are touched together, a pulse of current is observed to pass through the galvanometers in the reverse direction, as the plates are discharged.

For a given capacitor the charge stored is directly proportional to the applied p.d. This may be tested with the arrangement of Fig. 10.12. With the switch on contact X the capacitor is charged to the p.d. V indicated by the voltmeter; the switch is now moved quickly to the contact Y, discharging the capacitor through the galvanometer. The maximum deflection θ on the first swing of the coil is noted. According to the theory of the ballistic galvanometer, the first 'throw' of the instrument is proportional to the total charge passed through it (p. 132). A graph of θ against V should then be a straight line passing through the origin.

If an alternating supply is joined to the plates of a capacitor, charge flows onto and off the plates as the p.d. between them oscillates, and a considerable current may flow in the connecting wires. This could well be sufficient to light a large electric bulb placed in series (Fig. 10.13). One of the functions of a capacitor in an electric circuit is to 'block' the flow of direct current while allowing an alternating current to flow in the rest of the circuit. These processes are discussed in more detail in later chapters.

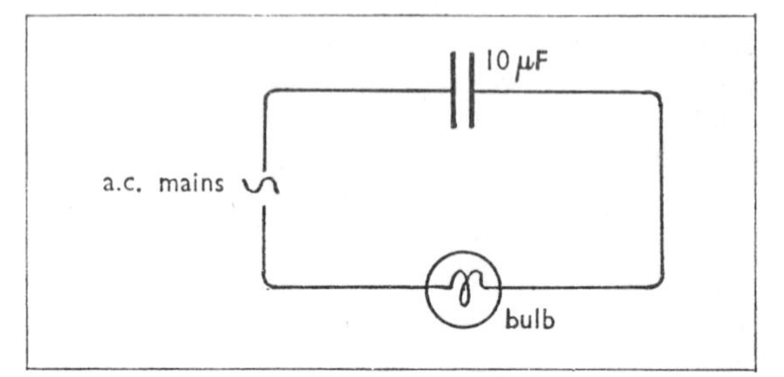

Fig. 10.13 The flow of alternating current 'through' a capacitor

10.4 Electric fields

Electric fields may be described by means of lines of force in much the same way as magnetic fields. The direction of *a line of force* at a point is taken as that of the force that would act on a *positive* charge placed there. A line of force therefore starts on a positive charge and ends on a negative one. It gives the direction in which current would flow (in the conventional sense) under the action of the field if a conducting substance occupied that region of space. Fig. 10.14 shows the nature of the field produced between two parallel plates joined to the terminals of a high-voltage battery. When the plates are first connected, there is a momentary flow of current conveying positive charge to one plate and negative charge to the other. The field between the plates will act on any charged particle

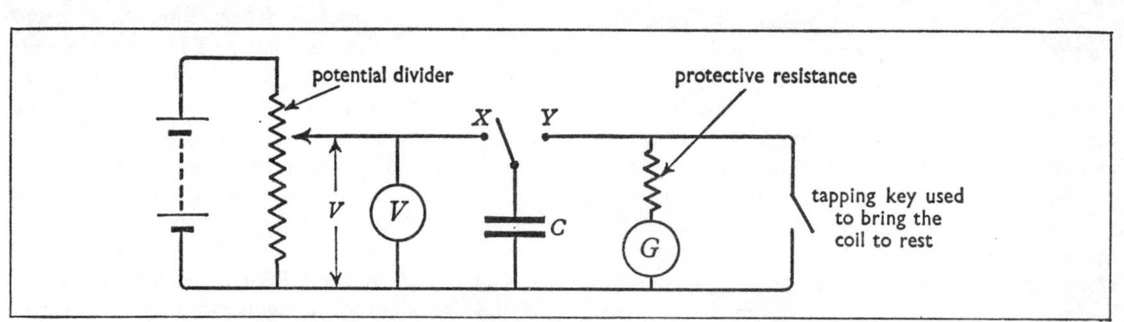

Fig. 10.12 Testing the proportionality of p.d. and charge for a capacitor

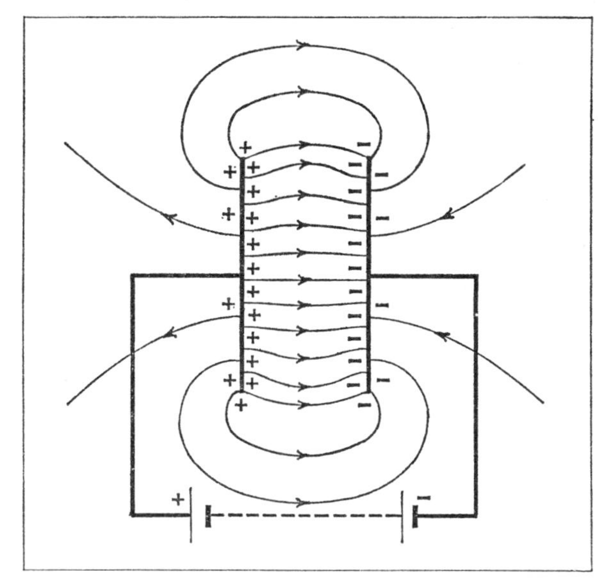

Fig. 10.14 The lines of force of the electric field between two parallel plates

impelling it in one direction or the other according to its sign. This may be demonstrated in a number of ways:

(i) A rod of insulating material is suspended on a thread and given a frictional charge. The end of it is inserted between the two plates. The forces acting on it will deflect it in the direction of the field if it carries a positive charge, and in the opposite direction if the charge is negative.

(ii) A table-tennis ball coated with conducting paint is hung on a nylon thread between the plates. As long as the ball remains uncharged it experiences no force. But as soon as it touches one of the plates (say, the positive one), it acquires from it a positive charge and is impelled towards the other plate; here it delivers up this charge, receiving a negative one instead. It then returns to the positive plate, and so the cycle is repeated indefinitely.

(iii) A mixture of powdered red lead and sulphur may be sprayed (dry) from a nozzle into the space between the plates. The frictional effects in the nozzle give the red lead particles a positive charge, while the sulphur ones become negative. The positive plate is therefore at once covered with the yellow sulphur and the negative plate with the red lead.

(iv) Two small metal plates on insulating handles (usually called *proof-planes*) are introduced into the field, carefully avoiding contact with either electrode; they are then brought into contact with one another face to face. The action of the field induces opposite charges on the two proof-planes, as shown. They are now separated (Fig. 10.15) while still in the field, and withdrawn. The induced charges remain insulated on them; and, using an electrometer, they may be shown to be equal in magnitude and of the signs shown.

(v) It is also possible to plot the lines of force of an electric field in a way rather similar to that used with magnetic fields. A light pivoted fibre tends to set itself along a line of force. If the fibre is conducting, equal and opposite charges are induced in the ends of it; and the electric forces will then align it with the field. Even with an insulating fibre there is some movement of the *bound* charges inside the atoms, and again the electric forces tend to align it with the field. The complete pattern of the lines of force can be shown up by scattering clippings of plastic fibres on an insulating plate in the field; the appearance is similar to that of iron filings scattered on a board placed over a magnet.

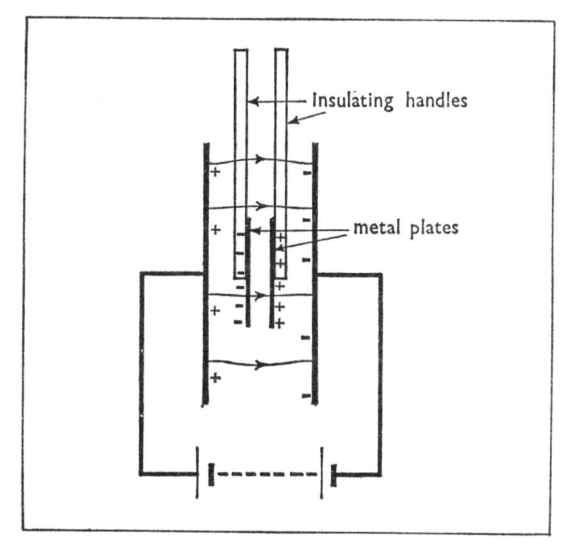

Fig. 10.15 Detecting the field between two parallel plates using a pair of proof-planes

For these demonstrations to be effective a p.d. of several thousand volts is needed between the plates. Modern electronic high-voltage generators are suitable for this purpose. It is also possible to

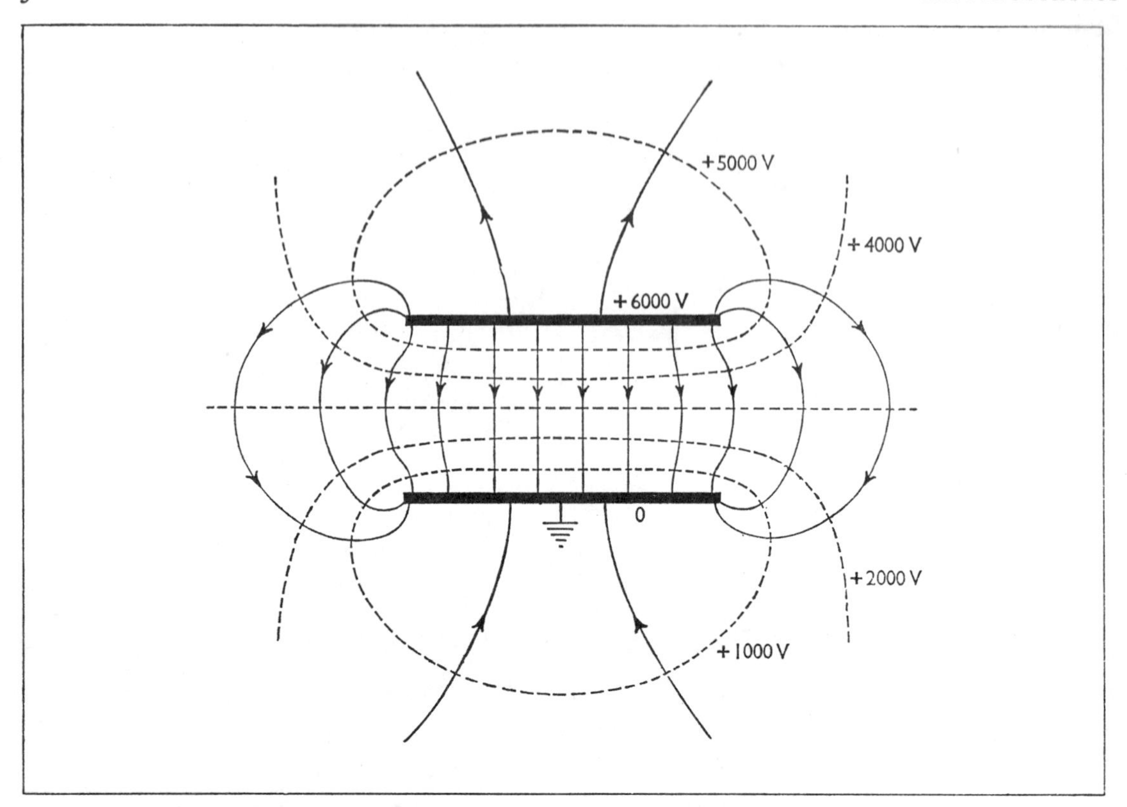

Fig. 10.16 The equipotential surfaces and lines of force between two charged plates

use an induction coil or an electrostatic machine (such as the Van de Graaff generator (p. 167) or the Wimshurst machine). The high p.d. can be dangerous unless connection is made to the plates through a very high resistance (high, that is, by the standards of current electricity). Two inches of coarse string has a suitable resistance for the purpose. Inadvertent contact of the observer with the plates then merely gives a slight tingling sensation!

The *intensity of an electric field*, denoted by E, may be defined in terms of the force that acts on a charged particle placed in it. It is assumed here and elsewhere that the presence of such a charge does not alter the field it is intended to measure.

The **intensity of an electric field** E *at a point is the force per unit charge acting on a positive electric charge placed there.*

From this definition the force F acting on a charge q in an electric field of intensity E is given by

$$F = qE$$

Equipotentials

An alternative way of describing an electric field is in terms of the potential at each point of it. The electric forces operate to drive any positively charged particle along a line of force, and the potential therefore decreases as we pass along one of these lines from a positive charge towards a negative one. We may in imagination connect together all points in the field at which the potential has a given value; the surface formed in this way is called an *equipotential surface*. A diagram showing equipotential surfaces provides a description of the field complementary to the line of force picture we have already used. In Fig. 10.16 the equipotentials are shown by dotted lines. The direction of the line of force at any point is that in which the potential changes most rapidly with distance; this means that the lines of force are always at right-angles to the equipotential surfaces. In the central region between the plates in Fig. 10.16 the field is uniform, i.e., the equipotentials are planes parallel to the plates, and the potential decreases uniformly through the air gap from one plate to the other.

The pattern of the equipotential surfaces round any system of electrodes may be plotted by using an *electrolytic tank model.* A scale model of the electrode system under investigation is immersed in a tank of electrolyte (e.g. weak copper sulphate solution). A low-frequency alternating supply (about 500 Hz) is connected to them. The current at any instant will flow along the lines of force; and the potential will be the same at all points on any given equipotential surface, just as it would if an insulator filled the space. Two metal probes on insulated handles are placed in the tank and connected to a pair of headphones. No sound is heard on the phones as long as the probes are on the same equipotential surface. One probe may then be fixed at a chosen point of the field, and the other is moved about to trace the shape of the equipotential surface through that point. From a map of the equipotentials we can then proceed, if we wish, to draw lines of force normal to them at every point.

The nature of the field between two parallel plates is not actually very difficult to find out by calculation, but with a complicated system of electrodes, such as that in the electron gun of a cathode-ray tube, an experimental approach by means of an electrolytic tank model may be very much simpler. On account of the small mass of an electron its path in an evacuated tube is to a first approximation along a line of electric force, though this is of course modified by the action of any magnetic fields present.

We now need a generalized version of our previous definition of potential (p. 10), so that the variation of potential from one point to another can be related to the intensity of the electric field.

> *The* **electrostatic potential** *at a point is the work done by the forces of the electric field per unit positive charge moved from that point to a place at zero potential.*

For most practical purposes the earth is taken as having zero potential; but for calculations on isolated systems a point at an infinite distance is taken as the zero of potential.

The relation between field intensity and potential may be derived as follows.

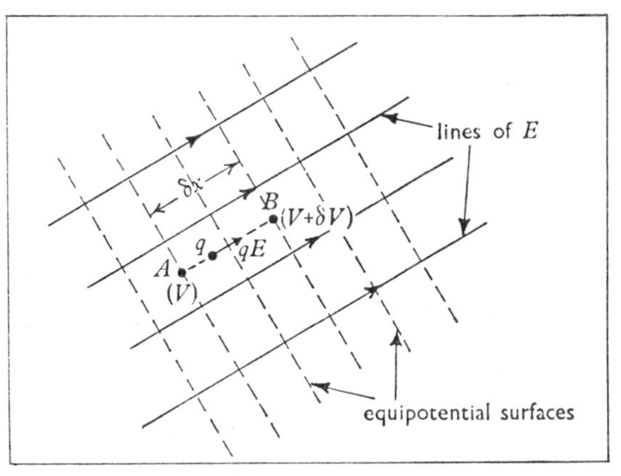

Fig. 10.17 The connection between the field intensity E and the potential gradient

Consider two points A and B a short distance δx apart in an electric field; let E be the component of the intensity of the field in the direction $\overrightarrow{AB}$ (Fig. 10.17). If a charge q moves from A to B, the work done by the forces of the field

$$= \text{force}.\delta x$$
$$= qE \times \delta x$$

If the potential difference between B and A $= \delta V$, we have

$$\delta V = -\frac{\text{work done by field}}{q} = -E\,\delta x$$

(The potential *decreases* in the direction of the field; hence the minus sign.)

In the limit, as $\delta x \to 0$,

$$E = -\frac{dV}{dx} = -(\text{potential gradient})$$

This enables us to calculate the field given the potential or vice versa. Because of this result, in the m.k.s. system, electric fields are usually expressed in *volts per metre*, though it would be equally correct to use *newtons per coulomb*. It is left to the student to show that these units are dimensionally equivalent.

The field between two parallel plates

Plotting the line of force diagram or the pattern of equipotential surfaces shows that the field between two parallel plates is a uniform one; i.e. the potential decreases uniformly as we pass across the gap from one plate to the other. We therefore have in this case

$$E = \frac{\text{p.d.}}{\text{distance apart}}$$

A knowledge of the potentials at two points enables us to calculate the total work done by the electric forces in moving a charged particle from one point to the other, without having to work out the field intensity at intermediate points. Suppose a particle of charge q is moved from a point at a potential V_1 to another at a potential V_2. From the definition of potential, the p.d. $(V_1 - V_2)$ is the work done by the field on unit charge moved from one point to the other.

$$\therefore \text{ work done on charge } q = q(V_1 - V_2)$$

If q is given in *coulombs* and the p.d. in *volts*, then the work done is obtained in *joules*.

Example. Calculate the velocity of an electron as it strikes the anode of a radio valve, if the p.d. between anode and cathode is 150 volts; assume that the velocity of the electron is negligible as it emerges from the cathode, and that its mass is $9{\cdot}1 \times 10^{-31}$ kg, and charge $-1{\cdot}60 \times 10^{-19}$ coulombs.

The work done on the electron by the field

$$= \text{charge} \times \text{p.d.} = 1{\cdot}6 \times 10^{-19} \times 150 \text{ joule}$$
$$= \text{the kinetic energy } (\tfrac{1}{2}mv^2) \text{ gained}$$
$$= \tfrac{1}{2} \times 9{\cdot}1 \times 10^{-31} \times v^2$$

where v is the velocity.

$$\therefore\ v^2 = \frac{2 \times 1{\cdot}6 \times 10^{-19} \times 150}{9{\cdot}1 \times 10^{-31}}$$

$$\therefore\ v = 7{\cdot}3 \times 10^6 \text{ m s}^{-1}$$

The electron-volt

The energy acquired by a charged particle accelerated by an electric field in a vacuum depends only on its charge and the p.d. through which it falls. A proton, having a larger mass than an electron, would acquire a smaller velocity with the same potential difference, but would gain the same kinetic energy. In any case in many atomic and nuclear calculations we are not much concerned to know the velocity; only the energy is important. It is therefore convenient in such applications to define a new unit of energy specially adapted for this purpose; it is called *the electron-volt (eV)*.

The ***electron-volt*** *is the quantity of energy gained by an electron in falling through a p.d. of 1 volt.*

Since the electronic charge $= 1{\cdot}6 \times 10^{-19}$ coulombs

$$1 \text{ eV} = 1{\cdot}6 \times 10^{-19} \text{ joule}$$

Some idea of the magnitude of the electron-volt may be gained if we say that the energy of vibration of an atom of a solid at room temperature (say, an atom on the page you are reading) is about 0·03 eV.

Example. Calculate the energy in eV of an α-particle (helium nucleus) accelerated through a p.d. of 4 million volts.

(The charge on an α-particle $= 2e$, where $-e$ is the charge of an electron.)

$$\text{The kinetic energy gained} = 2 \times 4 \times 10^6 \text{ eV}$$
$$= 8 \times 10^6 \text{ eV}$$
$$= 8 \text{ MeV}$$

10.5 The charge of the electron

Millikan's experiment

The principle of this experiment is to measure the electrostatic force acting on a charged oil drop in a known electric field, and so to find the charge it carries. The drops used are of microscopic size and the charges they carry are many orders of magnitude smaller than the smallest charge that could be measured by any other means.

Millikan found that the charges on such oil drops are always integral multiples of a certain smallest quantity of charge e; i.e. the charges obtained were

$$\pm e, \quad \pm 2e, \quad \pm 3e, \ldots \quad \text{and so on}$$

The charge was never $\frac{1}{2}e$, $1\frac{1}{2}e$ or any other fractional multiple. The experiment demonstrates conclusively the atomic nature of electric charge. Millikan assumed that e was the magnitude of the charge of one electron; and this is in agreement with the results of other methods of measuring the electronic charge. Millikan's experiment has not proved eventually to be the most precise method available. But other methods do no more than measure the *average* charge of electrons. It is the special achievement of Millikan's experiment that it demonstrates that all electrons (and presumably protons too) carry the *same* charge.

A simplified form of Millikan's apparatus is shown in Fig. 10.18. The electric field is produced by a steady p.d. applied between two horizontal plates P and Q; these are held exactly parallel by an accurately made spacer ring of insulating material. The ring has two windows let into it, as shown. The top plate has a small hole in the centre through which oil drops are allowed to fall from a spray. Frictional effects in the nozzle of the spray result in at least some of the oil drops being charged. A beam of light is concentrated into the space between the plates through one window; and a low-power microscope at the other is used to observe the drops by means of the light scattered from them. They are seen as sharp points of light against a relatively dark background. The microscope has a graduated scale in its eyepiece, by means of which distances in the object plane (where the drops are located) can be measured.

When the field is switched on between the plates, the motion of some of the slowly falling drops is reversed because of the electrostatic forces acting on them; a suitable drop is selected, and by switching off and on alternately it may be held near the centre of the space until all other drops have landed on one plate or the other. Measurements are now conducted on this single oil drop. The oil must be of the type used in vacuum apparatus; this has a very low vapour pressure so that evaporation of the drop is slow, and its weight remains practically constant for a considerable period.

First, with the electric field switched off (and the plates connected together) the drop is allowed to fall under the action of gravity. Using a stop-watch the time is measured for the spot of light to move between two selected divisions of the eyepiece scale; hence the velocity of the drop is obtained. The drop is so small that it reaches its terminal velocity almost at once; the viscous force due to the air is then exactly equal to the weight. According to Stokes's law the force F acting on a sphere of radius a moving with velocity v through a medium of viscosity η is given by

$$F = 6\pi a\eta v$$

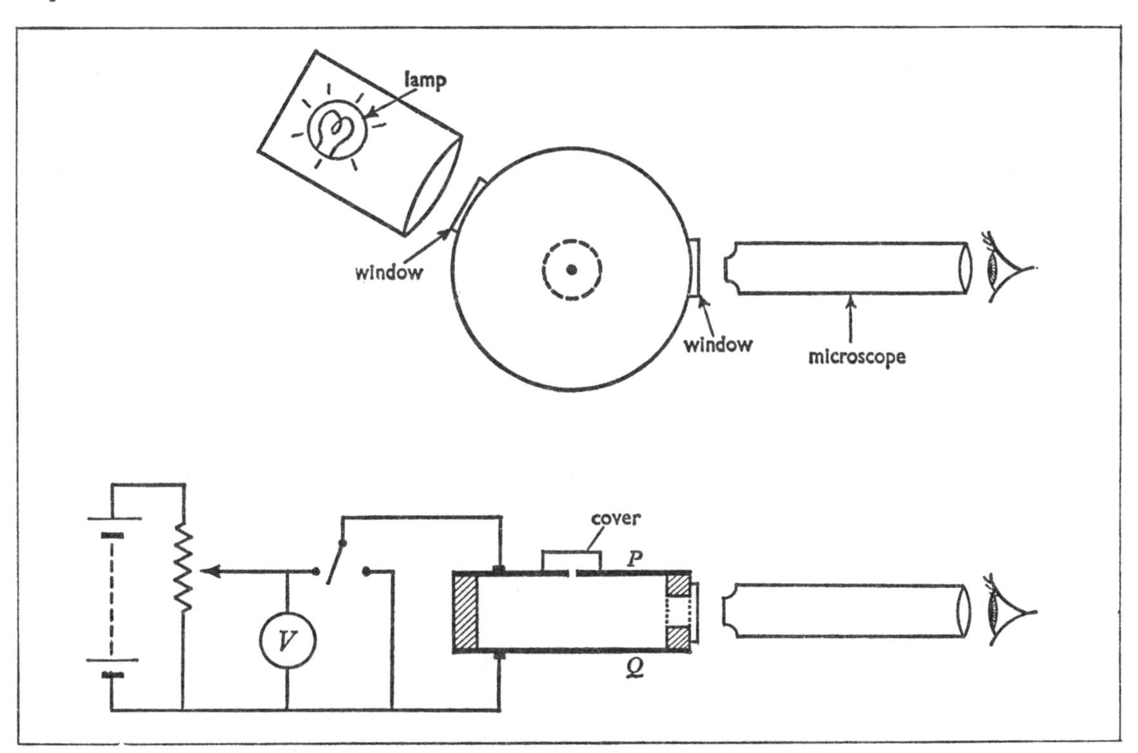

Fig. 10.18 A simplified modern form of Millikan's experiment

Now, the force of gravity on the drop $= \frac{4}{3}\pi a^3 \rho g$ where $\rho =$ the density of the oil.

The upthrust of the air on the drop

$$= \text{the weight of air displaced}$$
$$= \tfrac{4}{3}\pi a^3 \sigma g$$

where $\sigma =$ the density of the air. The apparent weight w of the drop is then given by

$$w = \tfrac{4}{3}\pi a^3 g(\rho - \sigma)$$

If the terminal velocity $= v_1$, we have

$$w = \tfrac{4}{3}\pi a^3 g(\rho - \sigma) = 6\pi a \eta v_1$$

$$\therefore \; a^2 = \frac{9\eta v_1}{2g(\rho - \sigma)}$$

knowing a, we may then also calculate w.

The field is now switched on in such a direction as to oppose the force of gravity on the drop. A first approximation to the charge q on the drop may be obtained by adjusting the p.d. between the plates until the drop is held exactly stationary. If the electric field intensity is E, we then have

$$\text{the electric force} = Eq = w = \tfrac{4}{3}\pi a^3 g(\rho - \sigma)$$

Now, $$E = \frac{V}{d}$$

where $V =$ the p.d., and $d =$ the separation of the plates.

$$\therefore \; q = \frac{4\pi d a^3 g(\rho - \sigma)}{3V}$$

However, this adjustment is not easy to make, and it is usually better to use a rather larger field and then measure the speed v_2 as the particle ascends between the plates again. Then

$$Eq - w = 6\pi a \eta v_2 = w . \frac{v_2}{v_1}$$

$$\therefore \; Eq = \frac{V}{d} . q = w . \frac{v_1 + v_2}{v_1}$$

$$\therefore \; q = \frac{wd}{V v_1}(v_1 + v_2)$$

The same drop is timed again and again as it falls and rises between the plates, and accurate values of v_1 and v_2 are obtained. The charge on the drop can then be changed by holding a radioactive source nearby. This slightly ionizes the air between the plates. The drop will soon collide with one or more ions and its charge is then changed, and with it the value of v_2. The measurements are repeated with many different quantities of charge on the drop—and again with other drops of differing sizes. It is found that, within the limits of error expected, the quantities of charge measured are always integral multiples of the smallest charge than can be obtained on a drop. This smallest quantity is taken to be the charge of the electron.

The accuracy of the result depends on the knowledge of the viscosity of air, which is difficult to measure with high accuracy. Also it can be shown that Stoke's law is not strictly applicable when the diameter of the sphere is comparable with the mean free path of the gas molecules.

The best values of e, the electronic charge, have been obtained indirectly from measurements of atomic quantities which theory shows to depend on e. One example has been given on p. 36. According to the ionic theory,

$$e = \frac{F}{N}$$

where $F =$ Faraday's constant, $N =$ Avogadro's constant. F may be found very precisely from electrolytic experiments. N is best found by X-ray diffraction experiments on suitable crystals (e.g. calcite); this gives a very precise measurement of the spacing of the atoms in the crystal. From this we can work out the number of atoms per unit volume and hence N, the number of atoms per kg-atom (p. 276). The presently accepted value is

$$= 1{\cdot}602 \times 10^{-19} \text{ coulomb}$$

10.6 The inverse square law

The forces between electric charges were first studied experimentally by the French scientist Coulomb. His measurements were made with a torsion balance, but we shall describe a similar technique using a simple spring balance.

A metallized pith ball is suspended by a nylon thread from a fine spring (e.g. a glass spring). A second similar pith ball is mounted on an insulating support vertically beneath (Fig. 10.19). Any force between the two pith balls can be measured by the extension of the spring, as shown. The scale reading is first taken with the pith balls uncharged. Then they are given suitable *opposite* charges; and the force F attracting them together is measured for different values of the separation d. By repeating the readings we can check that the charges have not altered during the course of the experiment.

It is found that the force obeys an *inverse square law*; i.e.

$$F \propto \frac{1}{d^2}$$

The method is not capable of high accuracy. For one thing each charge slightly alters the distribution of charge on the other pith ball, so that the effective centres of charge are not at the centres of the spheres. As with many other fundamental laws, indirect tests are more satisfactory than the direct test outlined above. The best methods have been those based on the experiments of Faraday (p. 164).

It may also be assumed that the forces between charges are additive; e.g. doubling one charge doubles the force. This means that the force should be proportional to *each* of the charges. These results may be summarised as follows.

The **law of force for electric charges:** *The force with which one electric charge acts on another is proportional to each of the charges and inversely proportional to the square of their distance apart.*

If the two charges are q_1 and q_2, the symbolic statement of this law is

$$F \propto \frac{q_1 q_2}{d^2}$$

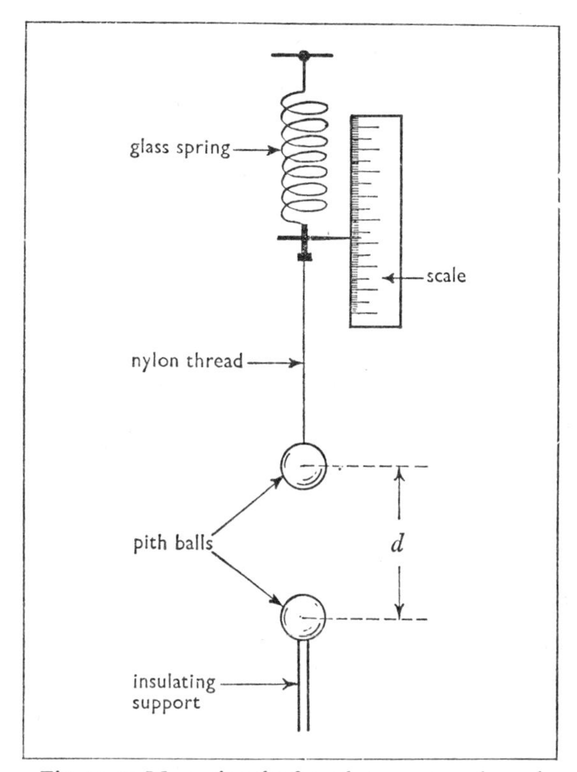

Fig. 10.19 Measuring the force between two charged pith balls

The effect of the insulating medium

In magnetism we found that most substances through which the field acts have little effect on its strength; only with the few ferromagnetic materials was any considerable effect produced. In electrostatics the position is quite different. All insulators produce a measurable effect on an electric field; this is small in gases, but in solids and liquids the field is usually *reduced* by a factor of between 2 and 10, and in a few materials by very much more than this.

In atomic terms this may be explained as follows. In an insulator the charges are *bound* in the sense that they cannot move away from the site in the crystal lattice to which they belong; but they are capable of small amounts of movement under the action of an electric field, the positive charges (atomic nuclei) in one direction and the negative ones (electrons) in the other. Suppose we have a slab of insulator between two parallel plates (Fig. 10.20*a*). When a p.d. is applied between them, the electric field acts on the bound charges in the insulator slightly displacing them. There is thus a small electric current in the insulator, called a *displacement current*, which flows momentarily while the electric field is changing; as soon as the field is steady, no further displacement of the charges takes place. The effect of this is to produce two layers of bound charges on the surfaces of the insulator; the field of these partly cancels out the electric field in the slab. The bound charge on the surface of the insulator is always less than that on the neighbouring metal plate, so that the field in the insulator is never completely cancelled. When the charges in an insulator are displaced in this way, we say that it is *polarized*.

Although a steady current cannot pass through an insulator, an electric field can; for this reason, an insulator is often referred to as a *dielectric*. By contrast, a piece of metal acts as a screen to an

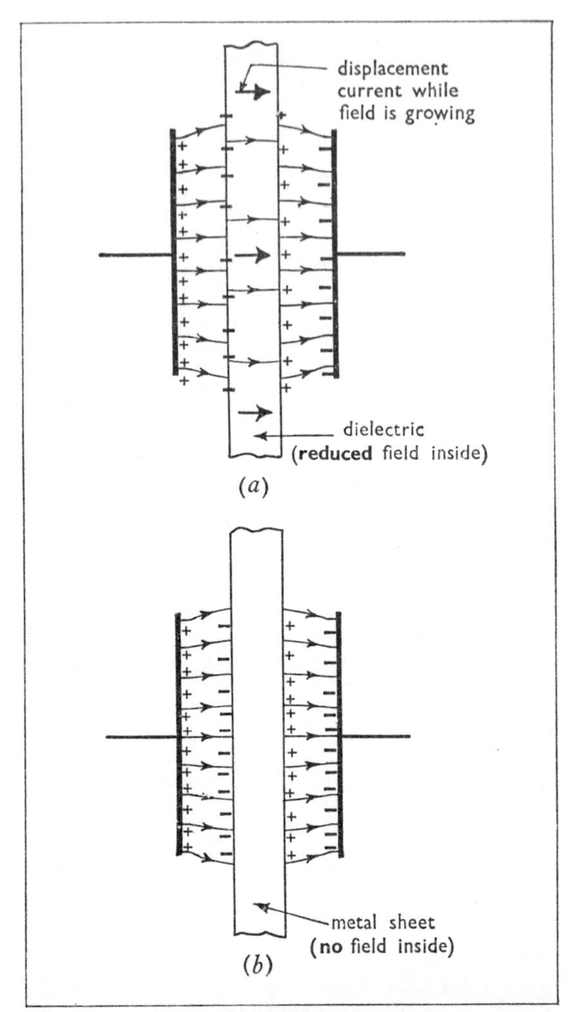

Fig. 10.20 (*a*) The polarization of a dielectric, leading to a *reduction* of the field inside it
(*b*) The separation of charges in a metal, producing *zero* field inside it

electric field. In a metal some of the charges (electrons) are completely free to move; under the action of an electric field they will continue in motion until they have taken up positions that reduce the field inside the metal everywhere to zero. Only while a current is flowing is there any electric field inside a conductor. When a sheet of metal replaces the slab of insulator between the plates of the capacitor (Fig. 10.20*b*), the charges induced on the surfaces of the sheet are *equal* to those on the neighbouring plates of the capacitor, thus reducing the field inside the sheet to zero.

The effect of an insulating medium around two charges on the force between them is always to *reduce* the force compared with its value in a vacuum. This is allowed for in calculations by introducing a property of the medium that we call its *permittivity* (usually denoted by the Greek letter ε). The permittivity of a vacuum is denoted by ε_0 (pronounced 'epsilon-nought'); it may be also called the *electric space constant.* It is convenient to define permittivity in such a way that a dielectric with high permittivity is one in which *large* polarization occurs for a given electric field; in such an insulator electric forces and fields are considerably *reduced* compared with a vacuum. To express this concept we need to include the permittivity ε in the *denominator* of the formula for the force F between two charges.

In this book we are using a 'rationalized' system of units (p. 69); this means that the value of ε_0 is adjusted so that π disappears from certain commonly used formulae involving *uniform* fields; it then re-appears instead in formulae for non-uniform fields. This is found to give a general simplification of the theory, particularly in advanced applications. Rationalizing is brought about by including a factor 4π with ε_0 in the denominator. (This has the effect of multiplying ε_0 by the factor $1/4\pi$ compared with the value it would have in an 'unrationalized' m.k.s. system.) We therefore write

$$F = \frac{q_1 q_2}{4\pi\varepsilon_0 d^2} \text{ in a vacuum}$$

In any other medium the force F is reduced by a factor ε_r (compared with its value in a vacuum), where ε_r is called the *relative permittivity* (or sometimes, the *dielectric constant*) of the medium. In the general case we thus have

$$F = \frac{q_1 q_2}{4\pi\varepsilon d^2} \text{ in any medium}$$

where $\varepsilon = \varepsilon_r \varepsilon_0$. To distinguish ε from ε_r we call the former the *absolute permittivity* of the medium. The relative permittivity ε_r is a dimensionless quantity, and is equal to 1 for a vacuum, by definition. The absolute permittivity ε has the dimensions of $q_1 q_2/F d^2$, and must be found by experiment. We already have units for charge, force and distance, and so in the m.k.s. system ε will be expressed in coulomb^2 newton^{-1} metre^{-2}. Any dimensionally equivalent combination of units may of course be used; the most common nowadays is one employing the unit of capacitance, the farad, enabling ε to be expressed in farad metre^{-1} (p. 177).

In principle the apparatus in Fig. 10.19 can be used to measure ε_0. The force F between the pith balls when they are a known distance d apart is found from the extension of the spring; and the charges q_1 and q_2 may be measured by suspending the pith balls in turn from glass springs in the known field between two parallel plates. Needless to say, this is a method of low accuracy, and can only give an order of magnitude for ε_0. (The experiment is performed in air instead of a vacuum; but this makes negligible difference, since the relative permittivity of air is almost unity.) A more precise experiment using a parallel-plate capacitor is described on p. 177. The result obtained is

$$\varepsilon_0 = 8{\cdot}84 \times 10^{-12} \text{ coulomb}^2 \text{ newton}^{-1} \text{ metre}^{-2}$$

The smallness of ε_0 is an indication of the enormous magnitude of the electrostatic forces that act between charged particles (ε_0 occurs in the *denominator* of the expression for F). It may be instructive to work out the force that would act between two isolated charges each of 1 coulomb placed 1 km apart in a vacuum.

Now 1 km = 10^3 m

$$\therefore\ F = \frac{1}{4\pi \times 8{\cdot}84 \times 10^{-12} \times (10^3)^2}$$
$$= 9 \times 10^3 \text{ newtons}$$
$$\simeq 1 \text{ ton-f!}$$

The field of an isolated charged particle

By symmetry the lines of force near an isolated small charge q must lie along radii, and the equipotential surfaces must form spheres concentric with the charged particle. The intensity of the electric field E is the force per unit charge acting on *another* small (positive) charge placed in the field. At a distance r from the particle we therefore have

$$E = \frac{q}{4\pi\varepsilon r^2}$$

This gives the field intensity in volts per metre when q is in coulombs and r in metres.

The potential V at a distance r from the particle may be calculated as follows. The connection between E and V is

$$E = -\frac{dV}{dr} \qquad \text{(p. 157)}$$

Integrating between the limits r and ∞ and taking the potential as zero at infinity, we have

$$V = \int_0^V dV = -\int_\infty^r E\,dx$$
$$= -\int_\infty^r \frac{q\,dx}{4\pi\varepsilon x^2} = \frac{q}{4\pi\varepsilon r}$$

In principle this last result enables the potential due to any system of charges to be calculated; but the mathematics involved is rarely straightforward. We do not often in practice have to deal with a system of *point* charges, but more usually with charges distributed over the surfaces of conductors. The calculation of potential then involves difficult integrations, except in those cases where there is some obvious symmetry in the system. None the less the calculation of potential is almost always simpler than the direct calculation of the field intensity E. The reason for this is that the intensity E is a *vector*, while the potential V is a *scalar*. Thus, the addition of the intensities due to a number of charges is a tedious process involving the parallelogram rule; but the potentials due to the charges may be compounded by simple addition.

10.7 The fields around conductors

Many of the features of electric fields may be understood without resort to detailed mathematics. In the absence of any source of e.m.f. producing a continuous supply of electrical energy, the charges in a system of conductors reach equilibrium under the action of the electric field in a very short time (less than 10^{-12} s in most cases). Without, at this stage, considering rigorous proofs, we can draw the following conclusions about the nature of the static electric fields in and near conductors.

(*i*) *The field is zero inside a conductor.* If not, the free charges in it will move under the action of the field until the distribution of charge is such as to make this so. It follows that the potential inside the material of the conductor is everywhere constant. In particular, the surface of the conductor is an equipotential surface.

(*ii*) *The field at the surface of a conductor is everywhere normal to the surface.* If not, it will have a component parallel to the surface and will set the free charges in the surface in motion. It follows that the equipotential surfaces very close to a conductor are everywhere parallel to the surface.

(*iii*) *The charge on a conductor resides entirely on its surface.* The distribution of charge must be such as to make the field zero everywhere inside the material of the conductor. If some of the charge on it came to rest somewhere inside the material, there would be lines of force running from this charge to some region containing charge of opposite sign; i.e. there would be an electric field in the material of the conductor, which is impossible under static conditions. It will be shown later (p. 171) that this result is a direct consequence of the inverse square law for the force between electric charges. What it amounts to is this: the description of a field in terms of lines of force is only fully consistent if the law of force in that field is an inverse square law. With a different law we might have to draw lines of force starting on positive charge and ending in empty space; or, we might have a region of positive charge in which no lines of force started at all. In actual fact, with a consistent line of force diagram, each line originates on a definite quantity of positive charge and ends on an equal negative charge. If the field is zero throughout a certain volume (such as the material of a conductor), then the charge contained is zero also.

(*iv*) *The field is zero inside a hollow closed conductor.* We have seen that the charge on a solid conductor resides entirely on its surface. Material might well be hollowed out from the interior of the conductor without affecting its surface, and therefore without altering the distribution of charge on its surface or the fields in and around it. The field should therefore be zero inside the hollow space—provided, of course, no charge is placed on an insulated support inside it. It follows that the potential inside such a hollow space is everywhere constant. This conclusion also can be shown to be a direct consequence of the inverse square law.

The prediction may be tested by standing an electroscope inside a closed metal box; the box has windows of copper gauze through which the instrument can be observed. Large electric fields may be produced near the box; or it may be stood on an insulating slab and raised to a high potential. In no case will any deflection be observed; there cannot therefore be any electric field inside the hollow space.

The result is of practical importance in the design of high-voltage apparatus. A hollow metal compartment may be raised to a high potential, and the apparatus and observers inside it are not in any way affected by the large fields outside. The ability of a closed metal screen to shield apparatus inside it from stray electric fields outside (and vice versa) is also made use of in high-frequency radio equipment. Many of the components—coils, valves, etc.—are mounted inside metal 'cans'; without these the coupling of the stray electric fields between different parts of the circuit would make its behaviour quite unpredictable. At high frequencies the cans also provide shielding from stray magnetic fields (p. 102).

(*v*) *There is no charge on the inside surface of a hollow conductor.* This follows from the previous result. If there is no field inside the hollow space and none inside the material of the conductor, there can be no charge anywhere except on the outside surface.

For the experimental test of this we use a deep metal can on an insulating support. Faraday did his original experiments with an ice-pail, which happened to be to hand at the time. So the cocoa tins, etc., used in doing this experiment are traditionally referred to as 'ice-pails'! The ice-pail is given a charge. We now make test of the charge on different parts of it with a proof-plane (Fig. 10.21). When placed on a conducting surface the proof-plane becomes for the time being electrically part of it, and collects a sample of charge from it. The

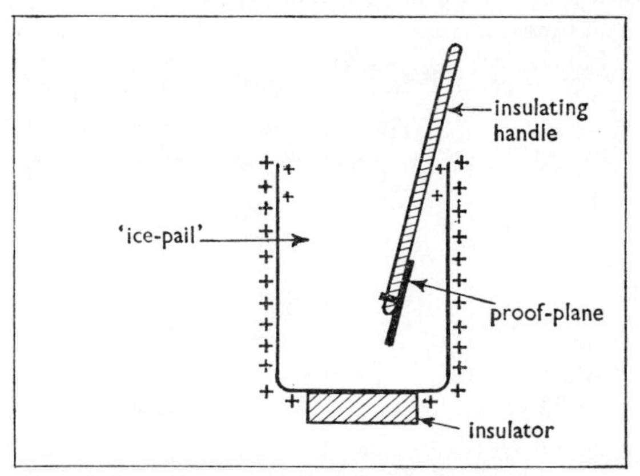

Fig. 10.21 No charge can be collected from the inside surface of a hollow charged conductor

charge may then be tested by touching the proof-plane on the cap of an electroscope. It will be found that no charge can be collected from the *inside* of the ice-pail, provided the proof-plane is not touched too near the lip of it. The charge will be found to reside entirely on the outside. A word of warning here: if there is any frictional charge on the handle of the proof-plane, this will induce an opposite charge on the metal disc itself, whenever it touches another metal object. Great care must therefore be taken to keep the handle free of frictional charges. If necessary it may be discharged by holding it for a few seconds near a small flame.

A useful experimental technique may be derived from this result: if we wish to communicate the *entire* charge on a proof-plane (or any other metal object) to a hollow conductor, this may be done by touching it on the *inside* surface of the latter; touching it on the outside surface would cause only part of its charge to be given up.

(*vi*) *The total induced charge is equal and opposite to the inducing charge.* An insulated positive charge induces a negative charge on its surroundings. If our line of force picture is consistent, we expect to find the lines of force starting on the positive charge ending on an equal quantity of negative charge.

To test this result experimentally the ice-pail is connected to the electroscope; this is often done by standing it on top of the cap. A charged metal object is then lowered on an insulating handle inside the ice-pail without touching it. This induces an opposite charge on the inside surface, and leaves a charge on the outside surface of the same sign as the inducing charge; a proportion of this charge is shared with the electroscope, and a deflection is observed. If the charged object is moved about inside the ice-pail (Fig. 10.22), no change is observed in the electroscope deflection, showing first that the induced charge is not affected by the position of the inducing charge, provided it is well inside the ice-pail. Now the charged object is touched on the inside of the ice-pail. We know from the previous experiment that the total charge inside the ice-pail is now zero, and careful observation shows that the deflection of the electroscope has not changed. It follows that the induced and inducing charges exactly cancelled one another at the moment of contact, and that therefore they were equal and opposite.

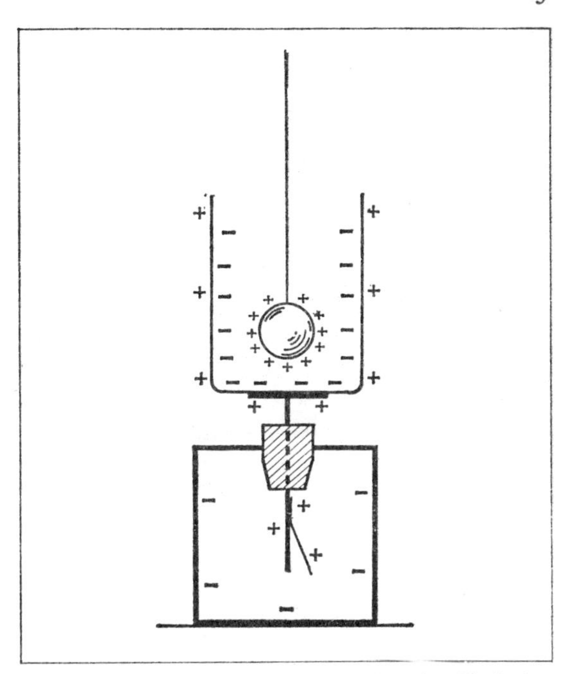

Fig. 10.22 Demonstrating that the induced and inducing charges are equal and opposite using ice-pail and electroscope

The experiments described so far in this section can all be interpreted as tests of the inverse square law for the force between electric charges. It can be shown that the results would not otherwise be as we have found them. *On any other law* we should find fields inside a hollow closed conductor and charges on its inside surface; and an induced charge would not always be equal and opposite to the inducing charge. Even with very simple apparatus these experiments provide a far more delicate test of the law than direct tests, such as Coulomb's.

(*vii*) *The field outside a charged conducting sphere is the same as though the charge on it were concentrated at its centre.* By symmetry both the conducting sphere and the equivalent point charge produce radial fields (Fig. 10.23). If, as we have assumed, a fixed number of lines of force originate on a given quantity of charge, then the number of lines of force is the same in the diagrams of both fields. Thus, according to the line of force picture the field outside the sphere is exactly the same as though the whole charge on it were concentrated at its centre. (Gauss's theorem provides a rigorous proof of this result—p. 172). Quoting the results

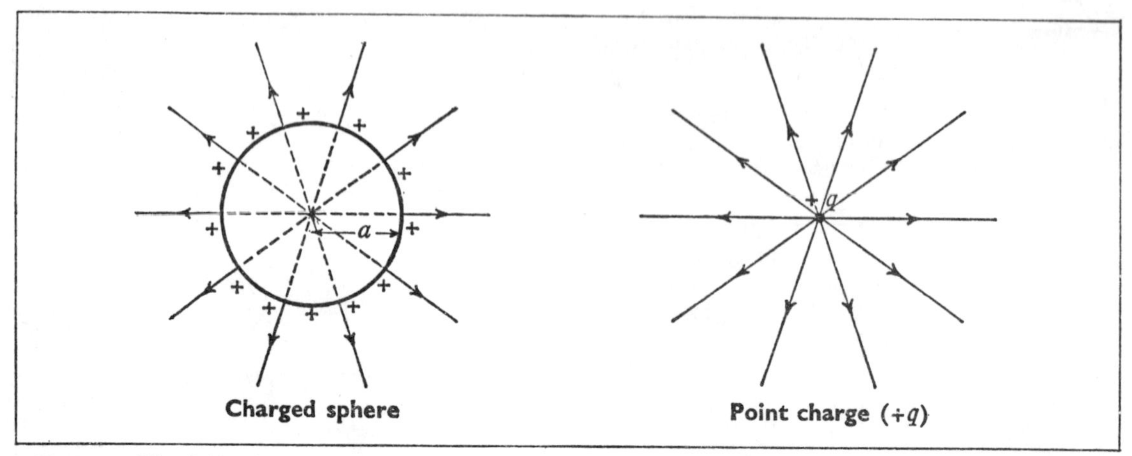

Fig. 10.23 The fields of a conducting sphere and of a point charge; outside the sphere the fields are identical

for a point charge (p. 163), the field intensity E and the potential V at a distance r from the centre of the sphere (outside it) are given by

$$E = \frac{q}{4\pi\varepsilon r^2}$$

and $$V = \frac{q}{4\pi\varepsilon r}$$

where $q =$ the total charge on the sphere.

If the radius of the sphere is a, then at its surface the field intensity E and potential V are given by

$$E = \frac{q}{4\pi\varepsilon a^2}$$

and $$V = \frac{q}{4\pi\varepsilon a}$$

The potential has this same value throughout the interior of the sphere (where the field is of course zero).

(*viii*) *The field at the surface of a conductor is proportional to the charge density on it.* The charge density σ on a surface is defined as the charge carried per unit area. Thus for a uniformly charged surface of area A carrying a charge q

$$\sigma = \frac{q}{A}$$

In a consistent line of force picture the strength of an electric field at a surface is represented by the number of lines of force emerging from it per unit area. Since each line is supposed to be associated with a definite quantity of charge, the field should be proportional to the charge per unit area on the surface, i.e. proportional to σ.

The relation between the field intensity E and σ can be derived by considering the special case of a charged conducting sphere. If the sphere is of radius a, the total charge q on its surface is given by

$$q = 4\pi a^2\sigma$$

$$\therefore\ E = \frac{q}{4\pi\varepsilon a^2} = \frac{4\pi a^2\sigma}{4\pi\varepsilon a^2}$$

$$\therefore\ E = \frac{\sigma}{\varepsilon}$$

(This result may be proved quite generally for a conductor of any shape by means of Gauss's theorem—p. 172.)

(*ix*) *The field and charge density are greatest at the most highly curved convex parts of a conductor's surface.* If a sphere of radius a is raised to a potential V, the charge q on its surface is given by

$$q = 4\pi\varepsilon aV$$

$$\therefore\ \sigma = \frac{4\pi\varepsilon aV}{4\pi a^2} = \frac{\varepsilon V}{a}$$

Thus the charge density is inversely proportional to the radius, and is greatest for a highly curved surface (for which the radius of curvature is small). The same is true of the field intensity E at the surface, since this is proportional to the charge density.

The result in fact applies quite generally to a conductor of arbitrary shape, whose curvature varies from one part of its surface to another. The charge is found to concentrate chiefly near corners and points, where the curvature of the surface is greatest. This may be tested experimentally with a proof-plane and electroscope. A conductor of irregular shape is stood on an insulating support and charged. Samples of the charge are then taken from various parts of its surface with the proof-plane. Each time the proof-plane is then touched on the *inside* of an ice-pail connected to the electroscope, so that we can be sure that the *entire* charge collected is given up. The deflection is noted, and is taken as a measure of this charge. It will be found that the charge is greatest when the proof-plane has been touched on the most highly curved convex parts of the surface, less on any flat parts, and least on any concave part (almost zero in any deep concavity).

The charge density also depends on the distribution of other conductors and charges in the neighbourhood. An earthed conductor (the hand) placed nearby greatly increases the charge density on the part of the surface nearest to it; and a charged object held close to the conductor may even produce a change of sign of the charge at some parts of the surface.

A consequence of this feature of the electric fields near conductors is that it is possible, without using very high potentials, to produce a very large field locally near a sharp point or a fine wire. The field in the immediate vicinity of the point or wire may then be sufficient to ionize the air there. Those ions which have the same sign of charge as the conductor are violently repelled away from it, giving rise to an appreciable electric 'wind'. This carries away the charge on the conductor into the surrounding air; it is called a *point discharge* or *corona discharge.*

Any object placed in the way of the 'wind' may collect some of the charge on its surface. This may be shown by holding a sharp needle connected to a high voltage source near the cap of an electroscope (Fig. 10.24*a*); a rapidly increasing deflection will be observed, as the charge blows onto the cap. The existence of the electric wind may also be demonstrated with the electric windmill shown in Fig. 10.24*b*. When this is raised to a high potential, the electric wind directed away from the points it sufficient to produce rapid rotation. Corona discharge has to be taken into account in the design of high-voltage equipment of all sorts:

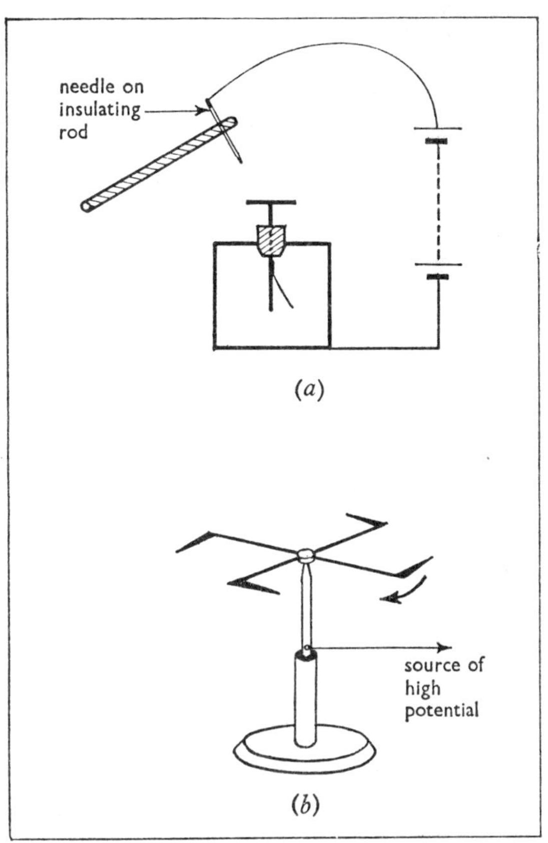

Fig. 10.24 (*a*) The corona discharge from a sharp needle
(*b*) A 'windmill' driven round by the electric wind from the points

(*a*) All parts of apparatus at a high potential must be gently curved; all sharp corners or protruding ends of wire must be eliminated. The high-voltage parts of a television set will be seen to be designed like this. With equipment working at millions of volts all high-potential parts are covered with large metal domes of uniform curvature.

(*b*) There is a maximum p.d. that can economically be used with a given diameter of cable for power transmission. If this is exceeded, the power loss into the air through corona discharge becomes significant. This provides an additional reason for making large power cables for use in air of aluminium (p. 19), since for a given resistance an aluminium cable has a larger diameter than a copper one and so has a less curved surface.

The Van de Graaff generator

This is a form of electrostatic machine that has

been used to provide the large p.d.'s needed for atomic particle accelerators; the largest models give p.d.'s up to 10 million volts. Fig. 10.25 shows the principle of the machine. Positive charge is sprayed by corona discharge from a row of points P on to a moving belt of insulating material—usually rubberized silk. This is carried up into the hollow conductor C and induces a negative charge on the inside of it. This negative charge is sprayed by the comb Q onto the belt, neutralizing the positive charge. The net result is the continuous transfer of positive charge from the points P to the belt and so to the upper conductor. A considerable tension is set up in the belt due to the repulsion between the charge on the upper conductor and the charge being carried up on the belt. The energy is supplied in the form of work done against this repulsion by the motor that drives the lower roller M. The maximum potential reached is decided chiefly by the quality of the insulation round the machine. For the highest p.d.'s the entire apparatus must be enclosed in an outer shell in which the gas pressure is raised.

In the simplified type of Van de Graaff machine often used in schools the row of points P is earthed,

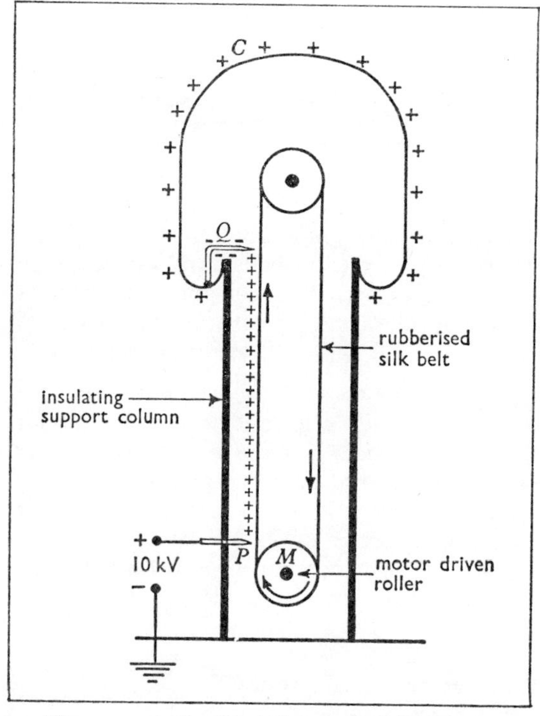

Fig. 10.25 A simplified Van de Graaff **generator**

and the action depends on the frictional charge developed on the inside of the belt by its contact with one roller (which is made of insulating material). This charge induces an opposite charge on the row of points P, and this is then sprayed onto the outside of the belt and carried up to the hollow conductor C.

10.8 The forces on charged surfaces

Consider a small element of area δA of the surface of a conductor carrying a charge density σ coulomb m^{-2} (Fig. 10.26). The field close to the surface can be regarded as the sum of two components:

(i) The field E_1 due to the charge $\sigma\,\delta A$ on the element considered; this must be directed away from the surface layer of charge in both directions.

(ii) The field E_2 due to the charge on the rest of the conductor; this component cannot change in the small distance between the two points A and B close to the surface on either side of it. E_2 must therefore act *outwards* through the surface at both points. We have seen that the distribution of charge over the surface of a conductor must be such as to make the resultant field zero everywhere inside it.

$$\therefore\ E_1 = E_2$$

Outside the surface the two components act in the same direction, and we have

$$E_1 + E_2 = \frac{\sigma}{\varepsilon}$$

$$\therefore\ E_1 = E_2 = \frac{\sigma}{2\varepsilon}$$

To calculate the force F on the element of area δA we need to find the product of the charge $\sigma\,\delta A$ and the field E_2 due to the *other* charges on the conductor; the field E_1 is due to the charge itself and cannot contribute to the force.

$$\therefore\ F = \sigma\,\delta A\,E_2 = \frac{\sigma^2 \delta A}{2\varepsilon}$$

$$\therefore\ \text{the force per unit area} = \frac{F}{\delta A} = \frac{\sigma^2}{2\varepsilon}$$

Whatever the sign of σ, this force acts *outwards* normal to the surface. In some respects it resem-

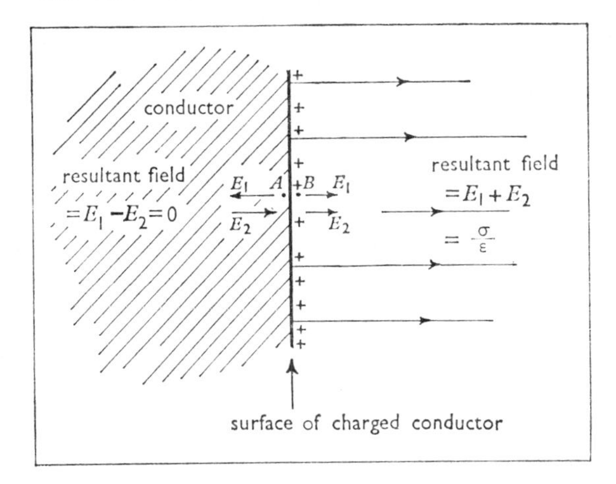

Fig. 10.26 The field near the surface of a charged conductor

bles a hydrostatic pressure acting 'inside' the surface.

When a potential difference is produced between two conductors, the surface forces act in such a way as to draw them together. This is the nature of the deflecting force in all electroscopes and electrometers. The field E between the fixed and moving electrodes in such an instrument is proportional to the p.d. V between them; it is also proportional to the charge density σ.

$$\therefore \sigma \propto V$$

The deflecting force F is proportional to σ^2.

$$\therefore F \propto V^2$$

One consequence of this behaviour is that the sensitivity of an electrometer increases with the p.d. These instruments are therefore insensitive for small p.d.'s, but are capable of measuring quite small *changes* in a large p.d.

The force F acts as an attraction between the electrodes whichever way round the p.d. is connected; electrometers are therefore suitable for measuring alternating p.d.'s as well as steady ones. Since the deflection depends on V^2, their readings indicate r.m.s. values (p. 128).

The attracted disc electrometer

This instrument measures p.d.'s by the force of attraction that acts between the plates of a parallel-plate capacitor. Two horizontal circular discs are used; the lower one is well insulated and can be raised and lowered by a micrometer screw, the

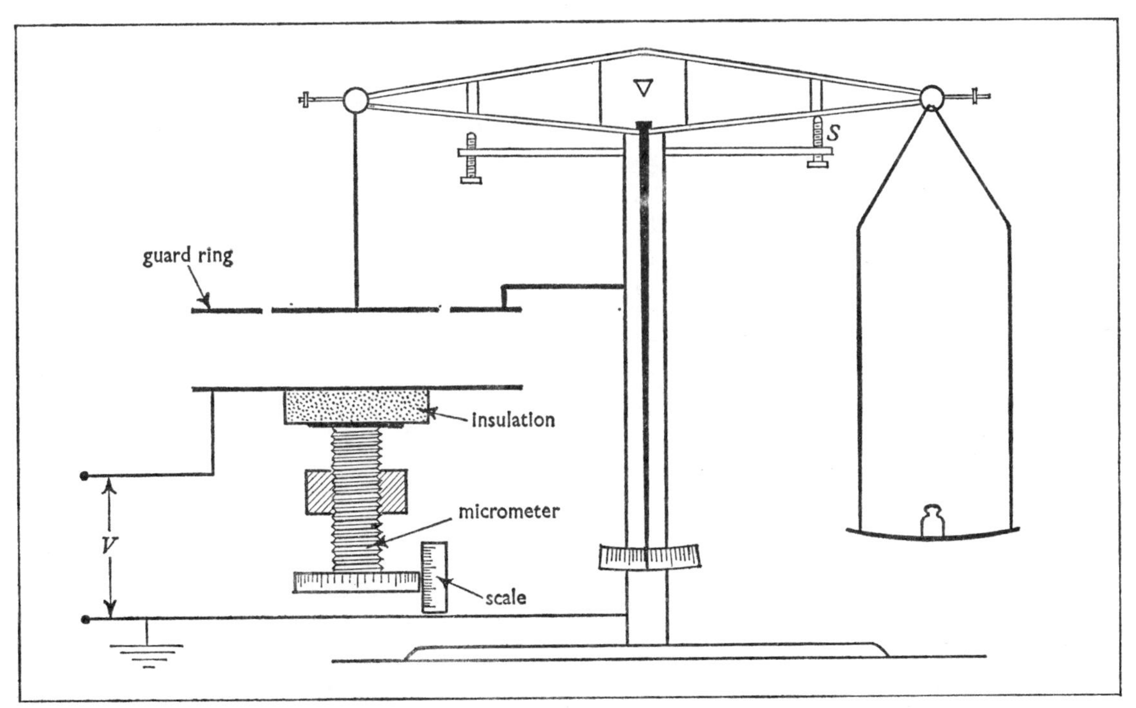

Fig. 10.27 An attracted disc electrometer

other hangs above it from one arm of a balance and is earthed (Fig. 10.27). In order to eliminate the uncertain edge effects of the field between two parallel plates, the suspended plate is surrounded by a fixed 'guard-ring', as shown; the plate moves freely through the circular hole in the guard-ring, but is electrically connected to it. The plate and guard-ring are arranged to be exactly coplanar when the balance arm is resting against the stop S. In this position the non-uniformity of field occurs only at the outside of the guard-ring, and the field below the suspended plate is truly uniform. If the separation of the plates is t and the p.d. between them V, then the field intensity E is given by

$$E = \frac{V}{t} = \frac{\sigma}{\varepsilon}$$

where σ = the charge density on the plates.

$$\therefore \sigma = \frac{\varepsilon V}{t}$$

If the area of the suspended plate is A, the force F acting on it is given by

$$F = \frac{\sigma^2 A}{2\varepsilon} = \frac{A}{2\varepsilon}\left(\frac{\varepsilon V}{t}\right)^2$$

$$\therefore F = \frac{\varepsilon A V^2}{2t^2}$$

This force may be balanced by the addition of a mass m to the other pan of the balance. Then, in equilibrium

$$F = mg = \frac{\varepsilon A V^2}{2t^2}$$

If the dimensions of the apparatus are measured in metres and the mass m in kg, then, taking

$$g = 9{\cdot}81 \text{ m s}^{-2} \quad \text{and} \quad \varepsilon_r = 1 \text{ (for air)}$$

so that

$$\varepsilon = \varepsilon_0 = 8{\cdot}84 \times 10^{-12} \text{ C}^2 \text{ N}^{-1} \text{ m}^{-2}$$

the potential V is obtained in volts. In practice the sensitivity used in this way is only sufficient for measuring very large p.d.'s (e.g. several kilovolts). But it may be adapted for measuring smaller p.d.'s by using it *differentially*; i.e. by connecting the p.d. v to be measured in series with a much larger constant p.d. V. With the large p.d. alone we have

$$V = \sqrt{\frac{2mg}{\varepsilon A}}\,.\,t_1$$

With the two p.d.'s in series the mass m is kept constant, and the separation of the plates is increased to t_2 with the micrometer screw to restore the balance. Then

$$V + v = \sqrt{\frac{2mg}{\varepsilon A}}\,.\,t_2$$

Subtracting,
$$v = \sqrt{\frac{2mg}{\varepsilon A}}(t_2 - t_1)$$

The change in separation $(t_2 - t_1)$ is found directly from the micrometer readings.

The chief difficulty in designing this type of electrometer is in preserving the stability of the balance. For a given mass m it is necessary to keep the separation t larger than a certain value for the equilibrium to be stable when the forces are balanced. In the original design of this instrument by Lord Kelvin a spring balance arrangement was used. In this case the requirements of stability make it necessary to use a rather stiff spring, and it is difficult then to achieve adequate sensitivity.

10.9 Gauss's theorem

The mathematical development of the theory of electric fields is in many respects very similar to that of magnetic fields. This similarity is of great assistance in suggesting new ways of analysing the two sorts of field.

We now continue the parallel development of electrical theory by defining the *flux density* of the electric field, usually denoted by the symbol D; it is sometimes called the *electric induction* and sometimes the *displacement.*

The **flux density** *(or electric induction) D of an electric field at a point is the product of the absolute permittivity ε of the medium and the electric intensity E at that point.*

Thus, $D = \varepsilon E$ (compare $B = \mu H$ in magnetism).

The *flux* of the electric field can then be defined in the same way as magnetic flux.

The **flux** *of an electric field through a small plane surface is the product of the area of the surface and the component of the flux density normal to it.*

Thus the flux ϕ of an electric field of flux density D through a small plane surface of area A to which

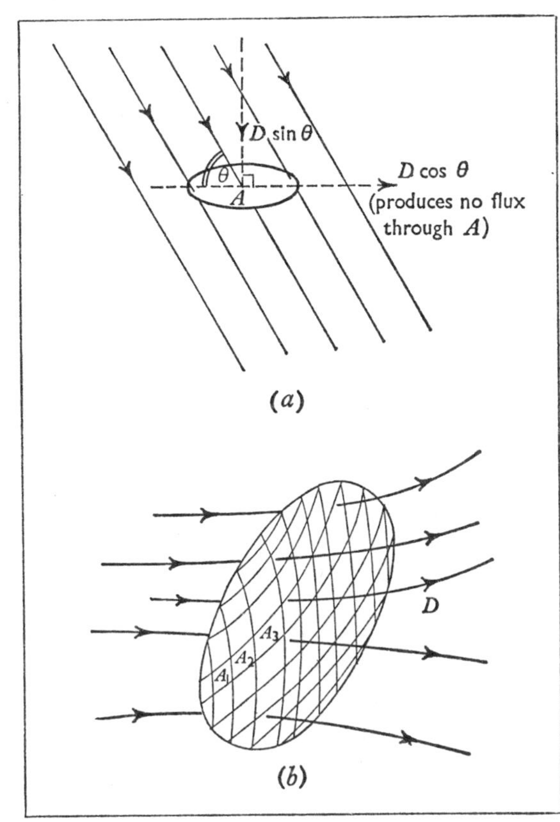

Fig. 10.28 The electric flux through a surface

the flux density is inclined at an angle θ (Fig. 10.28*a*) is given by

$$\phi = DA \sin \theta$$

To calculate the total flux through a large area (Fig. 10.28*b*), we imagine it divided up by a network of lines into small areas, A_1, A_2, A_3, etc., each of which is small enough to be treated as plane. The *total flux* is then found by summation.

The important result that follows from this way of describing an electric field is that the total flux emerging from a given charge is independent of the medium surrounding it. This is *Gauss's theorem.*

Gauss's theorem. *The total flux of the electric field outwards through any closed surface is equal to the total charge enclosed in the surface.*

It follows from this theorem that a fixed quantity of flux emerges from a given quantity of charge, since this amount will be found crossing *any* closed surface surrounding the charge. The use of a rationalized system of units ensures that this quantity of flux is numerically *equal* to the charge on which it arises. We therefore use the same unit for measuring both electric flux and electric charge—namely the *coulomb.* Thus the flux passing from an insulated conductor to its surroundings is equal to the charge (in coulombs) on the conductor —which is also equal (but opposite) to the charge on the surroundings. In particular the flux crossing the gap between the plates of a capacitor is equal to the charge stored on either plate.

The flux density D is the flux per unit area crossing any surface placed normal to the lines of force. The m.k.s. unit of flux density is therefore the coulomb metre^{-2}.

The formal proof of Gauss's theorem is deferred until the end of this chapter. But we shall now make use of it to give simple proofs of four results, which we have already discussed or accepted on experimental evidence.

(*i*) *Under static conditions charge resides entirely at the surface of a conductor.* Under static conditions the electric field is exactly zero everywhere inside the material of a conductor. Imagine a closed surface S described entirely inside the substance of the conductor. The flux density D is zero all over S, and therefore the flux out through it is zero. It follows from Gauss's theorem that the charge enclosed within S is also zero. Since the same applies to any surface of this kind, the charge on the conductor must be entirely on its surface.

(*ii*) *There is no charge inside a hollow closed conductor, provided no insulated charges are supported inside it.* Imagine a closed surface S drawn entirely inside the material of the hollow conductor, between its inside and outside surfaces (Fig. 10.29).

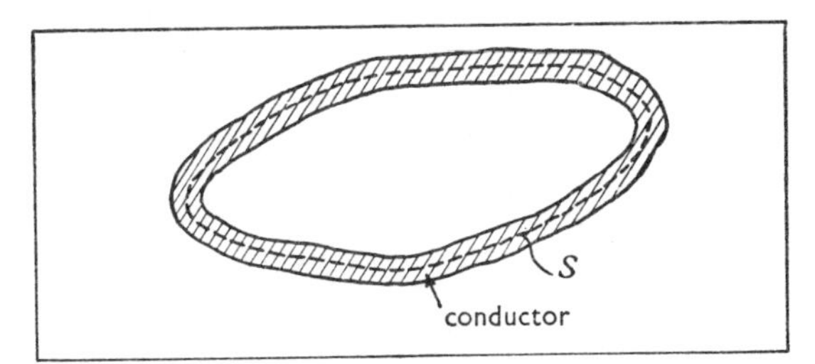

Fig. 10.29 Proof that the total charge is zero inside a hollow closed conductor

The field is necessarily zero at every point on S, and the flux through it is therefore zero. As before, the total charge enclosed in S must be zero. But there is nothing as yet to exclude the possibility of equal and opposite charges residing on different regions of the inside surface of the conductor. However, by Gauss's theorem, the flux q starting on any region of positive charge q must run through the hollow space to a region of negative charge; it cannot pass through the material of the conductor. Since the electric field is equal to the negative potential gradient (p. 157) this would require the regions of positive charge to be at a higher potential than those of negative charge. But this is impossible, since every point of a conductor must be at the same potential under static conditions. Therefore there can be no charge on the inside surface of the conductor, unless there are insulated charges supported inside it.

When the conductor contains an insulated charge, the same reasoning shows that the *total* charge inside it must still be zero; thus the charge induced on the inside surface must be equal and opposite to the inducing charge.

(*iii*) *The field outside a charged conducting sphere is the same as though the charge on it were concentrated at its centre.* Imagine a spherical surface of radius r concentric with the conductor. By symmetry the field is radial; suppose its intensity at the distance r from the centre is E. The flux ϕ out of the surface is given by

$$\phi = \varepsilon E . 4\pi r^2$$

If the total charge on the conductor is q, by Gauss's theorem we have

$$\varepsilon E . 4\pi r^2 = q$$

$$\therefore \; E = \frac{q}{4\pi\varepsilon r^2}$$

This is the same field intensity as would be produced by the charge q placed at the centre of the sphere.

(*iv*) *The relation between the field intensity E and the charge density σ at the surface of a conductor.* Imagine a small closed cylinder, partly inside the surface of the conductor and partly out, with its sides normal to the surface and its ends parallel with it (Fig. 10.30). The field is zero inside the conductor, and entirely normal to the surface outside it. The only flux out of the cylinder is therefore through the outside end of it. Let δA be the area of the surface enclosed by the cylinder

$$\therefore \text{ the enclosed charge} = \sigma\delta A$$

and the flux out through the end of the cylinder

$$= \varepsilon E\, \delta A$$

By Gauss's theorem, $\quad \varepsilon E\, \delta A = \sigma\, \delta A$

$$\therefore \; E = \frac{\sigma}{\varepsilon}$$

or

$$D = \sigma$$

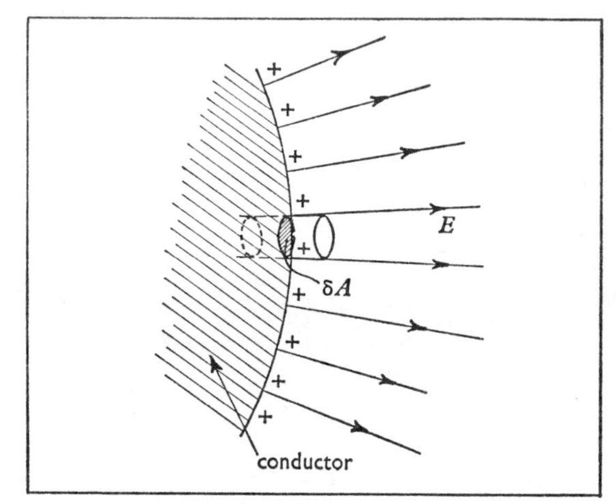

Fig. 10.30 The field at the surface of a conductor

★ The above result shows that the flux density D close to a conductor is equal to the charge density σ on its surface. This suggests a means of measuring flux density without first having to calculate or measure the intensity E. For instance, if we place a small proof-plane normal to the lines of force, the flux density close to it is equal to the charge density induced on its faces. The latter quantity may be measured by the technique of setting two proof-planes face to face and then separating them in the field—as described on p. 155 (Fig. 10.15). Indeed we might well define the flux density in terms of this experimental effect. Our two electric field vectors D and E would then be independently defined, just as B and H are in magnetism. There is no particular virtue in doing this as long as the relative permittivity ε_r of the dielectrics we employ are constants that do not vary with the polarization of the media. For the great majority of dielectrics this is the case. However, there are materials known as *ferroelectrics*, that exhibit hysteresis effects similar to those shown by ferromagnetics in magnetism. The relative permittivity ε of such a substance depends on the past history of polarization of the specimen, and is not a constant. We are then obliged to define and measure the two vectors by independent effects.

E.g. A parallel-sided piece of a ferroelectric may be sandwiched between the plates of a capacitor; D is then equal at any moment to the charge density σ on the plates and E is equal to the potential gradient between them. However, the consideration of ferroelectrics lies beyond the scope of this book, and we therefore retain the simpler definition of D in terms of E.

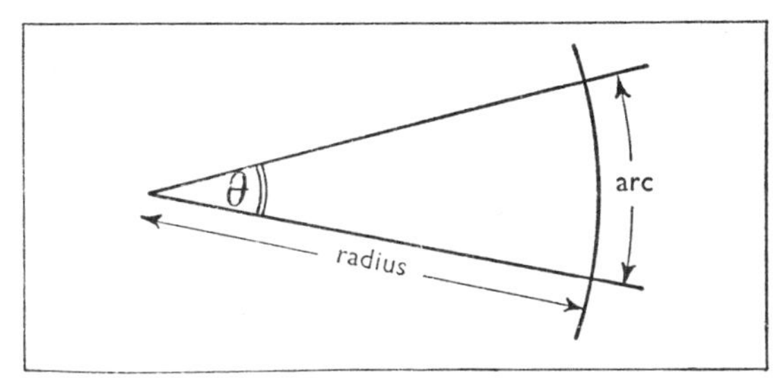

Fig. 10.31 The *angle* between two lines

Solid angle

The proof of Gauss's theorem requires an understanding of the concept of *solid angle*. In using two-dimensional figures drawn on a plane surface, we regard *angle* as a property of the junction of two lines. In Fig. 10.31 we find the value of the angle θ (in radians) by drawing a circle of any radius centred on the junction; then we measure the length of the arc cut off between the lines, and θ is given by

$$\theta = \frac{\text{arc}}{\text{radius}}$$

Angle is thus defined in such a way as to be dimensionless—that is, independent of the unit in which arc and radius are measured; it is also independent of the radius of the circle drawn for measuring it.

Similarly in three dimensions, we regard *solid angle* as a property of the apex of a cone (Fig. 10.32). We measure the solid angle ω by drawing a sphere of any radius centred on the apex; then we find the area of the surface of the sphere cut off by the cone, and ω is given by

$$\omega = \frac{\text{area cut off}}{(\text{radius})^2}$$

Again the definition is so framed as to make the quantity dimensionless and independent of the radius of the sphere. The unit of solid angle is sometimes called the *steradian*; but the usual practice is not to state the unit at all, since it is

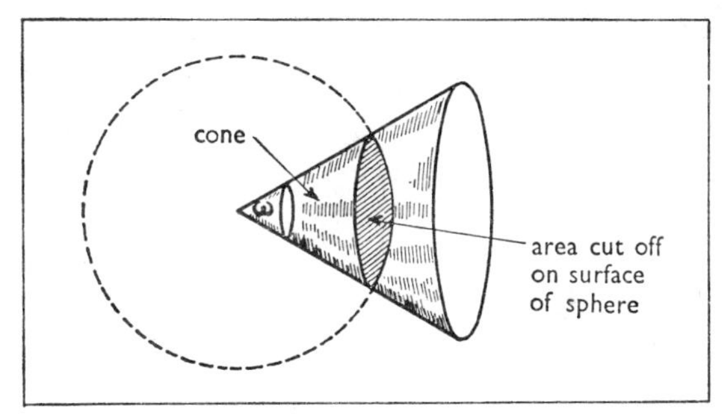

Fig. 10.32 The *solid angle* at the apex of a cone

always calculated in the way we have stated, and no confusion can arise.

The total solid angle round a point

$$= \frac{\text{the total surface area of a sphere}}{(\text{radius})^2} = \frac{4\pi r^2}{r^2} = 4\pi$$

Sometimes we need to calculate the solid angle 'subtended' by some irregular surface at a given point. It is then necessary to divide the surface into elements of small area, and the solid angle is obtained by summation or integration. Thus the solid angle $\delta\omega$ subtended at the point O by the element of area δA (Fig. 10.33) is the same as that subtended by the area $\delta A'$, the latter being the projection of δA on the surface of a sphere centre O of radius OP ($= r$). If θ is the angle between OP and the plane of the element, then

$$\delta A' = \delta A \sin\theta$$

$$\therefore\ \delta\omega = \frac{\delta A'}{r^2} = \frac{\delta A \sin\theta}{r^2}$$

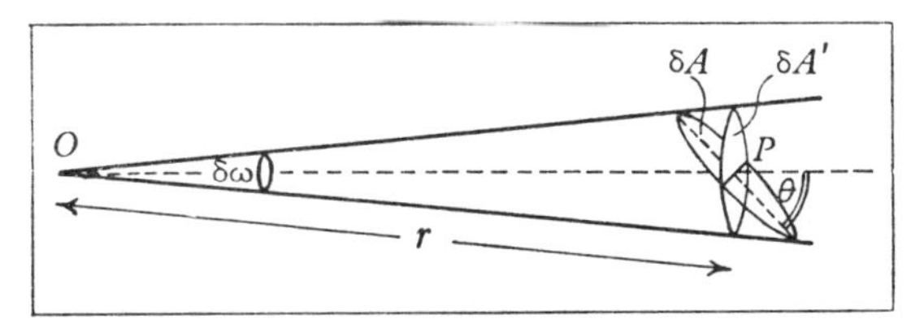

Fig. 10.33 The solid angle subtended by a small area δA at a point O

The proof of Gauss's theorem

Consider a point charge q inside a closed surface S. The flux $\delta\phi$ outwards through an element of the surface δA (Fig. 10.34*a*) is given by

$$\delta\phi = D\,\delta A \sin\theta$$

But $$D = \varepsilon E = \varepsilon \frac{q}{4\pi\varepsilon r^2} = \frac{q}{4\pi r^2}$$

$$\therefore\ \delta\phi = \frac{q\,\delta A \sin\theta}{4\pi r^2} = \frac{q\,\delta\omega}{4\pi}$$

where $\delta\omega$ is the solid angle subtended by the element δA at the charge q. The total flux ϕ outwards through the surface S is therefore given by

$$\phi = \frac{q}{4\pi} \times \text{(total solid angle subtended by S at O)}$$
$$= q$$

The same result can be derived separately for each charge that happens to be included *inside* S. However, any charge outside S contributes nothing to the total flux out through S. This can be seen from Fig. 10.34 *b*. The small cone, of solid angle $\delta\omega$, cuts off two elements of the surface S of areas δA_1 and δA_2, as shown; the quantities of flux passing *out of* S through these two elements are $-q\,\delta\omega/4\pi$ and $+q\,\delta\omega/4\pi$ respectively. Thus the *total flux* outwards through the two elements is zero, and the same applies to all small cones drawn with their apexes at O′ intersecting the surface S. Therefore the total flux ϕ outwards through S is given by

$$\phi = \text{total charge } \textit{inside} \text{ S}$$

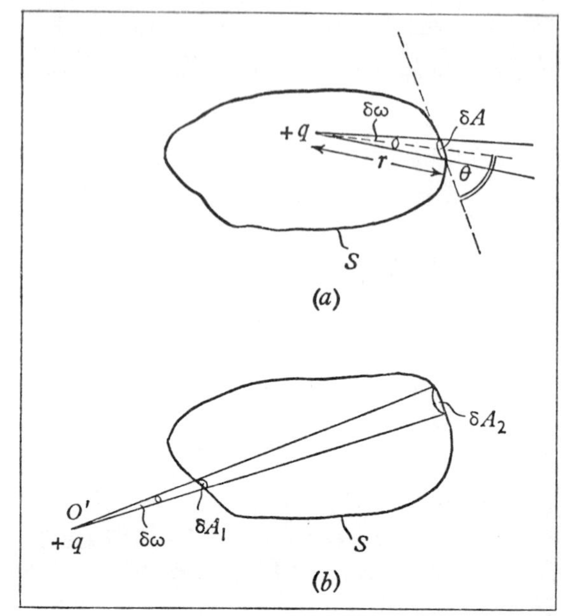

Fig. 10.34 The proof of Gauss's theorem

11: Capacitance

11.1 Types of capacitor

When a charge is placed on a conductor it induces an equal and opposite charge on other conductors in its neighbourhood, some of which may be earthed and some not. If the arrangement is such that effectively all the induced charge is on one other conductor, the combination of the pair of conductors and the insulator between them is called a *capacitor* (or *condenser*). An ideal capacitor would be one in

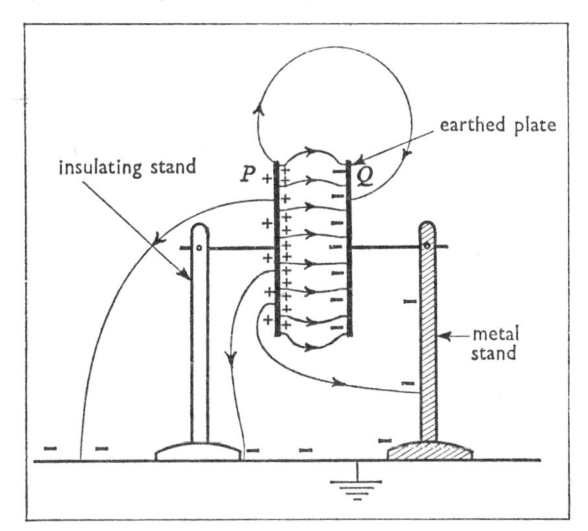

Fig. 11.1 The stray electric field of a parallel-plate capacitor

which the second conductor alone carried *all* the induced charge, but this could only be realized in practice if one conductor completely enclosed the other. But there are a number of ways of making a reasonable approximation to this condition. Even with two parallel plates a short distance apart a charge placed on one plate induces an almost equal charge on the other. Strictly in the arrangement shown in Fig. 11.1 some of the charge induced by the plate P is on the earthed table-top, but in practice this is negligible compared with that induced on the much closer earthed plate Q; and we can assume that the charges on P and Q are equal. The closer the plates are together the more exact the approximation is; it is further improved if the capacitor has a number of interleaved plates, as in Fig. 11.2. But even here there is a small 'leakage' of the electric field to other conductors in the neighbourhood, and there are occasions (particularly in high-frequency radio work) when this cannot be ignored.

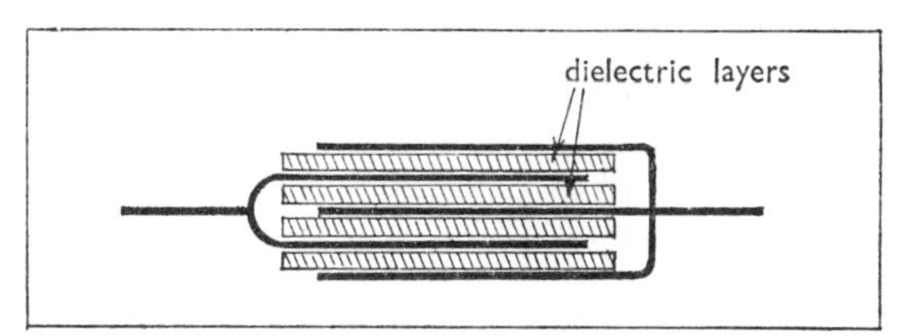

Fig. 11.2 A capacitor with interleaved plates

On theoretical grounds we should expect the p.d. V between the plates of an ideal capacitor to be directly proportional to the charge q on either of them; for the p.d. is proportional to the field between the plates, and this in turn is proportional to the surface density of charge on the plates and so to the total charge. The experimental test of this law is best made with a ballistic galvanometer by the method given on p. 154. Thus

$$V \propto q$$

We can therefore write

$$q/V = \text{a constant}$$

which is called the *capacitance* of the capacitor.

Capacitance. *The charge q on one plate of a capacitor is proportional to the p.d. V between the plates;*

$$\therefore\ q \propto V$$

or $$q = CV$$

where C is a constant for the capacitor known as its capacitance.

(The quantity q/V is still sometimes called the *capacity* of the *condenser*, but this use of words is becoming obsolete.)

★ The above analysis is only strictly applicable to an ideal capacitor, in which the charges on the two plates are exactly equal and opposite. The concept of capacitance may be extended to include non-ideal capacitors or indeed any arrangement of conductors; the following discussion indicates the way in which this can be done.

As long as we have to consider only a single unearthed conductor with no other insulated charges it remains true that the charge q on this is proportional to its potential V. Consider for instance a simple parallel-plate capacitor with one plate earthed. A charge $+q$ on the free plate induces a charge $-q$ on its surroundings; most of this appears on the other plate, but some of it is on the other earthed objects in the neighbourhood. We may regard this as equivalent to a pair of ideal capacitors as shown by the dotted lines in Fig. 11.3. The two capacitors are in parallel, since there is the same p.d. across both, but the charge q is shared between them, a fixed proportion of it appearing on each. The quantity q/V is thus the *total* capacitance of the free plate in its surroundings. If $-q_1$ is the charge induced on the second plate and $-q_2$ that on the other earthed objects, then

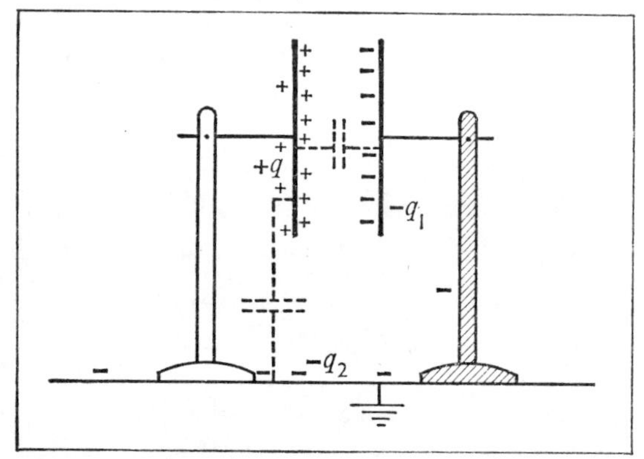

Fig. 11.3 The capacitance between two parallel plates and the stray capacitance to earth

$$q_1 + q_2 = q$$

$$\therefore \frac{q_1}{V} + \frac{q_2}{V} = \frac{q}{V}$$

That is, the total capacitance is the sum of the capacitance q_1/V of the parallel-plate combination and the *leakage* or *stray* capacitance q_2/V.

However, when there are two or more insulated conductors, the position is much more complicated. Each conductor then has a capacitance with every other conductor and with the earth. Needless to say, the analysis of such arrangements can be very complex; but in circuits containing only large capacitances we may usually ignore the effect of the many small stray capacitances associated with stray electric fields between different parts of the apparatus. Only in circuits in which small capacitances are used, such as in high-frequency radio and television equipment, do the leakage capacitances have a significant effect. Sometimes in these cases a change in the position of a component can radically alter the behaviour of the circuit, even though the circuit diagram appears to be unchanged; of course, if the stray capacitances were included in the circuit diagram, this behaviour would not appear so puzzling.

The m.k.s. unit of capacitance is called the *farad* (denoted by F).

The **farad** *is the capacitance of a capacitor on each plate of which there is a charge of* 1 *coulomb when a p.d. of* 1 *volt exists across it.*

The farad turns out to be an extremely large quantity. For example the capacitance of the earth itself, treated as an isolated conducting sphere, is only about 7×10^{-4} farad. Capacitances are therefore usually expressed in *micro-farads* (μF), i.e. 10^{-6} F, or *pico-farads* (pF), sometimes called micro-micro-farads ($\mu\mu$F), i.e. 10^{-12} F.

To calculate the capacitance C of a given capacitor it is necessary to work out the connection between the charge q on its plates and the p.d. V between them. Then

$$C = \frac{q}{V}$$

(*i*) *The parallel-plate capacitor.* To a first approximation we can take the field between the plates as uniform. There is bound to be a region at the edge in which the field is non-uniform (Fig. 11.4), but when the plates are close together, as with the mica-insulated or paper-insulated types (p. 179), the non-uniformity occupies only a negligible part of the space between them.

Suppose there is a charge $+q$ on one plate and $-q$ on the other, the charge density σ is then given by:

$$\sigma = \frac{q}{A}$$

where A = area of each plate. The field intensity E, being uniform, is the same as that at the surface of each plate.

$$\therefore E = \frac{\sigma}{\varepsilon} = \frac{q}{\varepsilon A} \qquad \text{(p. 166)}$$

where ε = the permittivity (absolute) of the insulator between the plates.

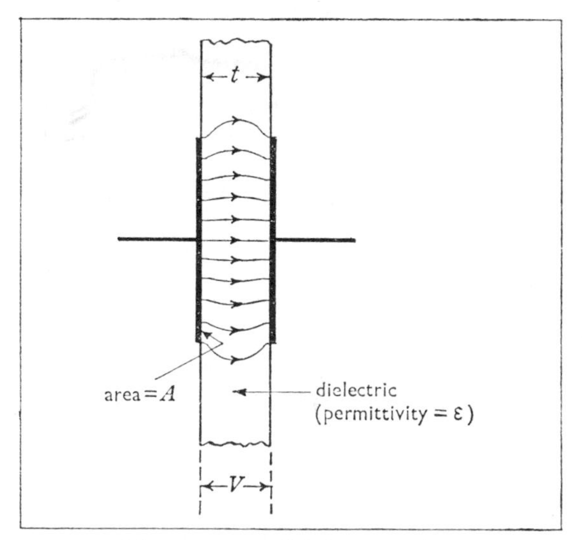

Fig. 11.4 A parallel-plate capacitor

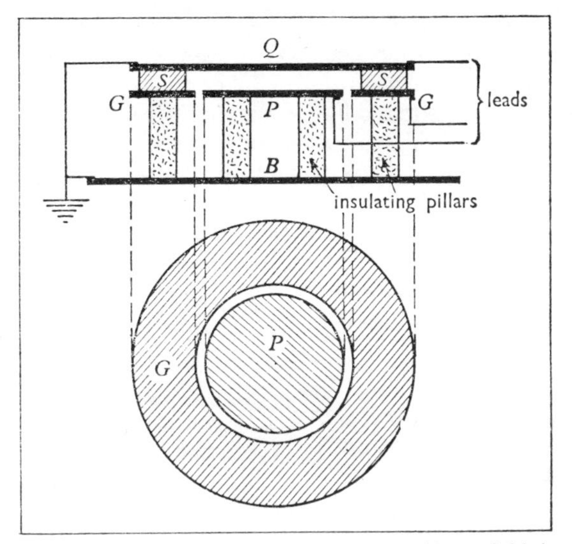

Fig. 11.5 A guard-ring capacitor; a uniform field is produced between P and Q

But $\quad E = \text{potential gradient} = \dfrac{V}{t}$

where $t =$ the separation of the plates.

$$\therefore \; \frac{V}{t} = \frac{q}{\varepsilon A}$$

$$\therefore \; C = \frac{q}{V} = \frac{\varepsilon A}{t}$$

The capacitance C is thus inversely proportional to t. This result may be tested by measuring C for a parallel-plate capacitor whose plate-separation t may be varied. Varying t also varies the extent of the edge region over which the field is non-uniform. To obtain reliable results it is necessary to employ a *guard-ring capacitor* as shown in Fig. 11.5. The lower plate P of this is surrounded by a coplanar guard-ring G; there is a narrow circular gap between P and G, and they are supported on separate insulating pillars. When the capacitor is charged, P and G are connected together momentarily, so that they acquire the same potential V. The non-uniformity of the field then occurs only at the outside edge of the guard-ring, and the field between P and the upper plate Q is accurately uniform. When the capacitor is discharged, the charge q on P is caused to pass through a suitable charge measuring instrument, while the guard-ring is separately discharged to earth. (The detailed methods used for this kind of measurement are discussed later in this chapter, p. 190.) The capacitance of P is then given by

$$C = \frac{q}{V}$$

This includes the stray capacitance between P and its other earthed surroundings, chiefly the base-plate B. Suppose this amounts to C_0. We then have

$$C = \frac{\varepsilon A}{t} + C_0$$

The plate separation t is adjusted to a series of pre-determined values by using sets of accurately ground insulating spacers S to support the upper plate Q; these rest on the guard-ring so that the space above P is occupied only by air. A straight-line graph may then be plotted of C against $1/t$, confirming the above theory. C_0 is given by the intercept on one axis; and the gradient of the line is equal to εA. The experiment thus yields a value for ε the absolute permittivity of air; this is negligibly different from the value ε_0 that would be obtained with a vacuum between the plates.

The expression for the capacitance C shows that the dimensions of ε_0 are those of

$$\left[\frac{\text{capacitance}}{\text{length}}\right]$$

and the unit in which its value may be expressed is therefore the farad metre^{-1}. Experiments such as the above give

$$\varepsilon_0 = 8{\cdot}84 \times 10^{-12}\ \text{F m}^{-1}$$

If the space between the plates of *any* capacitor

is filled with some insulator other than air, the capacitance is increased by the factor ε_r the *relative permittivity* of the insulator. The value of ε_r may therefore be found by comparing the capacitances measured with and without the insulator. As a practical matter, it is often rather difficult to fill the *whole* space between the plates with the insulator; and we may have to work with the insulator occupying only part of the gap. The calculation is then rather more elaborate (p. 192).

(*ii*) *An isolated conducting sphere.* In this case the other 'plate' of the capacitor is assumed to be another conductor completely surrounding the sphere at a very great distance from it. Taking the latter conductor as being at zero potential, the p.d. V across the capacitor is simply the potential of the sphere. If it carries a charge q, and is of radius a, then we have

$$V = \frac{q}{4\pi\varepsilon a} \qquad \text{(p. 166)}$$

$$\therefore\ C = \frac{q}{V} = 4\pi\varepsilon a$$

★ (*iii*) *A pair of concentric spheres.* A charge $+q$ on the inner sphere induces a charge $-q$ (exactly) on the inside of the outer one, which we shall suppose in this case to be earthed. The field in the region between the spheres is the same as that due to a point charge $+q$ at the centre of the system (Fig. 11.6). (This may be proved rigorously by using Gauss's theorem, p. 172.) Except for the region between the spheres the field is zero. The p.d. V is therefore given by

$$V = \int_b^a - E\,\mathrm{d}r = -\int_b^a \frac{q}{4\pi\varepsilon r^2}.\mathrm{d}r$$
$$= \frac{q}{4\pi\varepsilon}\left(\frac{1}{a} - \frac{1}{b}\right)$$

where a and b are the radii of the spheres. The capacitance C is therefore given by

$$C = \frac{q}{V} = \frac{4\pi\varepsilon}{(1/a - 1/b)} = \frac{4\pi\varepsilon ab}{b - a}$$

11.2 The construction of capacitors

A capacitor consists of two conducting sheets insulated a short distance apart by a suitable dielectric. To produce large values of capacitance it is necessary (*a*) to use sheets of large total area, (*b*) to use very thin layers of dielectric, (*c*) to choose a dielectric with a high dielectric constant. Breakdown of the insulation occurs in any given dielectric for electric fields above a certain critical value and there is therefore a maximum p.d. that can safely be applied across a given thickness of it. The thickness of the layer of dielectric is thus decided by the p.d. at which the capacitor is to operate. The choice of dielectric material generally depends on the frequency range of the alternating currents that will be flowing in the capacitor and on the stability of capacitance required.

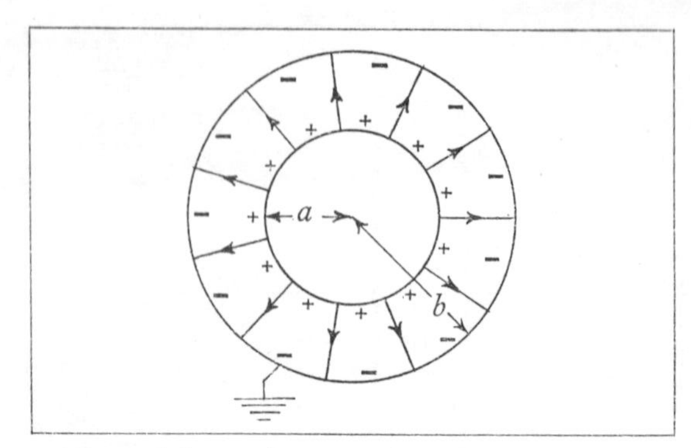

Fig. 11.6 A capacitor consisting of two concentric spheres

Paper capacitors

The electrodes consist of two long strips of metal foil. These are interleaved with two similar strips of paper, impregnated with paraffin wax or a suitable oil to improve its insulation qualities. The composite strip is then rolled up into a tight cylinder (Fig. 11.7). Connection is made to the two strips of foil by allowing them to project over the sides of the paper insulation, one strip in each direction; two metal discs are then pressed against the exposed edges. The whole arrangement is enclosed in a protective case of metal or waxed cardboard. With the larger values of capacitance the roll is packed into a rectangular metal box, and the connecting wires are brought out to terminals on top. Sometimes two such rolls are connected in series in the one box; this enables the working voltage to be doubled for a given type of paper insulation.

A more compact version of this type of capacitor is made with electrodes consisting of very thin layers of metal coated actually onto the surface of the dielectric. This type has the useful property of being self-healing after momentary breakdown of its insulation; if a spark passes through the paper

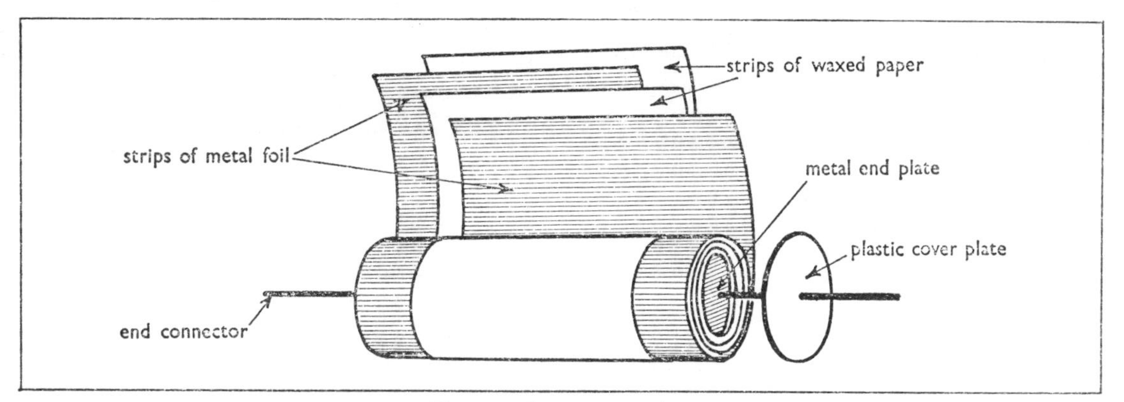

Fig. 11.7 A paper capacitor

at some point of weakness, the metal film is at once evaporated in the vicinity of the hole, restoring the original insulation strength.

The impregnated paper has a dielectric constant of about 5 and is suitable for low- and medium-frequency work (100 Hz to 1 MHz). Its insulation resistance is very high (often $>10^{10}$ ohms). The stability of this design of capacitor is not very good —10% variations are liable to occur with age. But within its proper frequency range its cheapness and relatively small size lead to its wide use for radio and general laboratory work. The values of capacitance made in this design cover the range from about 10^{-3} μF to 10 μF.

A similar method of construction is also used with a plastic strip (polystyrene or polythene) instead of the impregnated paper. These dielectrics are able to be used with very low loss at all frequencies from zero up to the highest encountered in radio work.

Mica capacitors

Mica is a mineral that has the useful property that it can be split into very thin uniform sheets—a thickness of about 0·05 mm is common in capacitors. The capacitor is built up of alternate sheets of metal foil and mica (Fig. 11.8). The odd-numbered sheets are taken out beyond the dielectric at one end and the even-numbered sheets at the other. The two groups of sheets form the two electrodes and are joined to connecting wires at each end of the stack. The whole capacitor is then bound tightly in a hermetically sealed plastic case.

In an alternative form of construction the metal foils are replaced by very thin layers of silvering on the surfaces of the mica sheets themselves. The result is a more compact unit; also in this way all air gaps are eliminated between the electrodes and the mica, which makes for greater stability of capacitance.

Mica has a dielectric constant of about 6 and produces very low losses for all frequencies up to about 300 MHz. The values of capacitance generally available cover the range from 5 pF up to 0·01 μF. The stability of capacitance is better than with the paper type. With careful design variations can be kept below 1%—sufficient for laboratory standards; special capacitors are made for this purpose with values up to about 1 μF.

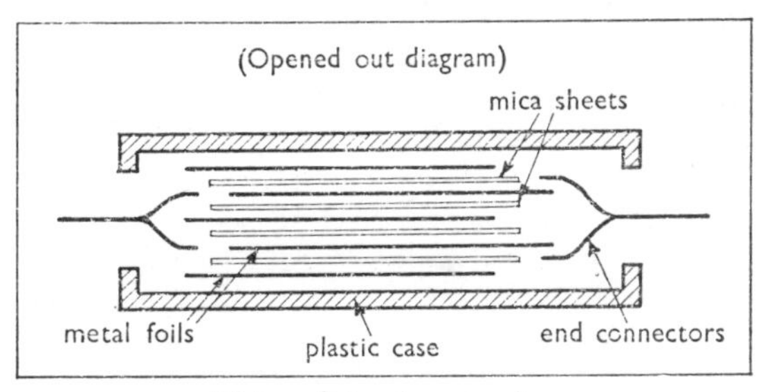

Fig. 11.8 A mica capacitor

Ceramic capacitors

There is a wide range of ceramic materials that can be used as low-loss dielectrics at all frequencies. The basis of the 'mix' is often steatite (talc), which has a dielectric constant of about 8. But by the addition of other materials, notably titania, the dielectric constant may be increased to 100 or more; and with barium titanate and related compounds dielectric constants up to 5000 have been realized!

These enable very compact capacitors to be made; but the stability is not good with the highest values of dielectric constant. A useful property of some of these materials is that they have a *negative* temperature coefficient of dielectric constant. Most forms of capacitor show an increase of capacitance with temperature. But by combining a suitable ceramic capacitor in parallel with another type the resulting capacitance can be made almost independent of temperature.

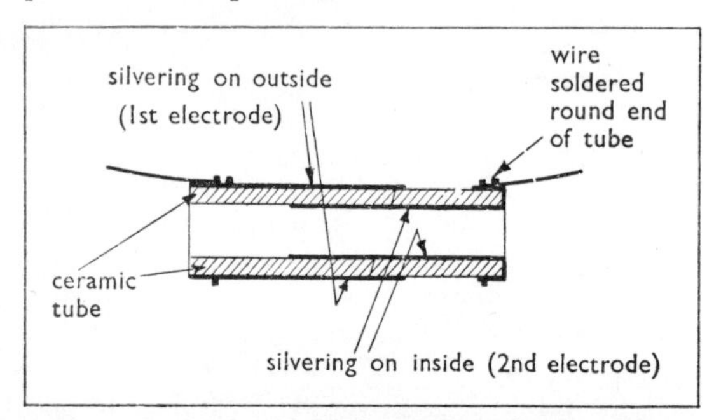

Fig. 11.9 A ceramic capacitor

A common form of construction is shown in Fig. 11.9. The purified material is pressed into a suitable shape, usually a small cylindrical tube. It is fired at a high enough temperature for vitrification to occur; this is necessary to make the ceramic non-porous. The cylinder is then silvered over parts of its inside and outside surfaces, as shown, thus forming the electrodes. Connection is made to these by means of wire rings soldered round the silvered layers at the ends of the tube. The whole tube is then given a coat of protective lacquer or is mounted inside another ceramic tube as a container.

Electrolytic capacitors

When an electric current is passed through a solution of ammonium borate using aluminium electrodes, a very thin film of oxide forms on the anode; this is an example of the process known as *anodic oxidation* or *anodizing*. The thickness of the film depends on the p.d. used and on the time of deposition. In the electrolytic capacitor the oxide layer is the dielectric; its thickness may be no more than 10^{-7} m, but it shows an insulation strength of about 10^9 volts per m—vastly greater than any other available dielectric. An extremely compact unit of high capacitance can thus be produced. The two 'plates' of the capacitor are the *anode* and the *electrolyte* (to which connection is made via the cathode). Electrolytic capacitors come into their own particularly in low-voltage applications since the thickness of the dielectric layer can be reduced almost indefinitely. The physical size of a 6-volt, 1000-μF capacitor may well be no more than that of a 350-volt, 1-μF, impregnated paper type. However, in order to maintain the electrolytic deposit it is necessary for a small 'leakage' current (about 0·1 to 10 mA) to flow continually through the capacitor in the right direction. This limits its use to circumstances in which an alternating p.d. is superimposed on a direct polarizing p.d. Care must always be taken to connect an electrolytic capacitor so that the polarizing current flows through it in the right direction; one terminal is normally coloured red indicating that this must be kept positive with respect to the other. The stability is very poor, since the thickness of the dielectric film depends on the average p.d. at which the capacitor is working. However, these drawbacks are of no

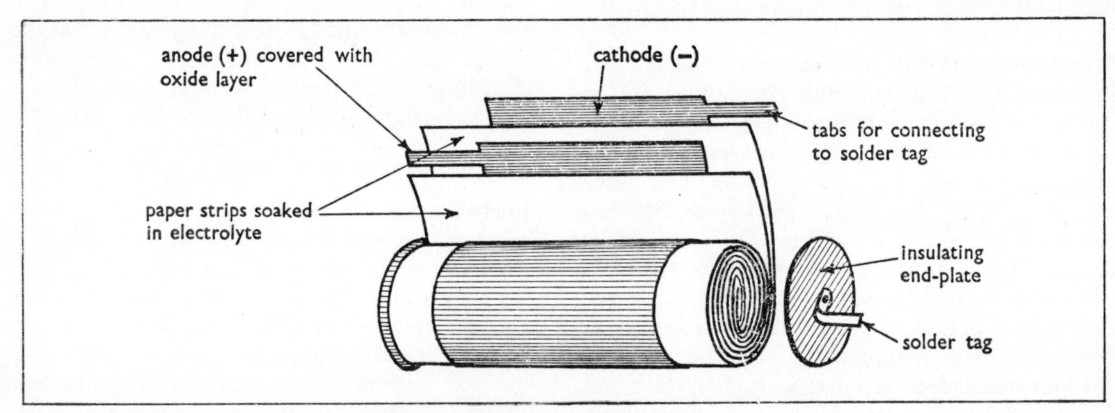

Fig. 11.10 An electrolytic capacitor

consequence in many applications, such as the storage and smoothing capacitors in rectifier circuits (p. 241). Their use is limited to low frequencies—up to about 10^4 Hz or 50 kHz in special cases.

A typical design is shown in Fig. 11.10. Two long strips of aluminium are interleaved with two strips of paper soaked in the electrolyte. The composite strip is rolled up into a cylinder. Tags are attached to the sides of the foil by which connection is made to the terminals of the capacitor. The whole arrangement is then sealed in a leak-proof container. Sometimes the case is made of aluminium, and is used as one terminal.

Even more compact capacitors can be produced by using *tantalum* foil instead of aluminium. In this case the polarizing current need only be a few micro-amps, so that its electrical properties compare favourably with those of an impregnated paper capacitor.

Air capacitors

Capacitors of this type have almost perfect electrical properties at all frequencies. Such losses as do occur arise chiefly in the insulation used in supporting the plates. However, air suffers from the disadvantage of having very low insulation strength—about 1/20 of the insulation strength of mica or impregnated paper. The spacing of the plates therefore has to be relatively large; also the plates themselves have to be thick enough to be mechanically rigid. Air capacitors are therefore extremely bulky, and *fixed* capacitance types are used only for laboratory standards.

However, air insulation provides an obvious means of constructing a *variable* capacitor (Fig. 11.11). It consists of two sets of interleaved parallel plates (or vanes). One set is fixed to the frame, the other to the central shaft. Rotation of this varies the area of the interleaved parts of the vanes, and therefore changes the capacitance. Miniature variable capacitors are also made with sheets of a plastic dielectric held loosely between the plates; this enables the plates to be thinner and closer together than with air insulation alone.

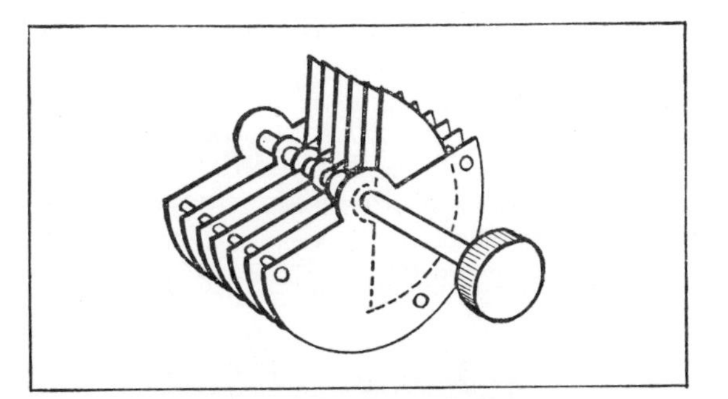

Fig. 11.11 A variable air capacitor

11.3 Energy; combinations of capacitors

The existence of a potential difference between the plates of a charged capacitor implies that the charge stored there carries electric potential energy. The quantity of energy may be calculated as follows.

Suppose the negative plate of a capacitor is earthed and the other at a potential v carries a charge $+q$. From the definition of potential this implies that the work done by the forces of the electric field in driving a quantity of charge δq from the positive plate to the other plate is $v\,\delta q$. This movement of charge would partially *discharge* the capacitor; the change of charge on the positive plate is therefore $-\delta q$. If the initial charge on this plate is Q, the total energy lost in completely discharging the capacitor is given by

$$\text{energy} = \int_Q^0 -v\,dq$$

But

$$v = \frac{q}{C}$$

where C = capacitance. Reversing the limits of integration,

$$\text{energy} = \int_0^Q \frac{q}{C}dq = \frac{1}{2}\cdot\frac{Q^2}{C}$$

If V = the initial p.d. between the plates, we have

$$C = \frac{Q}{V}$$

Substituting, we get three alternative forms:

$$\text{energy} = \frac{1}{2}\cdot\frac{Q^2}{C} = \tfrac{1}{2}CV^2 = \tfrac{1}{2}QV$$

If C is in farads, V in volts and Q in coulombs, then the energy is obtained in joules.

In order to calculate the potential at any point in a system of conductors we need to be able to find

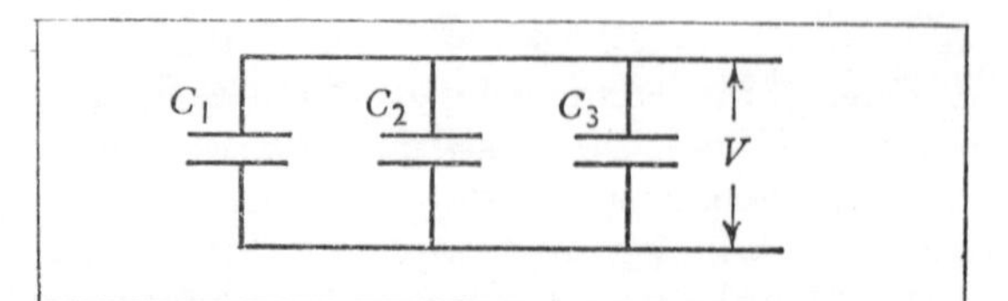

Fig. 11.12 Capacitors in parallel

the effective capacitance of various combinations of capacitors.

Capacitors in parallel

In this arrangement (Fig. 11.12) the p.d. V is the same across the capacitors C_1, C_2 and C_3, but the charges on them are different. The combined effective capacitance C is given by

$$C = \frac{Q}{V}$$

where Q is the total charge on the capacitors, i.e. the sum of their separate charges.

$$\therefore\ Q = C_1V + C_2V + C_3V = V(C_1 + C_2 + C_3)$$

$$\therefore\ C = C_1 + C_2 + C_3$$

Capacitors in series

Some caution is necessary before we pronounce a set of capacitors to be in series. Let us consider what happens when the capacitors in Fig. 11.13 are charged by applying a p.d. V between the points A and F. This will cause a charge $+Q$ to flow on to A, which will induce a charge $-Q$ on B. The plates B and C and the connecting link between them form a single insulated conductor, whose total charge must remain zero throughout the process. Therefore inducing a charge $-Q$ on **B** causes a charge $+Q$ to appear on C, which in turn induces a charge

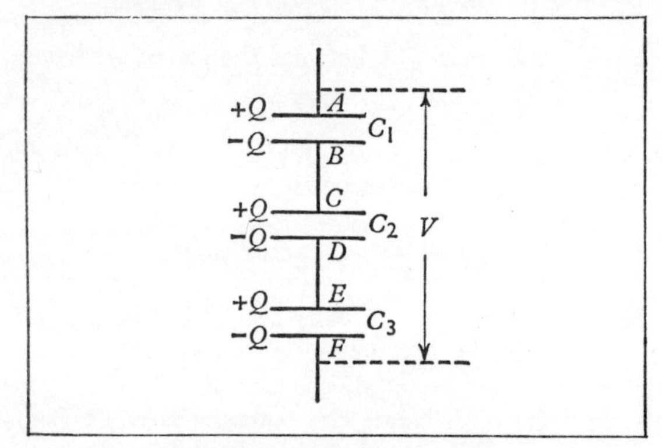

Fig. 11.13 Capacitors in series

$-Q$ on D, etc. Thus, with capacitors in series the *same charge* is stored in each capacitor. But it is quite possible to charge these capacitors in a different way. We might for instance join separate sources of e.m.f. of any magnitudes whatever across each capacitor individually; when the capacitors are charged, these sources of e.m.f. could then be removed so that the circuit diagram *looks* the same as in the first case. The p.d. across the combination is then the sum of the separate p.d.'s, but in general the charges on the capacitors will be *different*, and we cannot say they are in series, since this process of charging has involved also making temporary connections to the intermediate points of the chain.

However, in the case in which there is the same charge Q stored in each capacitor, the combined effective capacitance C is given by

$$C = \frac{Q}{V}$$

where V is the total p.d. across the combination.

$$\therefore\ V = \frac{Q}{C_1} + \frac{Q}{C_2} + \frac{Q}{C_3} = Q\left(\frac{1}{C_1} + \frac{1}{C_2} + \frac{1}{C_3}\right)$$

$$\therefore\ \frac{V}{Q} = \frac{1}{C} = \frac{1}{C_1} + \frac{1}{C_2} + \frac{1}{C_3}$$

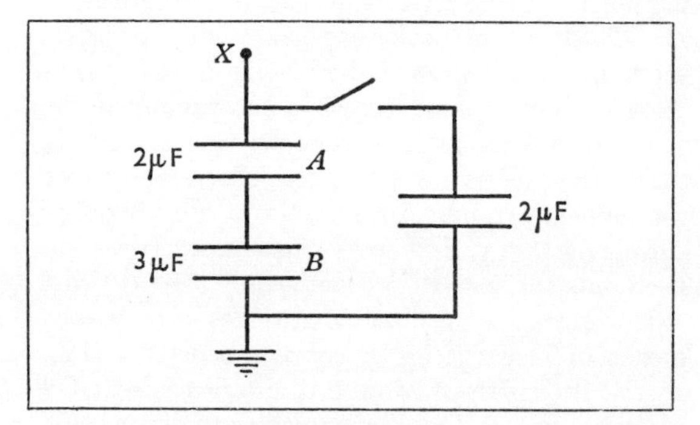

Fig. 11.14 Problem

Example. Three capacitors are connected as shown in Fig. 11.14. With the switch open the capacitors A and B are charged by momentarily connecting a 120-volt battery between the point X and earth. The switch is then closed. Calculate (*a*) the new potential of X, (*b*) the energy lost by closing the switch.

The combined effective capacitance C of A and B in series is given by

$$\frac{1}{C} = \frac{1}{2} + \frac{1}{3} = \frac{5}{6}$$

$$\therefore\ C = \frac{6}{5} = 1{\cdot}2\ \mu F = 1{\cdot}2 \times 10^{-6}\ F$$

When the switch is closed, this capacitance is in parallel with the 2 μF capacitor. The combined effective capacitance of all three capacitors is then $3{\cdot}2 \times 10^{-6}$ F.

The initial charge of A and B

$$= 1{\cdot}2 \times 10^{-6} \times 120 \text{ coulomb}$$

Assuming that no charge is lost on closing the switch, the p.d. V across the final combination

$$= \frac{\text{charge}}{\text{capacitance}}$$

$$= \frac{1{\cdot}2 \times 10^{-6} \times 120}{3{\cdot}2 \times 10^{-6}} \text{ volt}$$

$$= 45 \text{ volts}$$

[Or, more directly, we could have said that, for a given charge, p.d. is inversely proportional to capacitance

$$\therefore\ V = \frac{1{\cdot}2}{3{\cdot}2} \times 120 = 45 \text{ volts}]$$

The initial energy of A and B

$$= \tfrac{1}{2} \times 1{\cdot}2 \times 10^{-6} \times (120)^2$$
$$= 8{\cdot}64 \times 10^{-3} \text{ joule}$$

The final energy

$$= \tfrac{1}{2} \times 3{\cdot}2 \times 10^{-6} \times (45)^2$$
$$= 3{\cdot}24 \times 10^{-3} \text{ joule}$$

The difference between these is the energy lost (chiefly as heat in the connecting wires).

$$\therefore \text{ the loss} = 5{\cdot}4 \times 10^{-3} \text{ joule}$$

11.4 Charging and discharging processes

In general the charging and discharging of a capacitor is an oscillatory process. This happens because of the inevitable self-inductance of the circuit in which the capacitor is joined. Even when the wire of the circuit is not formed into a coil, the magnetic flux produced by the current in it links the circuit and gives rise to small self-inductive effects. Thus, if a capacitor is discharged by joining a short length of wire between its plates, the circuit diagram of the arrangement should be as shown in Fig. 11.15, where L represents the small self-inductance of the wire and R its resistance. The variation of the p.d. across the capacitor during the discharge process is shown in Fig. 11.16. At the point A the switch is closed, but the self-inductance delays the growth of current and the discharge only gradually gathers momentum. At B the capacitor is fully discharged, but the rate of discharge—i.e. the current—is a maximum; the effect of self-induction is now to maintain the current, thus charging up the capacitor in the reverse sense. As the p.d. grows in the reverse direction, the current slowly decreases until the point C when the p.d. across the capacitor is again momentarily steady. This process repeats itself many times over, giving rise to the oscillations shown in the figure. These are of decreasing amplitude since at each pulse of current some of the

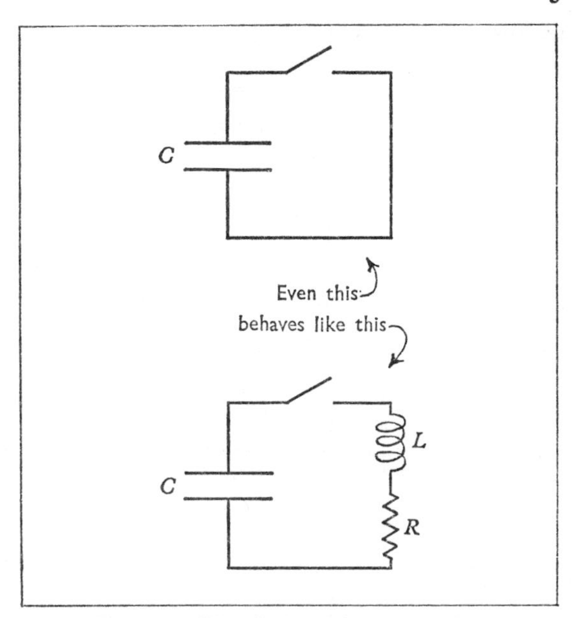

Fig. 11.15 The discharging of a capacitor

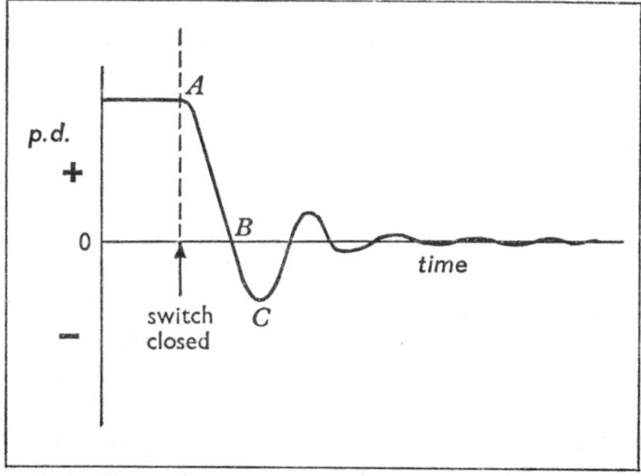

Fig. 11.16 The oscillatory discharge of a capacitor

energy originally stored in the capacitor is turned into heat in the resistance R. Another cause of energy loss from such a circuit is the emission of electromagnetic radiation; it is found that radio waves are emitted from any circuit in which a rapidly changing current flows (p. 263). The discharge of a capacitor will usually be found to make a 'crackle' in a nearby radio set. However, the loss of energy from this cause is almost always much less than the dissipation of heat in the resistance of the circuit.

Increasing the resistance R of such a circuit leads to heavier damping of the oscillations, and for values of R above a certain amount no oscillation occurs at all; the p.d. then falls asymptotically towards zero without ever reversing (Fig. 11.17).

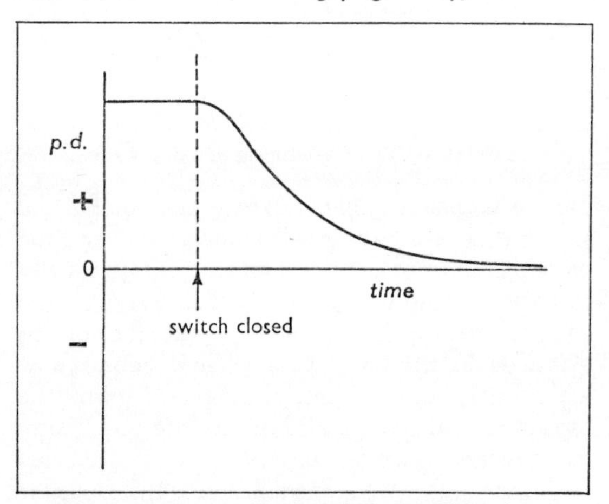

Fig. 11.17 Non-oscillatory discharge with a large resistance

Similar effects occur when a capacitor is charged by connecting it to a source of e.m.f. If the resistance of the circuit is low, the p.d. oscillates before settling down to its final value. But in a high-resistance circuit the p.d. rises without oscillation towards its final value equal to the e.m.f. of the source.

The loss of energy as heat, etc. in the charging process is the same as in the discharge of the capacitor. The source of e.m.f. must in fact supply twice as much energy as is eventually stored in the capacitor. Thus, if the e.m.f. is E and the charge stored Q, then

the energy stored in the capacitor $= \frac{1}{2}QE$

and the energy supplied by the source of e.m.f.

$$= \text{charge} \times \text{e.m.f.} = QE$$

$\therefore$ the energy lost as heat, etc.

$$= QE - \tfrac{1}{2}QE = \tfrac{1}{2}QE$$

This may be compared with the similar behaviour of a spring to which a weight is suddenly attached. The weight proceeds to oscillate about the position in which it will finally come to rest; the energy of oscillation is gradually dissipated as heat by frictional effects. If the weight is W and the final extension it produces X, then it may be shown that in the final rest position

the energy stored in the spring $= \frac{1}{2}WX$

and the potential energy lost by the weight

$$= WX$$

$\therefore$ the energy lost as heat $= WX - \frac{1}{2}WX = \frac{1}{2}WX$

Again, the weight must supply twice as much energy as is eventually stored in the spring, exactly half of it being lost as heat as the oscillation is brought to a stop. If the frictional forces are much larger (suppose the weight is hanging in treacle), then the weight will move to its final position without oscillation; but it will still be true that exactly half the potential energy given up by the weight will be dissipated as heat by friction. A detailed discussion shows that the behaviour of oscillating mechanical systems is closely analogous to that of oscillating electric circuits. The inductance L of the electric circuit occupies the same place in the equations as mass (inertia) in the mechanical system; the capacitance C fulfils the same function as the stretching of the spring (or whatever else is responsible for the storage of potential energy); the resistance R corresponds to the frictional forces that damp the oscillation. Cf. discussion on p. 132 ff.

The full analysis of the behaviour of oscillatory systems is beyond the scope of this book. But we shall consider here the special case of the discharge of a capacitor through a *very large resistance*. The charge q on the capacitor is given by

$$q = Cv$$

where v is the p.d. at any instant across the capacitor (Fig. 11.18). The *discharge* current i is therefore given by

$$i = -\frac{dq}{dt} = -C\frac{dv}{dt}$$

Also $$v = iR$$

$$\therefore \; v = -CR\frac{dv}{dt}$$

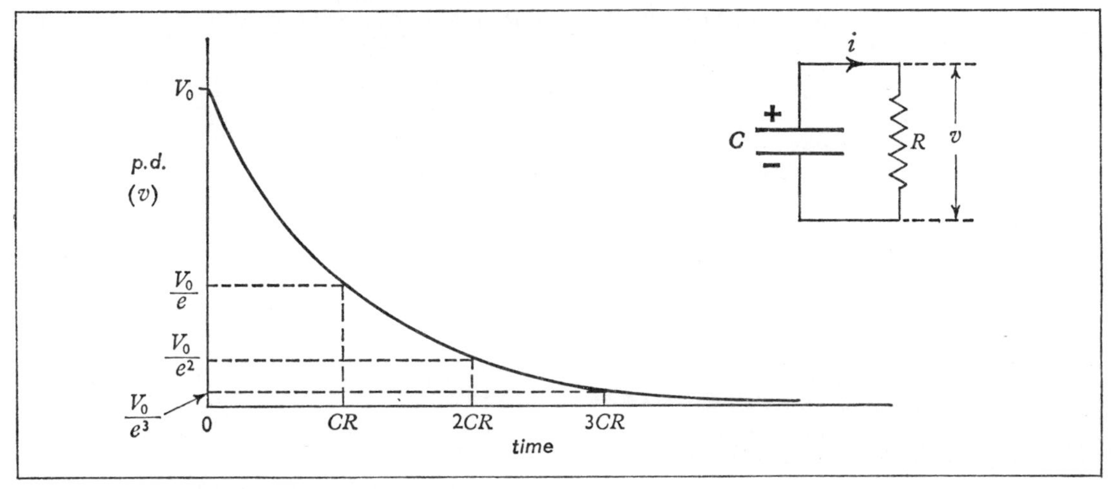

Fig. 11.18 The exponential decay of the p.d. across a capacitor when the inductance of the circuit is negligible

Re-arranging and integrating,

$$\frac{1}{CR}\int_0^t dt = -\int_{V_0}^{v} \frac{dv}{v}$$

where V_0 = the initial p.d.

$$\therefore\ \frac{t}{CR} = -\log_e\left(\frac{v}{V_0}\right)$$

$$\therefore\ v = V_0\,e^{-t/CR}$$

Thus the p.d. across the capacitor decays according to an *exponential law* (Fig. 11.18). The effective time taken by the decay process is measured by the product CR.

Thus, when $t = CR$, $v = \frac{V_0}{e} = 0{\cdot}37V_0$

and when $t = 2CR$, $v = \frac{V_0}{e^2} = 0{\cdot}14V_0$, etc.

The product CR has the dimensions of time and is called the *time constant* of the circuit. It is given in *seconds* when C is in *farads* and R in *ohms*.

A similar result is obtained when a capacitor is being *charged* through a large resistance by a source of e.m.f. E (Fig. 11.19). The differential equation in this case is

$$E = v + CR\frac{dv}{dt}$$

and the solution is

$$v = E(1 - e^{-t/CR})$$

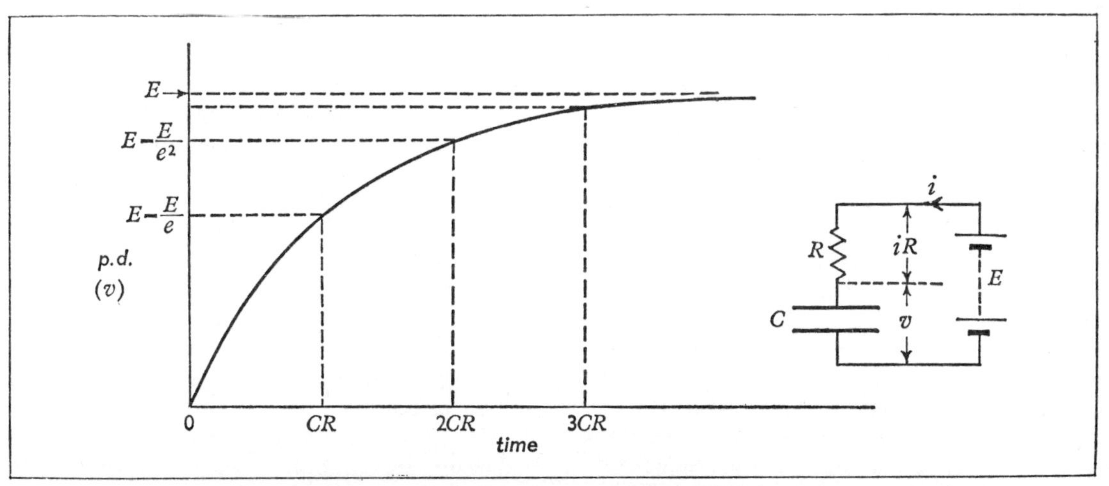

Fig. 11.19 The exponential rise of the p.d. across a capacitor charged through a large resistance

The p.d. therefore rises exponentially (with the same time constant CR) towards its final value equal to the e.m.f. E of the battery.

Measuring ionization currents

At low potentials and in the absence of any particular ionizing agency, the insulating property of air and other gases is almost perfect. However, many types of radiation are able to disrupt gas molecules into separate ions; the conduction of current through the gas can then go on as long as the radiation continues. These ionizing radiations are: (*a*) The shorter wavelength electromagnetic radiations, ultra-violet rays, X-rays and γ-rays;

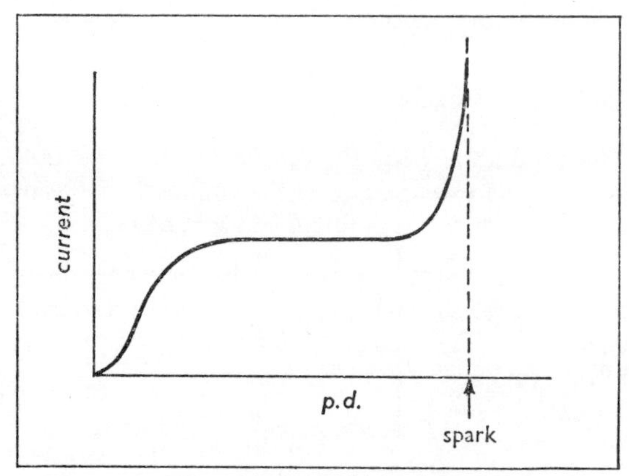

Fig. 11.20 The variation of ionization current with p.d. for an ionization chamber

(*b*) the rapid charged particles arising in nuclear processes, e.g. the α-particles and β-particles produced in radioactive disintegrations.

If ionizing radiation passes through the air between two electrodes, a small current will flow when a suitable p.d. is maintained between them. An arrangement of this sort is called an *ionization chamber*. At low p.d.'s a certain amount of recombination of the ions takes place before they are collected by the electrodes; under these conditions the ionization current increases with the p.d. However, if the p.d. exceeds a certain value, all the ions are collected before they can recombine; the current is then independent of the p.d. and is said to be *saturated* (Fig. 11.20). However, when the p.d. is made several times greater than this the electrons produced by the initial ionization are sufficiently accelerated by the field between collisions to cause further ionization. The number of ions in the gap then grows rapidly, producing a greatly increased current, and perhaps leading to a breakdown of the insulation and a spark. Measurement of the saturation current enables the rate of production of ions to be found.

The ionization currents usually obtained are far too small to give deflections in a moving-coil galvanometer; and instead an electrometer must be used to find the rate of collection of charge by the electrodes. The electrometer is first calibrated for two suitable p.d.'s, V_1 and V_2. These should be in the range of p.d.'s that give saturation currents. The electrometer and ionization chamber are then charged. The deflection slowly falls as the ionization current discharges the arrangement. The time t for the potential to fall from V_2 to V_1 is noted.

Now the charge q on the electrodes at any instant is given by

$$q = Cv$$

where C is the capacitance of the combination and v the p.d. across it. If the ionization current is i, we have

$$i = -\frac{dq}{dt} = -C\frac{dv}{dt} = \frac{C(V_2 - V_1)}{t}$$

since the current is constant. To obtain the actual value of i we need to know the capacitance C. But for comparison purposes we have

$$i \propto \frac{1}{t}$$

Techniques of this kind enable currents as small as 10^{-15} amp to be detected.

The pulse electrometer

This is an instrument specially adapted for the measurement of very small currents. It is also known as the *Wulf electrometer*. The principle may be demonstrated with an ordinary gold-leaf electroscope by mounting a metal plate inside the case a short distance in front of the sensitive leaf (Fig. 11.21); this is called the *counter electrode*. When a charge flows onto the cap, the leaf deflects as usual; but at a certain point its motion becomes unstable, and it flicks quickly across to the counter electrode and so discharges itself. If current continues to flow onto the cap, the movement is repeated again and again; each flick of the leaf corresponds to the transfer of a definite quantity of charge from the

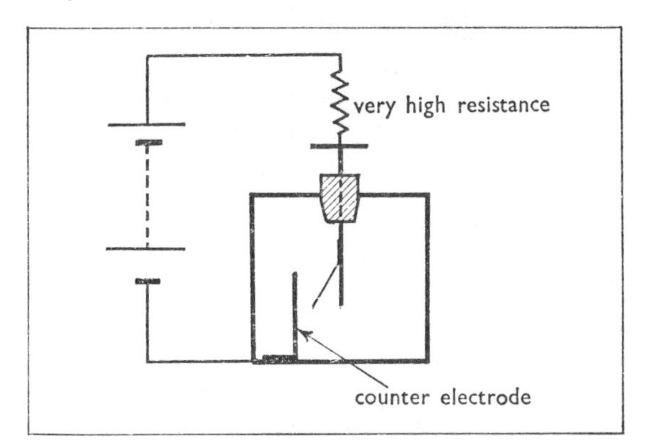

Fig. 11.21 The action of a counter electrode mounted inside a gold-leaf electroscope

cap to the case. The number of discharges per unit time is therefore a measure of the current flowing onto the cap and so through to the case.

In the gold-leaf electroscope the restoring force is provided by gravity, and the motion is rather sluggish. To increase the sensitivity of the device the moving part is made as small as possible and the restoring force is provided by an elastic fibre. It is then possible to register several discharges per second with complete reliability. Fig. 11.22 shows two different types of sensitive systems used in pulse electrometers. In (*a*) the foil resembles that used in a gold-leaf electroscope, but its motion is restrained by a flexible strip attached to its lower edge. In (*b*) the piece of foil is attached like a flag to the centre of a taut fibre, which twists when the leaf deflects towards the counter electrode.

Fig. 11.22*c* shows how the instrument may be connected to measure the current through a very high resistance. The same arrangement could be used to measure an ionization current with the resistor replaced by the ionization chamber. The sensitivity may be adjusted by moving the counter electrode. In its most sensitive condition (with the counter electrode about 1 mm away from the foil) it requires about 200 volts to deflect the foil across the gap; the charge transferred is then rather less than 10^{-9} coulomb. If we have the patience, we might time discharges occurring at the rate of one every 5 minutes; this corresponds to a current of about 10^{-12} amp. It is possible to measure the charge transferred per discharge, and so to calibrate the instrument to give currents in amps. But normally it is used for comparative measurements. The time t is found for a given number of discharges;

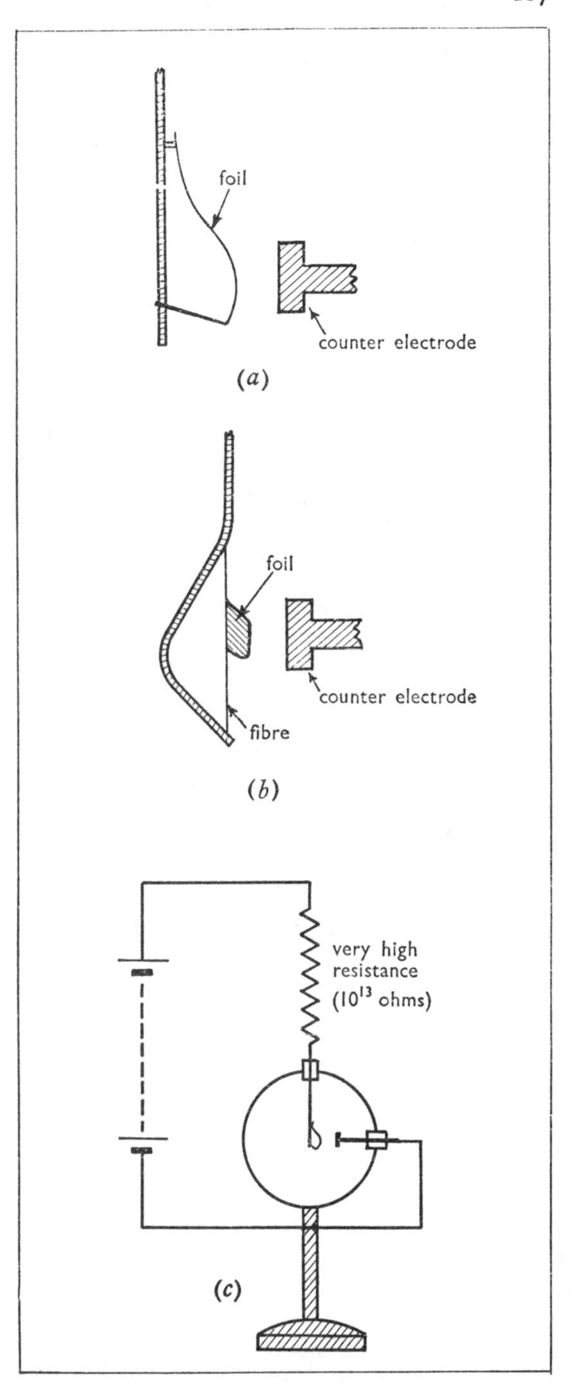

Fig. 11.22 (*a*) and (*b*) The sensitive systems of typical pulse electrometers

(*c*) The use of a pulse electrometer for measuring small currents

then, for a given setting of the counter-electrode, the current i is inversely proportional to t.

While it is operating, the p.d. across the electrometer fluctuates between zero and about 200 volts; but this is small compared with the p.d.'s of several thousand volts usually used in such circuits. It is therefore a reasonable approximation to assume that the entire applied p.d. exists steadily across the resistor or ionization chamber joined in series with the instrument.

The d.c. amplifier. This consists in principle of an electronic amplifier similar to those described in Chapters 14 and 17; however, *direct* connection between the stages is employed, the usual blocking capacitors being eliminated. The small current to be measured is passed through a very high resistance across the input terminals. The p.d. developed across this is amplified, and the current in the output stage is registered by a moving-coil meter. This may be calibrated to read the input current directly.

The instrument may also be adapted to give a direct measure of the total charge delivered to its input terminals. In this case a suitable capacitor is joined across the input and the charge to be measured is given to its plates. The output current now gives an indication of the p.d. across the capacitor and may be calibrated to show the total charge delivered. The capacitor slowly discharges through the high input resistance of the amplifier, but the time constant is made large enough so that the p.d. falls off to a negligible extent over a period of some minutes.

Used in this way the d.c. amplifier enables a number of elementary experiments in electrostatics to be done in a quantitative manner. For instance, an 'ice-pail' may be joined to the input, and the instrument then gives a direct measure of the charge transferred to it by, say, a proof-plane (p. 165).

11.5 The measurement of capacitance

Measurements of this kind are complicated by the irregular behaviour of some dielectrics. When a paper capacitor for instance is charged and then insulated, the p.d. falls for a short time afterwards and then becomes steady. It is evident that the polarization of the dielectric does not reach its final value until some time after the application of the electric field. Likewise, when the capacitor is discharged, the polarization of the dielectric does not immediately fall to zero, with the result that a second smaller discharge may be obtained a short while afterwards, and sometimes several more after that. This phenomenon is known as *dielectric absorption*. With mica capacitors the effect is not significant.

For this reason the results obtained with a paper capacitor vary to some extent with the time for which the capacitor remains charged. With a.c. measurements (in which the polarization is continually being reversed) the results are not affected by dielectric absorption except at very low frequencies, but they differ considerably from the results of d.c. measurements.

(*i*) *Using a ballistic galvanometer.* The capacitor is connected so that it may be alternately charged by a battery and then discharged through the ballistic galvanometer. This technique has already been described as a method of testing whether the charge Q on a capacitor is proportional to the p.d. V across it (p. 154, Fig. 10.12). If the instrument has been calibrated so that its ballistic sensitivity is known, then Q is obtained directly and the capacitance C is given by

$$C = \frac{Q}{V}$$

But usually the method is used only for comparison purposes. Thus, for a given p.d. the throw of the galvanometer is proportional to the capacitance. If the throws observed for two capacitances C_1 and C_2 are θ_1 and θ_2, respectively, then

$$\frac{C_1}{C_2} = \frac{\theta_1}{\theta_2}$$

Most ballistic galvanometers are suitable for measuring charges of about 1 micro-coulomb (10^{-6} coulomb). The p.d. V must therefore be chosen so that the charge Q is of this order of magnitude. For example, with a capacitance of 0·01 μF the p.d. needed is about 100 volts. It can be seen that the method is not suitable for capacitances very much less than this, if we are to avoid using excessive p.d.'s.

By allowing a delay between charging and discharging we may investigate the leakage resistance R of a capacitor. According to the analysis on p. 184, the time required for a given proportion of the charge on the capacitor to leak away depends on the time constant CR; the p.d. v after time t is given by

$$v = V_0 e^{-t/CR}$$

where $V_0 =$ the initial p.d.

$$\therefore \log_e \left(\frac{V_0}{v}\right) = \frac{t}{CR}$$

$$\therefore CR = \frac{t}{\log_e (V_0/v)} = \frac{t}{\log_e (\theta_0/\theta)}$$

where $\theta_0 =$ the initial galvanometer deflection obtained when the p.d. is V_0, and $\theta =$ the deflection obtained when the p.d. has fallen to v.

This technique is readily adapted for measuring a very high resistance. This may be joined as a 'leak' between the plates of the capacitor, and the time constant found as above. Resistances up to about 10^{10} ohms may be found by this method.

(ii) The vibrating reed method. As in the ballistic galvanometer method, the capacitance C is charged to the p.d. of the battery, but is then discharged through a microammeter. But this sequence of connections is performed many times a second by means of a metal strip (or reed) maintained vibrating at a known frequency. One way of doing this is to employ a device known as a *reed-switch*. In this kind of switch the contacts to be joined are carried on the ends of two steel strips fixed in the ends of a glass tube (Fig. 11.23*a*). When the strips are magnetized, unlike poles are formed on their adjacent ends at the centre of the tube, and the contacts are drawn together. The action is extremely rapid, allowing the switch to be closed and opened 400 times a second or more. For the purpose of this experiment a change-over form of the switch is used (Fig. 11.23*b*) in which the moving reed springs back in the unmagnetized condition on another (non-magnetic) contact. The reed switch is mounted inside a small coil which is joined in series with a germanium rectifier (p. 328) to an a.c. supply of known frequency (Fig. 11.24). The switch then closes in the conducting half-cycle of the rectifier, and springs back in the other half-cycle, when the rectifier is non-conducting. The number of discharges per second is thus equal to the frequency of the a.c. supply.

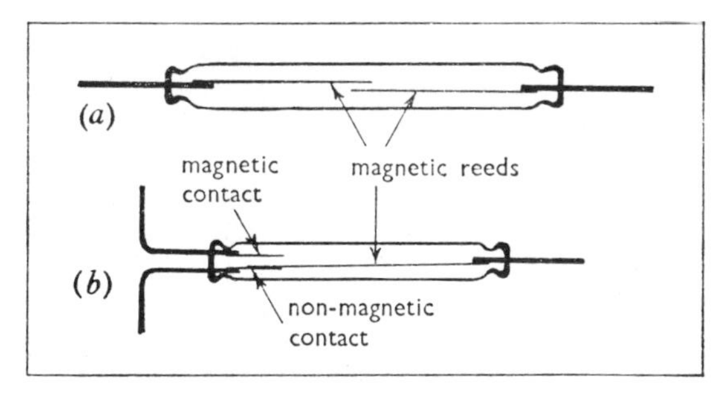

Fig. 11.23 Reed-switches: (*a*) a simple switch, (*b*) a change-over switch

If the pulses of current through the microammeter follow one another at high enough frequency the meter shows a steady deflection which records the *average* current I passing through it. Thus

$$I = \text{charge passing per second}$$

Now the charge Q stored in the capacitor at a potential V is given by

$$Q = CV$$
$$\therefore I = nCV$$

where $n =$ the frequency of vibration of the reed

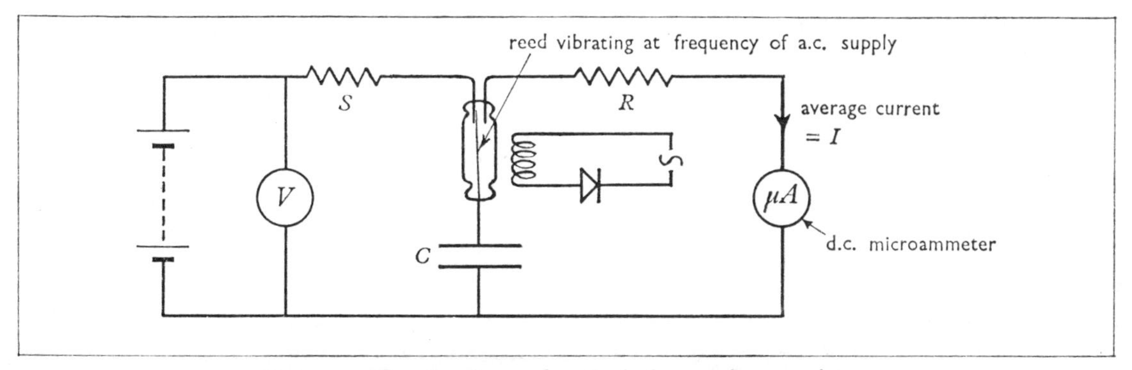

Fig. 11.24 The vibrating reed method of measuring capacitance

i.e. the number of discharges of the capacitor through the microammeter per second. By this means the capacitance C is obtained in terms of the two meter readings and the frequency of the a.c. supply driving the reed.

It is best to include resistances S and R, as shown, in the charging and discharging circuits to avoid excessive pulses of current as the contacts close. It is interesting to observe the effect of varying these resistances. Up to a certain value they have negligible effect on the average discharge current given by the microammeter. However, if R is increased until the time constant CR is comparable with the time during which the reed-switch contacts are closed, only partial discharging of the capacitor takes place, and the microammeter reading falls off. Likewise, if the other resistance S is made too large, the p.d. across the capacitor never rises to the full p.d. available from the battery. A cathode-ray oscilloscope (p. 236) may be joined across the capacitor to show the way in which the p.d. varies during the charging and discharging cycle; and the falling off of the recorded current may be correlated with the change in the waveform seen on the oscilloscope.

The vibrating reed method can be readily adapted for finding the capacitance of a guard-ring capacitor such as that discussed on p. 177 (Fig. 11.5). It may therefore be used as a method of measuring the electric space constant ε_0. In this case two reed-switches are used. One of these is joined to the central plate P in the usual way so that it discharges through the microammeter; the other is joined to the guard-ring and is alternately charged to the same p.d. as P, but is then discharged to *earth*. The two reed-switch coils are connected to the same a.c. supply so that they move in synchronism.

(*iii*) *Simple a.c. methods.* It is shown in the next chapter that the r.m.s. alternating current I 'through' a capacitance C and the r.m.s. p.d. V across it are connected by the relation

$$\frac{V}{I} = \frac{1}{\omega C}$$

where $\omega = 2\pi \times$ frequency.

The capacitance C may therefore be measured with a.c. ammeter and voltmeter as in Fig. 11.25*a*, using the a.c. mains as a supply of known frequency.

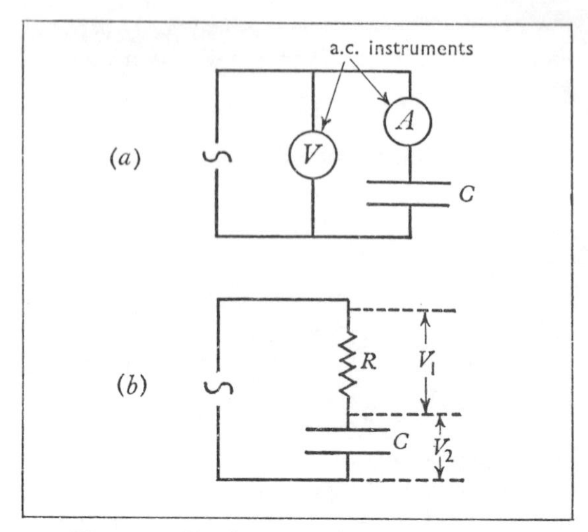

Fig. 11.25 Simple a.c. methods for measuring capacitance

The resistance of the ammeter to some extent affects the current; to avoid excessive error it is necessary for this to be small compared with $1/\omega C$.

Another arrangement is that shown in Fig. 11.25*b*. A known non-inductive resistance R is joined in series with the capacitor, and a cathode-ray oscilloscope is used as a high-impedance a.c. voltmeter (p. 236) to measure in turn the p.d.'s V_1 and V_2 across the resistor and capacitor, respectively. Then

$$V_1 = IR$$

and

$$V_2 = I\frac{1}{\omega C}$$

$$\therefore \frac{V_1}{V_2} = \omega CR$$

enabling C to be calculated.

To make a satisfactory measurement, V_1 and V_2

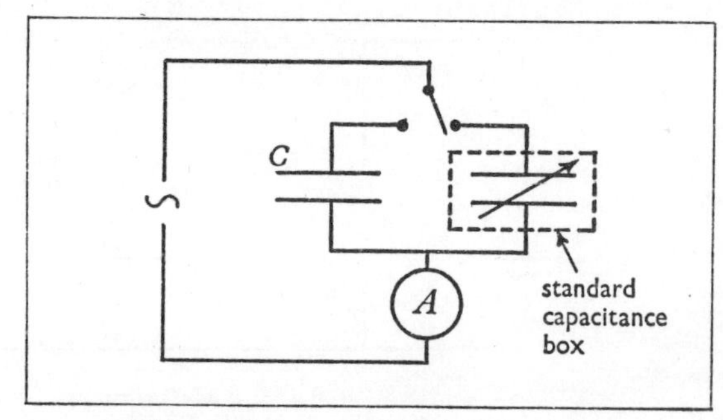

Fig. 11.26 An a.c. substitution method

must be of the same order of magnitude, and the values of R and of the frequency must be chosen accordingly.

There are many ways of devising substitution techniques for measuring capacitances, if a set of standard capacitors can be used. Capacitance 'boxes' are available which are designed so that any chosen value may be obtained by manipulating switches or plugs. One arrangement using the a.c. mains and an ammeter is shown in Fig. 11.26. In this case we do not depend on the calibration of the ammeter, nor does its resistance affect the result; nor does the frequency of the supply need to be known, as long as it is constant. The current reading is taken first with the unknown capacitance in circuit. The capacitance box is then substituted for this and its capacitance varied until the ammeter reading is brought back to its original value; and so the capacitances can be equated.

(*iv*) *A bridge method.* Many types of a.c. bridge circuits have been devised for measuring capacitance (and other quantities). One of the simplest, known as the De Sauty bridge, is shown in Fig. 11.27. It resembles the Wheatstone bridge, except that in two arms of the bridge the resistors are replaced by capacitors. C_1 is the unknown capacitance, and C_2 is a known fixed capacitance; R_1 is a known fixed resistance, and R_2 is an adjustable resistance box. A suitable alternating p.d. is connected across the diagonal of the bridge AC; and the sound in the headphones joined between B and D is adjusted to zero by varying the resistance R_2. When the bridge is balanced, no current flows through the headphones, and the current I_1 is the same 'through' both capacitors; likewise the current I_2 is the same through both resistors. Also at balance

$$\text{p.d. across } C_1 = \text{p.d. across } R_1$$

$$\therefore\ I_1\frac{1}{\omega C_1} = I_2R_1$$

Also $$\text{p.d. across } C_2 = \text{p.d. across } R_2$$

$$\therefore\ I_1\frac{1}{\omega C_2} = I_2R_2$$

Dividing these equations

$$\therefore\ \frac{C_2}{C_1} = \frac{R_1}{R_2}$$

The balance condition does not depend on the frequency of the supply. With this bridge it is

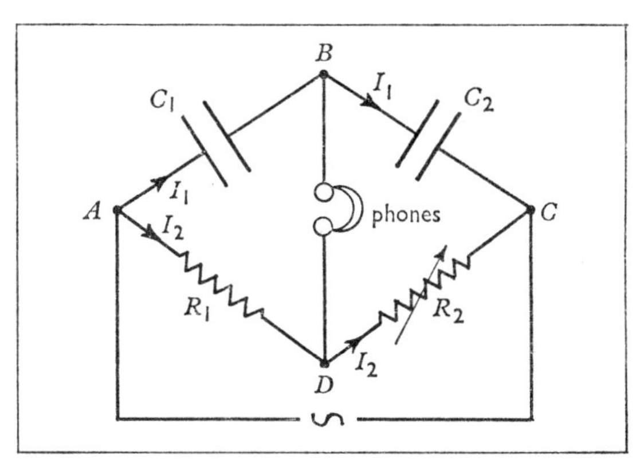

Fig. 11.27 The De Sauty bridge for comparing capacitances

therefore possible to use a mixture of frequencies such as would be produced in a circuit containing a simple make-and-break buzzer run from a battery. This is not the case with most types of a.c. bridge circuits; in general, the balance condition depends also on the frequency, and a source producing a pure sinusoidal waveform must be used.

In practice it is often impossible to reduce the sound in the headphones actually to zero. This is because of the stray capacitances between the experimenter (who is earthed) and the headphones and other parts of the apparatus. But for frequencies up to 1000 Hz* the effect of the stray capacitances may usually be reduced to manageable proportions by earthing a suitable point in the buzzer or other a.c. supply arrangement. The correct balance then produces a clear *minimum* of sound in the headphones, even though it is not strictly zero.

(*v*) *The method of sharing charges.* This is a d.c. method in which the capacitance of a calibrated electrometer is compared with another capacitance (Fig. 11.28). The electrometer is first given a charge, and the p.d. V_1 across it is given by its deflection. The capacitor under test is then joined in parallel with it so that the charge on the electrometer is shared between the two. The electrometer deflection decreases, and the new p.d. V_2 is noted. If the capacitance of the electrometer is C_1 and that of the other capacitor C_2, the charge Q stored in the combination is given by

$$Q = C_1V_1 = (C_1 + C_2)V_2$$

* I.e. 1000 c/s.

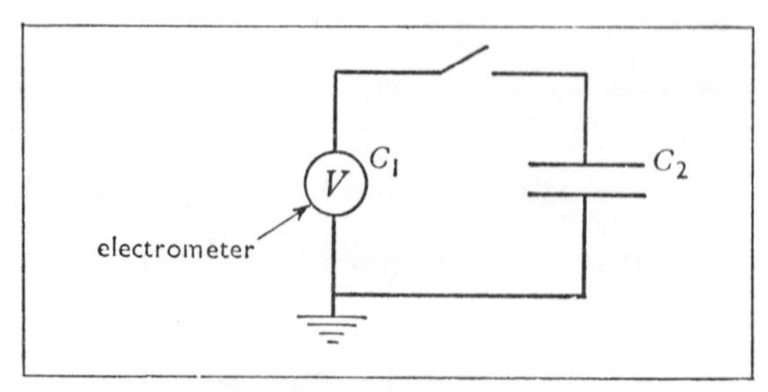

Fig. 11.28 The method of sharing charges for comparing capacitances

since the charge Q remains unaltered when C_2 is joined in parallel;

$$\therefore \frac{V_1}{V_2} = \frac{C_1 + C_2}{C_1} = 1 + \frac{C_2}{C_1}$$

$$\therefore \frac{C_2}{C_1} = \frac{V_1}{V_2} - 1$$

If C_2 is a known capacitance, the capacitance C_1 of the electrometer may thus be found in terms of it. Once C_1 is known, the procedure may be used to measure any other capacitance. It is clearly necessary for the two capacitances to be of the same order of magnitude. The electrometer capacitance is probably about 10 pF, so that the method is suitable for measuring very small capacitances. To use this technique for large values we can increase the capacitance of the electrometer artificially by joining another known capacitance in parallel with it.

11.6 Further calculations of capacitance

A parallel-plate capacitor partly filled with a layer of dielectric (Fig. 11.29).

It follows from Gauss's theorem that the electric flux is continuous across the capacitor from the positive charge on one plate to the negative charge on the other. The flux density D must therefore be the same throughout (ignoring edge effects).

Now $$D = \varepsilon E$$

Therefore the intensity E in any layer is inversely proportional to its permittivity ε. But E is also the potential gradient in the layer. Suppose the layer of dielectric is of thickness x and permittivity ε. The potential drop across it is

$$\frac{D}{\varepsilon}.x$$

This is the same as the potential drop across a layer of air (permittivity ε_0) of thickness

$$\frac{\varepsilon_0}{\varepsilon}.x = \frac{x}{\varepsilon_r}$$

where ε_r = the dielectric constant of the layer. If the combined thickness of the layers of air is b, the capacitance is the same as that of an air capacitor of total thickness

$$b + \frac{x}{\varepsilon_r}$$

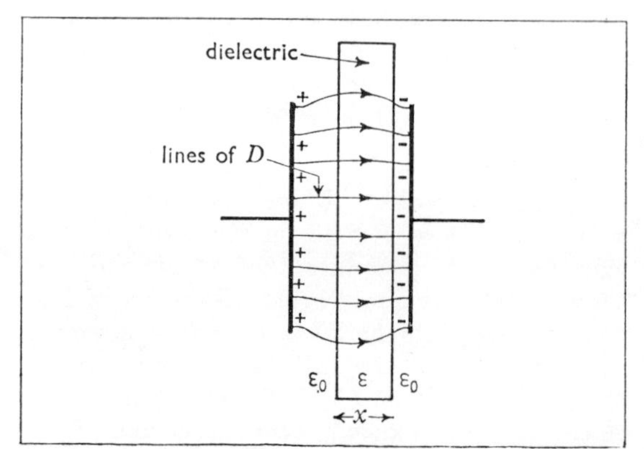

Fig. 11.29 A parallel-plate capacitor partly filled with a layer of dielectric

This result may be made the basis of a method of measuring dielectric constant. A capacitor consisting of a pair of parallel plates is connected across an electrometer. The earthed plate rests against the end of a micrometer screw so that the separation of the plates can be changed by a measurable amount. The capacitor is charged and the deflection of the electrometer noted. The slab of material of unknown dielectric constant ε_r and thickness x is now introduced between the plates, and the deflection falls. The deflection, and therefore the capacitance, are then restored to their original values by separating the plates by an additional amount y. If the original separation of the plates is t, the total thickness of the layers of *air* after moving the plates apart becomes

$$(t + y - x)$$

Using the result just proved, the final capacitance is equal to that of an air capacitor of thickness

$$(t + y - x) + \frac{x}{\varepsilon_r}$$

But the actual thickness of the air capacitor of this capacitance was t.

$$\therefore\ t = t + y - x + \frac{x}{\varepsilon}$$

$$\therefore\ \varepsilon_r = \frac{x}{x - y}$$

Concentric cylinders

This is an important practical case, since a coaxial cable is in effect a capacitor of this kind. If the cylinders are very long, end effects can be ignored, and the field must be everywhere radial (by symmetry). Imagine a closed cylindrical surface S, of length l and radius r, between the two conductors and concentric with them (Fig. 11.30). The flux through S is $\varepsilon E.2\pi rl$, where E is the intensity of the field at a distance r from the centre. If the charge on the inner conductor is q per unit length, then by Gauss's theorem we have

$$\varepsilon E.2\pi rl = ql$$

$$\therefore\ E = \frac{q}{2\pi\varepsilon r}$$

But E is also the potential gradient (p. 157).

$$\therefore\ E = -\frac{dV}{dr}$$

$$\therefore\ -\frac{dV}{dr} = \frac{q}{2\pi\varepsilon r} \qquad (1)$$

(The minus sign implies that with *positive* charge on the inner conductor the potential *falls* with increasing r.) Integrating, the p.d. V between the two conductors is given by

$$V = \frac{q}{2\pi\varepsilon}\int_a^b \frac{dr}{r} = \frac{q}{2\pi\varepsilon}\log_e\frac{b}{a}$$

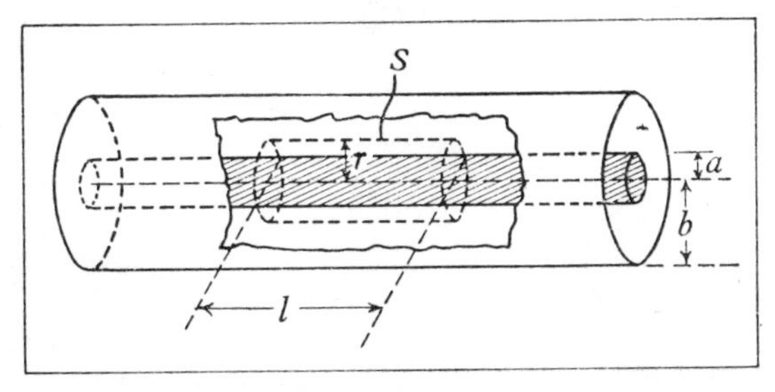

Fig. 11.30 A capacitor consisting of two concentric cylinders (or a coaxial cable)

where a and b are the radii of the surfaces of the two conductors. The capacitance C of the length l of the arrangement is given by

$$C = \frac{ql}{V} = \frac{2\pi\varepsilon l}{\log_e(b/a)}$$

$\therefore$ the capacitance per unit length

$$= \frac{C}{l} = \frac{2\pi\varepsilon}{\log_e(b/a)}$$

A typical polythene-insulated coaxial cable has a capacitance of about 75 pF per metre.

A pair of long, parallel wires

This also is an important practical case, since the transmission lines used for power distribution and for telephones are often of this kind. To simplify the analysis we shall suppose that the radius a of the conductors is small compared with their separation d. This enables us to assume that the charges $+q$ and $-q$ on unit length of the wires is uniformly distributed on their surfaces, so that we may use equation (1) above to calculate the potential due to the charges carried by each conductor. To preserve the symmetry of the arrangement we shall take the zero of potential as that of the centre plane between the conductors (Fig. 11.31).

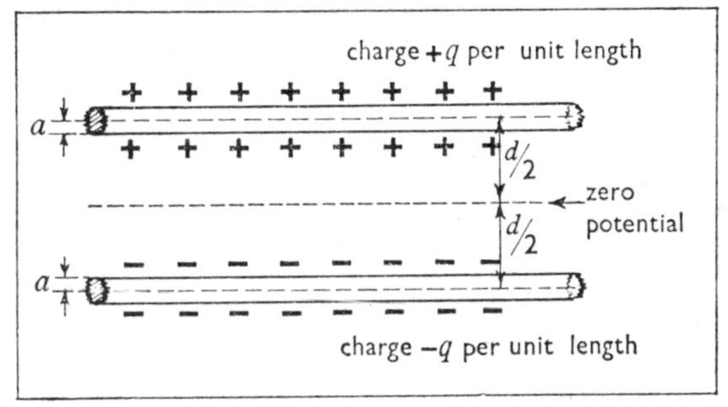

Fig. 11.31 The capacitance of two parallel wires

Let r_1 and r_2 be the distances of any point from the centres of the positive and negative conductors, respectively. Then the potential V_1 at the surface of the positive conductor *due to its own charge* is given by

$$V_1 = \int_{d/2}^{a} \frac{dV_1}{dr_1}.dr_1 = -\frac{q}{2\pi\varepsilon_0}\int_{d/2}^{a}\frac{dr_1}{r_1}$$

$$= \frac{q}{2\pi\varepsilon_0}\log_e\frac{d}{2a}$$

The potential V_2 at the same point *due to the*

charge on the other conductor is given similarly by

$$V_2 = \int_{d/2}^{d-a} \frac{dV_2}{dr_2} . dr_2 = -\frac{-q}{2\pi\varepsilon_0} \log_e \frac{d-a}{d/2}$$

$$= \frac{q}{2\pi\varepsilon_0} \log_e 2 \quad (\text{since } a \ll d)$$

The total potential at this point is then

$$V_1 + V_2 = \frac{q}{2\pi\varepsilon_0} \left(\log_e \frac{d}{2a} + \log_e 2 \right)$$

$$= \frac{q}{2\pi\varepsilon_0} \log_e \frac{d}{a}$$

The potential at the surface of the negative conductor lies the same amount below zero. The p.d. V between them is therefore given by

$$V = 2(V_1 + V_2) = \frac{q}{\pi\varepsilon_0} \log_e \frac{d}{a}$$

The capacitance C of a length l of the transmission line is given by

$$C = \frac{ql}{V} = \frac{\pi\varepsilon_0 l}{\log_e (d/a)}$$

∴ the capacitance per unit length is

$$\frac{C}{l} = \frac{\pi\varepsilon_0}{\log_e (d/a)}$$

The factor $\log_e (d/a)$ lies in almost all cases between 3 and 6. The capacitance of a parallel-wire transmission line therefore usually falls between 4 and 8 pF per metre.

12: Alternating Currents

12.1 Describing alternating quantities

Alternating currents and p.d.'s may have many different waveforms—saw-toothed, square-wave, sinusoidal, and many irregular forms; some examples are shown in Fig. 12.1. Each of these waveforms has its own special uses in electrical technology. But the simplest form, in terms of which all others may be analysed, is the *sinusoidal waveform* (Fig. 12.1*c*), expressed by

$$i = i_0 \sin \omega t \qquad \text{(p. 27)}$$

and $$v = v_0 \sin \omega t$$

where $\omega = 2\pi \times$ frequency. If the alternating current is being generated by a simple two-pole alternator (Fig. 9.3, p. 136), then ω is the angular velocity of the rotor in radians per second; i_0 and v_0 are the ***peak values*** of the current and p.d.

In an earlier chapter it was shown that, in many cases, the power produced in a.c. circuits may be calculated in the same way as for d.c. circuits (section 2.8, p. 27). This applies when the electric and magnetic fields of the circuit have negligible effect on the currents in it; that is, when capacitative and self-inductive effects are negligible and the currents are controlled almost entirely by the resistances of the circuit. In this case the current i in a resistance R and the p.d. v across it are proportional to one another at every instant, and

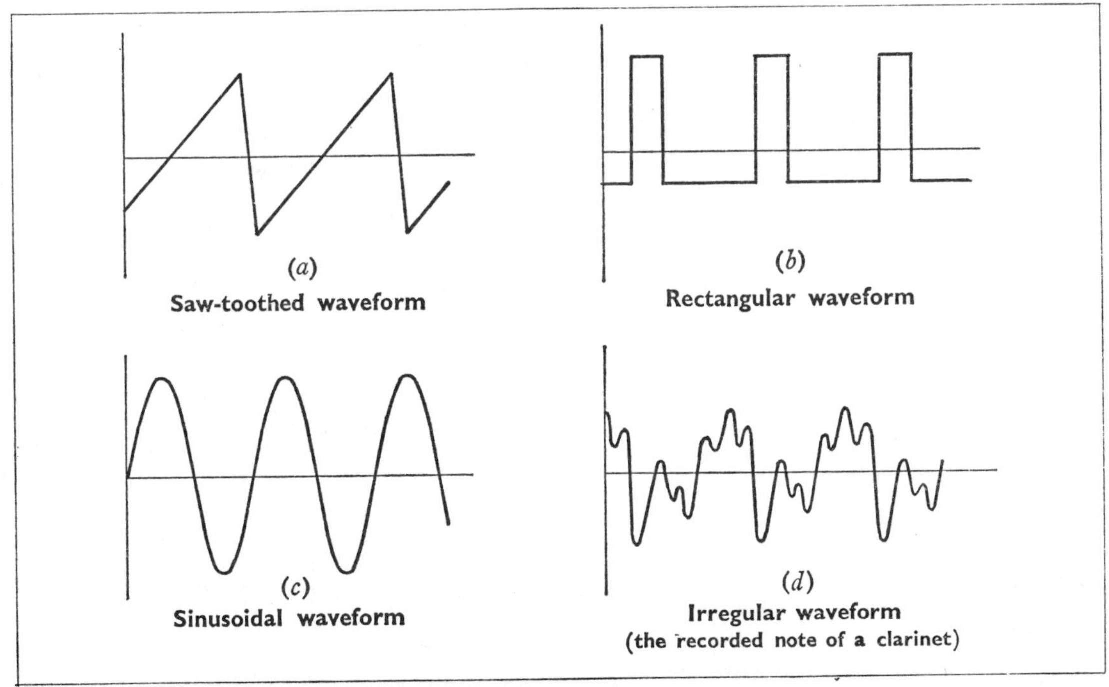

Fig. 12.1 Examples of a.c. waveforms

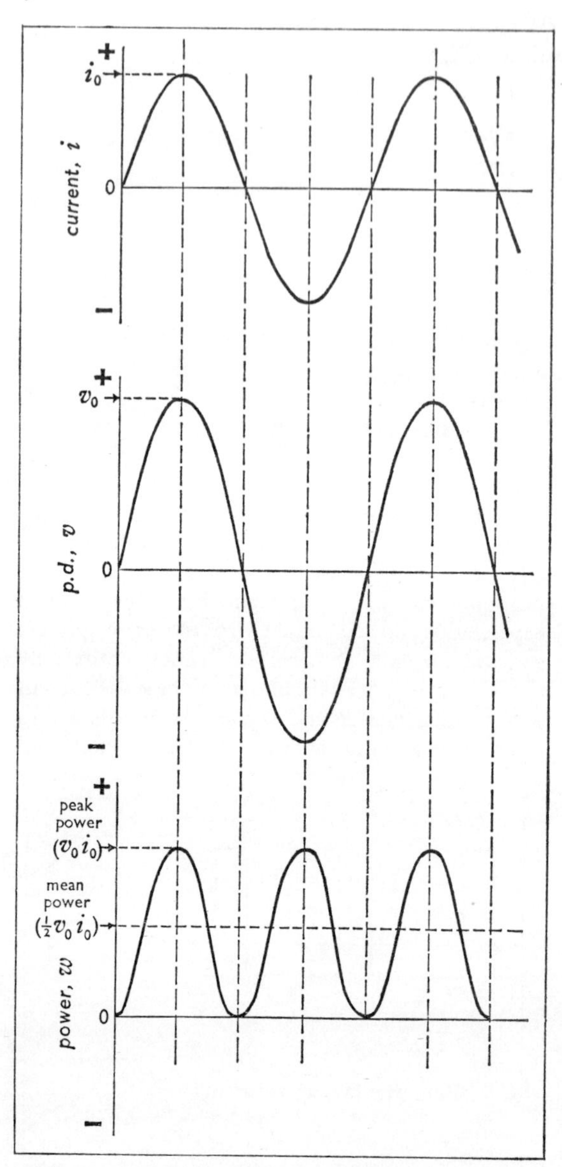

Fig. 12.2 Parallel graphs of current, p.d. and power dissipated, for a resistor

we have

$$\frac{v}{i} = R$$

The current and p.d. are therefore *in phase*—i.e. the peaks of current and p.d. occur at the same moments. The power w dissipated in the resistor *at any instant* is given by the three alternative formulae

$$w = vi = i^2R = \frac{v^2}{R}$$

and the *mean power* W by

$$W = VI = I^2R = \frac{V^2}{R}$$

where I and V are the r.m.s. values of current and p.d. In the special case of a sinusoidal waveform it was shown that

$$I = \frac{i_0}{\sqrt{2}} \quad \text{and} \quad V = \frac{v_0}{\sqrt{2}} \qquad \text{(p. 28)}$$

In terms of the peak values i_0 and v_0 the mean power W is then given by

$$W = \tfrac{1}{2}v_0 i_0 = \tfrac{1}{2}i_0{}^2R = \tfrac{1}{2}(v_0{}^2/R)$$

The relationship between these quantities is displayed in the parallel graphs of current, p.d. and power in Fig. 12.2.

But in general the behaviour of alternating current circuits is not as simple as this. When capacitative or inductive effects are appreciable, the currents are no longer controlled by the resistances only, nor are the currents and p.d.'s necessarily in phase—i.e. the peaks of current and p.d. may occur at different moments. The details of these processes are discussed later in the chapter; for the moment we are concerned with the way of describing these *phase differences*, as they are called.

Vector diagrams

Any sinusoidal alternating quantity may be represented by the following geometrical model (Fig 12.3):

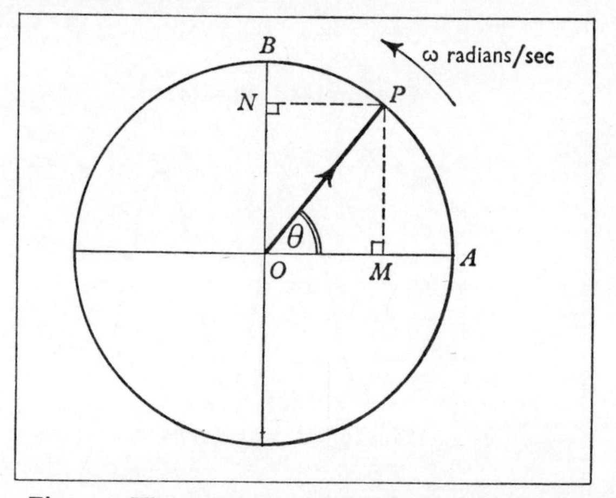

Fig. 12.3 The geometrical model of a rotating vector to represent an alternating quantity

OP is a line (or *vector*, as it is called), which we imagine to be rotating at a steady angular speed ω radians per second about the point O. If OP makes an angle θ with the line OA, then

$$\theta = \omega t$$

where t is the time (in seconds), measured from the point where $\theta = 0$. Then the projections of OP on OB and OA respectively are

$$ON = OP \sin \omega t \quad \text{and} \quad OM = OP \cos \omega t$$

In the language of vectors, we may say that $OP \sin \omega t$ and $OP \cos \omega t$ are the *components* of the vector OP in the directions OB and OA. Now, if the length of the vector OP is made proportional to the peak value of the alternating quantity to be represented, then the component of OP in a suitably chosen direction gives its instantaneous value.

The angle θ in this model is called the *phase* of the current or p.d. Thus, the full description of a sinusoidal alternating quantity requires the specification of

(*i*) its frequency
(*ii*) its peak value
(*iii*) its phase

The vector model provides a simple means of showing the relations between these quantities for two or more alternating currents or p.d.'s.

Fig. 12.4 shows the graphs of the current i through a piece of apparatus and the p.d. v across it; these are of the same frequency but are not in phase. Alongside is shown the vector diagram representing the same current and p.d.; the vectors are drawn for the moment X in the neighbouring graphs. As time advances, the whole vector diagram rotates at steady speed, but the angle between the two vectors remains constant; this angle is the ***phase difference*** ϕ between v and i. In algebraic terms i and v are given by

$$i = i_0 \sin \omega t$$

and

$$v = v_0 \sin (\omega t + \phi)$$

12.2 Capacitative circuits

When an alternating p.d. is connected across the plates of a capacitor, charge must be continually flowing onto and off the plates as the p.d. between them is changing. This means that an alternating current will flow in the wires connected to the capacitor, although there is no continuous conducting path by which a direct current could pass. In this sense we often speak of an alternating current flowing 'through' a capacitor, but this must be understood as a convenient scientific idiom. This process has already been outlined on p. 154, but we must now discuss it in more detail.

The behaviour of a capacitor in an alternating current circuit is quite different from that of a resistor. In both, the r.m.s. current that flows is directly proportional to the r.m.s. p.d. applied. But with the capacitor there is a phase difference of 90° between current and p.d.; i.e. the current reaches

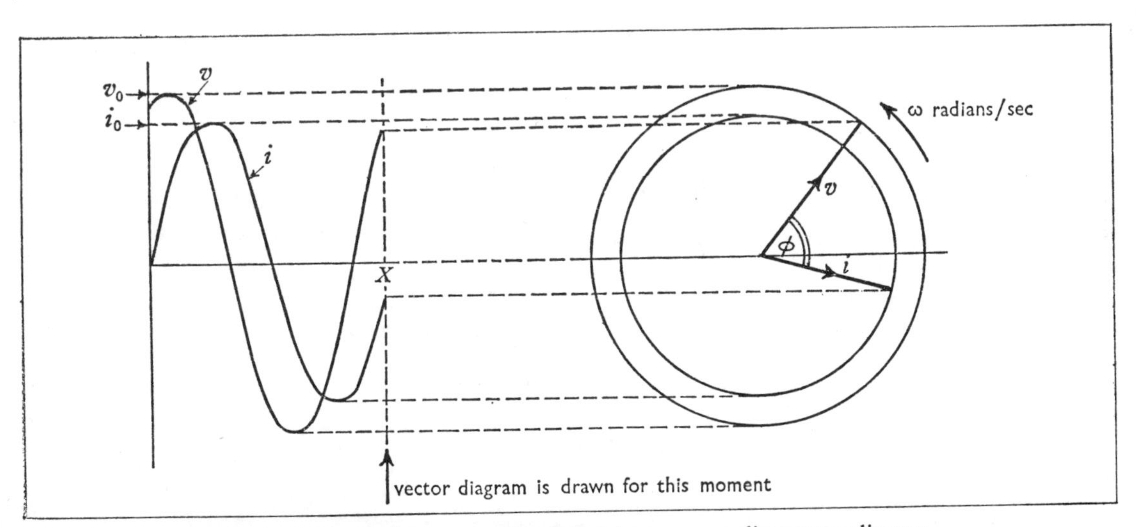

Fig. 12.4 Curves of current and p.d. and the corresponding vector diagram

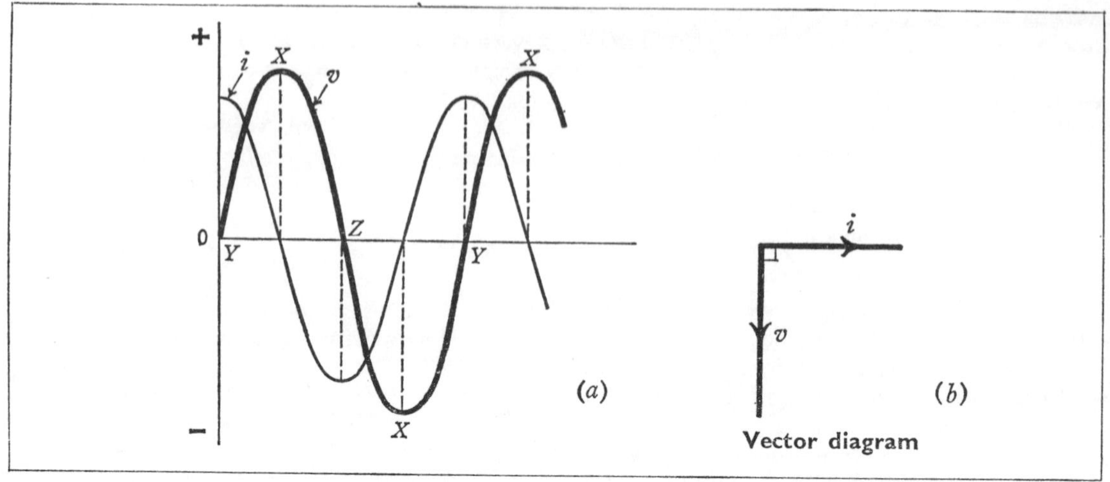

Fig. 12.5 Curves of current and p.d. for a capacitor, and the corresponding vector diagram

its maximum value at the instants when the applied p.d. is momentarily zero (Fig. 12.5*a*). This is often described by saying that the current and p.d. are *in quadrature*.

This behaviour can be explained by the following reasoning. The charge on the plates of a capacitor is directly proportional to the p.d. between them. At the moments marked X in Fig. 12.5*a* the applied p.d. is a maximum, but is momentarily unchanging; therefore the charge on the plates is momentarily constant, and the current flowing on to them (which is the rate of flow of charge) is zero. Now consider an instant such as Y, when the applied p.d. is momentarily zero, but is increasing at the maximum rate; the rate of flow of charge onto the plates (i.e. the current) is then also a maximum in the *positive* sense. Likewise, at an instant such as Z, the applied p.d. is again zero, but is now decreasing at the maximum rate; the current is again a maximum but in the *negative* sense. Thus the current and p.d. are 90° out of phase as shown. It will be noticed that each peak of current occurs *less far along the time axis* (i.e. earlier) than the corresponding peak of p.d. Thus the peaks of current are 90° ahead of the peaks of p.d. This is described by saying that '*the current leads* 90° *on the voltage*'. Fig. 12.5*b* shows the vector diagram representing this situation.

If we increase the frequency f of the supply while keeping the applied p.d. constant, the same charge as before has to flow onto and off the plates, but it now has to do this in a proportionately shorter time. The current I is therefore proportional to the frequency.

$$I \propto f$$

Increasing the capacitance C, while keeping the frequency and the applied p.d. constant, causes a larger charge to flow onto and off the plates in the same time, and so again increases the current. Thus

$$I \propto C$$

★ Using calculus methods, the reasoning of this section can be put rather more briefly and effectively. Let the applied p.d. v be given by

$$v = v_0 \sin \omega t$$

where $\omega = 2\pi \times$ (frequency). Then the charge q on the capacitor plates is given by

$$q = Cv = Cv_0 \sin \omega t$$

Now

$$\text{the current } i = \frac{dq}{dt} = \omega C v_0 \cos \omega t$$

$$= \omega C v_0 \sin (\omega t + \pi/2)$$

The current therefore leads 90° on the voltage, as represented in the curves of Fig. 12.5*a*. The expression shows also that the current is proportional to both the frequency and the capacitance. The peak current i_0 is given by

$$i_0 = \omega C v_0$$

$$\therefore \frac{v_0}{i_0} = \frac{1}{\omega C}$$

The same relationship also holds between V and I, the r.m.s. values of p.d. and current, since

$$V = \frac{v_0}{\sqrt{2}} \quad \text{and} \quad I = \frac{i_0}{\sqrt{2}}$$

$$\therefore \frac{V}{I} = \frac{1}{\omega C}$$

The quantity $(1/\omega C)$ is known as the *reactance* of the capacitor. It has the same dimensions as *resistance* and is therefore measured in *ohms*. But the two concepts should not be confused. In a resistance the current and p.d. are in phase, and heat is evolved. In a reactance the current and p.d. are in quadrature, and there is no net dissipation of energy (p. 201).

12.3 Inductive circuits

When an alternating current flows in an inductive coil (or *inductor*), it produces an alternating flux linkage with it. By Faraday's law of electromagnetic induction this gives rise to e.m.f.'s in the coil, and according to Lenz's law these must act in opposition to the applied p.d. The way in which the back e.m.f.'s can limit the current has already been discussed in general terms on p. 104; but we must now deal with the matter in greater detail. For the present we shall confine our attention to the case in which the self-inductance of the coil is large and its resistance small, so that the latter has negligible effect on the current.

The behaviour of an inductor connected to an alternating supply resembles that of a capacitor in that the current and p.d. are again in quadrature; however, with an inductor the current *lags* 90° behind the applied p.d. (Fig. 12.6*a*). This may be seen as follows. By Faraday's law, the back e.m.f. induced in the coil is proportional to the rate of change of the flux linkage of the magnetic field with it; for air-cored coils or those with 'ideal' magnetic cores, the flux linkage is proportional to the current. Therefore the back e.m.f. is proportional to the rate of change of current. At the moments marked X in Fig. 12.6*a* the current (and with it the flux linkage) is at a maximum value (in one direction or the other), but is momentarily unchanging; therefore the back e.m.f. is zero at these moments. Now consider an instant such as Y, when the current is momentarily zero but is increasing at the maximum rate; the back e.m.f. is therefore a maximum, but in the *negative* sense, since by Lenz's law it must act so as to tend to prevent the change of current. Likewise, at an instant such as Z, the current is again zero, but is now decreasing at the maximum rate; the back e.m.f. is therefore a maximum in the *positive* sense. Now, assuming the resistance of the coil to be small, the applied p.d. must at every instant be almost equal and opposite to the back e.m.f. to maintain the alternating current. It can be seen from Fig. 12.6*a* that the current and p.d. are 90° out of phase. Also the peaks of current occur further along the time axis than the corresponding peaks of p.d., and thus occur *after* them. This is described by saying that '*the current lags* 90° *on the voltage*'. This situation is represented by the vector diagram in Fig. 12.6*b*.

In this case an increase in the frequency f of the supply allows a shorter time for the changes of current to take place. But if the applied p.d. is to

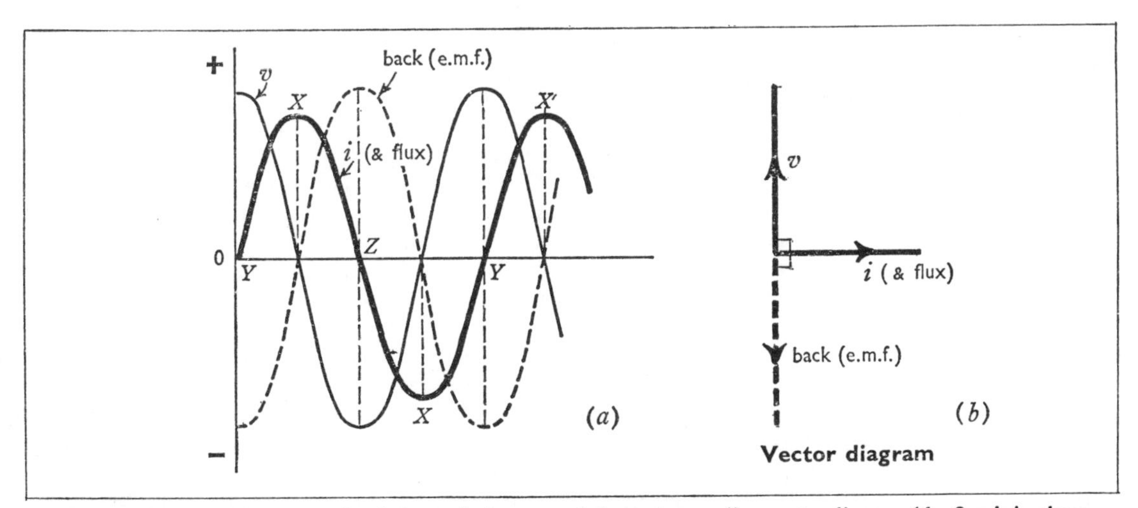

Fig. 12.6 Curves of current and p.d. for an inductor, and the corresponding vector diagram (the flux is in phase with the current)

remain constant, then the back e.m.f. must also be constant, and therefore so also must the rate of change of current. The current is therefore inversely proportional to the frequency.

$$I \propto \frac{1}{f}$$

If the self-inductance L of the inductor is increased, this again leads to an increase in back e.m.f. unless the current decreases in proportion. Therefore for constant applied p.d. the current varies inversely with the self-inductance.

$$I \propto \frac{1}{L}$$

★ Using calculus methods we may again greatly shorten this reasoning. Let the current flowing in the inductor be given by

$$i = i_0 \sin \omega t$$

where $\omega = 2\pi \times$ (frequency). Then by Neumann's law (p. 106), the back e.m.f. e is given by

$$e = -\frac{dN}{dt}$$

where N = the flux linkage of the magnetic field with the inductor.

Now

$$N = Li = Li_0 \sin \omega t$$

$$\therefore\ e = -\omega Li_0 \cos \omega t = -\omega Li_0 \sin (\omega t + \pi/2)$$

The applied p.d. v is given by

$$v = -e = +\omega Li_0 \sin (\omega t + \pi/2)$$

The current and p.d. are thus 90° out of phase, and the p.d. is leading (or the current lagging), as represented in the curves of Fig. 12.6*a*. The expression also shows that the current is inversely proportional to both the frequency and the self-inductance. The peak p.d. v_0 is given by

$$v_0 = \omega Li_0$$

$$\therefore\ \frac{v_0}{i_0} = \frac{V}{I} = \omega L$$

where V and I are the r.m.s. values of p.d. and current. The quantity ωL is known as the (inductive) ***reactance*** of the coil; like capacitative reactance it is measured in *ohms*.

12.4 Power calculations

Consider the general case of a circuit in which sinusoidal alternating current flows; suppose there is a phase difference ϕ between the current i and p.d. v. Then we can write

$$i = i_0 \sin \omega t$$

and $$v = v_0 \sin (\omega t + \phi)$$

The instantaneous power supplied w is given by

$$w = vi = v_0 i_0 \sin \omega t . \sin (\omega t + \phi)$$

The mean value of this is no longer $\frac{1}{2}v_0 i_0$, as with a purely resistive circuit, in which current and p.d. are in phase. There is a well known trigonometrical identity, which in its general form runs

$$\sin A \sin B = \tfrac{1}{2}[\cos (B - A) - \cos (B + A)]$$

Applying this to the above expression, we have

$$w = \tfrac{1}{2}v_0 i_0[\cos \phi - \cos (2\omega t + \phi)]$$

The first term in the bracket is constant, while the second term is an oscillating quantity whose mean value taken over a whole cycle must be zero. Therefore W, the mean power supplied, is given by

$$W = \tfrac{1}{2}v_0 i_0 \cos \phi = VI \cos \phi$$

where V and I are the r.m.s. values of p.d. and current.

The product VI is sometimes called the *apparent power*. It is equal to the true power W in the special case of v and i being in phase ($\phi = 0$); $\cos \phi$ is known as the power factor of the piece of apparatus. Thus

$$\textit{power factor} = \frac{\textit{true power}}{\textit{apparent power}}$$

This expression is taken as the definition of power factor in all cases, including those of non-sinusoidal currents and p.d.'s. It is only in the special case of sinusoidal waveforms that the power factor is equal to $\cos \phi$.

In specifying the maximum rating of many alternating-current devices, such as motors and transformers, it is common practice to state the maximum permissible apparent power rather than the true power. To avoid confusion it is usual then to quote apparent powers in volt-amps (VA) or kilo-volt-amps (kVA) rather than in watts. The reason for this practice may be seen from a numerical example:

A transformer designed for connection to the 200-V mains is rated at 2 kVA. If the load connected to its secondary is such as to make the whole circuit behave resistively ($\phi = 0$), the maximum power that could be taken from the mains is 2 kW. The primary current is then 2000/200 = 10 A, and the heat dissipated in the windings of the transformer is the maximum that can safely be allowed.

But suppose the secondary circuit is such as to make the power factor of the whole circuit 0·1. If we now allowed 2 kW to be taken from the mains, the apparent power would be 20 kVA, and the primary current would be 100 A, vastly more than the windings could tolerate without damage. Rather, we must ensure that the apparent power taken still does not exceed 2 kVA, which will keep the primary current below the allowed maximum of 10 A.

The special case of a phase difference of 90° between v and i is described by saying that the current and p.d. are *in quadrature*. In this case the

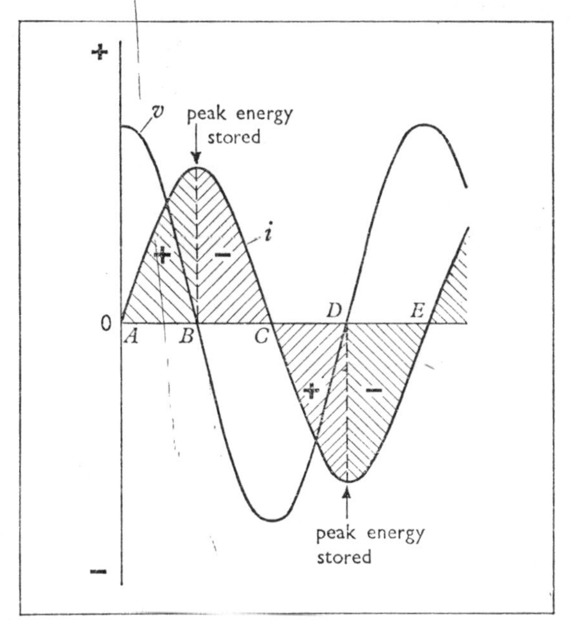

Fig. 12.7 The flow of energy when the current and p.d. are in quadrature

power factor is zero, and the mean power taken from the supply is also zero; this applies even though a large current may be flowing at a large potential difference. The details of the flow of energy in this case merit closer study (Fig. 12.7). During the first quarter-cycle from A to B both v and i are positive, and the power supplied w ($= vi$) is also positive. During the next quarter cycle from B to C the current i is still positive but v is negative; thus the power supplied is negative, and the energy that was supplied to the apparatus during the first quarter-cycle is returned again to the source of supply. This process recurs during the next half-cycle: from C to D energy is supplied to the apparatus (v and i both negative, and therefore w positive), and this returns to the source of supply from D to E. Although the mean power taken is zero, it is possible for considerable quantities of energy to be flowing into and out of the apparatus every quarter-cycle. In what form then is this energy stored in an entirely reversible manner? Now it can be seen from Fig. 12.7 that the *energy* stored in the apparatus reaches a maximum at the same moments as the *current*—namely, at the end of the interval AB, and again at the end of the interval CD. At these moments the magnetic field produced by the circuit is a maximum, and it is in the *magnetic field* that the energy is stored. This is in fact the situation that arises in an inductor.

If the curves for v and i are interchanged in Fig. 12.7, we have a similar situation; but now the *energy* stored reaches a maximum at the same moments as the p.d. This is the case in which the energy is stored in the *electric field* produced by the p.d., i.e. between the plates of a capacitor.

In the general case of a circuit containing resistance as well as capacitance or inductance there is a phase difference between the current and p.d. somewhere between 0 and 90°; the current leads on the p.d. in a capacitative circuit and lags in an inductive circuit. The vector diagram in Fig. 12.8

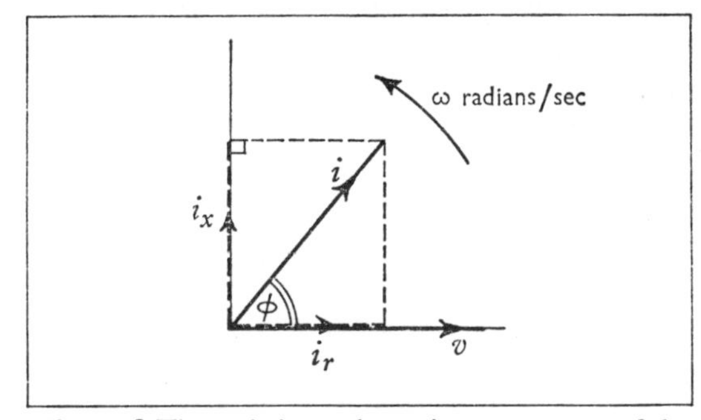

Fig. 12.8 The resistive and reactive components of the current in the general case

shows the way in which the movements of energy may be analysed in the general case. The current i may be separated into two components:

$$i_r = i \cos \phi$$

in phase with the p.d. v; and

$$i_x = i \sin \phi$$

in quadrature with the p.d. The quadrature component i_x contributes nothing to the mean power supplied; it is responsible for the reversible flow of energy between the source of supply and the electric and magnetic fields of the apparatus; it is sometimes known as the *wattless* or *reactive component* of the current. The in-phase component i_r is responsible for the irreversible supply of power w to the apparatus; i.e.

$$w = vi_r$$

whose mean value W is given by

$$W = \tfrac{1}{2}v_0 i_0 \cos\phi = VI\cos\phi$$

i_r is sometimes known as the *power* or *resistive component* of the current. The total current i is the *vector sum* of the resistive and reactive components.

12.5 The theory of the transformer

This consists of two coils wound on a laminated, closed iron core, as previously described on p. 102, Fig. 7.10. This design ensures that no free poles are produced when the core is magnetized; and there is very little demagnetizing effect. The maximum flux is therefore produced in the core for a given current. Also the magnetic coupling between the two coils is almost complete—i.e. nearly all the flux threads through all the turns of both coils. The a.c. supply is connected to one coil, called the *primary*. This is wound with a sufficient number of turns to reduce the no-load current I_0 in it to a low value.

The back e.m.f. E_p induced in the primary coil is nearly equal to the applied p.d. V_p.

$$E_p \simeq V_p$$

The alternating flux in the core produced by the small primary current induces an e.m.f. E_s in the secondary coil. The same flux in the coil threads equally through all the turns of both coils. Therefore the e.m.f. induced per turn is the same throughout the transformer.

$$\therefore\ \frac{E_s}{E_p} = \frac{n_s}{n_p}$$

where n_s and n_p are the numbers of turns in the secondary and primary respectively. By suitable choice of the turns ratio (n_s/n_p), we can convert an alternating supply at one p.d. to a supply of the same frequency at any other p.d.—up or down. When current is taken from the secondary, there is a drop in the p.d. V_s across the coil because of its resistance; but, provided the current is not too great, this drop is small, and we can write

$$E_s \simeq V_s$$

$$\therefore\ \frac{V_s}{V_p} \simeq \frac{n_s}{n_p}$$

provided the transformer is not overloaded.

When a current I_s is taken from the secondary, by Lenz's law the effect of it must be to reduce the magnitude of the alternating flux in the core that caused it. If nothing further happened, this would lead to a drop in the primary back e.m.f. E_p. However, the current in any low-resistance coil always adjusts itself to such a value that the back e.m.f. induced in it is nearly equal to the applied p.d. Therefore a large increase I_p' occurs in the primary current so as to restore the alternating flux in the core to its original value. The increase in current I_p' is in exact antiphase with the secondary current I_s, and their effects on the flux in the core are equal and opposite. Now the flux produced by a coil wound on a given core is proportional both to the current and the number of turns.

$$\therefore\ n_s I_s = n_p I_p'$$

$$\therefore\ \frac{I_s}{I_p'} = \frac{n_p}{n_s}$$

In normal use the no-load current I_0 is small compared with the total primary current I_p, so that we can write

$$I_p \simeq I_p'$$

$$\therefore\ \frac{I_s}{I_p} \simeq \frac{n_p}{n_s}$$

This is the reciprocal of the turns ratio. If the p.d. is transformed up in a given ratio, then the current is transformed down in the same ratio (approximately), and vice versa.

The small losses in a transformer are classified under two headings according to the site at which the loss arises, and from which accordingly the heat must be carried away.

(*i*) *Copper losses.* These are the ordinary heat losses that arise in the turns of any coil on account of its resistance. This loss (I^2R) can always be sufficiently reduced by using thick enough wire.

(*ii*) *Iron losses.* These are losses that occur in the iron core of the transformer; they arise from two different physical causes:

(*a*) *Eddy currents* (p. 101). In the presence of an alternating magnetic field eddy currents will be generated in any solid lump of metal. By making the core out of sufficiently thin laminations this loss can always be reduced to any desired extent.

(*b*) *Hysteresis* (p. 80). Some energy has to be expended in reversing the magnetization of the core. This appears in the iron as heat. The type of magnetic material for the core must be chosen accordingly.

★ The phase relations between the currents and p.d.'s for a loss-free transformer are shown in the vector diagram in Fig. 12.9. The common factor between primary and secondary circuits is the flux N in the core, which links both coils equally; we therefore draw the vector representing N first. When the secondary is on open circuit, the no-load current I_0 alone is responsible for the flux, and is therefore to be drawn in phase with N. Under these conditions the primary behaves simply as a high-inductance coil, and the applied p.d. V_p is therefore 90° ahead of I_0. The back e.m.f. E_p in the primary is nearly equal and opposite to V_p, and is therefore drawn in antiphase with it—i.e. lagging 90° behind the flux N. The same flux is also responsible for generating the secondary e.m.f. E_s, which is therefore in phase with E_p; the diagram in Fig. 12.9 is for a turns ratio of 1·5 : 1, so that $E_s = 1{\cdot}5 \times E_p$.

When a current I_s is taken from the secondary, the phase difference between E_s and I_s depends on the nature of the load—resistive, capacitative or inductive. Fig. 12.9 is drawn for an inductive load,

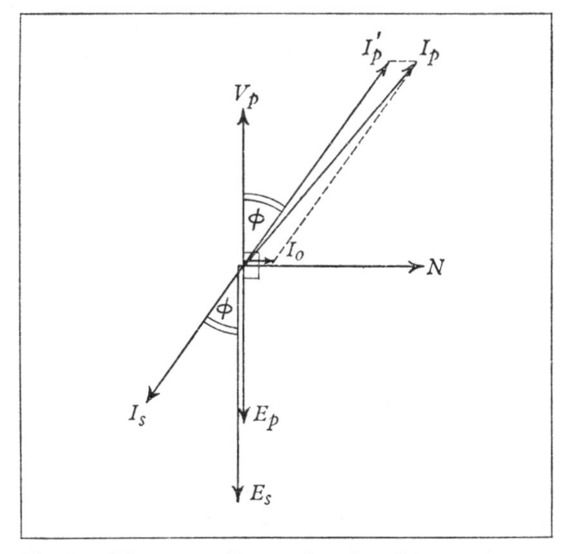

Fig. 12.9 The vector diagram for a loss-free transformer. The diagram is drawn for the case of an inductive secondary load. With a resistive load ϕ would be zero

the phase angle being ϕ. The primary current now grows by an amount I_p', so as to maintain the flux in the core at its initial value against the action of I_s. Thus I_p' is drawn in antiphase with I_s (and of 1·5 times the magnitude). The total primary current I_p is the vector sum of I_p' and the no-load current I_0. Under normal conditions I_0 is small, and I_p' and I_p are nearly equal both in magnitude and phase.

Since V_p is an antiphase with E_s and I_p nearly in antiphase with I_s, it follows that the phase angle ϕ between current and p.d. is nearly the same for the primary and secondary circuits. For example, if a capacitor is joined across the secondary, the primary coil behaves also like a capacitor in the circuit in which it is connected. In general the power factors of primary and secondary circuits are always approximately equal.

The most important practical case is that of a resistive secondary load (power factor 1). The primary coil then also behaves like a resistance, and V_p and I_p are nearly in phase. Suppose the load joined to the secondary is R_s, then

$$\frac{V_s}{I_s} = R_s$$

Now
$$V_p = V_s\frac{n_p}{n_s}$$

and
$$I_p = I_s\frac{n_s}{n_p}$$

The primary coil then behaves as a resistance of magnitude R_p given by

$$R_p = \frac{V_p}{I_p} = \frac{V_s}{I_s}\left(\frac{n_p}{n_s}\right)^2$$

$$\therefore\ R_p = \frac{R_s}{(\text{turns ratio})^2}$$

12.6 Power transmission

As long as electric power is to be generated and used locally, losses in transmission lines are not a serious matter, and there is little point in using alternating current rather than direct current. In the early days of public electric supply undertakings, each town (and in the country the individual house!) had its own generating station, and d.c. systems were the rule.

The virtue of a d.c. system was that it enabled batteries of accumulators to be used to maintain the supply during off-peak periods, when it was best to switch off the generators. At other times of the day the generators produced surplus current, which was used to re-charge the accumulators.

However, there are many economic advantages in increasing the scale of an electric supply network. The fluctuations of power at different times of the day are smaller; the industrial centre may be switching off at about the same time as the nearby 'dormitory town' is switching on, and so on. Also, it is much cheaper to generate power in some places than in others; in a large system the power stations can be sited at the most economic points in the network—near ports or in places where hydroelectric schemes are possible. In addition, it is found that the larger the amount of power being generated the more efficiently it can be done.

Under modern conditions, therefore, enormous quantities of electric power have to be transmitted continually from one part of the country to another, and reduction of the losses in the transmission lines becomes of prime importance. The following analysis shows that it then pays to use *high p.d.'s* for the connecting links between different parts of the system.

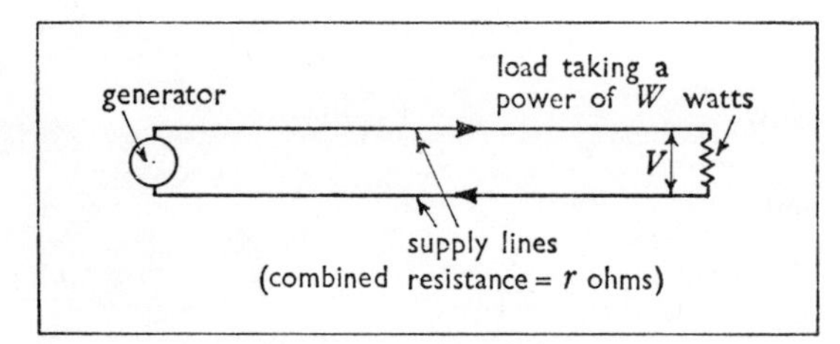

Fig. 12.10 Calculation of power loss in a pair of transmission lines

Consider a supply system delivering W watts of electric power at a p.d. of V volts; suppose the total resistance of the pair of cables connecting the generator to the load is r ohms (Fig. 12.10).

$$\therefore \text{ the current taken} = \frac{W}{V} \text{ amps}$$

$$\therefore \text{ the power loss in the cables} = \left(\frac{W}{V}\right)^2 . r \text{ watts}$$

$$\therefore \text{ the percentage power loss} = \frac{Wr}{V^2} \times 100$$

It is also important to keep the fluctuations of the supply p.d. small. When most of the load is switched off, the potential drop in the cables is suddenly reduced, and the supply p.d. may rise—with disastrous consequences in such equipment as is still switched on!

$$\text{the potential drop in the cables} = \left(\frac{W}{V}\right) \times r \text{ volts}$$

$\therefore$ the percentage fluctuation of p.d.

$$= \frac{Wr}{V^2} \times 100$$

This is the same as the expression for the percentage power loss. Therefore the same considerations apply in designing the system so as to keep both these quantities low. The maximum that can normally be tolerated is about a 5% power loss and fluctuation of p.d. Both quantities are inversely proportional to the square of the p.d., and it is therefore essential to use high p.d.'s.

However, against this advantage must be set the increased capital cost of the installation that ensues. Insulation problems are immensely increased. Indeed at a p.d. of several hundred kilovolts the loss from the surface of the cables by corona discharge (p. 167) becomes a significant factor. The power losses and fluctuations of p.d. can instead be reduced by increasing the diameter of the cable or by running several cables in parallel. The economic problem is one of achieving a nice balance between running costs and capital outlay for a given power W to be transmitted.

The advantage to be gained from the transmission of power at high p.d.'s makes the use of alternating current inevitable. The practicable limit for d.c. generators is a few thousand volts. Alternators function readily at much higher p.d.'s; and then the use of a.c. transformers enables the p.d. to be changed to any desired value. Transformers are such efficient devices that the losses in them are relatively unimportant, even when a supply link includes several—some for transforming the p.d. up for transmission, and some for transforming it down again to the p.d.'s at which it is used. It is only in certain rather unusual circumstances that the transmission of power by direct current is preferable; such a case, for instance, is the cross-channel power link that connects England and France.

For many industrial and domestic applications there is little to choose between d.c. and a.c. supplies; lighting, heating, etc. are equally efficient on either; and a.c. motors are almost as satisfactory as d.c. ones. It is only for electrolytic processes, electronic apparatus, and for one or two other applications that direct current must be employed. However, even here efficient rectifiers and a.c./d.c. converters have been developed, so that there is no serious disadvantage in using an a.c. main power supply.

The three-phase system

This is the name given to the system in which a.c. is generated, transmitted and consumed in three separate circuits simultaneously. In the alternator, one rotor sweeps past three sets of coils in succession (Fig. 9.5, p. 137), so that the phases of the currents in the three circuits are equally spaced—i.e. the phase differences are 120°. We have already seen that this makes the action of the alternator smoother and more efficient (p. 137). It also leads to considerable economies in transmission, since a common return lead can be used for all three circuits; if the three currents are then of equal amplitude, the current in the common wire is zero, and in principle it might be dispensed with altogether! Fig. 12.11 shows how this comes about. In the current/time graphs, the sum of the three currents at any instant is seen to be zero; thus, when one current is at its maximum value, the other two are each momentarily of half this magnitude in the opposite direction, etc. But the relationship may be proved easily from the vector diagram representing the three currents; it is clear from this that the vector sum of the currents is indeed zero. Even when the loads on the three circuits are not exactly equal, the vector sum of the three currents is not large; and only a small balancing current needs to flow in the common return wire, which can therefore be of much narrower gauge than the main cables. Also this wire is normally earthed, so that no expensive insulators have to be used to support it. The advantage for power transmission of the three-phase system is therefore apparent; three cables (and the small common neutral wire) can be used where six cables of the same kind would be needed in a single-phase system. The capital cost of the cables is therefore approximately halved.

The vector diagrams show that the same result holds also for any polyphase system; if the currents are equal, their vector sum is zero, and no current flows in the common return wire. It seems therefore that there is nothing to be gained by using a three-phase rather than a two-phase system. However, the action of alternators (p. 137) and polyphase a.c. motors (p. 144) is only really smooth when some multiple of three phases is used; the same would not apply for a two- or four-phase system.

The loads on the three circuits need to remain fairly closely 'balanced'; otherwise the losses rise seriously. Large industrial consumers are normally supplied with all three phases; most large motors are best designed as three-phase devices, and it is then not difficult to keep the phases balanced for any given factory. A private house is supplied from one phase only; but by arranging the numbers of domestic consumers joined to each of the phases to be approximately equal in any given locality, the three loads remain about equal, subject to statistical fluctuations. The larger the number of people

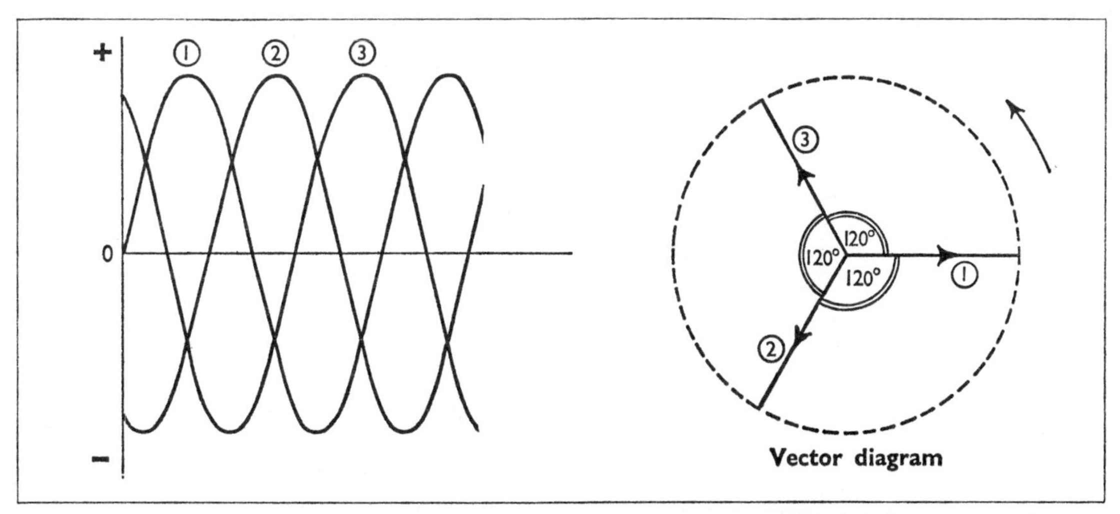

Fig. 12.11 Balanced currents in a three-phase system; the current in the common return wire is zero at every instant

using the supply, the smaller the percentage variations of current that occur between the phases.

12.7 The general a.c. circuit

Earlier in this chapter we have analysed the behaviour of single circuit 'elements'—capacitors, coils and resistors. In the general case the analysis of phase relations and the calculation of currents and p.d.'s is best done with vector diagrams. We shall in every case find that for a part of a circuit made up of capacitors, coils and resistors the r.m.s. values of current and p.d. are proportional to one another. Thus

$$\frac{V}{I} = Z$$

where Z is a constant known as the *impedance* of the arrangement; it has *resistive* and *reactive* 'components', and, like resistance and reactance, is measured in *ohms*. In the special case of a coil or capacitor in which there are negligible resistive losses the impedance is equal to the reactance; and for a 'pure' resistor the impedance is equal to the resistance. In drawing vector diagrams the first vector to draw should always be that which represents the quantity which is the same for all the circuit elements concerned—i.e. the *current* for a number of elements *in series*; or the *p.d.* for a number of elements *in parallel*. Conventionally, this first vector is drawn horizontal with the arrow pointing to the right. The other currents and p.d.'s are then drawn in relation to this.

(*i*) *A choke with appreciable resistance.* In terms of 'pure' circuit elements this may be represented as an inductance L and resistance R in series (Fig. 12.12). In the vector diagram we therefore draw the vector representing the current I first. The p.d. V_R across the resistance is in phase with I, and is given by

$$V_R = IR$$

The p.d. V_L across the inductance leads 90° on I, and we have

$$V_L = \omega LI$$

where $\omega = 2\pi \times$ frequency. The total p.d. V across the choke is the vector sum of V_R and V_L.

$$\therefore\; V = \sqrt{(V_R^2 + V_L^2)} = I\sqrt{(R^2 + \omega^2 L^2)}$$

The impedance Z of the choke is therefore given by

$$Z = \sqrt{(R^2 + \omega^2 L^2)}$$

The current lags behind the p.d., but not now by 90° as with a pure inductance. The phase angle ϕ is given by

$$\tan\phi = \frac{\omega L}{R}$$

At zero frequency (d.c.) the impedance is equal to the resistance; at high frequencies the resistance term in the impedance becomes negligible compared with the reactance term, and we then have

$$Z \simeq \omega L \quad (\text{i.e. } Z \propto \text{frequency})$$

Thus a coil of suitable design has the useful property of offering high impedance to alternating currents and low resistance to direct current; it can therefore be used as a filter to separate an a.c. 'ripple' from the direct current on which it is superimposed. A practical application is to be found in the smoothing circuit used to filter the output of a rectifier (p. 241).

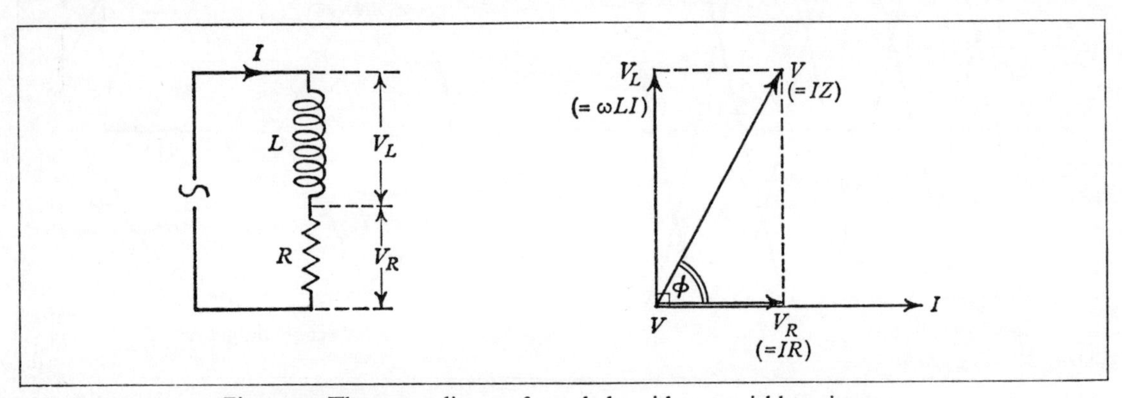

Fig. 12.12 The vector diagram for a choke with appreciable resistance

The impedance of a choke may be measured with the aid of an a.c. ammeter and a.c. voltmeter in much the same way as can be used for finding d.c. resistance (p. 49). If we can assume that the effective a.c. resistance is the same as that found for the coil by any d.c. method, then we have also the means of calculating the inductance L from the relationships given by the vector diagram. But this assumption is often not justified. Thus in a coil with an iron core the copper losses may well be no different. But to these must be added the iron losses (hysteresis and eddy current effects—p. 203), which increase the effective resistance of the coil even at low frequencies. At high frequencies the skin effect (p. 102) leads to big reductions in the effective cross-sectional area of the wire and the resistance of a coil is often many times greater than its d.c. value.

A method that avoids these uncertainties is shown in Fig. 12.13. A non-inductive resistance R is joined in series with the coil, and they are connected to an alternating supply of known frequency. A high-impedance voltmeter (e.g. a cathode-ray oscilloscope) is joined in turn across the coil, the resistance, and then the whole combination. If the three readings obtained are V', V_R and V, their relationship is shown in the vector diagram; from this the resistive and reactive components of V' can be worked out. In a practical measurement it is usually easiest to use graphical methods.

Example. A non-inductive resistance of 200 ohms is joined in series with a coil and connected to the 50 Hz a.c. mains. The Y-plates of a cathode-ray oscilloscope are joined in turn across the coil, the resistance, and then the combination of the two; the heights of the traces obtained are 8·3 cm, 5·0 cm, and 12·2 cm, respectively. Calculate the inductance and a.c. resistance of the coil.

The phase relations of the p.d.'s are of the general form shown in Fig. 12.13 below. Since the p.d.'s are proportional to the impedances, and are also measured by the heights of the C.R.O. traces, we may draw the lines in the vector diagram proportional to the C.R.O. measurements. Only half the parallelogram (i.e. a triangle) need be drawn to obtain all we want (Fig. 12.14). Choosing a scale of 1 inch to represent 2 cm (on the C.R.O. screen) we proceed as follows. Draw AB of length 2·5 inches; this can be taken to represent either the p.d. V_R or the 200-ohms resistance itself, since the two are proportional to one another. (The 'impedance scale' is then 1 inch to 80 ohms.) Then, with a compass, arcs of radii 6·1 inches and 4·15 inches are marked off centred on A and B, in order to find the point C. The line AC is now a measure of the impedance of the whole circuit (488 ohms); and BC is a measure of the impedance of the coil (332 ohms). Now produce AB, and drop the perpendicular CD onto it. The projection of BC on the resistance axis (i.e. BD), is a measure of the a.c. resistance of the coil; and its projection at right-angles to this (CD) is a measure of its reactance. By measurement on the diagram, we find

$$BD = 2{\cdot}75 \text{ inches} \qquad CD = 3{\cdot}11 \text{ inches}$$

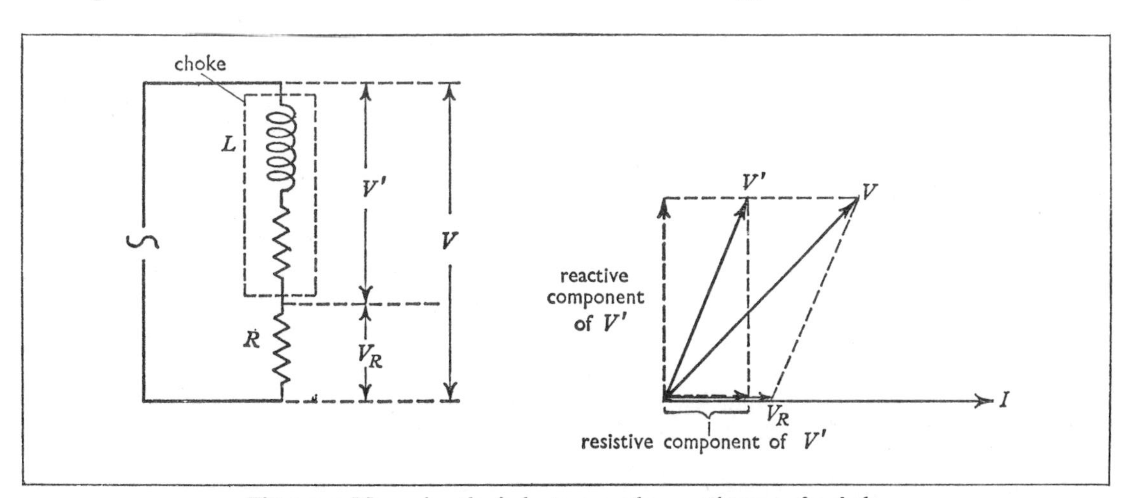

Fig. 12.13 Measuring the inductance and a.c. resistance of a choke

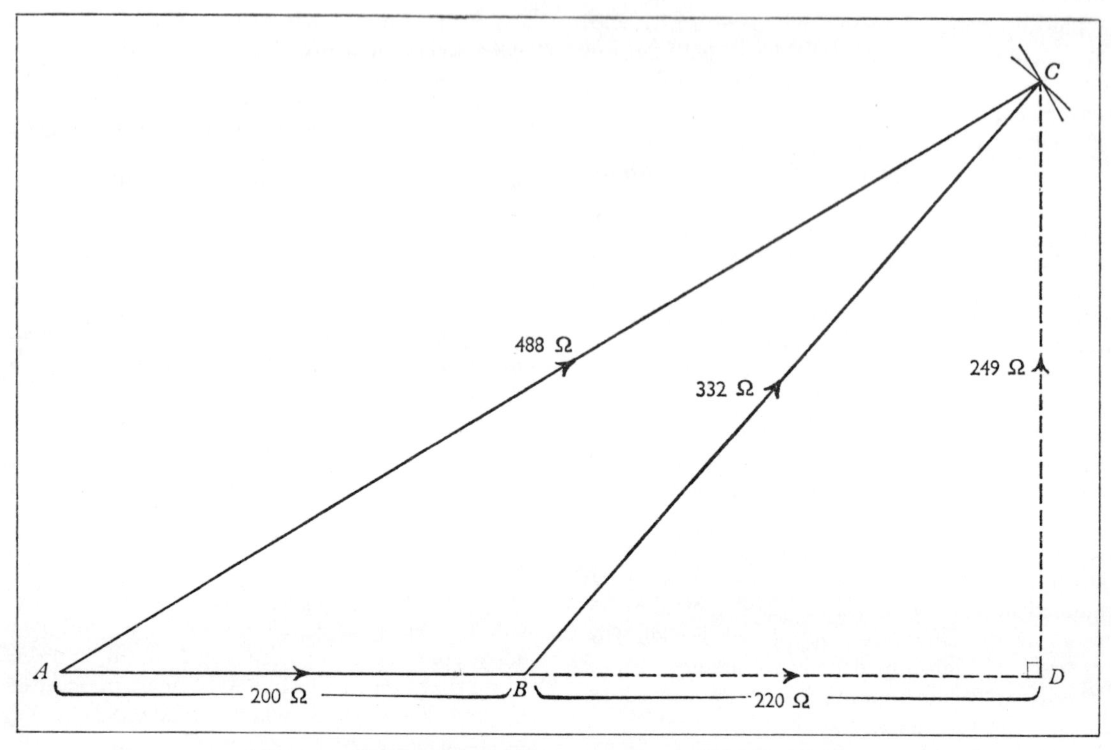

Fig. 12.14 The graphical solution of a vector diagram problem

$\therefore$ the a.c. resistance of the coil $= 2{\cdot}75 \times 80$
$= 220$ ohms
and its reactance $= 3{\cdot}11 \times 80$
$= 249$ ohms

But, inductive reactance $= \omega L$
$= 2\pi \times 50 \times L$

$$\therefore\ L = \frac{249}{2\pi \times 50} = \frac{2{\cdot}49}{\pi}$$
$$= 0{\cdot}79 \text{ henry}$$

(*ii*) *Capacitor and resistor in series* (Fig. 12.15). In the vector diagram we again draw the vector representing the current I first. $V_R (= IR)$ is in phase with this. The p.d. V_C across the capacitor lags 90° on I, and is given by

$$V_C = \frac{I}{\omega C}$$

The total p.d. V across the combination is the vector sum of V_R and V_C.

$$\therefore\ V = \sqrt{(V_R^2 + V_C^2)} = I\sqrt{\left(R^2 + \frac{1}{\omega^2 C^2}\right)}$$

The impedance Z of the combination is therefore given by

$$Z = \sqrt{\left(R^2 + \frac{1}{\omega^2 C^2}\right)}$$

and the phase angle ϕ by

$$\tan \phi = \frac{1}{\omega CR}$$

At zero frequency (d.c.) the impedance is infinite; at high frequencies it is almost equal to the resistance R.

This arrangement can be used for separating an alternating p.d. from the steady p.d. on which it is superimposed. The capacitor is then called a *blocking capacitor*. A practical application is to be found in the coupling between two stages of an R–C coupled amplifier (pp. 244, 335).

★ More exactly, a combination of capacitor and resistor in series can be regarded as a frequency filter. Thus the proportion of the applied alternating p.d. that appears across the resistor is

$$\frac{V_R}{V} = \cos \phi$$

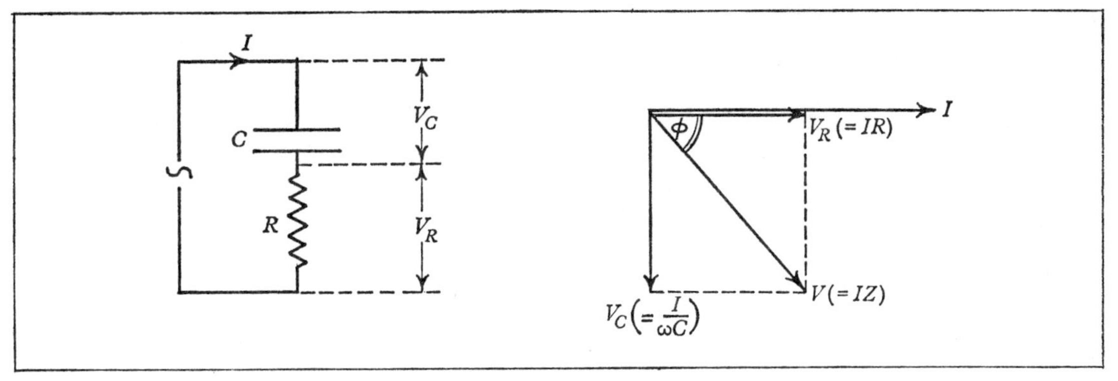

Fig. 12.15 The vector diagram for capacitor and resistor in series

At high frequencies this is nearly 1, and at low frequencies it drops off to zero (Fig. 12.16). The R–C coupling between the stages of an amplifier is then really a *high-pass* filter. The high-frequency components of the input pass the filter unaltered, while the low-frequency components are severely attenuated (and direct current is blocked altogether). There is no sharp cut-off frequency; but the 'knee' of the curve of Fig. 12.16 occurs roughly where

$$V_R = V_C$$

i.e. where $$R = \frac{1}{\omega C}$$

or $$\omega = \frac{1}{CR} = \frac{1}{\text{time constant of the filter}}$$

The time constant CR must therefore be adjusted so as to give little attenuation down to the lowest frequency that the amplifier is expected to handle.

Just as the high-frequency components of the input appear mostly across the resistor, so the low-frequency components appear mostly across the capacitor. Thus a capacitor and resistor in series constitute a *low-pass* filter if the output is taken from across the capacitor. Combinations of low-pass and high-pass filters of this type are used to provide the tone controls of record players and radios.

12.8 Resonant circuits

The series resonant circuit

Fig. 12.17 shows a circuit containing capacitance, inductance and resistance, all in series; the vector diagram is drawn alongside. In this case the p.d.'s V_L and V_C are in antiphase, and the vector sum V of all three p.d.'s is given by

$$V = [V_R^2 + (V_L - V_C)^2]^{\frac{1}{2}}$$
$$= I\left[R^2 + \left(\omega L - \frac{1}{\omega C}\right)^2\right]^{\frac{1}{2}}$$

The impedance Z is therefore given by

$$Z = \left[R^2 + \left(\omega L - \frac{1}{\omega C}\right)^2\right]^{\frac{1}{2}}$$

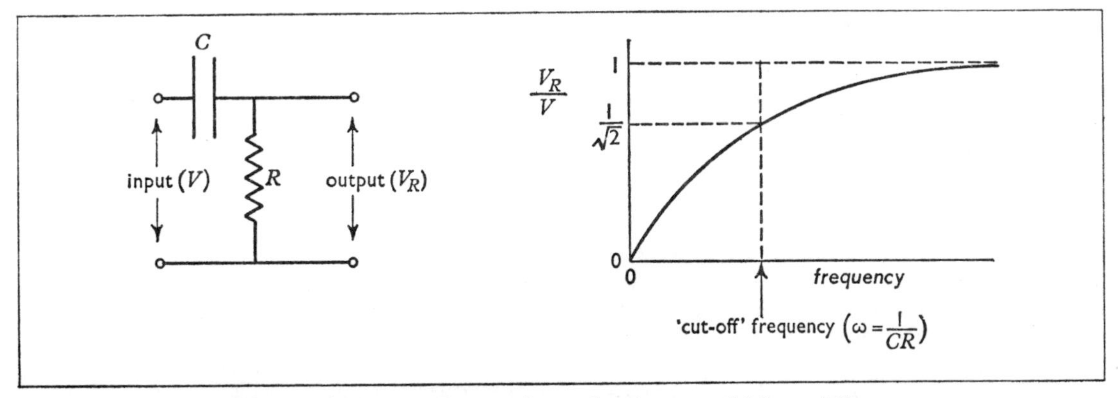

Fig. 12.16 The use of a capacitor and resistor as a 'high-pass' filter

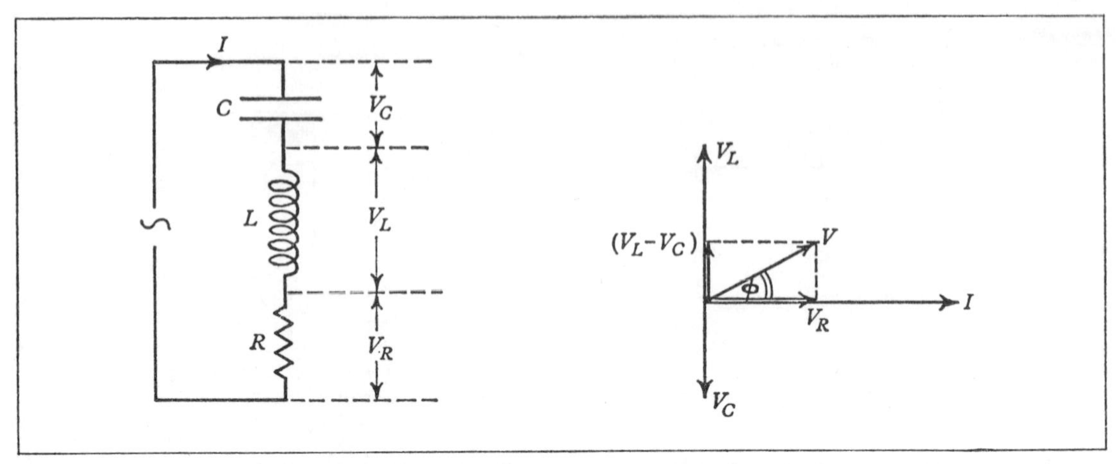

Fig. 12.17 The vector diagram of an L–C–R series circuit

and the phase angle ϕ by

$$\tan \phi = \frac{\omega L - (1/\omega C)}{R}$$

The impedance of the circuit drops to a minimum at the frequency for which the inductive and capacitative reactances are equal (and opposite); i.e. when

$$\omega L = \frac{1}{\omega C}$$

or $$\omega^2 LC = 1$$

At this frequency

$$Z = R \quad \text{and} \quad \phi = 0$$

The current I and p.d. V are then in phase, and the combination behaves like a pure resistance; if the losses (represented by R) are small, a very large current may flow. At any other frequency the impedance is greater (Fig. 12.18). At lower frequencies

$$\omega L < \frac{1}{\omega C}$$

and the circuit behaves capacitatively (ϕ negative—i.e. current leading on p.d.).

At higher frequencies

$$\omega L > \frac{1}{\omega C}$$

and the circuit behaves inductively (current lagging on p.d.). If a mixture of frequencies is applied to the circuit, the current only builds up to a large value for frequencies near the one to which the circuit is 'tuned', given by

$$\omega^2 LC = 1$$

This is an example of *electrical resonance*. Like other cases of resonant systems in mechanics, sound, etc. this circuit is able to act as a store of oscillatory energy. The process is worth considering in detail. At the resonant frequency the p.d.'s

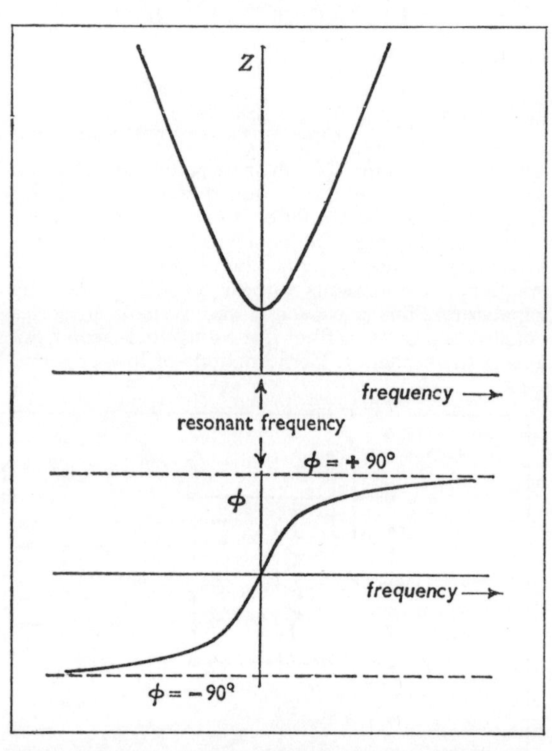

Fig. 12.18 The variation near the resonant frequency of the impedance Z and phase angle ϕ of a series resonant circuit

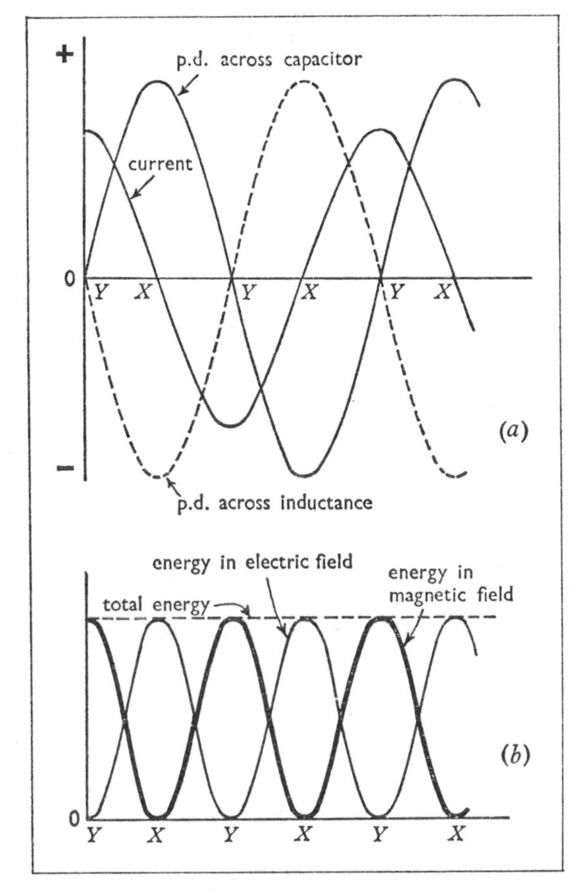

Fig. 12.19 The storage of energy in a series resonant circuit

across capacitor and inductance are equal in magnitude but in exact antiphase; the current is in quadrature with them (Fig. 12.19*a*). The energy stored in the electric field of the capacitor fluctuates with the p.d. across it; while the energy stored in the magnetic field of the inductance fluctuates with the current (Fig. 12.19*b*). At moments such as X the current is zero and the p.d. a maximum. There is then a maximum of energy stored in the electric field of the capacitor. At moments such as Y the p.d. is zero and the current a maximum. There is then a maximum of energy stored in the magnetic field of the inductance. At intermediate points the energy is partly in one and partly in the other. The total energy stored in the L–C system is constant, and is simply passed back and forth between the electric and magnetic fields. When the resonant current is first building up, this energy is drawn from the a.c. supply; but after that the supply need only make up the energy lost as heat in the resistance.

This may be compared with the similar behaviour of an oscillating mechanical system such as the pendulum of a clock. In this case the total energy is again constant, and is being continually changed from potential energy to kinetic energy and back again. The clock mechanism has only to supply the loss of energy (as heat) in the slight friction acting on the pendulum.

At the resonant frequency the p.d.'s across the two reactances build up to a value many times greater than the applied p.d. Thus at this frequency $Z = R$, and we have

$$I = \frac{V}{R}$$

$$\therefore\ V_L = \omega LI = \omega L\frac{V}{R}$$

$$\therefore\ \frac{V_L}{V} = \frac{\omega L}{R}$$

This ratio gives the ***magnification*** of p.d. produced by the resonance. Its value depends almost entirely on the design of the coil, since in practice the resistance of the circuit arises in this rather than in the capacitor. With a well-designed coil the magnification of p.d. at resonance can be 200 or more. It is known as the *Q-factor* of the coil (Q stands for 'quality'). Thus,

$$Q = \frac{\omega L}{R}$$

We can also regard Q as a measure of the extent to which the coil can be taken as a pure reactance. The phase angle ϕ between current and p.d. for a coil is given by

$$\tan \phi = \frac{\omega L}{R} = Q$$

For a perfect inductor, R would be zero, the Q-factor infinite, and ϕ exactly 90°.

The parallel resonant circuit

Consider now an arrangement of capacitor and inductor in parallel; the coil inevitably has both inductance L and resistance R (Fig. 12.20). Since this is a parallel arrangement, we draw first the vector representing the common p.d. V. The current through the capacitor I_C ($= \omega CV$) leads 90° on the applied p.d. The current through the

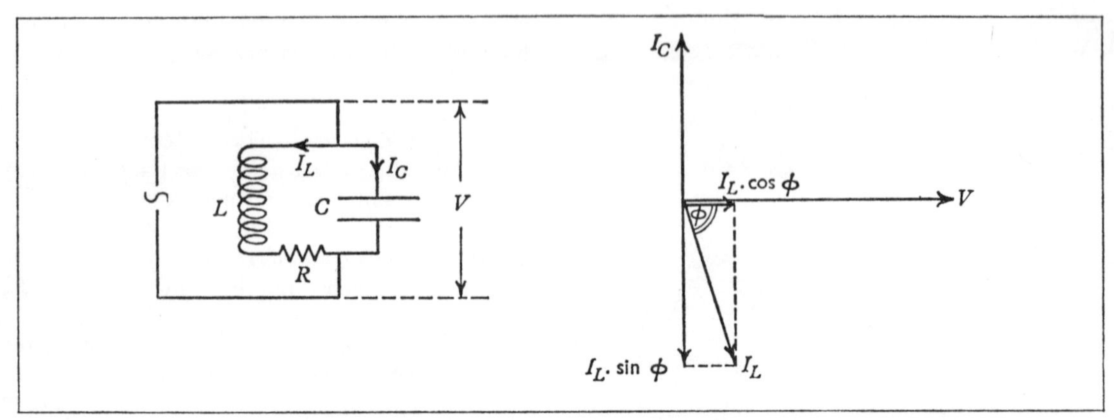

Fig. 12.20 The vector diagram of a parallel resonant circuit

inductor I_L lags on V, but by less than 90°, on account of the resistance R; the phase angle ϕ between I_L and V is given by

$$\tan\phi = \frac{\omega L}{R}$$

Also,
$$I_L = \frac{V}{\sqrt{(R^2 + \omega^2 L^2)}}$$

The reactive component of I_L ($= I_L \sin\phi$) is in exact antiphase with I_C. Resonance occurs when these two currents are equal in magnitude. Then

$$I_L \sin\phi = I_C$$

But
$$\sin\phi = \frac{\omega L}{\sqrt{(R^2 + \omega^2 L^2)}}$$

∴ at resonance

$$\frac{\omega L V}{(R^2 + \omega^2 L^2)} = \omega C V$$

Now,
$$Q = \frac{\omega L}{R}$$

$$\therefore R = \frac{\omega L}{Q}$$

Substituting for R and re-arranging,

$$\therefore \omega^2 LC\left(1 + \frac{1}{Q^2}\right) = 1$$

Q is usually a large quantity, so that for all practical purposes this reduces to the same condition as that for resonance with a series circuit with the same components:

$$\omega^2 LC = 1$$

We can regard the closed loop formed by the capacitor and inductor as a series resonant combination with the ends joined together. In a series circuit a large current flows for a given applied p.d.—i.e. the impedance of the circuit is very low; at the same time large p.d.'s are developed across the coil and capacitor. In the parallel circuit we still have a large current flowing in the coil and capacitor and a large p.d. developed across them; but the current now *circulates* within the closed loop formed by the two components. The only current that enters the loop under the action of the applied p.d. V is the small resistive component $I_L \cos\phi$ needed to make up the losses occurring within the loop (in the resistance R). At resonance the current entering the loop is small and is in phase with V; the impedance Z' is therefore very large and entirely resistive. It is given by

$$Z' = \frac{V}{I_L \cos\phi} = \frac{(R^2 + \omega^2 L^2)}{R}$$

$$= R + \frac{\omega^2 L^2}{R}$$

$$= \omega L\left(Q + \frac{1}{Q}\right)$$

$$\therefore Z' \simeq \omega L Q$$

since Q is normally large. At other frequencies the impedance is relatively small.

At lower frequencies

$$I_L \sin\phi > I_C$$

and the circuit behaves inductively (current lagging on p.d.).

At higher frequencies

$$I_L \sin\phi < I_C$$

and the circuit behaves capacitatively (current leading on p.d.).

The process of interchange of energy between the electric and magnetic fields is just as in the series resonant circuit. When the p.d. is a maximum, the current circulating in the loop is zero, and the stored energy is entirely in the electric field of the capacitor; and when the circulating current is a maximum, the p.d. is zero, and the energy is in the magnetic field of the inductor.

When a capacitor is charged or discharged through a low-resistance coil, the arrangement constitutes an L–C–R loop of just the kind we have been considering. The circuit therefore proceeds to 'ring' at its natural resonant frequency given by

$$\omega^2LC = 1$$

There is now no external source of energy to maintain the oscillation, and the alternating current in the loop dies away as the energy is dissipated in the resistance of the coil (p. 183).

Resonant circuits may be used as filters to select one particular frequency from a mixture of many others. One example is the aerial circuit of a radio set (Fig. 12.21). Currents of many frequencies are induced in the aerial by the radio waves impinging on it. These flow between the aerial and earth through the small coil. By mutual induction currents of the same frequencies are induced in the tuning coil L of the resonant circuit. At most of these frequencies no resonance occurs and the p.d.'s developed across the combination remain small. But if a current is induced at a frequency close to the resonant frequency, it builds up a large circulating current within the loop and an appreciable p.d. is developed. The magnification produced is proportional to the Q-factor of the coil. The Q-factor also decides the 'selectivity' of the tuned circuit—that is, the narrowness of the frequency band effectively magnified by the arrangement.

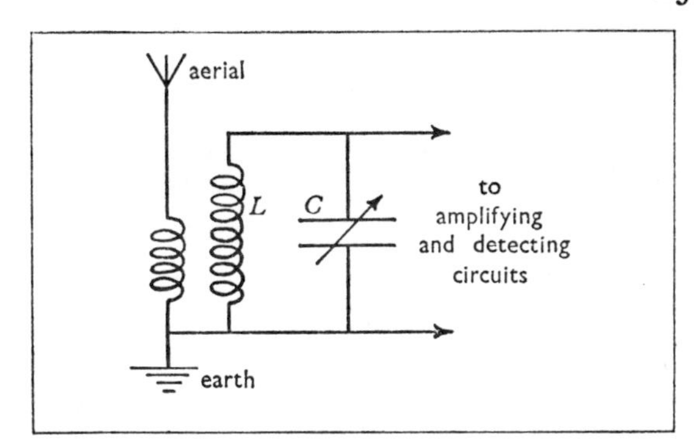

Fig. 12.21 The aerial circuit of a radio set

13: Electrons and Ions

13.1 The properties of electrons

Electrons were first discovered in electric discharges through gases at low pressures. The discoveries were in fact made possible by the technical development of efficient vacuum pumps. We shall in due course consider these phenomena (p. 226). But the experiments concerned are not always easy to interpret, nor are they very safe on account of the X-rays that may be emitted incidentally. We shall deal first with safer and more straightforward demonstrations depending on effects which, historically, came later.

The thermionic effect

When a piece of metal is heated to a high temperature, it is found that something carrying negative charge emerges from its surface. This is

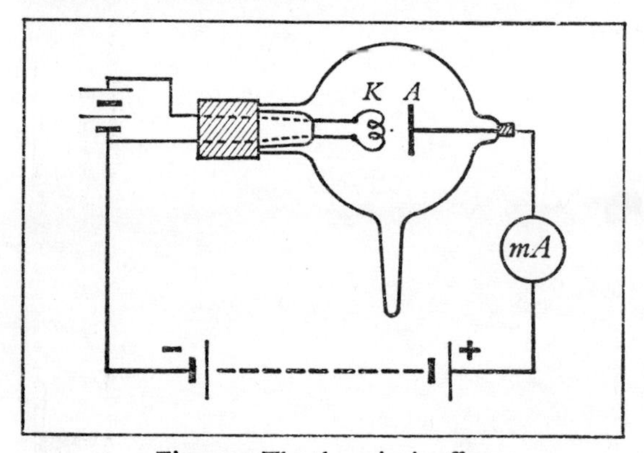

Fig. 13.1 The thermionic effect

known as the *thermionic effect*. It may be demonstrated with the apparatus in Fig. 13.1. A tungsten wire K is enclosed in a highly evacuated bulb; it can be heated by passing an electric current through it. Opposite the wire is a metal plate A. When the wire is cold, no current can pass through the vacuum from K to A. But when the wire is heated to incandescence a current can flow between the two electrodes—*in one direction only*. Presumably something carrying charge has emerged from the heated electrode into the empty space, and is now available to carry the current; the direction of this current is consistent with the idea that the charge emitted is negative. Thus, if A is made positive with respect to K, current flows, as shown; in this case the electric field in the space would act on the negative charge to attract it towards A and so complete the circuit. But if the p.d. between A and K is reversed, no current flows; the electric field in this case would draw any negative charge back again into K, and there is no complete circuit.

It is customary to label the two electrodes in the same way as in an electrolytic tank. The electrode from which the negative charge emerges (K) is called the *cathode*; that towards which this charge travels (A) is called the *anode*. However, unlike the phenomenon of electrolysis, no chemical change occurs in either the cathode or anode. Whatever it is that carries the negative charge out of the cathode and into the anode is apparently a constituent of the two metals, and is taken by the electric current through the electrodes and round the rest of the circuit. The materials used in the apparatus can be varied to some extent; only high-melting-point metals can be used for the cathode. But with this limitation the effect is unaltered by such changes. Likewise, the nature of the residual gas in the bulb does not affect the phenomenon, provided the pressure is less than about 10^{-5} mm of mercury (so that the mean free path is much greater than the dimensions of the bulb). It is reasonable to conclude that the agent responsible for carrying negative charge from cathode to anode is also that which carries an electric current through a piece of metal. It is presumably a universal constituent of all metals (and perhaps of all matter).

When the p.d. between the anode and cathode is large enough, the glass of the bulb opposite the cathode *fluoresces* with a green colour. The anode

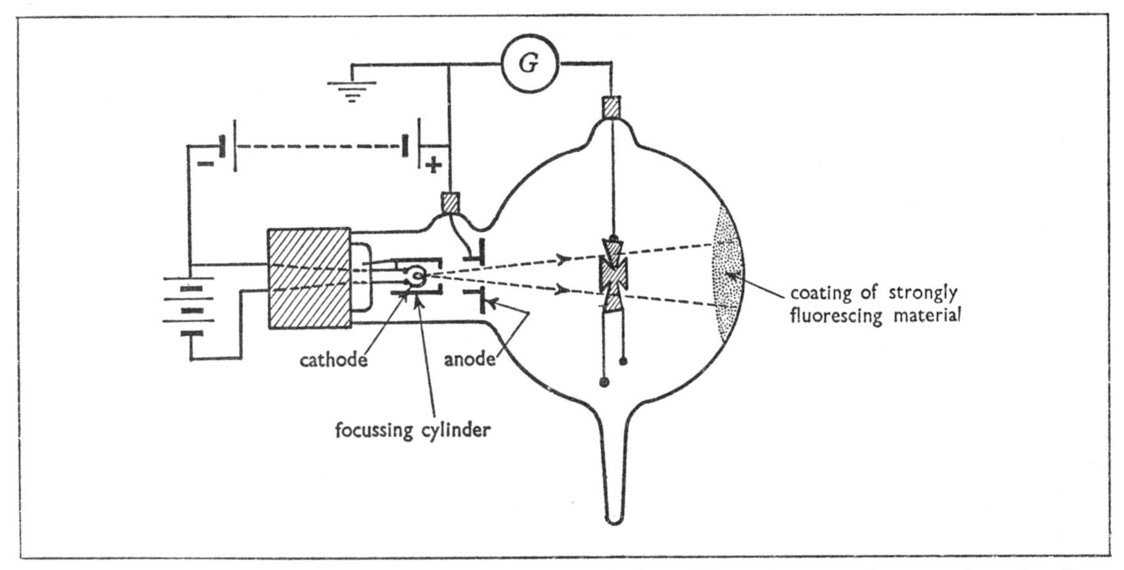

Fig. 13.2 The fluorescence produced by cathode rays; a shadow of the maltese cross is cast on the end wall

seems to cast a shadow in the middle of this fluorescence, and the impression is at once formed that the effect is due to some radiation emitted by the cathode. This is strikingly confirmed by using an electrode cut in a distinctive shape, like the maltese cross in Fig. 13.2. A shadow of the same shape appears in the fluorescence on the opposite wall of the tube; and its position and size clearly show that it is due to 'rays' of some kind travelling in approximately straight lines from the cathode. In the early development of the subject they were inevitably called *cathode rays*.

It is by no means certain at this stage of the investigation that the fluorescence is due to the same agent that carries the negative charge from the cathode to the anode, though that would be the simplest assumption to make. The emission of the cathode rays might be a separate effect *accompanying* the thermionic emission of negative charge. However, the point may be settled by showing that a beam of cathode rays does indeed consist of a stream of something carrying negative charge away from the cathode. This may be done by investigating the action of electric and magnetic fields on the beam of a 'cathode-ray tube' like that shown in Fig. 13.3. The anode A is a disc with a slit in it, through which a narrow beam of cathode rays passes. The beam passes obliquely over a sheet of fluorescent material, so that its path through the tube is made visible. When a p.d. is applied between the two metal plates P_1 and P_2, the electric field draws the cathode rays towards the positive plate and away from the negative one, and the beam is deflected accordingly. Likewise the cathode rays may be passed through a magnetic field—produced by a magnet or by a current flowing in suitably placed coils. The beam deflects as though the force acting on it is at right-angles to both the field and the line of the beam, just as we should expect for the magnetic force acting on an electric current. Thus, the beam could be deflected *downwards* in Fig. 13.3 by applying a magnetic field at right-angles to the plane of the diagram and directed away from the reader. If the beam carries a stream of *negative* charge, we must regard it as a current directed (in the conventional sense) *towards* the cathode. The direction of the force is then seen to be in agreement with Fleming's left-hand rule (p. 66).

Further evidence may be obtained by directing a beam of cathode rays onto an insulated electrode and collecting the charge carried. This may be done with the apparatus in Fig. 13.2. If a galvanometer or electrometer is joined between the maltese cross and the anode, a current is observed to flow in a direction corresponding with the collection of *negative* charge by the cross.

In experiments with cathode rays it is usual to work the tubes with the anode earthed. This

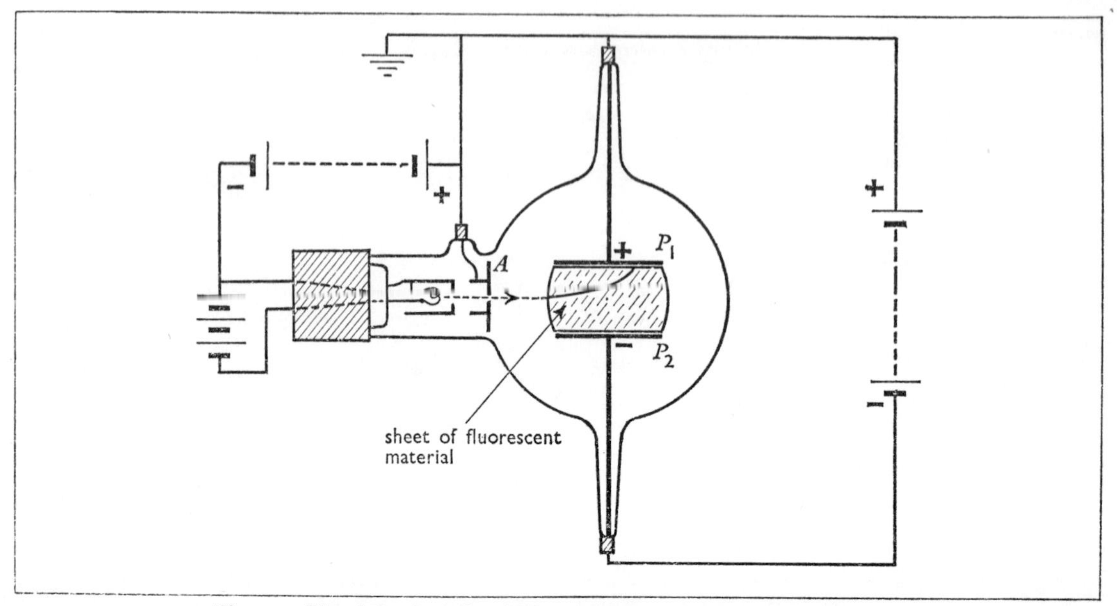

Fig. 13.3 The deflection of cathode rays by electric and magnetic fields

ensures that the region beyond the anode, in which the deflections are studied, is close to earth potential; the electric fields in the tube are not then much affected by movements of the observer's hand nearby. This procedure also makes for increased safety. But in any case such apparatus must always be handled with extreme caution, as the p.d.'s used can be lethal.

It is evident that a beam of cathode rays also carries energy. When it strikes the end wall of the tube, this energy is converted mostly into heat (the tube becomes warm); there is also light produced and a very small proportion of X-rays (p. 271). If the p.d. between anode and cathode is increased, the production of heat, light and X-rays goes up; at high p.d.'s the X-ray component also becomes more penetrating and is then dangerous. Some of the energy of the beam is always given up in ionizing the residual gas in the tube. If a cathode-ray tube is operated at a gas pressure of about 10^{-3} mm of mercury, the ionization makes the path of the beam visible in a well-darkened room. This provides another means of studying the deflections produced.

If we assume that cathode rays are particles that obey the known laws of mechanics, the measurement of their deflections in electric and magnetic fields enables us to calculate the velocity v and the ratio of the charge e to the mass m of the particles; the latter quantity (e/m) is known as the *specific charge* of the particles.

One of the simplest ways of doing this is to arrange the electric and magnetic fields to act in opposition on the beam. In Fig. 13.3 this may be done by applying a magnetic field perpendicular to the diagram directed away from the reader, in addition to the electric field shown. By adjusting the strength of one or both fields they can be balanced so that the beam is undeflected (at any rate in the centre where the fields are uniform).

The force F_E with which an electric field of intensity E acts on a particle of charge e is given by

$$F_E = eE \qquad \text{(p. 156)}$$

If the flux density of the magnetic field is B, the magnetic force F_M acting on the particle is given by

$$F_M = Bev \qquad \text{(p. 68)}$$

When the beam is undeflected

$$F_M = F_E$$

$$\therefore \; Bev = eE$$

$$\therefore \; v = \frac{E}{B}$$

The magnetic field B is produced by a current I flowing in a pair of coils arranged on either side of the tube. The distance between the coils is made equal to their radius, since this gives an extremely uniform field in the central region between them (the Helmholtz system of coils, p. 75). The value of B is proportional to the current.

$$\therefore \ B = kI$$

where k is a constant that can be calculated from the dimensions of the coils. The electric field E is the potential gradient between the plates P_1 and P_2.

$$\therefore \ E = \frac{V}{d}$$

where d is the separation of the plates and V the p.d. between them. With an accelerating p.d. between anode and cathode of 5000 volts the velocity of the cathode rays comes to about 4×10^7 m per second—about one-eighth of the speed of light. Since the cathode rays are observed to remain in a single concentrated beam in this experiment, it is clear that all the particles must have the *same velocity*.

Once the velocity is known, e/m may be calculated from the value of the accelerating p.d. between anode and cathode. The simplest way of using the apparatus in Fig. 13.3 is to join the cathode to the deflector plate P_2 and use the same source of e.m.f. (with the same voltmeter) both for deflection and acceleration; the accelerating p.d. is then V also. The kinetic energy gained by a particle in accelerating through this p.d. is given by

$$\tfrac{1}{2}mv^2 = eV \qquad \text{(p. 158)}$$

$$\therefore \ \frac{e}{m} = \frac{v^2}{2V}$$

If the electric field is switched off, and the magnetic field allowed to act alone, the force on the particle is entirely at right-angles to the motion at every point, and the beam is drawn into a circle (p. 68). The force F_M is now a centripetal force.

$$\therefore \ F_M = Bev = \frac{mv^2}{r}$$

where r is the radius of the circle.

$$\therefore \ r = \frac{mv}{Be}$$

An alternative way of finding e/m is then to measure the radius of curvature r of a cathode-ray beam; this also can be done with the apparatus described above, though the accuracy of the measurement is not high.

Each of the expressions we have deduced can be checked experimentally by performing the measurements with a range of different currents and p.d.'s. For instance the last result may be tested by verifying that the radius of curvature r is inversely proportional to the current I for a fixed accelerating p.d.

Notice again that the deflection produced by a given magnetic field is always the same for the entire beam; it is likely therefore that all the particles of the beam are identical. If it were not so, a 'spectrum' of deflections would be produced for the different values of e/m represented. There is thus little doubt that cathode rays consist of material particles of definite charge and mass. They are called *electrons*!

The currently accepted value of the specific charge of the electron is

$$\frac{e}{m} = 1{\cdot}759 \times 10^{11} \text{ coulomb kg}^{-1}$$

The specific charge of the electron can also be measured with the type of tube used in a cathode-ray oscilloscope (p. 235). In these only the point at which the beam strikes the end wall of the tube can be observed; and the theoretical analysis is rather involved. This is discussed further in section 13.5 below (p. 229). The best method for finding e/m with this kind of cathode-ray tube is to place it in an *axial* magnetic field; this causes the electrons to move in helical paths round the axis of the tube (p. 232), producing a rotation of the 'trace' on the screen.

The specific charge of the hydrogen ion is given by electrolytic experiments; let its mass be M. We assume it carries a charge of the same magnitude e. Then

$$\frac{e}{M} = 9{\cdot}58 \times 10^7 \text{ coulomb kg}^{-1} \qquad \text{(p. 37)}$$

Dividing,

$$\therefore \ \frac{M}{m} = 1836$$

The mass of the electron m is thus much less than the mass of even the smallest positive ion; and the mass of the electrons in any piece of matter is a minute proportion of the total. The value of m is best calculated from the known values of e/m and e (p. 160):

$$\therefore\ m = \frac{e}{e/m} = \frac{1{\cdot}602 \times 10^{-19}}{1{\cdot}759 \times 10^{11}} = 9{\cdot}11 \times 10^{-31}\ \text{kg}$$

13.2 The photoelectric effect

The emission of electrons from a metal may also be caused by illuminating its surface with light of sufficiently short wavelength—for most metals ultra-violet light is needed. This phenomenon is called the *photoelectric effect*. The photoelectric emission of electrons may be demonstrated with

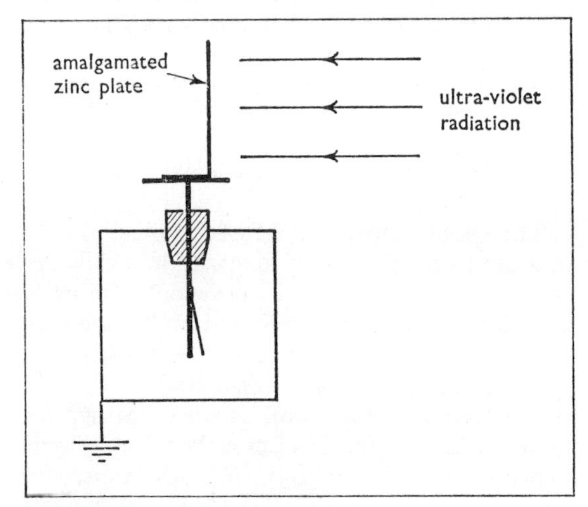

Fig. 13.4 Demonstrating the photoelectric effect with an electroscope

the apparatus in Fig. 13.4. A zinc plate is mounted on the cap of an electroscope so that it can be illuminated with ultra-violet light from a mercury-vapour lamp. The amount of emission varies considerably with small changes in the condition of the surface; it is found that the zinc surface for this experiment must either be freshly cleaned or else amalgamated with mercury. The zinc plate and electroscope are first given a *negative* charge; as soon as the light is directed at the plate the deflection starts to fall. The negative charge is evidently being lost from the illuminated surface. If the beam of light is intercepted with a sheet of ordinary glass (which absorbs ultra-violet light) the effect stops; while with an ultra-violet filter (opaque to visible light) the discharging of the electroscope continues unaffected. The rate of emission of electrons appears to be proportional to the intensity of the light. Thus, if the source of light is moved to half its original distance from the zinc plate, the illumination is increased four times; and the time for the deflection to fall a given amount is a quarter as much.

If now the electroscope is given a *positive* charge, no fall in the deflection occurs however bright the illumination of the zinc plate. We suppose that in this case the electrons emitted are at once drawn back to the plate, and there is no way by which the charge can escape to earth.

The electrons emitted by the zinc photocathode may be collected by another electrode, if this is at a high positive potential with respect to it. The current that flows between the electrodes may be detected with a pulse electrometer (p. 186). The arrangement used is shown in Fig. 13.5.

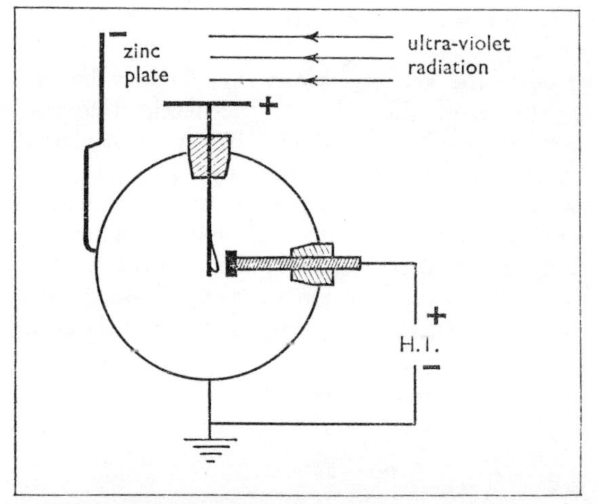

Fig. 13.5 Collecting a photoelectric current with a pulse electrometer

More extended experiments show that for a given metal emission of electrons occurs only for light whose wavelength is less than a certain critical value. The cut-off wavelength varies from one type of surface to another; for most metals it is in the ultra-violet region of the spectrum. But for the alkali metals—sodium, potassium, caesium, etc.—the cut-off is in the visible part of the spectrum; for specially treated surfaces it is even in the infra-red. However, the alkali metals can only be used in a vacuum or in a tube containing an inert gas.

Photocells

The photoelectric effect is used in a number of devices for controlling an electric circuit with light.

Fig. 13.6 shows a common design. The cathode is a semi-cylindrical plate coated with a photosensitive material. An alloy of antimony and caesium is satisfactory for the visible spectrum. For detecting the light from a tungsten filament lamp, a layer of caesium deposited on silver oxide is often used, since it responds well in the infra-red as well as in the visible spectrum. The anode consists of a single straight wire or loop, which is sufficient to collect the electrons emitted and does not much obstruct the incident light. The electrodes are enclosed in a highly evacuated glass bulb. To avoid damage to the photocathode the current is limited to very small values—about 10 μA maximum is

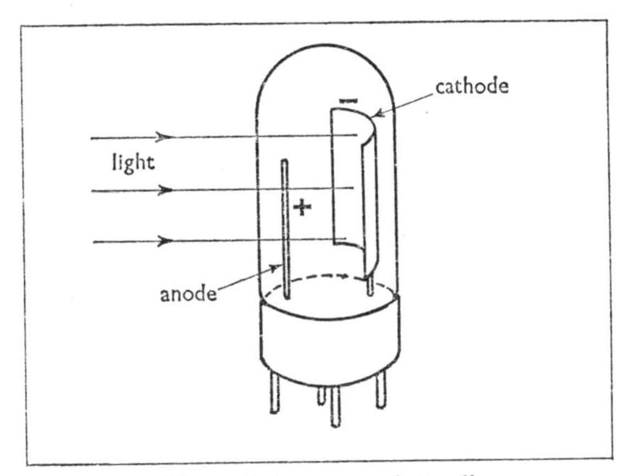

Fig. 13.6 A vacuum photocell

common. The sensitivity is about 30 μA per lumen of incident light; and the illumination of the cathode must therefore be controlled accordingly. The current is almost independent of the p.d. provided this exceeds a certain figure; it is also very closely proportional to the total incident flux of light. Fig. 13.7 shows the characteristic curves for such a photocell. The current may be increased by a factor of about 5 by filling the tube with an inert gas at low pressure. The electrons emitted produce some ionization as they are accelerated by the electric field, and a bigger current then flows. But the p.d. must be kept below about 100 volts, otherwise bombardment of the cathode by positive ions causes rapid deterioration. The gas-filled cell cannot be used for registering very rapid changes of illumination, since it takes about 0·1 milliseconds for the ions to recombine, and the current cannot follow the changes in a shorter time than this. However, it is well adapted for use in a film

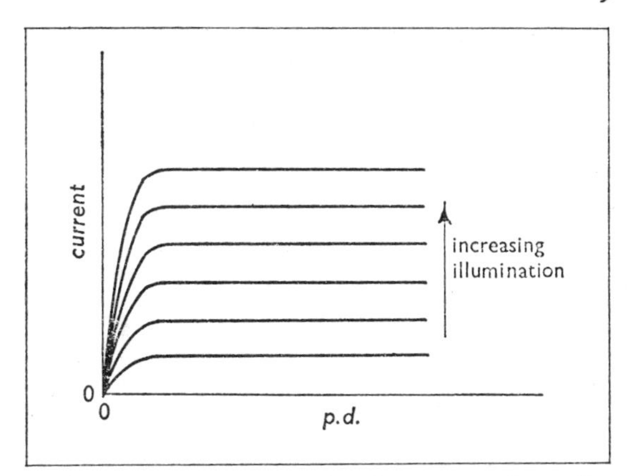

Fig. 13.7 The characteristics of a vacuum photocell

projector to follow the changes in light intensity from the sound track; the maximum frequency recorded in this case is probably about 5 kHz.

★ A very much higher sensitivity can be achieved by using the principle of *electron multiplication.* When an electron strikes a surface, it usually causes the emission of further electrons from it; this is known as *secondary emission.* By special treatment of the surface anything up to 4 electrons can be emitted for each incident particle. In the photomultiplier tube several intermediate electrodes (called *dynodes*) are inserted between the photocathode and anode so that the electrons must strike each of them in turn. By secondary emission the number of electrons in the beam is then multiplied by a factor of 4 at each impact. If there are 11 dynodes, the current in the beam is multiplied altogether by about 4^{11}, or nearly 5×10^6. The overall luminous sensitivity of the tube is thus about 10^8 μA per lumen, and it can detect very much weaker intensities than the human eye. Fig. 13.8

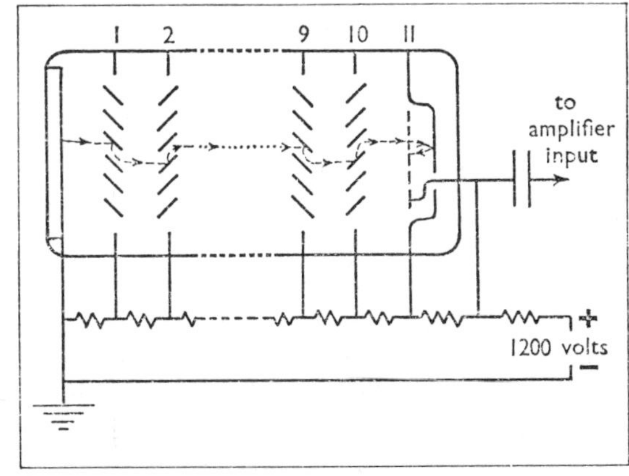

Fig. 13.8 A Venetian blind photomultiplier tube

shows the construction of the Venetian blind type of photomultiplier tube, so named because of the slatted form of the dynodes. To accelerate the electrons between the electrodes the potential must increase by about 100 volts from one dynode to the next through the tube. These potentials are supplied from a long resistance chain joined across a 1200-volt supply. The practical difficulties in operating a photomultiplier tube are considerable, and it is only used where sufficient sensitivity cannot be gained by other means.

A photocell of a very different type is shown in Fig. 13.9. Unlike the cells described above, this actually produces a small e.m.f.; it is therefore sometimes called a *photovoltaic cell*. It consists of an iron base-plate on which is fused a layer of selenium, which is a semiconductor (p. 20). To make contact with the front surface of the selenium a thin transparent film of gold or aluminium is deposited on it; connection with this is made by a metal ring clamped

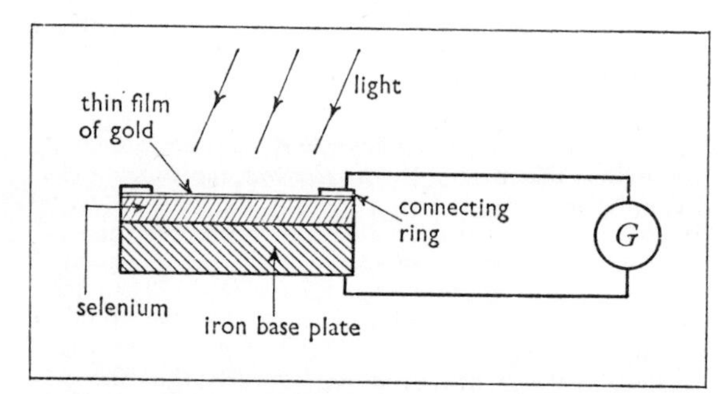

Fig. 13.9 A selenium photovoltaic cell

round the edge. The density of electrons is much higher in the metal film than in the underlying layer of selenium, and initially electrons diffuse through into the surface layer of selenium. This therefore becomes negative with respect to the metal film, i.e. a contact p.d. arises between the two (p. 45); this grows until the electric field developed across the boundary prevents further diffusion. No net e.m.f. acts in the circuit, since opposing contact p.d.'s arise at the other junctions in it.

When light falls on the cell, some of the electrons which are usually firmly bound to selenium atoms in the boundary layer are given sufficient energy to become effectively 'free', and are then available to conduct a current. The electric field at once drives them across the boundary into the metal film. The cell thus behaves as a source of e.m.f. with the metal film on the front as its negative terminal. Some of the electrons always diffuse straight back across the boundary layer. But if the resistance of the galvanometer is low, the majority of them take the external path back to the base-plate, and the current is then proportional to the illumination. Joined across a low-resistance microammeter the cell forms an illumination meter useful for photometric purposes. The sensitivity is about 500 μA per lumen, much greater than that of a photo-emissive cell—and no additional source of e.m.f. need be used. If the external resistance is high, the current is approximately proportional to the *logarithm* of the illumination; i.e. doubling the illumination produces approximately the same increase of current at all parts of the range. Joined across a high-resistance microammeter the cell is then useful as an exposure meter for photographic purposes, where illuminations are handled in the same way—i.e. changing the stop from one 'f-number' to the next alters the amount of light collected by a factor of 2; and so the usual scale of f-numbers is also a logarithmic one. An exposure meter can thus be calibrated with f-numbers, and the points are about evenly spaced across the dial. Photovoltaic cells cannot be used to record rapidly fluctuating illuminations, since the capacitance between the electrodes is quite high and acts for alternating currents as a low-impedance shunt across the output. Other types of semiconductor light-sensitive devices are described on pp. 331 and 339.

13.3 The quantum theory

When the photoelectric effect was discovered (towards the end of the nineteenth century) it had long been established that light was a form of wave motion. Particle theories of light had been suggested, and were found incapable of explaining the facts; and Maxwell's electromagnetic field theory predicted waves of just the right kind in the electromagnetic fields in space (p. 262 et seq.). The behaviour predicted for such waves had been found at every point to be in exact agreement with the observed behaviour of light; all the laws of geometrical optics and the phenomena of polarization, interference and diffraction were completely explained in every detail by Maxwell's wave theory. But there seemed no way of bringing the wave theory into line with the facts of the photoelectric effect; and before long physicists found themselves forced into a radical revision of their views of the nature of light and matter.

It is true that an emission of electrons from an illuminated metal might be expected on the electromagnetic theory. The oscillating field should cause oscillations of the charged particles in the metal, and an electron might in this way be given sufficient energy to escape from the surface. But, beyond this, the predictions are completely at variance with the facts. On the wave theory we should expect to find that no emission of electrons occurs for very weak illuminations, since the electromagnetic field would not in this case be sufficiently strong. But in fact there is no threshold illumination for the photoelectric effect; some emission of electrons occurs even for vanishingly small intensities of light. Again, the wave theory would suggest that the *velocity* of the emitted electrons should depend on the intensity of the illumination. But in fact the velocities of the electrons are not affected by the illumination, but depend only on the *frequency* of the light wave. The *illumination* decides the number of electrons emitted per second, i.e. the *current*.

The difficulties would be completely removed if we could regard a beam of light as a stream of particles. The intensity of the beam would now be measured by the number of particles arriving per second, and we should expect this to be proportional to the number of electrons emitted per second—which is what we observe. We should not now expect the velocity of the electrons to depend on the intensity of the beam, but rather on the amount of energy carried by each particle. If we suppose also that the position in the spectrum of a light 'particle' is decided by its energy, then the dependence of the velocity of the emitted electron on the 'frequency' of the light is readily understood. Also it is then to be expected that particles of light which belong to the 'low-frequency' end of the spectrum would not have sufficient energy to cause the emission of electrons. The only objection to all this is that a particle theory will not explain the other laws of optics, with which Maxwell's wave theory is so admirably successful!

At the turn of the century there was also one other effect that defied explanation in terms of the classical wave theory. This was concerned with the nature of the light emitted by a hot furnace. In 1901 an explanation of this was put forward by Planck based on the idea that light was emitted and absorbed by the walls of the furnace in indivisible packets of energy or *quanta*, as they were called. The energy E of each quantum he supposed to be proportional to the frequency ν of the light. Thus

$$E = h\nu$$

where h is a universal constant, which came to be called *Planck's constant*. Planck's theory was entirely successful; but there was still no need to suggest that the light, once emitted, consisted of anything other than the waves of Maxwell's theory.

It was left to Einstein (in 1905) to apply Planck's idea to explain the photoelectric effect. He suggested that the quantum of energy emitted by an atom continues as a concentrated, indivisible packet of energy until it is absorbed; it then gives up the entire amount to a single electron. Einstein supposed that in order to extract an electron from a given metal surface a minimum quantity of work W would have to be done on it; this he called the *work function* of the metal. The work done might well need to be more than this for some electrons (for instance those relatively far below the metal surface). The quantum of energy supplied by the light must be at least equal to the work function W, if there is to be any emission of electrons at all. The cut-off frequency ν_0 should therefore be given by

$$h\nu_0 = W$$

When light of higher frequency than this is used, the surplus energy is carried away as the kinetic energy of the electron. If v is the *maximum* velocity of the emitted electrons, then

$$\tfrac{1}{2}mv^2 = h\nu - W = h(\nu - \nu_0)$$

This prediction of Einstein's has been completely verified. It may be tested with a suitable photocell of the vacuum type. If the anode is made slightly negative with respect to the cathode, a retarding force acts on the emitted electrons; but some of them have sufficient kinetic energy to overcome this, and a small current continues to flow. If the reverse potential difference is V, the work done by the electron against the retarding forces of the electric field is Ve. The p.d. V is increased to the point at which the current is just stopped altogether. We then have

$$\text{maximum K.E. of the electrons} = Ve = \tfrac{1}{2}mv^2$$

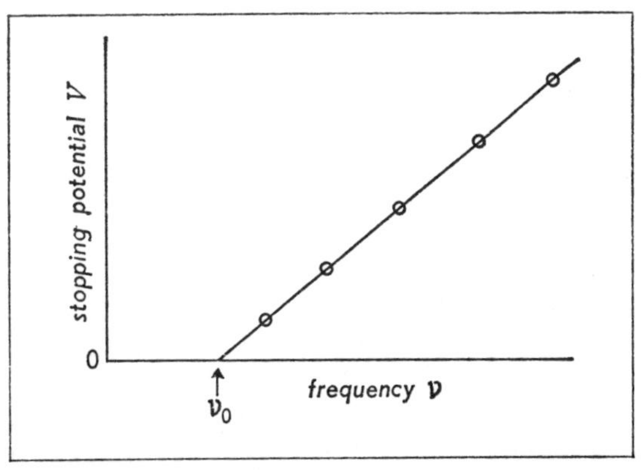

Fig. 13.10 Testing Einstein's photoelectric theory

$$\therefore \; Ve = h(\nu - \nu_0)$$

The stopping potential V is found for monochromatic light of a selected range of frequencies. A graph of V against ν should then be a straight line of the form shown in Fig. 13.10. The intercept on the frequency axis (where $V = 0$) gives the cut-off frequency ν_0; the gradient of the line is h/e. Since the electronic charge e is already known, we may use the experiment to measure Planck's constant h. The currently accepted value is

$$h = 6{\cdot}625 \times 10^{-34}\,\text{J s}$$

The wave–particle duality

The quantum of energy in Einstein's theory has all the trappings of a definite particle; it is often referred to as a *photon*. At the same time we do not deny the wave picture of the light, since the energy of the photon is expressed in terms of the *frequency* of the wave, which is still assumed to be of the electromagnetic type described by Maxwell's theory. It appears that we are expected to regard the light sometimes as a particle and sometimes as a wave—or even as both things simultaneously!

At first sight this appears to be nonsense. But closer reflection shows that this is only because of our preconceptions of what particles and waves should be like. If someone talks of a particle, our thoughts turn to billiard balls, bullets, etc.; and, if waves are mentioned, we think of ripples on a pond and the like.

On the large scale of everyday experience particles and waves seem quite distinct sorts of things. But there is no reason why the behaviour of light should be limited by the mental images we may be able to form of the processes involved. Indeed it is not difficult to see that the words *particle* and *wave* are bound to have a restricted meaning applied to light. If we are asked to make a record of the path of a billiard ball on a table, we might take a motion picture of the events. But there is no way of doing the same sort of thing for a photon. Suppose for instance we have a point source of light on the axis of a lens, which forms a real image of the source on a screen the other side of it. There is no doubt that a photon emitted by the source will fall on the screen in the image spot. But what path has it followed in between? This is a question which, in the very nature of things, cannot be answered. To decide whether the photon passed through a given point we should need to put something there (a photocell, say) to detect it. But this would absorb and destroy the photon, so that it never completes the journey at all. Likewise we might take a motion picture of the ripples on the surface of a pond, and we could thereby give an exact description of the progress of any given wavefront. But the same could not be done for a wavefront in the electromagnetic wave of a single quantum of light. The only way of deciding whether the wavefront is passing a given point is to detect it there; and this would absorb the quantum of energy concerned, so that the light no longer exists.

One interpretation of the wave-particle description of light is to regard the *intensity of the wave* at a given point as a measure of the *probability* of the photon appearing there. We cannot then predict the path of a photon with certainty; but we can state with certainty the probability of its following a given path. Whether or not it actually goes that way is only to be stated *after* the event.

As long as we are dealing with the vast numbers of photons emitted by ordinary sources of light the difficulty hardly arises, since statistical predictions become virtual certainties when the numbers are large enough. The intensity of the wave at a given point still gives the probability of photons turning up there; but this is now a 'certain' prediction of the number of photons arriving per second—i.e. of the intensity of the light. In such cases *wavefronts* and *rays* of light appear to be sharply defined concepts, in terms of which we may make exact predictions. But light interacts with atoms and electrons by the emission and absorption of single photons. In this context such phrases as 'the path of a single photon' or 'the shape of its wavefront' are essentially meaningless, since there is no conceivable way by which these lines could be traced. We must not then expect our mental images of waves and particles to mean very much.

Waves of probability and particles that can only be detected by making them vanish are certainly tenuous concepts! But this is not to say that we cannot calculate exactly what the behaviour of the light will be. We have found that Maxwell's wave theory is incomplete; to give a full description of the known behaviour of light we need to use also Einstein's quantum hypothesis. But the equations describing the effects observed can then be written down with great precision. Whether the human mind is able to form a valid mental picture of the processes involved is really of no consequence at all!

Optical spectra

A chemical element in the form of a gas or vapour may be made to emit light either by heating it, as in a flame, or by passing an electric current through it, as in a discharge tube (p. 227). If this light is examined with a spectrometer, the spectrum is found to consist of a number of sharply defined *lines*, whose wavelengths are characteristic of the element concerned. Now the kinetic theory of gases leads us to believe that the molecules of a gas or vapour are very small compared with the average distance between neighbouring molecules (p. 2); except actually at the moments of collisions a molecule in a gas must remain almost entirely unaffected by the presence of other molecules. It therefore seems likely that the *line spectrum* of an element arises from the properties of its individual atoms. Because of this the spectrometer is one of the most powerful instruments at our disposal for analysing the structure of atoms. However, even a cursory examination of a few spectra shows something of the immense difficulties involved in unravelling the evidence. For instance the spectrum of iron vapour (in an electric arc between iron electrodes) shows over 250 strong lines in the visible part of the spectrum alone, and many more in the ultra-violet and infra-red. The spectrum of sodium, which is about the simplest there is, appears at first sight to be deceptively straightforward; if a sodium salt is heated in a Bunsen flame, the spectrum consists of just one pair of lines very close together in the yellow region. But when the sodium atoms are more vigorously excited in an electric discharge tube, many other lines are revealed in addition to the predominant yellow ones; and there is no obvious regularity in the pattern.

Bohr's theory of the atom

In 1913 Bohr propounded a revolutionary extension of the quantum theory to account for atomic spectra. It was clear that the classical laws of Physics could not be valid in their usual form for the particles inside an atom. He was therefore prepared to look for behaviour not allowed for in the established laws of mechanics and electromagnetism.

The existence of sharply defined line spectra suggested that a given atom was capable of emitting or absorbing only certain definite parcels of energy. He therefore suggested that an electron in an atom could exist only in sharply defined states of fixed energy (E_1, E_2, etc.); the emission of light would then accompany the change from one such state to another, according to the established quantum relation

$$\text{energy emitted} = E_2 - E_1 = h\nu$$

The frequency ν and wavelength λ are connected by

$$c = \nu\lambda \qquad \text{(p. 258)}$$

where c = the velocity of light. It remained to find the correct way of describing these fixed states so that the behaviour of an atom could be predicted.

In 1911 Rutherford had put forward the model of the atom with which we are now familiar—a massive, positively charged nucleus round which circulate the appropriate number of negatively charged electrons to make the atom as a whole electrically neutral. He suggested that electrostatic forces alone were responsible for holding the atom together. The evidence had steadily accumulated that Rutherford's picture of the atom was fundamentally correct. However, it raised considerable theoretical difficulties. It had long been established that an *accelerating* electric charge should generate electromagnetic radiation. This is precisely what occurs when a high-frequency alternating current flows in a radio transmitting aerial; the free electrons in the aerial are accelerating as they oscillate to and fro in the wire, and radio waves are emitted (p. 263). Now the electrons circulating round the nucleus of an atom have an acceleration towards the centre, and should therefore be emitting radiation. According to the established laws of electromagnetism, the steady loss of energy in this way by the electron should cause it to spiral in towards the nucleus; and it was calculated that a Rutherford atom must collapse completely by this process in about 10^{-10} s!

Either Rutherford's picture of the nuclear atom was wrong or else the laws of classical mechanics and electromagnetism broke down within the atom. Bohr boldly chose the latter alternative. He accepted Rutherford's model, and suggested that the fixed states of the electrons that he was looking for consisted of particular 'allowed' orbits. To fit the observed pattern of the hydrogen spectrum he proposed that the *angular momentum* of an electron in its orbit could only be an integral multiple of $h/2\pi$. Thus

$$\text{angular momentum} = n\frac{h}{2\pi}$$

where n is an integer. All other orbits he supposed to be impossible. He then assumed that the electron does not in fact emit radiation when in an allowed orbit; but that the emission or absorption of radiation occurs when its motion changes from one allowed orbit to another.

These assumptions appeared highly arbitrary. But the theory developed from them agreed with experiment in a remarkably detailed manner. In general the working out of the orbits for an atom containing many electrons is a problem altogether too complex to be contemplated. However, when there is only a single circulating electron, the working is fairly simple. Detailed predictions with Bohr's theory could therefore be made only for the spectra of hydrogen, deuterium, and for heavier atoms which are ionized to the point at which only one electron remains near the nucleus. In these cases the predicted wavelengths agreed with the measured values to within 1 part in 40,000! The result was so striking that there could be little doubt of the fundamental correctness of Bohr's ideas.

It is beyond the scope of this book to discuss the detailed analysis of the hydrogen spectrum according to Bohr's theory. But the mathematics is not in fact very difficult, and the student with a fair knowledge of mechanics would do well to look it up in a more advanced work.

Fig. 13.11 is a diagram showing the allowed energy levels of the single electron in a hydrogen atom. The groups of transitions represented by the arrows each correspond to well-known series of lines in the hydrogen spectrum. Although the orbits of electrons cannot be worked out for anything more complicated than a hydrogen atom, the idea that the electrons in any atom have a limited number of allowed orbits and corresponding energy levels has turned out to be a universally consistent one. Even the most elaborate spectrum can be successfully analysed in terms of transitions between a relatively small number of atomic energy levels. Thus in the light of Bohr's theory the study of an apparently chaotic atomic spectrum becomes a means of measuring the energy levels of the atom; and energy level diagrams similar to Fig. 13.11 can now be constructed for all the elements.

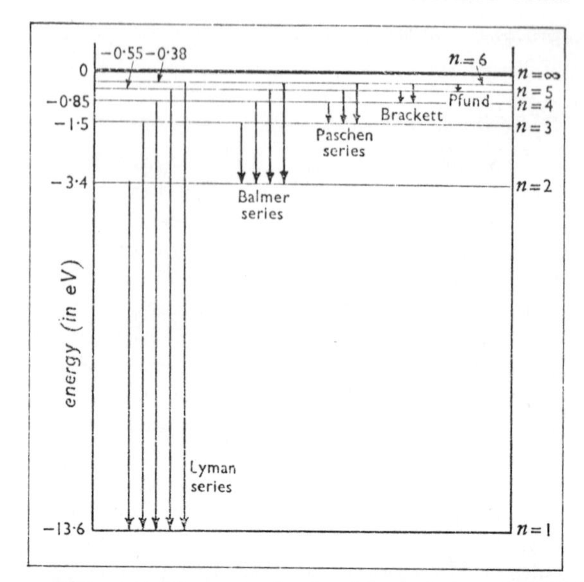

Fig. 13.11 The energy levels of the hydrogen atom

Wave mechanics

We have seen that the description of light as a wave motion can be regarded as a way of expressing the inevitable uncertainty about the path followed by an individual photon. Now there is a similar uncertainty about the movements of an electron. Any experiment we might perform to ascertain the exact position of an electron would deflect it so violently as to destroy all knowledge of its previous or subsequent motion. Only when we are dealing with vast numbers of electrons (as in a cathode-ray beam) can we make accurate predictions on a statistical basis. For a single electron all we can state is the probability of its following a given path. In particular, inside an atom there is no possible way of determining the actual position of an electron in its orbit; any experiment to do this would certainly change the orbit or ionize the atom completely.

In 1923 De Broglie made the suggestion that the behaviour of electrons might be more satisfactorily described by treating them also as waves. The intensity of the wave at a given point can then be taken as a measure of the probability of an electron being there. Just as with light, the wave picture can be regarded as the proper expression of the inherent uncertainty in our knowledge of the behaviour of an individual electron. De Broglie suggested that the wavelength λ of the electron *wave* would be connected with the momentum (mv) of the *particle* by

$$\lambda = \frac{h}{mv}$$

where h = Planck's constant.

The physical reality of the De Broglie waves was soon confirmed when it was shown that beams of electrons showed diffraction effects when incident on the surface of a crystal or when passing through a thin foil; these effects were very similar to those observed with beams of electromagnetic waves in the same circumstances (p. 273), and could only be explained by supposing a beam of electrons to be a train of waves. The wavelengths measured in these experiments agreed with those predicted by De Broglie's equation. Since that time it has been shown that not only electrons but also many other 'particles' must be regarded as waves for the purpose of predicting their paths. Diffraction effects essentially similar to those of electrons have been demonstrated for hydrogen and helium atoms, and even for hydrogen molecules. In each case the measured wavelengths are correctly given by De Broglie's equation. There is now little doubt that there is an inherent uncertainty about the behaviour of all particles, so that we must predict their paths on a probability basis by considering the movements of the associated De Broglie waves. The branch of Physics that deals with the behaviour and interpretation of these waves is known as *wave mechanics*.

There is no contradiction in treating even massive particles in this way. With a large 'particle', such as a tennis ball, the wave trains would be so sharply defined as virtually to coincide with the ball itself at every moment. This is merely another way of saying that there is a negligible chance of the ball following a path detectably different from that predicted by Newtonian mechanics. Even with atoms the differences are rarely significant. Thus wave mechanics turns out to be a refinement of classical mechanics—a refinement we need only consider when dealing with the fundamental particles of matter.

The equation to be obeyed by the electron waves was worked out by Schrödinger; and we can now use this to predict the behaviour of the waves, just as we can use the equations of Maxwell's electromagnetic theory to predict the behaviour of light.

An electron wave is refracted in passing near a nucleus, and there is the possibility of a wave system becoming trapped so that it revolves permanently round the nucleus. In wave mechanics this is how we describe a stable electron orbit. Schrödinger showed that the only possible sort of trapped wave would be a *standing wave system*. This occurs when the path of the wave round the nucleus contains a whole number of wavelengths, so that the wave pattern 'joins up' continuously round the loop.

If the radius of a wave path round the nucleus is r, a standing wave could occur, if

$$2\pi r = n\lambda = \frac{nh}{mv}$$

$$\therefore\ mvr = \frac{nh}{2\pi}$$

Now, mvr is the *angular momentum* of an electron moving in a circular orbit of this radius; and the above equation is therefore nothing more than Bohr's quantum condition for specifying the allowed electron orbits. Bohr's rather arbitrary rule is thus seen to be an inevitable consequence of the wave nature of the electron.

Scientists still tend to talk in terms of the Bohr picture of the atom, since this sort of language is more vivid than the 'hazy' wave mechanical picture. We refer to an electron 'rotating' in a certain 'orbit'; but this use of words can only be justified now if liberally sprinkled with inverted commas! It is really better to avoid talking about an electron in an atom as though it were a concentrated particle; the reality is rather the standing wave pattern, which in present terminology is referred to as the electron *orbital*. The position of the electron within the orbital can only be stated in terms of probabilities; and it is correct to regard its charge and mass as being 'smeared out' over the orbital in proportion to the intensity of the wave at each point.

The charge distribution is thus a stationary one. But we must not forget that a standing wave within an atom still represents an electron in motion round the nucleus; and the orbital may still have angular momentum and produce a magnetic field. In addition it is found that each electron has internal angular momentum, which is called its *spin*;

and this causes it to produce a magnetic field quite apart from its motion round the nucleus.

The modern quantum theory is able to give a very detailed description of the atom. It predicts that only two electrons can occupy each orbital, and these must have opposite spin. (This is known as the *Pauli exclusion principle.*) Such a pair of electrons has no resultant angular momentum or magnetic field. Imagine an atom being built up with the electrons added to it one by one; each electron will fall into the orbital of lowest energy that is still unoccupied. In this way the electron structure forms a series of *shells*, in each of which the orbitals have approximately the same energy. These shells are usually labelled in ascending order of energy with the letters *K*, *L*, *M*, etc. The *K*-shell is complete with only 2 electrons. The *L*-shell can hold 8 electrons, and the *M*-shell 18. However, the electron orbitals interact with one another; and it can happen that an orbital in an outer shell has lower energy than one in a still incomplete inner shell. This happens to some extent with most atoms whose atomic number is greater than 18 (i.e. with more than 18 electrons). Thus argon with 18 electrons has the *K*- and *L*-shells complete, and 8 electrons in the *M*-shell; but in potassium (with atomic number 19) the extra electron occupies an orbital in the *N*-shell, although there is room for 10 more electrons in the *M*-shell. Only in atoms with 29 or more electrons is the *M*-shell completely filled up. With the heaviest atoms there may be several incomplete shells; uranium with 92 electrons has 2 electrons in the *Q*-shell, while the *O*- and *P*-shells are very far from complete.

The spectrum and the chemical properties of an atom are found to depend to a large extent on the number of electrons in its outermost shell. Atoms with similar electron structures in this shell tend to have similar chemical properties and similar spectra. If the elements are arranged in the order of their atomic numbers, there is therefore a cyclic recurrence of atomic properties through the table; this feature of the chemical behaviour of the elements is displayed in the well-known periodic table. It is found that the number of electrons in the outermost shell can never exceed 8. Those atoms in which this shell is either complete or else contains exactly 8 electrons have particularly stable electron structures; these are the noble gases—*helium*, *neon*, *argon*, *krypton*, *xenon* and *radon*. Atoms which have one electron outside a complete shell (or a sub-shell of 8) likewise have similar properties; these are the alkali metals—*lithium*, *sodium*, *potassium*, *rubidium*, *caesium*. In the same sort of way the other groups of the periodic table arise from various recurring electron configurations.

The chemical bond

When two atoms are close together, the orbitals of their outermost shells become modified, and there exists the possibility of one or more orbitals being shared between the two atoms; these would represent electrons in motion round *both* nuclei. If the resulting electron configuration has a minimum total energy for a particular distance apart of the nuclei, then the atoms can be in equilibrium in this position, and a stable chemical bond can be formed between them.

In some cases the shared orbitals are so modified that one or more electrons are virtually transferred from one atom to another. One atom can thus acquire a negative charge, leaving a positive charge on the other. In this way ionic compounds are formed, in which the bond is largely due to the electrostatic force between the ions. But in general an orbital that provides a bond between two atoms is shared more or less equally between them, and there are no ionic forces to affect it. Such is the case, for instance, with the hydrogen molecule (H_2). Here, the orbitals are shared between the two nuclei; the energy of this configuration is much less than for the separate hydrogen atoms, and a very stable bond is formed.

13.4 Electric currents in gases

For small potential differences a gas is an almost perfect insulator. An electric current can only pass through it if some independent means is used to ionize the gas (e.g. X-rays), or in some other way to produce free electrons in it (e.g. by the photoelectric effect).

In such cases the current is limited by the rate at which ions and electrons are being produced in the space between the electrodes (p. 186). In Fig. 13.12 this is represented by the part of the curve between A and B. But if the applied p.d. is very much increased, a point is reached at which there is a considerable growth of current due to collision processes in the gas. The electrons initially present then gain enough energy between collisions to cause further ionization. The electrons so released are accelerated in their turn; and in this way a vast increase of ionization occurs. Each original ion pair leads to the creation of anything up to 10^5 fresh ion pairs, and the charge collected by the electrodes is increased by this factor. This form of magnified ionization current is known as a *Townsend discharge*, after its discoverer. It is represented in Fig. 13.12 by the part of the curve between B and C. However, the current is still dependent on the original source of ionization—X-rays, ultra-violet radiation, etc.; and without this it stops at once. This sort of discharge is non-luminous.

In order to maintain a self-sustaining discharge it is necessary for the production of ions to keep pace with their collection by the electrodes. This can occur if the p.d. is increased to the value represented by the point C in Fig. 13.12. At this

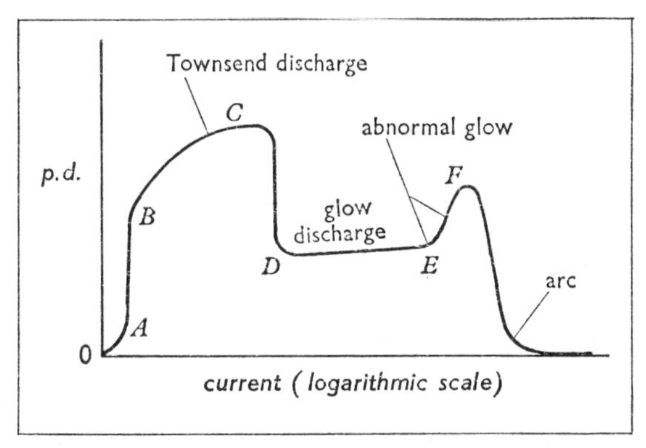

Fig. 13.12 The variation of current with p.d. in a typical gas discharge

point the current can grow indefinitely without any further increase of p.d.; indeed the p.d. needed to maintain the discharge falls off considerably as the current rises. The discharge is therefore highly unstable, and the current will only reach a steady value if it is limited by a sufficiently high resistance in the rest of the circuit; in the absence of this it increases until either a fuse blows or the equipment fails in some other way.

The actual nature of the discharge varies considerably with the configuration of the electrodes and the pressure of the gas. At n.t.p. it usually takes the form of a series of irregular sparks; though, if a large current is allowed to pass, it can degenerate into an electric arc. In this case the gas reaches a very high temperature and is almost completely ionized. The electrodes are rapidly vaporized; and the ionized vapour adds to the 'plasma' through which the current is passing. This form of discharge has a practical use as a concentrated and brilliant source of white light.

At lower pressures the sparking p.d. is reduced. This is because the mean free path of an electron in the gas becomes greater; it is therefore accelerated to a higher speed between collisions, and is more likely to cause ionization. At a pressure of a few millimetres of mercury the discharge takes on a quieter and more regular form; if the current is suitably limited by a series resistance, it settles down in the region between D and E (Fig. 13.12). The p.d. across the tube is now almost independent of the current, and a steady luminous discharge is observed, called the *glow discharge.* A characteristic of this form of discharge is that only a limited area of the cathode is used; if the current is increased (by decreasing the series resistance), the area in use increases in proportion so that the current flowing per unit area of the cathode remains constant. The reason for this behaviour is not properly understood. When the current has been raised to the point at which the whole cathode surface is in use, a further increase can only be obtained by once again raising the p.d. The discharge now becomes brighter, and is known as the *abnormal glow* (the region between E and F in Fig. 13.12). If the p.d. rises above the value at the point F, the discharge again becomes unstable; a high-temperature arc then develops and the p.d. drops off to a low value.

The glow discharge is employed in a number of important devices. Its essential characteristics may be investigated using a small neon lamp connected in the circuit of Fig. 13.13. These lamps are usually equipped with a built-in series resistance to limit the current to safe values; for this demonstration this must be disconnected. The potential divider across the d.c. supply is adjusted so that the p.d. applied to the lamp is slowly increased from zero. Up to a certain point no discharge is observed. (The Townsend discharge is non-luminous, and the current that might flow is far too small to be detected by the milliammeter.) But as soon as the p.d. reaches the striking value—usually about 180 volts—the glow discharge starts and the p.d. across the lamp at once falls to a lower value. The

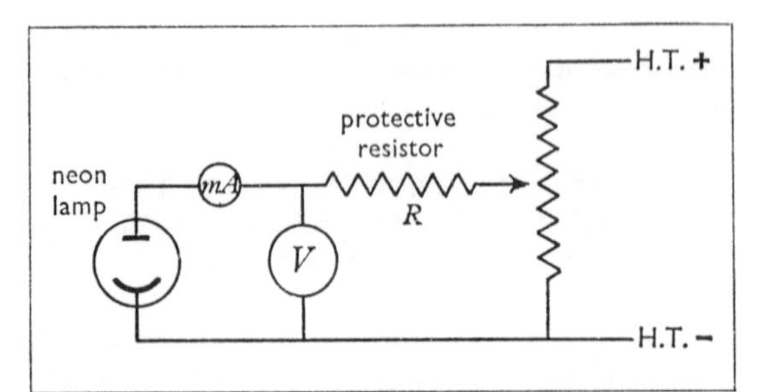

Fig. 13.13 Investigating the properties of the glow discharge

p.d. supplied to the lamp and its series resistor R can now be varied by moving the potential divider. But this only alters the current through the lamp, and leaves the p.d. across it at almost the same value. To stop the discharge the applied p.d. must be reduced below this.

The appearance of a gas discharge takes many different forms as the pressure in the tube is varied. Colourful and fascinating effects may be observed in a discharge tube connected to a vacuum pump and pressure gauge. At one time it was usual to operate such a tube from an induction coil. However, this is not now considered safe, since the p.d.'s may be sufficiently high to generate dangerous quantities of X-radiation. A tube about a foot long can be worked satisfactorily using an E.H.T. supply of about 6 kV. A typical appearance of the discharge for a pressure of about 0·1 mm is depicted in Fig. 13.14. Most of the tube is occupied by a bright luminous region called the *positive column*; this starts close to the anode, and is usually broken up into a series of bands or striations. A region of weaker luminosity (usually of a different colour) is observed a short distance from the cathode; this is known as the *negative glow*. On either side of this there are dark regions, called the *Faraday dark space* and the *Crookes dark space* after the men who first observed them. There is usually also a further small glowing patch to be seen covering the surface of the cathode, but separated from it by a very thin dark layer (*Aston's dark space*). The colours of the discharge depend on the gas used; for air, the positive column is a brilliant pink, and the negative glow deep blue.

The electrons emitted from the cathode are accelerated by the electric field in the Crookes dark space; the length of this is approximately equal to the mean free path of the electrons in the gas. The negative glow arises from the ionization caused by these electrons. Most of the applied p.d. exists across the Crookes dark space. The field here is very intense and accelerates some of the positive ions to strike the cathode; this in turn stimulates the emission of electrons on which the existence of the discharge depends. In the Faraday dark space the electrons are further accelerated, and the positive column is caused by the excitation of the atoms in the rest of the length of the tube.

If a greater length of tube is used, the cathode end of the discharge is unaltered (for the same gas pressure); and the positive column extends to occupy the extra length. The applied p.d. must then be increased to maintain the field. A neon advertising sign consists of a lengthy discharge tube bent into the tortuous shape required by the lettering, etc. It is filled with neon at a pressure of a few millimetres of mercury, and the positive column is arranged to occupy the entire visible portion. Owing to its length, p.d.'s of several thousand volts are normally needed to run the discharge.

As the pressure in a discharge tube is reduced, the mean free path of the electrons in the gas increases, and the Crookes and Faraday dark spaces and negative glow extend further down the tube; the positive column therefore contracts and becomes less intense. Eventually when the pressure is rather less than 0·01 mm of mercury, the negative

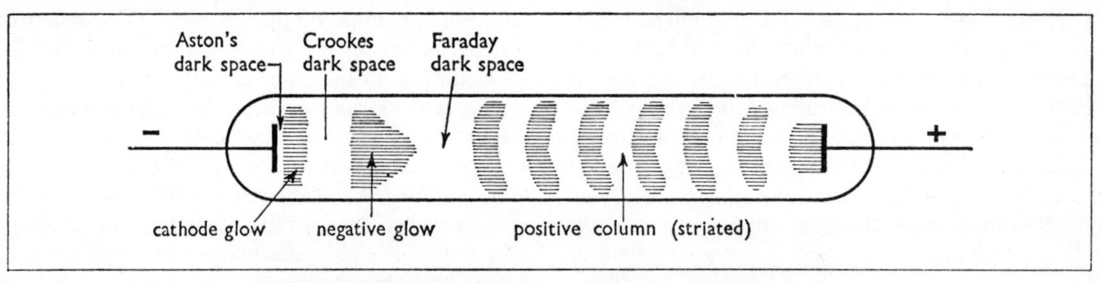

Fig. 13.14 The features of a low-pressure gas discharge

glow extends right to the anode and becomes much weaker. The walls of the tube are now seen to be emitting a green fluorescence arising from their bombardment with electrons emitted by the cathode. Indeed it was in discharge tubes at low pressures that cathode rays were first discovered. As the pressure is reduced still further, the p.d. needed to maintain the discharge rises rapidly; and at pressures below about 10^{-3} mm of mercury the tube usually becomes a good insulator again.

A glow discharge may be modified by using the thermionic effect to provide the necessary emission of electrons from the cathode. If this electrode consists of a heated tungsten wire, the negative glow disappears, and the p.d. across the tube is reduced by the amount that would normally exist across the Crookes dark space. The positive column then fills nearly the whole space. This sort of discharge is used in the fluorescent tubes which are now common in electric lighting installations. In these tubes *both* electrodes are in the form of heated tungsten wires; they are normally designed to work on alternating current, so that each electrode in turn becomes the cathode. The current is controlled by a choke joined in series (p. 105). Once the discharge has started, the current in the tube is sufficiently strong to keep the two 'cathodes' at the right operating temperature; but initially they must be heated by passing a current through them. This is done with a special *starter* connected as shown in Fig. 13.15. It consists of a small discharge tube

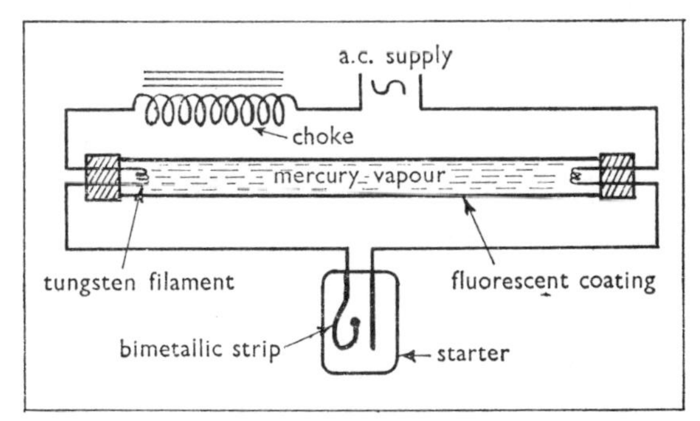

Fig. 13.15 The circuit of a fluorescent lighting tube

filled with argon. The electrodes are quite close together, and one of them is a small bimetallic strip. When the power is switched on, the discharge in the starter commences, and current therefore flows also through the two tungsten filaments joined in series. The heat generated in the starter causes the bimetallic strip to bend over; and within one or two seconds it makes contact with the other electrode. This short-circuits the starter, and the discharge in it stops. At once the bimetallic strip starts to cool, and the electrodes separate again. This switches off the current through the filaments and connects the full p.d. of the supply across the main tube. Provided the tungsten filaments have reached a high enough temperature the main discharge then starts, since the striking p.d. for the main tube is arranged to be less than that required for the starter discharge. If, however, the tungsten electrodes are still not hot enough to initiate a discharge, the starter strikes again; and the starting cycle is repeated until the lamp has lit up. The tube is filled with mercury vapour, which emits light of a blue-green colour. A bare mercury vapour lamp is sometimes used for street lighting; but even here its colour is far from satisfactory. It would be quite unacceptable in any other situation. However, it can be corrected by coating the inside of the tube with suitable fluorescent powders. The mercury vapour spectrum includes a considerable proportion of ultra-violet radiation. This is absorbed by the fluorescent coating and re-emitted as visible light, chiefly in the red part of the spectrum where mercury light is deficient. By suitable choice of the fluorescent materials the light can be made to have almost any required tone value.

13.5 The deflection of electron and ion beams

Electric fields

Over most of the space between a pair of deflector plates the field is nearly uniform, with lines of force parallel to one another and normal to the plates. The path of an electron in such a field is therefore very similar to that of a projectile in a uniform gravitational field. In the motion of a projectile (in the absence of air resistance) the horizontal component of velocity is unaltered by the gravitational forces. But in the direction of the field (i.e. vertically downwards) the projectile experiences a uniform acceleration. The resulting path is then part of a parabola. In the same way an electron moves in a parabolic path between a pair of

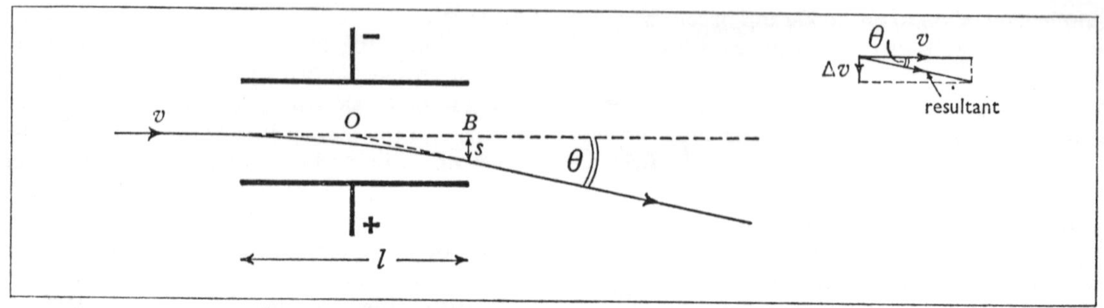

Fig. 13.16 The deflection of an electron beam in an electric field

deflector plates (Fig. 13.16). If its initial path is parallel to the plates, its velocity v in this direction remains unaffected. But at right-angles to this its acceleration a is given by

$$a = \frac{F_E}{m} = \frac{eE}{m}$$

The length l of the path in the field and the time t spent in it are related by

$$t = \frac{l}{v}$$

The component of velocity Δv acquired in the direction of the field is therefore given by

$$\Delta v = \frac{eE}{m}t = \frac{eEl}{mv}$$

The angular deflection θ of the electron beam is found from the parallelogram of vectors:

$$\tan\theta \ (= \theta \text{ in radians, for small angles})$$

$$= \frac{\Delta v}{v} = \frac{eEl}{mv^2} \qquad \text{(I)}$$

$$\therefore \ \frac{e}{mv^2} = \frac{\tan\theta}{El}$$

The lateral displacement s of the electron beam on emerging from the field is given by

$$s = \tfrac{1}{2}at^2 = \frac{eEl^2}{2mv^2}$$

The final direction of the beam is as though the deviation θ had occurred *at the point O* where the final path produced back intersects the initial path (Fig. 13.16).

Now, $$\tan\theta = \frac{s}{\text{OB}}$$

Substituting for s and $\tan\theta$, we find

$$\text{OB} = \tfrac{1}{2}l$$

The point O is thus at the exact mid-point between the plates.

In a cathode-ray oscilloscope (p. 235) we can therefore measure the deflection θ by dividing the displacement of the spot of light on the screen by the distance of the screen from the point O.

Equation (I) gives the value of θ in terms of the quantity e/mv^2. To find the specific charge e/m we have to perform a subsidiary experiment to give the velocity v of the electrons. For this purpose we need to study the magnetic deflection of the beam. It might be thought that v could be calculated from a knowledge of the p.d. through which the electrons have been accelerated (i.e. the p.d. between the cathode and final anode of the tube). But this calculation is only found to yield again the quantity e/mv^2. Thus, the kinetic energy of an electron that has fallen through the p.d. V between cathode and anode is given by

$$\tfrac{1}{2}mv^2 = eV$$

$$\therefore \ \frac{e}{mv^2} = \frac{1}{2V}$$

We can now substitute in equation (I) and eliminate e, m and v completely.

$$\therefore \ \tan\theta = \frac{El}{2V} \qquad \text{(II)}$$

The same deflection would therefore be obtained whatever the charge or mass of the particles in the beam. Indeed the study of electrostatic deflections in a cathode-ray tube provides no certain grounds for supposing the charge and mass of all electrons to be the same. To establish this point and to find the value of e/m we need to investigate also the magnetic deflection of the beam.

Equation (II) shows that the sensitivity of an

electrostatically deflected cathode-ray oscilloscope is inversely proportional to the accelerating p.d. V.

Magnetic fields

In general the path of an electron in a magnetic field is a *helix* (p. 68). But for the special case in which the motion is entirely at right-angles to the field the path is an arc of a circle; this is the arrangement normally used in deflection experiments. If the flux density of the magnetic field is B, then the radius of curvature r is given by

$$r = \frac{mv}{Be}$$

where v is the velocity of the particle (p. 217).

In a cathode-ray tube the deflection is usually produced by a pair of deflector coils mounted on either side of the neck of the tube. The magnetic field is not sharply bounded; but to a first approximation we may assume that it is uniform within the space enclosed by the coils and zero outside (Fig. 13.17). The beam follows a circular path in the field. It is clear from the geometry that the final direction of the beam is as though the deflection θ had occurred at the point C, the centre of the coil system. If the length of the path in the field is l, we have

$$\theta = \frac{l}{r}$$

For small deflections we may take l as equal to the diameter of the coils. As with electrostatic deflections, we measure θ by dividing the displacement of the spot of light on the screen by the distance of the screen from the point C.

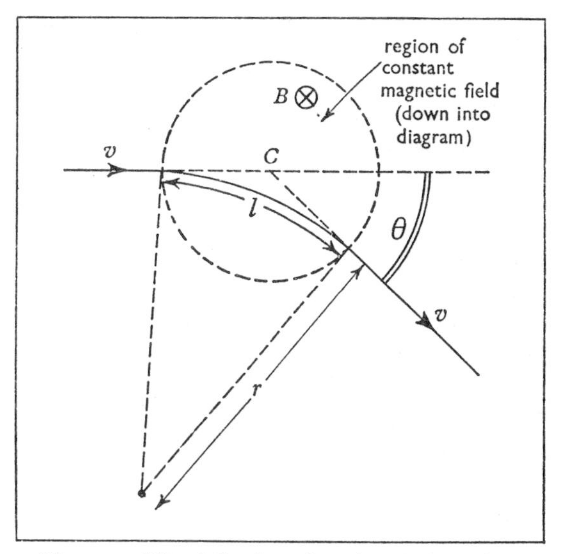

Fig. 13.17 The deflection of an electron beam in a magnetic field

The cathode-ray oscilloscope may be used to make an estimate of e/m by comparing the electric and magnetic deflections of the beam. The methods available are the same in principle as those described earlier in this chapter, except that we cannot now observe the curvature of the path in the magnetic field directly. From either the electric deflection or a knowledge of the accelerating p.d. we may deduce the value of e/mv^2; and from the magnetic deflection we obtain the value of e/mv. The two results may be combined to give the values of v and e/m.

However the method is not very satisfactory because of the difficulty of knowing the exact

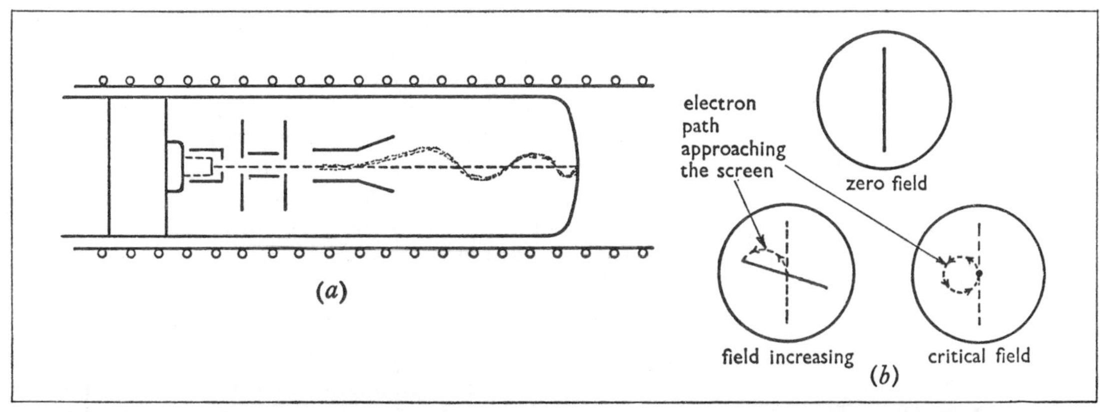

Fig. 13.18 The measurement of e/m with a cathode-ray tube in a longitudinal magnetic field

regions over which the fields act. Fig. 13.18 shows a method employing a different principle that avoids this uncertainty. In this case a *longitudinal* magnetic field is used. A small cathode-ray tube ($1\frac{1}{2}$ inch diam.) can be completely enclosed in a long solenoid so that the field is uniform and constant over the whole region between deflector plates and screen. As long as the beam is parallel to the magnetic field it is unaffected by it. But if a p.d. is applied to the Y-plates, the electrons acquire a component of velocity at right-angles to the magnetic field, and their subsequent path to the screen is a helix instead of a straight line. Viewed through the end of the tube, the electron beam would appear to be moving in a circular path as it approached the screen. If the magnetic field is adjusted so that the beam performs one complete revolution between deflector plates and screen, then the spot of light will be returned to the undeflected position. This adjustment is made by applying an alternating p.d. to the Y-plates sufficient to produce a straight-line trace covering the full height of the tube face; the frequency is of no importance. As the magnetic field is increased, the line rotates and gets shorter (Fig. 13.18*b*) until after a rotation of 180° it contracts to a point. The current in the solenoid required to do this is noted, and from this the magnetic field B is worked out. The velocity v of the beam depends on the p.d. V between final anode and cathode, and is given by

$$v^2 = \frac{2eV}{m} \qquad \text{(p. 217)}$$

The time t taken to travel the distance d from the centre of the Y-plates to the screen is given by

$$t = \frac{d}{v}$$

$$\therefore\ t^2 = \frac{d^2}{v^2} = \frac{md^2}{2eV}$$

Suppose the electrons acquire a component of velocity v' at right-angles to v in passing between the deflector plates; then the magnetic force Bev' is the centripetal force (mv'^2/r) that deflects them into the helical path (of radius r).

$$\therefore\ Bev' = \frac{mv'^2}{r}$$

If the electrons perform one complete revolution in the time t, then

$$v' = \frac{Ber}{m} = \frac{2\pi r}{t}$$

$$\therefore\ t = \frac{2\pi m}{Be}$$

(the same for all values of r)

$$\therefore\ t^2 = \frac{4\pi^2m^2}{B^2e^2} = \frac{md^2}{2eV}$$

Re-arranging this we get

$$\therefore\ \frac{e}{m} = \frac{8\pi^2 V}{B^2d^2}$$

If the magnetic field is increased beyond this point, the electrons perform more than one revolution, and the trace grows into a line again and rotates still further. It contracts into a point each time the field reaches an integral multiple of the value calculated above.

Refined modern techniques enable e/m to be measured with an uncertainty rather less than 1 part in 10^4. At these accuracies it is even possible to demonstrate the increase of mass of the electron with velocity. The theory of Relativity predicts that the mass m of a particle should vary with its speed v (relative to the observer) according to the relation

$$m = m_0\left(1 - \frac{v^2}{c^2}\right)^{-\frac{1}{2}}$$

where c = the velocity of light and m_0 = the mass at zero speed, called the *rest mass.* The mass should therefore *increase* and the specific charge *decrease* with increasing speed of the electron. The experimental results are in excellent agreement with this prediction at all speeds. The effect is by no means negligible even when quite moderate p.d.'s are used to accelerate the electrons. With the p.d.'s used in television tubes (16 to 20 kV) they reach about a quarter of the speed of light; and the corresponding increase of mass is then about 3%. With an accelerating p.d. of half a million volts (quite common in X-ray equipment) the mass of the electron is about doubled.

The mass spectrometer

In the course of the early experiments on gas discharges it was found that luminous 'rays' of some kind would emerge from the discharge

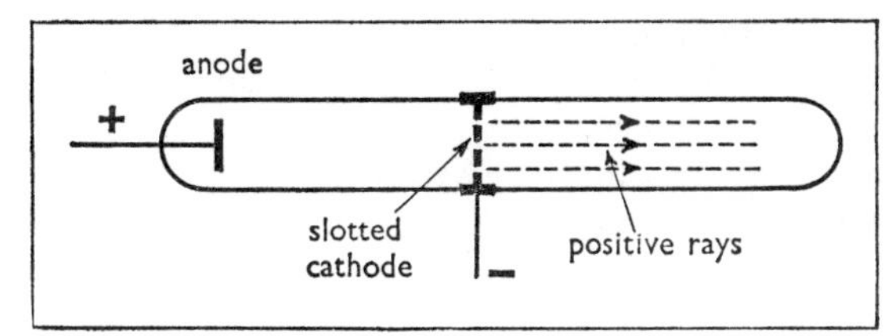

Fig. 13.19 Positive rays

through perforations in the cathode. They were known as *positive rays* (Fig. 13.19). It is not easy to demonstrate their deflection by magnetic fields. Their specific charges are thousands of times smaller than that of the electron, and the magnetic fields to deflect them must be proportionately greater. However, with suitable fields it is possible to show that they consist of streams of positively charged particles; and we now understand that they are the positive ions formed in the passage of the electrons through the gas.

By the combined use of electric and magnetic fields it is possible to separate out the many different sorts of ions present in a positive ray beam. In one type of apparatus the fields are arranged to 'focus' all the ions of any one specific charge into a single line on a photographic film. A different line shows up on the developed film for each type of ion. The *mass spectrograph*, as the instrument is called, makes possible the precise measurement of atomic masses.

Another type of instrument is equipped to detect the presence of particular ions and measure their relative abundances. This is called a *mass spectrometer*.

The principle is illustrated in Fig. 13.20. The material to be examined is introduced as a gas or vapour into the space A. An electron beam is directed across this space from the filament F, and ionizes some of the atoms there. The ions are drawn out of A by making the electrode Q slightly negative with respect to P, and are then accelerated by the much larger electric field between Q and R. Thus a narrow beam of ions emerges from the slit S_1 into the space D. If the p.d. between Q and R is V, the velocity v acquired by ions of charge e and mass M is given by

$$\tfrac{1}{2}Mv^2 = eV$$

$$\therefore\ v^2 = \frac{2eV}{M}$$

In the space D the particles move in semi-circular paths under the influence of a magnetic field of flux density B directed at right-angles to the plane of the diagram. The slit system, S_1–S_2–S_3, is arranged to select particles whose orbits are of one particular radius r.

$$\therefore\ r = \frac{Mv}{Be}$$

Eliminating v between these equations, the specific charge of the particles selected by the system is given by

$$\frac{e}{M} = \frac{2V}{B^2r^2}$$

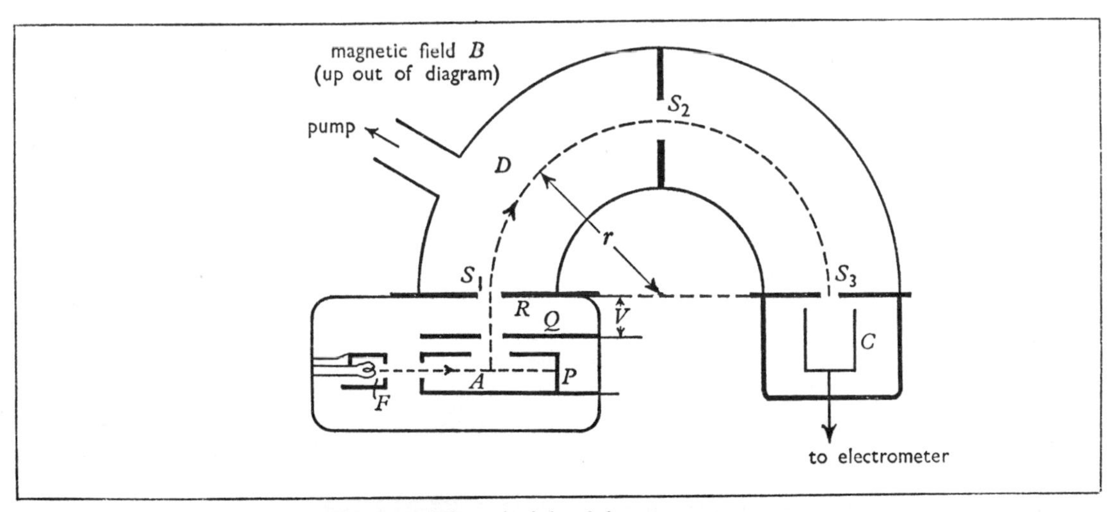

Fig. 13.20 The principle of the mass spectrometer

Any ions with this specific charge are collected by the electrode C, and the small current is registered by a sensitive electrometer. To make an analysis of the ions present in the apparatus, the ion current is measured as the p.d. V is varied, the magnetic field being kept constant. One type of ion after another is thus brought into the collector electrode C, and the relative sizes of the peaks of current indicates the proportions of different isotopes (and complex ions) present.

14: Electronics

14.1 The cathode-ray oscilloscope

The deflection of an electron beam by electric or magnetic fields can be used to study variations in the p.d.'s or currents by which these fields are produced. An instrument employing this principle is called a cathode-ray oscilloscope; the construction of a typical cathode-ray tube for this purpose is shown in Fig. 14.1. Electrons are emitted by the heated filament F. The anodes A_1, A_2, A_3 are maintained at positive potentials with respect to the filament. The electron beam is therefore accelerated down the axis of the tube (which is highly evacuated). The shapes and potentials of the anodes are chosen so that the electric fields between them converge the beam into a fine spot on the fluorescent screen of the tube at S. The filament F is surrounded by a cylindrical electrode G, which is kept at a negative potential with respect to F. This potential controls the proportion of the emitted electrons that reach the hole in the first anode A_1, and so controls the brightness of the spot of light on the screen. (Since the electrode G fulfils the same function as the grid in a triode valve (p. 242), it is usually called the *grid* of the cathode-ray tube.)

In modern tubes the side walls beyond the last anode are usually coated with a conducting layer of graphite, which is connected to the anode; the electron beam is then moving in an enclosure at a constant potential and is unaffected by external electric fields. The tube is usually operated with the final anode and tube wall earthed, so that movements of earthed objects (such as the observer's hand) near the tube face do not affect the beam. The current in the electron beam is of the order of 0·1 milliamp, and this must be allowed to return from the screen to the final anode so as to complete the electric circuit. This can be done by coating the inside of the fluorescent screen with a conducting layer; but small tubes usually rely on another mechanism to provide the return current. When an electron strikes any surface sufficiently rapidly, it causes the emission of further electrons from it; the process is called *secondary emission.* The secondary electrons ejected from the fluorescent screen accumulate as a weak negative 'space charge' in the regions behind it. This cloud of electrons then drifts back slowly through the tube towards the anode.

The deflection of the electron beam is effected by the electric fields between the pairs of deflector plates—X_1 and X_2 for horizontal deflection, and Y_1 and Y_2 for vertical deflection. One of each pair

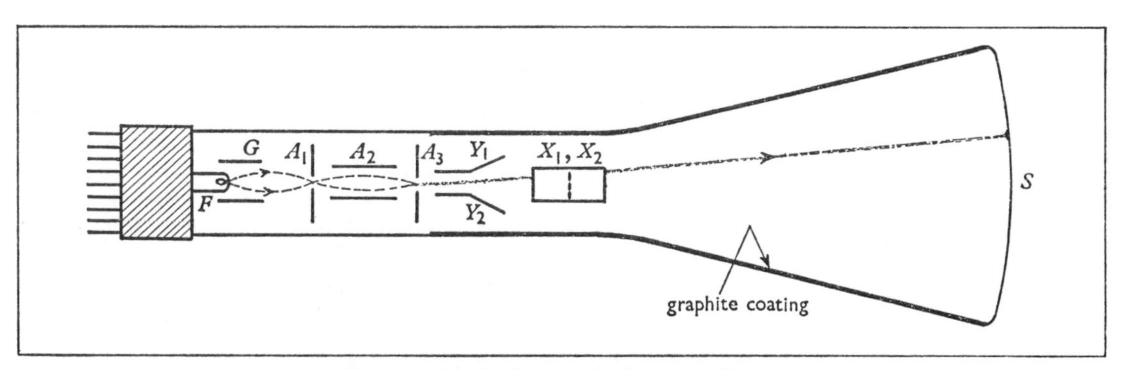

Fig. 14.1 A cathode-ray tube for an oscilloscope

of plates is connected to the final anode A_3 so as to avoid as far as possible stray electric fields between the deflector plates and the other parts of the tube. In a good modern tube the deflection of the spot of light is very closely proportional to the p.d. between a pair of deflector plates. The time of transit of the electrons through the tube is so small that the spot faithfully follows the variations of X and Y p.d.'s up to very high frequencies of oscillation. The beam may instead be deflected magnetically using currents flowing in coils mounted round the neck of the tube. But this is not usually done in tubes designed for instrument purposes. In fact such a tube is normally equipped with a mumetal screen to shield it from stray magnetic fields produced by transformers, etc. in the associated apparatus.

Some uses of a cathode-ray oscilloscope

(*i*) *To measure peak p.d.'s.* For this purpose only one pair of deflector plates is used—normally the *Y*-plates, since these, being further from the screen, produce bigger movement of the spot of light for a given p.d.; the *X*-plates are both earthed. The alternating p.d. joined to the *Y*-plates deflects the beam up and down in a vertical plane; and the trace observed is a straight vertical line whose length is proportional to the *peak p.d.* The instrument may be calibrated by first connecting a known p.d. to the plates.

(*ii*) *To study the waveforms of alternating p.d.'s.* For this purpose the *X*-plates are connected to an auxiliary circuit that generates a 'saw-toothed' p.d. (Fig. 14.2); such an arrangement is called a *time base.* The waveform to be investigated is joined to the *Y*-plates. During the interval from A to B the spot of light is drawn at a steady speed (ideally) in the *X*-direction across the screen, while in the *Y*-direction it follows the p.d. under study. The trace is therefore a graph of this p.d. against time. During the interval from B to C the spot flies back quickly to its starting point. If the pattern observed is to be a steady one, it is necessary for the second and subsequent traces to be exactly superimposed on the first. This will only occur if the frequency of the time base is an exact submultiple of the frequency of the p.d. under test; this p.d. will then complete exactly a fixed number of oscillations in the time of one oscillation of the time base. The time base frequency therefore needs to be continuously variable. Even then, drifting of one or both frequencies is almost inevitable, and to obtain a truly steady pattern it is necessary to lock the two oscillations together. This is done by ensuring that the fly-back of the time base is always synchronized to occur at a given point of the wave pattern under study.

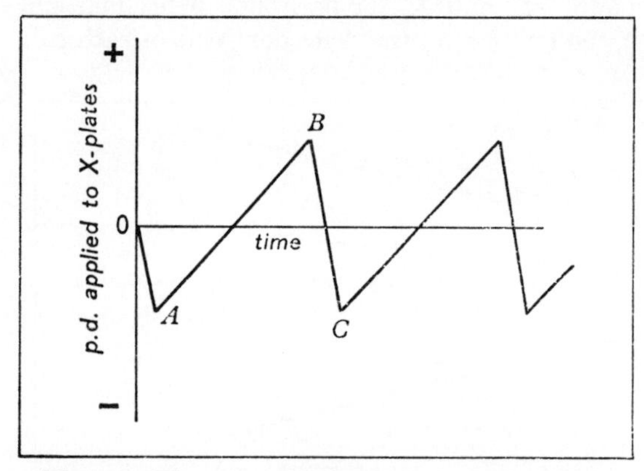

Fig. 14.2 The saw-toothed waveform required for a time base

It is often necessary to amplify the waveform to be examined before it is applied to the *Y*-plates, since a peak p.d. of at least 50 volts is likely to be needed to give an adequate height of trace. Needless to say, the amplifiers used must be of the highest quality so that they do not introduce distortions of any kind into the waveforms being used.

(*iii*) *To compare frequencies.* For this purpose the time base is switched off, and the two p.d.'s whose frequencies are to be compared are joined one to each pair of deflector plates. If the frequencies are in a simple ratio, a steady trace is observed of the kinds shown in Fig. 14.3. Patterns of this sort are known as *Lissajous figures*; the ratios of frequencies may be deduced from them as shown in the diagram. This technique is very useful when we need to conduct an experiment using alternating currents covering a range of known frequencies. A variable-frequency oscillator is used to supply the alternating current at the frequencies required; its output is joined to one pair of deflector plates. A standard source of known frequency is joined to the other pair. The oscillator is then used at the exactly specified frequencies that give stationary Lissajous figures.

(*iv*) *To measure phase differences.* If the two

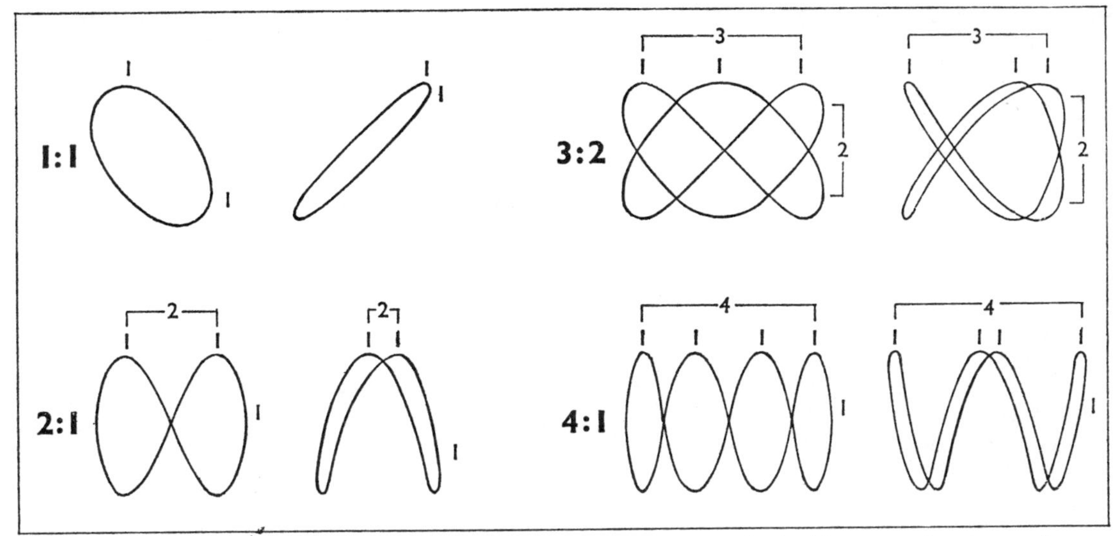

Fig. 14.3 Examples of Lissajous figures. The frequency ratio may be obtained by comparing the numbers of loops along the top and down one side

pairs of deflector plates are supplied with sinusoidal alternating p.d.'s of the same frequency, but differing in phase, the trace obtained will in general be an ellipse, as shown in Fig. 14.4. The shape of this figure may be used to give the phase difference. The variations of the x and y deflections

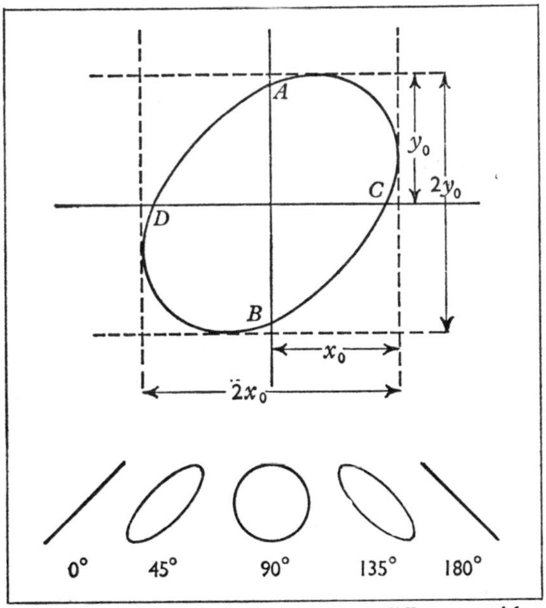

Fig. 14.4 The measurement of phase difference with a C.R.O.

with time t are given by

$$x = x_0 \sin \omega t$$

and $$y = y_0 \sin (\omega t + \phi) \qquad \text{(p. 197)}$$

where ϕ is the phase difference. The intercepts on the y-axis of the tube face are the values of y when $x = 0$; thus

$$\text{when } x = 0, \quad y = \pm y_0 \sin \phi$$

The distance between the two intercepts is therefore given by

$$\text{AB} = 2y_0 \sin \phi \qquad \text{(Fig. 14.4)}$$

The peak-to-peak height of the trace is $2y_0$. We therefore have

$$\sin \phi = \frac{\text{AB}}{2y_0}$$

Using the intercepts on the x-axis we have likewise

$$\sin \phi = \frac{\text{CD}}{2x_0}$$

In the special case when ϕ is exactly 0° or 180°, the ellipse collapses into a straight line inclined to the axes (Fig. 14.4). When the two p.d.'s are in quadrature (phase difference 90°), the trace can be made into a circle by adjusting the amplitudes to be equal.

14.2 Thermionic tubes

There are a number of devices that make use of the thermionic effect to conduct an electric current through a high-vacuum tube. Such tubes are made with two, three, four, five, etc. electrodes; and are referred to as *diodes*, *triodes*, *tetrodes*, *pentodes*, respectively (using prefixes based on Greek numerals to indicate the number of electrodes). The diode is the simplest of these; it contains a heated cathode, from which electrons are emitted, and an anode. Its essential property is that current can flow through it in one direction only, namely, by the flow of electrons from cathode to anode; and this can occur only if the anode is positive with respect to the cathode. When the anode is negative, the electric field in the tube drives the electrons back to the cathode again, and no current flows. It is this behaviour that has led to its being called a *valve*. Its most important use is in circuits for producing direct current from an alternating supply. This process is called *rectification*, and is discussed in detail below (p. 241).

The construction of a diode valve is shown in Fig. 14.5. It consists of a central cathode, surrounded by an approximately cylindrical anode. Two different types of cathode are used:

(*i*) *Directly heated.* In this type the cathode is a single fine wire or *filament*, usually zigzagged up and down the tube. It is heated by a direct current passing through it from an independent supply. A disadvantage of this construction is that the potential is necessarily higher at one end of the filament than the other; this means that the p.d. between anode and cathode must vary from one part of the cathode to another. The cut-off of the current through the tube when the anode is made negative is not therefore as sharp as it might be. Also, the cathode cannot normally be heated with alternating current, since this would impose a similar alternating component on the current flowing to the anode. Directly heated cathodes are now used only for specialized types of valve.

(*ii*) *Indirectly heated.* In this case the cathode consists of a sheath which is heated by a wire zigzagged up and down inside it. The heater wire is packed around with a refractory material (some form of clay) and is thus electrically well insulated from the cathode. The cathode is now at a uniform potential; and there is also no objection to using alternating current for the heater— which is a great practical convenience. The indirectly heated cathode suffers from the slight disadvantage that it takes an appreciable time to reach its operating temperature. But it is the form of cathode now used for the vast majority of small thermionic valves.

At one time the only material used for making thermionic cathodes was tungsten because of its high melting point. However, to obtain adequate electron emission a tungsten cathode must be run at about 2700°C; this is an inconveniently high

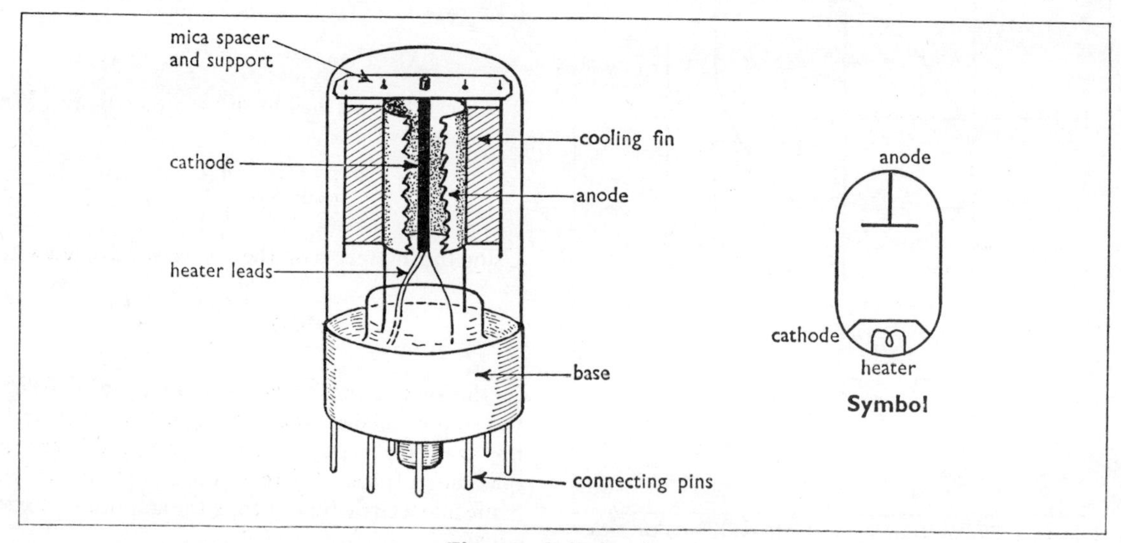

Fig. 14.5 A diode valve

temperature, requiring the dissipation of considerable amounts of power uselessly as heat. They are now used only in high-power valves, in which the bombardment of the cathode by positive ions from the residual gas would quickly destroy any other type of cathode.

In all other types of valve *oxide-coated* cathodes are now used. The emitting surface in this case is a specially prepared layer of barium and strontium oxides, which is built up on a suitable metal base (usually nickel or one of its alloys). This sort of cathode is normally run at a temperature of about 700°C, at which its electron emission is more than adequate for the needs of all types of small thermionic valves.

The other metal parts of the valve are generally made of nickel or a nickel alloy, since this substance is easy to handle in manufacture and is chemically stable. The electrodes are supported by the wires that pass through the glass base of the valve; accurately punched mica spacers are used to hold them rigidly in position.

For a valve to behave reliably it is necessary for a high vacuum to be maintained inside it—a pressure of less than 10^{-6} mm of mercury is usually required. It is found that the glass and metal parts of the valve can retain an appreciable layer of gas molecules more or less permanently stuck to their surfaces; the phenomenon is known as *adsorption*. However, when the valve is heated the adsorbed gas evaporates off, destroying the vacuum. It is therefore necessary for all parts of the valve to be heated above their normal working temperatures during the process of evacuation. The glass envelope is heated in an oven to a temperature near its softening point; and the metal parts are raised to an even higher temperature by eddy currents induced in them by a high-frequency alternating magnetic field (p. 101). The cathode is effectively de-gassed by passing a large current through the heater. Even with these precautions small amounts of chemically reactive gases and vapours would be left in the tube, which would quite quickly contaminate the cathode and reduce the electron emission. To cope with this a small quantity of a chemically active metal (barium, magnesium, etc.) is placed in a gauze container mounted near the base of the valve. During the eddy current heating of the metal parts this is vaporized and deposited as a thin film on the inside of the glass bulb. Here it remains to combine with any chemically active substances present in the tube; the vacuum is thus maintained and even improved after the valve has been sealed off. This device is called a *getter*. After sealing off, the glass bulb is mounted on a plastic base; and the copper leads are soldered into the connecting pins.

If the p.d. between cathode and anode is V and the current flowing between them is I, the power W dissipated as heat is given by

$$W = VI$$

The effect of the p.d. across the valve is to accelerate the electrons in the space between the electrodes. In the first instance the power W exists as kinetic energy of the electrons; only on impact with the anode does this energy get changed into heat. In addition, the anode receives most of the energy radiated by the hot cathode. The anode must therefore have a large external surface area, and is often provided with cooling fins to enable it to dissipate the heat generated without excessive rise of temperature.

14.3 Thermionic diodes

To study how the anode current of the valve depends on the anode potential and the cathode temperature, the circuit of Fig. 14.6 may be used.

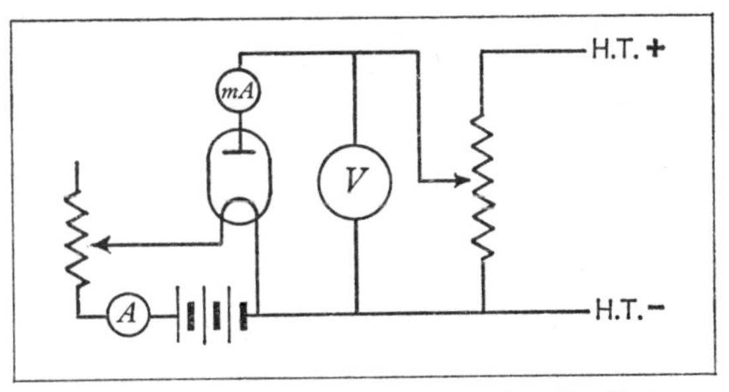

Fig. 14.6 Investigating the properties of a diode

The curves obtained are called the *characteristics* of the diode. For given temperature of the heater the variation of anode current with potential is as shown in Fig. 14.7. The current increases steadily with potential up to a certain value I_s (on curve A). But for p.d.'s greater than this there is no further increase; and the current is said to be *saturated*, all the electrons emitted by the cathode being collected at the anode. If the cathode

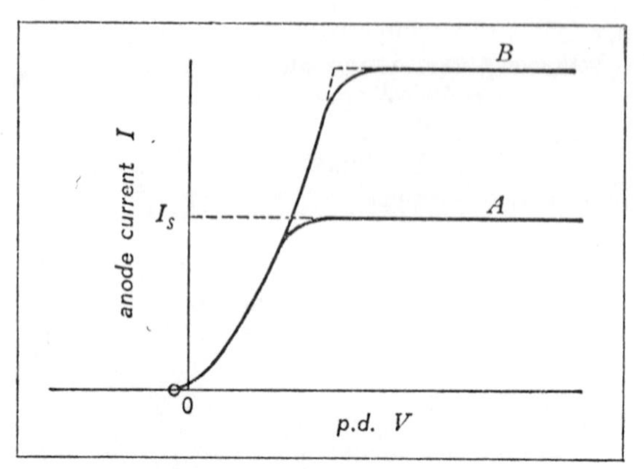

Fig. 14.7 The characteristics of a diode

temperature is raised (by increasing the heater current), the saturation current reaches a higher value (curve B). However, with oxide-coated cathodes at their normal working temperatures the emission of electrons is so copious that it is almost impossible to reach saturation without damaging the valve in some way; and this part of the characteristic is then irrelevant.

Below the saturation point only some of the electrons emitted are drawn away to the anode. The electrons emerge from the cathode with very small velocities and collect in the space between the electrodes as an almost stationary cloud of negative charge, called a *space-charge.* Most of the space-charge is concentrated quite close to the cathode; and in this region the electric field produced by it cancels out that due to the p.d. across the valve. For small anode potentials the space-charge exists in a dynamic equilibrium, in which the rate of return of the electrons to the cathode is almost equal to their rate of emission; a small proportion of them only is drawn away from the outer regions of the space-charge to the anode. As the anode potential is increased, the density of the space-charge is reduced and a higher proportion of the electrons reach the anode. On the curved part of the diode characteristic the current is therefore said to be *space-charge limited.* At the 'knee' of the curve the space-charge ceases to exist, and all the electrons emitted arrive at the anode.

A theoretical analysis of this process shows that the space-charge limited current I should be proportional to the *three-halves power* of the p.d. V, a result sometimes referred to as *Child's law.* Thus

$$I \propto V^{3/2}$$

or

$$I = k\,V^{3/2}$$

where k is a constant that depends on the cathode surface and temperature. However, in practice the behaviour of most diodes agrees only crudely with this law, since the temperature (and therefore the value of k) falls off drastically towards the ends of the cathode.

A small current is found to flow through a diode even when the anode is made slightly negative with respect to the cathode. This happens because some of the electrons are emitted with appreciable kinetic energy, and can reach the anode even against the action of a small opposing field. The

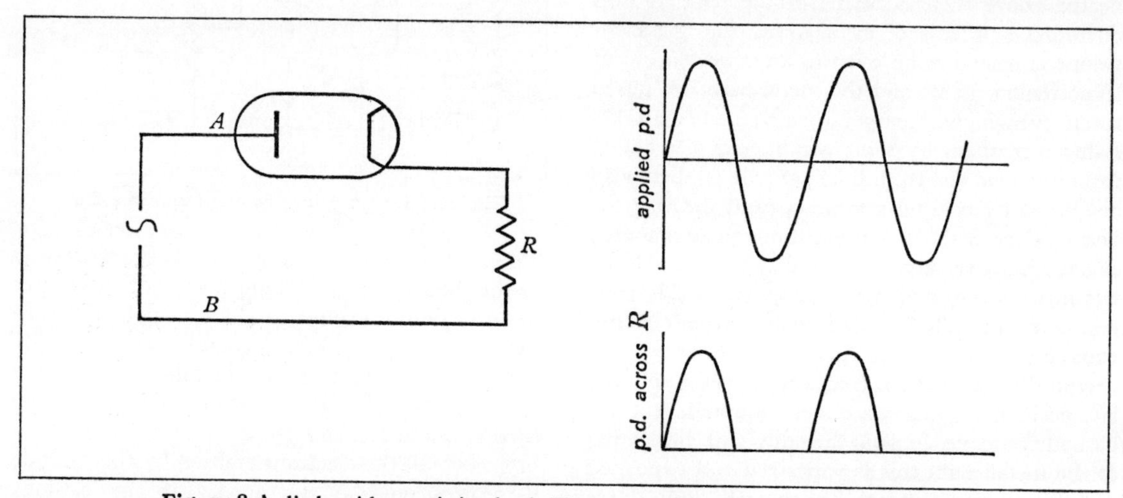

Fig. 14.8 A diode with a resistive load. The current is unidirectional but pulsating

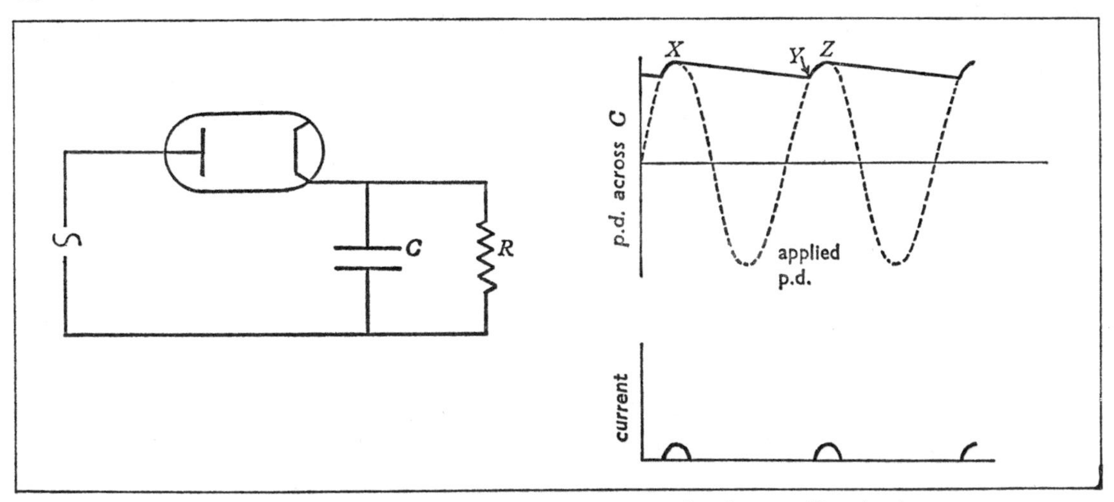

Fig. 14.9 The use of a *storage* or *reservoir* capacitor in a rectifier circuit

anode–cathode p.d. must be reduced to about $-\frac{1}{2}$ volt to prevent the flow of electrons completely.

Rectifier circuits

The ability of a diode to conduct current in one direction only enables it to be used as a rectifier for producing d.c. from an alternating supply. The simplest possible arrangement consists of a diode joined in series with the load (Fig. 14.8). (The word 'load' in this connection means that to which the rectifier is supplying power in the form of direct current; it is represented in Fig. 14.8 by the resistor.) The diode conducts only at those parts of the cycle when the point A is positive with respect to the point B. The current through the load is therefore uni-directional, but pulsating.

A better arrangement is that shown in Fig. 14.9. The connection of the *storage* or *reservoir capacitor* C in parallel with the load radically alters the behaviour of the circuit. Suppose first that the load R is disconnected, so that there is no way for the capacitor to discharge when the diode is non-conducting. Current flows through the diode during the positive half of the cycle charging up the capacitor with the upper plate positive. This continues in each cycle until the p.d. across the capacitor is equal to the *peak* p.d. of the supply. When this has occurred, the anode of the diode can no longer become positive with respect to the cathode, and no further current flows. With the load joined up, there is a continual steady discharge of the capacitor, and the output is of the form shown in Fig. 14.9. At the point X the capacitor is charged up to the peak p.d.; the diode then ceases to conduct and the p.d. across the capacitor falls slowly as it partially discharges through the load. At the point Y the anode again becomes positive with respect to the cathode and the diode again starts to conduct. A small pulse of current therefore flows through it between the points Y and Z recharging the capacitor once more. If the current taken through the load is small, the output is a nearly steady p.d. almost equal to the peak p.d. of the supply.

However, with increased current the average p.d. is less than this, and there is an a.c. 'ripple' superimposed on it. If further smoothing of the output is needed, a choke-capacitor filter can be used as in Fig. 14.10. The choke is chosen so as to have low d.c. resistance, but high a.c. impedance;

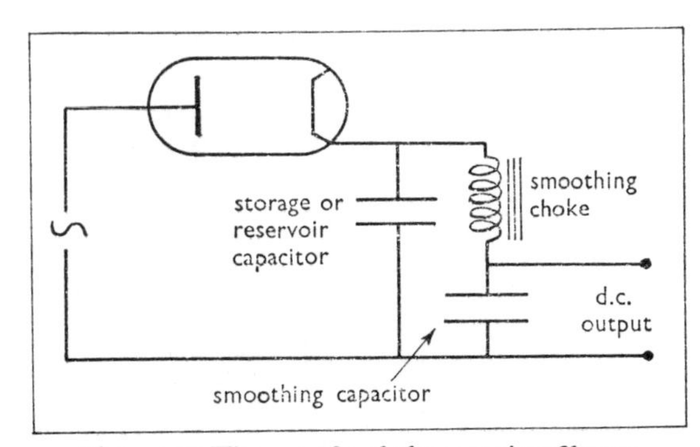

Fig. 14.10 The use of a choke-capacitor filter to obtain further smoothing

while the capacitor has very high d.c. resistance, but low a.c. impedance. Most of the 'ripple' p.d. therefore appears across the choke and very little of it across the capacitor where the output is taken. A small part of the steady p.d. across the storage capacitor is dropped in the resistance of the choke, but most of it is available across the load.

The valve voltmeter

A similar arrangement may be used to measure the peak value of an alternating p.d. In Fig. 14.11 the 'load' of the rectifier is a high-resistance d.c. voltmeter. The storage capacitor is chosen so that the

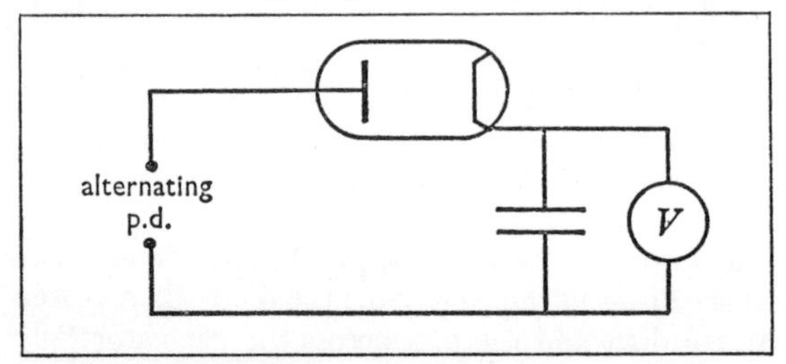

Fig. 14.11 A simple diode valve voltmeter reading peak values of p.d.

time constant CR is very long compared with the period of oscillation of the p.d. There is then a negligible ripple on the rectified output measured by the voltmeter; and the reading gives the *peak value* of the p.d. However, allowance must be made for the ability of the diode to conduct when its anode is slightly negative with respect to the cathode. The result of this is that the voltmeter reads about 0·5 volt even when the two input leads are joined together. This zero reading must therefore be subtracted from all subsequent readings of the instrument.

14.4 The triode

The addition of a third electrode to a thermionic valve provides another means of controlling the current flowing to the anode. This electrode takes the form of an open wire spiral supported between the other two electrodes, but much closer to the cathode than the anode; it is called the *grid*. Otherwise the construction of the valve is very similar to that of a diode (Fig. 14.12). The electrons leaving the cathode are subject to the action of two electric fields—(*i*) that between the cathode and anode, and (*ii*) that between the cathode and grid. Normally these fields are arranged to act in opposition, with the anode positive with respect to the cathode and the grid negative. Provided the anode field is the greater of the two, electrons will still be drawn to the anode through the gaps in the grid; but no electrons can be collected by the grid itself, if it is more than 0·5 volt negative with respect to the cathode. In this condition small changes in the grid potential cause considerable changes in the anode current. The triode valve can therefore be used as an amplifying device. It can

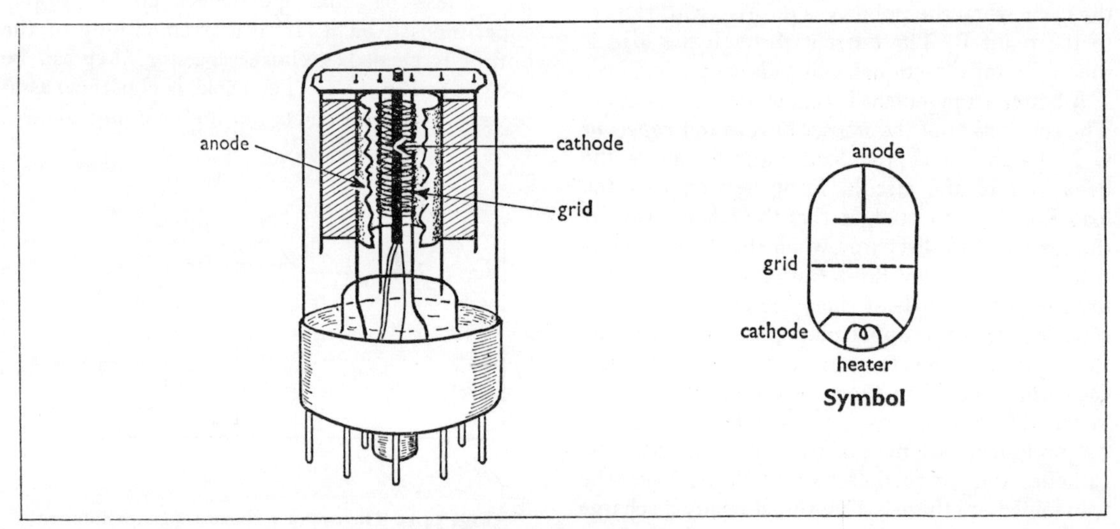

Fig. 14.12 A triode valve

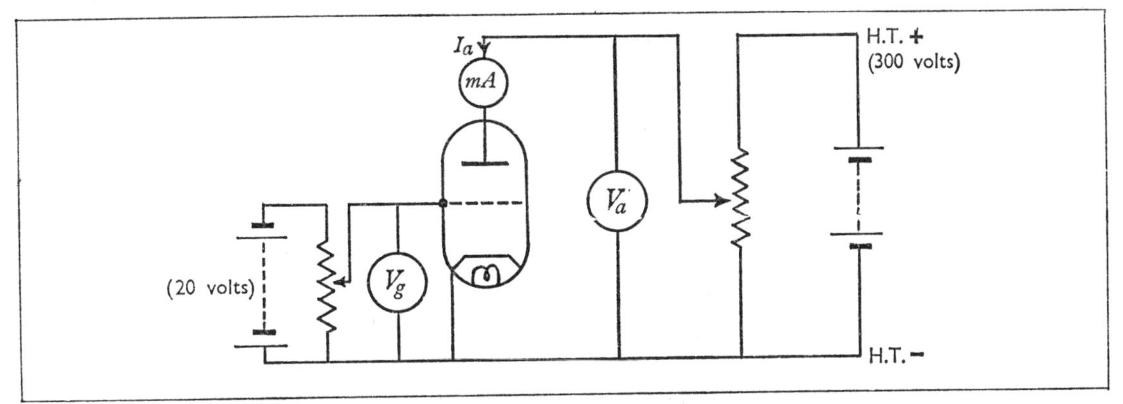

Fig. 14.13 Investigating the characteristics of a triode

be regarded as a generator of alternating current, in which the current at every instant follows exactly the variations of potential of the grid. But the power delivered comes from the H.T. supply rather than from the grid circuit. The alternating p.d. applied to the grid is superimposed on the steady negative grid potential or *bias*, as it is called, so that the grid remains always negative; no current flows in the grid circuit, and negligible power is therefore dissipated in it. But the corresponding alternating current in the anode circuit can supply appreciable quantities of power to a load connected in series.

The detailed *characteristics* of a triode may be investigated with the circuit shown in Fig. 14.13. Two batteries or other sources of p.d. are required; a low-voltage source (about 20 volts) for the grid bias, and a high-tension source (about 300 volts) for the anode. High-resistance potential dividers are joined across both sources so that the p.d.'s applied can be adjusted to any chosen values within these ranges. In this case there are three variables under consideration; the anode current I_a, the anode–cathode p.d. V_a and the grid–cathode p.d. V_g—which are measured by the three meters shown. The full representation of the valve characteristics would therefore require a three-dimensional graph! In practice we must be content with showing the variation of two of the quantities for selected values only of the third quantity. This may be done in either of the ways shown in Fig. 14.14. In (*a*) I_a is plotted against V_a for selected values of V_g. The curves obtained are known as the *anode characteristics* of the valve. For each value of V_g no anode current flows at all unless V_a exceeds a certain figure; but for greater values of V_a the curves become approximately straight and parallel to one another. In (*b*) I_a is plotted against V_g for selected values of V_a. These curves are called the *mutual characteristics* of the triode. Both anode and

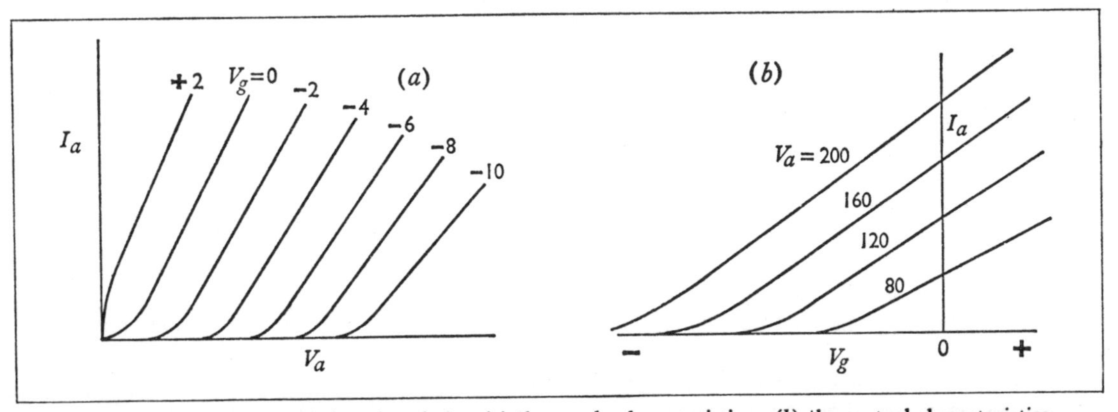

Fig. 14.14 The characteristics of a triode: (*a*) the anode characteristics; (*b*) the mutual characteristics

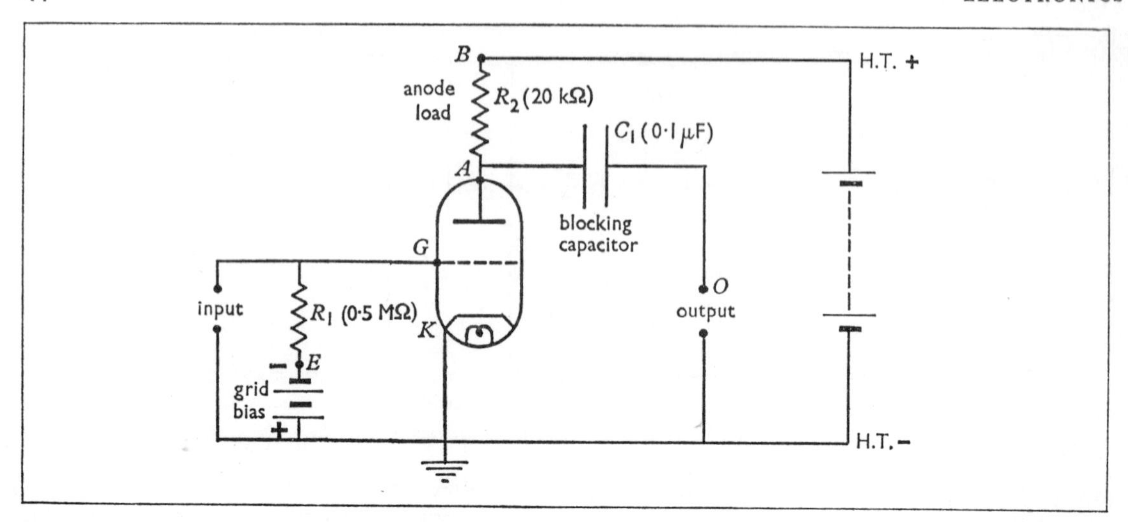

Fig. 14.15 The triode as a voltage amplifier

mutual characteristics may be plotted from the same set of readings; or one set of curves may be deduced from the other.

Voltage amplifiers

There are many familiar applications in which a small alternating p.d. must be amplified to drive some piece of equipment. For instance a microphone or gramophone pick-up may produce an alternating p.d. of only a few millivolts at negligible power; this signal must then be sufficiently amplified to actuate a loudspeaker or the recording head of a tape recorder. A triode can deliver appreciable quantities of power from a negligible power input; the power is *derived* from the H.T. supply, but is *controlled* by the potential of the grid. To develop enough power an alternating p.d. of several volts must usually be applied to the grid. A complete amplifier therefore consists first of several stages of *voltage amplification*; in these the small input p.d. is amplified by a factor of a thousand or more without necessarily generating large amounts of power. These are then followed by a single stage of *power amplification*, in which sufficient power is developed to drive the loudspeaker or recording head.

Fig. 14.15 shows how a triode may be used as a voltage amplifier. As long as there is no input 'signal' and the grid potential is constant, a steady current from the H.T. supply passes through R_2 and the valve in series. The anode A is therefore negative with respect to the point B. No current flows to the grid G, and there is therefore no p.d. across R_1; G is thus at the same potential as E, which is held negative with respect to the cathode K. When a small alternating p.d. is applied to the input terminals, this is superimposed on the steady grid bias and causes corresponding variations of anode current. Since the valve is to be used as a voltage amplifier, the alternating current in it must be passed through a suitable resistor, across which the output p.d. will be produced; this is called the *anode load resistor*—R_2 in Fig. 14.15. However, there is also a steady potential at A that is not wanted at the output terminal *O*. A *blocking capacitor* C_1 (p. 154) is therefore used to filter out the d.c. component of the p.d. across the valve; only the a.c. component then appears at the output.

To obtain distortionless amplification the valve must be operated on the linear part of its characteristics, and the values of the grid bias and R_2 are chosen accordingly. For example, suppose examination of the characteristics of a certain triode shows that the optimum operating conditions are: $I_a = 5$ mA; $V_g = -3$ V; $V_a = 150$ V. If the p.d. of the H.T. supply is about 250 V, the potential drop across R_2 must be about 100 V.

$$\therefore \; R_2 = \frac{100}{5 \times 10^{-3}} = 20{,}000 \text{ ohms}$$

If now a signal is applied to the input and the grid potential *rises* by (say) $\frac{1}{2}$ V, this might cause a 1-mA increase in anode current (i.e. from 5 mA to 6 mA). The p.d. across the load resistor therefore increases from 100 V to 120 V; i.e. the potential at A *falls* by 20 V. This is 40 times as great as the input, but is *in antiphase* with it.

In a complete amplifier several amplifying stages such as that shown in Fig. 14.15 are joined in cascade, the output terminals of one stage being joined to the input terminals of the next. The coupling capacitor C_1 then has the essential function of isolating the grid of the second valve from the high p.d. on the anode of the first. With an even number of stages the output and input are in phase; with an odd number they are in antiphase.

To achieve a big output signal the anode load R_2 needs to be as large as possible. However, if the valve is to operate under optimum conditions, R_2 can only be increased by using larger high-tension p.d.'s. Beyond a certain point it becomes uneconomic to increase the amplification by this means. In practice a voltage amplification of about 40 is all that can reasonably be obtained from a triode. The value of the grid resistance R_1 is not at all critical; its only function is to keep G at the same potential as E. However, it must be large enough to cause negligible drain of power from the input. A value of 0·5 megohm would probably be suitable. In some cases R_1 might be dispensed with altogether, if there is a direct connection between G and E through the input terminals. For instance the input might be fed through a transformer whose secondary coil would provide the necessary connection. The capacitance C_1 must be large enough so as to cause no loss of output at the lowest frequency used. In the amplifier of a record player 0·1 μF might be a suitable value.

★ Power amplifiers

For this purpose the valve needs to be able to pass relatively large currents. Power triodes are therefore designed with cathodes of larger surface area than usual. The same type of circuit is used, but the anode load is now usually a transformer (Fig. 14.16). A moving-coil loudspeaker or a recording head has an impedance of only a few ohms, and therefore requires a high current at low p.d.; whereas the valve supplies the power at a higher p.d. and a much smaller current. The transformer is therefore needed to 'match' the amplifier to the loudspeaker, so that power is transferred at maximum efficiency and with minimum distortion.

Fig. 14.16 shows also the method that is normally used in all valve circuits for supplying grid bias; this arrangement avoids the use of separate grid-bias batteries. The *bias resistor* R_3 is joined in series with the cathode, as shown. The anode current I_a flows also through this, and the cathode K is kept positive with respect to the point E. As before, no current flows in the grid resistor R_1, and the points G and E are therefore at the same potential. The p.d. across R_3 thus constitutes the grid bias. The function of the by-pass capacitor C_2 is to prevent the bias developed across R_3 from fluctuating with the anode current. Under steady conditions C_2 charges up to the p.d. across R_3, and no further current then flows onto its plates. But when the anode current is varying, the alternating component is 'by-passed'

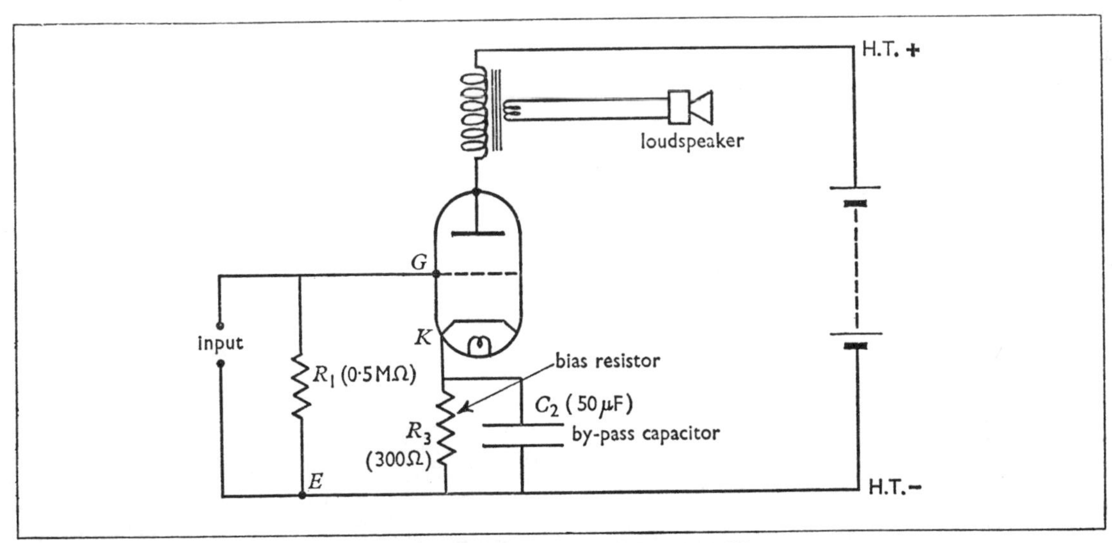

Fig. 14.16 The triode as a power amplifier

through C_2, which is of large capacitance (e.g. 50 μF) and therefore of low impedance at the frequencies concerned. An electrolytic capacitor would normally be used for this purpose, since in this circuit it has the necessary d.c. polarizing p.d. across it (p. 180).

14.5 Oscillators

Any amplifier can become a self-maintained generator of oscillations if some of the output is fed back to the input in the right phase. Indeed, quite unwanted oscillations can easily be generated in this way. In a public address system it sometimes happens that the sound fed back from the loudspeakers to the microphone is of greater intensity than the original sound to be amplified. The signal then builds up to a very large amplitude, and a piercing howl develops at some frequency to which the system is resonant. This is an example of acoustic feed-back. But the same effect may occur if there is an electrical connection or any other sort of link between output and input. Oscillations can start if there is enough feed-back *in phase* with the input; this is called *positive feed-back*. In designing an amplifier great care is taken to eliminate positive feed-back, since it causes distortion and instability, even when not sufficient to cause oscillations. (*Negative feed-back* is often applied to *reduce* distortion.) But the same effect may be used deliberately in circuits for generating alternating currents.

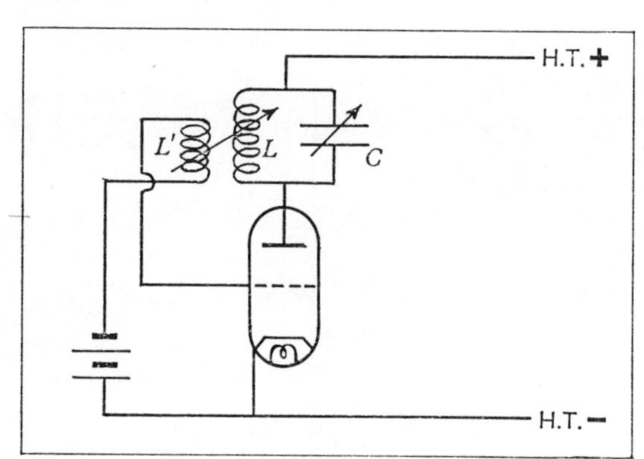

Fig. 14.17 A reaction oscillator

Fig. 14.17 shows one of many ways in which a triode can be used as an oscillator. The frequency of the oscillations is controlled by the resonant circuit consisting of the inductor L and capacitor C in parallel. Such a combination has a natural frequency of oscillation f given by

$$\omega^2 LC = 1 \qquad \text{(p. 212)}$$

where $$\omega = 2\pi f$$

If there were no losses, an alternating current of this frequency could circulate in the loop formed by L and C without ever decreasing. In practice the inevitable resistance of the circuit would cause the oscillations to die away rapidly. The function of the valve is to supply an alternating current to the resonant circuit to make up the losses so that the oscillations can continue indefinitely.

In the *reaction oscillator* in Fig. 14.17 the resonant circuit is in series with the anode. The small coil L' is inductively coupled to L so that the alternating current in the resonant circuit provides the alternating input p.d. for the grid. If L' is connected the right way round, the feed-back will be positive, and the anode current will be in the right phase to increase the alternating current in the resonant circuit. The oscillations then build up in amplitude until the power losses in the circuit are equal to the power that the valve can develop. It is not necessary to do anything to start the oscillations; the random motions of electrons in the coils are quite sufficient to provide the initial disturbance. Provided there is sufficient feed-back between anode and grid, the oscillations start as soon as the circuit is switched on. To control the feed-back the coupling between the two coils is made adjustable; this is represented by the arrow through the coils in the circuit diagram. Either the spacing of the coils is altered or a magnetic core linking them is moved in or out. This type of oscillator works very well provided the losses in the circuit are not too great, otherwise it may be impossible to provide sufficiently close coupling to give enough positive feed-back.

★ A rather better arrangement is shown in Fig. 14.18; this is known as the *Hartley oscillator*. In the reaction oscillator one end of the coil is kept at the fixed potential of the H.T. supply, and the potential of the other end oscillates about this. But in the Hartley circuit the H.T. supply is joined to a point B near the lower end of the coil. The potentials of the ends of the coil, O and I, are now both oscillating about the H.T. potential, and are necessarily in antiphase. This ensures that the feed-back will be positive. The capacitor C' serves to block the high H.T. potential from the grid; but

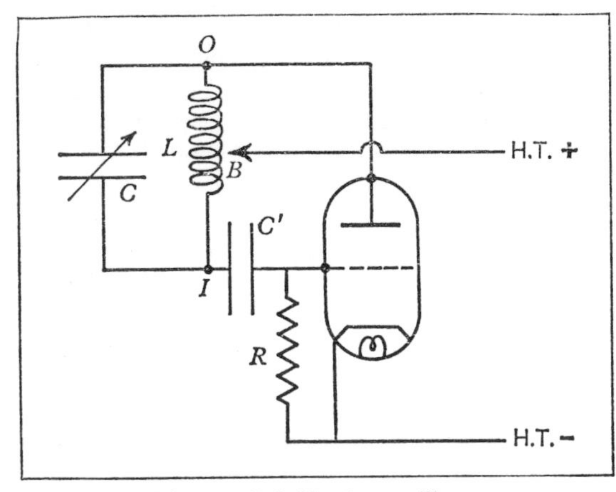

Fig. 14.18 A Hartley oscillator

the alternating potential at I is communicated to the grid through it. The resistor R controls the grid bias. The grid is inevitably driven positive with respect to the cathode during some part of each cycle of oscillation, and a steady current therefore flows through R making the grid negative. In fact the grid circuit behaves like a diode rectifier arrangement with C' as the storage capacitor and R the load (p. 241). The rectified p.d. constitutes the grid bias; superimposed on this is the alternating p.d. from the coil. The amplitude of oscillation and the grid bias normally settle down at such values that the grid just becomes positive at the peak of each cycle; i.e. the grid bias is approximately equal to the peak value of the alternating potential at I. The amount of feedback in the Hartley oscillator is altered by adjusting the position of the tapping point B on the coil; it can readily be made much greater than in the reaction oscillator. The Hartley circuit can therefore work at much higher frequencies, at which the losses are liable to be greater.

14.6 The theory of the triode

On the linear parts of the characteristics of a triode the variations of anode current I_a with anode potential V_a and grid potential V_g may be specified by the gradients of the appropriate curves. Three such parameters are commonly used:

(*i*) *The anode a.c. resistance*, r_a. This is defined as the ratio:

$$r_a = \frac{\text{increase in } V_a}{\text{increase in } I_a} \quad \text{for constant } V_g$$

It is thus the reciprocal of the gradient (or slope) of the anode characteristics. For this reason it is also known as the *anode slope resistance*. In Fig. 14.19*b*

$$r_a = \frac{\text{AC}}{\text{BC}}$$

The a.c. resistances of small triodes are mostly in the range from 1000 to 20,000 ohms. The types of valves with the smallest a.c. resistance are those used as power amplifiers. Valves with higher a.c. resistance are more suitable for voltage amplifiers.

(*ii*) *The mutual conductance*, g_m. This is defined as the ratio:

$$g_m = \frac{\text{increase in } I_a}{\text{increase in } V_g} \quad \text{for constant } V_a$$

It is therefore the gradient of the mutual characteristics. In Fig. 14.19*a*

$$g_m = \frac{\text{AB}}{\text{AC}}$$

It can equally well be calculated from the anode

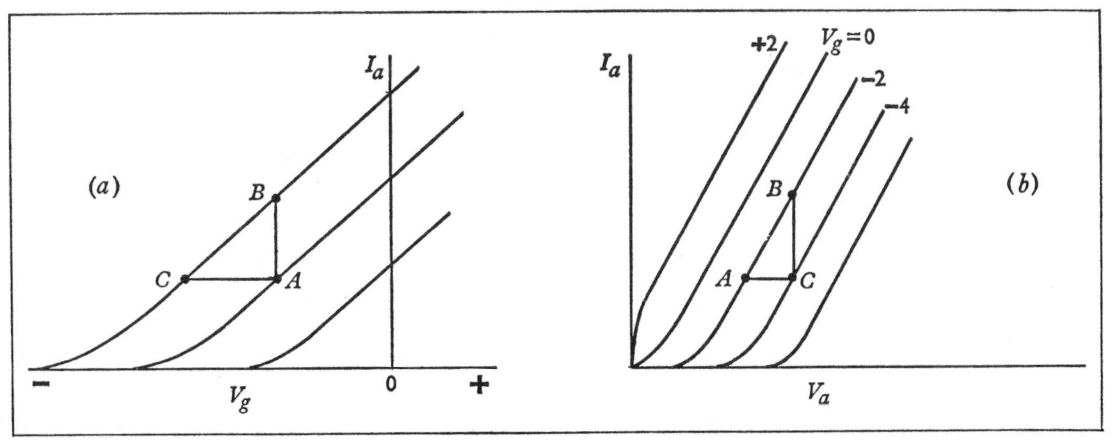

Fig. 14.19 Deducing the valve parameters: (*a*) from the mutual characteristics; (*b*) from the anode characteristics

characteristics. Thus, the anode potential is the same for the points B and C in (*b*). The change in grid potential between the two points is 2 volts; and the corresponding change in anode current may be read off from the appropriate axis.

The mutual conductance of a small triode is usually in the range from 1 to 6 mA/V. It is approximately proportional to the surface area of the cathode; the greater the surface area, the greater the total electron emission, and the greater is the effect on the anode current of a given change in grid potential.

(*iii*) *The amplification factor* μ. This is defined as the ratio:

$$\mu = \frac{\text{increase in } V_a}{\text{decrease in } V_g} \quad \text{for constant } I_a$$

It can be calculated from the anode characteristics by considering two points on the curves such as A and C, at which the anode current is the same. The increase in V_a may be read off from the appropriate axis and the corresponding decrease in V_g is 2 volts. And it can just as well be found by considering the points A and C on the mutual characteristics.

The amplification factor of a triode is usually in the range from 10 to 70. It is a purely geometrical property of the valve approximately equal to the ratio of the capacitances between its electrodes. Thus if the grid–cathode and anode–cathode capacitances are C_{gc} and C_{ac} respectively, then

$$\mu \simeq \frac{C_{gc}}{C_{ac}}$$

The amplification factor is therefore increased by close spacing between grid and cathode. It is also increased by winding the turns of the grid spiral close together.

The three valve parameters are clearly related to one another. Suppose an increase in anode current δI_a is brought about by an increase in V_a, keeping V_g constant. Then

$$\text{increase in } V_a = r_a\,\delta I_a$$

If now the anode current is brought back to its original value by a *decrease* in V_g keeping V_a constant, we have

$$\text{decrease in } V_g = \frac{\delta I_a}{g_m}$$

$$\therefore\ \mu = \frac{r_a\,\delta I_a}{\delta I_a/g_m}$$

$$\therefore\ \mu = g_m r_a$$

In practice the valve parameters vary to some extent with the anode and grid potentials—which is another way of saying that the valve characteristics are not really linear. The mutual conductance g_m always *increases* slightly with *decreasing* grid bias, corresponding with the slight upward curve of the mutual characteristics. In some valves this effect is deliberately accentuated so that the amplification can be altered by simply varying the grid bias. The amplification factor μ changes very little with grid bias, provided the valve is not taken near the cut-off point.

Voltage amplification

The three valve parameters are associated with changes of any *two* of the quantities I_a, V_a and V_g, while the remaining quantity is kept constant. But in general all three quantities change simultaneously. A change in grid potential δV_g by itself would cause a change in anode current of $g_m\,\delta V_g$. Likewise a change in anode potential δV_a would make the anode current change by $\delta V_a/r_a$. When both changes occur simultaneously, the increase in anode current δI_a is the sum of these two components.

$$\therefore\ \delta I_a = g_m\,\delta V_g + \frac{1}{r_a}\delta V_a \qquad (1)$$

Consider now a simple amplifier circuit such as that in Fig. 14.15, discussed previously. A rise in the anode current causes an increase in the potential drop across the anode load resistor R; and the anode potential then decreases by this amount. In this circuit δV_a is therefore given by

$$\delta V_a = -R\,\delta I_a \qquad (2)$$

Substituting for δI_a in equation (1), we have

$$-\delta V_a\left(\frac{1}{R} + \frac{1}{r_a}\right) = g_m\,\delta V_g$$

The voltage amplification A is therefore given by

$$A = \frac{\delta V_a}{\delta V_g} = -\frac{g_m}{(1/R) + (1/r_a)} = \frac{-g_m r_a R}{r_a + R}$$

But $$g_m r_a = \mu$$

$$\therefore\ A = \frac{-\mu R}{r_a + R}$$

The minus sign represents the phase reversal that occurs in a single amplifying stage.

The valve equivalent circuit

An alternative way of analysing the behaviour of a triode circuit is to treat the valve as a generator of alternating current.

Substituting for δV_a from equation (2) in equation (1) and multiplying through by r_a, we have

$$r_a\,\delta I_a = g_m r_a V_g - R\,\delta I_a$$

Re-arranging, and writing $\mu = g_m r_a$

$$\therefore\ \mu\,\delta V_g = \delta I_a(R + r_a)$$

The alternating current that flows in the anode circuit is therefore exactly the same as though the valve were a source of e.m.f. $\mu\,\delta V_g$ with internal resistance r_a. The equivalent circuit of the amplifier is shown in Fig. 14.20. The impedance of the H.T. supply is too small to be worth including. The equivalent circuit represents the behaviour of the triode for alternating currents only. In discussing its d.c. behaviour the characteristic curves must be consulted.

The voltage amplification A is very simply deduced using this equivalent circuit. The alternating current δI_a is given by

$$\delta I_a = \frac{\mu\,\delta V_g}{r_a + R}$$

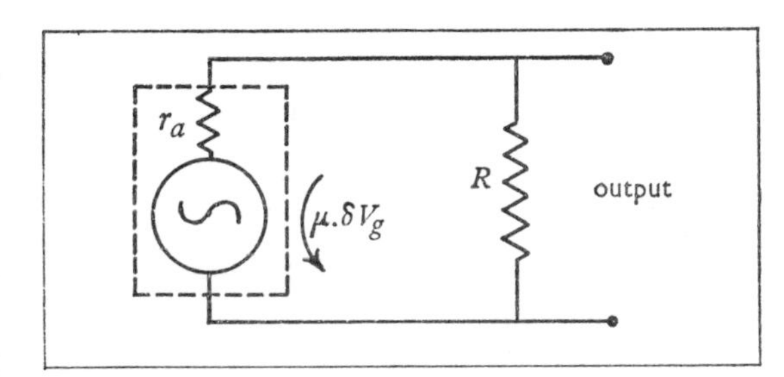

Fig. 14.20 The triode equivalent circuit

The output is the p.d. δV_a developed across the load R.

$$\therefore\ \delta V_a = \frac{\mu\,\delta V_g R}{r_a + R}$$

$$\therefore\ A = \frac{\mu R}{r_a + R}$$

Since the direction of the current through the valve (in the conventional sense) is *towards* the cathode, the upper end of the generator becomes negative when δV_g is positive. Thus the output and input are in antiphase.

The expression for A may be re-written

$$A = \frac{\mu}{1 + r_a/R}$$

The voltage amplification therefore increases with R, reaching a maximum value equal to μ when R is very large. In practical cases it is not economic to make A more than about two-thirds of this theoretical maximum.

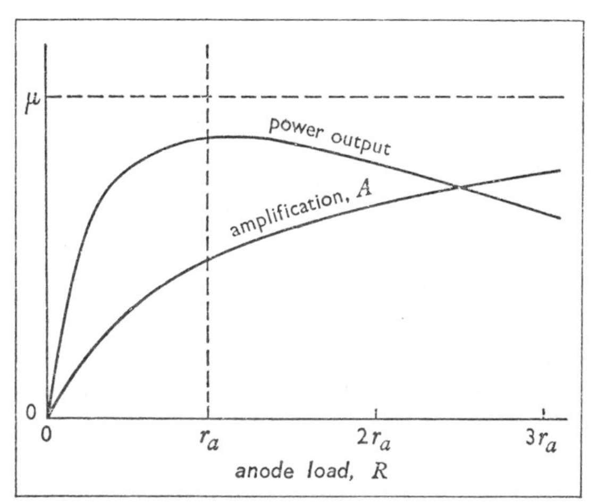

Fig. 14.21 The variation of amplification and power output with anode load for a triode amplifier

Fig. 14.21 shows how both the amplification and the power delivered vary with R. Maximum power is supplied to the load when its resistance is equal to the internal resistance of the source of e.m.f. (p. 24). A power amplifier should therefore be designed so that

$$R = r_a$$

However, it may be preferable in practice to use a different anode load to reduce distortion.

A loudspeaker or recording head usually has a very low impedance. To match this to a power amplifier a step-down transformer is used. Suppose the load joined to the secondary circuit is a resistance R_s, then the primary coil behaves in the anode circuit as a resistance R_p given by

$$R_p = \frac{R_s}{(\text{turns ratio})^2} \qquad \text{(p. 203)}$$

and the turns ratio (n_s/n_p) is chosen accordingly. For example, suppose the optimum load for a certain power amplifier is 2700 ohms; if it is being used to feed a 3-ohm loudspeaker, then

$$\frac{R_p}{R_s} = \left(\frac{n_p}{n_s}\right)^2 = 900$$

$$\therefore \frac{n_p}{n_s} = 30$$

i.e. a 30 : 1 step-down transformer is needed.

Load lines

The analysis of the behaviour of a triode circuit in terms of the three valve parameters is valid only in so far as the valve characteristics are truly linear. For small amplitudes of oscillation this is a fair approximation. But in the later stages of an amplifier where the signal is of relatively large amplitude this assumption is altogether too crude. The circuit must then be analysed by means of a *load line* drawn on the anode characteristics. Whatever the non-linearity of the characteristics, the anode load resistor may be assumed to obey Ohm's law; and the changes of anode current δI_a and anode potential δV_a are still correctly related by equation (2) above:

$$\delta V_a = -R\,\delta I_a$$

This is the equation of any straight line of gradient $-1/R$ drawn across the anode characteristics; the load line is the particular line of this gradient that passes through the initial operating point A of the valve (Fig. 14.22). The load line marks out all the possible combinations of V_a, I_a and V_g that could occur in that particular case. Thus in Fig. 14.22, if $\delta V_g = +2$ volts, the conditions in the valve shift to the point B on the characteristic curves. This means a reduction of V_a from 180 volts to 130 volts; i.e. $\delta V_a = -50$ volts. When $\delta V_a = -2$ volts, the valve conditions move to the point C, and $\delta V_a = +45$ volts. Clearly the circuit introduces appreciable distortion; a careful analysis in this manner allows the exact nature of the distortion to be computed.

In a numerical problem it is usually simplest to plot the load line by marking its intercepts on the axes. For instance the load line in Fig. 14.22 is for an amplifier running on a H.T. supply of 300 volts with an anode load resistance of 15,000 ohms. If the anode current were zero, there would be no p.d. across the anode load, and V_a would be equal to the full high-tension p.d. of 300 volts. Likewise, if V_a were zero, the p.d. across the load would be 300 volts; the current I_a would then be given by

$$I_a = \frac{300}{15{,}000}\ \text{amp} = 20\ \text{mA}$$

The intercepts on the axes are therefore the points (300 volts, 0) and (0, 20 mA); and the load line is joined connecting them.

14.7 Multigrid valves

One of the disadvantages of the triode valve is the appreciable capacitance that exists between its grid and anode. This amounts to about 10 pF,

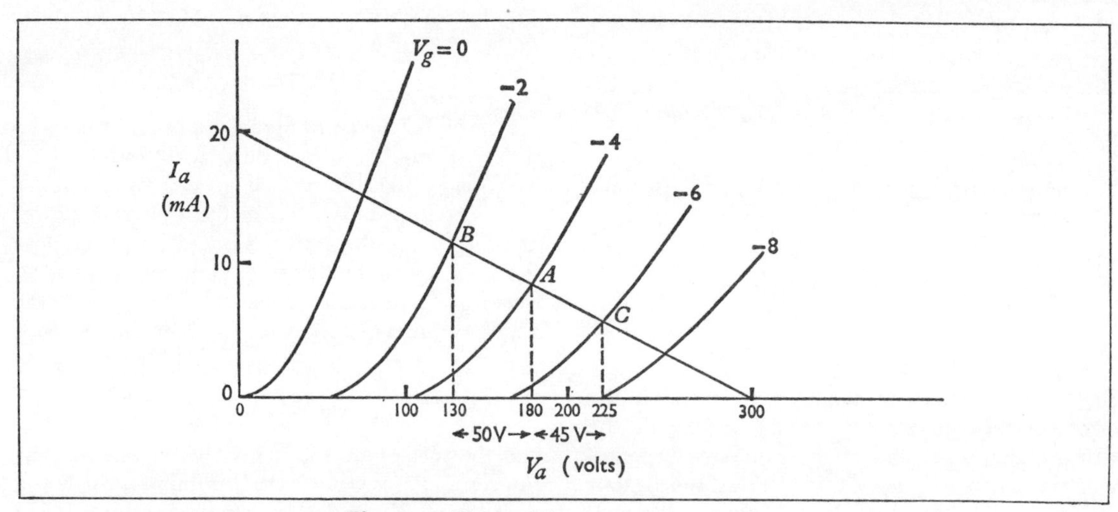

Fig. 14.22 The load line of a triode amplifier

which is sufficient at frequencies above 10 kHz* to feed back a significant fraction of the output p.d. into the grid circuit. Because of the phase reversal of the output, this generally means that the feed-back is negative, with the result that the amplification falls off seriously above this frequency. At radio frequencies the effect is even more serious. If the anode load is partly inductive, it is possible for the feed-back to be positive, and the circuit may then oscillate. To overcome these drawbacks the *tetrode* valve was introduced. This has two grids. The one nearest the cathode fulfils the same function as the grid of a triode, and is called the *control grid.* The other, called the *screen grid*, is maintained at a steady positive potential somewhat less than the anode potential. As its name implies, the screen grid acts as an electrostatic screen between control grid and anode, between which the capacitance may now be as little as 10^{-3} pF. The feed-back between anode and grid is thereby reduced to negligible proportions, and the valve may be used as an amplifier at high frequencies.

The screen grid also reduces the anode–cathode capacitance, with the result that small changes of anode potential now have very little effect on the anode current; in other words the anode a.c. resistance of a tetrode is very high, often as much as 1 megohm. But there is another less desirable effect that also influences the anode current. Whenever an electron strikes a surface it may cause the emission of one or more other electrons from it—the process known as *secondary emission* (p. 235). In the triode valve the secondary electrons emitted from the anode are immediately drawn back to it again; they are emitted at low velocity and are always collected by the most positive electrode in their vicinity. The anode current is not therefore affected by secondary emission. The same happens in the tetrode as long as the anode is more positive than the screen grid. But if the potential of the anode falls below that of the screen grid, many of the secondary electrons will be drawn to the latter. The anode current therefore falls, and the screen grid current rises by the same amount. In this region of the valve characteristics the curves are very irregular; linear amplification is only possible if the anode potential is kept substantially greater than the screen grid potential. This means the valve is limited to handling rather small signals in which the 'swing' of anode potential is not too large.

To extend the useful range of anode potentials over which the screen grid type of valve can be used the *pentode* valve was developed. A third grid, called the *suppressor grid*, is mounted between the screen grid and anode. This is maintained at a potential near to that of the cathode; in many valves the two electrodes are actually connected together internally. The electric field near the anode is then always in such a direction as to draw the secondary electrons back to the anode, and the kinks in the anode characteristics are completely eliminated (Fig. 14.23). The curves are almost straight and horizontal for all anode potentials above about 50 volts. Linear amplification is therefore possible for large signals as well as small.

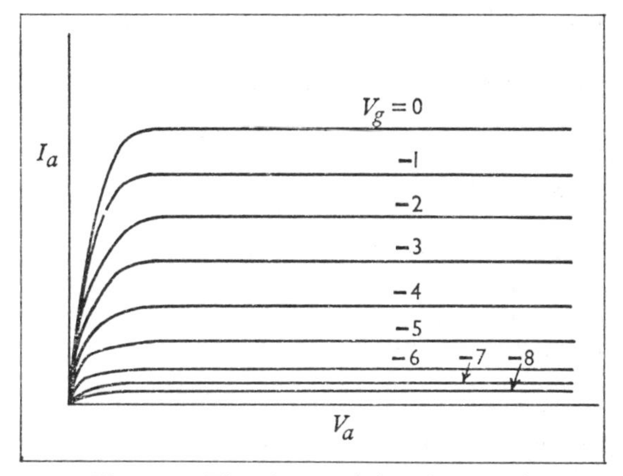

Fig. 14.23 The characteristics of a pentode

The amplification factor μ and anode a.c. resistance r_a are very large. Typical values are: $\mu = 4000$, $r_a = 1$ megohm. In fact the linear parts of the anode characteristics are so nearly horizontal that these two valve parameters are more or less unmeasurable and meaningless. Only the mutual conductance g_m is of significance. The expression previously obtained for the voltage amplification A of a triode applies also in this case (p. 248).

$$\therefore\ A = \frac{-g_m r_a R}{r_a + R} = \frac{-g_m R}{1 + R/r_a}$$

The load resistance R is usually much less than the anode a.c. resistance r_a, so that the expression can be simplified to

$$A \simeq -g_m R$$

* Or 10 kc/s.

Typical values might be:

$$g_m = 2 \text{ mA/volt}$$
$$R = 50{,}000 \text{ ohms}$$
giving $$A = 100$$

This is two or three times the voltage amplification that can reasonably be obtained with a triode.

The pentode can also be used as a power amplifier. In this respect it is more efficient than the triode, in the sense that a higher proportion of the power drawn from the H.T. supply is converted into a.c. power at the output. But it is harder to eliminate distortion, though with careful design this is not an insuperable difficulty.

Valves with even larger numbers of grids are also made for special purposes. The *hexode* for instance has four grids and is used for mixing two signals together. It has two control grids (the first and third). The second and fourth grids are screens to shield the control grids from each other and from the anode.

14.8 Sound reproduction

Sound consists of longitudinal wave motions transmitted through the air (or some other material medium). A microphone is essentially a device for producing an electrical signal whose time variation follows closely either the displacement or the pressure changes in the sound wave incident on it. The movements involved in a sound wave are extremely small. A microphone will commonly be expected to produce a satisfactory signal with air displacements of only 10^{-10} m—less than the diameter of one atom; the corresponding pressure changes amount to about 10^{-8} atmosphere—not much more than the *total* pressure inside a thermionic valve. The power available to vibrate the diaphragm of the microphone may not be more than 10^{-12} watt. It is not surprising therefore that the electrical signal produced is very small, requiring considerable amplification before it can be used to drive a loudspeaker or recording head. Furthermore the microphone should respond uniformly to all frequencies to which the ear is sensitive—from 20 Hz to 20 kHz.* But this is an ideal not readily attained in practice. We shall usually be content in high-quality work with a uniform frequency response from about 40 Hz to rather more than 10 kHz. Fortunately there is no great difficulty in designing amplifiers to work in the frequency range up to 20 kHz—usually known as the *audio frequency* range.

The carbon microphone (Fig. 14.24*a*)

This consists of a small box B loosely packed with carbon granules G. An electric current from a small battery is driven through the granules between the carbon block C at the back and the carbon button A attached to the light aluminium diaphragm D at the front. When the granules are compressed, the resistance of the unit decreases,

* 1 Hz = 1 c/s; 1 kHz = 1 kc/s = 10^3 c/s.

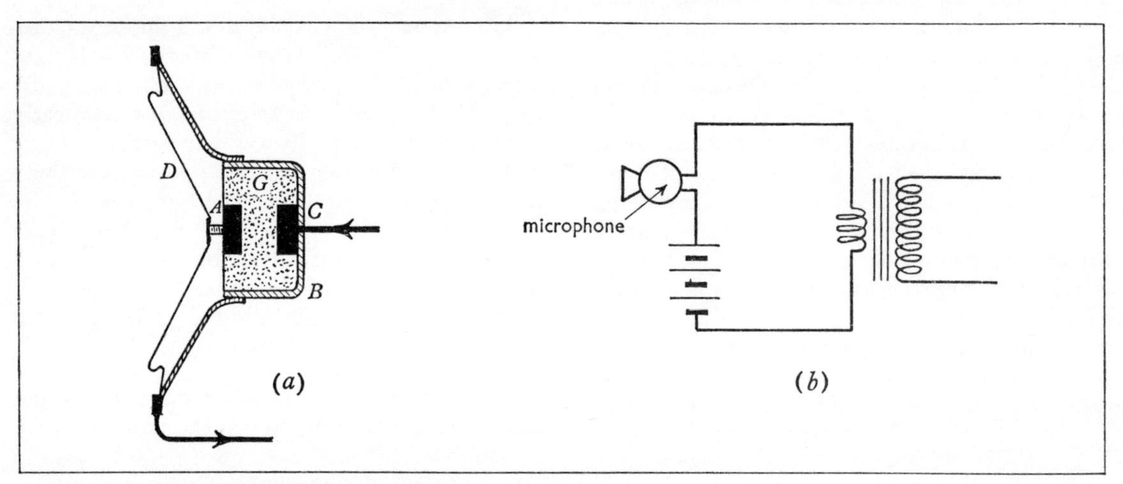

Fig. 14.24 (*a*) A carbon microphone; (*b*) the circuit of the microphone with battery and step-up transformer

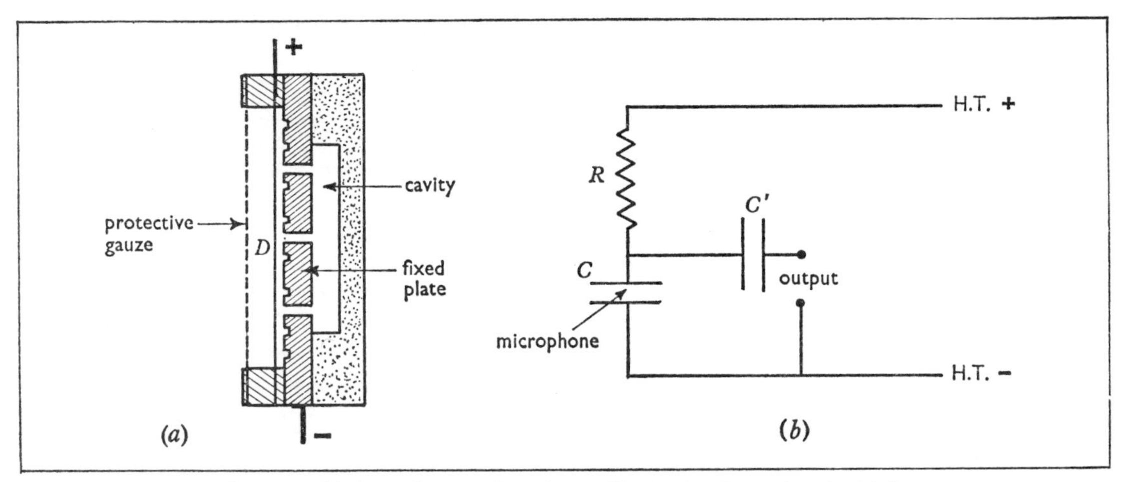

Fig. 14.25 (*a*) A condenser microphone; (*b*) the circuit employed with it

and the current in the circuit rises. Thus the oscillations of the diaphragm produced by the pressure changes in the incident sound wave cause corresponding oscillations of the current. The microphone and battery are joined in series with a high-ratio step-up transformer, so that the maximum alternating p.d. is produced at the output terminals (Fig. 14.24*b*). Although the signal generated in this way is controlled by the sound wave, its power is derived from the battery. The carbon microphone is thus an amplifying device, and is the only kind of microphone that can be used without additional electronic amplification. It is therefore very suitable for telephones. However, it is not satisfactory for high-quality sound reproduction because of the background 'hiss' produced by the loose contacts between the carbon granules.

The crystal microphone

In certain types of crystal, such as *quartz* and *Rochelle salt*, mechanical strains can produce electric polarization. Equal and opposite bound charges then appear on opposing faces of the crystal, and a p.d. is produced between two metal electrodes in contact with them. This is called the *piezo-electric effect.* A section from a single crystal can thus be used as a microphone if it is arranged so that the incident sound wave causes the appropriate type of strain in it.

The condenser microphone (Fig. 14.25*a*)

This consists essentially of an air capacitor, one plate of which is the flexible diaphragm D on which the sound waves act. Movements of the diaphragm produce corresponding changes of the capacitance. The microphone is connected in the circuit of Fig. 14.25*b*. Provided the resistance R is sufficiently large, the charge Q on the capacitor does not alter appreciably while the diaphragm oscillates. The p.d. V across it is given by

$$V = \frac{Q}{C}$$

Also the capacitance C is inversely proportional to the thickness t of the air film between the plates.

$$\therefore\ V \propto t$$

The blocking capacitor C' ensures that only the a.c. component of V passes to the grid of the first amplifying valve. The signal delivered to the amplifier is thus proportional to the displacement of the diaphragm.

The moving-coil or dynamic microphone

In this microphone a small coil is attached directly to the diaphragm. It is arranged to move in the strong radial field in an annular gap between the pole pieces of a permanent magnet (Fig. 14.26). When the diaphragm oscillates, the coil cuts the magnetic flux and an alternating e.m.f. is generated in it. The construction is very similar to that of a moving-coil loudspeaker (Fig. 14.28). In fact a moving-coil loudspeaker may be used as a micro-

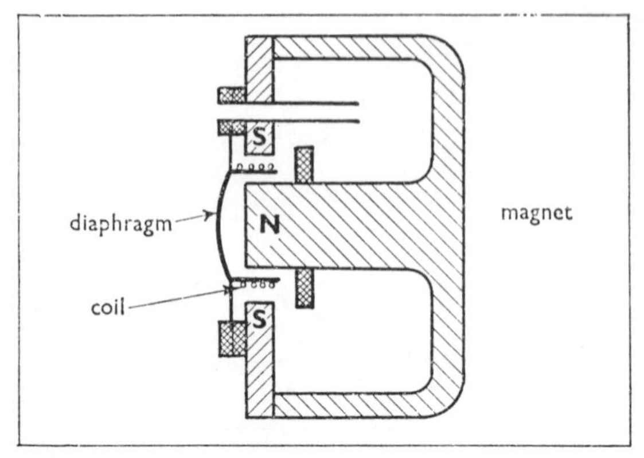

Fig. 14.26 A moving-coil (or dynamic) microphone

phone, though its bulk is a bit inconvenient for this purpose!

This microphone has a low impedance; and a relatively large current is generated in it at a small p.d. Since a valve amplifier is operated by variations of potential, the signal is applied to the amplifier input through a high-ratio step-up transformer.

The ribbon microphone (Fig. 14.27)

This consists of a corrugated duralumin ribbon suspended edgewise between the poles of a permanent magnet. When the ribbon vibrates, it cuts the magnetic flux, and an alternating e.m.f. is generated in it. Again the signal is applied to the input of a valve amplifier through a high-ratio step-up transformer.

Modern technology has reduced the design of microphones and their associated amplifiers to something near perfection. The electrical signal obtained at the output can (at a price) be an almost perfect replica of the variations in the original sound wave. The art of recording sounds on disc or magnetic tape has also been developed to the point at which it often leaves little to be desired. The weak link in the chain of sound reproduction is the loudspeaker. Even the best loudspeakers are not free from resonances of one kind or another, and no one has succeeded in designing a single unit that responds uniformly to the complete range of audio frequencies. High-fidelity systems make use of two or more speakers covering different frequency ranges—'woofers' for low frequencies and 'tweeters' for high. Even if a perfect loudspeaker system could be evolved, the quality of the reproduction could be spoilt by using it in a room with inadequate acoustics. The 'perfect' loudspeaker would need a room with perfect acoustics to go with it. Indeed, the ideal of reproducing a sound exactly as it was heard in the concert hall is plainly unattainable—unless, that is, we are prepared to design our living-rooms with the acoustics of an auditorium!

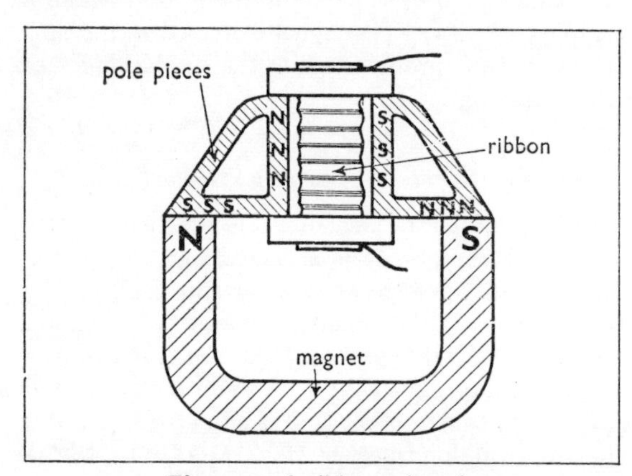

Fig. 14.27 A ribbon microphone

The moving-coil loudspeaker (Fig. 14.28)

The moving part consists of a conical diaphragm D made of stiff paper. This is supported round its circumference by a flexible ring of cloth and at the centre by a weak spring strip. Also attached to the centre is a single-layer coil located in the radial magnetic field in the annular gap between the pole pieces of a permanent magnet. The signal current flows in the coil, and the electromagnetic forces move it to and fro in the gap, carrying the diaphragm with it like a piston, and generating sound waves. *Two* sound waves are in fact produced—at the front and back of the cone; and these are bound to be in antiphase. If the acoustic path from back to front is much less than half a wavelength, there is almost complete destructive interference of the waves in all directions. The baffle-board or cabinet is designed to reduce this effect, and must be regarded as an essential part of the loudspeaker system.

At high frequencies, when the wavelength is comparable with the diameter of the speaker, there is a tendency for the sound to be emitted chiefly along the axis. The frequency balance therefore varies with the position of the observer. This is

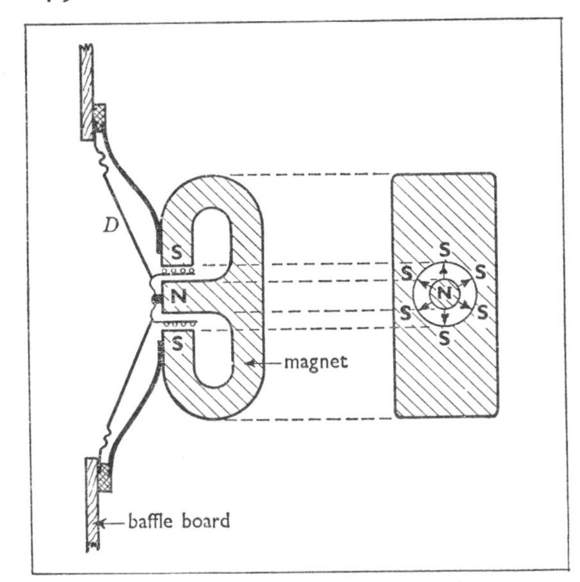

Fig. 14.28 A moving-coil loudspeaker

partly mitigated by the behaviour of the cone; at high frequencies instead of vibrating as a single unit, complicated standing wave patterns are set up on it by the oscillations of the coil in the centre. This reduces the effective radiating diameter, but creates new resonance problems on its own account. All in all it is fortunate that the average human ear is so uncritical of the sounds that fall upon it!

14.9 Waves on wires

When an e.m.f. is first generated in an electric circuit, it takes time for the electrical disturbance to be propagated round it. The charged particles in the conductors of the circuit move at quite slow speeds even when very large currents are flowing (p. 9). But the electric and magnetic fields they produce are propagated round the circuit at very great speeds (approaching the speed of light), and this ensures that the charged particles are set in motion almost simultaneously throughout the circuit. With direct currents and with low frequency alternating currents the transmission of the disturbance round a circuit is so rapid that we are not usually aware of the finite time taken. But at very high frequencies this time may become comparable with the period of oscillation of the current. We then have a situation in which the e.m.f. reverses before the original disturbance has travelled very far along the wire. In fact the high frequency e.m.f. is then sending *waves* of current and potential round the circuit; these are accompanied by waves of magnetic and electric fields in the space surrounding the conductors. The transmission of energy along a cable by an alternating current is thus accompanied by the passage of *electromagnetic waves,* as they are called, through the insulating medium surrounding the cable. In this section we shall study the nature of such electromagnetic waves in the neighbourhood of high-frequency transmission lines. After that we shall proceed to consider the still more important process of the propagation of electromagnetic waves in empty space far removed from any conductors.

There are many different kinds of wave system known to science, involving a great variety of physical processes. But they all tend to share certain common features, so that a clear understanding of one type of wave system is helpful in elucidating the functioning of another. Electromagnetic waves can be observed by their effects, as we shall see, but the detailed 'structure' of the waves has to be derived by theoretical reasoning and must be left largely to the imagination. The intangible nature of these waves means that they are by no means the most suitable type with which to *begin* a study of wave systems. It is therefore assumed below that the student already has a working knowledge of other kinds of waves. For instance, the properties of transverse waves on stretched cords and sound waves in tubes are in many ways closely analogous to those of electromagnetic waves on a transmission line; and we shall make use of these analogies in what follows.

Fig. 14.29 shows the nature of the wave 'pattern' for an electromagnetic wave travelling along a transmission line consisting of a pair of parallel wires. In this case the wave pattern moves steadily forward along the line, and it is therefore called a *progressive wave.*

At a given instant there are certain points (such as A and A′) at which the current and p.d. have their maximum values in one direction (call this the *positive* direction). Intermediate between these are other points (such as C) at which, at the same instant, the current and p.d. have maximum negative values. At B and D the current and p.d. are momentarily zero.

An associated pattern of electric and magnetic fields exists in the medium surrounding the

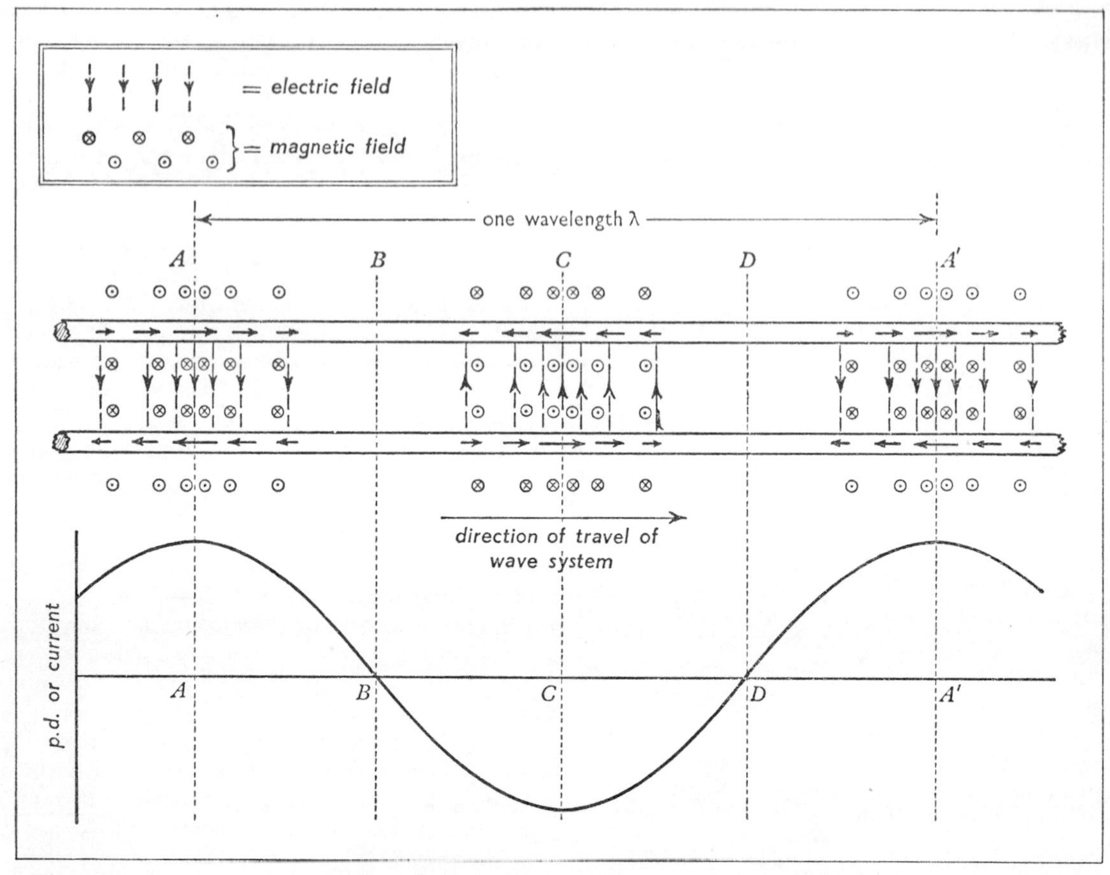

Fig. 14.29 A parallel-wire transmission line carrying a high-frequency electromagnetic wave, showing the distribution of currents and potentials at a given instant

conductors. The lines of force of the electric field run across from one wire to the other, the electric field strength in any section of cable being proportional to the p.d. between the wires. The lines of force of the magnetic field form closed loops round the wires, and are therefore at right-angles to the plane of the diagram in Fig. 14.29; the magnetic field strength in any section of cable is proportional to the current in the wires. Fig. 14.30 shows the nature of the field pattern in a plane at right-angles to the wires (at a position such as A or A′ in Fig. 14.29). Over a distance of one wavelength (e.g. from A to A′ in the figure) the p.d. and current vary over the complete range between their maximum positive and negative values; corresponding variations of the electric and magnetic fields occur over the same distance.

It may seem strange to find a situation in which the current can flow simultaneously in opposite directions in the same conductor, such as at A and C in Fig. 14.29. However, a little consideration will show that the pattern of currents in the figure is causing a flow of charge towards B in one conductor and away from it in the other, so charging up the capacitor formed by the part of the cable near B. A moment later this flow of charge will have caused the electric field to reach a maximum at B; likewise in the same time the flow of charge will have produced maximum electric field in the negative sense at D—and the electric fields at A and C will have died away to zero. Thus the system of currents shown causes a forward movement of the electric field pattern. Simultaneously the variation of potential along the wires is causing the pattern of *currents* to advance along the transmission line—and with it the

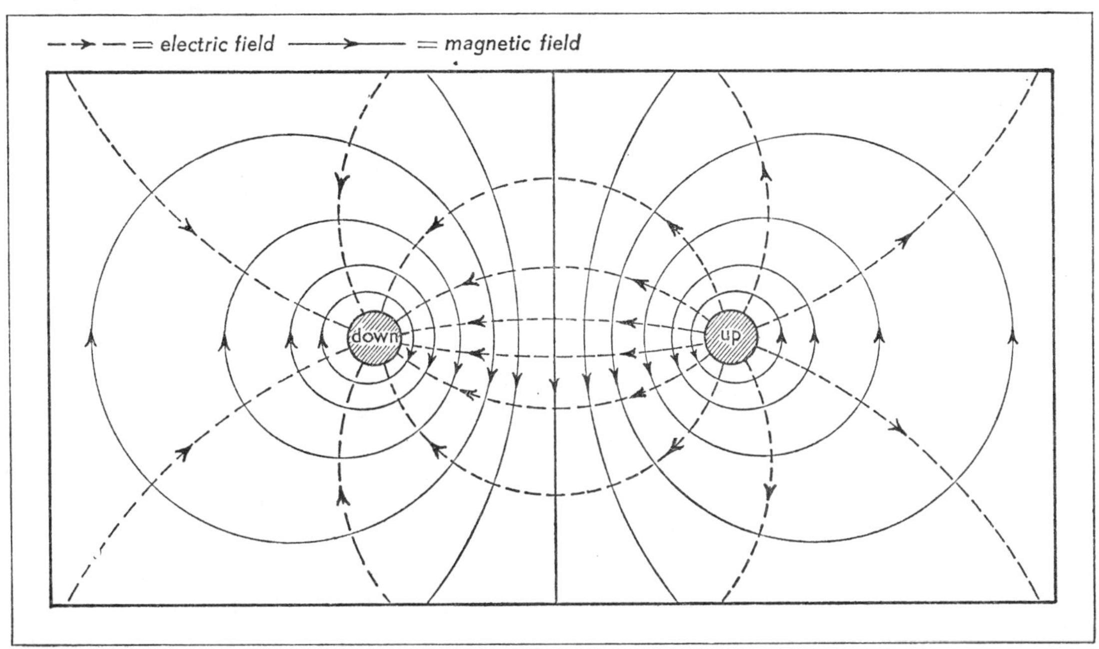

Fig. 14.30 The pattern of the magnetic and electric fields at a given instant in a plane at right-angles to a parallel-wire transmission line; the current is approaching the reader in the right-hand conductor, and receding in the left-hand one

associated pattern of magnetic field. For instance, at the moment depicted in Fig. 14.29, A is at a higher potential than C in the top conductor, and this is causing the current to grow in the section of cable between these points; the rate of growth of current near B depends on the inductance of this part of the transmission line. Likewise, C is at a lower potential than A′ (in the top conductor), and this is causing the current to grow in the negative sense in the part of the cable near D. In this way the current wave also is made to move forward along the transmission line. As the positive peak of current at A moves towards B, the current between B and C first falls off to zero and then increases in the positive sense; when the current at B reaches a maximum, that at C is momentarily zero. At the same time the negative peak of current at C moves forward to D, and the current at A′ falls momentarily to zero.

The energy of the wave is carried partly in the electric field produced by the p.d. between the conductors, and partly in the magnetic field associated with the current—in actual fact the energy is divided *equally* between these two forms. Thus, in Fig. 14.29, at A, C and A′ the p.d. is a maximum in one direction or the other and energy is stored temporarily in the capacitance of the cable near these points—i.e. in the electric field. This is a form of potential energy, since it depends on the potential of the charges in the conductors. At the same set of points the current also has its maximum positive or negative values and energy is stored temporarily in the inductance of the cable—i.e. in the magnetic field. This is a form of kinetic energy, since it depends on the motion of the charged particles in the conductors. The individual charged particles oscillate to and fro as the current alternates, but the concentrations of electric and magnetic field move steadily forward together along the cable and carry the pulses of energy with them.

A similar sharing of the energy between potential and kinetic forms occurs in other types of wave system. Consider for instance the propagation of a sound wave. The wave can be described either in terms of the pressure variations in the air or in terms of the velocities of the air particles. At the positions of maximum or minimum pressure energy is carried in potential form; at the same positions the velocities also have their maximum values in one direction or the other, and energy is carried in kinetic form. The individual particles of air vibrate to and fro along the line of propagation, but the regions of maximum pressure and velocity are transferred steadily forward, carrying the energy of the wave with them.

A detailed examination of the wave patterns in sound shows that the direction of propagation is that in which the air particles are moving at the points of *maximum* pressure. In a similar way it is clear from Fig. 14.29 that the electromagnetic wave is propagated in the direction in which the current flows at the positions of *maximum positive* potential.

The progressive electromagnetic wave we have described consists of two interlocking wave systems, the current wave and the potential wave; these remain always in phase at any given point, and each is associated with its own kind of field. Thus the patterns of the electric and magnetic fields also travel forward together and are in phase with one another at all points in any given plane at right-angles to the wires. The lines of force of the two kinds of field are in fact at right-angles to each other everywhere (Fig. 14.30) and are also at right-angles to the direction of travel of the waves along the cable. The associated electromagnetic wave is therefore of *transverse* type. (We are assuming in this section that the separation of the conductors is small compared with the wavelength; the situation is very much more complicated when this is not the case.)

The wavelength λ of the wave on the transmission line depends on its frequency f and velocity c. The relation between these quantities may be derived as follows. In the time taken for one complete oscillation of the current or p.d. the wave pattern advances a distance λ. In one second f oscillations take place and f complete waves pass any given point. The wave pattern therefore advances a distance $f\lambda$ per second.

$$\therefore c = f\lambda$$

One way of measuring the velocity c is to find the wavelength λ on the transmission line produced by an alternating e.m.f. of known frequency f; is then given by the above expression.

At the ends of a transmission line in general some reflection of the wave occurs. The simple, progressive wave described above is obtained only if special steps are taken to absorb the energy arriving at the end of the line so that none is reflected. If this is not done, the oncoming and reflected waves interfere with one another producing a *standing wave* system on the line. In such a wave system at certain points the two waves are permanently in antiphase and the resultant amplitude is a minimum; these points are called *nodes*. At other points the two waves are permanently in phase, and the amplitude is a maximum; these points are called *antinodes*. If the oncoming and reflected waves are of equal amplitude, they cancel completely at the nodes, and there is also no net transfer of energy in either direction along the line.

Since reflection is liable to occur at both ends of the line, there are usually only certain sharply defined lengths of transmission line for which large amplitudes of oscillation can be built up for any given frequency. The line is then *resonant* to the frequency in question.

Fig. 14.31 shows an arrangement that may be used for studying a standing wave system of this kind. When the 'coil' of one turn at one end of the line is placed near the coil of an oscillator (working at 100 MHz* or more), an e.m.f. is induced in it by mutual induction. A large current will flow in the coil only when the line is resonant to the frequency in use. Tuning may be effected by moving a short-circuiting strip along the line near its far end until the lamp indicates that a large current is flowing in the coil.

The current and potential waves behave in different ways when they are 'reflected' at the ends of the cable. The short-circuiting strip at A is of very low resistance. Although a large current may flow through it, the p.d. across it is always very small. The reflections therefore occur in such a way that the end A becomes an antinode of current, but a node of potential. Thus, although in a progressive wave system the waves of current and of potential are in phase at any given point, in a standing wave they are not, and their nodes and antinodes occur in different positions; the nodes of current coincide with the antinodes of potential and vice versa. In Fig. 14.31 the graphs drawn below the diagram show how the *amplitudes* of the oscillations in the standing wave system vary along the transmission line, the pattern repeating itself at intervals of one wavelength. At A′, one wave-

* I.e. 100 Mc/s, p. 27.

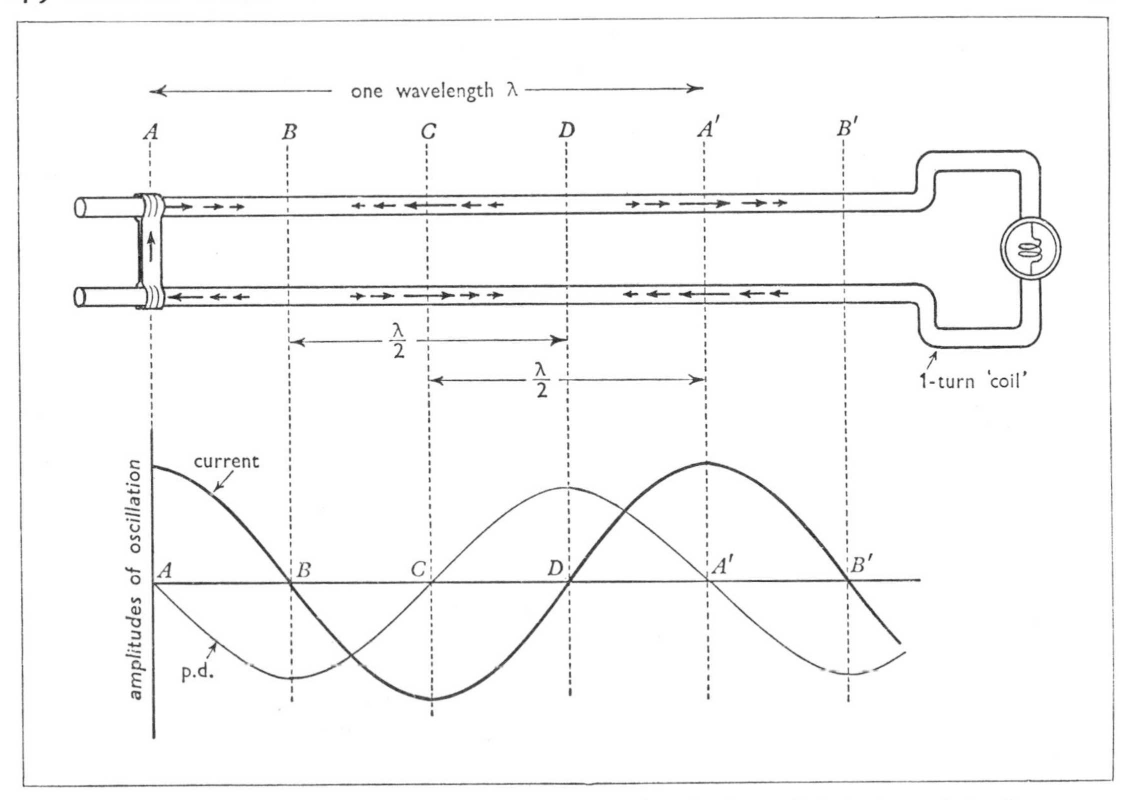

Fig. 14.31 The standing wave pattern on a resonant length of a parallel-wire transmission line

length from A, there is again an antinode of current, the current at A′ being in phase with that at A. Another antinode of current occurs at the intermediate point C; the negative amplitude indicates that the current here is in antiphase with that at A and A′. At A, C and A′ there are nodes of potential. At B, D and B′ there are nodes of current, but antinodes of potential, the oscillations of potential at B and B′ being in antiphase with that at D. Thus, as in all standing wave systems, the distance between adjacent nodes (or between adjacent antinodes) is *half a wavelength*; and the oscillations at adjacent antinodes are in antiphase with one another.

The standing waves of potential may be detected by using a miniature neon bulb. This passes a very small current, but only lights if the p.d. across it is more than a certain minimum, usually about 80 volts. As the neon bulb with its connecting clips is moved along the transmission line it reaches its maximum brightness at the positions of the potential antinodes, and does not light at all in the vicinity of the nodes. The wavelength is given by twice the separation of adjacent antinodes. If sufficiently high power is available from the oscillator, it is possible to observe the variations of the electric field strength between the wires using a large neon bulb without any leads attached to it. As the bulb is moved near an antinode of potential the electric field strength can be sufficient to ionize the low pressure gas inside the bulb, causing it to glow.

The standing wave of current may be detected with the aid of a small torch bulb attached to a pair of clips. The bulb is of low resistance and takes a relatively large current at a low p.d. It therefore acts in much the same way as the short-circuiting strip at A, which it may be used to replace. The bulb will be found to glow brightly at A, C and A′, where antinodes of current are located.

Energy is stored in a resonant transmission line in much the same way as in a resonant circuit consisting of an inductor and capacitor (p. 211). Thus the oscillations of current and p.d. in the line are *in quadrature* (p. 198). At one moment the currents are a maximum in (say) the directions shown by the arrows in Fig. 14.31; at this moment the p.d. between the conductors is everywhere zero. This flow of current causes charge to accumulate at B and D, and one quarter of an oscillation later the p.d.'s have reached their maximum values, and the current has fallen momentarily to zero. The energy in the line thus passes alternately from the magnetic field associated with the current to the electric field associated with the p.d., and back again.

The different behaviour of the current and potential waves at the short-circuited end comes about because the potential wave experiences a *phase reversal* at the reflection, whereas the current wave does not. Indeed a little thought will show that the phase of *one* of the two components of the wave system must be reversed if the wave pattern is to be propagated in the opposite direction. Thus at the point A the oncoming and reflected waves of potential are in antiphase and cancel; but the two waves of current remain in phase and produce maximum amplitude of oscillation.

A similar situation arises in many other kinds of wave systems. Consider, for instance, the waves that may be sent down a rope by waggling the end of it from side to side. If the rope is tied onto another cord of different mass per unit length, partial reflection of the wave occurs at the knot. The reflected wave suffers a phase reversal if it approaches the knot from the side of the less massive rope, but no phase reversal if approaching from the other side. Similarly it may be shown that a light wave travelling in air suffers a phase reversal when it is partially reflected at an air-glass boundary; but there is no phase reversal when it approaches the same boundary travelling in the glass.

Sound waves in a tube behave in a manner closely analogous to that of electromagnetic waves on a transmission line. In a progressive sound wave the velocity and pressure waves are in phase with one another, but in a standing wave system they are in quadrature—just as are the current and potential waves on a transmission line. Also, at a reflection, in both systems one component wave suffers a phase reversal where the other does not. Thus a sound wave reaching the open end of a tube is partially reflected back along it; in this case the pressure wave suffers a phase reversal, but the velocity wave does not. This means that the open end becomes a node of pressure but an antinode of velocity. The electromagnetic wave behaves in the same way at the short-circuited end of the transmission line—the potential is analogous to the pressure in the sound wave and the current to the velocity. The behaviour of the two components of the sound wave is interchanged if the tube is terminated in a rigid boundary. The velocity wave then experiences a phase reversal but not the pressure wave; this forms a node of velocity at the boundary (where the air particles *cannot* move longitudinally) and an antinode of pressure. A similar form of reflection occurs with the electromagnetic waves if the end of the transmission line is left open-circuited. No current can then flow at this point, which therefore becomes a node of current and an antinode of potential. At an open-circuited termination the current must therefore experience a phase reversal, but not the potential. This also may be demonstrated with the apparatus of Fig. 14.31 with the shorting strip removed. To tune the transmission line we must now find a way of varying its length, or else the frequency of the oscillator must be adjusted, until resonance is achieved.

It is interesting to note the paradoxical fashion in which a quarter-wave length of line behaves. Consider, for instance, the section between B and A in Fig. 14.31. This is short-circuited at A, but behaves at B as though it were an open-circuit. Although there is maximum current flowing through the shorting strip at A, no current flows at B, in spite of the large alternating p.d. that exists at this point. Conversely an open-circuited quarter-wave length of line behaves like a short-circuit. At the open end the p.d. is a maximum and no current flows. But a quarter of a wavelength back along the line the current reaches its maximum value, although there is a potential node at this point.

It may be shown (see below) that, provided the resistance losses of a transmission line are small, the velocity c of the electrical waves on it is given by

$$c = \frac{1}{\sqrt{\varepsilon_0 \mu_0}}$$

where ε_0 and μ_0 are the electric and magnetic space constants (p. 162 and p. 69). Substituting

$$\varepsilon_0 = 8{\cdot}84 \times 10^{-12}\ \mathrm{F\,m^{-1}}$$

and $\mu_0 = 4\pi \times 10^{-7}\ \mathrm{H\,m^{-1}}$

we obtain $c = 3{\cdot}00 \times 10^{8}\ \mathrm{m\,s^{-1}}$

This is the same as the velocity of light, and provides a strong indication that light is an electromagnetic wave process. We shall show in the next section how it is possible for an electromagnetic wave to be propagated without the aid of the currents that accompany it in the case of a transmission line. The velocity of the waves turns out to be the same in both cases.

Once the theory of electromagnetic waves is satisfactorily confirmed by experiment, the measurement of their velocity (e.g. the velocity of light) provides the most accurate means of finding the electric space constant ε_0. In the m.k.s. system of units the magnetic space constant μ_0 is chosen arbitrarily in the course of defining the ampere (p. 70); ε_0 is then given by the theoretical relationship

$$\varepsilon_0 = \frac{1}{\mu_0 c^2}$$

★ *The velocity of waves on transmission lines*

A transmission line behaves as it does because of the inductance and capacitance that exists distributed along its length. It may thus be represented by the chain of inductors and capacitors shown in Fig. 14.32 (ignoring the resistance of the wires).

Over the length δx of the line from P to Q the total distributed inductance is δL, half of it in each wire. Let the capacitance distributed over the same length be δC. Quoting results derived in earlier chapters, the inductance and capacitance per unit length are given by

$$\frac{\delta L}{\delta x} = \frac{\mu_0}{\pi}.\log_e (d/a) \qquad \text{(p. 114)} \qquad (1)$$

and

$$\frac{\delta C}{\delta x} = \frac{\pi\varepsilon_0}{\log_e (d/a)} \qquad \text{(p. 194)} \qquad (2)$$

where d and a are the separation and radius of the conductors, respectively.

In the section of transmission line considered let the p.d. between the conductors be v_1 at P, and v_2 at Q, as shown. Taking into account the fall in potential along the distributed inductance between P and Q, we have

$$v_1 - v_2 = \delta L \frac{di}{dt}$$

where i = the average current at any instan tin the section between P and Q. The time taken for the wave to travel at speed c along the transmission line from P to Q is $\delta x/c$. In this time interval the p.d. at Q increases from v_2 to v_1, the value that previously existed at P. The rate of increase of the p.d. at Q (dv/dt) is therefore given by

$$\frac{dv}{dt} = \frac{v_1 - v_2}{\delta x/c} = \frac{\delta L(di/dt)}{\delta x/c} = c\left(\frac{\delta L}{\delta x}\right)\frac{di}{dt} \qquad (3)$$

Likewise the current flowing into the distributed capacitance δC between P and Q is $\delta C.(dv/dt)$. If the currents entering and leaving the section are i_1 at P, and i_2 at Q, as shown, we have

$$i_1 - i_2 = \delta C \frac{dv}{dt}$$

The current at Q must increase by this amount in the time $\delta x/c$ taken by the wave to travel from P to Q along the line. The rate of increase of current at Q (di/dt) is therefore given by

$$\frac{di}{dt} = \frac{i_1 - i_2}{\delta x/c} = \frac{\delta C(dv/dt)}{\delta x/c} = c\left(\frac{\delta C}{\delta x}\right)\frac{dv}{dt} \qquad (4)$$

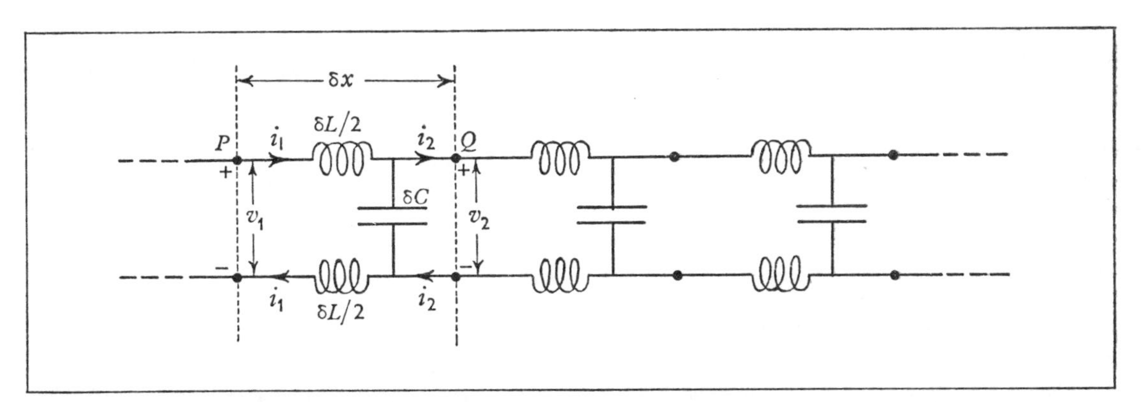

Fig. 14.32 Deducing the velocity of the waves on a transmission line

We may eliminate dv/dt and di/dt between equations 3 and 4 (by multiplying them together) giving

$$c^2 \frac{\delta L}{\delta x} \frac{\delta C}{\delta x} = 1$$

Substituting from equations 1 and 2 we find the important result

$$c^2 \varepsilon_0 \mu_0 = 1$$

$$\therefore c = \frac{1}{\sqrt{\varepsilon_0 \mu_0}}$$

A similar relation is found to exist between the inductance and capacitance per unit length of *any* air-insulated type of transmission line—for instance a coaxial cable with air insulation. A fuller analysis shows that the resistance losses in the line always have the effect of slightly reducing the speed of the wave. But the amount of this reduction is rarely significant; and we may normally assume that the velocity with which an electrical disturbance is propagated along a cable is equal to the velocity of light, $3{\cdot}00 \times 10^8$ m s^{-1}.

14.10 Electromagnetic waves without wires

In the previous section we have discussed how an electromagnetic wave accompanies the propagation of energy by an alternating current in a cable. In that case the electric field was produced by the alternating p.d. and the magnetic field by the alternating current. The possibility of electromagnetic waves being propagated through empty space far distant from any conductors was first mooted by Maxwell in 1865. He showed that if this is to happen there must exist mechanisms whereby

(*i*) an electric field can be produced by an alternating magnetic field; and

(*ii*) a magnetic field can be produced by an alternating electric field.

The first of these is nothing more than the familiar process of electromagnetic induction. When the magnetic field in a coil changes, an e.m.f. is induced in it. This e.m.f. is evidence that an electric field has been produced acting round the circumference of the coil and moving the free charges in it (p. 10); but the electric field exists even in the absence of any conductor for it to act in. Thus, a changing magnetic field produces an electric field at right-angles to it, with lines of force running in closed loops. The other mechanism is less familiar but occurs in much the same way—namely the production of a magnetic field by a changing electric field. When a capacitor is being charged (Fig. 14.33), the dielectric between its plates polarizes, i.e. the bound charges within it are displaced slightly from their normal positions (p. 161). While they are in motion these charges produce a magnetic field just like any other moving charge; but the displacement current only exists as long as the electric field in the dielectric is changing. Some magnetic field is of course produced by the pulse of current flowing in the connecting leads to the capacitor. But there is no doubt that the changing electric field in the dielectric also contributes to it; again the lines of force run in closed loops and are at right-angles to the electric field.

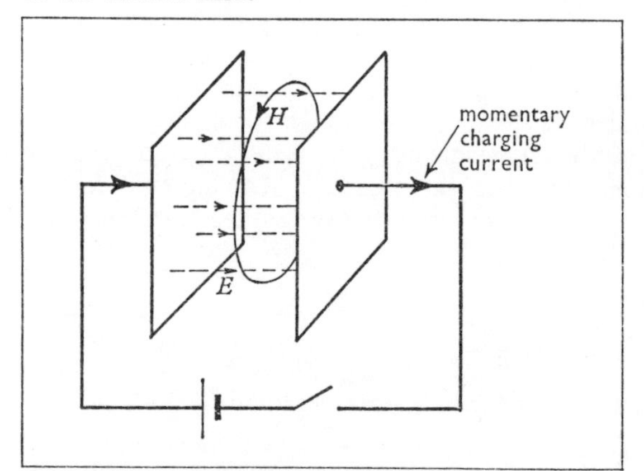

Fig. 14.33 A magnetic field is produced by a changing electric field between the plates of a capacitor

The two effects are exactly complementary. But there seems at first sight to be an important difference. Electromagnetic induction takes place with or without the presence of matter. A ferromagnetic core for the coil increases the effect, but it still happens in a vacuum. But in the second case the effect, as discussed so far, depends on having a material dielectric in the capacitor. In a vacuum there would be no bound charges and no reason to expect the 'dielectric' to contribute to the magnetic field.

Maxwell made the bold suggestion that there can be a displacement 'current' in a vacuum between two capacitor plates, producing a magnetic field just like any other form of current, though without any actual movement of charge. By this mechanism a changing electric field would produce a magnetic field even in a vacuum.

He could find no direct experimental test of his hypothesis, but it introduced a beautiful symmetry into the theory of electric and magnetic fields; and, if it were true, electromagnetic waves in a vacuum were a possibility. The theory showed that the waves should be of transverse type, and should therefore show polarization effects. It also predicted that the velocity c of the waves should be given by

$$c = \frac{1}{\sqrt{\varepsilon_0 \mu_0}}$$
$$= 3{\cdot}00 \times 10^8 \text{ m sec}^{-1}$$

This figure was remarkably close to the known velocity of light, which had already been shown to consist of transverse waves of some kind. Maxwell therefore suggested that light was an example of an electromagnetic wave motion.

It remained to show that electromagnetic waves could be produced by electrical means. The theory predicted that the waves should be emitted by any *accelerating electric charge*, since this would generate the necessary pattern of *changing* electric and magnetic fields. Some of the lines of force of the fields would then be formed into closed loops that would travel out into space carrying away some of the kinetic energy of the charge. To generate waves of given frequency the charge must perform oscillations at that frequency. In practical terms this means that a high-frequency alternating current must be made to flow in a suitably placed conductor—an *aerial* or *antenna*, as we now call it.

Possibly a high-frequency transmission line like that considered in the previous section might radiate away some of the energy it carries. But the currents in the two wires of a transmission line are necessarily in opposite directions at every instant, and the waves they radiate would be in antiphase; they would therefore almost entirely destroy one another by interference, and no radiation would be detected. Only if the conductors could be placed far apart compared with the wavelength was there any prospect of their producing electromagnetic waves. Thus, the length of the aerial and its distance from other conductors (such as the ground) needed to be comparable with the wavelength of the waves.

To detect the radiation it had to be shown that an alternating e.m.f. was produced in a 'receiving' aerial (or antenna) placed near the 'transmitter'. For, when the radiation flowed past the second aerial, the alternating electric and magnetic fields should generate corresponding oscillations in it.

The electrical technology of 1865 could not provide the high-frequency generators that were needed for this demonstration. A frequency of perhaps 1 kHz* might have been produced by mechanical means; but the wavelength would then be

$$\frac{3 \times 10^8}{10^3} = 3 \times 10^5 \text{ m} = 300 \text{ km}$$

and the aerials for this were unthinkable!

Eventually in 1888 Hertz produced very high-frequency currents in the oscillatory discharge of a capacitor (p. 183), and used these to generate electromagnetic waves. Maxwell's brilliant prediction was then confirmed in every detail. Hertz's apparatus is shown in Fig. 14.34. The 'aerial' consisted of two square metal sheets A_1 and A_2 arranged one above the other. These were joined to small metal spheres K_1 and K_2 connected to an induction coil. The two halves of the aerial thus formed a sort of capacitor. Each pulse of high p.d. from the induction coil charged up the capacitor until the insulation in the gap broke down; the ionized air then provided a low-resistance path through which the capacitor was rapidly discharged. Because of the inevitable small inductance of the plates and connecting rods each discharge took the form of a damped oscillatory current of very high frequency. According to Maxwell's theory this arrangement should generate trains of electromagnetic waves with the electric vector parallel to the gap and magnetic vector at right-angles to it (i.e. with the lines of force in circles concentric with the connecting rods). This was duly detected by Hertz's 'receiver', which consisted of nothing more than a large loop of wire with a narrow gap at one point. When this was held in a vertical plane in line with the aerial, the alternating magnetic field in the radiation linked with the loop, generating an e.m.f. in it. A fine train of sparks was then observed crossing the receiver gap. The sparks ceased when the plane of the loop was turned through a right-angle, showing that the radiation was polarized in the way predicted. All the familiar properties of waves

* I.e. 1 kc/s, p. 27.

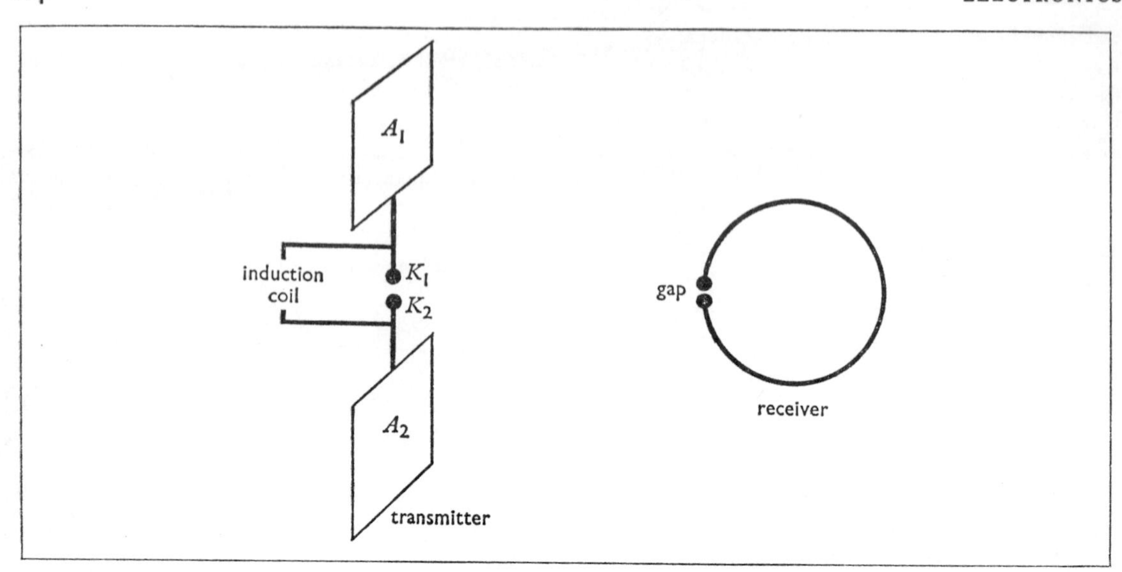

Fig. 14.34 Hertz's spark 'transmitter' and 'receiver'

were demonstrated with apparatus of this kind. They were reflected from metal sheets and refracted by various materials, such as paraffin wax and pitch; and interference effects and standing wave patterns were readily produced. The wavelength used was found to be about 6 metres, so that the frequency of oscillation in Hertz's transmitter must have been about 50 MHz.

Spark transmitters are now strictly illegal because of the extensive interference with radio reception they cause—a point that did not matter in Hertz's day. However, similar experiments may now be performed with more efficient transmitters and receivers that do not produce interference outside their own frequency limits; and in any case the experiments can be done in a laboratory at such low power that the radiation would be undetectable outside the building. But before embarking on any such tests the appropriate regulations should be consulted; for some purposes licences are required.

The valve oscillator provides the ideal means of generating the high-frequency currents required in a radio transmitter. Fig. 14.35 shows a very simple arrangement. The alternating current in the tuned circuit of the Hartley oscillator induces an e.m.f. in the aerial coil coupled with it, and so feeds a high-frequency current into the aerial.

Waves of current and p.d. therefore travel along the aerial in much the same way as along a transmission line. The waves are reflected at the far end of the aerial, and a standing wave system is therefore set up on it with a node of current (and an antinode of potential) at the far end. Maximum current flows and maximum power is radiated when the aerial is resonant to the frequency in use. The shortest resonant length is a quarter of a wavelength. For maximum efficiency either the

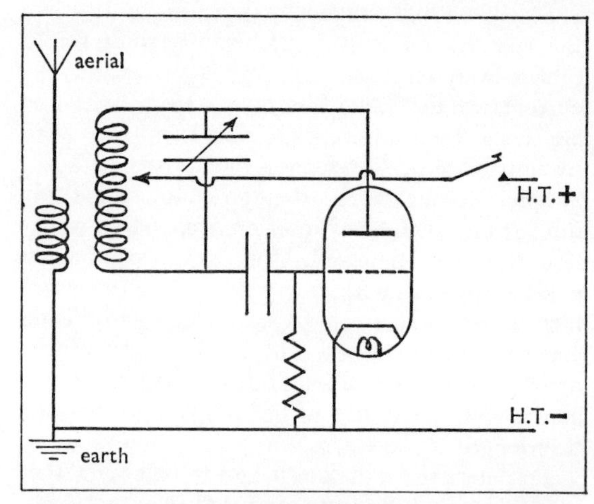

Fig. 14.35 A simple radio transmitter using a Hartley oscillator

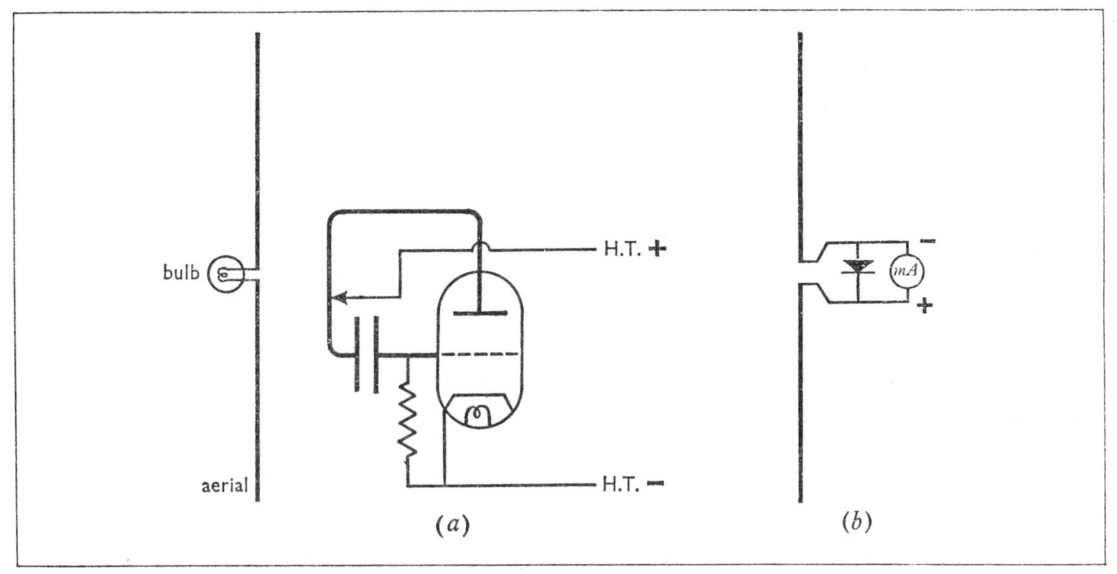

Fig. 14.36 A v.h.f. transmitter (*a*) and receiver (*b*)

length of the aerial or the inductance of the coupling coil must be adjusted so that this condition is effectively fulfilled.

By suitable choice of the tuning coil and capacitor this simple transmitter may be made to operate at any frequency up to several hundred megahertz. At the highest frequencies the tuning circuit need be no more than a single-turn coil together with the anode-grid capacitance of the valve. The apparatus then takes the form shown in Fig. 14.36*a*. The single vertical rod fulfils the functions of both coupling 'coil' and aerial. Its length is arranged to be half a wavelength (i.e. it is two quarter-wave aerials back to back). Such an aerial is called a *half-wave dipole.* The current flowing up and down through its centre may be demonstrated by joining the two halves through a low-current torch bulb. The receiver consists of a similar dipole with the appropriate means of detecting the current at its centre (Fig. 14.36*b*). Close enough to the transmitter a torch bulb will suffice for this also. But greater sensitivity is achieved by using a small semiconductor diode (p. 329) and milliammeter as shown. Current flows through the diode only in the forward direction, so that the high frequency current in the aerial develops a steady p.d. between the two halves of the dipole. This is registered by the milliammeter, and is a measure of the intensity of the incident wave. Many of Hertz's experiments may be simulated with this apparatus.

★ *The 'plane of polarization'*

A notorious difficulty arises at this point in describing plane polarized electromagnetic waves, whether radio waves or light. The plane of polarization may be taken as the plane of either the electric or the magnetic vector. We know that the two vectors are at right-angles, and it is simply a matter of convention which we choose. Unfortunately there is no general agreement on which it should be. When polarization of light was first demonstrated, it was not known what kind of wave it might consist of; and the phenomenon of polarization by reflection was used to define the plane of polarization. When electromagnetic waves are reflected at the boundary between two 'transparent' media, the reflected waves are partially plane polarized; the plane containing the incident and reflected 'rays' was originally taken as the plane of polarization of the reflected waves. It turns out that this is in fact the plane of the magnetic vector. (The matter may be demonstrated by studying the reflection of very short wavelength radio waves from the surface of a block of paraffin wax.)

However, it appears that the electric vector is really the more significant of the two when we consider the interaction of electromagnetic waves with matter; and many writers have taken this vector as defining the plane of polarization. Indeed this is quite general in radio engineering—e.g. television aerials have to be aligned in the plane of the electric vector, and this is called the plane of polarization of the waves. Other writers (in textbooks on Light) have sought to avoid ambiguity by calling the plane of the electric vector the *plane of the vibrations*; and they do not then use the phrase 'plane of polarization' at all.

In the Author's opinion the student does well to avoid using both these ambiguous phrases, and to describe plane polarized waves by specifying the direction of either the electric or the magnetic vector. Thus Hertz's apparatus generates waves polarized 'with the electric vector in a vertical plane'; and light waves reflected from a glass plate are partially plane polarized 'with the magnetic vector in the plane of the incident and reflected rays'. This procedure leaves no loophole for ambiguity.

The velocity of electromagnetic waves in a vacuum

It is beyond the scope of this book to consider the detailed analysis by which Maxwell predicted the existence of electromagnetic waves. But we give here a deduction of the expression for the velocity c of the waves on the assumption that they do exist in the form that Maxwell predicted.

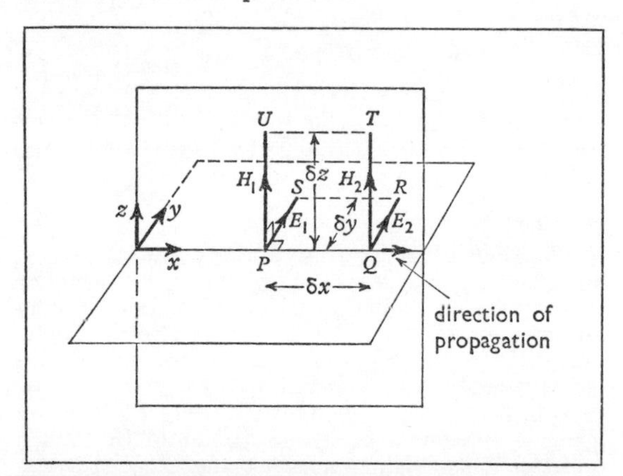

Fig. 14.37 Deducing the velocity of plane electromagnetic waves in a vacuum

Fig. 14.37 shows the instantaneous pattern of the fields in a plane electromagnetic wave travelling in the x-direction. The wave is taken to be polarized so that the electric field lies in the y-direction and the magnetic field at right-angles to it in the z-direction. Neither field has any component in the direction of travel of the wave. Consider two points P and Q a short distance δx apart; let the two fields have intensities E_1 and H_1 at P, and E_2 and H_2 at Q, as shown.

We shall apply first the law of electromagnetic induction (p. 106) to the rectangular path PQRS, which is of width δy. Now the electric field intensity E is equal to the potential gradient. The p.d. between P and S is therefore $E_1\,\delta y$; and that between R and Q is $-E_2\,\delta y$. There is no potential gradient in the x-direction. The e.m.f. acting round the path in the sense $\overrightarrow{\text{PSRQP}}$ is therefore $(E_1 - E_2)\delta y$. (I.e. the e.m.f. is the *line integral* of the electric field intensity round the path, p. 88.) This is equal to the rate of change of the magnetic flux linking the rectangle. If the mean value of the magnetic field intensity over the rectangle is H, the mean magnetic flux density B is given by

$$B = \mu_0 H$$

The magnetic flux linking the rectangle is $B.\delta x\,\delta y$.

$$\therefore\ \text{e.m.f.} = (E_1 - E_2)\delta y = \frac{\mathrm{d}B}{\mathrm{d}t}.\delta x\,\delta y$$

$$\therefore\ E_1 - E_2 = \frac{\mathrm{d}B}{\mathrm{d}t}.\delta x$$

The time taken by the wave to travel from P to Q is $\delta x/c$. In this time interval the electric field intensity at Q increases by the amount $(E_1 - E_2)$. The rate of increase $\mathrm{d}E/\mathrm{d}t$ is therefore given by

$$\frac{\mathrm{d}E}{\mathrm{d}t} = \frac{E_1 - E_2}{\delta x/c} = c.\frac{\mathrm{d}B}{\mathrm{d}t} = c\mu_0.\frac{\mathrm{d}H}{\mathrm{d}t}$$

Similarly the *magnetomotive force* or *m.m.f.* (p. 88) acting round the rectangular path $\overrightarrow{\text{PUTQP}}$ is $(H_1 - H_2)\delta z$. According to Ampère's rule (p. 87) this should be equal to the electric current linked with the path. Since we suppose that the wave is travelling in a vacuum, there is no current in the ordinary sense anywhere in the neighbourhood. It was at this point that Maxwell introduced his hypothesis of the *displacement current*; without it no electromagnetic wave propagation is possible. To complete the symmetry of the equations we suppose that the m.m.f. acting round the closed path is equal to the rate of change of the total electric flux linking the rectangle. If the mean electric flux density is D ($= \varepsilon_0 E$, see p. 170), the electric flux linking the rectangle PQTU is $D.\delta x\,\delta z$.

$$\therefore\ \text{m.m.f.} = (H_1 - H_2)\delta z = \frac{\mathrm{d}D}{\mathrm{d}t}.\delta x\,\delta z$$

$$\therefore\ H_1 - H_2 = \frac{\mathrm{d}D}{\mathrm{d}t}.\delta x$$

As before, the magnetic field intensity at Q increases by this amount in the time $\delta x/c$ taken by the wave to advance from P to Q.

$$\therefore\ \frac{\mathrm{d}H}{\mathrm{d}t} = \frac{H_1 - H_2}{\delta x/c} = c.\frac{\mathrm{d}D}{\mathrm{d}t} = c\varepsilon_0.\frac{\mathrm{d}E}{\mathrm{d}t}$$

Thus if the pattern of the electric and magnetic fields is to remain constant as it moves forward in the x-direction, the rates of change of E and H must be connected by the two symmetrical relations derived above, which may be written

$$\frac{\mathrm{d}E}{\mathrm{d}t} = c\mu_0.\frac{\mathrm{d}H}{\mathrm{d}t}$$

and

$$c\varepsilon_0.\frac{\mathrm{d}E}{\mathrm{d}t} = \frac{\mathrm{d}H}{\mathrm{d}t}$$

Dividing and re-arranging we obtain the expression for c.

$$\therefore\ c^2\varepsilon_0\mu_0 = 1$$

$$\therefore\ c = \frac{1}{\sqrt{\varepsilon_0\mu_0}} = 3{\cdot}00 \times 10^8\ \mathrm{m\ s^{-1}}$$

14.11 Radio

In order to send messages with radio waves it is necessary to vary or *modulate* the wave in some way. An unmodulated *carrier wave*, as it is called, conveys no information except that the transmitter is working! The simplest form of modulation is to switch the transmitter on and off in time with the dots and dashes of the Morse code. A tapping key may be arranged so that the waves are only sent out when it is pressed down. For instance in the simple transmitter of Fig. 14.35 the tapping key may be inserted in the lead to the H.T. supply. To transmit speech or music the modulation must take the form of varying some feature of the wave in time with the electrical signal from the microphone. Either the amplitude or the frequency of the wave may be varied.

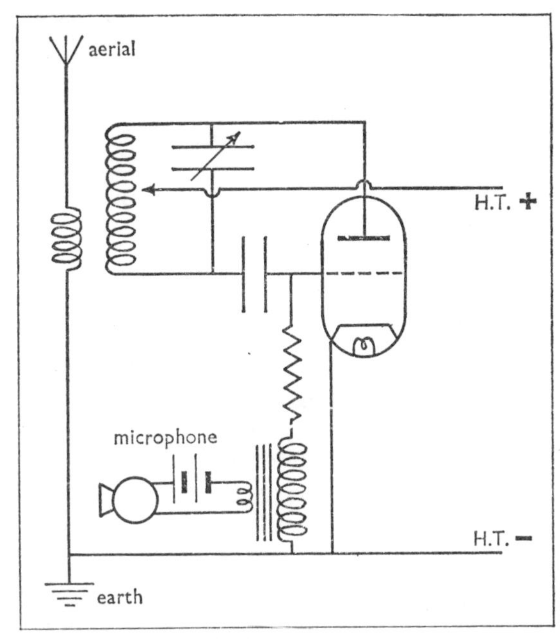

Fig. 14.39 A simple amplitude modulated transmitter

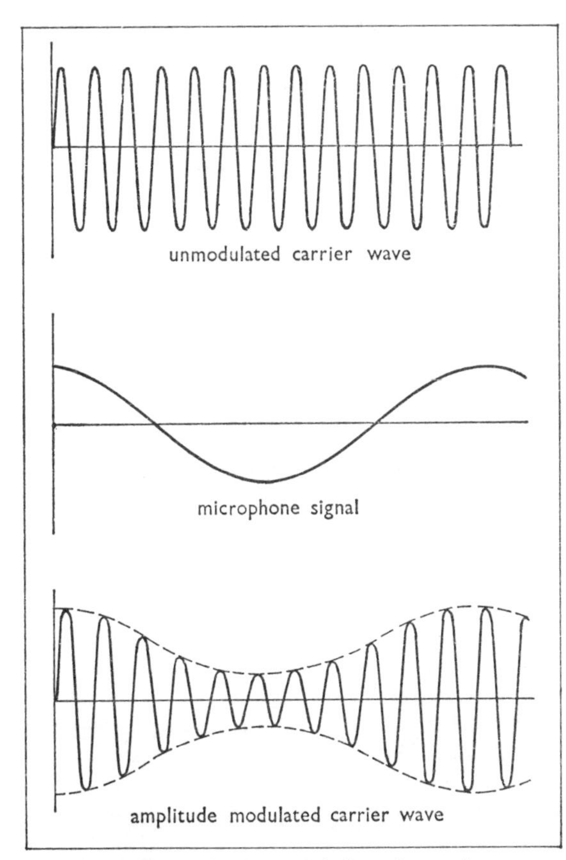

Fig. 14.38 Amplitude modulation of a carrier wave

Amplitude modulation is the simplest to transmit and receive and is used for most broadcasting and communication (Fig. 14.38). Frequency modulation gives higher quality reception, but can only be used at very high frequencies (v.h.f.).

Fig. 14.39 shows how a simple Hartley oscillator may be adapted to transmit an amplitude modulated carrier wave. The speech or music signal is produced by the carbon microphone, which is joined into the circuit as usual through a transformer. The secondary of the transformer is connected in series with the grid resistor. The audio frequency signal is thus superimposed on the steady negative bias on the grid. Because of the curvature of the valve characteristics a change in the grid potential alters the amplification, and so varies the amplitude of the high-frequency oscillations in time with the signal. Like all very simple methods this technique of modulating the wave is not very satisfactory. Just because the valve is being used on the non-linear parts of its characteristics, appreciable distortion is introduced. Also varying the valve amplification tends to affect the frequency of the transmission as well as its amplitude. A better method is to employ a master oscillator, which generates oscillations of constant frequency and amplitude. Several stages of power amplification are then used before the signal is fed to the aerial. The modulation is applied to one of these stages in such a way as to vary its amplification.

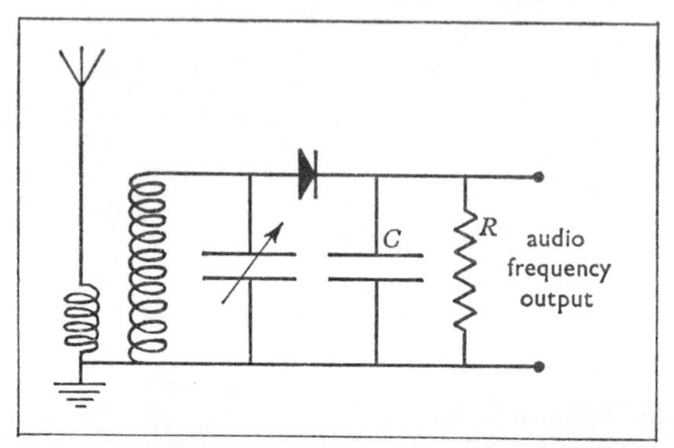

Fig. 14.40 A diode detector circuit

At the receiving end the passage of the radio waves past the aerial generates small alternating currents in it. The audio frequency signal must now be recovered from the variations in amplitude of this current. Circuits for doing this are called *demodulators* or *detectors*. Fig. 14.40 shows a simple diode detector; either a semiconductor or thermionic diode may be used. The circuit should be compared with the power rectifier arrangement of Fig. 14.9 (p. 241). The parallel resonant circuit is tuned to the frequency to be received, and a magnified alternating p.d. is developed across it; this is applied to the diode and storage capacitor C in series. The time constant CR is made to be about 5×10^{-5} s. This is very long compared with the period of oscillation of the aerial current, and the storage capacitor therefore charges up to the peak p.d. across the tuned circuit. However, the time constant is short compared with the period of the audio frequency modulation. The p.d. across C therefore follows faithfully the variations of amplitude of the aerial signal.

To give distortionless detection the diode must be able to pass sufficient current for all amplitudes of signal. Unfortunately, very little current flows in any kind of diode for p.d.'s less than about 0·4 volt. This method of detection is not therefore suitable for small signals. However, with alternating p.d.'s of 10 volts or more it is the best method available; it is regularly used in modern radio sets, in which there are several stages of high-frequency amplification before the detector stage. The load R in the detector could consist of a pair of high-resistance headphones, but this takes a rather large drain of power from the tuned circuit and so reduces its Q-factor. It is better to develop the audio frequency signal across a high-resistance load and apply it to the grid of an amplifying valve.

Fig. 14.41 shows a way of combining detection with amplification all in one triode valve. At the same time the efficiency of the tuned circuit is improved by providing a certain amount of positive feed-back, or *reaction*. Detection takes place just as before in the 'diode' formed by the grid and cathode of the valve. Sufficient grid current flows on the peaks of the signal to charge up the storage capacitor C to the peak p.d. The net signal actually applied to the grid is then the sum of the audio frequency p.d. across C and the radio frequency p.d. across the tuned circuit, and both of these affect the anode current. The reaction coil L_3 is magnetically coupled with the tuning coil L_2, and by this means some of the high-frequency signal in the anode circuit is fed back in phase with the input. Excessive feed-back makes the circuit oscillate; and the coupling between the coils is adjusted so that this does not quite happen. In effect the positive feed-back is making up for some of the losses in the tuned circuit, and so increasing enormously its Q-factor. The audio frequency signal passes through the headphones, the high-frequency part of the current being by-passed through a small capacitor. In practice it is inconvenient to have to adjust the feed-back by varying the coupling between the coils, and the same result is achieved rather by other means. One method is to join a variable capacitor across the reaction coil L_3, so that some of the high-frequency current is by-passed through it. The effective increase in Q-factor in the reaction detector not only magnifies the signal applied to the grid, but also increases the selectivity of the receiver, since frequencies other than that to which it is tuned are not magnified to the same extent (p. 213). This type of circuit can be a source of much pleasure to the amateur radio enthusiast who is prepared to master the fine art of adjusting it to maximum efficiency.

One cause of poor radio reception is the rapid

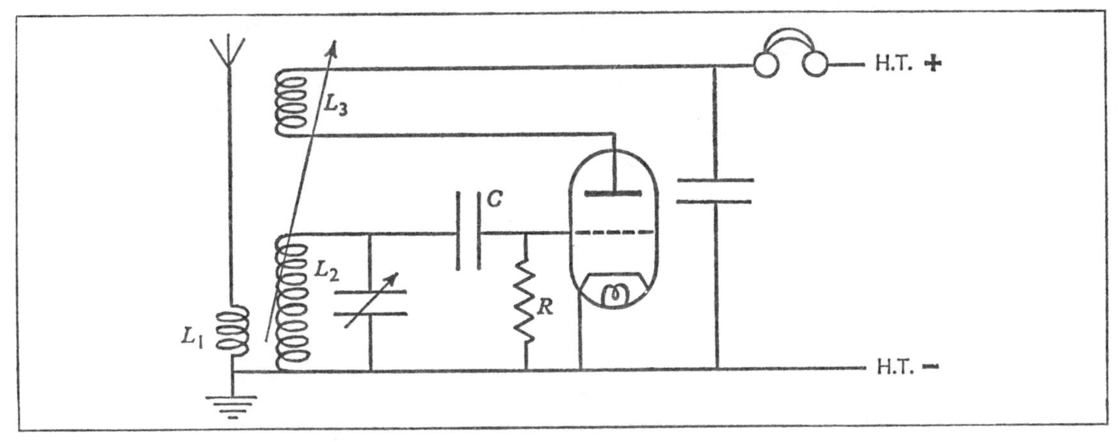

Fig. 14.41 A reaction detector

changes of signal strength that are liable to occur; this is known as *fading*. It arises because the waves can reach the receiving aerial by two different paths. One wave travels directly over the surface of the ground; diffraction effects enable it to follow the curvature of the earth and to be little affected by obstacles such as hills and buildings. Another wave travels up into the air and is reflected from layers of ionized particles in the upper atmosphere, called the *ionosphere*. Close to the transmitter the *ground wave* predominates, but at distances of 100 miles or so the *sky wave* may be of about the same amplitude; and interference then occurs between the two beams. Conditions in the ionosphere are changing continually. At one moment the receiving aerial may be at a maximum in the interference pattern and a strong signal is received. At another the resultant signal may drop off almost to nothing. To some extent fading may be countered by using Automatic Gain Control (A.G.C.) in the receiver; in this arrangement the average strength of the signal is made to control the amplification (or gain) of the receiver. When the signal increases, the gain is reduced; and vice versa.

The ionization in the upper atmosphere is produced by radiations of various kinds from the sun. It occurs in a number of separate layers at heights between 80 and 500 km; the layers fluctuate continually both in position and in density of ionization. In the medium waveband (500 to 1500 kHz) the conditions are such that the sky wave is of negligible intensity during the day time; fading does not then occur, and reception of the ground wave is possible within a range of a few hundred miles. But when darkness falls the sky wave grows in strength, and fading and interference from adjacent channels are at their worst. Beyond the range of the ground wave reception is only possible by means of the sky wave. Even then fading can be serious, because the waves may reach the aerial by two or more different paths within the ionosphere. Sometimes the waves can reach the receiver after several reflections between the ionosphere and the ground, since the earth's surface, whether land or sea, is a fair conductor, and reflects electromagnetic waves strongly.

In the short waveband (from 3 to 30 MHz) conditions are rather different. The ground wave is of very short range (about 50 miles). But the sky wave is strongly reflected, at any rate for oblique incidence on the ionosphere; waves that strike it nearly normally pass straight through and are not reflected. Up to a certain range (known as the *skip distance*) no sky wave can therefore be received; but beyond this good reception is possible. By choosing the frequency of the transmitter to match the ionospheric conditions at a given moment reasonably reliable communication is possible with any part of the world. To increase the efficiency of short-wave communication and broadcasting it is common practice to *beam* the transmission towards the part of the world concerned. This is done by using several aerials in such a way that interference between the waves emitted concentrates the radiation at the required bearing and elevation. Many of these aerial systems are analogous to devices for producing interference in optics. Indeed some interference effects occur with almost all aerial systems. Even a simple half-wave aerial behaves like a Lloyd's mirror arrangement because of reflection by the ground; the radiation is therefore concentrated at particular elevations at which the waves from the aerial and its reflected 'image' in the ground happen to be in phase. A number of aerials side by side produce an interference pattern resembling that of a diffraction grating in optics; and by such means very narrow beams of radiation may be transmitted.

At still higher frequencies (>30 MHz) the ionosphere does not reflect the waves at all, even at oblique incidence; and the range is limited by the ground wave. However, this also limits the range from which interference can arise, and high-quality reception is possible.

15: X-rays

15.1 The production of X-rays

In 1895 Röntgen discovered that a very penetrating radiation was being emitted by a discharge tube in which cathode rays (p. 228) were passing at a high p.d. The radiation produced fluorescence on screens several metres away, and fogged photographic plates nearby, even when these were securely wrapped in dark paper. It was known that cathode rays themselves could not penetrate more than a few millimetres into air at n.t.p.; and it was therefore clear that a new kind of radiation was involved, to which Röntgen gave the name *X-rays.*

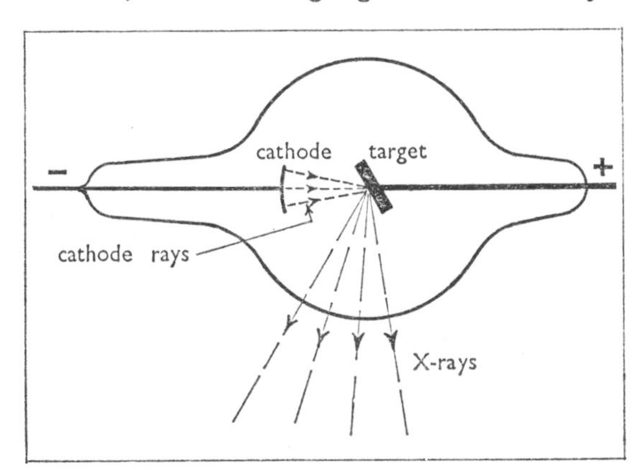

Fig. 15.1 A gas-filled X-ray tube

X-rays are produced whenever cathode rays (electrons) strike a piece of matter and so are brought to rest. In a discharge tube they are therefore emitted at the anode and from all parts of the glass wall reached by the cathode rays. In order to produce a localized source of X-rays it is necessary to focus the cathode rays into a small spot on the *target.* In the early days gas-filled tubes were used (at a pressure of about 10^{-3} mm of mercury). The *cathode* was a concave plate whose surface was arranged to be concentric with the target spot on the *anode* or *anti-cathode* (Fig. 15.1). In this way some focusing of the cathode rays is achieved since they tend to leave the cathode along lines normal to its surface. However, the p.d. across such a tube and the current through it depend critically on the gas pressure, and this tends to alter as the tube is used; it is therefore difficult to achieve reliable control.

A modern X-ray tube produces the cathode rays by thermionic emission from a heated tungsten cathode (Fig. 15.2). A high vacuum is used (less than 10^{-6} mm of mercury), and the beam is focused by means of an electrode arrangement similar to that used in the electron gun of a cathode-ray tube (p. 237). The efficiency of the tube is increased by using a target material of high atomic number. Even so, only about 0·2% of the energy of the cathode-ray beam is converted into X-rays; the rest is changed into heat. The target is therefore made of a plug of tungsten (which has a high melting point); this is set in a substantial copper bar with external cooling fins. The maximum emission of X-rays occurs in the directions shown—approximately at right-angles to the cathode-ray beam. In all other directions the tube is heavily shielded with lead glass and sheets of lead. The *penetrating power* of the X-rays may be adjusted by varying the p.d. across the tube; the higher the p.d., the more penetrating the radiation. P.d.'s up to 100 kV are commonly employed; and for special purposes p.d.'s over a million volts are used. The *intensity* of the beam depends on the number of electrons striking the target per second—i.e. it depends on the current through the tube. This may be adjusted by varying the filament current, which controls the cathode emission.

One way of providing the high p.d. required to operate an X-ray tube is simply to connect it across the secondary of a high-voltage a.c. transformer—T_1 in Fig. 15.3. The tube is itself a thermionic diode and conducts only on the half-cycles when the anode is positive with respect to the cathode.

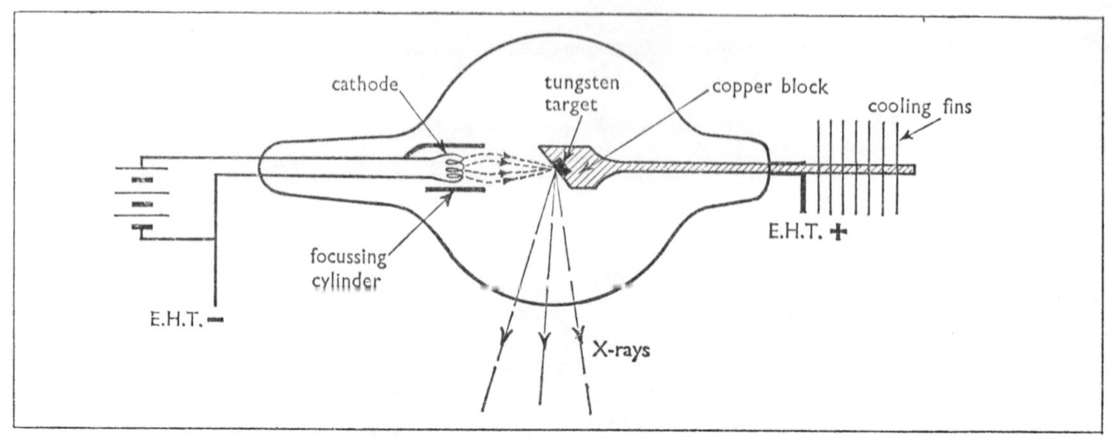

Fig. 15.2 A hard vacuum X-ray tube

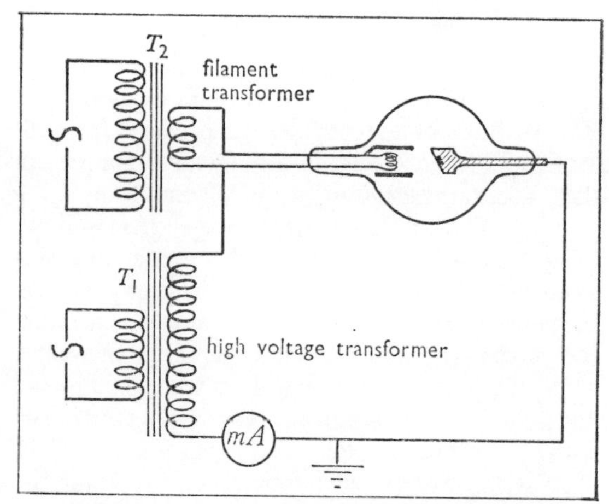

Fig. 15.3 A simple self-rectifying circuit for an X-ray tube

On the other half-cycles no current flows and no X-rays are emitted. The anode may need to be water-cooled, and it is usually convenient to earth this part of the apparatus. This means that the cathode and the whole secondary winding of the filament transformer T_2 are operating at very high potentials, and their insulation must be designed accordingly. This arrangement is satisfactory for some purposes; but when the tube must be operated at a steady known p.d., a separate rectifier employing a diode and storage capacitor must be used (p. 241).

X-rays are to some extent absorbed by all matter through which they pass; but their penetrating power is very great. They are therefore to be treated with the greatest possible respect; and adequate, well-tested safety precautions must always be observed. Their effects on the human body are only partly understood; but they are known to produce deep-seated burns, destruction of living cells and unpredictable chemical changes. Most insidious of all is their tendency to produce serious diseases, sometimes many years after the time of irradiation; and they also produce genetic changes which may only become apparent in subsequent generations. All unnecessary exposure to X-radiation should be rigorously avoided; and the student is strongly advised never to use X-ray shoe-fitting equipment and similar apparatus. Discharge tubes operating at p.d.'s above 6 kV should never be used without adequate shielding. Television tubes (using p.d.'s up to 20 kV) should never be used without the screens designed by the manufacturers to go with them; however, often the front wall of the tube itself is made of glass that will give sufficient protection.

When X-rays pass through matter, their immediate effect is to produce ionization. Electrons may be ejected from all levels of the atoms, and the energy given to them is then dispersed through the rest of the matter by collisions. By this means the energy of an X-ray beam is mostly absorbed as heat. For a given thickness of screen the absorption is approximately proportional to the total number of electrons per unit volume. It is therefore greatest for dense substances containing elements of high atomic number.

The X-ray photography of the human body is

made possible by this effect. The bones contain much denser material than the soft organs, and therefore absorb the X-rays passing through them to a much greater extent. An X-ray picture is simply a *shadow photograph*, which is made sharp enough by using a very small source of X-rays and placing the photographic film as close as possible behind the part of the body under investigation. When the film is developed, all the tissues show up relatively pale against the darkened background; and the bones show up particularly white. Apart from the bones, the differences of density between the parts of the body are very slight, and it is not normally possible to distinguish the details of the soft organs —stomach, liver, brain, blood vessels, etc. However, in a number of cases doctors are able to introduce into an organ a harmless compound containing a dense element. For instance, in order to observe the stomach and intestinal tract, the patient is given a 'barium meal', consisting of a paste of barium sulphate and water. Each part of the alimentary canal in turn then shows up clearly as the barium meal reaches it. By techniques of this kind the radiographer can now take satisfactory X-ray pictures of many organs that could not otherwise be photographed.

When X-ray photography must be used for medical purposes, care is always taken to ensure that the dose is far below that which is definitely known to be harmful. Sometimes the destructive power of X-radiation is deliberately used for therapeutic purposes. It so happens that rapidly dividing cells are more readily damaged by X-rays than stable ones. A carefully controlled dose can thus destroy a growing tumour without doing irreparable harm to the surrounding tissue. Similar techniques are also used with the radiations from radioactive substances, considered in Chapter 16.

15.2 The nature of X-rays

When X-rays were first investigated, the conviction rapidly grew among physicists that they were a form of *electromagnetic radiation*—essentially the same as radio waves, light, etc., but of much shorter wavelength even than ultra-violet radiation. Several of the properties of X-rays suggest this conclusion:

(*i*) They are not deflected by electric or magnetic fields. We can therefore dismiss the possibility of their being charged particles.

(*ii*) They are apparently produced by the deceleration of electric charges; and theory leads us to expect the production of electromagnetic waves in such a case (p. 263).

(*iii*) They affect a photographic emulsion just like light and ultra-violet radiation. However, other things besides electromagnetic waves can do the same; for instance, beams of electrons and other charged particles are known to produce darkening of a film.

(*iv*) They produce ionization of the air through which they pass. It may be simply demonstrated by passing a beam of X-rays through the air inside or near a charged electroscope. Whatever the sign of the charge on the electroscope, there is a rapid fall of the deflection. The rate of discharge is proportional to the intensity of the X-ray beam, and is sometimes used as a means of measuring it (p. 186).

However, none of this evidence was conclusive; and it was many years before a way was found of demonstrating a distinctive wave effect that would confirm the suggestion that X-rays were electromagnetic waves. They seemed to show no reflection or refraction like other sorts of waves, and the early attempts to demonstrate diffraction or interference of X-rays were quite inconclusive.

Eventually, in 1912, von Laue suggested that the spacing of the atoms in a crystal was probably of the same order of magnitude as the wavelength of X-rays. Therefore diffraction effects should be observed when a beam of X-rays was passed through a crystal. Von Laue calculated that diffracted beams should be observed only in certain sharply defined directions in relation to the crystal structure. The process is in some ways similar to the behaviour of light in passing through a diffraction grating, except that in this case the grating (i.e. the crystal) has a repetitive structure that extends in *three* dimensions. The predictions were duly tested with the apparatus shown in Fig. 15.4*a*. A very fine beam of X-rays was allowed to pass through the pin-holes in the two lead screens S_1 and S_2. It then impinged on the crystal C, behind which was fixed a photographic plate. The developed plate revealed a pattern of fine spots (Fig. 15.4*b*), which was in exact agreement with von Laue's predictions. This clearly established the wave nature of X-rays, and showed that the wavelengths used were approximately in the range 10^{-11} to 10^{-10} m—about

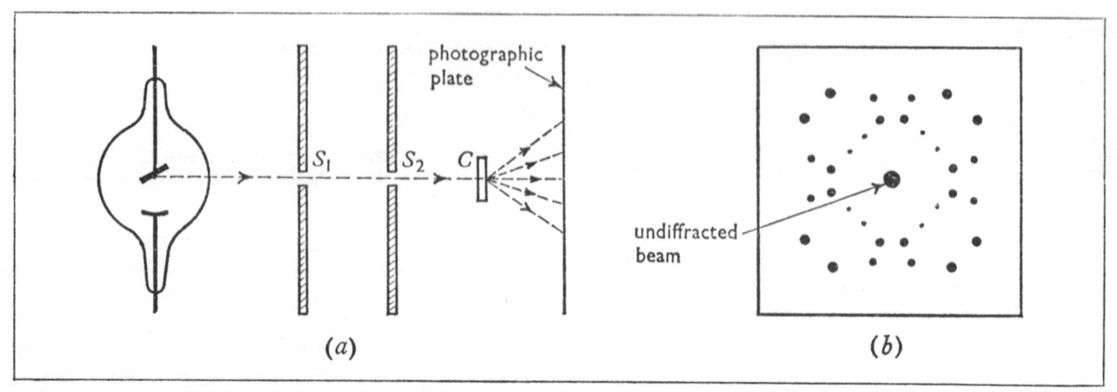

Fig. 15.4 The diffraction of X-rays by a crystal: (*a*) Von Laue's apparatus; (*b*) the diffraction spots obtained

1/10,000 of the wavelength of visible light. It was not surprising that previous experiments with ordinary diffraction gratings had not revealed anything.

Bragg's X-ray spectrometer

Soon afterwards Sir William Bragg suggested a much simpler way of analysing the diffraction of X-rays by a crystal lattice. When a beam of X-rays is incident on a single layer of atoms in one plane of a crystal, each atom scatters a minute proportion of the beam, i.e. it becomes the source of a small spherical wavelet of X-rays (Fig. 15.5). In most directions the wavelets destroy one another by interference. But in the direction for which the angle of incidence is equal to the angle of 'reflection' they reinforce one another. In other words the X-rays behave as though they were weakly reflected by the layer of atoms. The regularity of the arrangement of atoms in a crystal ensures that there are many such layers, parallel and equally spaced. In general therefore the beams reflected from successive layers will again destroy one another by

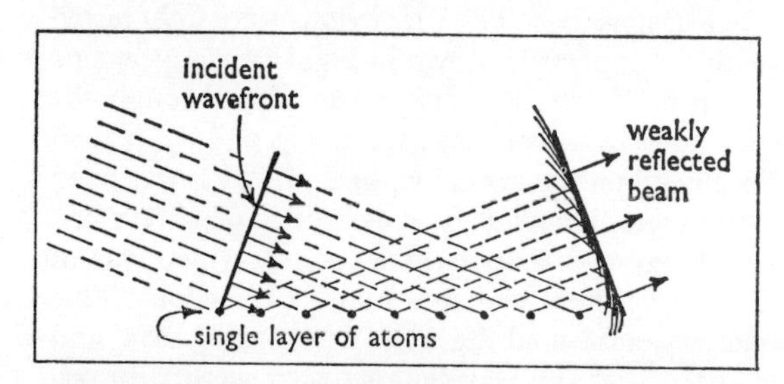

Fig. 15.5 The 'reflection' of X-rays by a single layer of atoms

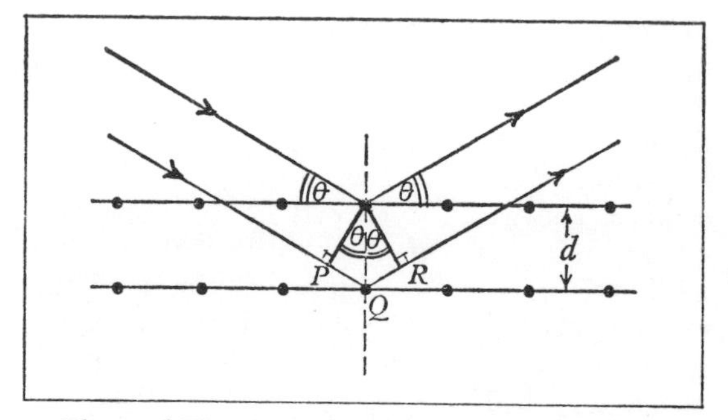

Fig. 15.6 The deduction of Bragg's law for X-ray reflection

interference, and only in certain sharply defined directions will constructive interference take place. These directions are those for which the path difference for the waves reflected from adjacent layers of atoms is a whole number of wavelengths. Referring to Fig. 15.6, this path difference is equal to PQ + QR. Strong reflection therefore occurs when

$$PQ + QR = n\lambda$$

where λ = the wavelength, and n is an integer.

If d is the separation of the layers of atoms and θ is the *glancing angle*, then there is strong reflection when

$$2d \sin \theta = n\lambda$$

a result known as *Bragg's law*. Notice that Bragg found it convenient to make his measurements in terms of the glancing angle θ, and not in terms of the *angle of incidence*—which would be measured from the normal and is equal to $(90° - \theta)$. For any given wavelength and crystal spacing Bragg expected to find several angles at which reflection occurred, corresponding to n = 1, 2, 3, etc. These

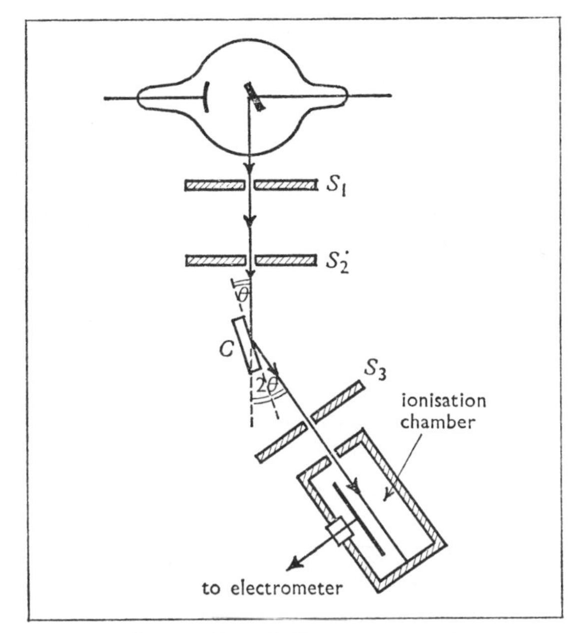

Fig. 15.7 Bragg's X-ray spectrometer

he described as the *first-order reflection*, *second-order reflection*, etc.

Bragg's analysis was fully confirmed by experiment, and provided the means of measuring not only the spacing of atoms in crystals, but also the wavelengths of the X-rays emitted by any given source. The instrument that Bragg developed is therefore rightly called an *X-ray spectrometer*. The principle is shown in Fig. 15.7. A very narrow beam of X-rays was defined by the slits (less than 0·1 mm wide) in the two lead screens S_1 and S_2. The crystal C was mounted on a table that could be rotated about a vertical axis and carried a graduated scale, so that the glancing angle θ of the beam on its faces could be measured. The reflected beam (deviated through an angle 2θ, as shown) passed through a slit in another screen S_3 and into an ionization chamber with lead walls. This was connected to a suitable source of p.d. and an electrometer to measure the ionization current, which was taken as a measure of the strength of the reflected beam. The arrangement was rather similar to the familiar prism spectrometer used for studying optical spectra. The ionization chamber and screen S_3 were mounted on an arm that could rotate about the same central axis as the crystal table—just like the telescope and prism table of an optical spectrometer; while the slits in the screens S_1 and S_2 took the place of the collimator.

The ionization currents obtained for varying values of the glancing angle θ were as shown in Fig. 15.8. These measurements revealed two significant features of X-ray spectra. Firstly, the spectrum emitted by a given source contained a pattern of sharply defined wavelengths which were characteristic of the element used as the target of the X-ray tube. This can be compared with the characteristic line spectra emitted by gases and vapours in the visible region. For a given target element the wavelengths of the X-ray 'lines' were unaffected by varying any other factors, such as the p.d. in the X-ray tube—except that when the p.d. was too low a particular line of the spectrum might not be excited at all. Secondly, a continuous background spectrum was apparent containing all possible wavelengths within a certain range. This can be compared with the continuous spectrum emitted in the visible region by a hot solid (such as a lamp filament). The character of the continuous spectrum was unaffected by changing the target material, but was found to vary in a characteristic way with the p.d. used in the tube.

The same wavelength can appear in a number of different peaks in the diffraction pattern. Thus, peaks Nos. 3 and 4 (in Fig. 15.8) are due to the same pair of wavelengths as peaks Nos. 1 and 2; but 1 and 2 are the first-order reflection, while 3 and 4 are the second-order. This may be proved by showing that the values of sin θ for the second pair are twice those for the first—in agreement with Bragg's law.

When the experiment is repeated with different

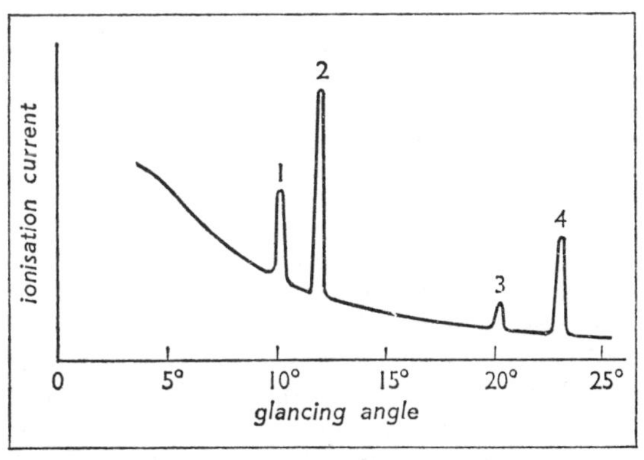

Fig. 15.8 The variation of ionization current with glancing angle

crystals, the same pattern is found in every case, but the values of θ are shifted in accordance with the lattice spacing d of the particular crystal. This enables us to compare the lattice spacings of different crystals. Thus, for a particular wavelength in the first-order spectrum we have

$$d \propto \frac{1}{\sin \theta}$$

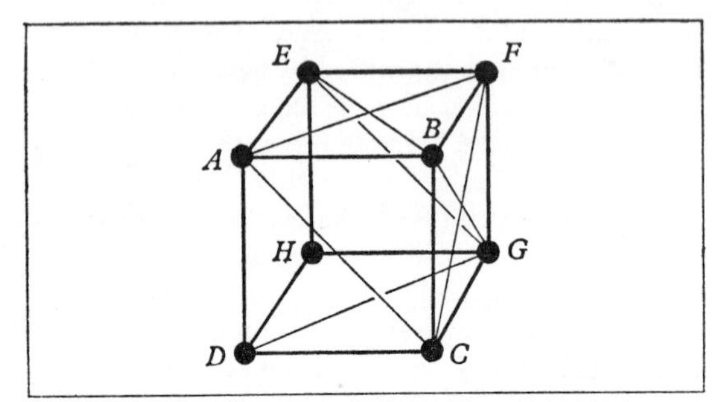

Fig. 15.9 Part of a simple cubic crystal

Even within a single crystal there are several sets of planes which are rich in atoms. Fig. 15.9 shows the positions of the centres of the atoms in a simple cubic crystal lattice. First of all there are planes such as ABCD or AEFB parallel to the faces of the cube. Then there are planes such as ADGF or AEGC inclined at 45° to the cube faces. Other planes rich in atoms are those parallel to ACF or EBG. The separation of adjacent planes can be calculated in each case in terms of the distance between neighbouring atoms. For the three types of planes in a simple cubic lattice, the separations are in the ratio

$$1 : \frac{1}{\sqrt{2}} : \frac{1}{\sqrt{3}}$$

Other forms of crystal lattice give different ratios for their plane separations. By studying the diffraction patterns for different orientations of a crystal we can measure these ratios and establish exactly which type of crystal structure we are dealing with. If we also know the wavelength of the X-rays used, we can go further and actually measure the spacing of the atoms (or ions) in the crystal. Conversely, if we know the atomic spacing in any particular crystal, we can use this to measure the value of some X-ray wavelength. Once the wavelength of one X-ray line is known, other wavelengths can be found by comparison with it, using Bragg's law.

To fix the wavelength scale we must have an independent means of calculating the atomic spacing for some suitable crystal. Bragg turned his attention first to the crystals of halogen salts—potassium chloride and sodium chloride (rock salt). He soon established by the method outlined above that these substances crystallize in a cubic system. Now imagine the material divided up by sets of planes parallel to the cube faces and passing midway between the atoms—as in Fig. 15.10. If the distance

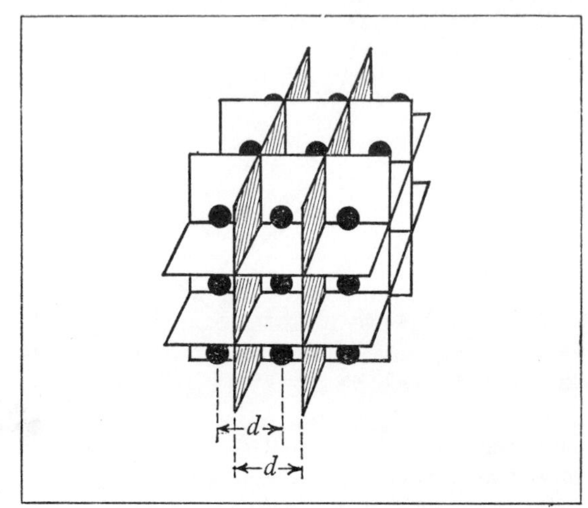

Fig. 15.10 Calculating the number of atoms per unit volume of a simple cubic crystal

between adjacent atoms (or, more strictly, ions) is d, then each cubical box in the figure is of volume d^3 and contains just *one* ion. Suppose the density of the material is ρ we then have

$$\rho = \frac{\text{mass}}{\text{volume}} = \frac{\text{average mass of one ion}}{d^3}$$

One kilogram-atom of any element contains N atoms, where N is Avogadro's constant (p. 37). If the atomic masses of the two elements in the salt are A_1 and A_2, then $(A_1 + A_2)$ kg of the salt contain $2N$ ions altogether.

$$\therefore \text{ the average mass of one ion} = \frac{A_1 + A_2}{2N}$$

$$\therefore \rho = \frac{A_1 + A_2}{2Nd^3}$$

$$\therefore d = \sqrt[3]{\frac{A_1 + A_2}{2N\rho}}$$

Taking the following values for rock salt:

$$\rho = 2{\cdot}18 \text{ g cm}^{-3} = 2{\cdot}18 \times 10^3 \text{ kg m}^{-3}$$

A_1 (sodium) = 23 a.m.u.
A_2 (chlorine) = 35·5 a.m.u.
and $N = 6{\cdot}02 \times 10^{26}$ per kg-atom

we get $$d = \sqrt[3]{\frac{58{\cdot}5}{2 \times 6{\cdot}02 \times 10^{26} \times 2{\cdot}18 \times 10^3}}$$
$$= 2{\cdot}81 \times 10^{-10}\ \text{m}$$

There are a number of simple crystals in which the atomic or ionic spacing can be calculated in the same sort of way. The calculated spacings are always entirely consistent with the ratios found from the diffraction patterns using Bragg's law.

X-ray spectroscopy was thus established on a firm footing, and has proved to be one of the most powerful techniques of modern science. It has made possible the analysis of the arrangement of atoms in even the most complicated molecules. It has also provided an additional weapon for investigating further the inner structure of the atom. Incidentally the technique has removed all doubt that X-rays are indeed very short wavelength electromagnetic waves. The wavelengths emitted by X-ray tubes are mostly in the range 10^{-8} m to 10^{-13} m—a range that overlaps appreciably with the wavelengths of ultra-violet radiation produced in discharge tubes. In the overlapping region the properties of the two types of radiation are found to be identical.

15.3 X-ray spectra

(i) The continuous spectrum. The distribution of energy in the continuous background spectrum of an X-ray tube is found to depend only on the p.d. across the tube. The results obtained for different values of the p.d. are as shown in Fig. 15.11. As the p.d. is raised, the total intensity increases (for a given tube current), and X-rays of shorter wavelength are emitted—which are therefore more penetrating.

An important feature of these curves is that for any given p.d. there is a sharply defined *minimum wavelength* emitted by the tube. According to classical electromagnetic theory any accelerating (or decelerating) charged particle should generate electromagnetic waves (p. 263). It is therefore to be expected that radiation would be emitted by electrons brought suddenly to rest on striking the target of an X-ray tube. However, the classical

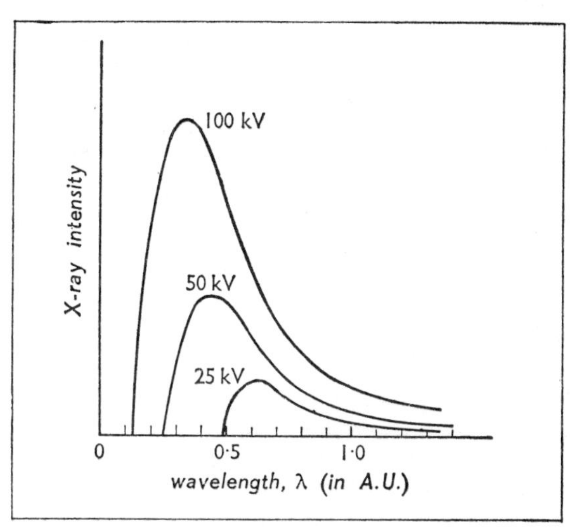

Fig. 15.11 The continuous X-ray spectrum for various p.d.'s

theory is quite unable to explain why there should be any *sharp* cut-off wavelength of the kind observed. Once again the quantum theory (p. 220) comes into its own in explaining this effect. According to this theory the emission or absorption of electromagnetic radiation can take place only in packets of energy (or *photons*) of amount given by

energy = $h\nu$ (p. 221)
where h = Planck's constant,
and ν = the frequency of the waves.

Now, the electrons in an X-ray tube are all accelerated through the same p.d. and therefore all reach the target with practically the same energy, given by

$$\text{electron energy} = eV$$

where V = the p.d. across the tube, and e = the charge on an electron.

This is also the maximum energy that can be imparted to a photon, if we assume that each photon arises in the deceleration of a single electron. We should therefore expect to find a *maximum* X-ray frequency ν_{max} given by

$$h\nu_{max} = eV$$

and a minimum wavlength λ_{min} given by

$$c = \nu_{max}\lambda_{min}$$

where c = velocity of light.

$$\therefore \lambda_{min} = \frac{hc}{Ve}$$

Thus the quantum theory predicts that the cut-off wavelength λ_{min} should be inversely proportional to the p.d. V across the X-ray tube; and this is in good agreement with experiment. Most of the electrons that generate photons give up only part of their energy in this way, and most of the X-radiation is therefore of longer wavelength than λ_{min}.

The cut-off wavelength λ_{min} can be measured with great precision with an X-ray spectrometer. Since c and e are already known to high accuracy, this measurement provides one of the best ways of finding the value of Planck's constant h. The result is in excellent agreement with the value found from the photoelectric effect.

(*ii*) *The line spectrum.* Like the optical spectra of gases and vapours the line spectrum of an X-ray tube is characteristic of the element used for the target. However, unlike optical spectra, which are of great complexity, the X-ray spectra of the elements are all strikingly similar. A typical pattern for an element of medium atomic weight is shown in Fig. 15.12*a*. The group of lines of shortest wavelength is known as the *K*-series. Usually two lines only of this series can be detected (though there are more); these are known as the K_α and K_β lines, in order of decreasing wavelength. The K_α line is always the strongest in the spectrum. At longer wavelengths there is another group of lines known as the *L*-series (L_α, L_β, L_γ, etc.). The pattern is essentially the same for all elements. But as we proceed through the periodic table the whole pattern shifts progressively towards shorter wavelengths with increasing atomic weight. For the heaviest elements a third series, the *M*-series, can be detected; while for the lightest elements the *L*-series is beyond the range of our instruments, and only the *K*-series can be observed.

In 1913 Moseley measured the wavelengths of the *K* and *L* lines for a large range of elements; he used an improved design of spectrometer, in which the spectral lines were recorded on a strip of photographic film. Fig. 15.12*b* shows the results obtained for the *K*-series of a selection of neighbouring elements in the periodic table. It is apparent that the elements can be arranged in a definite sequence on the basis of the wavelengths of their *K* lines. (The same sequence is obtained using the wavelengths of other lines in the X-ray spectra.) Now this order coincides exactly with that obtained on the basis of the chemical properties of the elements—expressed in the periodic table. To some extent the same order is given by arranging the atoms in order of ascending atomic weights. But this sequence differs in detail from that given by the X-ray wavelengths or the chemical evidence. Fig. 15.12*b* shows one example of this. On the basis of their X-ray spectra the order should clearly be:

chlorine–argon–potassium

and this agrees with the sequence in the periodic table. But in order of atomic weights the sequence is:

chlorine (35·5)–potassium (39·1)–argon (39·9)

When Moseley numbered the elements according to their positions in the periodic table, he found there was a linear relation between the *atomic number* (as he called it) and the *square root of the frequency* of the K_α line. It was thus established that both the chemical properties and the X-ray spectrum of an element were decided by its atomic number rather than its atomic weight. It remained to be settled what feature of the structure of an atom the atomic number described.

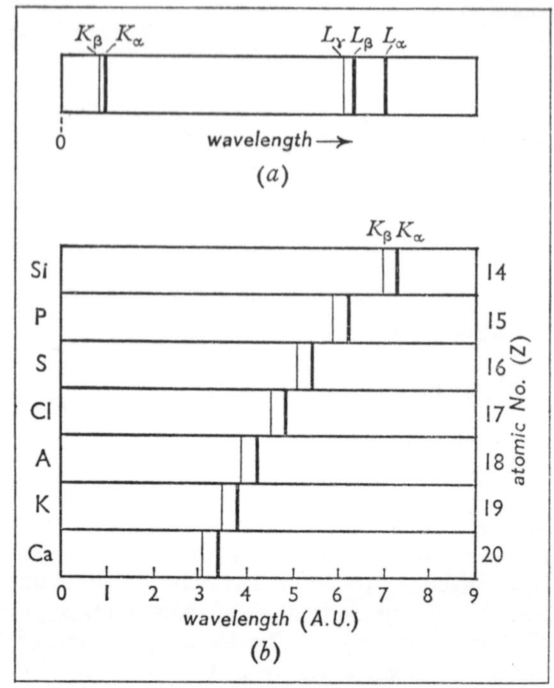

Fig. 15.12 (*a*) The characteristic X-ray spectrum of an element
(*b*) The K-series of a set of neighbouring elements of the periodic table

Bohr's theory of the atom (p. 223) was propounded in the same year in which Moseley did his work on X-ray spectra. In the first instance the theory was put forward to account for the features of optical spectra; but it was also immediately successful in explaining Moseley's results. The theory showed that the *optical* spectrum of an element arises from changes in the outermost electron orbitals of the atom. Thus, if one of the outermost electrons of the atom is raised into an orbital of higher energy than normal, it subsequently returns to its ground state, emitting the surplus energy as a quantum of radiation (light). The frequency ν of the radiation is given by Planck's relation

$$h\nu = E_1 - E_2 \qquad \text{(p. 223)}$$

where E_1 and E_2 are the energies of the initial and final electron orbitals. The optical line spectrum of the element contains the frequencies corresponding to the possible transitions of the outermost electrons of the atom. The energy jumps are in all cases relatively small; and so the frequencies are low and the wavelengths long—in fact, in or near the visible region of the spectrum.

In an X-ray tube we suppose that the energy of the incident electrons is so high that they are able to ionize even the innermost shells of the atoms of the target. When this happens a vacancy is created in one of these shells, which can then be filled by a transition from a shell of higher energy. Because the innermost shells are very tightly bound to the nucleus, the energy differences between them are considerable. The photon emitted in such a transition is therefore of high frequency and short wavelength—in fact, in the X-ray region of the spectrum.

Fig. 15.13 shows the energy level diagram for the inner shells of an atom. When there is a vacancy in the K-shell, transitions are possible from all the other shells. A transition from the L-shell produces the K_α line; that from the M-shell gives the K_β line. The K_β transition involves a larger change of energy, and this line is therefore of shorter wave-

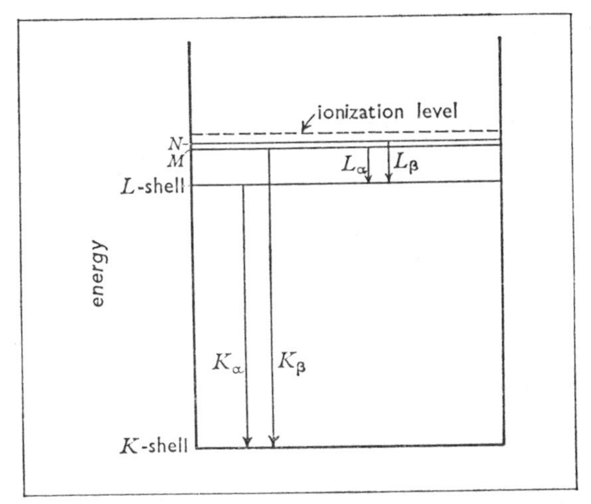

Fig. 15.13 The energy levels and transitions involved in X-ray spectra

length than the K_α line. However, it is a less probable transition; and the K_β line is therefore weaker than the K_α. Likewise the L-series arises from transitions following ionization of the L-shell. These are of lower energy than the K-series transitions, and are thus of much longer wavelength.

It is now clear why all X-ray spectra show the same pattern. The innermost shell structure of all atoms is the same, since in all cases these shells are completely filled in the ground state. When electrons are ejected from the inner shells, the same pattern of transitions is possible whatever the type of atom involved. However, if one atom carries a larger charge on its nucleus than another, its shells are more tightly bound and are therefore of lower energy. The corresponding changes of energy are then of larger amount, giving a spectrum in which all the lines are of shorter wavelength. It therefore seemed that the atomic number, which Moseley showed to be of such significance in fixing the properties of the atom, was to be identified with the *total charge* on the nucleus. It is equal to the number of orbiting electrons in the neutral atom.

16: Nuclear Physics

16.1 Radioactivity

In 1896 the French scientist Becquerel discovered a hitherto unknown type of radiation which was emitted spontaneously by any sample of the element uranium or its compounds; the phenomenon was called *radioactivity*. Becquerel had been investigating an effect in which certain substances emit light weakly for a time after being exposed to a strong beam of light; this is known as *phosphorescence*. It occurred to him that there might be a connection between this effect and the *fluorescence* observed on the walls of a cathode-ray tube, which was known to be accompanied by the emission of X-rays. Accordingly he set out to find whether X-rays were emitted by phosphorescent materials. This he did by wrapping photographic plates in black paper and standing on them a mineral that was phosphorescing through being exposed to sunlight. The results obtained were negative until phosphorescent compounds of uranium were used. The developed photographic plates then showed intense blackening. However, it soon became clear that the effect had nothing to do with the phosphorescence, the investigation of which had prompted the experiment. The blackening of the plates was found to continue even when the uranium compound had been kept strictly in the dark; also it occurred with compounds of uranium that did not exhibit phosphorescence and for metallic uranium as well.

At first it appeared that the new radiation was similar to the X-radiation discovered by Röntgen. Thus, it was very penetrating; also it was strongly absorbed by a sheet of metal. A metal screen placed between the uranium compound and the photographic plate cast a 'shadow' on the plate, just as would occur with X-rays. However, fuller investigation showed that the effect was a complex one, in which several different kinds of radiation were involved; some of these were found to have properties very different from those of X-rays.

Becquerel's discovery led to an intensive search by many scientists for other radioactive materials. Madame Curie and others found that *thorium* and its compounds were also radioactive. She then discovered that the mineral known as *pitchblende*, which contains uranium and thorium, showed about four times as much activity as could be explained by its content of these elements. None of the other elements known to be present in pitchblende had been found to be radioactive. Madame Curie therefore deduced that the mineral must contain one or more hitherto undetected substances; since they were presumably present in very small quantities, it was clear that they must be vastly more radioactive than uranium or thorium. After years of tedious chemical separations she and her husband succeeded in identifying two new radioactive elements, which they called *polonium* and *radium*. The latter is more than a million times as radioactive per unit mass as uranium. Within a few years a long list of radioactive elements had been discovered by various scientists; many of these turned out to be isotopes of already known elements, and nearly all of them were of large atomic mass. In recent years the list has been enormously extended following the discovery of techniques for making artificial isotopes not found in nature; most of these are radioactive. Indeed it is now possible to find radioactive isotopes of almost every element in the periodic table, light elements as well as heavy ones. The same techniques have also given rise to new radioactive *elements* with atomic masses greater than that of uranium, the heaviest element found in nature.

Radium is still one of the most useful radioactive sources for laboratory purposes, since a small quantity of it strongly emits radiations of several different kinds simultaneously. But it is sometimes convenient to have a source that provides a single kind of radiation only. For this purpose certain artificial radioactive isotopes are particularly suitable; those usually encountered in school laboratories are *plutonium* (a 'transuranic' element), radioactive *strontium*, radioactive *cobalt* and radioactive *caesium*.

The chief effect produced by the radiations from radioactive substances is the ionization of the matter through which they pass. This may be demonstrated by placing a radioactive source inside or near a charged electroscope, when the deflection at once starts to fall. The rate of decrease of the deflection is a measure of the ionization current (p. 186), and for a given type of source may be taken as an indication of the strength of the radiation emitted. A more detailed investigation of the ionizing effect of the radiations may be conducted using an ionization chamber connected to a pulse electrometer (Fig. 16.1*a*). For this purpose the outer metal cylinder of the chamber can form one electrode and the cap of the electrometer the other. A p.d. of about 3 kV joined between them ensures that all ions produced in the space will be collected. Any ionizing radiation passing through the chamber then causes the leaf to strike the counter electrode at regular intervals; the number of discharges per minute is taken as a measure of the ionization current.

α-rays and β-rays

The most strongly ionizing component in the radiations emitted by radium has a very short range —only a few centimetres in air, and proportionately less in a denser substance. One way of demonstrating this is to equip the ionization chamber with a gauze cover (Fig. 16.1*a*). When a radium source is held a short distance above the gauze, a steady ionization current is collected; this can only be due to that part of the radiation which has passed through the gauze and caused ionization in the electric field inside. When the source is raised, the current decreases, and finally falls to a very small value when the radium is about 6 cm from the gauze. If the radium is placed once more close to the gauze, it will be found that even a single thick sheet of paper is sufficient to absorb the radiation almost completely.

A more detailed investigation may be made by mounting the radioactive source *inside* the ionization chamber (Fig. 16.1*b*). A higher proportion of the radiation then passes through the space of the chamber, and, with a weak source, larger ionization currents are collected. The absorption of the radiation by matter may be studied by placing thin foils of aluminium on top of the source. A foil 10^{-2} mm thick reduces the current to about a quarter of its initial value. With three or four such foils the current falls to about 1% of its original value. However, it then becomes apparent that the radium is also emitting a second kind of radiation, which is both more penetrating and also produces less ionization. The thickness of absorbing foil over the source may be doubled or trebled without much affecting the residual 1% of ionization current; and

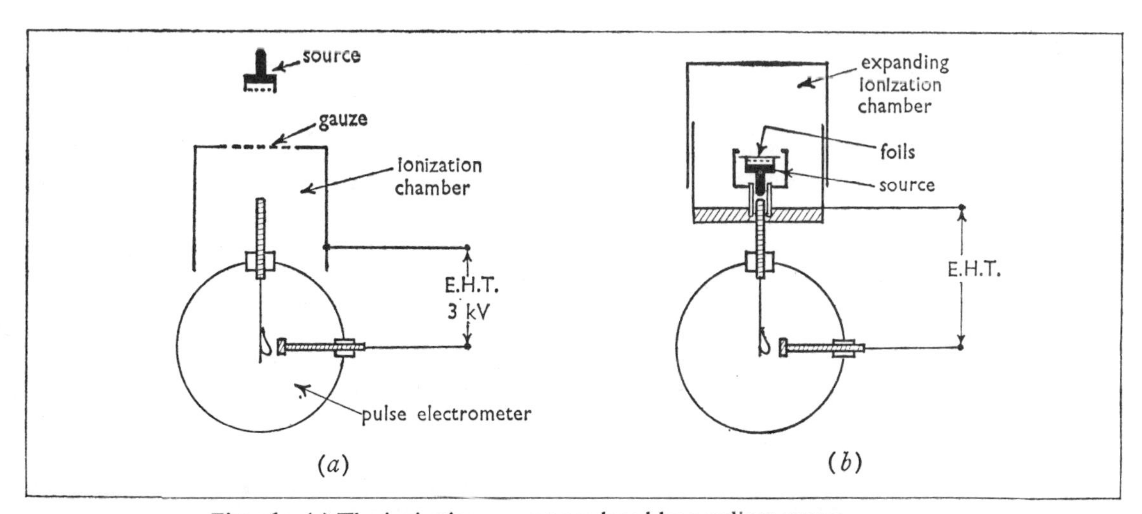

Fig. 16.1 (*a*) The ionization current produced by a radium source
(*b*) Measuring the range and absorption of the radiations from radium

a sheet of aluminium 1 mm thick is barely sufficient to halve it. These two kinds of radiation are called *α-rays* and *β-rays*. The α-rays are stopped by a few centimetres of air or a thin film of solid material; but within this short range they cause heavy ionization. To stop the β-rays requires a thickness of about half a millimetre of lead or about half a centimetre of wood; their range is about 10 times that of the α-rays and the ionization they produce in a given distance is about a tenth as much.

The range of α-rays in air may be found by providing the chamber with an expanding top (Fig. 16.1*b*). When the top of the chamber is close to the source, only a small part of the ionization produced by the α-rays occurs in the air of the chamber; as the chamber is expanded, the ionization current steadily increases, but reaches a maximum when the entire path of the α-radiation is confined to the space of the chamber—i.e. when the top is about 6 cm above the source. Expanding the chamber beyond this point produces no further detectable increase of current. When a thin sheet of metal foil is mounted above the source, the current is reduced, and the range of the α-rays in the air above the foil is also less. By such measurements we can compare the stopping powers of different materials. It is found that the stopping power of a slab of material is approximately proportional to its mass per unit area of cross-section. Thus

$$\text{the density of air} = 1{\cdot}3 \times 10^{-3}\ \text{g cm}^{-3}$$

and

$$\text{the density of aluminium} = 2{\cdot}7\ \text{g cm}^{-3}$$

$$\therefore\ \text{the ratio of densities} = \frac{2{\cdot}7}{1{\cdot}3 \times 10^{-3}} \simeq 2 \times 10^{3}$$

and this is also the ratio of the stopping powers of the two materials for a given thickness. For the most energetic α-rays emitted by radium,

$$\text{range in air} = 6\ \text{cm}$$

$$\therefore\ \text{range in aluminium} = \frac{6}{2 \times 10^{3}} = 3 \times 10^{-3}\ \text{cm}$$

The mass per unit area required to stop these rays

$$= 6 \times 1{\cdot}3 \times 10^{-3}\ \text{g/cm}^2 \simeq 8\ \text{mg/cm}^2$$

A graph of ionization current against effective thickness of stopping foil is of the form shown in Fig. 16.2. (Since the current covers a large range of values, it is convenient to plot this sort of graph using a logarithmic scale of current.) At the point where there is a total thickness equivalent to 8 mg/cm² there is a pronounced discontinuity in the graph. At this point the α-rays have been completely absorbed and the small remaining ionization current is caused by the β-rays only. The absorption of β-rays may be studied in the same way, though we must now use a very much thicker range of foils. The ionization current due to the β-rays is approximately halved for an additional absorbing thickness of about 300 mg/cm²; and the absorption is complete when the thickness approaches 1 g/cm².

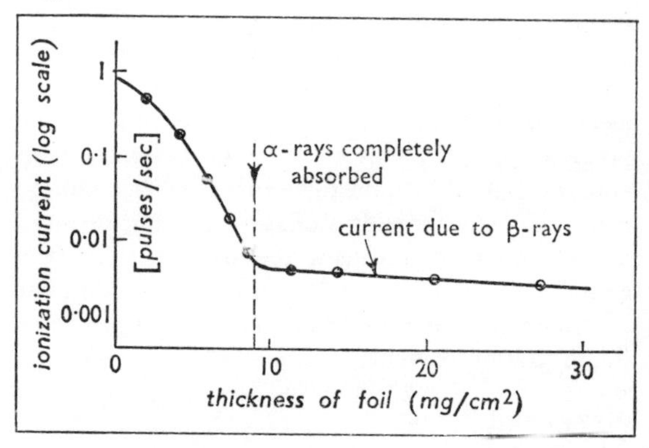

Fig. 16.2 Absorption curve showing the existence of at least two kinds of radiation

γ-rays

Using even thicker absorbing screens than these it is possible to show that a radium source emits yet a third kind of radiation, called *γ-rays*. These are even more penetrating than β-rays, and produce much less ionization—about 1/1000 of that produced by β-rays in the same distance. To obtain satisfactory ionization currents from γ-rays requires radioactive sources at least 100 times as strong as can be permitted for school use. The consideration of ways of demonstrating the properties of γ-rays must therefore be deferred until we have dealt with more sensitive devices for detecting ionization (e.g. the Geiger–Müller tube, p. 287).

Safety precautions

The effects on living organisms of the radiations from radioactive substances are in many ways similar to those of X-rays (p. 272). They can cause burns and destruction of living cells, and are capable of initiating serious diseases and even genetic changes. Also the effects are to some extent cumulative, so that a very small daily dose of radiation may add up in the course of a year or so to a total dose sufficient to cause irreparable harm. Radioactive sources must therefore be treated with the greatest possible care—particularly because, unlike an X-ray tube, they are never 'switched off', and it is all too easy to forget their presence.

The damage caused by the radiations arises in the first instance from the ionization they produce. This in turn leads to many different kinds of chemical changes that can affect the functioning of living cells. *α-rays* produce intense ionization, but their range is so short that they would not penetrate even the surface layers of skin. There is therefore little danger from α-active materials provided they are not absorbed into the system through the stomach or lungs. In particular, radium sources must be carefully sealed to prevent any escape of the α-active gas *radon* that is generated continually by radium. *β-rays* are less strongly ionizing, but their penetration is rather greater. However, a layer of wood or Perspex $\frac{1}{4}$ inch thick gives complete protection against external β-radiation. The main hazard from β-active materials is again the possibility of absorbing them into the system through the stomach or lungs. *γ-rays* constitute the chief danger in handling radioactive materials, owing to their great penetrating power. Even a layer of lead 1 cm thick is only sufficient to absorb about 50% of the γ-radiation incident on it, and so it is difficult and expensive to provide adequate screening. All γ-sources must therefore be as weak as possible, and must be kept at a sufficient distance from people in the laboratory to reduce the radiation dose well below the acceptable limit—allowing for whatever lead screening is in use. It is usual to insist on the following rules being observed in a school laboratory in which experiments of the types described in this chapter are performed:

(*i*) All sources must be weak (less than 10 μC—see p. 307), and must be sealed in foil to prevent the escape of radioactive material.

(*ii*) All sources must be stored in a cupboard in a little-frequented part of the laboratory; they must be kept and transported in a suitable lead container.

(*iii*) Sources must only be handled with forceps, and should always be held well away from the body.

(*iv*) No eating, drinking or smoking must take place in a laboratory where radioactive materials are in use; also the licking of labels or sucking of pipettes should be avoided.

Having said all this, it is only fair to point out that by observing these rules the dangers are indeed reduced to quite negligible proportions. To give further confidence it is always possible to carry a pocket dosimeter (p. 150), which gives a record of the total dose received by the experimenter. These instruments are usually calibrated in *röntgens* (R), a special unit devised for measuring the total dose of X- or γ-radiation by means of the ionization produced. The röntgen was originally defined in terms of the quantity of positive and negative charge produced per unit volume of air at n.t.p. through which the radiation passed. However, it is now redefined in terms of the amount of energy liberated in the ionization process.

The **röntgen** *is the quantity of X- or γ-radiation that liberates by ionization* $8{\cdot}38 \times 10^{-3}$ *joules of energy per kilogram of air at n.t.p.*

Other units used for measuring an absorbed dose are the *rad* and the *rem*. With the types of radiation likely to be encountered in schools these can both be taken as equal to the röntgen. The total dose received by anyone in one year should not normally exceed 50 mR (50 milliröntgens); and the dose rate at any given time should not exceed 0·25 mR per hour. Actual destruction of living tissue requires a total dose of several thousand röntgens.

16.2 Detecting individual particles

The spark counter

The ionization produced in the air by α-rays is sufficiently intense to trigger off a spark discharge in an electric field that would not otherwise be strong enough for a spark to pass. This provides a useful means of detecting α-rays. One form of spark counter consists of a fine wire stretched about 1 mm above the surface of a metal plate, and carefully insulated from it (Fig. 16.3). The p.d. between wire and plate is adjusted to a value slightly less than that required to produce sparking. At this p.d. there is usually a weak corona discharge from the wire, which may be slightly luminous in the dark. When a radium source is held close to the

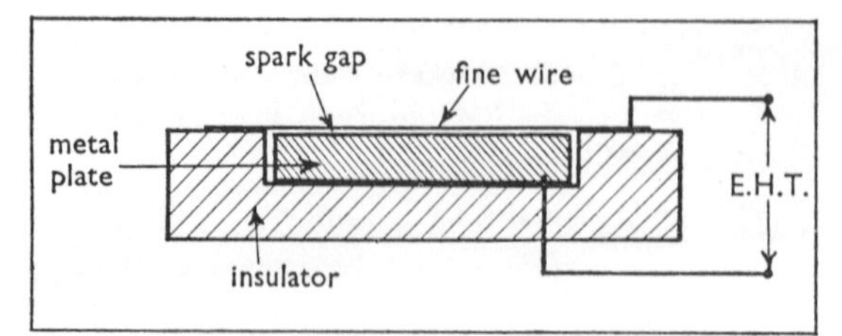

Fig. 16.3 A simple spark counter

wire, sparks are seen to pass between the two electrodes. The sparks occur at irregular intervals and appear to be distributed at random along the wire. They are clearly due to the α-rays only, since no sparks occur when the source is moved to more than 6 cm from the wire or if a sheet of paper is placed to intercept the α-rays. The β-rays evidently do not produce sufficient ionization to initiate a spark. By watching the spark counter we get the impression that the α-rays do not form a steady stream of radiation but rather a random 'rain' of particles. This impression, as we shall see, is confirmed by other means available for detecting α-rays.

We thus have the means of *counting* the individual α-particles (as we shall call them) that arrive in the vicinity of the wire. It is possible to count the particles visually by watching for the sparks. But this is rather tedious and not very reliable. The pulse of current that passes in a spark is large enough to be detected by a sensitive electrometer —the pulse electrometer (with the counter electrode screwed back) can well be used for this purpose. Each time a spark passes a clearly visible flicker of the sensitive leaf occurs. Alternatively, the momentary changes in the p.d. across the spark counter may be fed into an electronic *scaler* (p. 289), which provides an exact record of the number of pulses occurring in a given interval of time.

By testing a radioactive source first with a spark counter and then with an ionization chamber and pulse electrometer we can analyse the nature of the emissions it gives. A plutonium source produces discharges in the spark counter, and must therefore emit α-particles. However, if it is placed in the ionization chamber and covered with sufficient foil to stop the α-particles altogether, no ionization current can be detected; we therefore assume that in this case there is no emission of β-rays. Using the same technique we can show that a radioactive strontium source emits β-rays but not α-rays. A cobalt-60 source appears to emit neither α-rays nor β-rays; we shall see later that it produces γ-rays. (Actually the cobalt also emits β-rays, but these are of such low energy that they are absorbed completely by the piece of foil enclosing the source.)

Scintillation counters

The radiations from radioactive substances can

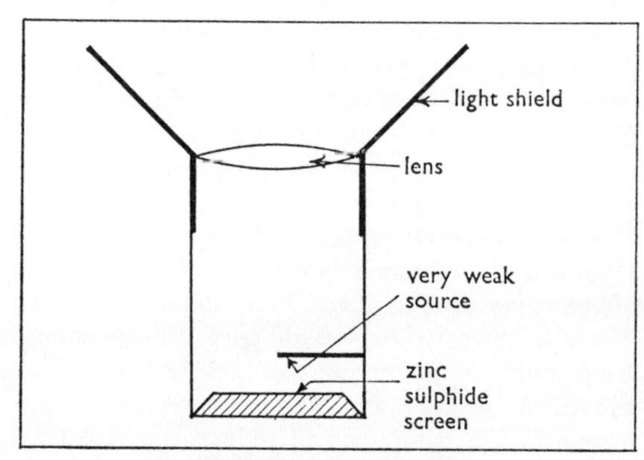

Fig. 16.4 A spinthariscope

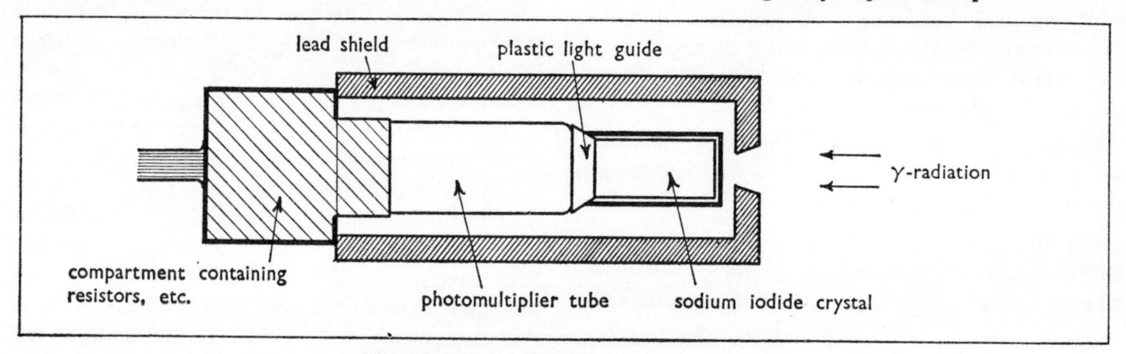

Fig. 16.5 A scintillation counter for γ-rays

also be detected by the fluorescence they produce in certain types of materials known as *phosphors*. This effect may be observed with an instrument called a *spinthariscope* (Fig. 16.4). It consists of a screen covered with a suitable phosphor, such as zinc sulphide, with a speck of radium mounted on a needle a short distance from it. The screen is viewed through a magnifying glass. When the eye has got adapted to the darkness inside the instrument, minute flashes of light, called *scintillations*, are seen coming from points on the screen. It is difficult again to avoid the conclusion that we are observing the impact of individual particles ejected from the source in an irregular and random way. The scintillations cease if the source is covered by a thin absorbing sheet, and it is therefore clear that they are due to the impact of α-particles.

It is now known that any ionizing radiation can produce scintillations in a suitable phosphor, though with β-rays and γ-rays the flashes of light are too weak to be detected by the human eye. They are, however, within the range of a photomultiplier tube (p. 219). The combination of phosphor and photomultiplier tube is called a *scintillation counter*; it has become one of the most useful instruments for detecting ionizing radiations. It has the special merit that the strength of the flash of light depends on the type and energy of the particle that caused it. Employed with suitable electronic circuitry a scintillation counter can be used to discriminate between different sorts of ionizing particle. To detect α-particles a zinc sulphide phosphor is used. This is deposited on a Perspex plate mounted on the end of the photomultiplier tube; the phosphor is covered with a thin aluminium foil, which allows α-rays to pass through, but serves to exclude light from the system. For β-particles a crystal of anthracene is commonly employed as a phosphor. Fig. 16.5 shows a typical design of scintillation counter for detecting γ-rays. The phosphor in this case is a crystal of sodium iodide containing a trace of thallium. The tube and phosphor are enclosed in a light-tight case (which must also be air-tight, since the crystal would deliquesce in contact with moist air). The crystal is a dense material, and a high proportion of the incident γ-rays are absorbed and detected by this instrument.

The scintillations caused by α-particles in zinc sulphide are used to provide luminous dials for clocks and watches. The numbers and other markings on the dial are made of a mixture of the phosphor with a suitable radioactive substance. The luminosity is the combined effect of thousands of scintillations. This is clearly seen if part of the dial is viewed in the dark through a low-power microscope.

The cloud chamber

In this device the actual tracks of the α-rays are made visible, thus providing an even more vivid demonstration of their particulate nature. A common form of cloud chamber is shown in Fig. 16.6. The walls and top of the chamber are made of a transparent plastic. Round the top of the walls is fixed a strip of felt which is kept soaked with a suitable volatile liquid, such as methylated spirits. The air in the chamber is therefore saturated with the alcohol vapour. The floor of the chamber

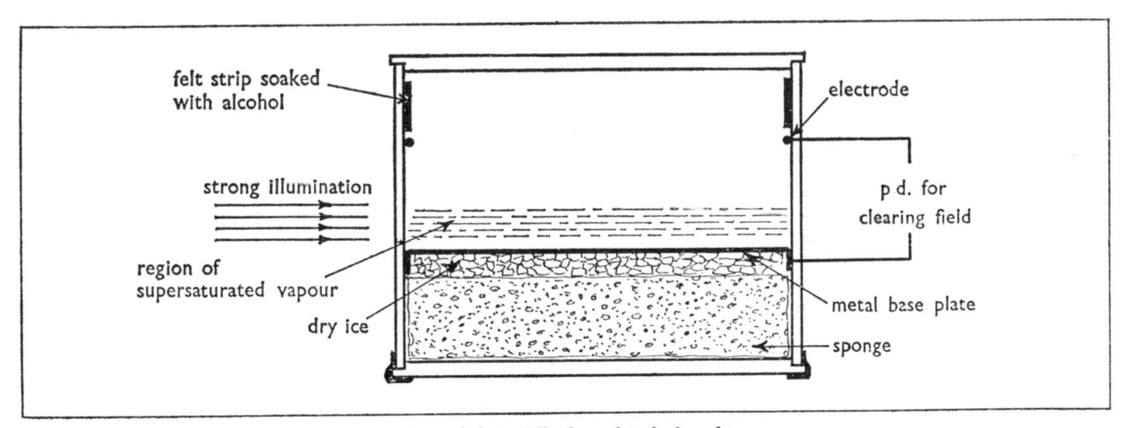

Fig. 16.6 A diffusion cloud chamber

consists of a metal plate, which is cooled by a layer of 'dry ice' (solid carbon dioxide) packed beneath it. The alcohol condenses on the plate, and there is thus a steady diffusion of vapour from the top to the bottom of the chamber. A short distance above the metal plate the vapour passes through a narrow region in which it becomes supersaturated. In this condition it will condense on any ions that happen to be present. A string of droplets therefore forms along any track in which ionization occurs, and in this way the paths of any ionizing particles are made visible. To see the tracks clearly it is necessary to illuminate brightly the active region in the chamber, and the base-plate needs to be painted black to cut out background reflections. Also the ions must be swept out of the chamber by an electric field so that the space is continually being cleared for the detection of fresh ionizing particles. The clearing field may be provided by joining a suitable p.d. between the base-plate and another electrode mounted near the top of the chamber. Alternatively, the plastic top may be rubbed with a duster, when the static charge often produces a sufficient clearing field.

In another design of cloud chamber the supersaturation is produced by a sudden reduction of pressure. To effect this the chamber is connected to a cylinder and piston (a bicycle pump with the valve reversed is often used). When the piston is pulled back, the sudden expansion causes a fall in temperature, and the vapour becomes supersaturated. In this design of cloud chamber the tracks are only visible for an instant just after the expansion.

When a suitable radioactive source is mounted in the chamber, the tracks of α-particles are strikingly shown up. Their range in air and the manner in which they are stopped by thin foils shows that the tracks are indeed those of α-particles such as we have already detected by other means. The ionization produced by an α-particle is always very heavy; detailed measurements show that each particle produces about 30,000 ion pairs per cm of its path in air at n.t.p. It is therefore not surprising that the energy of the particle is rapidly dissipated, bringing it to rest in the short distance that we observe. The paths of the particles are seen to be almost exactly straight, though small deflections of about 1° are fairly frequent (Fig. 16.7*a*). The patient observer, who watches a chamber for several minutes, may be rewarded by seeing an α-particle deflected through a much larger angle (90° or more). The study of the large-angle deflections of α-particles provided Rutherford with the evidence he needed to propound the nuclear theory of the atom (p. 298; see also Plate A.1).

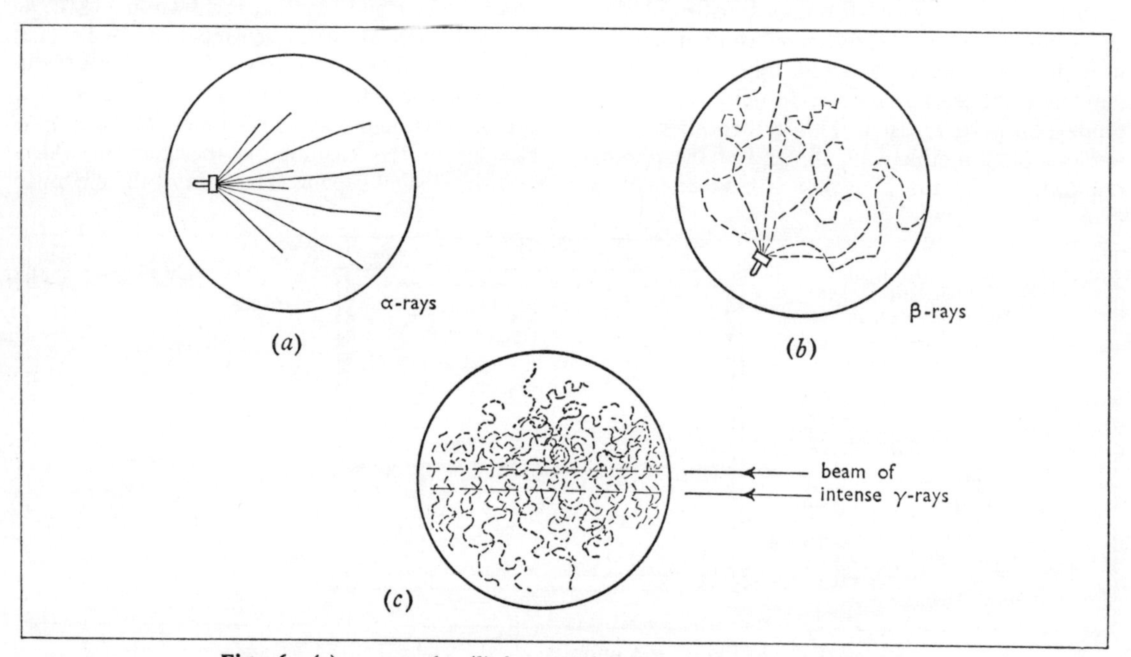

Fig. 16.7 (*a*) α-ray tracks: (*b*) β-ray tracks; (*c*) tracks produced by γ-rays

It is also possible to use a cloud chamber to observe β-rays. Again we find a series of clearly marked tracks, and it seems that β-rays also must be regarded as particles. The ionization along the track of a β-particle is much less heavy than for an α-particle, and the path is only thinly marked out by a line of droplets; it vanishes within a fraction of a second of being formed. But with a powerful magnifying glass focused on the right region of the chamber an alert observer will just manage to see the occasional track. Satisfactory observation is really only possible by photographic means. The appearance of β-particle tracks is shown in Fig. 16.7*b*. The paths of the particles are very far from straight; they seem to suffer frequent small deflections and occasional large deflections of 90° or more. The deflections become more and more frequent as the speed of the particle falls; and the end of the track is usually very tortuous.

An *intense* beam of γ-rays shows up in a cloud chamber as in Fig. 16.7*c*. In this case there is no clear line of droplets marking the path of the beam; but a number of short tracks are observed, resembling those of β-particles, each track originating in the line of the beam. This is quite different from the tracks of α- or β-particles, which only rarely show 'whiskers' of the kind that characterize the paths of γ-rays. With the strengths of γ-ray sources available in schools it is very difficult to observe these tracks (see Plate A.2).

Detection by photographic films

The tracks of α, β and γ-rays may also be studied in special photographic films. The emulsion reduces the range of α-particles to about 0·02 mm, and even β-particle tracks are only about 1 mm in length. A microscope must therefore be used to observe them. With ordinary photographic plates this method is not satisfactory, since the grains of silver bromide are not sufficiently close together to show up the tracks clearly. But special 'nuclear plates' are obtainable, in which the emulsion is thicker and contains a higher density of silver bromide grains. These have been used extensively in the investigation of radiations of various kinds. The method has the advantage of automatically making a permanent record of the 'events' studied. Also the tracks are so small that an enormous number of events can be recorded on one plate—sufficient sometimes to keep an experimenter busy at the eyepiece of a microscope for months on end!

16.3 The Geiger–Müller counter

This instrument is probably the most versatile and useful of the devices available for detecting radiations from radioactive substances. Like the ionization chamber and the spark counter it is actuated by the ionization of the gas it contains; but it is far more sensitive for this purpose than either of the other devices. It is essentially a form of discharge tube, containing gas at a pressure of about 10 cm of mercury; it is operated at a p.d. somewhat less than that which would produce a continuous discharge in it. A typical design is shown in Fig. 16.8*a*. The anode consists of a fine wire which runs along the axis of the cylindrical cathode. A large electric field is therefore produced in the immediate vicinity of the anode; and in this region any free electrons are sufficiently accelerated to cause further ionization. The process is cumulative, and a small amount of initial ionization can give rise to a considerable 'avalanche' of electrons. The electrons, being very light, are collected almost at once by the anode, leaving behind a space-charge formed by the more massive and slow-moving positive ions. In a short time ($\sim 10^{-6}$ s) the space-charge becomes sufficiently dense to cancel the electric field round the anode; the ionization process then ceases, and the positive ions are drawn away by the field to the cathode. Thus any ionization of the gas in the tube triggers off an appreciable pulse of current. When the right p.d. is used, the charge collected by the electrodes is independent of the amount of the original ionization—a single ion pair may be sufficient to initiate a full-scale pulse.

The G–M tube (as it is usually called) is connected in the circuit shown in Fig. 16.8*b*. When an ionizing particle enters the tube, the resulting pulse of current causes a corresponding pulse of p.d. across the resistance R in series with it. This is amplified and registered by a suitable detecting device.

It is obviously important that only one pulse should be registered for each ionizing particle entering the tube. Unfortunately, the positive ions formed in the gas may well reach the cathode with sufficient energy to eject secondary electrons from it; these then cause further avalanches of electrons, and a whole train of pulses may be triggered off by only one original particle. One way of preventing this is to make the resistance R in series with the

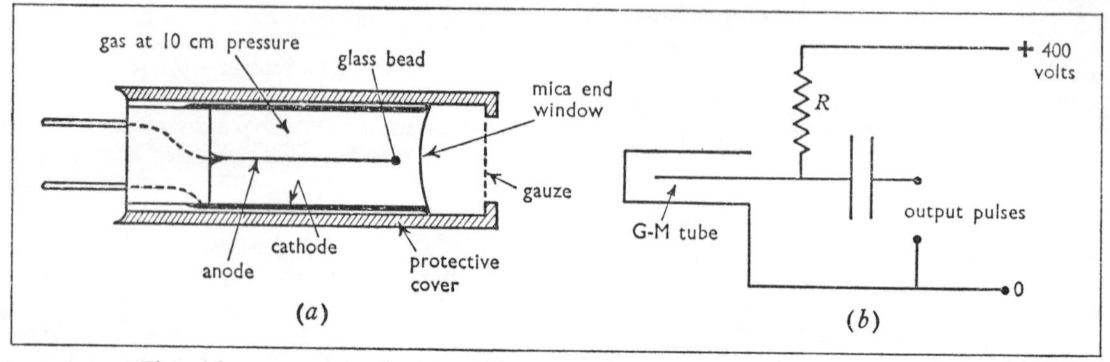

Fig. 16.8 (*a*) An end-window Geiger–Müller tube; (*b*) the circuit in which the tube is used

tube very large (10^9 ohms, say); the time taken for the p.d. across the tube to return to its original value after a pulse of current is then relatively long (10^{-2} s)—long enough for all the ions to be collected by the cathode while the p.d. is still at a low value. However, this technique is not very satisfactory, as the tube remains insensitive for such a long time after each pulse; it is therefore only suitable for low rates of counting. A better method is to include in the filling of the tube a small quantity of a suitable vapour that will absorb the energy of the positive ions before they can cause secondary electron emission; such substances are known as *quenching agents*. Various organic compounds have been used—e.g. ethyl alcohol. These are slowly destroyed by the action of the tube, and the quenching action becomes ineffective after some 10^8 counts. Such counters require operating p.d.'s of about 1500 volts. Another method is to include a small quantity of a halogen vapour as quenching agent—e.g. chlorine or bromine at a pressure of about 1 mm of mercury. A counter of this kind works on a p.d. of about 400 volts, and there seems to be no definite limit to the number of counts it can make. The interval during which these tubes are insensitive to the arrival of further particles is about 0·1 milliseconds, a quantity known as the *dead time* of the counter. However, the time taken for the tube to recover completely from a discharge is usually about 0·3 milliseconds. This is known as the *recovery time*; during this interval any pulses delivered by the tube are likely to be of reduced amplitude and may not be registered by the recording equipment. This means that there is always a proportion of the particles passing through the tube that are not counted. At low rates of counting this effect is insignificant; but when the counting rate is more than a few thousand per minute a correction for the recovery time should be made to the readings.

It is possible to design a Geiger–Müller tube to detect α-particles, but this is not often done, because of the extreme delicacy of the window that must be provided for the particles to enter. An effective thickness of about 2 mg/cm^2 is all that can be allowed; and even supported by a metal grill it is very easily damaged. In any case we already have satisfactory means of detecting α-particles—ionization chamber, cloud chamber or spark counter. To detect β-rays a rather thicker window can be used—20 to 30 mg/cm^2 is a common figure. It is usually made of mica or glass, and is still obviously thin enough to be treated with great care.

A modified design of G–M tube is often used when it is required to detect γ-rays only. The thin window can be dispensed with, since γ-rays penetrate readily through any part of the tube. Indeed the cathode and the walls of the tube are often deliberately made thicker than necessary in order to provide the maximum chance of a γ-ray ejecting an electron to initiate the discharge. Even so the γ-rays are so weakly ionizing that they have only a small chance of producing any ionization at all inside the tube. In most cases not much more than 1% of the γ-rays passing through the tube are actually counted. A scintillation counter has a very much better performance in this respect.

Recording pulses from a G–M tube

(*i*) *With a pulse electrometer.* In order to avoid insulation difficulties it is usually best to mount the

G–M tube directly on top of the electrometer. The instrument is operated with the counter electrode screwed back; and a flutter of the leaf is then observed each time a discharge passes in the G–M tube. The method is obviously only suitable for very low counting rates.

(*ii*) *With headphones.* The G–M tube may be connected to an amplifier and a pair of headphones or loudspeaker. Each pulse is then detected as an audible click. At very low counting rates it is possible to note the individual clicks; but at higher speeds they merge into a continuous roar, and the method is then only of use for qualitative observations—we can at least detect the presence or absence of radiations by this means.

(*iii*) *With a scaler.* The exact number of pulses occurring in a given time interval may be recorded by using an electronic counting device, known as a *scaler*. Many different forms of circuit have been used for this purpose; but one of the most commonly encountered is that employing a special gas-filled discharge tube, known as a *decatron* (Fig. 16.9). The anode of the tube is the circular metal plate in the centre of its end face. This is surrounded by 10 separate cathodes, each consisting of a single metal wire. The current through the tube is limited to such a value that the discharge can cover only one of the cathodes at any one time (p. 227). The glow in the tube shows clearly which cathode is in use. The counting process consists in causing the discharge to transfer from one cathode to the next for each pulse delivered to the apparatus. This is effected by applying the pulse to sets of intermediate cathodes, called *guide electrodes*. The minimum transfer time is about 0·1 milliseconds, and this limits the speed of operation; but the circuit works reliably at rather more than a thousand counts per second. The 'zero' cathode is brought out to a separate terminal. When the discharge reaches this, a pulse of p.d. is produced, which can be used to operate a second decatron unit. One pulse is thus passed on to the second decatron for every 10 pulses delivered to the first. At the output of the second decatron the pulses are usually coming sufficiently slowly to operate an electro-mechanical register (with a scale resembling the milometer of a car); and this is more economical than using further sequences of decatrons. The first two digits of the total count are then read off from the positions of the glows in the decatrons, and the remaining digits (hundreds, thousands, etc.) from the mechanical register (Fig. 16.9).

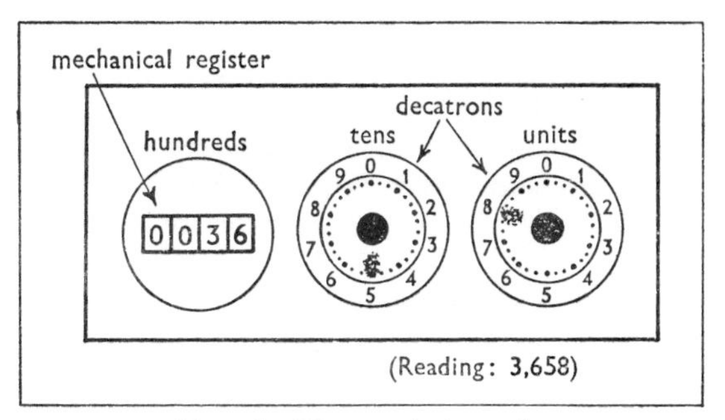

Fig. 16.9 The panel of a scaler

(*iv*) *With a ratemeter.* Sometimes it is only necessary to know the *average rate* at which pulses are delivered by the G–M tube; and a ratemeter is designed to give a direct reading of this figure. The pulses are first amplified, and an electronic device is employed to ensure that the pulses at the output of the amplifier are all of the same amplitude; these are then fed to a capacitor–resistor combination known as an *integrating circuit* (Fig. 16.10). Each pulse delivers a fixed quantity of charge to the capacitor C, and this then leaks away slowly through the resistor R. The time constant of C and R is arranged to be large compared with the probable interval between pulses. The p.d. across the capacitor then builds up to a steady value which is proportional to the number of pulses arriving per unit time. Thus, if there are n pulses per second, each delivering a quantity of charge q to the capacitor, then the average current I registered by the meter in series with R is given by

$$I = nq$$

It is usual to calibrate the meter to read the average number of counts per minute directly. Because of

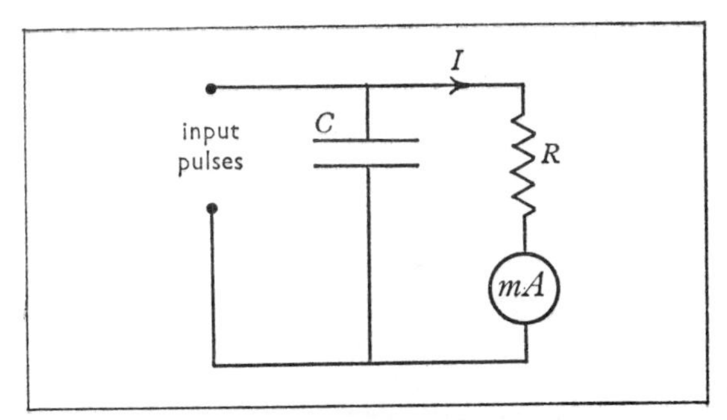

Fig. 16.10 The principle of a ratemeter

the random nature of the counting process there is always a certain amount of 'wandering' of the meter, due to statistical fluctuations of the count rate. This is more pronounced when the time constant (*CR*) is low (1 second, say). Steadier readings are obtained by increasing the time constant; but it is then more tedious taking the readings, since we have to wait correspondingly longer for the meter to settle down to its equilibrium reading.

16.4 Some experiments with a Geiger counter

The characteristic curve of a G–M tube

It is important to operate a Geiger–Müller tube at the correct p.d. This is found experimentally as follows. The tube is mounted at a fixed distance from a suitable radioactive source; we can then assume that a constant number of particles enters the tube per minute (subject to statistical fluctuations). The count rate is then recorded for different values of the applied p.d. Up to a certain value of the p.d. (known as the *threshold p.d.*) no counts are recorded at all, since the amount of gas amplification in the tube is not enough to give pulses of sufficient magnitude to be detected. Just above the threshold the count rate steadily increases with applied p.d. In this region of p.d.'s (between A and B in Fig. 16.11) the magnitude of the pulse developed in the tube depends on the initial ionization, and this varies from one incident particle to another; only some of the particles give pulses of

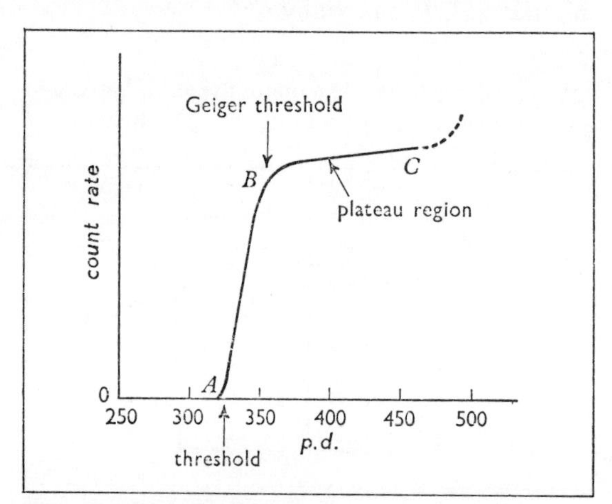

Fig. 16.11 The characteristic curve of a G–M tube

sufficient magnitude to be recorded by the scaler or ratemeter. However, beyond a certain point (the *Geiger threshold*) there is a range of p.d.'s in which the count rate is almost constant; this is known as the *plateau region* (between B and C in Fig. 16.11). The pulses are now all of the same magnitude, and every incident particle that produces any ionization at all is being registered. This is the right range of p.d.'s in which to operate the G–M tube. If the p.d. is raised above the plateau region, the quenching of the pulses is no longer fully effective, and one incident particle may start a whole train of pulses. The count rate therefore increases abruptly. The p.d. should not be raised much above this point, or there is a danger of a continuous glow discharge starting, which could damage the tube permanently. In a halogen quenched tube the plateau region usually extends over about 100 volts, so that any operating p.d. in the middle of this range may be used. Actually there is a slight rise of count rate through the plateau region—usually between 5 and 10%; this is apparently due to effects arising in the non-uniform fields at the ends of the anode wire. It is therefore important to maintain the tube p.d. constant during any particular experiment, although the exact value is not critical.

The background count

Even without any special radioactive source in the neighbourhood a stream of background radiation is recorded by a Geiger–Müller counter. This is due partly to radioactive contamination of the apparatus and its surroundings and partly to cosmic radiation entering the earth's atmosphere from outer space. The background count usually amounts to between 20 and 50 particles per minute, though with adequate lead shielding it can be reduced to about 10 per minute. When we are investigating the emissions from a radioactive source, it is necessary first to record the background count, and then to subtract this from all the subsequent readings. To obtain satisfactory results we must ensure that the total count in a given interval of time is substantially greater than the background count in that time. At high rates of counting (over 5000 per minute, say) the contribution from the background radiation is obviously negligible, and this correction can be ignored.

Statistical fluctuations

Because of the random nature of the emission of

particles by a radioactive source, the count recorded by a Geiger counter for a given time interval fluctuates from one reading to another. The amount of these fluctuations is described in statistical theory by the *standard deviation* of the readings. This is a figure that indicates the deviation from the average count to be expected in any one reading. It is shown in works on statistics that for a random process such as we are considering the standard deviation σ in the count obtained in a given time interval is given by

$$\sigma = \sqrt{N}$$

where N is the average count obtained in that time. This result is readily confirmed by experiment, and serves to establish that the emission of the particles is indeed a random process. Thus, suppose the background count on some occasion averages 50 per minute. The standard deviation for a count of 50 is expected to be $\sqrt{50}$, i.e. about 7. Thus the actual values obtained in a series of 1-minute counts will mostly lie between 43 and 57, though about one-third of the readings will lie outside this range. However, a deviation more than twice the standard value will be quite rare—it should occur for only about 1 in 20 of the readings. If now a radioactive source is held near the tube so that the count rate averages 10,000 per minute, the standard deviation for 1-minute counts should now be $\sqrt{10{,}000}$ or 100; and most of the readings will now be found to lie between 9900 and 10,100. If enough 1-minute counts are made in each case, the correct statistical procedure may be applied to calculate the standard deviation; and the results will be found to agree with the predicted values.

These inevitable statistical fluctuations limit the accuracy that can be obtained in any observation of the count rate. Thus, if the total count recorded is only 100, we must expect a standard deviation of 10; and the accuracy of the measurement cannot exceed 10%. To obtain an accuracy of 1% the count must be continued long enough for the total count to reach at least 10,000. An accuracy of 0·1% can only be achieved by extending the total count to 1,000,000. In general

$$\text{fractional error} \simeq \frac{\text{standard deviation}}{\text{total count}}$$

$$= \frac{\sqrt{N}}{N} = \frac{1}{\sqrt{N}}$$

β- and γ-absorption curves

Using a G–M tube and a suitable recording device we now have the means of making further investigations of the types of radiation emitted by radioactive substances. A radium source mounted near the tube causes a rapid count. This is clearly not due to α-particles, since the count continues with the source more than 6 cm from the tube—outside the range of this sort of radiation. Presumably it is chiefly due to the β-radiation, and the assumption is soon confirmed by studying the absorption characteristics of the radiation. A graph of the observed count rate against the thickness of absorbing screens introduced between source and G–M tube is of the form shown in Fig. 16.12. The thickness needed to produce a given percentage reduction of the reading is the same as for β-radiation in an ionization chamber—i.e. the count rate is reduced by a half for a total thickness of about 300 mg/cm² (p. 282). It is therefore a fair assumption that we are dealing with the same kind of radiation in both cases. The same experiment also reveals the γ-radiation emitted by the radium source. Thus, even when the thickness of absorbing screen is much more than is needed to absorb the β-radiation completely, some form of ionizing radiation is still reaching the G–M tube. It is definitely distinguished from β-radiation by its absorption characteristics. To reduce the intensity of the γ-radiation by a half requires a thickness of absorber of nearly 10 g/cm² (about 1 cm thickness of lead); this is about 30 times as much as is required to produce the same reduction for β-radiation.

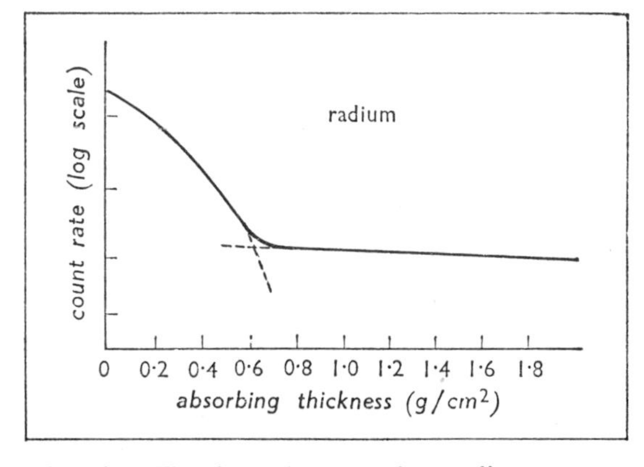

Fig. 16.12 The absorption curve for a radium source

The behaviour of a Geiger–Müller counter provides clear evidence that both β-rays and γ-rays are emitted by a radioactive source in an irregular and random manner—rather than as steady streams of radiation. Just as with α-rays it is difficult to avoid the conclusion that we are again dealing in both cases with *particles* ejected at random by the source.

Fig. 16.13 shows the absorption curves obtained with other types of source. A strontium-90 source evidently emits β-rays only, while only γ-rays emerge from a cobalt-60 source. Sources that emit both β-rays and γ-rays—such as radium—show a discontinuity in their absorption curves, where the radiation detected changes from the mixture of β- and γ-radiation to γ-radiation only.

There is also an important difference in the *character* of the absorption curves for β-rays and γ-rays shown in Fig. 16.13. The β-rays are *completely* absorbed in all cases for a total thickness of about 1 g/cm$_2$. But γ-rays are never completely absorbed no matter how thick the layer of absorbing material may be. There is no maximum 'range' for γ-rays as there is for α- and β-radiation. We find instead that a given thickness of absorbing material always reduces the intensity of the γ-radiation by a certain fixed proportion. Thus the intensity of the cobalt γ-radiation is reduced by one-half by a layer of lead 1·25 cm thick. Twice this thickness reduces it to a quarter of the original intensity; three times to one-eighth, etc. This is another way of saying that the intensity I decreases *exponentially* through the layer of absorbing material, and may be expressed by the equation

$$I = I_0 \, e^{-\mu x}$$

where I_0 = the initial intensity, x = the thickness, and μ = a constant known as the *linear absorption coefficient* for the given γ-radiation. This coefficient is approximately proportional to the density of the material; it is therefore usually convenient to express the 'thickness' x in g per cm^2 of material, and the value of μ so obtained is then called the *mass absorption coefficient.*

Taking logarithms on both sides of the above equation, we have

$$\log_{10} I = \log_{10} I_0 - \mu x \log_{10} e$$

A graph of $\log I$ against x should therefore be a straight line, which indeed is confirmed by experiment (Fig. 16.13*b*). The value of μ can be calculated from the gradient of the line. It is found that μ varies with the type of γ-radiation; for a high-energy γ-ray μ is smaller (i.e. the ray is more penetrating) than for a low-energy γ-ray.

The absorption of β-rays does not strictly follow an exponential law; and a graph of $\log I$ against absorbing thickness is not a straight line. However, the central part of such a curve is usually approximately straight, and we may use this to find an absorption coefficient for β-rays also. The value of μ is not in this case an exactly defined quantity; but this measurement sometimes provides a useful means of comparing the absorption of β-rays in different cases.

Once the absorption characteristics of the β-particles from a given source are known, we may use the results to measure the thickness of an absorbing sheet placed between source and detector. This is the principle of the *β-particle thickness gauge*, which is widely used in factories for monitoring the thickness of thin metal or plastic sheets. The source is placed on one side of the sheet of material emerging

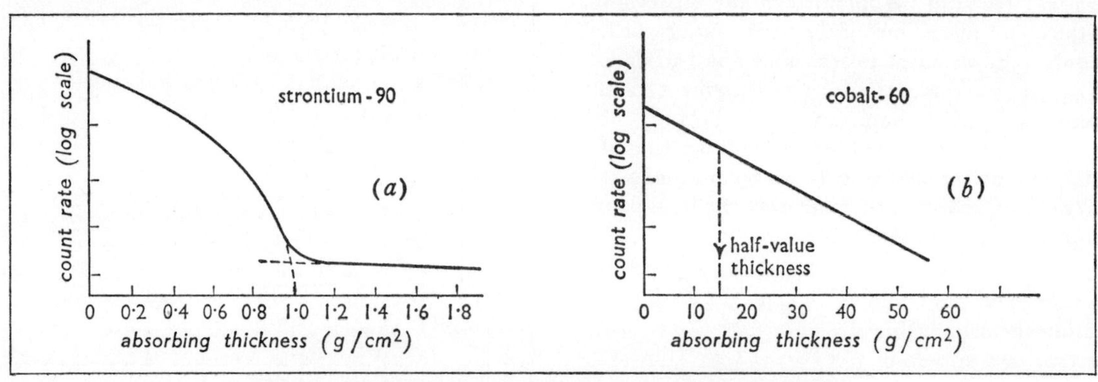

Fig. 16.13 Absorption curves for (*a*) a β-source, (*b*) a γ-source

from the mill and a Geiger counter and ratemeter on the other. The ratemeter may be calibrated to read thickness directly. A continuous record of the output thickness is thus obtained without any part of the apparatus actually coming in contact with the material. Similar instruments have also been designed using γ-radiation; these can be used with metal sheets several inches thick.

The inverse square law

It is found that the intensity of γ-radiation near a small source varies inversely with the square of the distance from the source. This may be tested by measuring the γ count rate I with a G–M tube at varying distances d from the source. (As always, the background count must be subtracted from each reading to give the true γ count rate.) Thus, we expect to find

$$I \propto \frac{1}{d^2}$$

and a graph of I against $1/d^2$ should be a straight line through the origin. However, there is always some uncertainty in the exact value of d; the position of the source in its holder is not readily ascertained, nor can we be sure of the point in the G–M tube at which a γ-ray is detected. But in spite of this, *changes* in thevalue of d can be measured with precision. It is therefore best to re-arrange the equation—

$$\frac{1}{\sqrt{I}} \propto d$$

and plot a graph of $1/\sqrt{I}$ against d. Uncertainty in the value of d then merely has the effect of shifting the points parallel to the d-axis so that the graph no longer passes through the origin—but a straight line should still be obtained.

It is of some interest to consider why the inverse square law holds for γ-radiation. A similar law holds also for some other emission processes, e.g. light, radio waves and sound, provided the distribution of the 'radiation' is not modified by reflection, refraction, etc. A little thought shows why this must be so. Consider a point source O emitting W units of radiation per unit time uniformly in all directions (Fig. 16.14). At a distance r from the source this radiation is spread over the surface of a sphere of this radius (provided there has been no absorption). The intensity I is the rate of incidence of the radiation per unit area. Therefore at the distance r

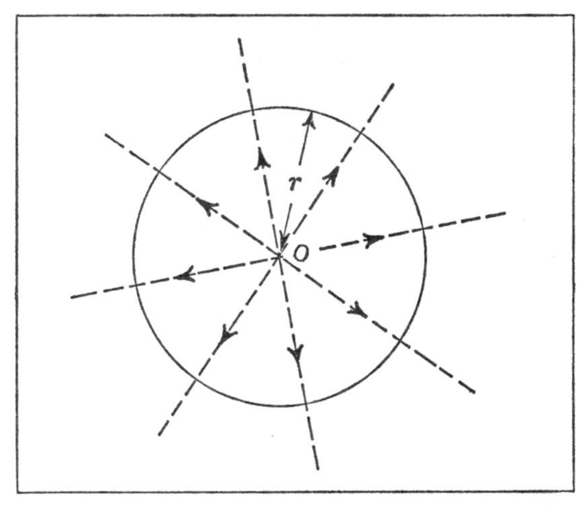

Fig. 16.14 The basis of the inverse square law for the intensity of some forms of radiation

$$I = \frac{W}{4\pi r^2}$$

i.e.

$$I \propto \frac{1}{r^2}$$

which is the inverse square law.

The result depends upon three conditions being satisfied:

(*i*) The source must be small compared with the distances involved.

(*ii*) There must be negligible absorption of the energy in the intervening medium.

(*iii*) There must be rectilinear propagation of the 'radiation'—i.e. the 'rays' must proceed in straight lines. (The law could still hold for radiation that diffused outwards following random paths, but only if we were dealing with a source radiating uniformly in all directions. If the radiation is confined to a narrow beam, it must travel radially, otherwise the energy will not remain in the beam; and the inverse square law breaks down.)

For example, the law holds fairly precisely for a small source of light in air (away from reflecting objects). The result may be tested with a small bulb and photocell. But it breaks down if the light has been refracted in a lens or scattered from a white ceiling, so that the paths of the rays are deviated. Likewise the law breaks down close to a large source of light such as a fluorescent lighting tube; and certainly it does not apply in a strongly absorbing medium, such as a fog!

The verification of the inverse square law for narrow beams of γ-radiation therefore depends on the γ-rays being very weakly absorbed, which we know to be the case; but it also establishes the rectilinear propagation of γ-rays. The law also holds for α-rays at points close enough to the source, since they travel in nearly straight lines and are not absorbed at all before the end of their range. But for β-rays in air it does not hold at all. This is partly because the tracks of these particles deviate considerably from straight lines; and partly because there is a large spread of energies represented in a β-source, and the slower particles are completely absorbed close to the source.

In a vacuum the inverse square law holds for β-radiation also. Both the absorption of β-particles and the tortuous tracks they normally follow arise from their interactions with matter. In a vacuum the tracks are straight and there is no absorption; the β-rays do not just come to a certain point and peter out. The deflections they normally suffer, and their eventual absorption are due to collisions with the molecules of the air.

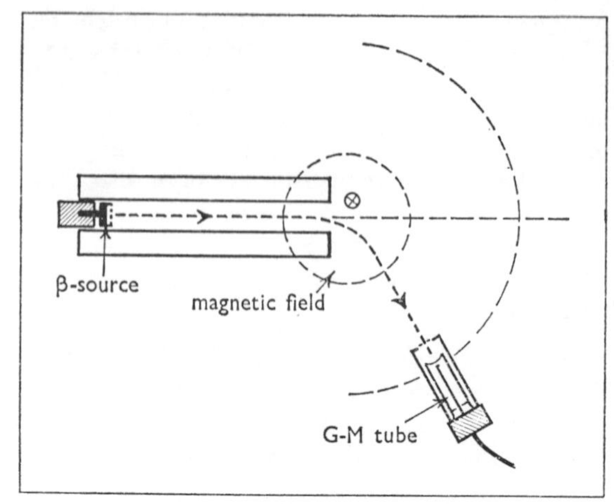

Fig. 16.15 The magnetic deflection of β-rays

16.5 Identifying the particles

We have seen how the three kinds of radiation from radioactive substances may be distinguished from one another by the different ways in which they are absorbed in passing through matter. All three kinds give the impression of being particles. In order to establish what these particles consist of we need to investigate more of their properties; and in particular we must find out how they are deflected in magnetic and electric fields.

β-rays

A preliminary experiment shows that β-rays are readily deflected by a magnetic field. One way of doing this is with the arrangement in Fig. 16.15. A β-emitting source (Sr–90) is mounted at the end of a tube whose walls are thick enough to stop the β-radiation completely. A Geiger–Müller tube is supported in a clamp so that it can be moved to various positions along an arc, as shown. The β-radiation is thus virtually confined to a narrow beam near the axis of the tube—which may be confirmed by observing the count rate for various positions of the G–M tube. Outside the beam the count rate falls to a low value about equal to the background count. However, when a magnetic field is applied at the mouth of the tube (at right-angles to the plane of the diagram), the beam is deflected to one side; the count rate falls to a low value in the central position and the β-particles may now be detected by moving the tube to the right position on one side or the other. The direction of the deflection suggests that the β-rays are *negatively* charged particles. Thus in Fig. 16.15, if the magnetic field is *down* into the paper, the beam is deflected towards the *lower* edge of the diagram. Fleming's left-hand rule (p. 66) shows that the beam must constitute an electric current directed (in the conventional sense) *towards* the source; the particles emerging from the source must therefore be negatively charged. Quite moderate magnetic fields are found to produce large deflections. A flux density of 10^{-2} tesla is quite sufficient for this demonstration (see also Plate B.1).

The β-particles may also be deflected by an electric field, though this is harder to demonstrate, since rather large fields must be used. Again the deflection is consistent with the particles being negatively charged. Thus, when the beam is passed between two metal plates maintained at a large p.d., the deflection is towards the *positive* plate.

It is also possible simply to collect the β-particles on a suitable conductor and show that this has acquired a negative charge. But to do this the conductor must be completely embedded in an insulating material so that the air ionized by the β-radiation does not come in contact with it. With strong radioactive sources the charge collected can be identified with a suitable electrometer.

In these experiments there is not much point in making precise measurements of the deflections, since the β-particles are being slowed down and scattered by collisions with air molecules along their paths; and little useful information is gained. But using vacuum apparatus such measurements enable the velocities and specific charges of the particles to be found. We shall not describe the methods here, since they are not readily repeated in a school laboratory; and in any case the principles involved are the same as those used in studying cathode rays (p. 216) and mass spectra (p. 233). The results show that:

(*i*) The β-particles from a given source have all possible velocities from zero up to a definite maximum.

(*ii*) The specific charge of the β-particles is found to *decrease* slightly with increasing velocity. This variation is entirely consistent with the assumption that the charge of the particles remains constant, while their masses *increase* with velocity in the manner predicted by the theory of Relativity (p. 232). At low velocities the specific charge is identical with that of the electrons produced in the thermionic or photoelectric effects. There is therefore little doubt that β-particles are *electrons* emitted with speeds up to two-thirds the velocity of light. Their energies may be anything up to several million electron-volts.

α-*rays*

When radioactivity was first discovered, it was concluded that the least penetrating component of the radiation—the α-rays—could not be deviated by electric or magnetic fields. Fields sufficient to produce large deflections of cathode-rays and β-rays could give no detectable deflection of the α-rays. However, Rutherford eventually succeeded in devising very sensitive techniques for observing the minute deflections that do actually occur, and thus managed to show that α-particles are relatively massive and carry a *positive* electric charge. One piece of apparatus he used is shown in Fig. 16.16. A layer of radium was placed at the bottom of a cavity in the base of a gold-leaf electroscope. Above the radium was fixed a stack of parallel metal plates, so that the α-rays could enter the electroscope only through the narrow gaps (about 1 mm wide) between them. The rate of fall of the leaf was observed, and could be taken as a measure of the ionization current produced by the α-rays. A strong magnetic field was then applied at right-angles to the plane of the diagram, and it was found that the ionization current decreased; the greater the magnetic field, the more slowly the gold leaf collapsed. Presumably, many of the particles that had previously passed through the gaps were now being deflected to strike the plates. To determine the direction of the deflections the tops of the gaps were each masked by a small projection on one side, as shown. It was then found that a much greater reduction in ionization current occurred with the magnetic field directed *down* into the diagram (in Fig. 16.16) than in the reverse direction. By Fleming's left-hand rule (p. 66) the stream of particles emerging from the source must be *positively* charged (see also Plate B.2).

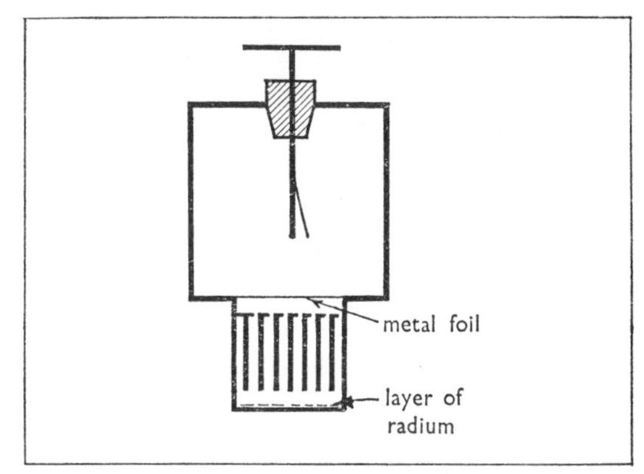

Fig. 16.16 The magnetic and electric deflection of α-rays

The same apparatus was modified to detect the deflection of the particles by electric fields. Alternate plates were connected together like an air capacitor, and a p.d. was applied between the two sets of plates. The electric field so produced would certainly have interfered with the functioning of the electroscope. To prevent this the floor of the case above the plates was covered with a fine metal foil, which acted as an electrostatic screen; but this was thin enough to allow the α-particles to pass through. Again a reduction of the ionization current was observed when the electric field was applied.

Knowing the magnitudes of the fields used in these experiments and the spacing of the plates, Rutherford was able to make preliminary calculations of the specific charge and velocity of the α-particles. These enabled him to design apparatus to work in a vacuum for measuring more precisely the deflections of the particles. In principle the methods were similar to those employed for finding the specific charge of cathode rays (p. 216); but because of the small deflections of the α-particles the technical difficulties were considerable (see Q. 24 on p. 385). We shall not consider these methods further here. The experiments yielded two important results:

(*i*) With some radioactive sources *all* the α-particles were found to be emitted with the same velocity; and even when this was not the case, the α-particles always fell into one of a few groups, each having a sharply defined velocity. This is sometimes described by saying that the α-particle energies form a *line spectrum.*

(*ii*) The specific charge of the α-particles from all sources was found to be about 5×10^7 coulomb kg^{-1}. This is just half the value of the specific charge of the hydrogen ion obtained from experiments in electrolysis (p. 37). Rutherford therefore concluded that the α-particle was either a singly charged atom of atomic mass 2 or else a doubly charged atom of helium (of atomic mass 4).

To decide between these possibilities it was necessary to find some way of measuring the actual charge carried by an α-particle. This was done by Rutherford and Geiger using the two pieces of apparatus shown in Fig. 16.17. The tube T in (*a*) was an early form of Geiger counter, the prototype of the modern G–M tube. The α-particles from the source S entered the counter through a small aperture A covered with a thin film of mica. The count rate with their equipment had to be limited to a few pulses per minute, but knowing the area of the aperture A and its distance from the source the measurements enabled the total number of α-particles emitted into the space above the source to be calculated. This figure is half the total number of particles emitted, since the same number presumably pass downwards into the supporting plate. The same source was then mounted opposite the collecting plate C in an evacuated enclosure (Fig. 16.17*b*). A magnetic field was applied across the apparatus sufficient to deflect any β-particles back into the source. The only charge collected by the plate C was therefore that carried by the α-particles, and this was measured by a sensitive electrometer. The charge q carried by one α-particle was then given by

$$q = \frac{\text{charge collected per second}}{\text{No. of particles emitted per second}}$$

The value obtained was $3{\cdot}1 \times 10^{-19}$ coulomb, which is almost double the charge on a hydrogen ion. Rutherford therefore decided that the α-par-

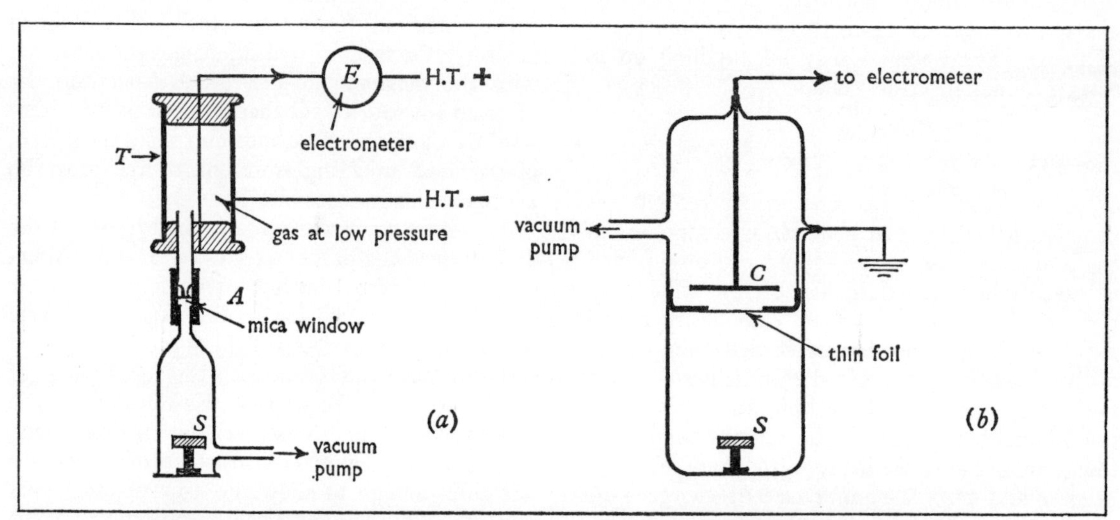

Fig. 16.17 Measuring the charge on the α-particle: (*a*) Counting the particles emitted
(*b*) Measuring the total charge carried

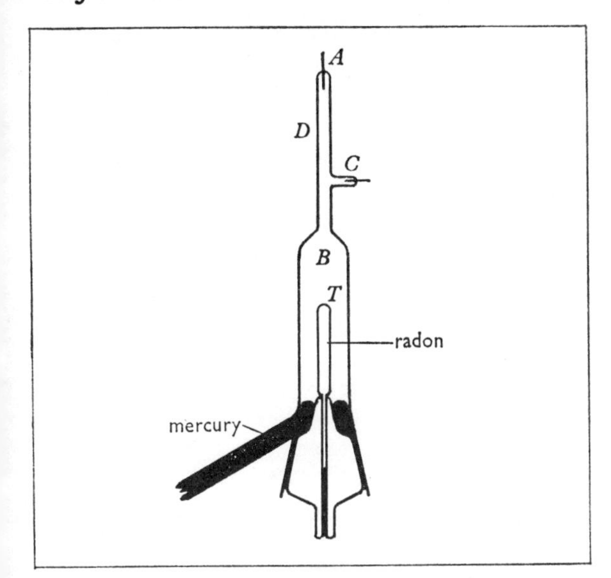

Fig. 16.18 Rutherford and Royd's experiment to show that α-particles are helium nuclei

ticle must be a doubly charged helium atom. The matter was settled by the celebrated experiment of Rutherford and Royd depicted in Fig. 16.18. A small quantity of the radioactive gas *radon* was placed in the central tube T; this had sufficiently thin walls to allow the α-particles emitted by the radon to pass through into the evacuated outer compartment B. Any gas which collected in this space could be compressed up into the discharge tube D by raising the level of the mercury. After six days enough gas had collected in B to enable a discharge to be passed between the electrodes A and C. The light from this was examined with a spectrometer, and the usual spectrum of helium was revealed. There remained the possibility that the helium was present already mixed with the radon in the tube T and had diffused through into the outer compartment. However, when the tube T was filled with helium instead of radon, no helium could be detected in the discharge tube even when several days had been allowed for possible diffusion to occur.

γ-rays

Early in the investigation of radioactivity it was suggested that γ-rays were identical with short wavelength X-rays; and this supposition has been fully confirmed by later work. The difference between γ-rays and X-rays seems to be one of origin rather than of nature; γ-rays arise in the nucleus, whereas X-rays are generated in the inner shells of the electron structure outside the nucleus. But the properties of the two forms of radiation appear in every respect to be identical:

(*i*) Neither form of radiation shows any deflection by electric or magnetic fields; nor is there any other evidence suggesting that they carry electric charge.

(*ii*) Both forms of radiation cause ionization of the air through which they pass; and in both cases the ionization is relatively slight compared with that produced by α-rays or β-rays.

(*iii*) The diffuse tracks they cause in a cloud chamber or a nuclear photographic plate are similar in character (p. 286; see also Plate A.2).

(*iv*) Both forms of radiation are very penetrating and show the same kinds of absorption characteristics in passing through matter.

(*v*) Both X-rays and γ-rays may be diffracted in the same way by crystals (p. 273); and their wavelengths may be found by this means.

There is thus very little room for doubt that γ-rays, like X-rays, are very short wavelength electromagnetic waves. While γ-rays are generally of shorter wavelength than X-rays, there is in fact a considerable overlap between the X-ray and γ-ray parts of the spectrum. From a given γ-source only certain sharply defined wavelengths are observed, which are characteristic of the type of nucleus concerned. In other words, like the visible and X-ray spectra of the *atom*, the γ-radiation from the *nucleus* forms a 'line' spectrum. This implies that the particles in the nucleus, like the electrons in the atom, can exist only in certain particular states, each of well-defined energy.

As with optical and X-ray spectra, the energy and frequency ν of the radiation are connected by

$$\text{energy} = h\nu$$

where h = Planck's constant (p. 221). The quantities of energy liberated in nuclear processes are generally far larger than those arising from rearrangements of electrons in the outer parts of an atom; a γ-ray photon is therefore of very high frequency (and so of very short wavelength). The energy carried by the photon is quite sufficient for it to be detected individually by a Geiger–Müller tube (although it may pass right through the tube without causing any ionization). The behaviour of a Geiger counter in a beam of γ-radiation is therefore a convincing demonstration of the particulate nature of electromagnetic radiation (discussed on p. 220 *et seq.*).

Like other forms of electromagnetic radiation, a γ-ray photon is able to give up its energy to a charged particle in its path; and it is this process that causes the ionization of the matter through which the radiation passes. By analogy with the equivalent optical effect this process is called *photoelectric emission.* The quanta of energy in γ-radiation are large enough to eject electrons from any level of an atom with a considerable amount to spare. An ejected electron then behaves just like a β-particle. This is shown up in a cloud chamber through which the γ-radiation is passing. The tracks revealed are those of β-particles each originating in the path of the beam. Most of these tracks represent the *complete* absorption of a γ-ray photon (see Plate B.3).

This may be contrasted with the process by which α-particles and β-particles are brought to rest. An α-particle has a relatively high probability of ejecting an electron from an atom, but loses only a minute proportion of its energy in the process; it may produce 10^5 ion pairs before its speed becomes too small to cause further ionization; and there is little variation in the lengths of the tracks of α-particles of given energy. A β-particle has a lower probability of ejecting an electron, but the events are sufficiently frequent for the line of droplets that marks its path in a cloud chamber to appear almost continuous. However, the path of a γ-ray is not marked at all—until the point at which it is absorbed by an electron. We can then infer its path in a straight line from the source to the start of the observed β-particle track.

The γ-ray is very penetrating, i.e. the probability of its being absorbed is very low. We can describe this by specifying the thickness of matter in which a γ-ray photon has an even chance of being absorbed; on average, half the photons emerge through this thickness. Thus, the intensity of the γ-rays from cobalt-60 is reduced to a half in passing through about 1 cm of lead; 2 cm of lead reduce it to a quarter, and so on. An α-particle has a certain maximum range because it is extremely unlikely not to make the 10^5 collisions that will bring it to rest within this distance. But there is no maximum range for γ-rays; all we can say is that a given thickness of matter will reduce the intensity in a certain proportion.

X-rays are produced when fast electrons are brought to rest in a target. It is therefore to be expected that β-particles will generate X-rays in any matter through which they pass. These X-rays are of such high energy that we may well class them as γ-rays. The effect is apparent when we are investigating the absorption characteristics of β-rays from a given source. There is a pronounced 'tail' at the end of the curve (Fig. 16.13*a*—p. 292). This is due to the 'γ-rays' generated by the β-particles within the absorbing material. To obtain the true β-absorption curve it is necessary to produce back the tail of the curve and subtract the part of the total count rate caused by these γ-rays.

16.6 The nucleus

It is known that atoms are of such a size that in a piece of solid material they are more or less tightly packed together. When therefore an α-particle traverses a thin piece of metal foil, it must actually pass right through the interior of a large number of atoms. Any changes of direction that occur in the process give some indication of the nature of the forces that act on a charged particle *inside* the atom. In general the tracks of α-particles in cloud chambers and nuclear emulsions are very nearly straight; but a close inspection shows that they suffer frequent small deflections of 1° or less —particularly towards the ends of their paths, where their speed is much reduced. These deflections, small though they are, cannot be accounted for by collisions with the electrons in the atoms. The α-particle is more than 7000 times as massive as the electron, and the maximum possible deflection in such a collision is less than half a minute of angle. The deflections observed are therefore evidence of the interaction of the α-particle with the positively charged part of the atom, which is presumably associated with most of its mass. Acting under Rutherford's guidance, Geiger and Marsden therefore embarked on a detailed investigation of the deflection of α-particles in passing through thin metal foils. Most of the particles were scattered through angles of about 1°; but a small proportion (about 1 in 8000) were deflected through angles of more than 90°—in fact they appeared to be *scat-*

tered back from the metal foil. In due course the same effect was found also in cloud chamber tracks. The same small proportion of α-particle tracks showed sharp deviations of 90° or more in collisions with atoms of gas in the chamber.

If the force acting on an α-particle was electrostatic in nature, it was evident that the field in at least some part of the atom was an exceedingly intense one. At that time (1910) it was generally supposed that the atom consisted of a sphere of uniformly distributed positive charge, which carried most of the mass of the atom; the electrons were supposed to be embedded within this. It was thought that the electrons must be stationary, since otherwise they would radiate energy until they came to rest (p. 223). This was affectionately called the 'plum pudding model' of the atom; and there was a certain amount of experimental evidence to support it. However, it was quite incapable of explaining the large deflections observed by Geiger and Marsden. The maximum possible electric field in the plum pudding atom would occur at the surface of the positively charged sphere; and this would only be sufficient to account for a deflection of about 1°. The chance of 100 deflections of this magnitude all occurring in the same direction in succession is inconceivably remote. Besides, cloud chamber photographs show that the large-angle deflections occur as single events, not as the result of multiple small-angle scattering. Rutherford therefore suggested that the positive charge of the atom and most of its mass must be concentrated in a small central nucleus. Assuming that the inverse square law for electrostatic forces holds for such short distances, the electric field close to the nucleus could then be of sufficient magnitude to account for the deflections observed. Rutherford worked out the consequences of his hypothesis, and deduced a mathematical formula predicting the proportion of α-particles that should be scattered through any given angle. Geiger and Marsden then set out to test the formula with the apparatus in Fig. 16.19. The radioactive source A was enclosed in a container with a slit S in the side; the emerging α-particles were thus confined to a narrow beam that fell normally on the metal foil F. The particles deflected through any given angle θ were detected by looking for the scintillations on the zinc sulphide screen Z attached to the microscope M. The box B was evacuated through the tube T so that the α-particles would not be affected

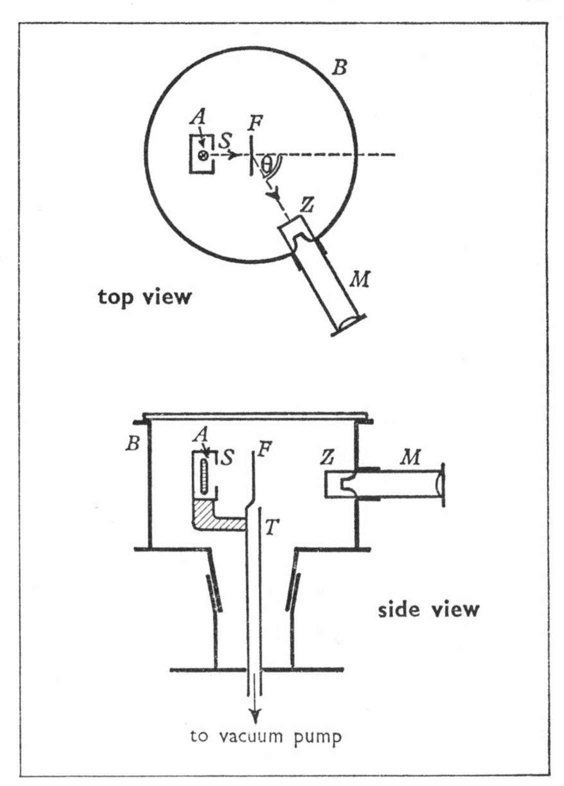

Fig. 16.19 Geiger and Marsden's investigation of the large-angle scattering of α-particles

by collisions with molecules of air. The source and foil were fixed to the central tube T, while the box B and the microscope could be rotated in an air-tight joint so as to vary the angle θ. The results of the measurements agreed in every detail with Rutherford's theory; and there was therefore very little doubt that the nuclear model of the atom was essentially correct.

One of the quantities on which the proportion of scattered particles depends is the charge carried by the nucleus. Once the validity of the theory had been established the experiment became a means of finding this quantity directly. Geiger and Marsden's experiment yielded only low accuracy in this measurement; but later experiments on the same lines showed that the charge on the nucleus, expressed in electronic charge units, was exactly equal to the *atomic number*. This number had been introduced in the first place to specify the position of an element in the periodic table, and it is thus a means of indicating the chemical properties of the element. Moseley's work had shown that the X-ray spectrum of an element also depended on its atomic number (p. 278). Rutherford's discoveries had finally demonstrated that the atomic number was to be identified with the *charge on the nucleus*. This charge presumably controlled the number and distribution of electrons in the atom, and in this way fixed its spectrum and chemical properties.

The scattering of β-particles

It is not easy to investigate the scattering of α-particles in a school laboratory. But the equivalent effect for β-particles is readily demonstrated with the arrangement in Fig. 16.20. When the metal plate is in the position shown, a substantial proportion of the β-particles emitted by the source are scattered back, and the G–M tube registers a greatly increased count. The amount of back scattering increases with the thickness of the metal plate up to a certain maximum (equal to about half the range of the β-particles in the metal). The use of thicker plates than this does not then produce any further increase.

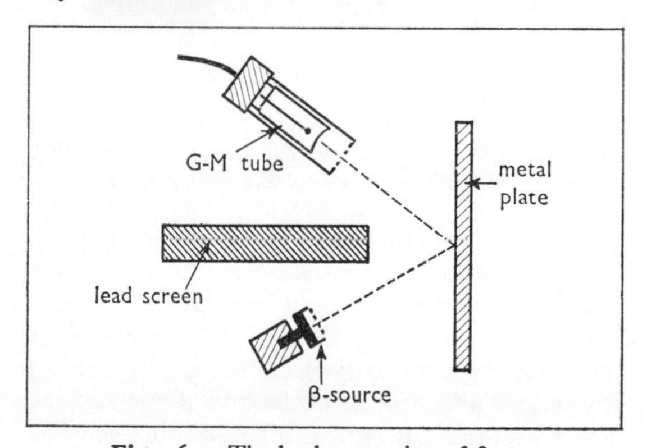

Fig. 16.20 The back-scattering of β-rays

As with α-particles, the back scattering of β-particles can be shown to be due almost entirely to their interactions with atomic nuclei; their collisions with electrons in the metal are relatively infrequent and do not much affect the phenomenon. We should therefore expect the proportion of particles back-scattered to increase with the atomic number of the metal used. With a thick aluminium plate the proportion is about 13%; for copper about 30%; and for gold 50%. The proportions being as high as this, it is clear that the majority of the β-particles must experience more than one large deflection. The tortuous nature of β-particle tracks observed in a cloud chamber supports this conclusion. It is therefore impossible to make any exact prediction relating the amount of back scattering to the nuclear charge; and this phenomenon adds little to the information about atomic structure provided by α-particle scattering experiments.

The back scattering thickness gauge

The scattering of β-particles is used in an instrument for measuring the thickness of a thin sheet, when it is impossible to have access to both sides of it—e.g. the wall of a tube or a layer of paint. One design is shown in Fig. 16.21. The β-emitting

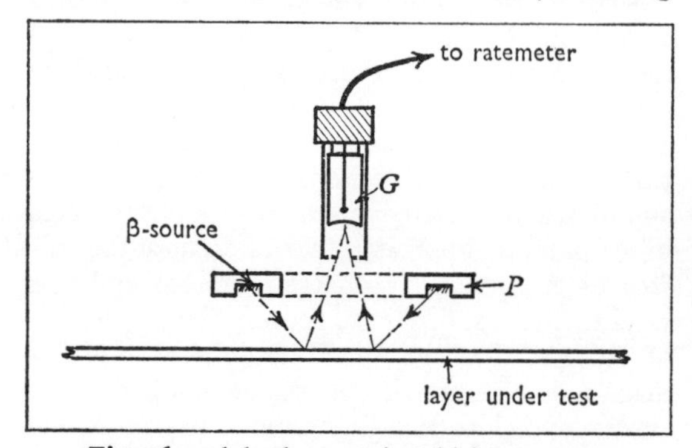

Fig. 16.21 A back-scattering thickness gauge

material is arranged in an annular groove in the brass plate P. The G–M tube G is mounted centrally to collect the particles scattered along the normal from the layer under test. A ratemeter is used to register the count rate, and its scale can be calibrated to read thicknesses direct.

16.7 Nuclear transformations

With the picture of the nuclear atom clearly established, we are in a position to see what the process of radioactive emission of particles consists of. It seems likely that the source of the particles is actually the nucleus of the radioactive atom. The β-particles could conceivably arise from the electron structure of the atom, external to the nucleus; but the α-particles, which are helium nuclei, could not possibly do so. However, the point can be settled by considering the transmutations that should accompany the emission of these particles from a nucleus. If a nucleus ejects an α-particle, which carries a charge of 2 electronic units, its atomic number should *decrease* by 2; at the same time its mass number should *decrease* by 4, since that is the mass number of helium. Likewise, the emission of a negatively charged β-particle should *increase* the positive charge on the nucleus by one unit, and so raise its atomic number by one; but the mass of the β-particle is much less than 1 atomic mass unit, and so leaves the mass number of the atom *unchanged*. Thus a radioactive element should be continually transmuting its atoms into those of another element, according to the rules outlined above. There is abundant evidence that this is exactly what happens.

We shall consider how this works out by analysing the radioactive processes that occur in a radium source. The atomic number of radium is 88, and the mass number of its principal isotope is 226. Using the usual chemical symbol for radium, this information can be expressed symbolically:

$$_{88}Ra^{226}$$

The subscript to the left indicates the atomic number, and the superscript to the right the mass number.* Since an α-particle is a helium nucleus, it can be described in the same notation by

$$_{2}He^{4}$$

Strictly speaking the information carried by the subscript is redundant; if the chemical element is known, a copy of the periodic table suffices to tell us also the atomic number. To specify a nucleus it is only necessary to state the element and its atomic mass number. Sometimes therefore a different notation is used, in which the above two isotopes are described as radium-226 and helium-4. But the fuller notation is always used in writing nuclear equations.

When a nucleus of radium-226 emits an α-particle, we expect its atomic number to fall to 86 and its mass number to 222. It should therefore form a nucleus of the noble gas, *radon*. The decay process of the radium can thus be expressed symbolically by the nuclear 'equation'

$$_{88}Ra^{226} \rightarrow {}_{86}Rn^{222} + {}_{2}He^{4}$$

In all known processes of this kind the total nuclear charge is always conserved. It follows that the subscripts on the right of the equation must add up to the same total as the subscripts on the left. Likewise the total mass is conserved; therefore the superscripts also must 'balance' on both sides of the equation. [Strictly, the masses are not exactly conserved, since some of the mass is changed in the process into the kinetic energy carried away by the α-particle (p. 311); but the loss of mass is always much less than 1 atomic mass unit, so that to the nearest atomic mass unit the total mass of the particles remains unaltered.]

The evolution of radon from a quantity of radium is readily demonstrated, though the experiment is not safe enough to attempt in a school laboratory. If a solution of a radium salt is stored in a sealed vessel, within a few days a small bubble of radon appears at the top. This may be collected and analysed (by observing its spectrum in a discharge tube). Subsequently the radium salt will generate a fresh supply of radon. In fact the radium could be used as a source of radon for as long as it lasts—i.e. for some thousands of years.

Radon itself is a radioactive element which emits an α-particle, and so changes into an isotope of *polonium*—one of the radioactive substances discovered by the Curies.

$$_{86}Rn^{222} \rightarrow {}_{84}Po^{218} + {}_{2}He^{4}$$

The polonium in its turn emits an α-particle and decays to form a radioactive isotope of *lead*.

$$_{84}Po^{218} \rightarrow {}_{82}Pb^{214} + {}_{2}He^{4}$$

This lead isotope is a β-emitting substance, which therefore increases its atomic number by one when it decays, forming bismuth-214. The original radium-226 is thus the start of a sequence of radioactive changes, involving 8 isotopes and terminating eventually on the stable isotope lead-206. A sample of originally pure radium, left for a long enough time, is found to contain all of these elements. Several of them are present in sufficient quantity to be detected by chemical means, though the existence of the shorter-lived substances can only be inferred from the radiations they emit.

The nature of the radiations observed from a radium source depends on whether the first decay product, radon, remains trapped in it or not. In practice care is taken to seal a laboratory source, as radon-222 and its decay products would be most harmful if they escaped into the air of the laboratory. Usually a radium source consists of a piece of foil of a radium–silver alloy. Almost all the radon then remains trapped in the foil. What

* Some scientists prefer to write the mass number on the left, thus:

$$^{226}_{88}Ra$$

we call a 'radium source' is really a mixture of isotopes of nine elements, including the parent radium. Between them they emit α-, β- and γ-radiations of many different energies. A freshly purified sample of radium emits α-particles of one range only, together with some γ-radiation. But if it is kept with its decay products it soon produces β-particles also, and its α- and γ-emission increases by a factor of about 4 as well.

Since a γ-ray consists of electromagnetic radiation, it does not alter the charge of the nucleus from which it comes and affects its mass very little (much less than 1 atomic mass unit). It therefore leaves the composition of the nucleus unaltered. A γ-ray is found always to *follow* the emission of either an α-particle or a β-particle. When one of these particles emerges, it may leave the resulting nucleus in an excited state. Shortly afterwards the nucleus emits the surplus energy as a γ-ray. In a few cases it happens that the emission of an α- or β-particle takes away *all* the available energy, so that the resulting nucleus is left already in its ground state. There is then no emission of γ-radiation. But this is exceptional.

Artificial transmutations

The close agreement between the results of Rutherford's scattering experiments and his predictions on the basis of the nuclear model of the atom made the acceptance of the nuclear theory inevitable. However, when such experiments were performed with materials of low atomic number, discrepancies were revealed. It appeared that in a sufficiently close collision between an α-particle and a light nucleus the inverse square law for the electrostatic force between them broke down. Rutherford supposed that in such a collision the α-particle actually penetrated the nucleus; and so there was the possibility that an artificial nuclear transmutation might be brought about by this means.

The first effect of this kind was discovered in the bombardment of nitrogen atoms by α-particles. It was found that the passage of α-particles through nitrogen generated a new kind of 'radiation' which was about four times as penetrating as the α-particles themselves. The 'radiation' caused scintillations on a fluorescent screen, though the flashes of light were noticeably weaker than those produced by α-particles. Radiation with similar properties had previously been observed when α-particles were passed through hydrogen. In this case there was little doubt that it consisted of *protons* (hydrogen nuclei) knocked on by the impact of the α-particles. Rutherford therefore concluded that some of the incident α-particles had reacted with nitrogen nuclei, forming a new nucleus and causing the ejection of a proton. Such a reaction must be represented by the equation

$$_7N^{14} + {_2He^4} \rightarrow {_8O^{17}} + {_1H^1}$$

The requirement of balancing the subscripts and superscripts leads us to expect a nucleus of oxygen-17 to be formed in the process—an isotope that is known to form a small part of naturally occurring oxygen. Cloud chamber photographs confirmed Rutherford's supposition. With nitrogen in the chamber about one α-particle track in 50,000 was found to be branched as in Fig. 16.22. The α-particle track comes to an end at the fork. Of the two tracks emerging from this point, one is more thinly marked than an α-particle track, and is assumed to be that of a proton; the other is short and thick, and is taken to be that of the residual oxygen nucleus (see Plates A.3 and A.4).

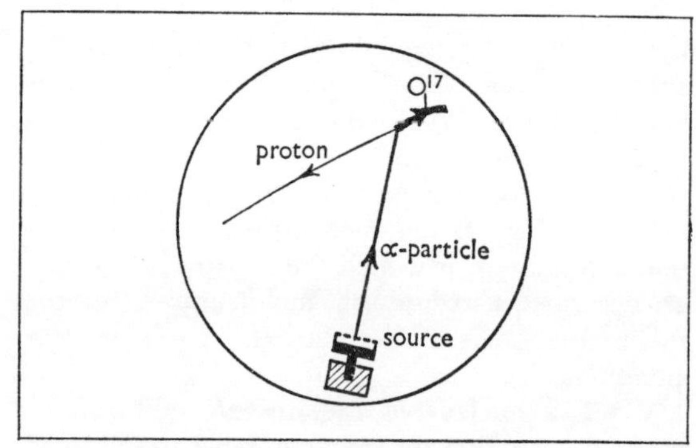

Fig. 16.22 A nuclear transmutation of a nitrogen-14 nucleus into an oxygen-17 nucleus

Many other examples were subsequently discovered of the transmutation of nuclei by bombardment with α-particles. In some cases the residual nuclei turned out to be radioactive isotopes not found in nature.

The neutron

Up to this point in the development of the subject the composition of the nucleus had remained something of a mystery. It seemed almost certain

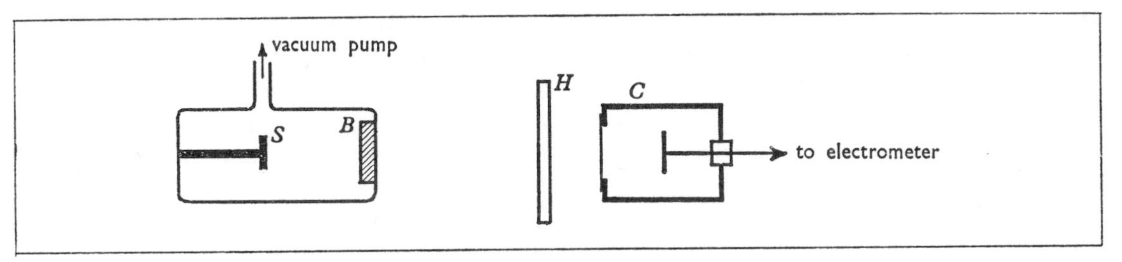

Fig. 16.23 The discovery of neutrons by Chadwick

that the *proton* was one ingredient; the presence of this particle in the nucleus presumably accounted for its positive charge. Since, however, the mass number of a nucleus is usually more than double its atomic number, it must contain other particles besides protons. Rutherford suggested the electron as another ingredient. This seemed reasonable enough, since electrons were known to be emitted in the process of β-decay. On this basis an α-particle would consist of 4 protons and 2 electrons; its net positive charge would thus be 2 units, while its mass would be 4 units. However, this theory ran into difficulties.

The uncertainty was eventually cleared up with the discovery of yet another kind of 'radiation'; this was generated in the element *beryllium* by bombardment with α-particles. The nature of this effect was elucidated by Chadwick in 1932 with the apparatus shown in Fig. 16.23. The α-emitting source S and the beryllium target B were mounted in an evacuated enclosure. The beryllium disc was thick enough to stop the α-particles completely. In the absence of any other sheets of matter between the beryllium and the ionization chamber C a very small ionization current was registered. The amount of ionization was comparable with that producedby γ-radiation; but the insertion of absorbing sheets showed that the new radiation was even more penetrating than normal γ-radiation. It was not likely to consist of charged particles of any kind, since these would cause much greater ionization than was observed. The real surprise came, however, when a sheet of paraffin wax H (or any other material containing a large amount of hydrogen) was inserted in the path of the beam. The ionization current was then found to increase by a factor of two or more. It was soon established that the increased current was due to the ejection of protons from the paraffin by the new radiation. It seemed that the radiation had no interaction with the electrons in the matter through which it passed but only with the atomic nuclei. This was confirmed by cloud chamber photographs. No electron tracks were observed in a chamber through which the radiation passed. Instead, occasional heavy straight tracks were found that were interpreted as those of nuclei projected forward by the impact of some particle.

Chadwick assumed that the action of α-particles on beryllium caused the emission of a hitherto undetected particle, which he supposed to be uncharged. It was named the *neutron*. He suggested the following nuclear reaction for the process:

$$_{4}Be^{9} + {}_{2}He^{4} \rightarrow {}_{6}C^{12} + {}_{0}n^{1}$$

The symbol 'n' is used to represent the neutron in such equations.

By careful measurement of the energies of nuclei struck by the neutrons Chadwick estimated that the mass of the neutron was about equal to that of the proton. It is now known that the neutron is in fact slightly more massive.

Mass of the neutron = 1·00867 a.m.u.
Mass of the proton = 1·00728 a.m.u.

The discovery of the new particle settled the problem of the composition of the nucleus. The previous difficulties were removed if it could be assumed that the number of protons in the nucleus was equal to the atomic number, and that the additional mass was made up by neutrons. On this basis the α-particle must consist of just 2 protons together with 2 neutrons. However, there were new difficulties in explaining the process of β-emission. On the neutron theory this must consist of the transformation of a neutron into a proton and the creation of an electron to carry away the surplus negative charge. At the time this seemed rather unlikely; but subsequently this very process has been observed in neutrons outside the nucleus. In the free state the neutron itself is a *radioactive particle*, which decays into a proton by β-emission; its half-life is 12·8 minutes (p. 305). In fact according to modern understanding protons and neutions in the nucleus are continually changing one into the other, so that it is not really appropriate to talk about them as though they could be distinguished from one another (p. 3). In this connection they are both referred to as *nucleons*; only when one of them emerges from the nucleus can we say whether it is a proton or a neutron.

We can now see more precisely the reason for balancing the superscripts on both sides of a nuclear equation. The masses of the nucleons usually change very slightly in a nuclear reaction, but the *total number of nucleons* does not alter, although protons may change into neutrons and vice versa. The superscript therefore represents the *number of nucleons* in a nucleus, rather than their exact mass.

By using beams of neutrons many new kinds of nuclear reaction can be produced. A charged particle entering a nucleus must have sufficient energy to overcome the intense electrostatic repulsion. Protons and α-particles must have large energies to do this. But the neutron, being uncharged, can pass readily *through* a nucleus, whatever its energy; and it may in the process be absorbed by it. Many of the artificial radioactive isotopes in current use are produced by irradiation of suitable substances with the intense beams of neutrons obtained in nuclear reactors (p. 319).

One such reaction provides one of the most convenient modern methods of detecting neutrons:

$$_5B^{10} + {_0n^1} \rightarrow {_3Li^7} + {_2He^4}$$

The boron-10 nucleus absorbs neutrons very readily, and then disintegrates into an α-particle and a lithium nucleus—both of which are heavily ionizing particles. The ionization produced by neutrons in most gases is so slight that an ordinary Geiger counter scarcely responds to them at all. But by filling the tube with boron trifluoride gas a high proportion of the incident neutrons may be detected.

Because of their high penetrating power neutrons can constitute a serious health hazard. The most effective substances for stopping them are those which contain as many light nuclei as possible per unit volume. This means that protective screens round neutron sources are best made of compounds of hydrogen and other light nuclei, e.g. water or paraffin wax, rather than lead.

16.8 Radioactive decay

For many radioactive substances the decay rate is so slow that there is no perceptible reduction in the quantity of the material within the lifetime of one observer! But in all cases that can be investigated in a reasonable time the decay is found to follow the same pattern. For instance a sample of

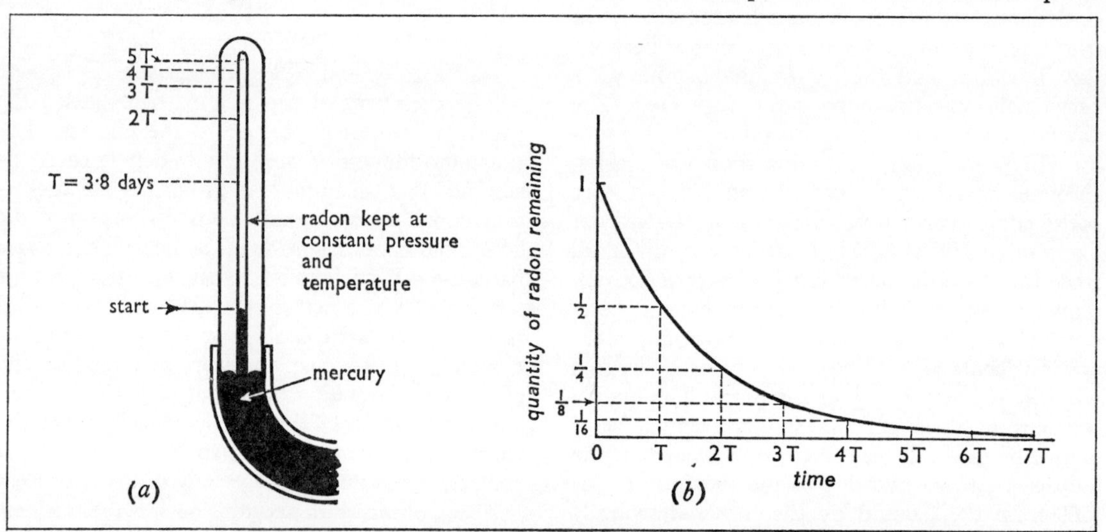

Fig. 16.24 (*a*) The radioactive decay of a sample of radon-222; (*b*) the exponential decay curve

radon-222 may be collected over mercury in a capillary tube (Fig. 16.24*a*). The radon may be kept at constant pressure and temperature while its volume is measured at intervals during the next few days. Its decay products are solid substances; so the volume of the gas decreases as the radon decays. This is found to happen according to an *exponential law*, as shown in Fig. 16.24*b*. In 3·8 days half the radon has decayed; this is called the *half-life* of the isotope. In another 3·8 days half of what remains decays; after a third period of 3·8 days only $\frac{1}{8}$ of the original sample remains; and in about seven times the half-life the radon has virtually disappeared. The walls of the tube are then lined with the decay products—which may be detected by the radiations they emit. Similar exponential decay curves are found in all cases that have been investigated. All such processes are completely described by stating the half-life of the radioactive isotope concerned.

The **half-life** *of a radioactive isotope is the time taken for any quantity of the isotope to decay by one-half of the quantity of it initially present.*

An exponential curve of the type shown in Fig. 16.24 may be expressed by the equation

$$m = m_0\,e^{-\lambda t}$$

where m_0 is the mass of the isotope initially present, and m is the mass remaining after the passage of time t. The constant λ is a measure of the decay rate, and is known as the *decay constant*. The half-life T may be expressed in terms of λ as follows:

From the definition of the half-life T,

$$\text{when } t = T$$
$$m = \tfrac{1}{2}m_0$$
$$\therefore\ e^{-\lambda T} = \tfrac{1}{2}$$
$$\text{or}\quad e^{\lambda T} = 2$$
$$\therefore\ \lambda T = \log_e 2 = 0{\cdot}693$$
$$\therefore\ T = \frac{0{\cdot}693}{\lambda}$$

The radon-222 from the decay of natural radium has harmful and relatively long-lived decay products. However, another isotope (radon-220), formed as one of the decay products of thorium-232, can safely be handled in a school laboratory. About 25 g of thorium hydroxide form a convenient source of this isotope. The thorium reaches approximate equilibrium with its decay products in about 12 years, and may then be used to produce radon-220 (often called *thoron*). The 'vintage' thorium hydroxide is kept in a polythene bottle that can be connected to an ionization chamber by a flexible tube. The neck of the bottle must be plugged with a suitable filter to prevent the escape of the fine powder. Radon-220 is formed continually, and may be transferred to the ionization chamber by simply squeezing the bottle. The mass of radon expelled from the bottle is probably less than 10^{-19} kg—quite undetectable by chemical means. However, some 10^5 atoms may enter the ionization chamber, and these will decay

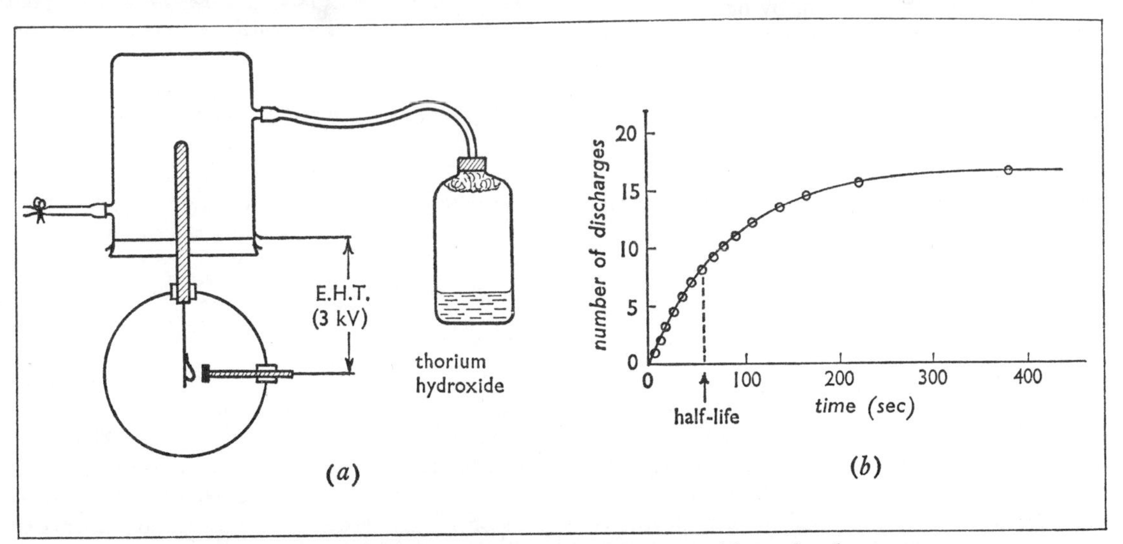

Fig. 16.25 Measuring the decay time of radon-220 with a pulse electrometer

initially at a rate of about 600 per second, which provides an ionization current that may be measured with a pulse electrometer (Fig. 16.25*a*).

The ionization current at any instant is proportional to the quantity of radon still remaining in the chamber, and can therefore be used as a measure of this quantity. The current is inversely proportional to the time interval between successive discharges of the electrometer; and a graph of the number of discharges against time is thus of exponential form (Fig. 16.25*b*), and the half-life of radon-220 may be calculated from it.

Another decay process that may conveniently be observed in a school laboratory is that of the isotope of protactinium (Pa^{234}) formed as one of the decay products of uranium-238. Any compound of uranium contains trace amounts of protactinium-234, which may be extracted from it by chemical means. The protactinium decays by β-emission into another long-lived isotope of uranium (U^{234}) which is itself α-emitting. The β-activity at any instant of the extracted solution can therefore be used as a measure of the quantity of protactinium still present in it. The solution is placed in intimate contact with a G–M tube and ratemeter, and the count rate is recorded at 20-second intervals. A graph of count rate against time is of the usual exponential form.

An exponential decay law is best tested by plotting the variables in such a way as to obtain a straight-line graph. Thus the exponential decay of the count rate I above is represented by

$$I = I_0 e^{-\lambda t}$$

where λ = the decay constant.

Taking logs (to the base e), we have

$$\log I = \log I_0 - \lambda t$$

and a graph of $\log I$ against t should thus be a straight line whose gradient is $-\lambda$. From this the half-life may be calculated.

Radioactive tracers

In both the above experiments we use quantities of radioactive isotopes which, by other standards, would be regarded as vanishingly small. However, the numbers of atoms in each sample are large enough to ensure a measurable decay rate, and this is what matters. This shows up a feature of work with radioactive isotopes. It is possible to follow an isotope through a physical or chemical process, and its presence at any point can be detected by the radiations it emits; but, it is only necessary to use sufficient material to ensure that the count rate obtained is significantly greater than the background count.

Numerous practical applications of such techniques have been developed. For instance water mains may be tested for leaks by filling them with water containing minute quantities of a radioactive isotope of sodium (Na^{24}), usually in the form of sodium bicarbonate. If there are any leaks in the pipe, the 'labelled' solution flows out into the surrounding soil. The pipe is then emptied and washed through with plain water. Sodium-24 emits high-energy γ-radiation, which is readily detected near a leak by a G–M tube drawn along the inside of the pipe or lowered into small holes made in the ground beside suspected joints. The half-life of the isotope is only 15 hours, so that it soon vanishes from the soil and leaves no hazard for the public. In any case the quantities used are too small to present any danger, even if someone should mistakenly drink the water with which the pipe was initially filled. In spite of this, the method is sensitive enough to detect leaks as little as 10 ml per hour!

There are many branches of pure and applied science in which the use of *radioactive tracers* has become an indispensable tool of research. For instance it has become possible by this means to unravel some of the complex sequences of chemical reactions that take place in living organisms. The organism may be 'fed' with a compound labelled with a suitable radioactive isotope. The subsequent growth of radioactivity in other compounds present in the organism reveals the pattern of chemical reactions into which the original 'food' enters.

It is outside the scope of this book to go further into details of the many ingenious methods that have been developed for using radioactive tracers. The student who is interested should consult books of a more specialized nature.

The decay of long-lived isotopes

The half-lives of the known radioactive substances cover an enormous range—from less than 1 microsecond to more than 10^{15} years. The direct measurement of the decay rate by (say) chemical analysis is unthinkable in most cases. The half-life of radium (1620 years) is trifling compared with that of uranium ($4{\cdot}5 \times 10^9$ years). Even so the amount of radium in a given sample would have decreased by only 4% during the time since this element was discovered by the Curies; and this cannot be made the basis of a measurement of its half-life. However, each particle that emerges from a source represents the decay of one nucleus; and in principle it is possible by counting the emitted particles to estimate the number of nuclei decaying per second. If we can also estimate the total number of atoms in the source, we can then calculate the half-life of the specimen. Thus, if N_0 is the number of atoms initially present in a source, and N is the number present after time t, then

$$N = N_0 e^{-\lambda t}$$

where λ = the decay constant.

Differentiating,

$$\frac{dN}{dt} = -\lambda N_0 e^{-\lambda t} = -\lambda N$$

This can be interpreted to mean:

$$\text{no. of disintegrations per second} = \lambda N$$

$$\therefore \lambda = \frac{\text{no. of particles emitted per sec}}{\text{no. of atoms in the source}}$$

Avogadro's constant ($6{\cdot}02 \times 10^{26}$) is the number of atoms in 1 kg-atom of any element (p. 37). The denominator of the above expression may therefore be calculated if the mass of active isotope in the source is known. Then λ and the half-life may be found.

Example. A radioactive source contains 10^{-6} g of plutonium (mass number 239). The source is estimated to emit a total of 2300 α-particles per second in all directions. Calculate the half-life of plutonium.

239 kg of plutonium contain $6{\cdot}02 \times 10^{26}$ atoms.

$$\therefore 10^{-9}\,\text{kg contains} \frac{10^{-9} \times 6{\cdot}02 \times 10^{26}}{239} \text{atoms}$$

$$= \frac{6{\cdot}02}{239} \times 10^{17} \text{ atoms}$$

Therefore the decay constant λ is given by

$$\lambda = \frac{2300 \times 239}{6{\cdot}02 \times 10^{17}} = 9{\cdot}15 \times 10^{-13}\,\text{s}^{-1}$$

$$\therefore \text{half-life} = \frac{0{\cdot}693}{\lambda}$$

$$= \frac{0{\cdot}693}{9{\cdot}15} \times 10^{13}\,\text{s}$$

$$= 7{\cdot}57 \times 10^{11}\,\text{s}$$

$$= \frac{7{\cdot}57 \times 10^{11}}{3600 \times 24 \times 365} \text{ years}$$

$$= 24{,}000 \text{ years}$$

The curie

In using a radioactive source we are not generally concerned so much with the actual mass of active material present as with the rate of disintegration taking place within it. It is therefore useful to have a unit of *disintegration rate*; it is called the *curie*. Originally this was fixed as the disintegration rate of 1 g of radium. But it is now redefined on a numerical basis.

The **curie** *is a radioactive disintegration rate of* $3{\cdot}7 \times 10^{10}$ *nuclei per second.*

In school laboratories 10 microcuries (10 μCi) is usually considered the maximum safe activity for any one source; and for most purposes the sources can well be much weaker than this. The strength of the plutonium source appearing in the example above is therefore

$$\frac{2300}{3{\cdot}7 \times 10^4} = 6{\cdot}2 \times 10^{-2}\,\mu\text{Ci}$$

For more advanced work sources in the millicurie range are used; safety precautions must then be of a much more stringent kind than those we have described. For medical purposes radium sources of a few curies are sometimes used; and γ-emitting sources of cobalt-60 of more than 2000 curies have been used for deep therapy. Such sources could well give lethal doses of radiation in quite a short space of time.

It is found that no physical or chemical process is able in any way to affect the rate of disintegration of a given isotope. It may be exposed to extremes of temperature or placed in large electric and magnetic fields. It may be fixed in any number of chemical compounds—gaseous or solid—or it may be ionized in a high-temperature discharge. But in no case has any change in the disintegration rate been detected. In other words no action *outside* the nucleus is able to affect the moment at which any particular nucleus disintegrates. Since we cannot know what is going on inside the nucleus, we can only describe the behaviour of the nucleus by stating the *probability* of its decay in a given interval of time.

When we are dealing with very large numbers of atoms, the probability P of decay of a particular atom within an interval of one second is equal on average to the fraction of the atoms that do in fact decay in this time.

$$\therefore P = \frac{\text{no. of disintegrations per second}}{\text{total no. of atoms in the source}}$$

$$= \frac{\lambda N}{N} = \lambda$$

The true significance of the decay constant λ is thus apparent; it is equal to the probability of a given atom decaying in one second, and is a constant for the atom that is unaffected by any external conditions. The universal validity of the exponential type of decay law shows also that λ does not depend on the *age* of the atom. Any variation of the probabilities with time would yield a different decay curve.

16.9 Radioactive equilibrium

The radioactive decay of radium is the start of a series of disintegrations, ending eventually with a stable isotope of lead. Several such radioactive series are known. Indeed the series produced from radium is itself only part of a much longer series starting with the most abundant isotope of uranium (U^{238}). The complete uranium series is displayed in the table.

The half-life of radium is very short compared with the age of the earth, so that any radium originally present in the earth must long since have decayed away. The radium we now have is that which has arisen as a daughter product of uranium and has not yet decayed into radon, etc. Radium and the other members of the uranium series are therefore found naturally in uranium-bearing ores (such as pitchblende). Given a sufficient period of time the quantity of each isotope in the ore builds up to the point at which its rate of decay is equal to the rate at which it is formed from the decay of the previous member of the series. When this condition is reached, the proportions of the radioactive isotopes in the rock do not change with time, and we say that the uranium is in *radioactive equilibrium* with its daughter products. The half-life of uranium is so long ($4{\cdot}5 \times 10^9$ years) that it takes about 50 million years for the quantity of it in the rock to change by even 1%. Its disintegration rate is therefore almost constant, and this is equal to the disintegration rate of each member of the series in equilibrium with it. It is also the rate at which the stable end-product (Pb^{206}) is being formed. The net result is therefore the conversion of U^{238} to Pb^{206} at a steady rate, while the quantities of the intermediate members of the series remain constant.

The amounts of daughter products in equilibrium with a given amount of the parent substance may be calculated from a knowledge of their half-

The uranium series

Isotope	*Symbol*	*Emissions*	*Half-life*
Uranium-238 ↓	${}_{92}U^{238}$	α, γ	$4{\cdot}5 \times 10^9$ y
Thorium-234 ↓	${}_{90}Th^{234}$	β, γ	24 d
Protactinium-234 ↓	${}_{91}Pa^{234}$	β, γ	1·2 m
Uranium-234 ↓	${}_{92}U^{234}$	α, γ	$2{\cdot}5 \times 10^5$ y
Thorium-230 ↓	${}_{90}Th^{230}$	α, γ	$8{\cdot}0 \times 10^4$ y
Radium-226 ↓	${}_{88}Ra^{226}$	α, γ	1620 y
Radon-222 ↓	${}_{86}Rn^{222}$	α	3·8 d
Polonium-218* ↓	${}_{84}Po^{218}$	α	3·1 m
Lead-214 ↓	${}_{82}Pb^{214}$	β, γ	27 m
Bismuth-214* ↓	${}_{83}Bi^{214}$	β, γ	20 m
Polonium-214 ↓	${}_{84}Po^{214}$	α	$1{\cdot}6 \times 10^{-4}$ s
Lead-210 ↓	${}_{82}Pb^{210}$	β, γ	19 y
Bismuth-210* ↓	${}_{83}Bi^{210}$	β	5·0 d
Polonium-210 ↓	${}_{84}Po^{210}$	α	138 d
Lead-206	${}_{82}Pb^{206}$	Stable	—

* In these isotopes alternative modes of decay occur in a small proportion of cases. For example

(*i*) Polonium-218 may decay by β-emission to astatine-218; this then decays by α-emission to bismuth-214.
(*ii*) Bismuth-214 may decay by α-emission to thallium-210 which decays by β-emission to lead-210.
(*iii*) Bismuth-210 may decay by α-emission to thallium-206, this decays by β-emission to lead-206.

lives. Suppose the decay constants of the members of the series are λ_1, λ_2, λ_3, etc. Let the numbers of atoms of each isotope present be N_1, N_2, N_3, etc. The corresponding rates of disintegration are $\lambda_1 N_1$, $\lambda_2 N_2$, $\lambda_3 N_3$, . . .; and in equilibrium these rates are all equal.

$$\therefore \lambda_1 N_1 = \lambda_2 N_2 = \lambda_3 N_3 = \ldots$$

i.e.

$$N \propto \frac{1}{\lambda}$$

for a given series.

Now the half-life T of an isotope is given by

$$T = \frac{\log_e 2}{\lambda}$$

$$\therefore N \propto T \quad \text{for the series}$$

The mass m of any isotope present is proportional to its atomic mass A and the number of atoms N.

$$\therefore m \propto AT$$

for a given radioactive series.

Example (*i*). A quantity of ore is found to contain 1·0 kg of uranium. Estimate the mass of radium in the ore. The half life of uranium-238 is $4\cdot5 \times 10^9$ years and that of radium-226 is 1620 years.

Let the mass of radium in the ore be x kg. Using the relation deduced above (assuming the ore to be in radioactive equilibrium),

$$\frac{x}{1} = \frac{226 \times 1620}{238 \times 4\cdot5 \times 10^9}$$

$$\therefore x = \frac{2\cdot26 \times 1\cdot62}{2\cdot38 \times 4\cdot5} \times 10^{-6}\ \text{kg}$$

$$= 0\cdot34\ \text{mg}$$

Example (*ii*). One of the decay products of uranium-238 is lead-210 which is a β-emitting substance. The lead-210 extracted from a quantity of uranium ore is found to emit $1\cdot5 \times 10^5$ β-particles per second. Calculate the mass of uranium-238 in the ore. (The half-life of uranium-238 is $4\cdot5 \times 10^9$ years; Avogadro's constant $= 6\cdot0 \times 10^{26}$ per kg-atom.)

The decay constant λ of uranium is given by

$$\lambda = \frac{0\cdot693}{4\cdot5 \times 10^9 \times 365 \times 24 \times 3600}$$

$$= 2\cdot78 \times 10^{-18}\ \text{s}^{-1}$$

Assuming that the lead-210 is in equilibrium with the uranium in the ore, the disintegration rate of the uranium is the same as that of the lead.

$\therefore$ the no. of atoms of uranium present

$$= \frac{\text{the disintegration rate}}{\lambda}$$

$$= \frac{1\cdot5 \times 10^5}{2\cdot78 \times 10^{-18}}$$

$\therefore$ the no. of kg-atoms of uranium present

$$= \frac{1\cdot5 \times 10^5}{2\cdot78 \times 10^{-18} \times 6\cdot0 \times 10^{26}}$$

Now 1 kg-atom of uranium consists of 238 kg.

$\therefore$ the mass of uranium in the ore

$$= \frac{238 \times 1\cdot5 \times 10^5}{2\cdot78 \times 10^{18} \times 6\cdot0 \times 10^{26}}$$

$$= 21\cdot3 \times 10^{-2}\ \text{kg}$$

$$= 21\ \text{g} \quad \text{(to 2 significant figures)}$$

Radioactive dating

The existence of radioactive substances with long half-lives provides us with a means of measuring time intervals on a geological scale. If a radioactive isotope and its decay products remain trapped in a rock, the measurement of the proportion of the parent isotope that has decayed enables the age of the rock to be determined. This can be done for instance with uranium-bearing rocks. Uranium-238 reaches equilibrium with its decay products in about 2 million years. After this time the net result of the disintegration processes is the production of Pb^{206} from U^{238}, while the intermediate products remain in constant concentrations. The measurement of the ratio of Pb^{206} to U^{238} enables the age of the rock (since it first solidified) to be established. By this means ages of rocks between 50 million years and 4000 million years have been measured with fair reliability. Some assumption must be made about the proportion of the Pb^{206} that was in the rock when it was first formed, and this introduces an uncertainty. It is usually desirable to check the measurement by several similar methods. For instance the less common isotope of uranium (U^{235}) is also the start of a radioactive series that terminates with Pb^{207}; and the same sort of measurement can be performed for these two isotopes. Sometimes a check can also be provided by analysing the proportions of lighter elements that have radioactive isotopes. For instance about 30% of the element rubidium exists as a radioactive isotope, rubidium-87. This decays by β-emission into strontium-87, which is a stable isotope. The half-life of this decay process is $4{\cdot}3 \times 10^{10}$ years. The measurement of the relative concentrations of these isotopes in a rock thus provides another means of estimating its age. Measurements on the oldest rocks have established that the age of the earth must be about 4500 million years. The ages of meteorites are sometimes found to be much greater than this, showing that many of these pieces of matter must have solidified long before the earth was formed.

Carbon-14 dating

The half-lives of the naturally occurring radioactive substances discussed above are far too great to be of much value for dating archaeological finds, whose ages range up to a few thousand years only. Fortunately it has been established that a radioactive isotope of carbon (C^{14}) is formed continually by the action of cosmic radiation in the earth's atmosphere. The cosmic rays cause some nuclear disintegrations from which neutrons emerge. Some of the neutrons react with nitrogen nuclei forming carbon-14

$${}_{7}N^{14} + {}_{0}n^{1} \rightarrow {}_{6}C^{14} + {}_{1}H^{1}$$

Because of this a small proportion of the carbon in the atmosphere is in this radioactive form. Provided the intensity of cosmic radiation has not changed during the last few thousand years the proportion can be assumed to have been constant during this period. There is a continual interchange of carbon (in the form of carbon dioxide) between plants and the atmosphere; and animals (including man) feed on the plants and on each other. All living creatures therefore contain the same small proportion of C^{14}. However, when a creature dies, the interchange of carbon with the atmosphere ceases, and the C^{14} content of the remains starts to decay. The half-life is 5600 years, which is ideal for archaeological purposes. Also carbon is an ingredient of many materials of archaeological interest—wood, paper, peat, household refuse, etc.

The technical difficulties of dating objects by this means are enormous. Very often only small quantities of material can be spared for the measurement. In any case even in contemporary specimens only about 1 part in 10^{12} of the carbon is in the form of the active isotope. This is equivalent to about 15 disintegrations per minute in 1 g of carbon. Also the β-particles emitted by C^{14} are of low energy; so the thickness of the specimen must be kept to a minimum. In fact it is usual to convert the carbon into a suitable gas (carbon dioxide or acetylene) and introduce it into the counter itself. Even so, the count may not greatly exceed the normal background count of a Geiger–Müller tube. Elaborate precautions must therefore be taken to reduce the background radiation as much as possible. Thick lead shielding is employed; and as far as possible all radioactive substances are eliminated from the materials employed in constructing the apparatus. For instance potassium must be excluded from the glass of the tube, since a small proportion of natural potassium is in the form of the radioactive isotope potassium-40.

This technique has enabled ages of suitable finds to be measured from 600 years up to more than 10,000 years old; often this can be done with an uncertainty of less than 100 years. It has been possible on a number of occasions to check the method by measurements on objects whose age can be independently established from historical records. In every case the dates agree within the stated limits of error.

16.10 Mass and energy

It is convenient to express the energies of particles emitted by radioactive sources in *electron-volts* (p. 158). For example, the α-particle emitted by radium has an energy of 4·8 MeV. This means that it has the same energy as that acquired by an electron accelerated through a p.d. of $4{\cdot}8 \times 10^{6}$ volts. It is also useful to be able to express the energies of γ-ray photons in the same unit.

Example. Calculate the frequency and wavelength of a 2-MeV γ-ray.

Taking the charge on an electron as $1{\cdot}6 \times 10^{-19}$ coulomb, the energy gained by an electron accelerated through a p.d. of 2×10^{6} volts is

$$1{\cdot}6 \times 10^{-19} \times 2 \times 10^{6} \text{ joule} = 3{\cdot}2 \times 10^{-13} \text{ joule}$$

and this is the energy of the photon. Taking

Planck's constant h as $6{\cdot}6 \times 10^{-34}$ joule s, we have

$$\text{frequency of } \gamma\text{-ray} = \frac{3{\cdot}2 \times 10^{-13}}{6{\cdot}6 \times 10^{-34}}$$
$$= 4{\cdot}85 \times 10^{20}\ \text{Hz}$$
$$\text{the velocity of light} = 3{\cdot}0 \times 10^{8}\ \text{m s}^{-1}$$
$$\therefore \text{the wavelength} = \frac{3{\cdot}0 \times 10^{8}}{4{\cdot}85 \times 10^{20}}$$
$$= 6{\cdot}2 \times 10^{-13}\ \text{m}$$

One of the most important conclusions of the Special theory of Relativity is that energy and mass are equivalent. The theory predicts that a body gaining an amount of energy E thereby increases in mass by an amount m given by

$$E = mc^2$$

where c = the velocity of light. In this equation E is in joules, when m is in kg and c in m s^{-1}. The quantities of energy handled in everyday life are far too small to involve detectable changes of mass. Thus 1 kg of water absorbs $4{\cdot}2 \times 10^5$ joules of energy to raise its temperature from 0°C to 100°C. The theory of Relativity therefore predicts that its mass will increase accordingly by an amount m, given by

$$m = \frac{4{\cdot}2 \times 10^5}{9 \times 10^{16}} \simeq 5 \times 10^{-12}\ \text{kg!}$$

In chemical reactions also the changes of energy are of this order of magnitude, involving at the most a few electron-volts of energy per atom; and the associated changes of mass are again quite undetectable.

However, in nuclear processes the quantities of energy released are of the order of a few MeV, and are sufficient to produce significant changes in the masses of the particles concerned. The mass equivalent of 1 MeV of energy may be calculated as follows.

$$1\ \text{MeV} = 1{\cdot}6 \times 10^{-19} \times 10^{6}\ \text{joule}$$
$$= 1{\cdot}6 \times 10^{-13}\ \text{joule}$$
$$\text{mass equivalent} = \frac{1{\cdot}6 \times 10^{-13}}{9 \times 10^{16}}$$
$$= 1{\cdot}78 \times 10^{-30}\ \text{kg}$$

This is almost exactly twice the mass of an electron; i.e. a 1-MeV γ-ray carries this quantity of mass away from the nucleus that emits it. In the same way a 1-MeV β-particle has a total mass *three* times the rest mass of a stationary electron (its rest mass + the mass of its kinetic energy).

Thus the processes of radioactive decay involve appreciable changes in the total masses of the nuclei and particles concerned; these changes can readily be measured by modern techniques of mass spectroscopy. For instance the atomic mass of Ra^{226} is 226·0254 a.m.u. This decays by emission of an α-particle (i.e. a helium nucleus of atomic mass 4·0026 a.m.u.) into Rn^{222}, whose atomic mass is 222·0175 a.m.u. The combined mass of the radon and helium is thus 226·0201 a.m.u. This is less than the mass of the parent radium atom by 0·0053 atomic mass units—a quantity known as the ***mass defect*** of the reaction. The energy equivalent of this quantity of mass appears as the kinetic energy of the emitted particle. Thus

$$1\ \text{atomic mass unit} = 1{\cdot}66 \times 10^{-27}\ \text{kg}$$
$$\therefore \text{the surplus energy} = \frac{0{\cdot}0053 \times 1{\cdot}66 \times 10^{-27}}{1{\cdot}78 \times 10^{-30}}$$
$$= 4{\cdot}95\ \text{MeV}$$

The α-particle carries away 4·8 MeV of this energy, the remainder being emitted as a γ-ray shortly afterwards. In every case of radioactive decay the large quantities of energy released are found to be provided at the expense of the masses of the nuclei and particles involved, as predicted by the Special theory of Relativity. Indeed we may say that a given nuclear change could not take place unless the combined masses of the daughter products were less than the mass of the parent nucleus, so that a surplus of energy can be made available to carry away the decay products. In general, the greater the mass defect of any possible nuclear change, the more probable the change becomes. This is another way of saying that a radioactive substance emitting a very high-energy particle will be very short lived.

★ *Kinetic energy in the Special theory of Relativity*

In Newtonian mechanics the kinetic energy of a particle of mass m moving with velocity v is given by

$$\text{K.E.} = \tfrac{1}{2}mv^2$$

The Special theory of Relativity approaches the calculation of kinetic energy from a different point of view. The total energy E of the particle is given in all circumstances by

$$E = mc^2$$

This includes the rest mass energy that the particle has even when at rest. When the particle is moving with velocity v, its mass increases according to the relation

$$m = \frac{m_0}{\sqrt{1 - v^2/c^2}}$$

where m_0 = the rest mass.

The kinetic energy of the particle is accounted for by the increase in its mass $(m - m_0)$, and is therefore given by

$$\text{K.E.} = (m - m_0)c^2$$

Substituting for m, and expanding the bracket in powers of v^2/c^2, we have

$$\text{K.E.} = m_0c^2\left[\left(1 - \frac{v^2}{c^2}\right)^{-\frac{1}{2}} - 1\right]$$

$$= m_0c^2\left[1 + \frac{1}{2}\frac{v^2}{c^2} + \frac{3}{8}\frac{v^4}{c^4} + \ldots - 1\right]$$

$$= \tfrac{1}{2}m_0v^2\left[1 + \frac{3}{4}\frac{v^2}{c^2} + \ldots\right]$$

For sufficiently small velocities (i.e. $v^2 \ll c^2$), this expression agrees closely with the Newtonian formula, (K.E. $= \frac{1}{2}m_0v^2$). But at speeds approaching that of light the Newtonian theory breaks down completely.

For example, according to the Newtonian formula the K.E. of a particle moving at half the speed of light is

$$\tfrac{1}{2}m_0\frac{c^2}{4} = \tfrac{1}{8}m_0c^2 = 0{\cdot}125m_0c^2$$

According to the Special theory of Relativity the mass of the particle at this speed becomes

$$m = \frac{m_0}{\sqrt{1 - 1/4}} = \frac{2}{\sqrt{3}}m_0$$

$$\therefore \text{K.E.} = (m - m_0)c^2 = \left(\frac{2}{\sqrt{3}} - 1\right)m_0c^2$$

$$= 0{\cdot}155m_0c^2$$

In this case the Newtonian calculation gives an error of about 25%.

16.11 High-energy accelerators

Most of our knowledge of the nucleus has come from collision experiments, in which we study the effects of firing high-speed particles at a piece of matter. Much of it was gained using as projectiles the α-particles from radioactive substances. But the maximum energy obtainable in this way is less than 10 MeV, and there is no control over the direction of emission of the particles. To obtain a reasonably strong collimated beam a dangerously large source must be used with all its attendant difficulties.

Cosmic rays

Nature provides also another source of high-speed particles in the cosmic radiation that enters the earth's atmosphere from outer space. Here the energies available are of quite a different order of magnitude. The majority of primary particles have energies of about 10^4 MeV; but energies have been observed up to a maximum of about 10^{13} MeV. The primary radiation consists mostly of protons with a small proportion of other light nuclei and of electrons. However, all these particles are lost in collisions with nuclei in the upper atmosphere, and only the secondary products (mostly electrons) can be observed at ground level. To experiment with the primary radiation the equipment must be flown in a balloon high in the atmosphere. In spite of the difficulties several vital discoveries have been made with this technique. But the intensity of the radiation is very low—fortunately for us—and the gathering of detailed statistical data about the behaviour of the particles is an impossibly lengthy business. It has become clear since the 1930's that progress depends on the development of artificial means of producing very high-energy particles (see Plate D.1).

Electrostatic accelerators

The most obvious way of accelerating a charged particle is to use a very high p.d. between two electrodes in an evacuated tube. Several techniques for doing this were developed in the 1930's. Cockroft and Walton adapted conventional rectifier circuits to produce high p.d.'s from an a.c. input. Eventually they reached p.d.'s up to 2 million volts by this means. At the same time Van de Graaff was developing his electrostatic generator (p. 167). In both machines the maximum p.d. depends on the quality of the insulation surrounding the charged electrode. By encasing the whole apparatus in a pressure tank containing *freon* gas Van de Graaff generators have been made to work at p.d.'s up to 7 million volts.

The ions to be accelerated are produced in an ion source contained inside the charged electrode; a hydrogen tube gives protons, heavy hydrogen deuterons, helium α-particles, etc. Some of the ions then pass through a small hole into the main accelerating tube. This is lined with a number of cylindrical electrodes maintained at intermediate p.d.'s (Fig. 16.26). The electric field is thus con-

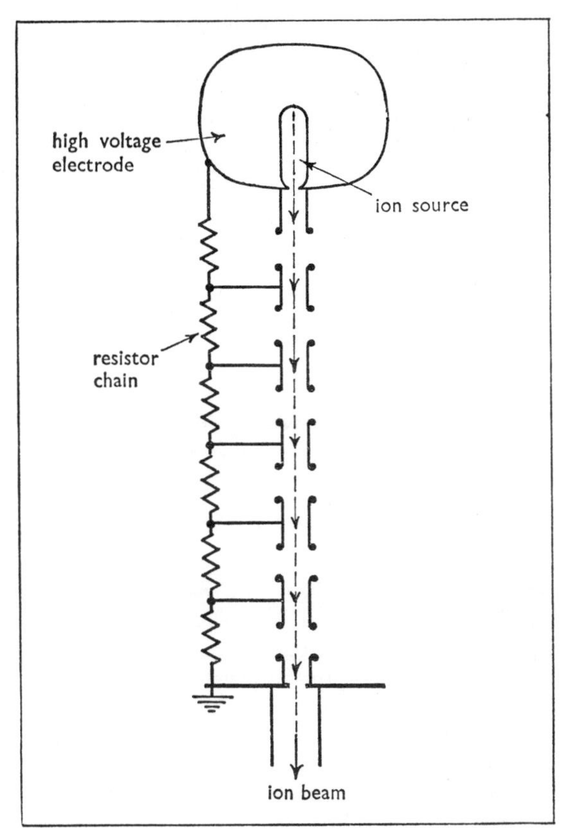

Fig. 16.26 An electrostatic accelerator tube for use with a high-voltage generator

centrated in the gaps between the cylinders, and has the effect of focusing the ions into a narrow beam.

The energies achieved were less than those of natural α-particles; but the strong, collimated beams of exactly known energy enabled a wide range of new experiments to be performed. Also for the first time it was possible to observe artificial nuclear reactions produced by protons and deuterons.

The linear accelerator

In this machine high energies are produced without employing particularly high p.d.'s by using what is known as the principle of *synchronous acceleration.* The electrode system consists of a series of coaxial cylinders which increase in length towards the target at the far end (Fig. 16.27). A high-frequency alternating p.d. is applied to the electrodes, as shown, the odd-numbered cylinders being joined to one terminal of the supply and the even-numbered ones to the other. Any positive ions that reach the gap between cylinders A and B at a moment when B is at a high negative potential with respect to A will be strongly accelerated. But once inside the hollow electrode B they are travelling in a field-free region and are unaffected by the changing potential of the electrode. The length of each cylinder is so chosen that the time taken by the ions to move through it is equal to half the period of oscillation. The electric field in the next gap (between B and C) will then be in the right phase to accelerate the ions again as they pass through it; and the same applies at each gap right down the tube. Suppose the peak p.d. of the supply is 200 kV. Then at each gap the ions acquire about 200 keV of energy; and the final energy is limited only by the number of electrodes that can be accommodated inside a tube of reasonable length. Very high frequencies are employed so that the electrodes can be as short as possible.

This machine was developed first for accelerating heavy ions. But it has also been adapted for electrons. In this case the electrode system is in some ways rather simpler. No particle can travel faster than the speed of light; and at 2 MeV an electron has already reached 98% of this speed. Scarcely any acceleration is then possible, and any

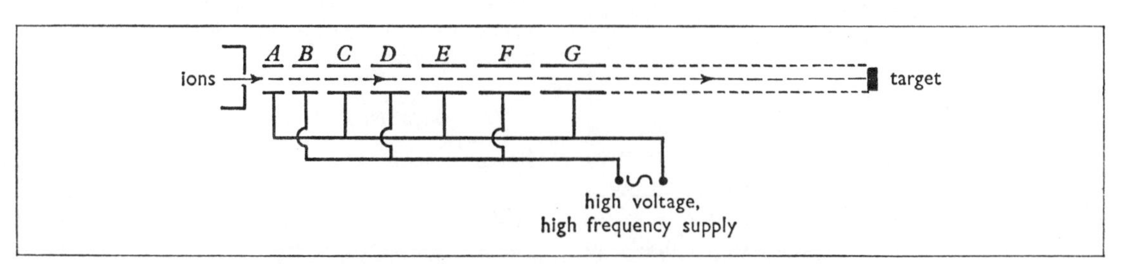

Fig. 16.27 The principle of a linear accelerator

gain in energy comes rather from the increase in mass of the particle. Thus after passing through the first few electrodes the electron is virtually moving at constant speed, and all subsequent electrodes can be of equal length. However, the technical problems of getting the p.d. to each gap at the right moment in the right phase are very considerable.

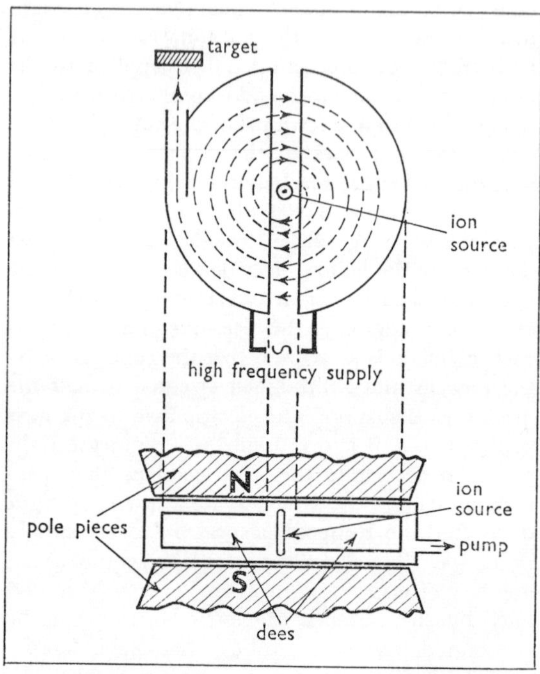

Fig. 16.28 The principle of the cyclotron

The cyclotron

This machine also employs the principle of synchronous acceleration; but excessive length is avoided by using a magnetic field to wrap the paths of the particles into tight spirals. The high frequency p.d. is applied to two hollow D-shaped electrodes (Fig. 16.28). The whole system is placed in an evacuated box between the poles of a very large magnet. The positive ions are injected near the centre and are accelerated each time they cross the gap between the D's. Inside the D's there is no electric field, and they move in semicircular orbits. As a particle gains speed, the radius of its orbit in the magnetic field increases proportionately, but its period of revolution remains constant. If this is also the period of oscillation of the alternating p.d., the field in the gap between the D's is automatically in synchronism with the motion of the particles. The frequency f required for this may be worked out as follows. Suppose the flux density of the magnetic field is B teslas, and that the particles are of charge e and mass M. Then the radius r of the orbit of a particle of velocity v is given by

$$Bev = \frac{Mv^2}{r} \qquad \text{(p. 217)}$$

$$\therefore \frac{r}{v} = \frac{M}{Be}$$

The period of revolution T is given by

$$T = \frac{2\pi r}{v} = \frac{2\pi M}{Be}$$

$$\therefore f = \frac{1}{T} = \frac{Be}{2\pi M}$$

After performing about a hundred revolutions the particles reach the edge of the system and enter a subsidiary electric field that deflects them out of the circle to strike the target.

In the earliest forms of cyclotron the energy attainable was limited by the relativistic increase of mass of the particles. At 20 MeV the mass of a proton is about 2% greater than its rest mass, and beyond this point it becomes very difficult to preserve the synchronism between the motion of the particles and the alternating p.d. However, much higher energies than this have been produced by varying the frequency during the acceleration process. A high frequency is used at the moment of injection; but as the particles gain energy and move out towards the edge of the system the frequency is steadily reduced to maintain the synchronism. Machines employing this principle have produced energies up to 600 MeV. They are known as *synchrocyclotrons*. The maximum energy is decided by the strength of the magnet and the radius of its pole pieces. Magnets weighing hundreds of tons have been used in the largest machines, and the economic limit has undoubtedly been reached.

The synchrotron

In this machine the particles are accelerated inside a large annular ring (Fig. 16.29). The magnetic field is then only required over a limited region, and a ring-shaped magnet can be used to hold the particles in the channel. Since the acceleration is to

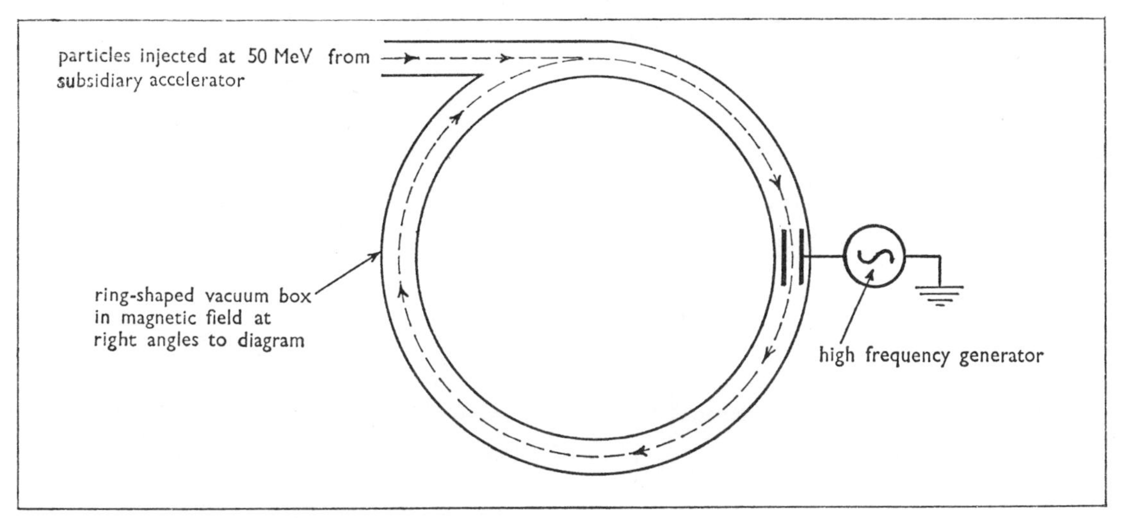

Fig. 16.29 The principle of the synchrotron

take place at fixed radius, the magnetic field must be steadily increased as the particles gain energy. At one point of the ring the particles pass through a hollow cylindrical electrode connected to a high-frequency generator, and are accelerated both on entering and leaving it. For this to happen they must enter the electrode at the moment when it is at a large negative potential (with respect to the earthed frame of the apparatus), and they must leave it after precisely half an oscillation of the potential. The frequency must therefore be steadily increased as the particles speed up, and this increase must be kept in step with the variations of magnetic field.

There seems to be no limit in principle to the energy that might be achieved with machines of this kind, except that the cost soon becomes too much for the resources of any one nation. The largest machine in operation at the time of writing is the internationally owned one at Geneva, which can accelerate protons to energies of 30,000 MeV.

To keep a true perspective in these matters it is well to remember that the highest energies so far produced by man-made accelerating machines are of the same order of magnitude as the *average* energy of cosmic-ray particles. The highest energies recorded in cosmic rays are greater than this by a factor of about 10^9, but experiments with these rare particles can only be conducted above the protective blanket of the earth's atmosphere.

16.12 More particles

The years since 1930 have seen the discovery of an ever-increasing number of 'fundamental' particles. At the time of writing over 100 are known, and the pattern revealed is one of great complexity. The scope of this book does not allow anything beyond a sketch of the more important discoveries.

The positron

In 1930 Dirac produced a theory of the electron which seemed to show that there should be a positive electron as well as the familiar negative one; and a few years later the new particle, called the *positron*, was duly discovered. It was found that a very high energy γ-ray passing near an atomic nucleus may give rise to an electron–positron pair (see Plate C). The discovery of *pair production* provided one of the most convincing demonstrations of the equivalence of mass and energy implied in the theory of Relativity. On this theory the minimum energy of a γ-ray that can cause pair production should be $2mc^2$, where m is the mass of an electron (or positron). This is equal to 1·02 MeV, which is in good agreement with the experimental results. At very high energies pair production accounts for most of the absorption of γ-rays by matter. A positron has a short life, since it soon annihilates itself in combination with an electron. The energy liberated is carried away by two γ-ray photons; each of these is of energy 0·51 MeV.

In 1934 the Joliots discovered a positron form of β-disintegration. When aluminium is bombarded with α-particles, some of the nuclear collisions result in the formation of phosphorus-30.

$$_{13}Al^{27} + {_2He^4} \longrightarrow {_{15}P^{30}} + {_0n^1}$$

The phosphorus^{30} then decays by β^+-emission with a half-life of 3 minutes.

$$_{15}P^{30} \longrightarrow {_{14}Si^{30}} + {_{+1}e^0}$$

This was the first example to be discovered of artificially induced radioactivity. Since then numerous other β^+-emitting substances have been found. Generally they are produced by bombarding stable nuclei with *protons*, since this tends to give nuclei with too many protons for stability. Bombardment by *neutrons* tends to produce nuclei with excess neutrons; and these usually revert to the more stable isotopes by β^--emission, which in effect changes one of the neutrons into a proton.

K-capture

In some radioactive nuclei, as an alternative to β^+-emission, the nucleus may capture one of the electrons from the *K*-shell of the atom. This is called *K-capture*. No particles are emitted; and the only evidence for the process is the appearance of the daughter product and the emission of the characteristic *K*-lines of its X-ray spectrum as another electron falls into the *K*-shell. Sometimes also the nucleus is left in an excited state, and a γ-ray is emitted. An example of this process is

$$_4Be^7 + \underset{(K\text{-shell})}{_{-1}e^0} \longrightarrow {_3Li^7}$$

It now appears that Dirac's theory applies more widely than just to electrons and positrons. There seems to be complete symmetry in the table of fundamental particles; to every particle there is an *anti-particle* of opposite charge (if any), but otherwise with similar properties. A particle and its anti-particle are mutually annihilated when they combine together, and the energy liberated appears in some other form. To produce a nucleon–anti-nucleon pair requires about 2000 times as much energy as for an electron–positron pair; and the full confirmation of the theory therefore had to await the development of giant accelerators in the 1950's. In a few cases a particle and its anti-particle are identical; this applies for instance to the photon. But generally, even with uncharged particles, the two particles are distinct and capable of mutual annihilation.

The neutrino

In the 1930's an unexplained feature of the process of β-decay came to light. The energy spectrum of β-particles from the nucleus was known to be a *continuous* one (p. 295). With α-particles and γ-rays only a limited number of sharply defined energies are found; and this is what the modern atomic theory leads us to expect. A decay process involves a definite reduction in the mass of the nucleus, and by the mass/energy relationship a fixed quantity of energy should be released. This applies to β-emission as much as to any other process; and yet the β-particles may actually be emitted with *any* energy up to a certain maximum. To account for this Pauli made the suggestion in 1933 that a second particle was emitted at the same time as the β-particle; the total energy carried away by the two particles could then be constant, but might be divided between them in any way whatever. The conservation of charge required the new particle to be uncharged, and it was labelled the *neutrino*.

It is now believed that its anti-particle, the *anti-neutrino*, also exists; in fact this particle is thought to be the one that emerges in β^--decay, while the *neutrino* is emitted in conjunction with a positron in β^+-decay. In modern nuclear theory electrons and neutrinos are grouped together in a class of light particles, called *leptons*. It is believed that the total number of leptons remains unaltered in any nuclear reaction. (For this purpose the number of anti-particles is counted as negative.) In nuclear equations the neutrino is represented by the Greek letter ν, and the antineutrino by the same letter with a bar over it. Thus the β^--decay of lead-210 is written

$$_{82}Pb^{210} \longrightarrow {_{83}Bi^{210}} + {_{-1}e^0} + {_0\bar{\nu}^0}$$

Counting the antineutrino as 'minus one' particle, the total number of leptons is zero before and after the reaction. The same happens in the disintegration of the neutron.

$$_0n^1 \longrightarrow {_1H^1} + {_{-1}e^0} + {_0\bar{\nu}^0}$$

But in the β^+-decay of phosphorus-30 a neutrino

is emitted; this also happens in the process of K-capture

$$_{15}P^{30} \longrightarrow {}_{14}Si^{30} + {}_{+1}e^{0} + {}_{0}\nu^{0} \cdot$$

and $$_{4}Be^{7} + \underset{(K\text{-shell})}{{}_{-1}e^{0}} \longrightarrow {}_{3}Li^{7} + {}_{0}\nu^{0}$$

The neutrino is thought to have *zero rest mass.* The theory of Relativity requires that such a particle should only exist at all if it is travelling with the speed of light! In this respect it resembles the photon, which also has zero rest mass. But unlike the photon, the neutrino has very little interaction with matter—so little that it can pass right through the earth with only a slight chance of interacting with any of the particles in its path. But in spite of this its interaction with matter has been directly observed. This was achieved in 1956 by Cowan and Reines with the intense stream of radiations emerging from a nuclear reactor. Using a large quantity of an organic liquid near the reactor, they were able to demonstrate a number of cases of the simultaneous production of a neutron and positron in the following reaction

$$_{0}\bar{\nu}^{0} + {}_{1}H^{1} \longrightarrow {}_{0}n^{1} + {}_{+1}e^{0}$$

Mesons

In 1935 the Japanese scientist Yukawa sought to explain the very large forces that operate between nucleons by postulating the existence of yet another group of particles, which came to be called *mesons.* He supposed that the circulation of mesons round the nucleons provided the means of binding the latter particles together. This way of accounting for forces is a familiar one from other parts of atomic theory. For instance the chemical bond between atoms arises from the sharing of electron orbitals between them (p. 226). The mesons were supposed to fulfil the same function for the nuclear forces. To explain the observed nature of the forces it had to be assumed that the mass of the meson was about 200 times that of the electron. The theory predicted that the meson would be unstable and have a half-life of the order of 10^{-6} s.

Not long afterwards a particle was discovered in the cosmic radiation that appeared at first to fit this description. Its mass was 207 electronic masses, and it was found both in positive and negative forms. It was unstable, and decayed (with a half-life of 2×10^{-6} s) into an electron (or positron) and *two* neutrinos. It later came to be known as the μ-meson or ***muon.*** However, the muon was found to be very penetrating, and could pass through great thicknesses of matter without any interactions with atomic nuclei. Yukawa's particle by its very nature would need to be a strongly interacting particle.

Eventually in 1947 another particle was discovered by Powell in nuclear emulsions exposed to cosmic radiation at high altitudes; this appeared to have the required strong interaction with matter. It was produced in collisions of the primary cosmic ray particles with atomic nuclei in the upper atmosphere (or in Powell's emulsions); and it could cause disintegration in another nucleus that it then struck. It has come to be called the π-meson or ***pion.*** Its mass is about 270 electronic masses, greater than that of the muon. Both positive and negative forms were discovered, and these were shown by Powell to decay into muons (and neutrinos) with half-lives of about $2{\cdot}5 \times 10^{-8}$ s (see Plate D.2).

$$\pi^{+} \longrightarrow \mu^{+} + \nu$$
$$\pi^{-} \longrightarrow \mu^{-} + \bar{\nu}$$

A neutral pion was also found. This has a shorter existence than the other two, having a half-life of about 10^{-15} s. It decays into two γ-rays.

Before very long evidence began to accumulate for yet another kind of meson, less frequently produced than the pion. This became known as the K-meson or ***kaon.*** It has a mass of 966 electronic masses. Like the other mesons it decays with a very short half-life; its modes of decay are varied and complex. Thus Yukawa's theory earned the distinction of having provoked the search for new types of particle, where none was previously suspected. But the pattern then revealed by experiment has turned out to be vastly more complex than could have been predicted.

Hyperons

The giant accelerating machines have made experiments possible with particles of ever-increasing energies. One by one whole new *families* of particles have been discovered in the range of masses heavier than the proton or neutron. Their modes of decay and their interactions with other matter are even more elaborate than those of the mesons. At the time of writing a theoretical analysis of all these particles is just beginning to emerge.

16.13 Nuclear energy

Fission

In 1934 Fermi discovered that the bombardment of uranium with neutrons produced radioactive substances with half-lives that could not be matched with those of any known isotopes of neighbouring atomic number. He concluded that he had produced new elements of atomic number greater than that of uranium. This interpretation was partly correct. It is now known that such a reaction does occur in uranium-238.

$$_{92}U^{238} + {}_0n^1 \rightarrow {}_{92}U^{239}$$

U^{239} is a β-emitting substance that decays with a half-life of 24 minutes into the first transuranic element, *neptunium*. This in turn decays by β-emission with a half-life of 2 days into *plutonium* (p. 281).

$$_{92}U^{239} \rightarrow {}_{93}Np^{239} + {}_{-1}e^0 + {}_0\bar{\nu}^0$$
$$_{93}Np^{239} \rightarrow {}_{94}Pu^{239} + {}_{-1}e^0 + {}_0\bar{\nu}^0$$

Then in 1939 Hahn and Strassmann found by chemical analysis that a number of elements of medium atomic mass were also formed when uranium was bombarded with neutrons. They suggested that it might be possible for a uranium nucleus to break up into two or more fragments when it absorbs a neutron, a process now known as *fission*. The hypothesis was soon confirmed in cloud chamber photographs. It was found that the energy released in the fission of one uranium nucleus was about 200 MeV! The nucleus may break up in many different ways; usually the two main fragments are unequal. Most of the immediate fission products are radioactive.

Just before the second world war it was discovered that, in addition to the two main fragments, several high-speed neutrons were emitted at the moment of fission. There was thus the possibility of starting a chain reaction in a piece of uranium. The fission of one nucleus might produce enough neutrons to cause the disintegration of several more; these in their turn could trigger off the fission of an even larger number; and so on. In this way it might be possible to tap some of the vast amount of energy stored in the material (200 MeV per nucleus).

Fission has been found to occur in a number of heavy elements. Both of the chief uranium isotopes show the effect. In U^{238} it can only happen with high-energy neutrons (>1 MeV); at lower energies the most probable result is the absorption of the neutron to form neptunium-239. As most of the neutrons evolved at the moment of fission are of lower energy than 1 MeV, there is no possibility of a chain reaction developing in this isotope (over 99% of natural uranium). But in U^{235} fission is caused chiefly by slow neutrons; and the same applies to plutonium-239. Both these materials can support chain reactions. All that is necessary to start the reaction is to bring together enough of the material in one confined space. With a small quantity of fissile material too many of the neutrons escape from the surface before colliding with another nucleus. But there is a certain critical size above which enough neutrons remain in the material to support the reaction. The *atomic bomb* consists essentially of an arrangement for shooting together suddenly two subcritical masses of fissile material. There are always a few neutrons present to initiate the reaction, which then spreads with great rapidity through a substantial fraction of the material—with effects that need no description.

In the *nuclear reactor* the chain reaction is made to proceed at a steady controlled rate. This is achieved by absorbing a proportion of the neutrons in some non-fissile material, so that on average *one* neutron from each fission event remains in the reactor to produce further fission. Any increase in the number of neutrons beyond this point causes the reaction rate to rise. In one kind of reactor natural uranium is used as the 'fuel', though only the small proportion of U^{235} is actually consumed. To prevent most of the neutrons being absorbed by the heavier isotope the fuel is arranged in the form of slender rods which are embedded in a material of low atomic mass known as a *moderator*. Most of the neutrons escape at once from the rods into the moderator, where collisions quickly reduce their energy to that of the atoms around them (about 0·03 eV). When they re-enter the fuel rods they are travelling at speeds at which they are very strongly absorbed by U^{235} and scarcely at all by the heavier isotope. By this means the more abundant isotope is prevented from interfering with the reaction. However, a certain proportion of the neutrons are inevitably absorbed by the U^{238}, so that there is a steady production of plutonium in the reactor. At intervals the fuel rods are removed, and the plutonium is extracted from them by chemical means. The reactor can therefore be used as a source of pure fissile material (plutonium). This is an alternative to the lengthy process of separating the two isotopes of uranium to obtain pure U^{235}; this can be done, but the chemical separation of plutonium from uranium is a much simpler technique. The reaction rate is adjusted by means of control rods made of materials that absorb neutrons very strongly; cadmium and boron are found to be suitable for this. The control rods lie in channels running through the reactor core. By withdrawing them slightly the absorption of neutrons is decreased, and the reaction rate goes up. Automatic devices are of course provided to move the rods deeper into the core if an excessive temperature rise occurs. The moderator must be of a material that does not itself absorb neutrons to any extent. Carbon (in the form of graphite) and heavy water have both been used successfully.

One of the most important uses of nuclear reactors has been the production of radioisotopes. Many of these are formed as the fission products of the uranium, and are extracted when the fuel rods are changed. Caesium-137 and strontium-90 are familiar examples. Others are produced by irradiation with neutrons. To do this a suitable compound is drawn into the centre of the core, where the flux of thermal neutrons may well be more than 10^{16} per m^2 per second. Sodium-24 (p. 306) is manufactured in this way.

$$_{11}Na^{23} + {}_0n^1 \rightarrow {}_{11}Na^{24}$$

Neutron irradiation has also been used as a means of non-destructive chemical analysis—a technique of great value to archaeologists. The constituents of the object to be analysed react in different ways with neutrons and can be identified by the radiations they subsequently emit. In this way the compositions of ancient coins have been measured with sufficient precision to establish the mine from which the metal was taken.

Most of the energy liberated in the reactor appears in the form of heat, which must be carried away by a cooling fluid. In a nuclear power station this heat is used to generate steam which drives turbines and alternators of conventional design. The great merit of this type of power station is the small fuel consumption. A few pounds of uranium are sufficient to produce hundreds of megawatts of power for a whole day. Fuel transport costs are therefore trivial, and nuclear power stations can be sited in remote places where a conventional power station would be uneconomic. A serious practical problem encountered is the safe disposal of the large quantities of radioactive 'waste'.

Once a sufficient stockpile of plutonium-239 is available, it is possible to operate what is known as a *fast breeder reactor*. No moderator is used, and the core consists of a set of plutonium rods. They are enclosed in a container of natural uranium. The fission proceeds in the core for fast and slow neutrons alike. The neutrons that escape from the core are absorbed by the U^{238} of the containing walls, thereby generating further supplies of plutonium in it. The reactor therefore 'breeds' more fuel for itself in the containing walls while it is using up the fuel in the core. It is even possible that reactors using this principle might generate more fuel than they use—which would be very satisfactory!

Fusion

In the fission process energy is released because the fragments of the explosion are of much lower mass than the original nucleus. But there is also a considerable reduction of mass when light nuclei are fused together. Such reactions are readily observed in experiments with high-energy beams of protons and deuterons. For instance

$$\underset{\text{(protons)}}{{}_1H^1 + {}_1H^1} \rightarrow \underset{\text{(deuteron)}}{{}_1H^2} + {}_{+1}e^0 + {}_0\nu_0$$

and $${}_1H^2 + {}_1H^2 \rightarrow {}_2He^3 + {}_0n^1$$

In each case the products are of lower mass than the original particles, and considerable quantities of energy are released (as kinetic energy of the particles). Numerous such reactions have been discovered. In each case one condition for the reaction to proceed is that the two reacting particles should approach one another at sufficient speed to overcome the coulomb repulsion between them. To achieve this on a large scale it is only necessary to raise the temperature of the ingredients to the point at which some of the particles will have the required speed. For this reason these are known as *thermonuclear reactions*; unfortunately temperatures of about 10^8 °C are needed to bring them about.

The energy of the sun and stars is evolved by cycles of thermonuclear reactions, whose net effect is the conversion of hydrogen into helium. The formation of each helium nucleus liberates 27 MeV of energy, more than 0·8% of the mass of the original particles being lost in the process. The sun loses some 4 million tons of matter per second in this way; and there seems no reason why it should not continue in this profligate style for the next 10^{10} years! Paradoxically a larger star than the sun uses up its greater resources of hydrogen fuel in a shorter time; and the largest stars may effectively consume all their hydrogen in about 1 million years.

The world contains almost unlimited resources of hydrogen in the oceans. But so far, mankind's only successful attempt to tap this store of energy has been the *hydrogen bomb*. In this weapon an ordinary fission bomb is used as a detonator to bring the other ingredients to the temperature at which thermonuclear reactions start. There seems to be no theoretical limit to the size of such a bomb or the devastation it might produce. If on the other hand a way could be found of maintaining a controlled thermonuclear reaction, a source of cheap and abundant energy would be available—to the very great benefit of the human race. The resources of several nations have for many years been directed towards this goal, but at the time of writing the prize still remains tantalizingly out of reach.

A.1 A typical fan of α-ray tracks in a cloud chamber. The tracks are almost all straight, and heavy ionization is produced along them (p. 286). (C. T. R. Wilson)

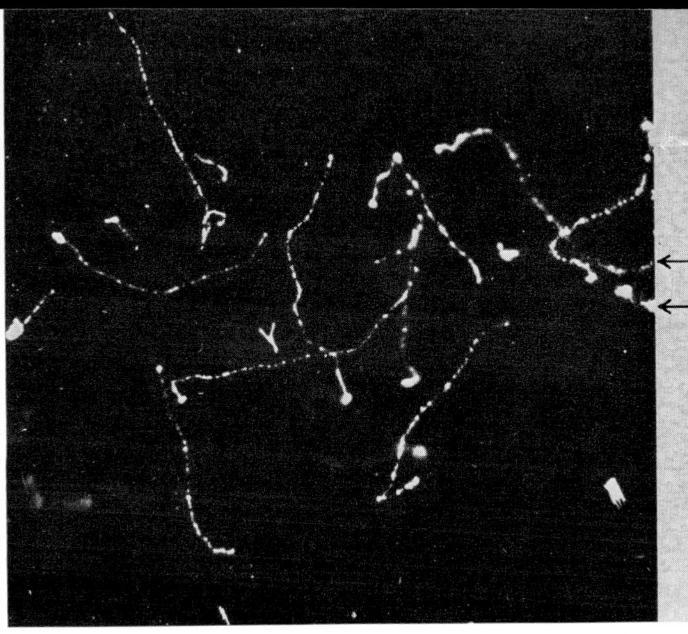

A.2 Cloud chamber tracks produced by a beam of X-rays entering the chamber from the right. The tracks start in the path of the beam and are identical with those of weak β-rays. Similar tracks are caused by γ-rays. Their lengths depend on the wavelength of the radiation involved. In this case the K-series X-rays from silver were used, and the tracks are about 1·5 cm long (energy 24 keV). (C. T. R. Wilson; print prepared by W. H. Andrews)

A.3 A classical cloud chamber photograph, showing the reaction of an α-particle with a nitrogen nucleus. The long thin track across the picture is that of a proton emerging from the collision; the α-particle is absorbed in the process forming a nucleus of oxygen-17 (p. 302). (P. M. S. Blackett & D. S. Lees (1932), *Proc. Roy. Soc. A*, **136**, 325)

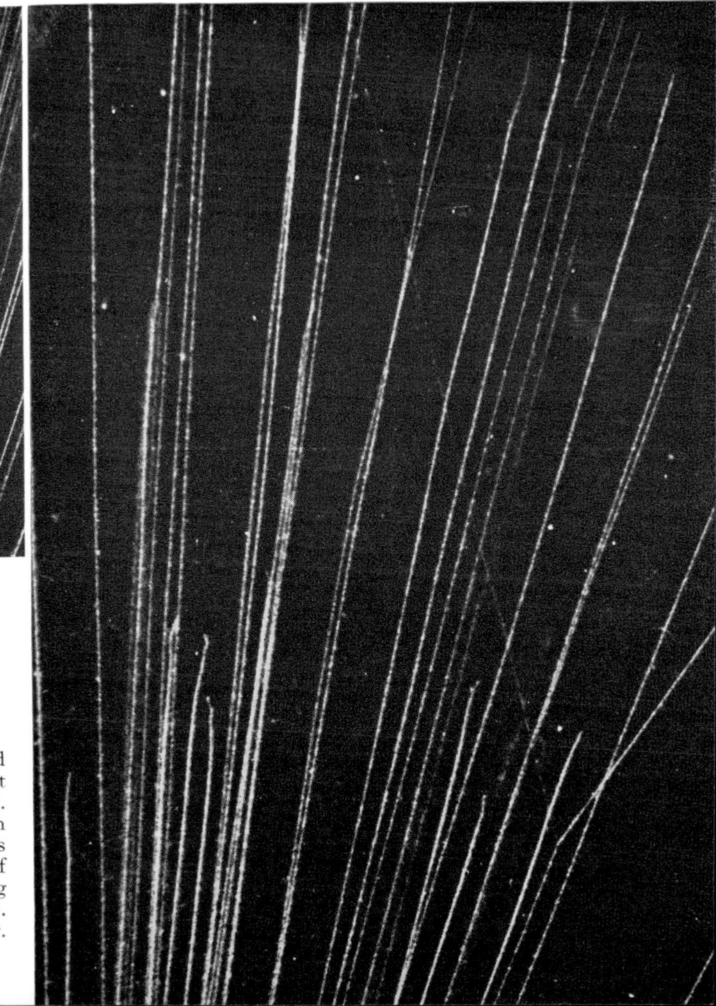

A.4 The cloud chamber in this case was filled with hydrogen. The picture shows the impact of an α-particle on a hydrogen nucleus (proton). The α-particle is deflected and the proton projected forward. Measurement of the angles involved in the collision shows that the ratio of masses of the particles is close to 4, confirming this analysis of the event (p. 302). (P. M. S. Blackett & D. S. Lees (1932), *Proc. Roy. Soc. A*, **136**, 325)

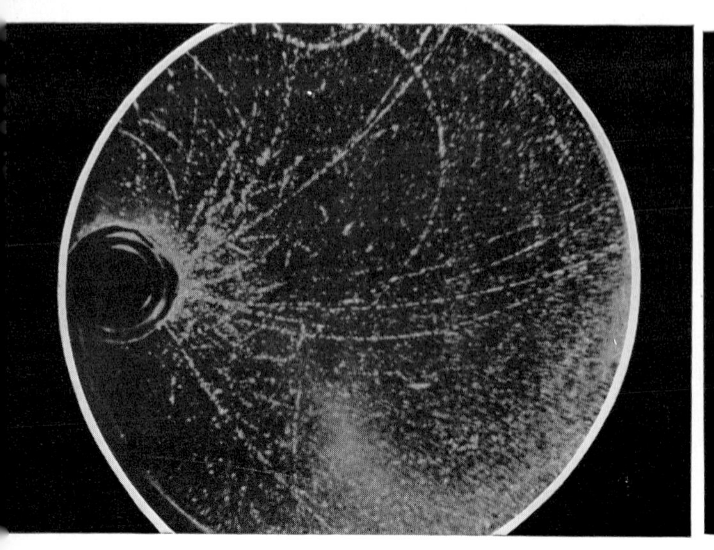

B.1 A cloud chamber photograph of the β-rays from phosphorus-32. The whole chamber is placed in a magnetic field (directed up out of the diagram), and the tracks are bent into arcs, the fastest particles being least deflected. There is clearly a considerable range of speeds represented in the source (pp. 294–5). (Kurie, Richardson & Paxton (1936), *Phys. Rev.*, **49,** 368)

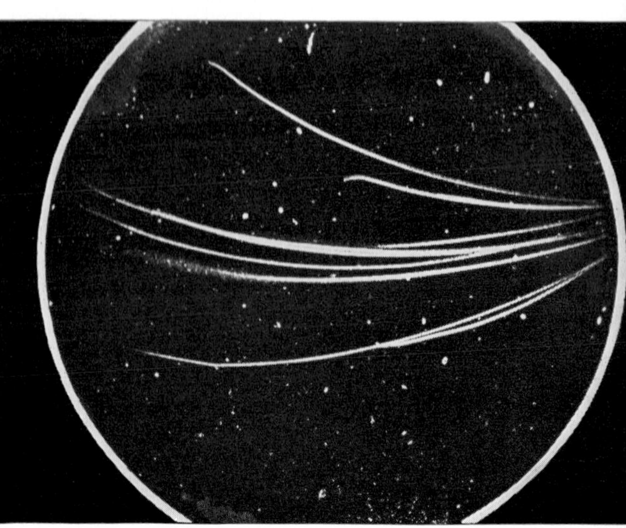

B.2 α-rays are not readily deflected by magnetic fields. For this cloud chamber photograph a flux density of over 40,000 gauss was used (directed up out of the diagram). Notice that the tracks of the group of particles of shorter range are more sharply curved than those of longer range. One of the tracks shows a large-angle deflection near its end, presumably the result of a close nuclear encounter. (P. Kapitza (1924), *Proc. Roy. Soc. A*, **106,** 622; print prepared by W. H. Andrews)

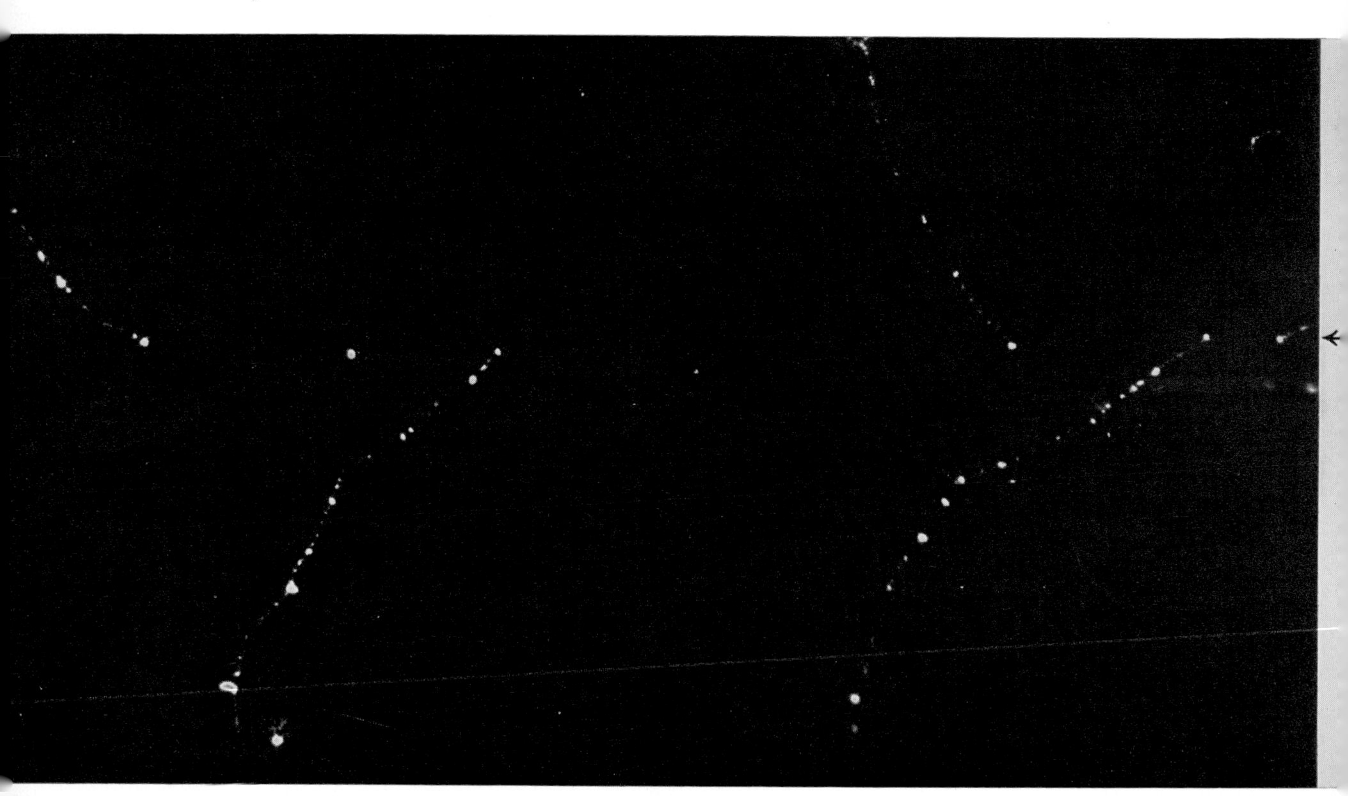

B.3 In this photograph a very narrow beam of X-rays enters the chamber from the right, as indicated. The tracks observed are those of electrons ejected from the molecules of the air in the path of the beam. There is nothing to mark the path of an X-ray photon between the source and the point at which it gives up its energy to a single electron (p. 298). (C. T. R. Wilson (1923), *Proc. Roy. Soc. A*, **104,** plate 5)

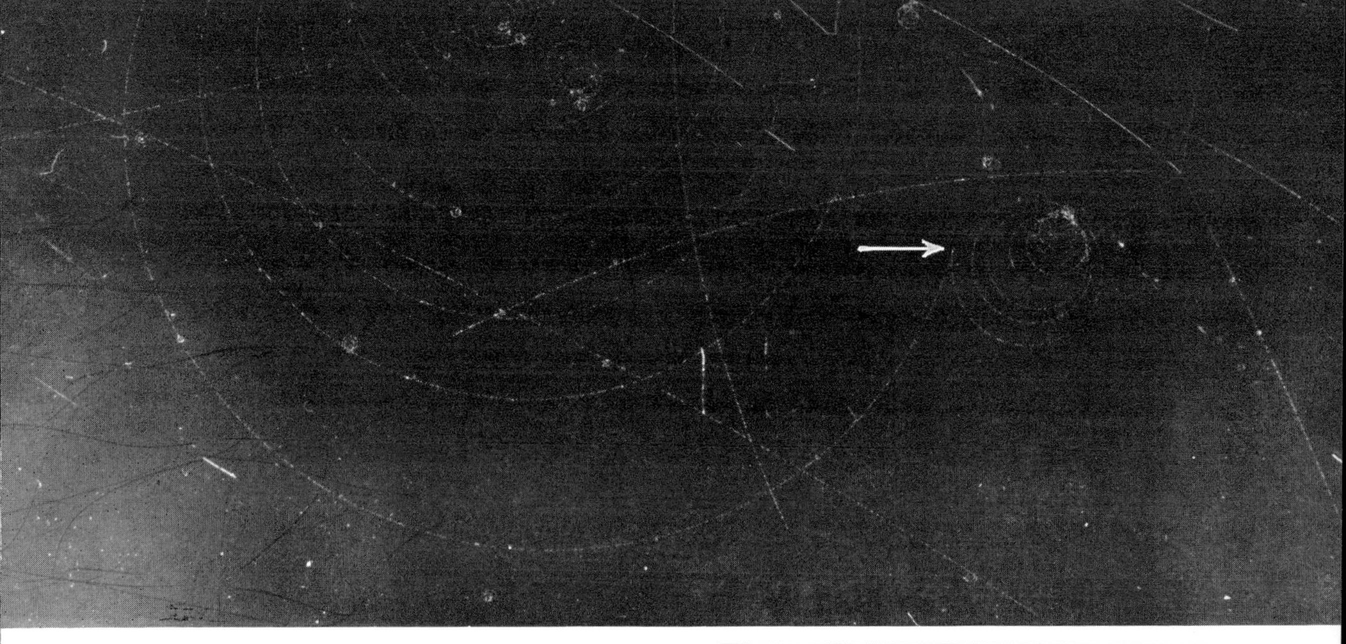

C.1 An interesting example of pair production photographed in a *bubble chamber*. In this method of detection the particles pass through a tank of liquid hydrogen. The pressure on the liquid is then suddenly reduced, and the hydrogen is brought to an unstable condition in which bubbles of vapour form along the paths of any charged particles. In the present instance a beam of anti-protons (which are negatively charged) was directed into the chamber from the top left of the picture. Most of these suffer annihilation with protons in the chamber, the energy released being carried away by pions and γ-rays. One of the γ-ray photons forthwith changes into an electron–positron pair. Since the whole chamber was in a magnetic field at right-angles to the plane of the photograph, the two particles curve in opposite directions and move in tightening spirals as they gradually come to rest (p. 315). (Cavendish Laboratory; print prepared by R. Marsh)

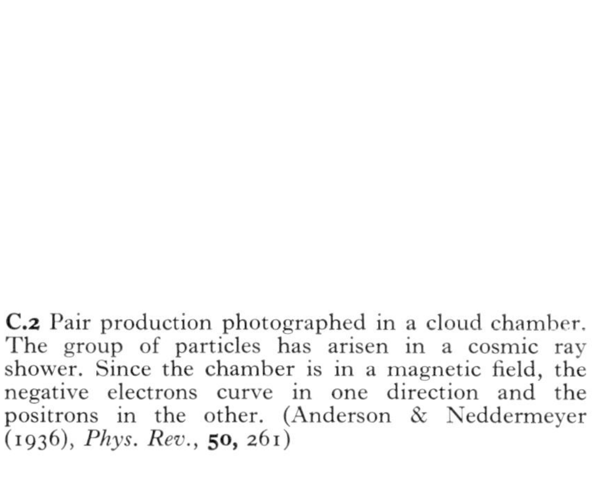

C.2 Pair production photographed in a cloud chamber. The group of particles has arisen in a cosmic ray shower. Since the chamber is in a magnetic field, the negative electrons curve in one direction and the positrons in the other. (Anderson & Neddermeyer (1936), *Phys. Rev.*, **50,** 261)

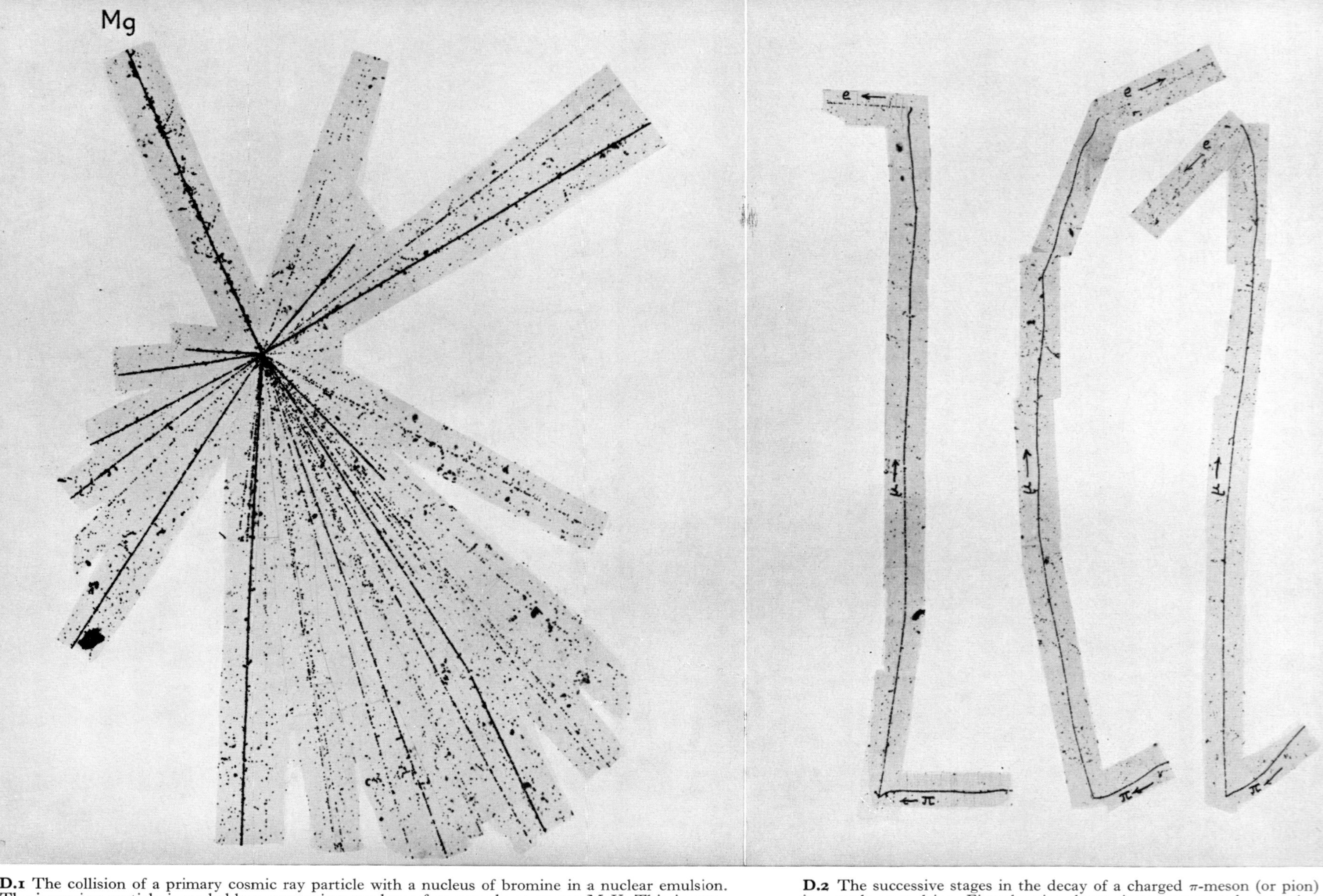

D.1 The collision of a primary cosmic ray particle with a nucleus of bromine in a nuclear emulsion. The incoming particle is probably a magnesium nucleus of energy about 15,000 MeV. This is completely dissolved into its component nucleons, and the bromine nucleus is shattered into several fragments. At still higher energies collisions of this sort produce π-mesons (or pions). (Dainton & Kent, Bristol University, 1950)

D.2 The successive stages in the decay of a charged π-meson (or pion) in a nuclear emulsion. First the pion decays into a muon and a neutrino (which, being uncharged, leaves no track). Then the muon decays into an electron and two further neutrinos (p. 317). (Bristol University, 1949)

17: Semiconductors

17.1 The electronic theory of solids

The characteristic property of a conducting substance is that it contains 'free' electrons, i.e. electrons that are not fixed at any particular site in the crystal lattice, but are free to wander through it in any direction. Under equilibrium conditions these electrons diffuse at random rapidly in all directions and there is no net movement of electric charge. They form in fact a kind of 'electron gas' confined within the crystal structure. When the substance is placed in an electric field, the electrons gain momentum from the field and a slow drift of the electron gas through the material takes place; this is superimposed on the rapid random diffusion and constitutes an electric current. The average velocity of the charge carriers in the direction of the field is always quite small (p. 9), though the disturbance in the electric field that accompanies their motion is propagated through the material at almost the speed of light (p. 255), so that the charge is set in motion almost simultaneously at all points.

The conductivity of a material depends on two factors:

(*i*) the freedom of movement (or *mobility*, p. 20) of the charge carriers in it; i.e. the average velocity of drift acquired per unit electric field;

(*ii*) the number of charge carriers per unit volume (p. 21).

The latter quantity varies over an enormous range for different types of materials. In a metal about one valence electron per atom is effectively free to move through the crystal lattice and conduct electric currents; the number of conduction electrons does not change much with temperature. In an insulator there are virtually no free electrons—at least the number is less than in a metal by a factor of 10^{20} or more. Intermediate between these extremes lies the class of materials known as *semiconductors*. In a pure semiconductor at very low temperatures there are no free electrons and it behaves as an insulator. But as the temperature rises the material exhibits increasing conductivity as electrons are liberated by the thermal vibrations of the crystal. But even at room temperature only about one atom in 10^{10} has parted with an electron in this way.

We need now to consider how exactly an electron becomes 'free' in a crystal lattice, and why this spontaneous 'ionization' of the atoms occurs readily in some materials and not in others; while in semi-conductors it is dependent on energy being available in the thermal vibrations of the lattice.

In an isolated atom the valence electrons can exist in a number of distinct orbitals each of sharply defined energy (p. 223 *et seq.*). Normally it exists in the *ground state*, which is the orbital of lowest possible energy; only when the atom absorbs energy from radiation or from a collision with another particle will the valence electron change to an orbital of higher energy—from which it quickly reverts to the ground state, with the emission of radiation. When two atoms come close together, the possible orbitals of their valence electrons are modified by the change in the electric field in which they move, and the energy levels are altered. If a stable chemical bond is formed (p. 226), the pattern of available energy levels is inevitably more complex than for the constituent atoms on their own; a large molecule has a very large number of closely spaced energy levels in any of which the valence electrons may exist, though they will always tend to occupy chiefly the states of lowest energy. An assembly of atoms in a crystal must be regarded as a single chemical unit, and in this case the energy levels become spread out into a series of bands. Each band consists of a very large number of levels (one pair of levels for each atom in the crystal); but these overlap to form in effect a continuous band of available energies. At sufficiently low temperatures (and in the absence of an external supply of energy such as radiation) all the electrons will exist in the states of lowest possible energy, and will therefore fill or partially fill the lowest available energy band. This is called the *valence band*. This situation represents a balanced electronic configuration in which there is no net movement of electrons in any particular direction, although the orbital of each electron is effectively spread out over a sizeable region of the crystal lattice. In order to conduct an electric current some of the valence electrons must be in an energy band that is only partly filled. This can happen either by some of the electrons being given sufficient energy to raise them into the next lowest energy band—called the *conduction band*—or by the valence band being only partly filled, as in the alkali metals that have only one valence electron per atom. If an electron is in a partly filled band, it is possible for its energy to be increased steadily in infinitesimal steps—in other words the electron can be accelerated by an applied electric field, thus gaining energy from it. In this way it can take part in the conduction of an electric current, and we may call it a 'free' electron.

The differences between the properties of the various classes of materials arise from the nature of the valence and conduction bands and the width of the energy gap between them.

(*i*) *Metals*. The majority of pure elements in crystalline form have metallic properties. In these the gap between the valence and conduction bands is non-existent, and in many cases the two energy bands actually overlap (Fig. 17.1); alternatively, the valence band is only partly filled. A metal therefore has relatively high conductivity at all temperatures. The proportion of valence electrons that are free to be accelerated increases slightly with temperature, and this tends to increase the conductivity. However, this effect is more than counteracted by the increasing interaction between the electrons and the crystal lattice as the temperature rises, which reduces the mobility of the electrons. The conductivity of a metal therefore falls with rising temperature (i.e. the resistivity rises, p. 21).

The freedom of movement of the valence electrons is also responsible for the high *thermal* conductivity of metals. Since the number of effectively free electrons increases (slightly) with temperature,

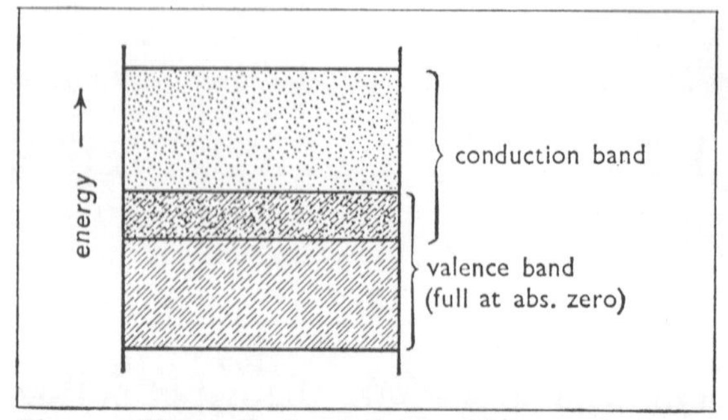

Fig. 17.1 The energy band diagram for a metal in which the conduction and valence bands overlap

a temperature gradient through the material produces a gradient in the concentration of the free electrons in it. The rapid diffusion of the electrons then tends to equalize the electron concentration and conveys the surplus energy between the regions of high and low temperature. Electrical conduction consists of the diffusion of electric charge through the material under conditions of non-uniform electric potential; while thermal conduction consists of the diffusion of thermal energy under conditions of non-uniform temperature. In a metal both processes take place essentially by the same mechanism, and some connection between the two coefficients of conductivity is therefore to be expected. An examination of tables of physical properties shows that the ratio of thermal and electrical conductivities at a given temperature is approximately constant for a large range of metals. This is known as the *Wiedemann-Franz law*; it provides strong confirmation of the electronic theory of metals. The law does not hold, even approximately, for non-metals.

In addition to the conduction of heat through a metal by the free electrons in it there is also a transference of thermal energy by the vibrations of the ions of the crystal lattice. The rate of conduction of heat on this account is comparable with that occurring in poor (non-metallic) conductors of heat, and it is normally swamped by the much more rapid process of electronic conduction. It therefore has little effect on the validity of the Wiedemann-Franz law.

(*ii*) *Insulators*. In this class of materials the valence band is completely filled and there exists a large gap between the valence and conduction bands (Fig. 17.2). In good insulators the gap is

5 electron-volts or more. (This means that an electron must gain an amount of energy equivalent to being accelerated through a p.d. of 5 volts in order to be transferred to the conduction band.) This is about 100 times as great as the average energy of vibration of the atoms of the crystal lattice, and the chance of an electron being set 'free' for conduction is therefore very remote. However, if a sufficiently strong electric field is applied, the electrons may gain the requisite energy from it, and the insulation breaks down. The insulation strength therefore depends (amongst other things) on the width of the *forbidden zone* between the two energy bands.

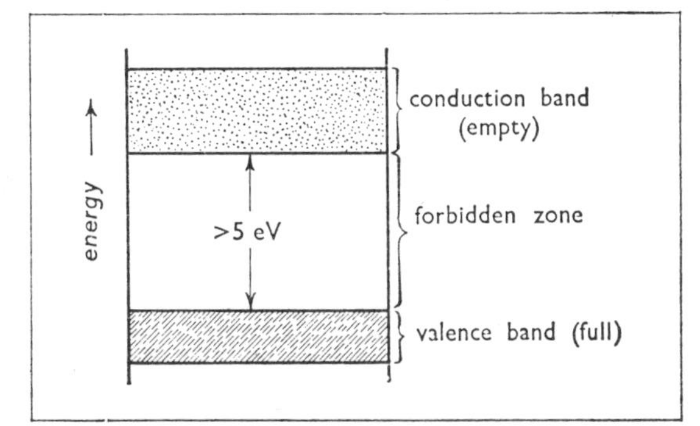

Fig. 17.2 The energy band diagram for an insulator in which there is a large forbidden zone between the valence and conduction bands

The chance of electrons appearing in the conduction band increases when the temperature is raised. The conductivity therefore increases with temperature, and may even become relatively high. For instance, if a glass rod is heated to a temperature just below its melting point, a large enough current may be passed through it to maintain its temperature and even to melt it.

(*iii*) *Semiconductors.* In semiconductors, as in insulators, the valence band is completely filled when the electrons are all in their ground states. However, in this case the width of the forbidden zone between the valence and conduction bands is of the order of 1 electron-volt (Fig. 17.3). This means that at low temperatures there will be virtually no electrons in the conduction band and the conductivity will be very low. But at normal room temperatures there is a small but significant chance of an electron gaining sufficient energy from the vibrations of the crystal lattice to raise it

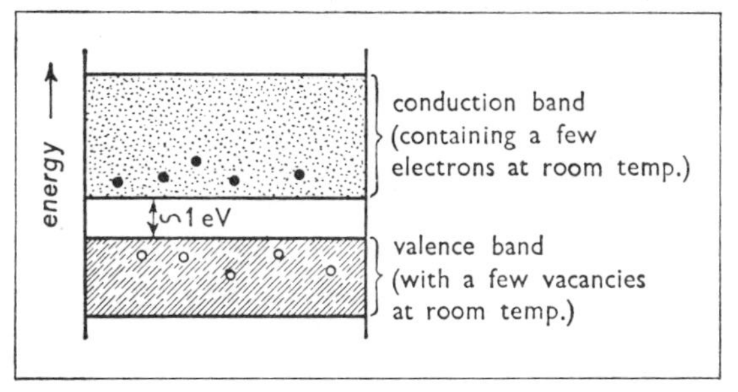

Fig. 17.3 The energy band diagram for a semiconductor in which there is a small energy gap between the valence and conduction bands

into the conduction band; and a very small proportion (typically about 1 in 10^{10}) of the atoms will be ionized in this way at any instant. The most commonly used semiconductors at present are *germanium* and *silicon.* In germanium the width of the forbidden zone is 0·72 eV; in silicon it is 1·12 eV. At a given temperature therefore silicon has fewer electrons in the conduction band, and its conductivity is considerably less than that of germanium.

Because of the low level of 'ionization' in a semiconductor the conduction of a current through it can take place by two distinct mechanisms; Fig. 17.4 shows how this comes about. The black dots represent the few conduction electrons. Each of these has escaped from some atom, leaving behind it a positively charged *hole* in the electron structure of the crystal. If a p.d. is applied to the specimen, the free electrons move under the action of the electric field, as we should expect; and the current is partly accounted for in this way. But it is also possible for a *bound* electron to move into a neighbouring hole; the electron still remains a bound one (in its new site), but the hole is transferred to the atom just vacated. Part of the conduction is thus by the movement of positive holes

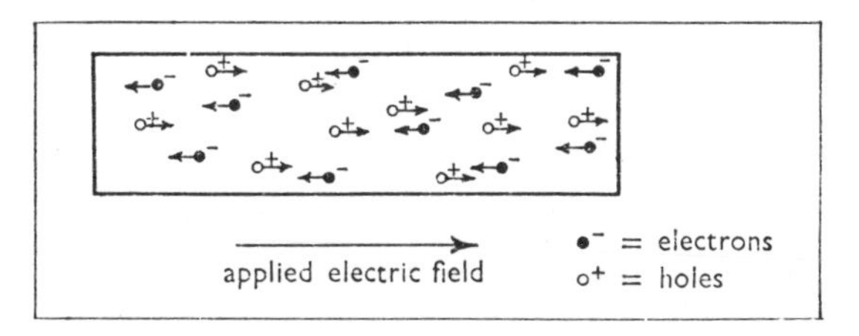

Fig. 17.4 The flow of electrons and positive holes in an intrinsic semiconductor

through the material; and it is found that these behave very much as though they were positively charged electrons.

In terms of the energy level structure of the material the effect may be explained as follows. The existence of a hole in the electron structure implies that there is an unoccupied energy level in the valence band. The valence band now possesses an *unbalanced* electronic configuration which can represent a net movement of charge in one direction. Under the action of an electric field the unoccupied energy level of the hole can be filled by a bound electron from a lower energy level in the valence band. In this way the energy represented by the hole is increased by the electric field; and this can be regarded as an acceleration of the positive charge of the hole in the direction of the field. Thus an applied p.d. causes the electrons in the largely unoccupied conduction band to drift in one direction, while the holes in the almost fully occupied valence band drift in the opposite direction.

The behaviour of electron and holes in a semiconductor is analogous to the movement of water in a tilted tube. If a few drops of water are introduced, they will run down to the lower end of the tube gaining energy from the gravitational field. If the tube is sealed at both ends and is completely full of water, no net movement of water takes place when the tube is tilted. But if now a few 'holes' are introduced in the form of air bubbles, these will move *upwards* in the gravitational field. The movement of the bubbles upwards accompanies a net downward displacement of the water; but we can analyse the process by treating the bubbles as though they were droplets of negative weight (but, of course, of positive mass). Likewise the movement of holes in the valence band of a semiconductor is most conveniently analysed by treating them as positively charged electrons. In the tilted tube the resistance to the motion of the water droplets would doubtless be very different from that of the bubbles. So, likewise, the mobilities of electron and holes in a semiconductor are usually quite different. In germanium the average velocity of the electrons in a given electric field is about twice as great as that of the holes; and in silicon the electrons have more than four times the mobility of the holes.

Some 'recombination' of electrons and holes is taking place continually in a semi-conductor. The energy so released goes to increase the thermal vibration of the crystal lattice. At the same time there is a small chance of the lattice providing sufficient energy to raise an electron to the conduction band, and 'ionization' (i.e. pair production of electron and holes) therefore occurs also at a small but steady rate. The number of current carriers is maintained at such a level that recombination and pair production proceed at equal rates. A rise in temperature increases the rate of production of electron-hole pairs and so increases the number of current carriers. On this account the conductivities of semiconductors rise very rapidly with temperature (p. 21).

A material in which there are equal numbers of electrons and holes is known as an *intrinsic semiconductor*. This only happens in an extremely pure material. Impurities tend to produce additional electrons or holes in the crystal structure. Since only 1 in 10^{10} of the atoms contributes a current carrier to the crystal in a pure semiconductor, an impurity level of 1 in 10^7 can be enough to increase the conductivity a thousandfold and to modify drastically the other electrical characteristics. Semiconductors acquire their useful properties from the controlled addition during manufacture of minute traces of impurities.

Germanium and silicon both crystallize in the same type of structure—the same in fact as that of diamond. Each atom has four valence electrons and is bonded to four neighbouring atoms in a tetrahedral arrangement. A double bond consisting of one pair of shared electrons thus exists between every pair of adjacent atoms. When an impurity atom of about the same size is incorporated in the crystal it simply occupies the site that would otherwise be taken by a germanium or silicon atom without affecting the structure of the lattice around it. Its effect on the electrical properties of the crystal depends on its valency.

Suppose an impurity is added having five valence electrons per atom, such as antimony or arsenic. In this case one electron per impurity atom will be left over after all the valence bonds with neighbouring atoms have been satisfied, and this will be free to move through the crystal with the other conduction electrons. These are known as *donor* impurities, and have the effect of greatly increasing the number of conduction electrons. Since the conduction is now mainly by the *negative* current

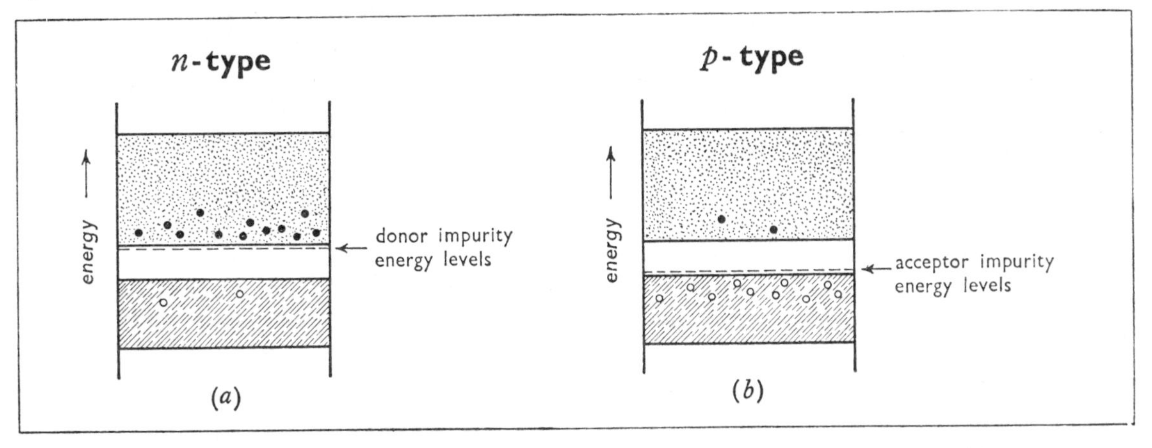

Fig. 17.5 Energy band diagrams for the two types of impurity semiconductor, (*a*) *n*-type, (*b*) *p*-type

carriers, the semiconductor is said to be *n-type.* In fact a small amount of energy is required to separate the spare electron from each donor atom; but unless the temperature is very low the available thermal energy of the lattice is sufficient to ensure that most of the donor atoms are ionized. From the point of view of the energy level diagram the donor impurity creates extra localized energy levels in the forbidden zone just *below* the conduction band (Fig. 17.5*a*). The gap between the impurity energy level and the conduction band is less than 0·1 eV, comparable with the thermal energy of an ion of the lattice at ordinary room temperatures.

It is also possible to add an impurity having three valence electrons per atom, such as *indium* or *gallium.* In order to complete the electron structure these atoms will tend to acquire an extaa electron each from their neighbours. Each one thus becomes a centre of bound negative charge, while a positive hole is created near by. These are known as *acceptor* impurities, and have the effect of greatly increasing the number of holes in the crystal structure. Since the conduction is now mainly by the *positive* current carriers, the semiconductor is said to be *p-type.* Again a small amount of energy is needed to complete the electron structure round the impurity atom and so create the extra hole; but at room temperature this is readily available from the thermal energy of the lattice. On the energy level diagram an acceptor impurity creates an extra localized energy level in the forbidden zone just *above* the valence band (Fig. 17.5*b*).

In each type of semiconductor the addition of the impurity not only provides a large number of extra current carriers—the *majority carriers,* as they are called—but it also largely suppresses the carriers of opposite type—the *minority carriers.* Thus the addition of a trace of donor impurity might, say, increase the number of conduction electrons tenfold. But it also thereby increases the chance of recombination of electrons and holes by the same factor; and the net result is a tenfold reduction in the number of holes in the lattice. The minority carriers, holes in this case, then play a very small part in the conduction of current. Likewise, a trace of acceptor impurity produces an enormous preponderance of holes in the material. However, the proportion of minority carriers increases rapidly if the temperature rises because of the increased rate of production of electron-hole pairs; and this tends to set a limit to the current carrying capacity of semiconductor devices whose functioning depends on maintaining the preponderance of majority carriers.

The Hall effect

When a conductor carrying an electric current is placed in a magnetic field, it is found that a p.d. is developed between the sides of the conductor in a direction perpendicular to the field. This is called the *Hall effect.* Thus in Fig. 17.6, if a current I is passed through the slab of conductor between the edges P and Q, and a magnetic field is applied at right-angles to the faces of the slab, a p.d. appears between the points R and S on the sides. The Hall p.d. V_H is found to be proportional to the flux density B and to the current density j (i.e. the current per unit area of cross-section); it is also proportional to the breadth b of the slab at right-angles to current and field. Thus

$$V_H \propto Bjb$$

where $$j = \frac{I}{bd} \qquad (d = \text{thickness})$$

$$\therefore V_H = hBjb \qquad \text{(I)}$$

or $$V_H = \frac{hBI}{d}$$

where h is a constant for the material known as the *Hall coefficient*.

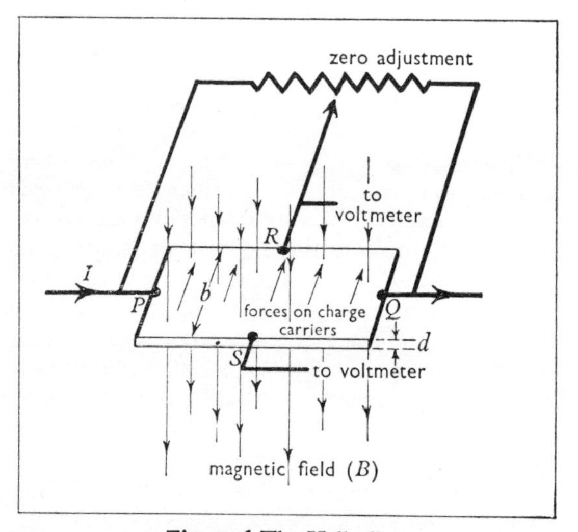

Fig. 17.6 The Hall effect

The effect is readily explained in terms of the forces acting on the moving charge carriers in the conductor. In Fig. 17.6, with the directions of current and field shown, according to Fleming's left-hand rule (p. 66) the forces on the current carriers act across the slab from S towards R, and the concentration of current carriers should increase in this direction. This will manifest itself as a difference of potential between the terminals R and S. The direction of the force is the same whichever sign the charge carriers happen to have, so that the sign of the Hall p.d. may be used to show whether n- or p-type conduction predominates in the material. The Hall coefficient h is taken as positive if the p.d. is in the direction caused by positive current carriers—i.e. for p-type material—and negative for n-type material.

With metals the Hall effect is very small; the coefficient h is of the order of 10^{-6} m^3 coulomb^{-1}, and with the electromagnets usually available in schools the Hall p.d. is unlikely to exceed 1 millivolt even with very thin specimens. But with semiconductors the effect is many orders of magnitude greater, and p.d.'s of the order of 0·1V are obtainable quite easily. The chief practical difficulty in demonstrating the effect is in the positioning of the Hall electrodes R and S. It is necessary to ensure that, in the absence of a magnetic field, R and S are at exactly the same potential. Since the Hall p.d. may well be much less than the p.d. applied between P and Q, this adjustment needs to be done with precision. Fig. 17.6 shows one way of doing it. One of the side electrodes R is connected to the slider of a high resistance potential divider joined between the ends of the specimen. Through this electrode a small current can be made to enter or leave the slab, as may be necessary to bring R to the same potential as S. With no magnetic field the slider is positioned to give zero reading on a voltmeter connected between R and S. The Hall p.d. may then be read directly from the meter when a magnetic field is applied.

The measurement of the Hall coefficient enables not only the sign but also the concentration of the majority carriers to be found; and it has therefore provided one of the most powerful techniques for studying the properties of semiconductors. The following approximate analysis shows how this may be done.

When the moving charge carriers migrate across the slab of conductor under the action of the magnetic field, they give rise to an electric field that opposes this lateral movement. The lateral separation of charge continues until the magnetic and electric forces on the charge carriers are equal and opposite. We have already shown (p. 9) that the average velocity of drift v of the charge carriers is given by

$$v = \frac{I}{nAe} = \frac{j}{ne}$$

where j ($= I/A$) is the current density, and n is the number of charge carriers per unit volume.

$\therefore$ the magnetic force on each charge carrier

$$= Bev = \frac{Bej}{ne} = \frac{Bj}{n}$$

and the electric force

$$= e \times (\text{potential gradient})$$

$$= \frac{eV_H}{b}$$

Equating the two forces, we have

$$V_H = \frac{1}{ne}.Bjb$$

This expression is in agreement with that obtained experimentally (equation I above); comparing the two equations we see that the Hall coefficient h is given by

$$h = \frac{1}{ne}$$

Since the electronic charge e is known, this enables the concentration n of charge carriers to be found. This result shows why the maximum Hall effect is found in materials having a low concentration of charge carriers—i.e. in semiconductors rather than in metals.

17.2 Junction diodes

There are a number of devices besides the thermionic diode that conduct current chiefly in one direction. These employ the special characteristics of junctions between two sorts of semiconductor or between a semiconductor and a metal. Because they have two terminals and fulfil a similar function in electric circuits to the thermionic diode, these semiconductor devices are commonly referred to as *diodes* also.

Such a diode may be made by forming regions of n-type and p-type semiconductor *within the same crystal* of germanium or silicon. At the junction electrons diffuse across from the n-type side and holes from the p-type side (Fig. 17.7). Recombination of these electrons and holes produces on either side of the boundary a narrow *depletion layer* (or *barrier layer*) relatively free of current carriers and therefore of high resistance. This leaves the n-type material positive with respect to the p-type material; the diffusion continues until the electric field developed across the depletion layer is sufficient to prevent further movement of charge carriers.

If now a p.d. is applied to the crystal making the p-type region positive (or less negative) with respect to the n-type region, the electric field in

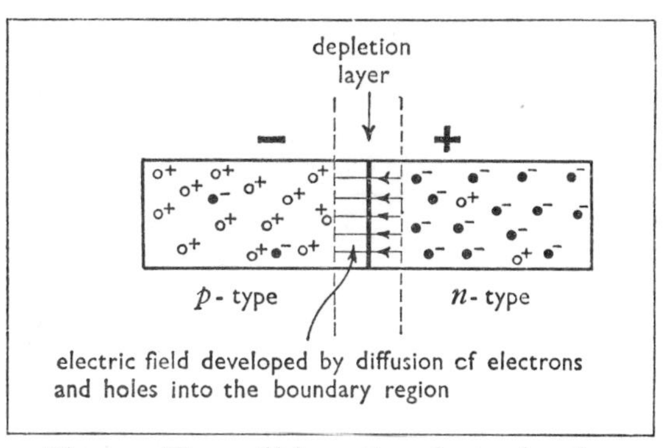

Fig. 17.7 The rectifying action of a p–n junction

the depletion layer is reduced and a current can flow across the junction. But a p.d. applied the other way round increases the field in the depletion layer, and the flow of majority carriers across the junction is further inhibited. The only current is then that due to the very small proportion of minority carriers (electrons from the left, holes from the right, in Fig. 17.7). The junction thus conducts current readily in one direction (the *forward* direction, it is called), but has a high resistance in the *reverse* direction.

To complete the diode it is of course necessary to provide metallic contacts to the n- and p-type regions of the crystal; and these junctions must themselves be of non-rectifying types. A junction between a metal and a semiconductor can form a barrier layer in much the same way as one between p- and n-type semiconductors, and this must be avoided where the connecting leads are joined on. This may be done by alloying the metallic contact with impurities of the same kind as in the adjoining region of the semiconductor—or by using solders that contain these. The impurities then diffuse into the immediately adjacent layer of semiconductor and produce there such a high concentration of current carriers that no depletion layer is able to form near the junction. This is called an *ohmic* contact. In general there will still be a contact p.d. (p. 45) between the metal and semiconductor arising from their different concentrations of charge carriers; but there is no barrier layer and the resistance is low for both directions of current flow through the junction.

One form of germanium rectifier is shown in Fig. 17.8. A wafer of n-type germanium is soldered onto a metal base, and a button of indium is then

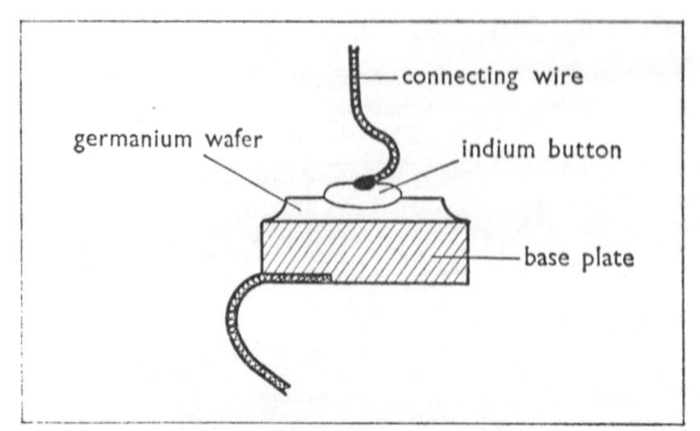

Fig. 17.8 A germanium diode

fused onto its top surface. The indium and germanium diffuse together in the boundary region, and the germanium there is thus converted into *p*-type. Connecting wires are soldered to the indium button and the base, and the whole unit is hermetically sealed in a light-proof capsule.

The electrical characteristics of a germanium diode are shown in Fig. 17.9. In the *forward* direction the current increases approximately exponentially with p.d. Very little current flows until the forward p.d. rises above about 0·2 V; it then increases very rapidly. The curve for the reverse direction is plotted separately in Fig. 17.9, using different scales on the axes, since the ranges of current and p.d. involved are quite different. For a large range of reverse p.d.'s the leakage current due to the minority carriers remains quite small. However, at a sufficiently high p.d. (the *turnover voltage*) breakdown of the barrier layer takes place, and beyond this point the diode is likely to suffer damage unless the current is limited by some means. When breakdown occurs the electric field in the barrier layer is high enough for electrons to be lifted straight into the conduction band. Thus, as the field reaches this critical value there

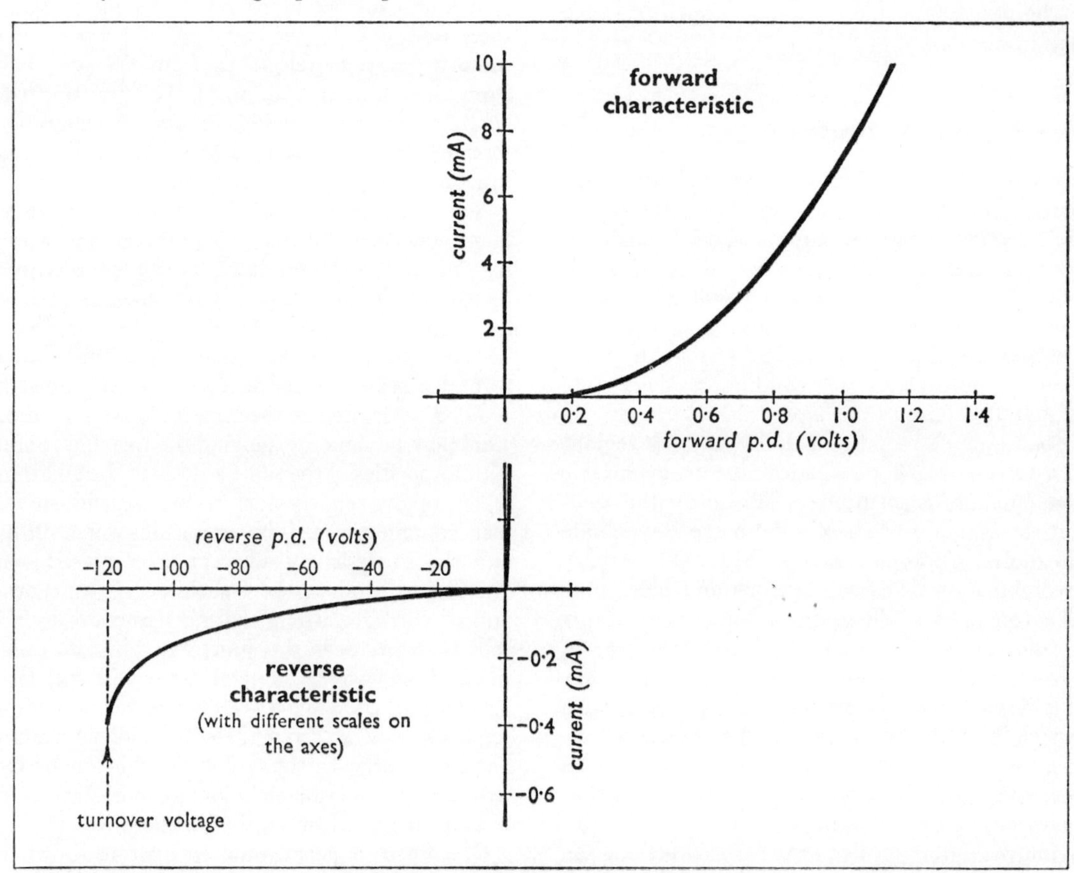

Fig. 17.9 The characteristics of a germanium diode. The forward and reverse characteristics are plotted on different scales, the latter covering a larger range of p.d.'s and a smaller range of currents

is a sudden evolution of electron-hole pairs and the current rises abruptly; this is called the *Zener* or *avalanche effect.*

In the reverse direction a junction diode behaves like a leaky capacitor between its *p*- and *n*-type regions, the barrier layer being the dielectric. This means that the efficiency of the diode as a rectifier is greatly reduced at high frequencies, since it is effectively shunted by its own interelectrode capacitance. For rapidly varying currents it is usual therefore to employ an older design of rectifier known as a *point-contact diode.* This consists of a single *n*-type crystal of germanium on which a springy metal wire (called traditionally a *cat's whisker*) is made to bear. During manufacture the diode is 'formed' by passing a relatively large current between the whisker and the crystal. This produces a thin layer of *p*-type material under the point of the wire, and the rectifying action occurs between this and the rest of the crystal. The reverse capacitance of this type of diode is sufficiently low for it to be usable up to the highest frequencies; but of course it is less robust than the usual junction type of diode.

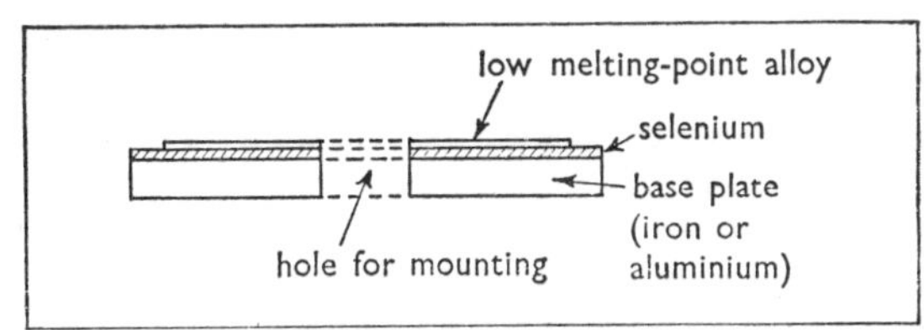

Fig. 17.10 A selenium rectifier

For small-scale power rectification the *selenium rectifier* (often called a *metal rectifier*) is commonly used (Fig. 17.10). It consists of a layer of selenium (a semiconductor) coated on a base-plate of iron or aluminium; connection is made to the front surface through a film of a low-melting-point alloy. The construction differs little from that of the selenium photovoltaic cell (p. 220), except that in the rectifier the front film of metal does not need to be transparent. Light must in fact be excluded from it, and it is usually protected by a thick coat of paint. The treatment received during manufacture forms on the front surface of the selenium a barrier layer of very low conductivity. It is thin enough for electrons from the metal film to diffuse through it into the main layer of selenium. This leaves the metal film positive with respect to the selenium, and the process continues until the electric field developed across the layer is sufficient to prevent further movement of electrons. As in the *p–n* junction diode a p.d. applied so as to decrease the electric field in the barrier layer (i.e. making the base positive with respect to the alloy film) causes current to flow. But in the reverse direction the electric field is increased and the resistance is high.

A typical small unit ($\frac{1}{2}$ inch diameter) can pass a current of 300 mA or more in the forward direction, the drop in p.d. across the unit being about 2 volts. In the reverse direction the leakage current is rather less than 50 μA, and breakdown occurs for p.d.'s above about 25 volts. To stand higher p.d.'s several discs may be stacked together in series. The current-carrying capacity can be raised by using larger diameters and by fitting cooling fins to the discs.

These properties compare unfavourably with those of a *p–n* junction diode. The reverse current in a germanium diode is much smaller, and peak reverse p.d.'s can well be 100 volts or more. The current-carrying capacity for a given area of junction is hundreds of times greater than in a selenium rectifier, and the forward drop in p.d. is much smaller for a given current. The silicon diode is even better in some respects; it can be designed to stand peak reverse p.d.'s of about 1000 volts, and the reverse currents are even smaller than in a germanium diode.

The chief problem in designing germanium and silicon rectifiers to handle high powers is in arranging for the heat evolved in them to be conducted away rapidly enough. If the temperature of a germanium diode is allowed to rise much above 100°C, irreversible changes take place in the crystal structure of the junction, and the diode is destroyed. It is therefore often necessary to bolt the diode to a *heat sink* consisting of a thick metal plate with cooling fins. The temperature of a silicon diode can be allowed to rise to about 200°C, and the problem of arranging for heat dissipation is not as acute as with germanium.

Semiconductor diodes are now employed in many of the applications in which thermionic diodes have formerly been used. For instance any of the rectifying circuits described in section 14.3 (p. 241) may be adapted to use germanium or silicon diodes. However, because of the small size and simplicity of semiconductor diodes it is often

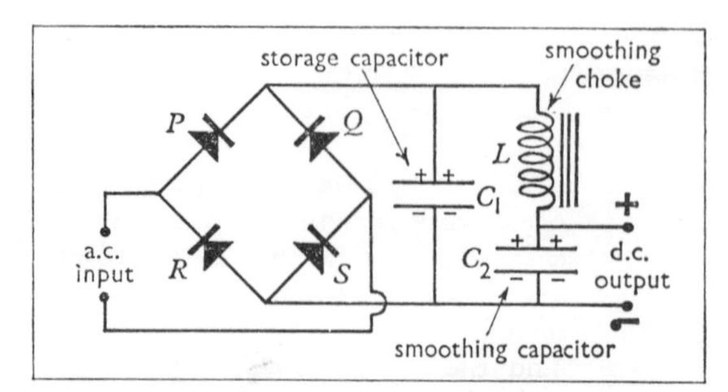

Fig. 17.11 A full-wave bridge type rectifier with storage capacitor and choke-capacitor smoothing

preferable to use a full-wave type of rectifier circuit employing four diodes joined in a bridge (Fig. 17.11). On one half-cycle current passes through diodes P and S, charging the storage capacitor C_1 to near the peak p.d. of the supply; on the other half-cycle P and S remain non-conducting and current passes instead through Q and R. If the load current is very small, the p.d. across C_1 remains nearly constant and equal to the peak supply voltage. When an appreciable load current is being taken, C_1 partly discharges during part of each half-cycle; and the average p.d. across C_1 is then less than the peak p.d. and an a.c. ripple is superimposed on it. Further smoothing to reduce the a.c. ripple can be provided by a choke-capacitor filter consisting of the inductor L and smoothing capacitor C_2 (cf. Fig. 14.10, p. 241). Silicon diodes are usually preferable for high-voltage and high-power applications at low frequencies; germanium diodes are rather more efficient in low voltage power units because their forward drop in p.d. for a given current is less than for silicon diodes. Germanium diodes are also particularly suitable for detector circuits in radio receivers (p. 268, Fig. 14.40); normally the point-contact type would be used for this purpose.

The Zener diode

When an increasing p.d. is applied to a diode in the reverse direction, there is a sudden rise in current as the p.d. reaches the turnover point. This is particularly marked in silicon diodes. By careful adjustment of the impurity level near the junction the characteristic beyond the turnover point becomes almost a vertical line (Fig. 17.12). Thus, in this region of its characteristic the p.d. across the

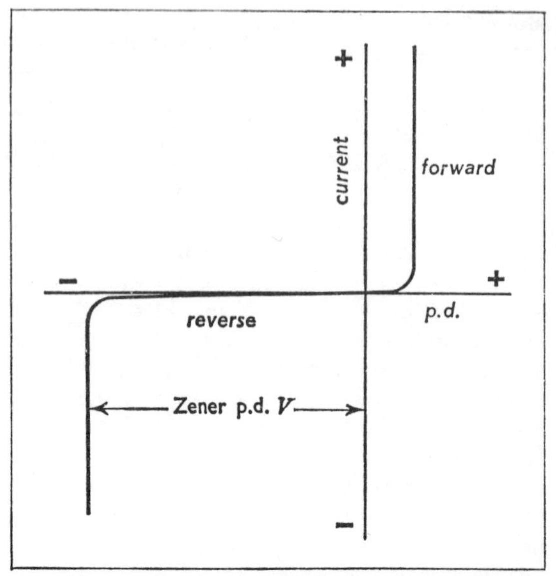

Fig. 17.12 The characteristic of a Zener diode

diode remains almost constant for a large range of currents. It may therefore be used as a *voltage reference* device for stabilizing a p.d. to a predetermined value. Diodes designed for this function are called *Zener diodes*. At present they can be manufactured to operate at any p.d. between about 4 V and 100 V, but their current-carrying capacity is limited to a fraction of an amp.

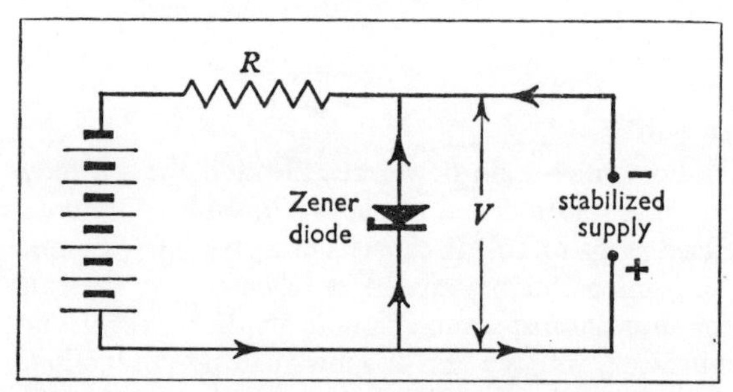

Fig. 17.13 A simple arrangement for stabilizing a battery supply using a Zener diode

Fig. 17.13 shows a simple arrangement for stabilizing the supply p.d. of, say, a transistor radio. The series resistance R is chosen so that, with a fresh battery and no current taken by the load, the maximum safe current (at the Zener p.d. V) flows in the diode. When the load draws a current, that through the diode falls by the same amount, so that the drop in p.d. across R remains almost constant. Thus variations in the load current affect the p.d. V supplied to it very little. The arrangement is also stabilized to some extent against variations in the battery p.d., provided this remains greater than the Zener p.d. of the diode and the current to the load is not excessive.

The photodiode

When a diode is biased in the reverse direction, the normal small leakage current is caused by the diffusion of minority carriers across the depletion layer; these carriers arise from the spontaneous formation of electron-hole pairs in and near the junction. If the junction is illuminated by light of sufficiently short wavelength, the energy of the photons (p. 222) can be sufficient to create extra electron-hole pairs. An increase in the reverse current thus occurs which is proportional to the flux of light falling on the sensitive area. A diode adapted to be used in this way is called a *photodiode*. If no bias is applied, a small e.m.f. is generated in it by the light in the same manner as in the selenium photocell (p. 220). But it is usually more satisfactory to operate it biased in the reverse direction by an external p.d. The incidence of light then causes an apparent fall in the resistance of the diode, and a current flows that may be used, for instance, to operate a relay or some other form of detector.

The solid-state particle detector

A suitable reverse-biased diode may also be used to detect other kinds of ionizing radiations, such as α- and β-particles. The passage of one of these particles through the sensitive region round the junction of the diode produces many extra electron-hole pairs, and causes a pulse of current through the detector. The pulses may then be counted with a scaler or ratemeter (p. 289). Unlike the G-M tube the solid-state detector produces pulses whose amplitude depends on the amount of ionization produced by the particle, and it may be used to discriminate between particles of different type. To detect α-particles the thickness of material on one side of the junction must of course be reduced to a minimum (p. 288). With β-particles and γ-rays it is best to increase the reverse bias as much as possible; this increases the thickness of the sensitive depletion layer round the junction, and gives a better chance of detecting weakly ionizing radiations.

17.3 Transistors

A transistor consists of a thin layer of *n*- or *p*-type semiconductor sandwiched between two layers of semiconductor of opposite type—shown diagrammatically in Fig. 17.14*a* and *b*. The circuit symbols for the two types are shown alongside, *c* and *d*. (The symbols derive from the construction of an early form of transistor, consisting of two fine wires bearing on a wafer of semiconductor—a sort of double point-contact diode.) The three layers are not just in electrical contact but must be formed within a single crystal.

A number of different methods of construction are now employed. One of these is the *alloy-junction* technique (Fig. 17.14*e*). The central region in this type of transistor is a wafer of *n*-type germanium about 0·2 mm thick. Two buttons of indium are fused into the sides of this, and sufficient indium dissolves in the neighbouring layers of germanium to convert these into *p*-type semiconductor. The indium buttons are then the means of conveying the current into the *p*-type layers formed on their surfaces. Fig. 17.14*f* shows another type of transistor produced by the *grown-junction* technique. In this type all three layers are formed within a single crystal as it is grown from molten germanium. A 'seed' crystal of germanium is lowered into molten germanium and then slowly withdrawn whilst being rotated, so that a rod of germanium is formed consisting of a single crystal. The type of semiconductor at each point of the rod can be controlled by adding the appropriate impurities to the melt. To start with a trace of arsenic is added so that *n*-type material is formed. Then enough gallium is added to the melt to convert it into *p*-type germanium. Almost immediately afterwards the melt is heavily doped again with arsenic so that it reverts to *n*-type material. The bar is then sawn into strips, which are shaped to form the small unit shown in the figure. Its length is about 3 mm, and at the centre its cross-section measures about 1 mm $\times$ 0·2 mm. Connecting wires are bonded onto the three regions. The central *p*-type layer may be only 10^{-2} mm thick, so that the fine gold wire that makes connection with it must be positioned with

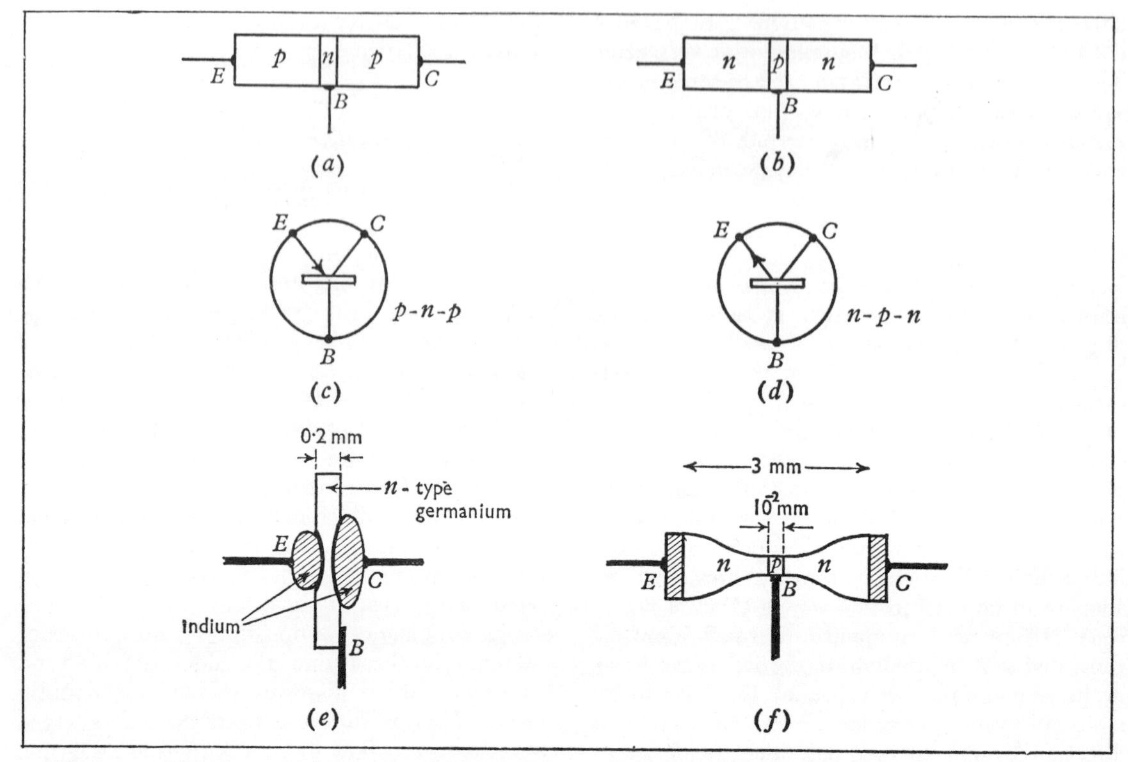

Fig. 17.14 Transistors—(*a*) *p–n–p* type; (*b*) *n–p–n* type; (*c*) and (*d*) circuit symbols; (*e*) an alloy junction transistor; (*f*) a grown junction transistor

some care! Finally the transistor is enclosed in a hermetically sealed, light-proof container. All operations on hot germanium have to be performed in an inert atmosphere (e.g. argon), and contamination has to be rigorously excluded at all stages of manufacture. Considering also the small scale and close tolerances of the work, it is clear that the production of transistors demands manufacturing techniques of the highest sophistication.

These and many similar processes have been used to produce both *p–n–p* and *n–p–n* types of transistor. The two types differ little in their properties, except for the directions of flow of current in them. We shall describe the action of the *p–n–p* type, which is rather more common.

The *p–n–p* transistor can be regarded as two *p–n* junction diodes joined back to back. It is operated with one junction (that between E and B) biased in the forward direction and the other in the reverse direction (Fig. 17.15). The left-hand layer E (called the *emitter*) is a very heavily doped *p*-type region, so that the current between E and B

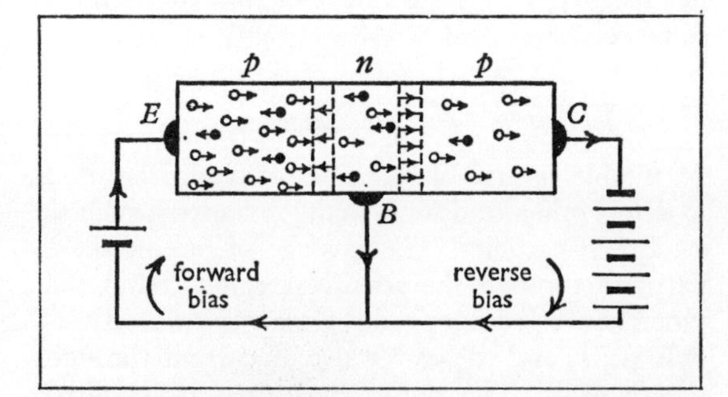

Fig. 17.15 The biasing of the junctions of a transistor. The current at the base-collector junction is almost entirely by means of holes that have diffused from the emitter-base junction

is almost entirely by the movement of positive holes from E into B. The holes entering the central region (called the *base*) diffuse across it; and, if it were thick, they would soon be neutralized by the electrons there. But by making the base layer very thin a high proportion of the holes will diffuse to the far side of it. They then enter the electric field at the junction between B and C and are quickly swept into the third region C (called the *collector*). In practice over 95% of the positive holes from the emitter reach the collector; and in many cases the proportion is more than 99%. The remainder recombine with electrons in the base layer and cause a small current to flow in the base lead. [In the *n–p–n* type of transistor the conduction process is similar, but takes place instead by migration of *electrons* from emitter to collector, a few of which recombine with *holes* in the base. The battery connections must of course then be the other way round to give the correct bias.]

In a typical small transistor (e.g. the Mullard OC71, which is of *p–n–p* type) a current of about 5 mA flows through the emitter-base junction when the p.d. across it is 0·2 V. Only about 2% of this current (0·1 mA) passes to the base lead, leaving 4·9 mA to flow out through the collector lead. The collector current is scarcely affected by changes in the collector potential and bears a nearly constant ratio to the base current (about 50 in this case). Variations in the base current may be produced by causing small changes in the emitter-base p.d.; these cause proportionately larger variations in the main current that flows through the transistor from emitter to collector.

If a small alternating current is fed into the base lead (superimposed on the steady bias current), it causes an amplified alternating current to flow from emitter to collector (superimposed on the steady collector current). The transistor may thus be used as a *current amplifier*. The input current flows from emitter to base, and the amplified output current from emitter to collector.

The characteristics of a *p–n–p* transistor may be investigated with the circuit shown in Fig. 17.16. The object is to find how the collector current I_c varies with the base current I_b and the p.d. V_c between collector and emitter. To represent the variation of all three of these quantities on one graph we can plot a set of curves showing the variation of I_c with V_c for a selected range of values of I_b. These curves are called the *collector characteristics*. The base current I_b may be adjusted to the required value by varying the p.d. V_b applied between base and emitter by means of the potential divider P_1. The resistance R is chosen so as to limit the base current to the maximum allowed value with the slider of P_1 set for maximum p.d. Thus, if the maximum current is to be 200 μA and a 2-volt cell is used, a value of 10 kΩ would be suitable for R. [If it is required to measure V_b, an instrument must be used that takes a current small compared with I_b—e.g. a d.c. amplifier (p. 188). Alternatively, the voltmeter can be joined to the left of the microammeter, but correction must then be made for the p.d. across the microammeter. The variation of I_b with V_b resembles that obtained with a *p–n* junction diode, since that is what the

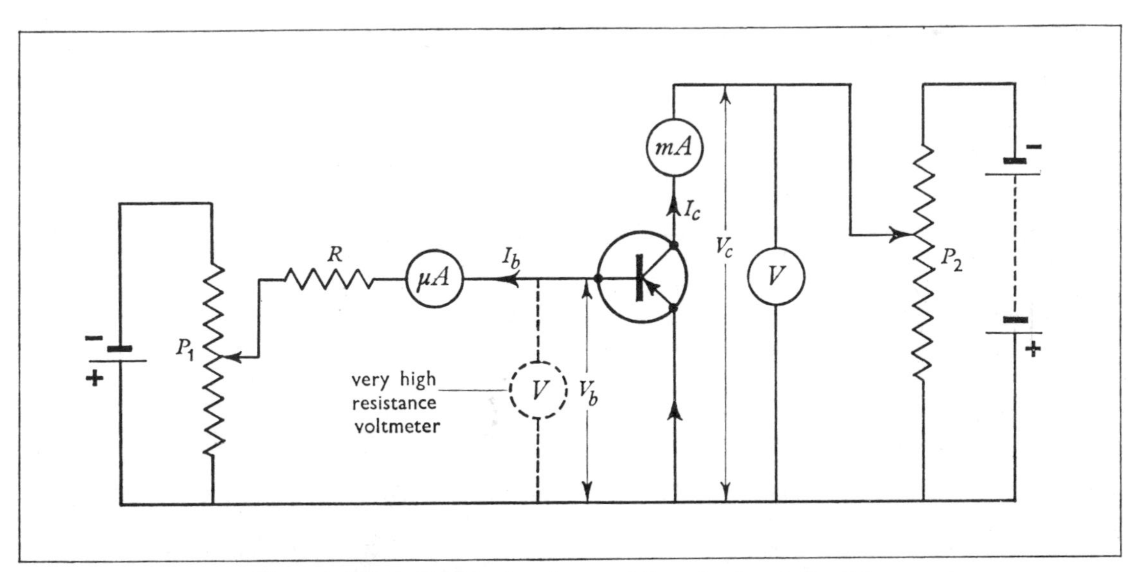

Fig. 17.16 Investigating the characteristics of a *p–n–p* transistor

emitter-base junction in fact consists of. With small transistors V_b is always less than 0·25 V, and to a first approximation we can often assume that it is negligible compared with the other p.d.'s in the circuit—i.e. we can assume that the base and emitter are approximately at the same potential.]

Having selected a suitable value for I_b, the p.d. V_c applied between collector and emitter is varied with the other potential divider P_2, and the collector current I_c is read on the milliammeter. The base current I_b is slightly affected by variations in V_c and must therefore be re-adjusted to the selected value before each observation. Fig. 17.17

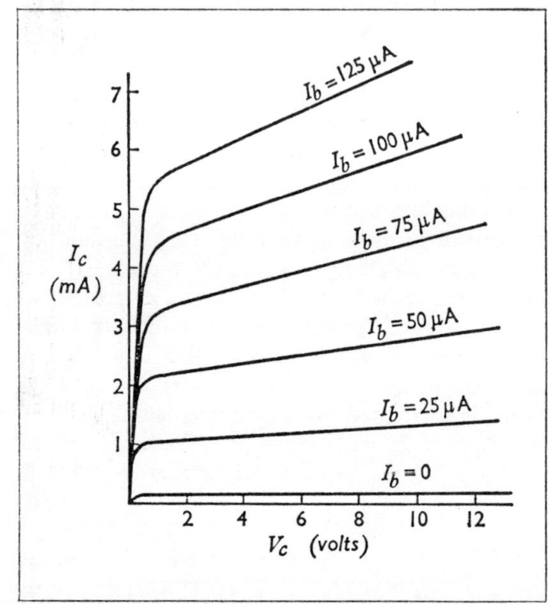

Fig. 17.17 The collector characteristics of a small transistor (such as an OC71)

shows the collector characteristics of a low-power transistor, such as the OC71. For very small p.d.'s (comparable with that between emitter and base) I_c varies rapidly with V_c. But for values of V_c above about 0·3 V the collector current I_c is decided almost entirely by I_b, and V_c has relatively little effect on it. In this region there is approximately a linear relation between I_c and I_b. This is shown up by plotting the characteristics as in Fig. 17.18, to reveal the variation of I_c with I_b for selected values of V_c. These are called the *transfer characteristics*. For a given value of V_c the characteristic is almost a straight line; but for different values of V_c the gradient varies over a range

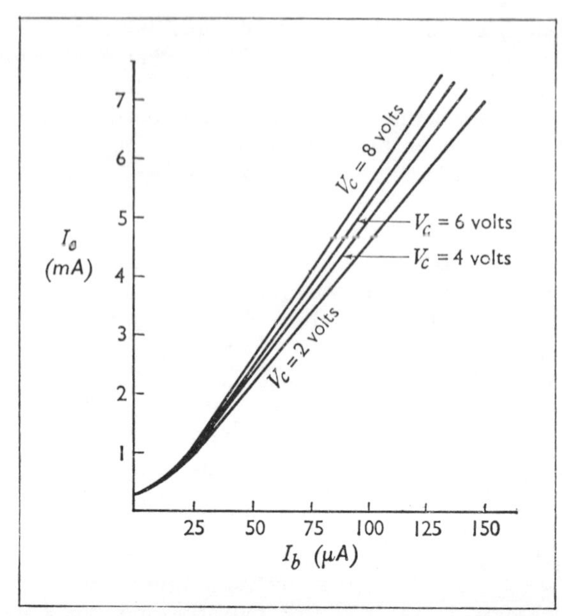

Fig. 17.18 The transfer characteristics of a small transistor (e.g. an OC71); the average gradient is called the current amplification factor β

of about $\pm 10\%$. The average value of the gradient is known as the *current amplification factor* of the transistor, and is denoted by β. For a typical OC71 it is about 50, although for individual transistors of this kind it may lie anywhere between 30 and 75. The transfer characteristics do not pass exactly through the origin, since a small leakage current flows across the collector-base junction even when I_b is zero. This leakage current varies rapidly with temperature; a rise in temperature of 8°C can be sufficient to double it. The increased leakage current is then added to the current that flows by transistor action, and has the effect of displacing the characteristics upwards (parallel to the axis of I_c). It is therefore important always to measure and record the temperature at which the characteristics are plotted.

Transistor amplifiers

Several different forms of amplifying circuit are possible with transistors; that in Fig. 17.19 is one of the most frequently used. It is known as the *common-emitter* circuit, since the emitter terminal is common to the input and output. In the absence of any input 'signal' the steady base current is controlled by the resistance R_1. Since there is only

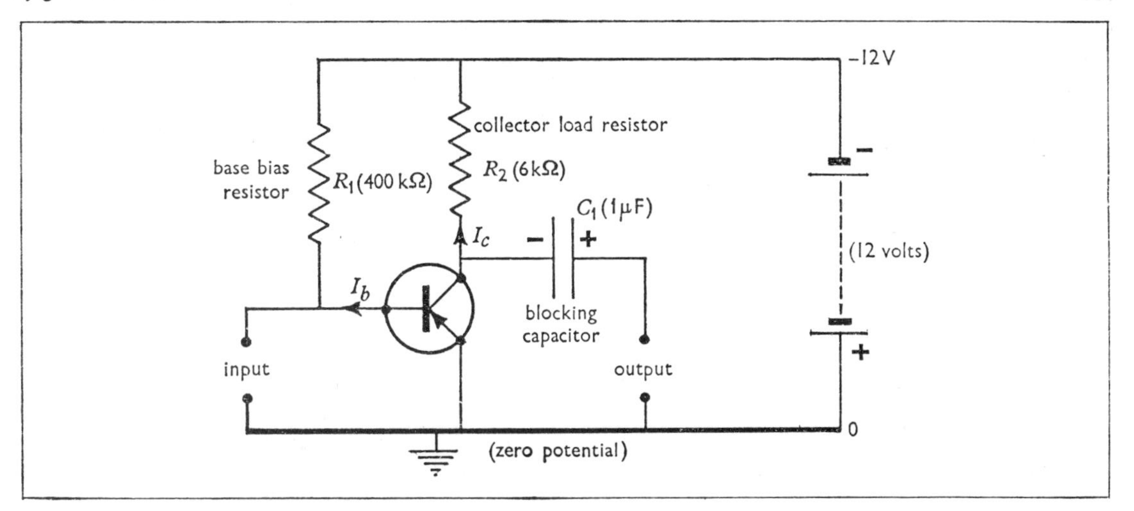

Fig. 17.19 A common-emitter amplifier, using a *p–n–p* transistor

a small p.d. between base and emitter, the p.d. across R_1 is nearly equal to the supply p.d. V of the battery.

$$\therefore\ I_b \simeq \frac{V}{R_1}$$

The collector current I_c is then given by

$$I_c \simeq \beta I_b$$

(The exact value of I_c can be read off from the characteristics.)

If a small alternating current is supplied to the base through the input terminals, it will be superimposed on the steady bias current that flows out of the base through R_1, and will cause corresponding variations of collector current (β times as great as the variations of base current). The collector current passes through a resistor R_2 known as the *collector load resistor*, across which the output p.d. is produced. The p.d. across R_2 thus contains an alternating component that varies in the same way as the input; this is superimposed on the steady p.d. that exists across R_2 in the absence of a signal. A *blocking capacitor* C_1 (p. 154) is therefore used to filter out the steady component so that only the alternating part appears at the output terminals. As in all amplifying devices, the power of the output is derived from the supply battery but is controlled by the very small power available at the input terminals.

A numerical example may perhaps make the working of the circuit clearer. Suppose it is intended to operate with a 12-volt supply battery, and that the transistor is to be used on the part of its characteristics where

$$I_c = 1{\cdot}5 \text{ mA} \quad \text{and} \quad V_c = 3{\cdot}0 \text{ volts}$$

The steady p.d. across the load resistor R_2 mus then be 9·0 volts, and the required value of R_2 is therefore given by

$$R_2 = \frac{9}{1{\cdot}5 \times 10^{-3}} = 6{,}000 \text{ ohms}$$

Taking the current amplification factor β as 50, the base current is given approximately by

$$I_b = \frac{1{\cdot}5}{50} \text{ mA} = 30\ \mu\text{A}$$

The base bias resistance R_1 has nearly the full supply p.d. across it.

$$\therefore\ R_1 = \frac{12}{30 \times 10^{-6}} = 400{,}000 \text{ ohms}$$

With this current the p.d. between base and emitter is probably about 150 mV, but it will vary to some extent with the temperature. If we take the emitter as the zero of potential (i.e. we imagine the emitter earthed), the steady potentials of base and collector are thus −150 mV and −3 V, respectively.

Suppose now a positive signal is applied to the input terminals causing a current of (say) 8 μA to flow *into* the top terminal (in Fig. 17.19). This causes the base potential to become slightly more positive (by perhaps 10 mV), so that the base potential changes from −150 mV to −140 mV and the base current falls. The total current in R_1 is very little affected by these small changes in base potential, and the fall in base current must therefore be equal to the input current of 8 μA (i.e. I_b falls to 22 μA). The corresponding decrease in I_c is 50 times as great as this, namely 400 μA (0·4 mA). Thus I_c falls to 1 1 mA, and the p.d. across the load resistor R_2 *decreases* by 2·4 volts—i.e. the potential of the collector falls to −5·4 volts. The change in the output p.d. is thus 240 times as great a the change in the input p.d. and is *in antiphase* with it.

A complete amplifier will usually consist of several such stages joined in cascade, the output terminals of one stage being connected to the input terminals of the next. The output of the complete amplifier will be in phase with the input if there is an even number of such stages and in antiphase with an odd number. The coupling capacitors (C_1) between the stages must be large enough to avoid loss of output at the lowest frequencies used. For an audio frequency amplifier designed as above a value of 1 μF might be suitable. For the sake of compactness an electrolytic capacitor would normally be used at this point; to provide the correct polarizing p.d. its negative terminal would need to be joined to the collector.

In the output stage of an amplifier, if more than a fraction of a watt is to be delivered, a special type of transistor is used that is able to pass currents of several amps. Such 'power' transistors are made by the same alloy-junction technique; but the collector pellet, in which most of the heat is dissipated, is made in the form of a flat disc that is fused onto a thick metal mounting plate forming one face of the unit. The transistor must be bolted down with the mounting plate in close thermal contact with a heavy metal sheet to act as a heat sink (p. 329).

At first sight there appear to be many similarities between thermionic valve and transistor amplifying circuits; compare, for instance, the transistor circuit given above with the circuit in Fig. 14.15, p. 244. In the thermionic triode the cathode emits electrons which pass through the grid to reach the anode; the p.d. between grid and cathode controls the anode current. So likewise in the *p–n–p* transistor the emitter is a source of positive holes, which pass *mostly* through the base to the collector; and the p.d. between base and emitter controls the collector current. However, although the functionings of the circuits are similar, the physical processes involved are very different. In the thermionic triode the current is controlled by the electric fields between the electrodes, that between grid and cathode being the most significant. The current therefore depends on the p.d.'s between the electrodes and the changes in current are approximately proportional to the changes in grid potential. In the transistor the collector current depends on the *current* that is caused to pass through the base-emitter junction; and the changes in collector current are proportional to the changes in the base-emitter current (a fixed fraction of which emerges through the base lead). For this reason the transistor is sometimes described as a *current-operated* device This term is in some ways a misleading one, since changes in base current are caused by changes in the base-emitter p.d., and the collector current can just as well be regarded as a function of the base potential. The point is that Ohm's law does not apply to the base-emitter diode; changes in the collector current are proportional to changes in base current, but neither of these is even approximately proportional to the changes of potential of the base. The transistor is therefore essentially a current amplifier, producing approximately distortionless amplification of alternating *current* fed into its base terminal. If it is required to amplify the alternating e.m.f. produced by, say, the pick-up of a record player, we must ensure that the base *current* is proportional to the e.m.f. This will only be so if the internal resistance of the source of e.m.f. is large compared with the effective input resistance of the transistor (which is typically in the neighbourhood of 1000 ohms). The resistance of the complete input circuit will then remain nearly constant, even though the input resistance of the transistor varies with the applied p.d. In the example quoted above the input resistance with a 10 mV signal might change by as much as ±200 ohms as the base potential fluctuates. A source resistance of at least 2000 ohms is therefore needed to keep the total resistance of the input circuit constant to within 10%. When amplifying stages are connected in cascade, each stage can be regarded as an a.c. generator connected to the input terminals of the following stage. Fortunately the internal resistance of the 'generator' formed by a common-emitter amplifier is fairly high (more than 6000 ohms in the example above) so that reasonably distortionless amplification is possible.

By contrast a thermionic triode can accurately be referred to as a *voltage-operated* device. The input resistance is to be measured in megohms, and the grid current is negligible—and is in any case irrelevant to the functioning of the amplifier. It is thus essentially a voltage amplifier, giving approximately distortionless amplification of an alternating *potential* applied to the grid.

The transistor is in many respects much superior to the thermionic valve. It is small and robust. It requires no heater current, and operates on low-voltage supplies. Very little power is dissipated in it, and very compact circuits can therefore be made with it. It is obviously ideal for portable equipment. Transistors also suffer from a number of disadvantages. They tend to create more background noise than thermionic valves. They cannot yet be manufactured for very high-power applications, such as radio transmitters, so that valves must still be used for this purpose. Also there is a limit to the frequency at which a given transistor can be used for amplification; this is because of the finite time taken by the charge carriers to move through the base. A similar effect occurs in thermionic valves, but the speeds of the electrons are much higher in this case, and their *transit time* between cathode and anode is of the order of 10^{-9} s; valves of conventional kinds can therefore be used at frequencies up to 1000 MHz or more. However, in a transistor the charge carriers *diffuse* through the base at relatively slow speeds, and simple alloy-junction transistors (such as the OC71) cannot be used in common-emitter amplifiers outside the audio frequency range (i.e. above 20 kHz). New methods of manufacture have had to be devised to produce transistors for use at high frequencies.

In one design both the base and emitter layers are formed by allowing the appropriate impurities to diffuse into the wafer of *p*-type germanium that constitutes the collector. The base and emitter leads go to two small metal pellets placed very close together on the same side of the collector wafer. The base pellet contains donor impurities only, while the emitter pellet contains both donor and acceptor impurities; however, the two kinds of impurities are selected so that the donor impurity will diffuse more rapidly into the germanium than the acceptor impurity. The assembly is heated for a carefully controlled time to a suitable temperature at which diffusion takes place. The result is shown in diagrammatic form in Fig. 17.20 (the layers are really extremely thin). It is called the *alloy-diffused* type of transistor. The donor impurities diffuse rapidly out of both pellets, and produce beneath them a continuous layer of *n*-type germanium, which becomes the base of the transistor.

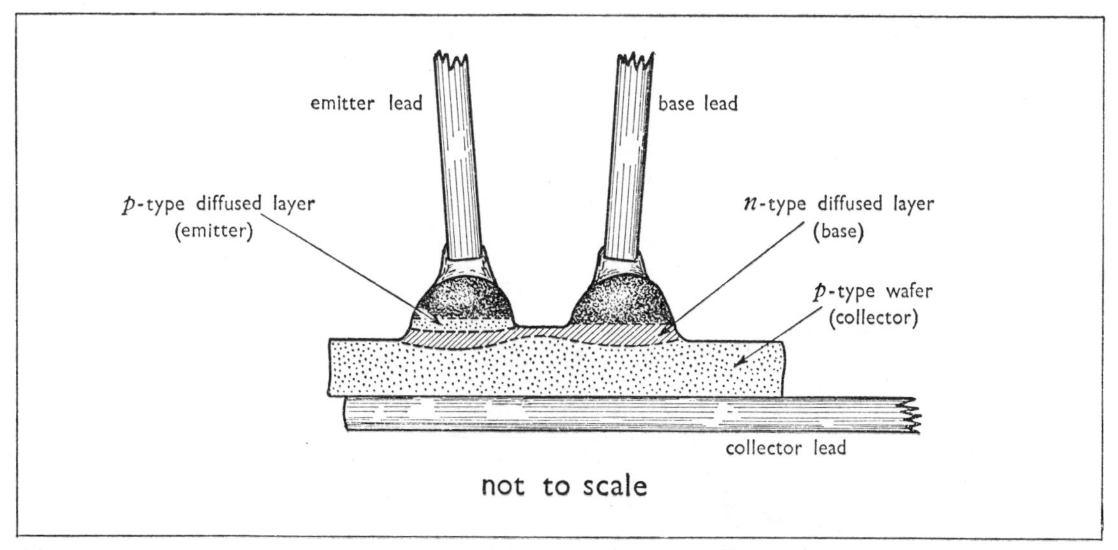

Fig. 17.20 The construction of an alloy-diffused type of transistor; the diagram is not to scale—in reality the emitter and base layers are very thin (less than 10^{-2} mm)

The acceptor impurity diffuses out of the emitter pellet more slowly, and forms a layer of *p*-type germanium between it and the base layer; this constitutes the emitter. The thickness of the base layer produced by this means can be less than 5×10^{-3} mm (to be compared with 0·1 mm for the alloy-junction type of transistor). Also the diffusion process leads to the concentration of donor impurity varying through the base layer in such a way as to produce an electric field in the base that accelerates the holes on their passage towards the collector. The 'transit time' of the charge carriers through the base layer is thus greatly reduced, and this enables this kind of transistor to be used at frequencies above 1000 Mc/s.

There seems little doubt that before very long technical developments will make it possible to use transistors in almost every application in which valves have hitherto been employed.

17.4 The theory of transistor circuits

Leakage current

There are three leads into a transistor. Normally current I_e enters through the emitter lead (with a *p–n–p* type) and leaves through the base and collector leads (I_b and I_c respectively). Taking the currents in the directions shown in Fig. 17.21*a*, we must therefore have in all circumstances

$$I_e = I_b + I_c$$

Ideally the only current crossing the base-collector junction would be that caused by the diffusion of holes from the emitter, a small, fixed proportion of which flow out through the base lead. In such an ideal transistor zero base current would imply zero collector current. We should then have

$$I_c = \beta I_b$$
$$\therefore\ I_e = (\beta + 1)I_b$$

where β is the current amplification factor of the transistor in a common emitter circuit.

However, in germanium transistors the leakage current across the reverse-biased base-collector junction cannot be ignored. Its value does not vary much with the collector potential provided this exceeds about 0·3 volt. It is, however, very strongly temperature dependent. At 20°C its value for a germanium transistor is typically about 3 μA; but a rise in temperature of 8°C can be sufficient to double it. We shall denote this collector leakage current by I_{co}. (In silicon transistors I_{co} is much less than 0·1 μA at 20°C, and can be ignored.) If the transistor is used in such a way that $I_b = 0$, i.e. with the base open-circuited, I_{co} is amplified by the usual transistor action. The way in which this comes about can be seen from Fig. 17.21*b*. The leakage current I_{co} passes from base to collector. Since no current is allowed to flow in the base lead, an equal balancing current must flow into the base through the emitter-base junction; in other words the potential of the base must adjust itself so that the base-emitter p.d. is of such a value as to allow current I_{co} to flow into the base. However, only a small proportion of the holes that cross the emitter-base junction are able to remain in the base layer; the majority of them diffuse

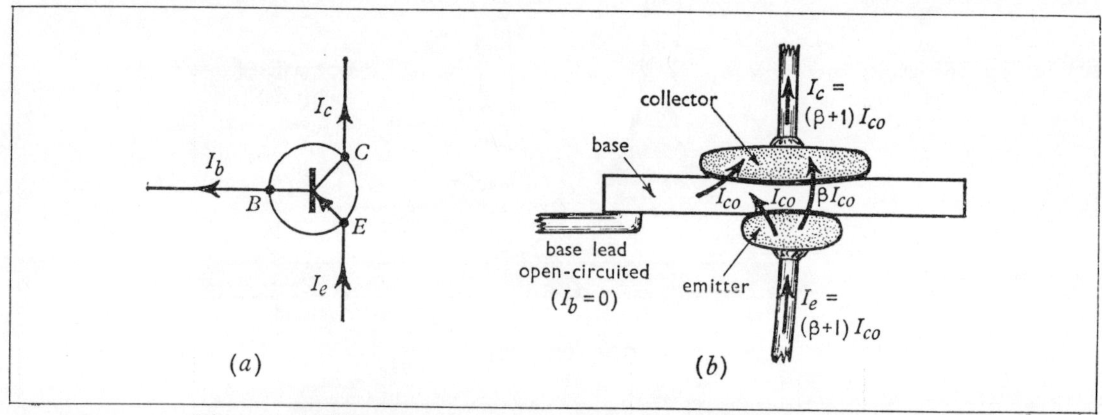

Fig. 17.21 (*a*) The currents entering and leaving a transistor
(*b*) The leakage currents in a transistor with the base open-circuited

across into the collector. The part of the current that consequently flows straight across from emitter to collector is βI_{co}. The total current leaving the emitter or entering the collector is therefore $(\beta + 1)I_{co}$; and this is the effective leakage current through the transistor when it is being used in a common-emitter type of circuit.

When a current I_b flows out of the base lead, the collector current is *increased* by an amount βI_b, which flows in addition to the leakage current $(\beta + 1)I_{co}$. We then have

$$I_c = \beta I_b + (\beta + 1)I_{co}$$

and

$$I_e = (\beta + 1)(I_b + I_{co})$$

It is quite possible for the collector current to be made less than $(\beta + 1)I_{co}$ by causing current to flow in the reverse direction through the base lead —i.e. *into* the base. This happens, for instance, if the base is joined to the emitter through a low resistance. This reduces the p.d. between base and emitter below the value it has when the base is open-circuited; a reduced current then flows into the base from the emitter, and a small balancing current enters the base through the base lead. If the resistance from base to emitter could be reduced to zero, we should have

$$I_b = -I_{co}$$
$$I_c = I_{co}$$
$$I_e = 0$$

But in practice the internal resistances of the transistor prevent the collector current falling to as low a value as this.

The phototransistor

This is an ordinary *p–n–p* alloy-junction transistor with a transparent case. Light falling on the base region creates electron-hole pairs in it; and, if these appear within about 0·5 mm of the collector junction, the holes have a high chance of diffusing to the junction and being swept into the collector. The light thus causes a flow of current *out of* the base region. If the base lead is open-circuited, an equal current must flow *into* the base from the emitter, and by the usual transistor action a current β times as great accompanies it flowing straight across from the emitter to the collector. The total current is thus $(\beta + 1)$ times as great as the original photoelectric current. In fact the phototransistor can be regarded as a photodiode (p. 3[illegible]) with built-in current amplification. The sensitivity is about 300 mA per lumen as against about 8 mA per lumen for a photodiode and about 0·5 mA per lumen for a selenium cell (p. 220). In practice it pays to use the phototransistor with a resistance joined between base and emitter, since this reduces the dark (leakage) current and gives greater stability of current if the temperature changes.

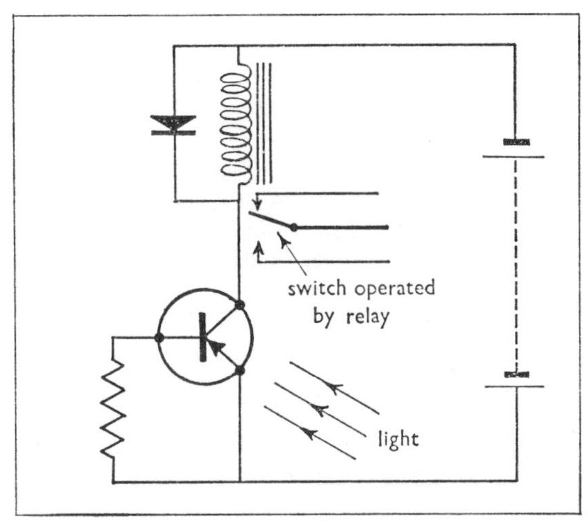

Fig. 17.22 A light-operated switch using a phototransistor

Fig. 17.22 shows a simple arrangement for a light-operated relay using a phototransistor. The increase of current produced by the light magnetizes the iron core of the relay; this acts on a small soft-iron armature and closes or opens a pair of contacts. In this way the light can be used to switch on or off some device that takes a very much greater current than that flowing in the transistor. This circuit also demonstrates an interesting use of a semiconductor diode. If the circuit were used without the diode joined across the relay coil, there would be a risk of damaging the transistor when the light falling on it is suddenly interrupted. The sudden reduction of current in the relay coil would cause a large e.m.f. of self-induction to appear in it, which would act so as to tend to maintain the current. This would make the lower end of the relay (in the diagram) highly negative with respect to the top end, and might momentarily exceed the voltage rating of the transistor. However, the diode provides a low resistance path across the coil if the lower end of it should ever become more negative than the top end, and thus it effectively prevents the p.d. across the transistor rising above that of the supply. At other times the lower end of the coil is positive with respect to the top end, and the diode remains non-conducting.

Stabilization

The rapid variation of I_{co} with temperature means that circuits employing transistors are apt to vary in behaviour with the temperature of their surroundings. Similar changes in behaviour occur as the circuit warms up after switching on. The warming up leads to increases in the leakage currents of the transistors, and this in turn causes their temperatures to rise even further. Indeed it is quite possible for the rise in temperature to get out of control, so that the transistors are destroyed. This is known as *thermal runaway*; because of the small thermal capacity of a transistor it can take place in a mere fraction of a second! It is therefore important to find means of stabilizing a circuit to some extent against the effects of changes of temperature.

Some measure of stabilization is automatically provided if it can be arranged that less than half the total supply p.d. appears across the transistor. This is the case, for instance, in the common-emitter amplifier considered on p. 335. The supply p.d. V was taken as 12 V, and 9 V of this was arranged to be dropped across the load resistor R_2. If $V_c < \frac{1}{2}V$, as in this case, it is easily shown that a rise in the collector current causes a *fall* in the heat dissipation in the transistor. Thus, the p.d. V_c between emitter and collector is given by

$$V_c = V - I_c R_2$$

Hence the power W dissipated in the transistor is given approximately by

$$W = V_c I_c = V I_c - I_c^2 R_2$$

$$\therefore \quad \frac{dW}{dI_c} = V - 2I_c R_2 = 2V_c - V$$

This is positive if $V_c > \frac{1}{2}V$; but negative if $V_c < \frac{1}{2}V$, so that W then falls as the current grows. This compensates for the rise in temperature that occurs when the circuit is first switched on; and, provided the transistor is not being overrun in the first instance, there is then no danger of sudden thermal runaway. Even so, the behaviour of the circuit is bound to depend to some extent on the temperature of the surroundings. For instance in the amplifier on p. 335, if the collector current were to rise at any time to 2 mA, the full supply p.d. of 12 volts would be dropped across the load resistance of 6000 ohms, and the p.d. across the transistor would fall to near zero—a condition known as *bottoming*. Further change of collector potential in the positive direction is then impossible and serious distortion of the signal is bound to occur.

What we must do is to ensure that the collector current is largely independent of the properties of the transistor, in particular of its leakage current. Fig. 17.23 shows one of the most commonly used methods of doing this. In this circuit the base potential is fixed by the potential divider (R_1 and R_4) across the supply. If the currents through these two resistors are much greater than the base current, the steady (no signal) value of the base potential will be very little affected by changes in I_b. The emitter current then automatically adjusts itself so that the emitter is slightly more positive than the base—i.e. so that the p.d. across R_3 is slightly less than that across R_4. If the collector leakage current increases, this scarcely affects the emitter current; but the base current is reduced by an amount equal to the increase in I_{co}. If the leakage current increases sufficiently, it may happen that the base current reverses. Normally the base current flows *out of* the base and then along R_1 to the negative terminal of the supply. But if the current in the base reverses it will flow from the positive end of the potential divider up through R_4 *into* the base. Either way the potential of the base is very little affected, and the emitter current remains almost constant. When an alternating current is supplied to the input terminals it is partly shunted through R_1 and R_4, and these therefore absorb some of the input power. In choosing the values of these resistances we therefore have to achieve a compromise between making them large and so absorbing little power from the input, and making them small and so ensuring constancy of the base potential under changing temperatures. In practice there is little difficulty in choosing values that enable the circuit to function satisfactorily over a large range of temperatures. A *by-pass* capacitor C_2 is joined across the emitter resistor R_3 to prevent the emitter potential fluctuating at the signal frequency. C_2 is of large capacitance (100 μF perhaps) so that it has low impedance at the frequencies concerned; an electrolytic type would normally be used at this point. The rapidly varying part of the current flows onto and off the

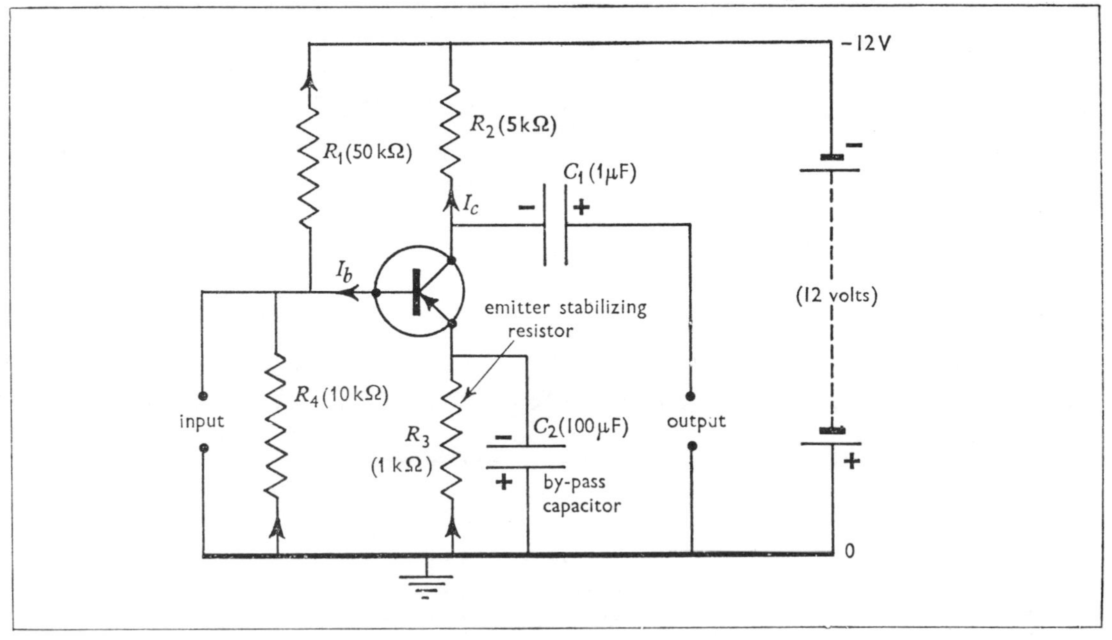

Fig. 17.23 A fully stabilized common-emitter amplifier

plates of C_2 without significantly affecting the p.d. between them.

Suppose the transistor is to operate (as before) from a supply of 12 volts and on the part of its characteristics where $I_c = 1{\cdot}5$ mA, and $V_c = 3{\cdot}0$ volts. A suitable value for R_3 might be 1000 ohms; the current through it is only slightly greater than 15· mA so that the p.d. across it would be 1·5 volts. The p.d. across R_4 must be slightly greater than this (about 1·7 volts). Suppose we choose a value for this resistance of 10 kΩ (which is much greater than the input resistance of the amplifier and therefore shunts only a small part of the input power). The current through R_4 would then be 170 μA. If the current amplification factor β is 50, the base current is approximately 30 μA. The total current through R_1 is therefore 200 μA and the p.d. across it is to be $(12 - 1{\cdot}7) = 10{\cdot}3$ volts.

$$\therefore\ R_1 = \frac{10{\cdot}3}{200 \times 10^{-6}} \simeq 50{,}000 \text{ ohms}$$

Notice that R_1 and R_4 have been chosen so that the currents through them are several times greater than the base current. Even if the leakage current increased to the point where I_b dropped to zero, the base potential would only rise from 1·7 volts to 2·0 volts, and the emitter current would not rise above about 1·8 mA.

Because some of the supply p.d. is now dropped across the emitter resistor R_3 the value of the collector load resistor R_2 must be less than in the unstabilized circuit by 1000 ohms (the value of R_3). So we take $R_2 = 5000$ ohms. The voltage amplification is therefore slightly reduced. The performance of the amplifier has thus been stabilized against changes of temperature at the expense of (*i*) 'wasting' some of the supply p.d. across the emitter resistor, (*ii*) slightly reducing the amplification, (*iii*) shunting the input with the two resistances R_1 and R_4.

Load lines

To some extent it is possible to analyse the behaviour of a transistor circuit, as we have done so far, by supposing that its current amplification factor β and internal resistances are constants. A cursory examination of the transistor characteristics is enough, however, to reveal that this is only approximately true. In general we must study the behaviour in relation to the measured characteristics themselves. In any given circuit there is only

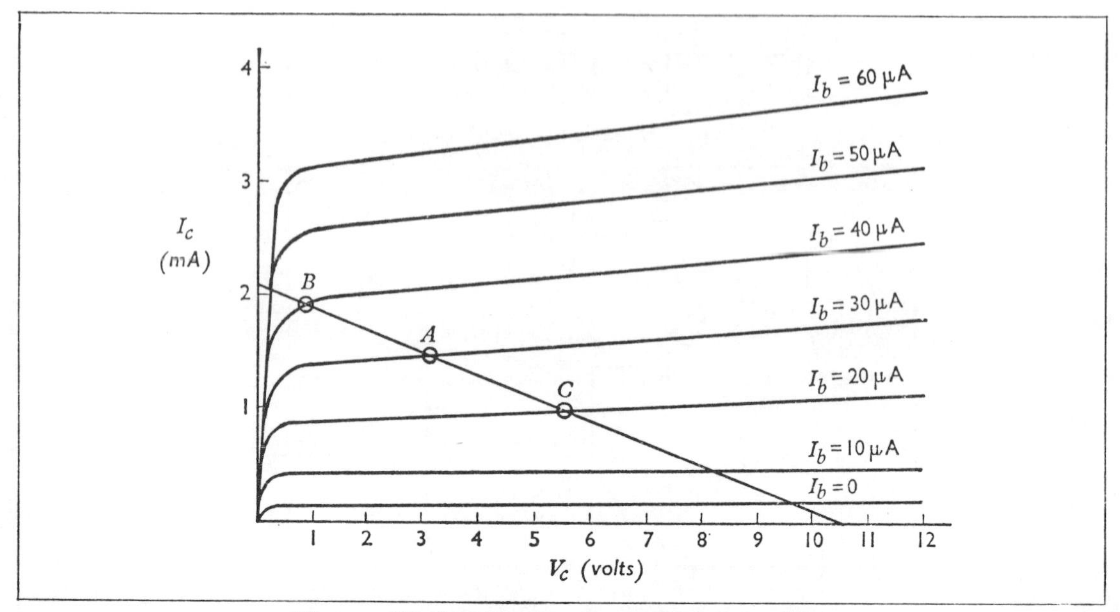

Fig. 17.24 The load line for a common-emitter amplifier

a limited range of possible values for I_c, V_c and I_b. For instance, if the collector load is a pure resistance, the *changes* in the collector potential δV_c and the corresponding *changes* in collector current δI_c must be related by

$$\delta V_c = -R_2\,\delta I_c$$

(V_c falls when I_c increases; hence the minus sign.) This shows that the values of V_c and I_c that are possible with a given collector load R_2 lie on a straight line across the collector characteristics of gradient $-1/R_2$. The line must in fact cut the axis of V_c at a value equal to the supply p.d., since, when $I_c = 0$, there is no p.d. across R_2. (But remember that the effective supply p.d. is reduced by the amount of the p.d. across the emitter resistor, if one is being used.) This line is known as the *load line*. It is drawn in Fig. 17.24 for the stabilized common-emitter amplifier circuit of Fig. 17.23. The working point to which the current and p.d. return when there is no signal is at A, where $I_c = 1{\cdot}5$ mA, and $V_c = 3{\cdot}0$ volts. The line marks out the possible ways in which I_b, I_c and V_c can vary with one another when a signal current is injected into the base in addition to the steady base current. B and C are the points on the characteristics corresponding to a rise and fall of the base current of 10 μA. In this way it is possible to study just how the output potential will vary with the input current, and so to estimate to what extent the amplifier introduces distortion. (See also the discussion of load lines on p. 250.)

Input resistance

The graph in Fig. 17.25 shows the variation of I_b with the p.d. V_b between base and emitter, for a fixed value of V_c; this is known as the *input characteristic*. The d.c. resistance of the base-emitter junction is V_b/I_b and varies between infinity and a value of a few hundred ohms. But the response of the amplifier to the input signal depends on the manner in which the base current *changes* when the base potential is *changed*. The effective input resistance for a small signal is therefore dV_b/dI_b, and is equal to the reciprocal of the gradient of the input characteristic. In Fig. 17.25 this quantity has been written beside the curve for a number of points along it. The input resistance can be taken as constant only as long as the changes in I_b produced by the signal are very small indeed. In general allowance must be made for the non-linear variations of current with p.d. in assessing the performance of the amplifier.

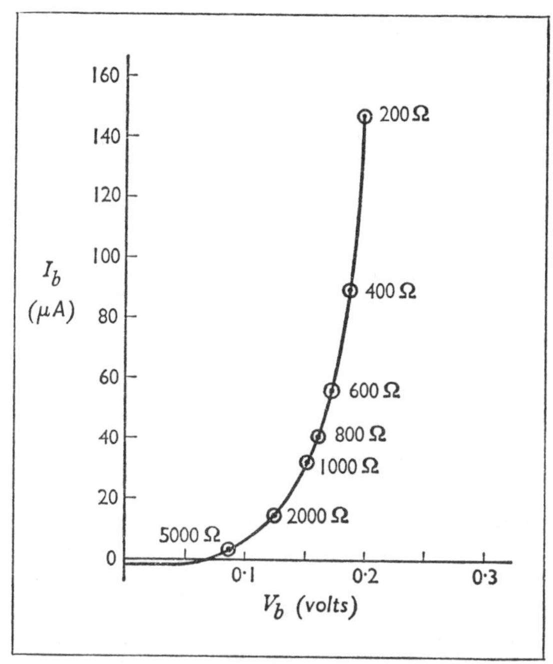

Fig. 17.25 The input characteristic of a small transistor (e.g. an OC71), showing how the effective input resistance of the transistor varies with the base current

17.5 Transistor oscillators

In principle a generator of alternating current (i.e. an oscillator) can be made by feeding some of the output of an amplifier back to its input. If there is sufficient feedback *in phase* with the input, the circuit is unstable and oscillations build up at some frequency to which the system is resonant. The simplest way to control the frequency is to include in the circuit a resonant combination of inductance L and capacitance C; the natural frequency of the oscillator will then be close to the resonant frequency of this combination. The methods of designing oscillators using thermionic triodes have already been discussed (p. 246). Similar techniques can be used with transistors, and the student should have no difficulty in explaining the functioning of the reaction and Hartley oscillators in Fig. 17.26 by comparing them with the corresponding triode circuits on p. 246 and p. 247. In both the transistor circuits the average base potential is fixed by a potential divider (R_1 and R_4) across the supply; together with an emitter resistor R_3 (and by-pass capacitor C_2) this arrangement gives the circuit stability against changes of temperature. In the reaction oscillator (Fig. 17.26*a*) a capacitor must be joined across R_4 so as to provide a low impedance path for the alternating component of the base-emitter current. If this were not included, the

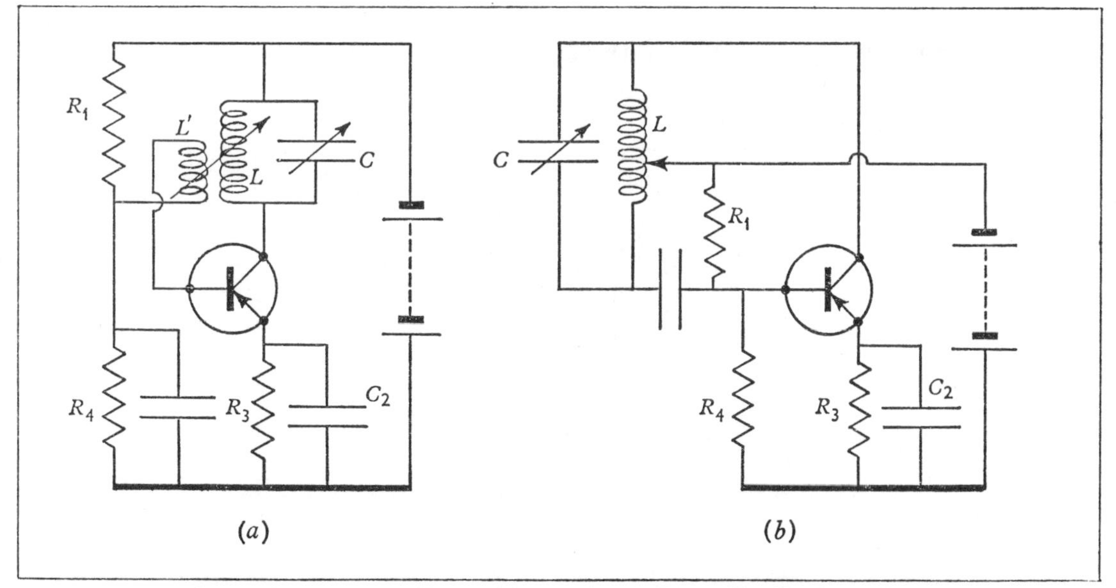

Fig. 17.26 Transistor oscillators: (*a*) a reaction oscillator (*b*) a Hartley oscillator

e.m.f. induced in the reaction coil L′ by mutual induction would not be able to drive a large enough alternating base current to maintain the oscillations.

The multivibrator

Fig. 17.27 shows an oscillator circuit of quite a different kind, in which the frequency is controlled by the times of discharge of two capacitors. It consists of a two-stage common-emitter amplifier with the output fed directly back to the input. Each stage produces both amplification and phase reversal, so that overall there is considerable positive feedback. The circuit is therefore highly unstable. If the current is *increasing* in one transistor, the p.d. across its collector load resistor will increase, and this will cause the current in the other transistor to *decrease*. This in turn acts to increase still further the current in the first transistor. When the circuit is switched on, it therefore goes rapidly into a condition in which one transistor is non-conducting and the other carries the maximum possible current. Which way round this happens initially is a matter of chance; suppose the conducting transistor is A. The maximum current in it is decided by the value of its load resistance R_4. When the current in R_4 has grown to the point at which almost the entire supply p.d. exists across it, the collector potential of A will be nearly zero and further growth of current will be impossible. In this condition the transistor is said to be *bottomed*. The instability of the circuit will have driven transistor A very suddenly into the bottomed condition and therefore will have caused a sudden change of its collector potential from a negative value up to zero. The p.d. across a capacitor cannot change suddenly (p. 184), and therefore this *change* of potential is communicated through the coupling capacitor C_1 to the base of B and holds this transistor effectively non-conducting (apart from the small leakage current). The collector potential of B is therefore almost −6 V. The circuit seems now to have reached a state of stability; however, this is only short-lived, because the p.d. across C_1 cannot remain constant. Its left-hand plate (in the diagram) is at about zero potential and its right-hand plate is positive; and current can flow to its plates through transistor A and resistance R_1. It therefore starts to discharge, and, if nothing further happened in transistor B, it would eventually charge up the other way round with its right-hand plate at the negative potential of the top supply rail. However, as soon as the right-hand plate reaches a small negative potential it causes transistor B to start conducting again. The potential of the collector of B therefore rises (towards zero). This is communicated to the base of A through the capacitor C_2 and starts to switch off this transistor. The collector potential of A therefore falls towards that of the negative supply rail, and this acts to switch on B still more rapidly. Thus the regenerative action of the circuit quickly switches it into the opposite state in which A is off and B is bottomed. The base of A is now at a

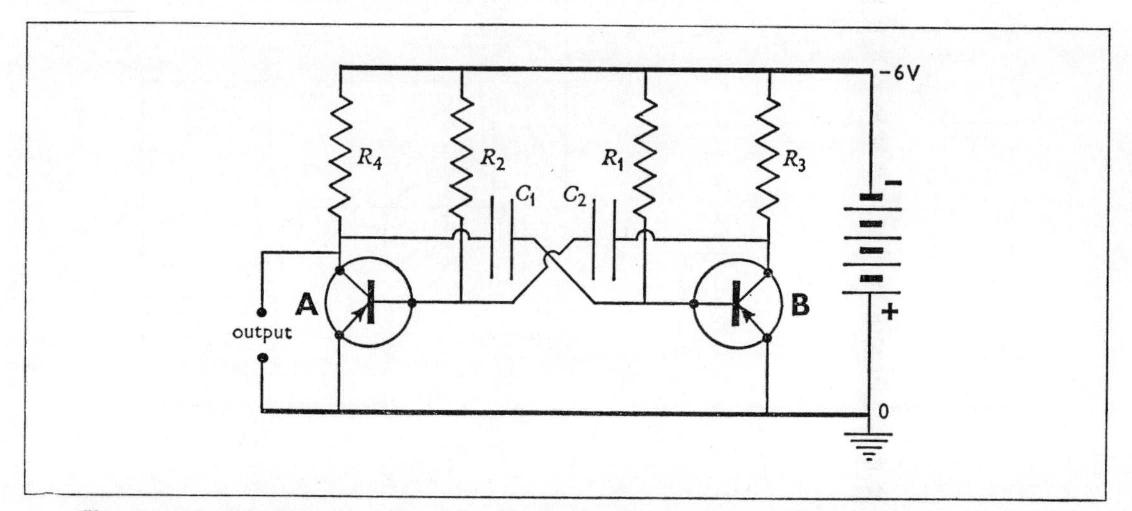

Fig. 17.27 A multivibrator—a type of oscillator producing an approximately rectangular waveform

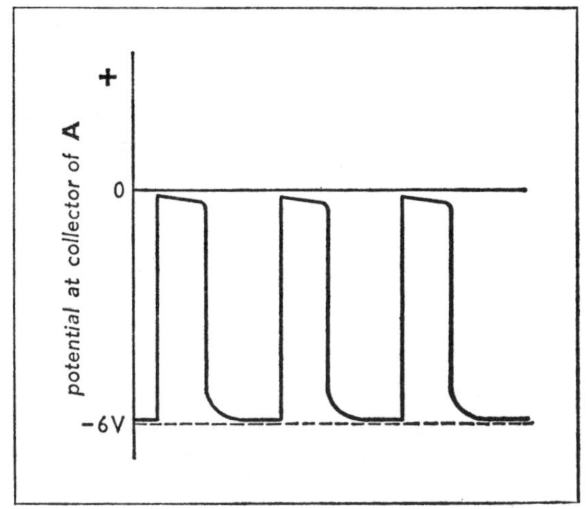

Fig. 17.28 The waveform of the oscillations produced by a multivibrator

large positive potential, and the capacitor C_2 discharges through R_2, until the switching action is repeated when the base of A once more reaches a small negative potential. The period of oscillation is decided chiefly by the time constants of the two capacitor-resistor combinations and is approximately proportional to $(C_1R_1 + C_2R_2)$.

The output of the oscillator may be taken from the collector of either transistor. The waveform is approximately a *rectangular* one in which the potential switches alternately between the two values 0 and -6 V (Fig. 17.28). This sort of waveform can be analysed as a mixture of sinusoidal frequencies (harmonics) that are multiples of the fundamental frequency. In fact it generates a very wide range of harmonics; it was for this reason that the name *multivibrator* was coined for it. If the circuit is made entirely symmetrical, with $C_1 = C_2$, $R_2 = R_1$, etc., it becomes a square-wave generator, and the two parts of the waveform are of equal duration.

17.6 The transistor as a switch

The essence of the action of a transistor is that the collector current can be controlled by a small change in the base current. It may therefore be used in the same way as a relay switch (p. 339), in which the input current causes the switch contacts to close and so controls a much larger current in another circuit.

Fig. 17.29 shows how changes in potential at the base of a transistor can be used to switch on and off the current in the collector circuit. When the point A is connected to earth (zero potential), the potential divider formed by R_1 and R_2 makes the

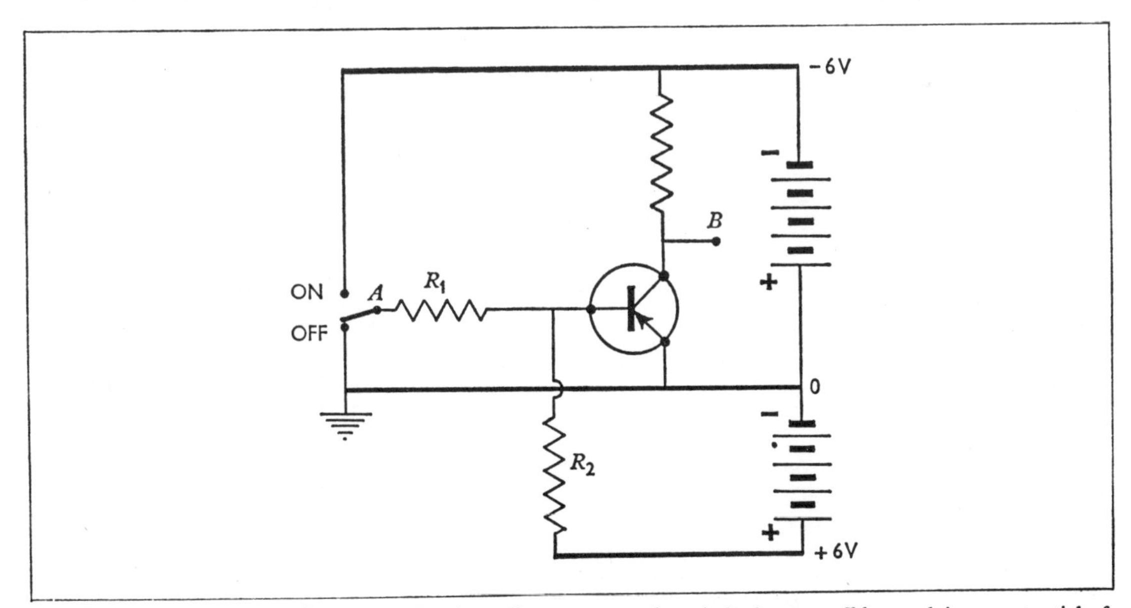

Fig. 17.29 A transistor used as a switch; the collector current is switched on or off by applying a potential of -6 V or 0 at A

base of the transistor slightly positive and so renders it non-conducting. The point B will then be at the potential of the negative supply rail (−6 V). If A is joined to the negative rail and R_1 is appreciably less than R_2, the base potential will become negative and the transistor will conduct. Usually it can be arranged so that the current is sufficient to bottom the transistor; the potential at B is then nearly zero. In this simple arrangement the output potential at B takes one of two values depending on the input potential applied at A; when A is at −6 V, B is at zero potential, and vice versa. The virtue of using a transistor to achieve this result is that the input potential can be provided by another electronic circuit, and so the 'position' of the switch can be made dependent on potentials derived from other points in the apparatus. Furthermore electronic switches of this kind can be 'reversed' millions of times a second, and so enable interlinked switching operations to be carried out with great rapidity. It is this function of the transistor that has made possible the construction of high-speed digital computers.

Any process of 'computation' consists in performing a sequence of operations on the data of the problem; at each stage in the computation the nature of the operation to be performed is decided partly by the pre-determined programme and partly by the outcome of earlier stages in the process. We therefore need 'switches' with multiple inputs, whose outputs are determined in specified ways by the condition of their inputs. These arrangements are known as *gates*, and mostly they are elaborations of the simple transistor switch considered above.

The gate in Fig. 17.30 is designed so that it will 'open' if any one (or more) of the inputs A_1, A_2, A_3 is switched 'on' (i.e. to the potential of the negative rail). It thus performs a function equivalent to the logical connective 'OR'. When all the inputs are at zero potential, the transistor is switched off, and the potential at B is −6 V. But if any one of the inputs is connected to the negative rail it is sufficient to make the base negative and bottom the transistor; B then has nearly zero potential. The output is of course of opposite polarity to the input that caused it. The circuit is therefore strictly to be described as a NOR gate. At the bottom of Fig. 17.30 is shown the symbol used to represent this kind of gate in diagrams of the logical intercon-

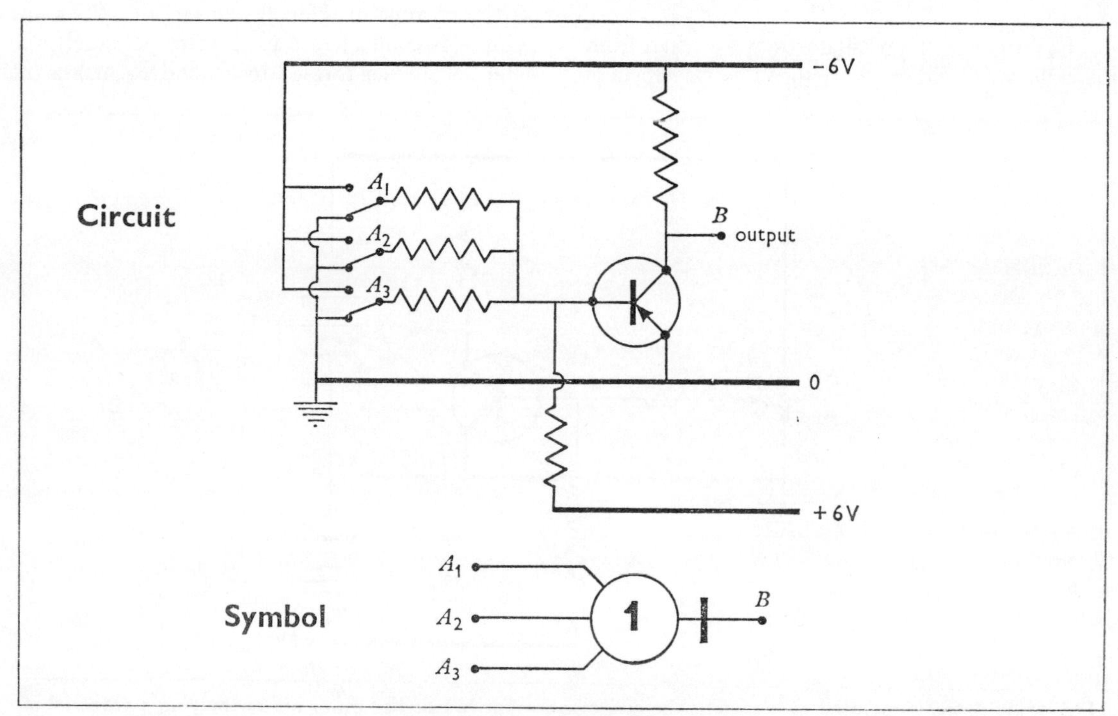

Fig. 17.30 A three-input NOR gate

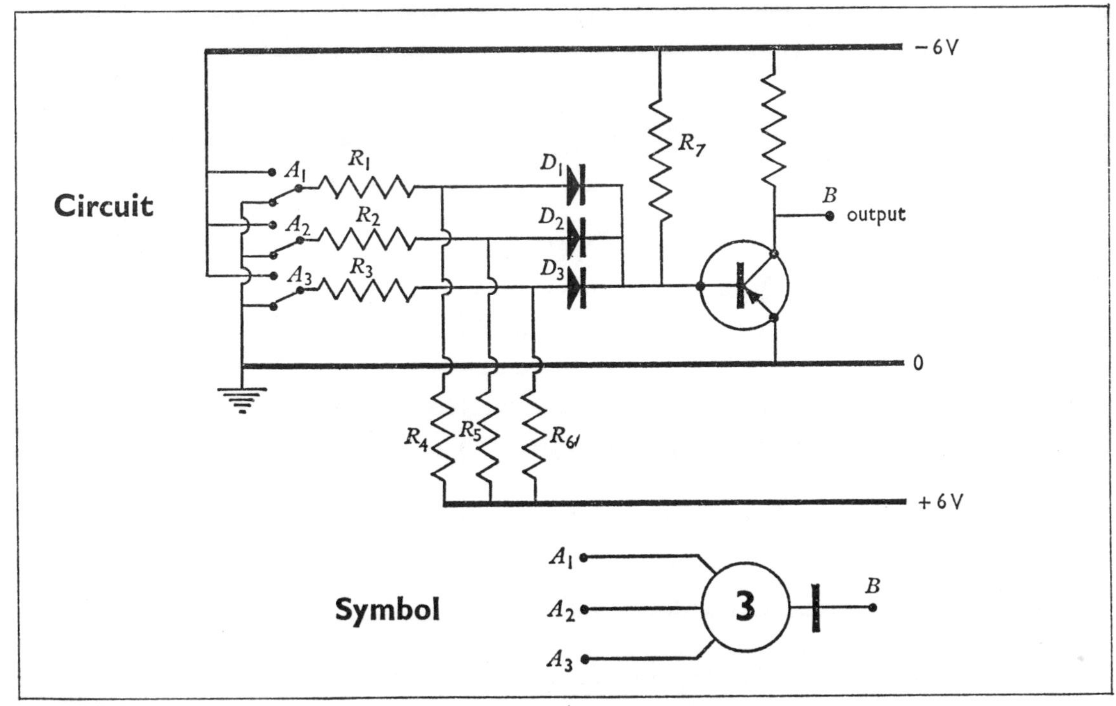

Fig. 17.31 A three-input NAND gate

nections in a computer. The **1** indicates that the gate opens if only *one* of the inputs is 'on' (i.e. switch *up*). The bar across the output line shows that the pulses in it are of opposite polarity to those at the input. The simple switch in Fig. 17.29 can be regarded as a NOR gate with only one input; or we can refer to it as a NOT gate. The triple input NOR gate in Fig. 17.30 could be converted into an OR gate just by connecting a NOT gate to its output.

The gate in Fig. 17.31 is designed to open only if *all* the inputs are switched 'on' (i.e. to the potential of the negative rail). It therefore performs a function equivalent to the logical connective 'AND'. If any one of the inputs (say A_1) is at earth potential, the potential divider formed by R_1 and R_4 maintains the diode D_1 biased in the forward direction. Sufficient current therefore flows through R_7 to ensure that the base potential is kept slightly positive and the transistor non-conducting. If the other inputs are also at earth potential, an even larger current flows through R_7; in any case the base potential of the transistor remains positive. If A_1 is connected to the negative rail at −6 V, D_1 will be biased in the reverse direction and no current flows through it to R_7. But only if all the inputs are negative ('on') will it be possible for the base to become negative so that the transistor conducts and bottoms. The output at B is again of opposite polarity to the inputs that caused it; and this arrangement is therefore described as a NAND gate. The symbol for it is shown at the bottom of Fig. 17.31. The **3** indicates that *all three* of the inputs must be 'on' simultaneously for the gate to open. It can be turned into a 3-input AND gate by joining a NOT gate to its output.

A binary scale counter

Arithmetical operations in a digital computer are normally carried out using a binary scale of arithmetic. This is because only two digits, 0 and 1, are required in this system, and these can be represented by the two possible 'positions' of a switch or gate. A counting circuit can be built up out of units each consisting essentially of two NOT gates joined together so that the output of each is the input for the other. Thus when one transistor is

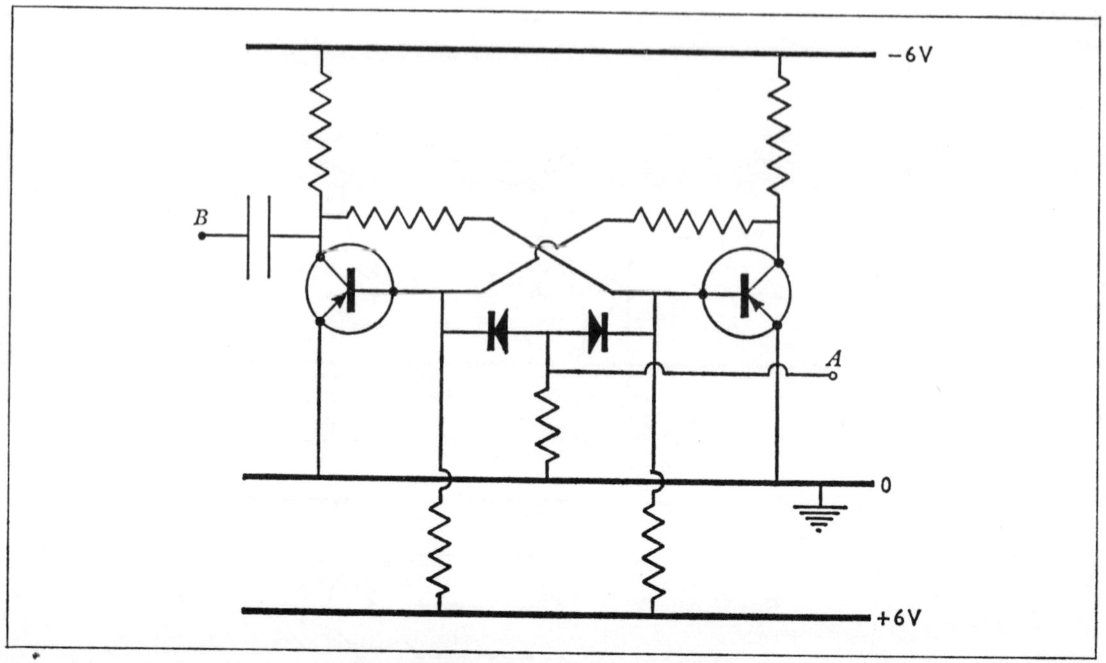

Fig. 17.32 One stage of a binary counter

bottomed it holds the other off, and vice versa. The circuit is in fact closely related to the multivibrator in Fig. 17.27 (p. 344), except that it has direct coupling between the transistors and so is permanently stable in either of its two possible states. One of these states can represent the digit 0 and the other the digit 1.

To make it perform its counting function we must arrange for it to change over from one state to the other when a suitable electrical pulse is fed into it. This can be done by applying a short positive pulse of potential to the base of the conducting transistor. This switches it off and therefore causes the other transistor to switch on. Fig. 17.32 shows a practical form of the circuit. A detailed analysis of its functioning is left as an exercise for the student. The two diodes are connected so that a positive pulse applied at A only affects the transistor that is conducting (and whose base is therefore slightly negative); the other diode is automatically reverse-biased, and so is non-conducting. When a succession of short positive pulses is applied at A, the circuit changes over each time from one state to the other. A negative pulse does not affect it. At the output terminal B positive and negative pulses will appear alternately. This output may be applied to the input of another similar circuit; the latter will then change over its state on every other pulse applied at the first input. A third such circuit will change over once in every four pulses arriving at A, and so on. A complete binary scale counter is made by joining up a sufficient number of such units, as in Fig. 17.33. Suppose we decide that the digit 0 is to be represented by the left-hand transistor conducting in each case, and 1 by the opposite state. The initial condition of the counter will then be as in the top diagram; the conducting transistor is shown shaded in each case. If we now feed in (say) 19 pulses at the input, the counter will reach the condition shown in the lower diagram, which represents 19 on the binary scale.

$$(1\times 2^4+0\times 2^3+0\times 2^2+1\times 2^1+1\times 2^0=19.)$$

A digital computer contains many such counters, the condition of each one representing a number involved in the computation. Arithmetical operations of all kinds may be effected by performing the right sequence of logical processes on the data represented by the potentials in the counters. For this purpose gating circuits, such as those described above, are used.

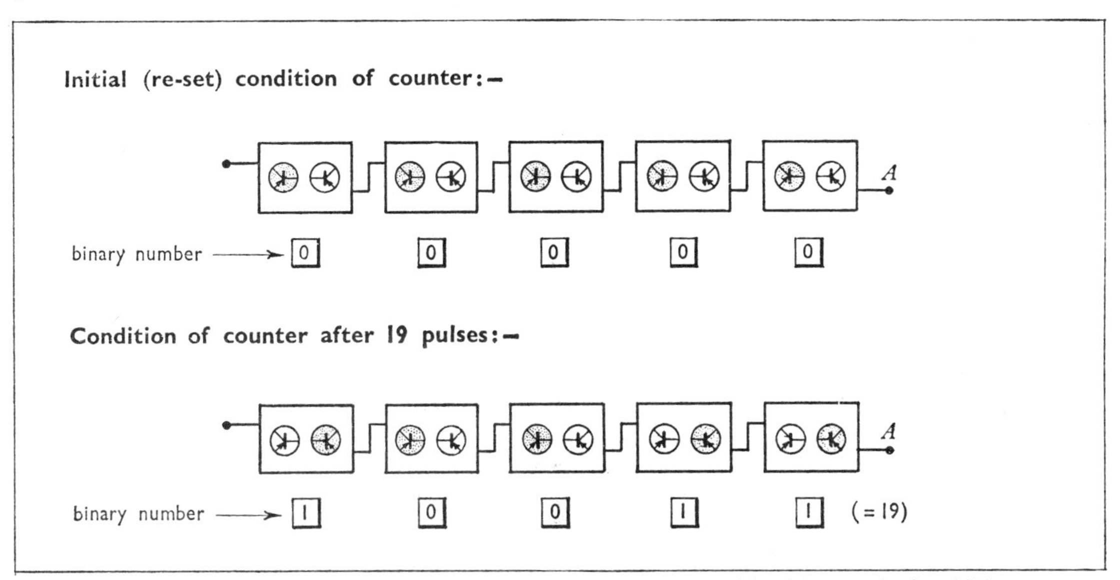

Fig. 17.33 A complete binary counter consists of many counter stages joined in cascade; in addition some arrangement is needed for showing which transistor in each stage is conducting

A counter can also be regarded as a temporary 'memory' for the number that has been fed into it. As long as no further pulses arrive its condition provides a record of the original number of pulses concerned. However, a computer needs to store in its memory many thousands of numbers (and other coded data) besides those on which arithmetical operations are being performed at any given instant; and it would be far too costly to use counters of this kind throughout. A variety of other techniques are therefore used to provide the different parts of the computer memory. One method is to use sets of small ferrite loops. These can be magnetized by pulses of current flowing in wires threaded through them. The two digits 0 and 1 are represented by the two possible directions of magnetization of a loop. Another method used when much longer access-time is permissible is to record the pulses representing the binary number on a magnetic tape; the technique is simply a refinement of the process used in an ordinary tape recorder.

It is beyond the scope of this book to consider any further the fascinating details of modern computer technology. Enough perhaps has been given to enable the student to perform experiments with transistors for himself and to follow more detailed technical books on the subject. The advent of the high-speed digital computer has revolutionized many aspects of both pure and applied science, to say nothing of its use in commerce for high-speed book-keeping. In applied science and engineering it enables us to shorten the processes of design by performing calculations for which it would never be economic to employ human beings. In industry it makes possible the automatic control of machinery on a scale that has never before been contemplated. In pure science it provides us with the means of analysing complex data, and so has equipped us to conduct scientific investigations which previously would have entailed unthinkably lengthy computations. Indeed, it is hard to estimate what the effect of the digital computer on human civilization may prove to be by the time its potentialities have been fully realized.

Questions

In some cases abbreviations have been standardized to conform with those used throughout the text.

Chapter 1 Atoms and electric currents

Additional data required:
The charge of 1 electron $= -1{\cdot}60 \times 10^{-19}$ *coulomb*
The mass of 1 electron $= 9{\cdot}1 \times 10^{-31}$ *kg*
The specific heat of water $= 4.2$ *joules* $°C^{-1}\,g^{-1}$
(*i.e.* 1 *calorie* $= 4{\cdot}2$ *joules*)

A

1. The current taken by a small torch bulb is 0·25 amp. What total charge passes a point in the circuit in 15 minutes? How many electrons pass the point in this time?

2. The e.m.f. of a small cell is 1·5 volts. How much energy does it produce when it drives a total charge of 160 coulombs round a circuit? What energy would be produced if the same charge was driven by a battery of three such cells joined (*a*) in series, (*b*) in parallel?

3. How much energy (in joules) is produced by a 60-watt electric light bulb in 5 minutes?

4. A p.d. of 12 volts is maintained between the ends of a wire. What quantity of electric charge must flow through it to supply 900 joules of energy? If it takes 1 minute for this charge to pass, what is the current?

Other questions: Ch. 3: 6 Ch. 13: 4, 6 Ch. 15: 3 Ch. 16: 6

B

5. A charged thunder cloud is producing rain drops of average diameter 3 mm, each carrying a charge of 3×10^{-15} coulomb. How heavy must be the rainfall (in cm per hour) for the electric current density to the earth to be 5×10^{-11} amp/m^2? (*L*)

6. The electron beam in a television picture tube travels a total distance of 50 cm in the evacuated space of the tube. If the speed of the electrons is $8{\cdot}0 \times 10^7$ m s^{-1} and the beam current is 2·0 milliamps, calculate the number of electrons in the beam at any one instant.

7. It is usually reckoned that the maximum safe current for a piece of copper wire is 10^7 amps per m^2 of cross-sectional area. If there are $1{\cdot}0 \times 10^{29}$ 'free' electrons per m^3 of copper, calculate the mean drift velocity of the electrons when the current reaches this value.

8. It is estimated that the average quantity of electric charge transported in a lightning flash is 30 coulombs. If the energy liberated is 2×10^{10} joules, what is the p.d. involved? In a typical thunderstorm lightning flashes strike the ground at intervals of about 3 minutes. Over the whole surface of the earth the current carried in this way between the atmosphere and the ground amounts to 1800 amps. Calculate the average number of thunderstorms taking place at any one instant over the whole earth.

9. An electric kettle (rated accurately at $2\frac{1}{2}$ kW) is used to heat 3 litres of water from 15°C to boiling point. It takes $9\frac{1}{2}$ minutes. How much heat has been lost? Outline briefly the explanation of this loss of heat. (*S, part qn*)

10. An immersion heater is placed in 160 g of water, and a current of 4·0 amps is passed through it for 2 minutes. The temperature is found to rise by 5·0°C. What is the p.d. across the heater?

11. An aluminium electric kettle weighing 500 grams contains 1 kilogram of water at 10°C. It is fitted with a 1-kilowatt heating element. How long will it take the kettle to come to the boil, and how much longer will it take to boil dry? Assume that there is no heat loss.
(Specific heat of aluminium = 0·22 cal g^{-1} °C^{-1}
Latent heat of steam = 540 cal g^{-1}.) (*Ox. Schol.*)

Other questions: Ch. 2: 25 Ch. 13: 18

C

12. Find the momentum acquired by the electrons in 10 cm of wire when a current of 1 amp starts to flow. (*Cam. Schol.*)

13. Given that one gram of copper contains approximately 10^{22} atoms, and that the density of copper is 9 g cm^{-3}, make a very rough estimate of the diameter of the copper atom, stating any assumptions you make. (*O & C*)

14. Copper is an element of *atomic weight* 63·5 and *atomic number* 29. It has two naturally occurring *isotopes* of *mass numbers* 63 and 65. Give the meanings of the terms printed in italics and estimate the relative abundance of each isotope. (*N*)

Chapter 2 Resistance

Assume:
1 calorie = 4·2 joules

A

1. (*a*) The current in a 10-ohm resistance must not exceed 0·4 amp. What is the maximum p.d. that may be applied to it? What is then the power dissipated?

(*b*) A p.d. of 5 kV is applied to a resistance of 20 megohm. What current flows? What power is dissipated?

(*c*) What is the resistance of (i) a 240-volt, 60-watt bulb, (ii) a 220-volt, 100-watt bulb?

(*d*) Two cells each of e.m.f. 1·5 volts and internal resistance 0·4 ohm are connected (i) in series, (ii) in parallel. What is the e.m.f. and internal resistance of the battery formed in each of these ways? What current would be driven by each of these batteries through a resistance of 1·6 ohms?

2. A small torch bulb is marked '2·5 V, 0·3 A'. (*a*) What is its resistance at its normal working temperature? (*b*) What resistance would you join in series with it to run it from a 6-volt battery? (*c*) What resistance would you join in parallel with it so that the combination could be run in series with a 200-volt, 100-watt bulb from a suitable mains supply (with both bulbs at normal brightness)? (*d*) When the torch bulb is run directly from a battery of e.m.f. 3·0 volts, the correct p.d. of 2·5 volts is produced across it. What is the internal resistance of the cell?

3. A nominal 2-ohm resistor is found on test to have an actual resistance of 2·040 ohms. What length of constantan wire of resistance 8·5 ohms per metre should be joined in parallel with the resistor so that the combined resistance is 2·000 ohms?

4. Equal lengths of iron and copper wire of the same diameter are connected first in parallel and then in series. If a potential difference applied between the ends of each arrangement in turn is gradually increased from a small value, it is found that in the first instance the copper begins to glow before the iron; in the second instance the reverse occurs. Explain this. (*N*)

5. An electric power line has a total resistance of 5 ohms. If the power input is 20,000 kilowatts at 100,000 volts, find (*a*) the percentage loss of power in transmission, (*b*) the voltage drop on the line.

If the figure of 100,000 volts refers to the r.m.s. value of a sinusoidal a.c. what will be the maximum voltage for which the line must be insulated? (*O & C*)

6. A power cable consists of one strand of steel and six of aluminium, each of diameter 3·0 mm. Calculate its resistance per kilometre length.

(Resistivity of steel = $9{\cdot}0 \times 10^{-8}$ ohm m; resistivity of aluminium = $2{\cdot}7 \times 10^{-8}$ ohm m.) (*Cam. Forces*)

7. An electric hot-plate has two coils of manganin wire, each 20 metres in length, and 0·23 sq. mm cross-sectional area. Show that it will be possible to arrange for three different rates of heating, and calculate the wattage in each case, when the heater is supplied from 200-volt mains. The specific resistance of manganin is $4{\cdot}6 \times 10^{-5}$ ohm cm. (*O & C*)

8. An electric heating element to dissipate 450 watts on 250-volt mains, is to be made from nichrome ribbon of width 1 mm and thickness 0·05 mm. Calculate the length of ribbon required.

(The resistivity of nichrome is 110×10^{-6} ohm cm.) (*L*)

9. Calculate the resistance between opposite edges of a square film of carbon deposited on an insulating plate; the film is $5{\cdot}0 \times 10^{-7}$ m thick, and the resistivity of the carbon is $4{\cdot}0 \times 10^{-5}$ ohm m. If a similar film is deposited on the outside of an insulating rod 3 mm in diameter, what length of film on the rod will have a resistance of 100 ohms?

10. The heating 'element' of an electric radiator is rated as 1 kilowatt on 240-volt supply, and the temperature of the element may be taken as 1000°C. When the element is at 0°C its resistance is 50·0 ohms. Calculate the mean temperature coefficient of resistance of the metal over the range 0°–1000°C. (*L*)

11. An alternating current passes through a wire of resistance 8·4 ohms immersed in oil and it is found that the temperature rises at the rate of 3·0 deg. C per minute when the temperature of the calorimeter and its contents is equal to that of the surroundings. If the thermal capacity of the calorimeter and its contents is 200 cal deg.$^{-1}$ C, calculate a value for the root-mean-square current. (*L*)

12. The heating element of an electric kettle is rated at 2·2 kW; it takes 6 minutes to bring its contents to the boil. If electrical energy costs $7\frac{1}{2}$d. per kilowatt-hour, how many kettles full of boiling water can be produced for one shilling?

13. An undergraduate left his electric fire (230 volts, 5 amperes) switched on all night (9 hours). The college fined him 5s. With electricity at 1d. per unit did this cover the loss to the college? (*Ox. Schol.*)

Other questions: Ch. 1: 3, 4 Ch. 4: 1, 2, 4 Ch. 8: 4 Ch. 9: 1 Ch. 11: 5, 14(*b*) Ch. 14: 2 Ch. 15: 2

B

14. A circuit consists of a battery and two 1-ohm resistances. When the resistances are connected in parallel the battery delivers 1·5 amps and when they are in series the current through them is 0·6 amp. Calculate the e.m.f. of the battery and its internal resistance. (*Ox. Schol.*)

15. A cell has an e.m.f. of 1·5 volts and a resistance of 1·8 ohms. It is connected in series with a rheostat R and a milliammeter of resistance 3 ohms. A 2-ohm coil is then joined in parallel with the meter. What value of R will cause the meter to read 30 milliamps? What is the potential difference between the terminals of the cell in these circumstances? (*L*)

16. A lead accumulator of e.m.f. 2·0 volts and negligible internal resistance is connected in parallel with a battery of Daniell cells, of e.m.f. 3·3 volts and internal resistance 9 ohms. Find the current flowing through a 1-ohm resistance connected across the combination, and the contribution of the accumulator to this current. (*O*)

17. An ammeter, accurately calibrated, indicates 12·2 amps when it is inserted in a circuit. When an identical ammeter is inserted in series with the first one, the reading of each becomes 11·8 amps. Find the current in the circuit before the first ammeter was inserted. (*L*)

18. Current from 200-volt mains is sent through resistances of 400 ohms and 600 ohms connected in series. What will be the potential difference between the ends of the 400-ohm coil? If an accurate voltmeter of resistance 1000 ohms is connected in parallel with the 400-ohm coil, how many volts will it read? (*O & C*)

19. Current is passed from a battery having negligible internal resistance through resistors of 5000 ohms and 3000 ohms in series. An accurate voltmeter of resistance 2000 ohms reads 6 volts when joined across the 3000-ohm coil. What is the e.m.f. of the battery, and what would the voltmeter read if joined across the 5000-ohm coil? (*Cam. Forces*)

20. A letter H is formed of five equal resistances of 10 ohms each. Across the two ends of the top is connected a 2-volt cell of negligible resistance. Across the bottom is connected an ammeter of 5 ohms internal resistance. Calculate the current passing through the ammeter. (*Cam. Schol.*)

21. Six resistors, AB, BC, CD, DE, EF, FA each having resistance 1000 ohms, are joined to form a closed circuit, and the points A and E are joined by a battery having e.m.f. 20 volts and negligible internal resistance. What will be the reading of a voltmeter having resistance 2000 ohms joined between (*a*) A and C, (*b*) A and F, (*c*) F and C? (*Cam. Forces*)

22. Two buildings, A and B, one mile apart, are connected by two telephone wires each of which has a resistance of 15 ohms. During tests, a small piece of equipment, known to have a resistance of 90 ohms, is accidentally left connected between the wires at the top of one of the telegraph poles supporting them. In order to locate the fault, the wires at B are disconnected from the telephone and joined together; the resistance in the circuit as measured at A is then found to be 29 ohms. Where will the fault be found? (*S, part qn*)

23. A certain wire has resistance 27·30 ohms when free from tension. What will its resistance become if tension is applied so that the length increases by 1% and the diameter decreases by 0·4%, if the resistivity is unaltered? (*C*)

24. The resistance of a piece of steel wire of diameter 0·6 mm is reduced to a third of its value by coating it with copper. What is the thickness of the copper?
(Resistivity of copper $= 1{\cdot}8 \times 10^{-6}$ ohm cm
Resistivity of steel $= 1{\cdot}98 \times 10^{-5}$ ohm cm.) (*S, part qn*)

25. Copper contains 10^{29} 'free' electrons per m^3 and its resistivity is $1{\cdot}72 \times 10^{-8}$ ohm m. A potential difference of 10 millivolts is set up between two points 10 cm apart on a uniform wire. Calculate the average drift velocity with which electrons will move through the wire, given that the charge on an electron is $1{\cdot}60 \times 10^{-19}$ coulomb. (*O & C*)

26. A wire has a resistance of 10·24 ohms at 20·5°C and 14·65 ohms at 99·8°C. Obtain a value for its temperature coefficient of resistance. (*L*)

27. Define the *volt* and deduce an expression for the heat H dissipated in time T by a conductor of resistance R in which a current I is maintained. Write down a similar expression in terms of T, R and the p.d. V across the conductor, and explain why it appears in one expression that the heat is directly proportional to the resistance, and in the other, that it is inversely proportional.

A current of 0·75 amp passes through a lamp filament of diameter 0·012 cm. Find the rate of heat loss per unit area of surface of the filament.
(Resistivity of filament material $= 1{\cdot}8 \times 10^{-6}$ ohm cm.) (*Cam. Overseas*)

28. A coil of platinum wire is immersed in a calorimeter containing water. A current of 2·0 amps is passed through the coil and the temperature rises from 17°C to 23°C in the first minute. Assuming that the thermal capacity of the system remains constant and that steps are taken to prevent loss of heat, calculate the rate of rise of temperature of the system at a temperature of 80°C when the current through the coil is 1·5 amps. The temperature coefficient of resistance of platinum is 39×10^{-4} deg.$^{-1}$ C. (*N*)

29. Five batteries, each of which has the same e.m.f. and the same internal resistance of 0·5 ohm, are connected in parallel and used to supply current to a heating coil whose resistance is 2 ohms. By what factor would the heat dissipated in the coil be increased if the five batteries were connected in series? (*S, part qn*)

30. A surge suppressor is made of a material whose conducting properties are such that the current passing through is directly proportional to the fourth power of the applied voltage. If the suppressor dissipates energy at a rate of 6·0 watts when the potential difference across it is 240 volts, estimate the power dissipated when the potential difference rises to 1200 volts. (*L*)

31. A heating coil is made of a material of which

the resistance is proportional to the absolute temperature. Assuming that Newton's law of cooling holds, calculate the ratio of the currents needed to maintain the coil at the two temperatures 77°C and 127°C when the room temperature is 27°C. (*S, part qn*)

32. Derive an expression for the rate of energy dissipation when an electric current i flows through a resistance r, stating in what units the quantities concerned are measured.

A wire of resistance R is connected in series with a battery of e.m.f. V volts whose internal resistance is r ohms. Find the rate of energy dissipation in the resistance R, and the condition for this to be a maximum.

Find the wattage and voltage of the brightest lamp which can be used with a $4\frac{1}{2}$-volt battery of resistance 2 ohms. (*O & C*)

33. The resistance of the nichrome element of an electric fire is found to be 39·23 ohms when the current passing through it is very small, the room temperature being 20°C. If, in use, the current passing is found to be 4·92 amps with a potential difference across the element of 220 volts, calculate (*a*) the rate of heat production in the element, (*b*) the steady temperature reached by it.

(The temperature coefficient of resistance of nichrome may be assumed to have the constant value $1{\cdot}70 \times 10^{-4}\,°C^{-1}$ over the range of temperature involved.) (*C*)

34. A liquid flows at a steady rate along a tube containing an electric heating element. When the power supplied to this heater is 10 watts and the rate of flow of the liquid is 50 g per min, the outflow temperature of the liquid is 5 deg. C higher than the inflow temperature. When the rate of flow is doubled, the power supplied to the heater must be increased to 19 watts for the outflow temperature to be unchanged. If the inflow temperature is the same in each experiment, calculate a value for the specific heat of the liquid. (*N*)

35. Electrical energy is supplied at a rate of 55 watts to a heating coil immersed in a liquid, causing the liquid to boil. The vapour is condensed and collected at a steady rate of 8·52 g min^{-1}. When the power is changed to 34 watts, the steady rate at which the liquid is collected becomes 4·97 g min^{-1}. Calculate the latent heat of vaporization of the liquid and the heat loss per min. Discuss the advantages and disadvantages of this experimental method. (*N*)

36. A small electrical heating coil is immersed in 100 cm^3 of oil (density 0·80 g cm^{-3}) contained in a calorimeter of thermal capacity 8·0 cal deg.$^{-1}$ C. When electrical energy is supplied to the heating coil at a rate of 4·3 watts the temperature of the calorimeter and oil rises to a steady value of about 50°C. When the current is switched off this temperature commences to fall at a rate of 1·20 deg. C min^{-1}. Calculate the specific heat of oil. If the experiment is repeated with all the data and conditions the same except that the oil is replaced by an equal volume of water, at what rate will the temperature fall? (*N*)

37. Given that the resistivity of aluminium is twice that of copper, and that the density of aluminium is one-third that of copper, find the ratio of the masses of aluminium and copper conductors of equal length and equal resistance. (*N*)

Other questions: Ch. 1: 9, 10, 11 Ch. 3: 10 Ch. 4: 7, 9, 10, 13 Ch. 5: 7 Ch. 8: 10, 11, 12 Ch. 9: 5 Ch. 10: 20 Ch. 11: 19 Ch. 12: 15, 17, 22 Ch. 14: 11

C

38. The current (in amps) through a rod of a certain metallic oxide is given by $i = 0{\cdot}20\, V^3$, where V is the potential difference (in volts) across it. The rod is connected in series with a resistance to a 6·0-volt battery of negligible internal resistance. What values should the series resistance have so that (i) the current in the circuit shall be 0·40 amp, (ii) the power dissipated in the rod shall be twice that dissipated in the resistance? (*C*)

39. In the circuit shown the cells A and B have e.m.f.'s E_1 and E_2 and internal resistances r_1 and r_2 respectively. Derive the condition for no current to flow in R_3. Suggest a method based on this result for finding the ratio of the e.m.f.'s of two cells having unknown, but not negligible, internal resistances. (*L*)

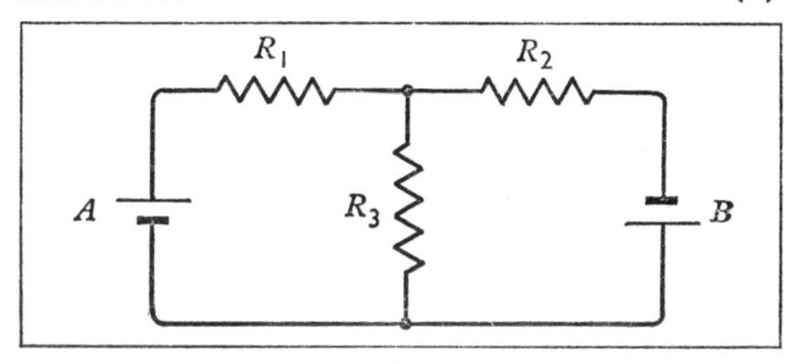

Fig. Q 2.39

40. Discuss the factors which govern the final steady temperature attained by a conductor carrying a current. How is the heating effect applied in the case of a fuse wire? A fuse wire breaks a circuit when a current of 5 amps is passed through it. If the diameter of the fuse wire is doubled, all other factors remaining the same, what will now be the maximum current that can be passed? Show that your result does not depend upon the assumption of any particular law for the way in which the heat lost depends upon the excess temperature. (*W*)

41. Find the ratio of the lengths of two wires of the same metal, but of different radii, which are heated to the same steady temperature when the same potential difference is applied in the same circumstances. Find also the ratio of the powers dissipated.

A wire is heated to whiteness *in vacuo* by an electric current, the surroundings being kept at a constant temperature. In course of time, owing to the evaporation of the metal, the wire gets thinner. Show that its temperature will remain constant if $V^3 i$ is kept constant, where V is the potential difference to the wire and i the current passing through it. (*N*)

42. A wire of length l, radius r and resistivity ρ is heated *in vacuo* by a current i which raises it to a steady temperature of T°K. Assuming that ρ is proportional to T, that the rate of radiation of energy per unit surface area of the wire is proportional to T^4, and that heat losses by conduction through the supports are negligible, find an expression for T in terms of the other quantities involved.

Compare (*a*) the radii, (*b*) the lengths of the filaments of a 100-volt 60-watt tungsten lamp and a 230-volt 100-watt tungsten lamp, assuming that the operating temperatures of the two filaments are the same. (*O & C*)

43. The temperature attained by a lamp filament is 1200°C when the power supply is 60 watts at 230 volts. Assuming that the filament radiates as a black body and that it loses heat only by radiation, calculate its temperature when the power supply is increased to 100 watts.

If the resistance of the filament is directly proportional to its absolute temperature, find the voltage needed to dissipate 100 watts in it. (*O*)

44. A long thin copper strip of width 5·0 mm is sandwiched between two sheets of asbestos 1·0 mm thick; the outside of the asbestos is maintained at 0°C. By calculating the temperature of the strip as a function of the current flowing through it, show that at a certain current the temperature will increase indefinitely; calculate this current. Assume that the flow of heat is normal to the strip.

(Resistance per unit length of the copper strip at T°C = $2{\cdot}2 \times 10^{-4}(1 + 0{\cdot}0043\,T)$ ohm cm^{-1}; thermal conductivity of asbestos = $1{\cdot}3 \times 10^{-3}$ watt cm^{-1} °C^{-1}.) (*C*)

45. State Ohm's law and define *electrical resistance*. Discuss the application of your definition to (*a*) a metal wire, (*b*) a battery.

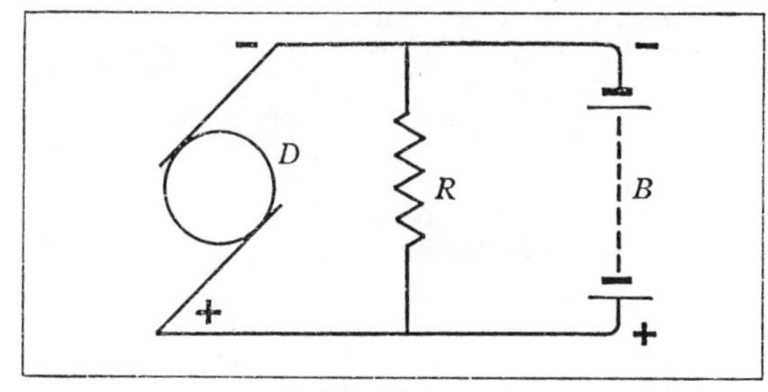

Fig. Q 2.45

A circuit is joined as shown. D is a dynamo, B a battery, and R a load resistance, the following being the relevant data:

Dynamo: e.m.f. = 17·0 volts, and internal resistance = 0·50 ohm.
Battery: e.m.f. = 12·6 volts, and internal resistance = 0·10 ohm.
Load resistance = 2·00 ohms.

Calculate the current through the load resistance and that through the battery, and state whether the latter is being charged or discharged. (*Cam. Forces*)

46. Two resistances are joined in parallel and a specified current flows through the combination. Show that of all the conceivable ways in which the current might divide between the two paths, the one actually adopted involves the least rate of energy dissipation. (*S, part qn*)

47. Two Daniell cells, A and B, each of e.m.f. 1·08 volts, are connected in parallel across a 5-ohm resistance. The internal resistance of A is 1·0 ohm and that of B, 0·5 ohm. Calculate the current through each cell and the potential difference across the 5-ohm coil. (*L*)

Other questions: Ch. 4: 25, 27 Ch. 9: 17 Ch. 12: 36 Ch. 14: 20 ,21

Chapter 3 Electrolysis and thermoelectricity

Additional data required:
Faraday's constant = $9{\cdot}65 \times 10^7$ *C kg-equiv.*$^{-1}$
The volume of a kg-mole of any gas at n.t.p. = $22{\cdot}4\ m^3$
E.c.e. of copper = $3{\cdot}3 \times 10^{-7}$ *kg C*$^{-1}$
1 *calorie* = 4·2 *joules*

A

1. A steady current is passed for 50 minutes through a solution of copper sulphate and 5·5 g of copper are deposited on the cathode. Find the value of the current.

2. An article of total surface area 200 cm^2 is to be plated to a depth of 0·020 mm with silver. How long will this take using a plating current of 3·5 amps?
(Atomic weight of silver = 108
Density of silver = 10·5 g cm^{-3}.)

3. Two copper voltameters in parallel are joined in series with a silver voltameter and a d.c. supply. After a certain time the mass of silver deposited is 5·40 g and the mass of copper in one of the voltameters is 0·636 g. What mass of copper is deposited in the other voltameter?
(The relative atomic masses (atomic weights) of copper and silver are 63·6 and 108 respectively; copper is divalent and silver monovalent.) (*L*)

4. Define *electro-chemical equivalent*, and draw a labelled diagram of the circuit you would use to determine the electro-chemical equivalent of copper. In such an experiment the following measurements were recorded:

Weight of electrode before experiment = 10·50 g
Weight of electrode after current has passed = 10·59 g
Current through voltameter = 0·88 amp
Time for which current passed = 5 min. 30 s.

Calculate the electro-chemical equivalent of copper, and comment on the expected accuracy of the result. How could this accuracy have been improved? (*C*)

5. One kilogram of copper is purified by electrolysis, the p.d. across the electrolytic cell being 8·0 volts. Calculate:

(*a*) the total quantity of electric charge that must be passed;
(*b*) the energy used (in joules);
(*c*) the cost, if the price of electrical energy is 2d. per kilowatt-hour.
(The atomic weight of copper = 63·6 The valency of copper = 2)

6. The specific heat of nitrogen at constant volume is 0·733 joule $°C^{-1}$ g^{-1}, and its molecular weight is 28·0. Calculate the average energy gained by a nitrogen molecule when the temperature of the gas is raised from 0°C to 100°C.
(Avogadro's constant = $6{\cdot}02 \times 10^{26}$ kg-mole^{-1}.)

7. A 12-volt 40-ampere-hour accumulator is used to operate the 2-h.p. starter motor of a car. Assuming the arrangement is completely efficient, find the current taken. If the starter is used for an average period of 10 s on each occasion, and the accumulator is not recharged in the intervals, how many times can the car be started before the accumulator is discharged?
(Take 1 h.p. to be 746 watts.) (*O*)

8. A battery of forty 2-volt accumulators, each with internal resistance 0·1 ohm, is connected in series to be charged from a 100-volt supply. If the charging current is 2 amps, what resistance should be connected in the circuit? (*Ox. Schol.*)

Other questions: Ch. 4: 1 Ch. 11: 14(*c*) Ch. 16: 4

B

9. State Faraday's laws of electrolysis.

A current of 0·50 amp is passed through a voltameter containing acidulated water. After 20 minutes, 76 cm^3 of dry hydrogen have collected in one limb of the voltameter, the pressure being 74 cm of mercury and the temperature 20°C. Calculate the electro-chemical equivalent of hydrogen. The density of hydrogen is 0·090 g litre^{-1} at s.t.p.

Describe how you would examine the dependence of the current through the voltameter on the potential across it when platinum electrodes are used. Show by means of a graph the result you would expect to obtain. (*N*)

10. After a constant current has been passed for 30 min through a copper voltameter and a 150-ohm resistor connected in series, it is found that one of the electrodes in the voltameter has increased in weight by 1·05 g. Calculate the heat produced in the resistor during the 30 min. (*Cam. Overseas*)

11. An electric lamp takes 60 watts on a 240-volt circuit. How many dry cells, each of e.m.f. 1·45 volts and internal resistance 1·0 ohm, would be required to light the lamp? How much zinc would be consumed by the battery in 1 hour?
(Equivalent weight of zinc = 32·5.) (*O & C*)

12. An accumulator of e.m.f. 12 volts and of internal resistance 1 ohm is to be charged from a 112-volt d.c. supply.

(i) Draw a diagram of the circuit you would use.
(ii) Calculate the charging current had the accumulator been connected directly across the supply.
(iii) Calculate the value of the series resistor necessary to limit the charging current to 10 amps.
(iv) With this resistance in the circuit, calculate the potential difference between the terminals of the accumulator.
(v) If the accumulator consists of plates of total mass 2 kg and of specific heat 0·03 cal g^{-1} $°C^{-1}$ immersed in one litre of acid of specific heat 0·9 cal g^{-1} $°C^{-1}$ and of density 1·1 g cm^{-3}, find the rate of rise of temperature of the accumulator.

State any assumptions you make in your calculations. (*N*)

13. A battery of 24 accumulators in series is charged from 100-volt d.c. mains. If the internal resistance of the *battery* is 2 ohms and the e.m.f. of each *cell* during charging is 2·5 volts, what series resistance will be required to give a charging current of 10 amps? What percentage of the energy taken from the mains will be converted into heat? (*O & C*)

14. A box containing electrical equipment has two external terminals. One terminal is connected to one side of a 100-volt electric supply. The other terminal is connected to the other side of the supply through a fixed resistance. An ammeter shows a constant current of 0·5 amp through the box, and a voltmeter connected across the box terminals shows a voltage of 75 volts. The dissipation of power in the box is 25 watts. Suggest what is in the box, and find the value of the external resistance. (*W*)

15. A d.c. dynamo of e.m.f. 12 volts and of 1 ohm internal resistance is charging a battery of e.m.f. 10 volts and of 2 ohm internal resistance. Draw a diagram of the circuit and calculate (i) the current, (ii) the capacity, of the battery if it takes 12 hours to charge under these conditions.

A resistance, *R* ohms, is placed across the battery while it is still connected with the dynamo. Calculate *R* if no current flows through the battery. (*S, part qn*)

16. Two identical copper wires AB, CD are joined together by a uniform wire BC of different material so as to form a perfectly symmetrical circuit with two junctions at B and C. The positive terminal of a battery is connected to A and the negative terminal to D, and a large current is allowed to flow for about a minute. The battery is then disconnected and a sensitive galvanometer immediately connected between A and D. A small current is observed to flow through the galvanometer, gradually decreasing to zero.

Explain these observations. In which direction would you expect the galvanometer current to flow, and why? *(O & C)*

Other questions: Ch. 4: 18 Ch. 9: 6 Ch. 11: 15

C

17. Find the volume of oxygen at n.t.p. liberated per second in a water voltameter carrying a current of 1 amp. Atomic weight of oxygen = 16. *(Ox. Schol.)*

Chapter 4 Circuit measurements

A

1. A certain thermocouple, which has a total resistance of 10 ohms, has one junction in melting ice and the other in steam. The e.m.f. between its ends, measured with a potentiometer, is 4·2 millivolts. What would be its reading when it is connected to a millivoltmeter which has a resistance of 50 ohms? *(O & C)*

2. The e.m.f. of a cell is balanced by the fall of potential along 150 cm of a potentiometer wire. When the cell is shunted by a resistance of 16 ohms, the length required is 120 cm. What is the internal resistance of the cell? *(S, part qn)*

3. A cell is balanced against the fall of potential along 120 cm of a potentiometer wire. When the terminals of the cell are connected by an accurate voltmeter of resistance 100 ohms, the voltmeter reads 1·21 volts and the cell balances against 110 cm of the wire. Account for the difference in the balancing lengths and find the internal resistance and e.m.f. of the cell. *(L)*

4. Explain how to use a potentiometer to measure the e.m.f. of a cell. What advantages does this method possess over a moving-coil voltmeter?

Two resistors, one of 800 ohms and the other of 1200 ohms, are joined in series and a current is passed through them from a cell of e.m.f. 2·00 volts and negligible internal resistance. What value will be obtained for the potential difference across the 800-ohm coil (*a*) using a correctly calibrated voltmeter of resistance 1000 ohms, and (*b*) using a potentiometer? *(Cam. Forces)*

5. Explain the principle of the potentiometer. Draw circuit diagrams to show how you would use a potentiometer (*a*) to compare the e.m.f.'s of two cells; (*b*) to compare the resistances of two coils; (*c*) to measure an electric current.

An electric current is passed through an ammeter and a standard coil of resistance 0·100 ohm, connected in series, and is adjusted until the ammeter reads 5·00 amps. The p.d. between the ends of the standard coil is balanced by the p.d. across 128·8 cm of a potentiometer wire. A standard Weston cell of e.m.f. 1·018 volts is found to require 254·5 cm of the same potentiometer wire for a balance. What is the error in the ammeter reading? *(O & C)*

6. The resistance of the coil of a platinum thermometer is 3·020 ohms at 0°C, 4·172 ohms at 100°C, and 13·20 ohms in a furnace. Deduce the temperature of the furnace as given by this thermometer. *(S, part qn)*

B

7. Derive the relation between the resistances of the resistors forming a balanced Wheatstone bridge.

Four resistors AB, BC, CD and DA are joined with a cell and galvanometer to form a Wheatstone bridge. At temperature 0°C each has resistance 10 ohms. AB and CD are made of iron wire, while BC and DA are made of constantan wire. What change must be made in the resistance of BC in order to preserve the balance if the temperature of the whole rises to 100°C?

(Temperature coefficient of resistance of iron = 0·005 $°C^{-1}$; temperature coefficient of resistance of constantan is negligible.) *(Cam. Forces)*

8. Four 10-ohm coils AB, BC, CD, DA have a cell of e.m.f. 1 volt and negligible internal resistance connected across AC.

(*a*) Calculate the change in the p.d. between B and D that occurs if an addition of 0·1 ohm is made to CD.

(*b*) The p.d. between B and D can be restored to the old value by connecting a resistor in parallel with one of AB, BC or DA. Calculate the resistance of this resistor and state the position in which it must be connected. *(Cam. Forces)*

9. The resistivity of iron is 9×10^{-6} ohm cm at 0°C, and its temperature coefficient of resistance is $6{\cdot}5 \times 10^{-3}$ per C degree. A coil is made up of 2·5 metres of iron wire of diameter 0·5 mm, and this is connected in the left-hand gap of a metre-bridge, while a standard ohm fills the right-hand gap. Find, to the nearest millimetre, where the balance point will be if a determination is made at 25°C. *(O)*

10. The left-hand and right-hand gaps of a simple metre bridge are closed by coils having resistances 2 ohms and 1 ohm respectively. When the 2-ohm coil is shunted by a length of wire the balance point is found to be at 55·2 cm from the left-hand end of the bridge wire. What is the resistance of the shunt? If the shunt is 126 cm long and 0·234 mm in diameter, obtain a value for the resistivity of its material. *(L)*

11. An accumulator of e.m.f. 2·0 volts and negligible internal resistance is connected across a uniform wire of length 100 cm and resistance 5·0 ohms. The appropriate terminal of a cell of e.m.f. 1·5 volts and internal resistance 0·90 ohm is connected to one end of the wire, and the other terminal of the cell is connected through a sensitive galvanometer to a slider on the wire. What length of the wire will be required to produce zero deflection of the galvanometer?

How will the balancing length change when a coil of resistance 1·0 ohm is placed (*a*) in series with the accumulator, (*b*) in parallel with the cell of e.m.f. 1·5 volts? (*C*)

12. Explain how you would use a potentiometer to compare two small resistances. Why is this method preferable to a Wheatstone bridge method?

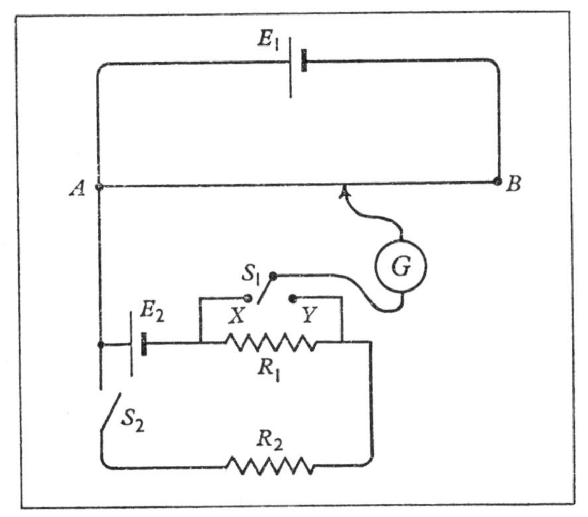

Fig. Q 4.12

In the above circuit AB is a uniform potentiometer wire 100 cm long. E_1 is an accumulator of e.m.f. 2·0 volts and negligible internal resistance. E_2 is a cell of e.m.f. 1·1 volts and internal resistance 1·0 ohm. The values of the resistors R_1 and R_2 are 1·0 and 2·0 ohm respectively. Switch S_1 enables the galvanometer G to be connected to either X or Y. Calculate the balance lengths obtained for each of the two positions of S_1 when switch S_2 is (*a*) open, (*b*) closed. (*C*)

13. A steady current (of about 2 amps) is passed through a 0·0200-ohm standard resistance R and a long uniform copper wire of diameter 1·22 mm connected in series. The potential differences across R and across 160 cm of the wire are compared using a wire potentiometer, the respective lengths on the potentiometer wire being 530 mm and 628 mm. Draw a circuit diagram of the arrangement of the experiment, and find (i) the resistance per metre of the copper wire, (ii) the resistivity of the copper. (*N*)

14. Describe, giving full experimental details, how you would compare the values of two unknown resistances each of the order of 0·1 ohm using a potentiometer. Draw a circuit diagram and give the theory of the method.

Suppose that, having set up the circuit to carry out this experiment, you found that no balance point could be obtained along the potentiometer. Discuss *three* possible reasons for this fault and the procedure you would adopt in order to trace it. (*N*)

15. Explain the principle of the potentiometer.

If a slide wire potentiometer of total resistance 5·0 ohms and a Weston standard cell of e.m.f. 1·0187 volts were available, together with the necessary auxiliary apparatus, describe in detail how you would (*a*) determine the variation of the e.m.f. of a thermocouple with the temperature difference between its junctions (maximum e.m.f. 3 millivolts), (*b*) compare the values of two resistances nominally 0·010 ohm and 0·005 ohm. (*L*)

16. Explain in detail how you would measure a small e.m.f. such as that of a thermocouple using a potentiometer method.

A 2-volt cell is connected in series with a resistance *R* ohms and a uniform wire AB of length 100 cm and resistance 4 ohms. One junction of a thermocouple is connected to A, and the other through a galvanometer to a tapping key. No current flows in the galvanometer when the key makes contact with the mid-point of the wire. If the e.m.f. of the couple was 4 millivolts what was the value of *R*? If the resistance of *R* is now increased by 4 ohms, by how much would the balance point change? (*S*)

17. A potentiometer of length 200 cm and resistance 4·00 ohms is joined to an accumulator through a coil of resistance 1000 ohms and a variable resistance. The variable resistance is adjusted until the potential difference across the 1000-ohm coil is exactly equal to the e.m.f. of a standard cell, 1·018 volts. When this is so the e.m.f. of a thermocouple is balanced against the potential difference across a length of the wire, and it is found that the detecting galvanometer shows no deflection for lengths between 124·0 cm and 126·0 cm. What is the e.m.f. of the thermocouple? (*C*)

18. Describe and explain how to determine by experiment the relation between the e.m.f., *E*, of a thermocouple and the temperature, θ°C, of its hot junction, the cold junction being maintained at 0°C.

The relation is sometimes given as (*a*) $E = \alpha\theta + \beta\theta^2$ and sometimes as (*b*) $E = k\theta^n$, where α, β, k and n are constant. Indicate how you would plot your observations to test these equations and to determine values for the constants. (*L*)

19. Write a short account of the variation with temperature of the electrical resistance of pure metals.

Give a brief description of a platinum resistance thermometer, and explain the principle of the electrical circuit in which it is normally used.

The resistance of the tungsten filament of an electric lamp measured by means of a Wheatstone bridge was found to be 44·0 ohms at 0°C and 65·7 ohms at 100°C. When the lamp was running normally on 230-volt mains it took 100 watts. Calculate the temperature of the hot filament measured on the tungsten resistance scale. *(O & C)*

20. If the resistance R_t of the element of a resistance thermometer at a temperature of t°C on the deal gas scale is given by $R_t = R_o(1 + At + Bt^2)$, where R_o is the resistance at 0°C and A and B are constants such that $A = -6{\cdot}50 \times 10^3$ B, what will be the temperature on the scale of the resistance thermometer when $t = 50{\cdot}0$ °C? *(L)*

21. What is meant by the statement that the *temperature coefficient of resistance* of platinum is $3{\cdot}8 \times 10^{-3}$ °C^{-1} in the range 0°C to 100°C? Describe a practical form of resistance thermometer, and draw a circuit suitable for measuring its resistance.

The resistance of such a thermometer is 10·54 ohms when placed in melting ice, and 14·69 ohms when placed in boiling water. Calculate (*a*) the temperature coefficient of resistance of the metal of the coil, (*b*) the temperature of a liquid in which the thermometer has a resistance of 11·30 ohms, (*c*) the minimum detectable change of temperature if the least change of resistance that can be detected is 0·01 ohm. *(C)*

Other question: Ch. 2: 21

C

22. One arm of a Wheatstone bridge network of equal resistances of 1000 ohms is replaced by a 1-volt cell. The galvanometer resistance is 50 ohms and the driver cell is 2 volts. The positive terminals of both cells are connected to the same point. If neither of the cells has appreciable internal resistance, calculate the current through the galvanometer and each of the cells. *(Ox. Schol.)*

23. P, Q, R and S are resistances, taken in cyclic order, in the four arms of a Wheatstone bridge. P and Q are the ratio arms and each is approximately 1 ohm; S is a standard resistance, 1·00150 ohms. To balance the bridge it is necessary to shunt Q with 2000 ohms. When R and S are interchanged and the shunt across Q removed, balance is restored when P is shunted with 1000 ohms. Obtain a value for the resistance of R. *(L)*

24. A Wheatstone bridge is made up from four resistances each of 100 ohms, a galvanometer also of resistance 100 ohms and a 2-volt battery of negligible internal resistance. If one of the resistances is used as a thermometer and has a temperature coefficient of resistance of 0·004°C^{-1}, find the least change of temperature detectable if the galvanometer is sensitive to a current of 10^{-6} amp. *(Cam. Schol.)*

25. An approximately balanced Wheatstone bridge network is constructed as follows. Two equal resistances of 100 ohms AB, BC, are connected in series, and another two, AD, DC, of 200 and 201 ohms respectively, connected in series with each other but as a whole in parallel with ABC. A 2-volt accumulator of negligible resistance is connected to A, C, and a galvanometer of resistance 50 ohms across BD. Find the current through the galvanometer. *(O & C)*

26. When one junction of a thermocouple is in melting ice and the other in steam at standard atmospheric pressure an e.m.f. of slightly less than 4 millivolts is produced. With the aid of diagrams describe and explain the procedure you would adopt to measure the temperature of a liquid within the range 100°C to 120°C using this thermocouple in conjunction with:

(*a*) a voltmeter, of resistance 100 ohms with a linear scale 10 cm long reading from 0 to 1·0 millivolt and a resistance box with steps of 100 ohms;

(*b*) a galvanometer, of resistance 40 ohms with a linear scale 50 cm long, reading from 0 to 60 microamps, and a resistance box with steps of 10 ohms;

(*c*) a uniform potentiometer wire of length 100 cm and resistance 5·00 ohms with any other necessary apparatus.

If the error in reading the scales in (*a*) and (*b*) and in finding the balance point in (*c*) is 1·0 mm, estimate the error in the temperature as determined by each method.

If the e.m.f. E produced by a thermocouple, when one junction is at 0°C and the other at a temperature t°C measured on the ideal gas scale, is given by $E = At - Bt^2$ where $A/B = 6{\cdot}00 \times 10^2$ deg. C, calculate the temperature which would be observed on the thermoelectric scale of the thermocouple when $t = 50$°C. *(N)*

27. In measuring the e.m.f. of an unknown cell (whose e.m.f. is in fact 1·1 volts and whose internal resistance is 1 ohm), an experimenter taps in at the centre point of a 50-ohm potentiometer wire supplied by a 4-volt accumulator of low internal resistance. If he uses a 14-ohm galvanometer without any protection, what current will flow through it? *(O & C)*

Chapter 5 Magnetic effects of a current

The permeability of a vacuum $\mu_0 = 4\pi \times 10^{-7}$ NA^{-2} $(H\,m^{-1})$
Charge of an electron $= -1{\cdot}60 \times 10^{-19}$ *coulomb*
Acceleration due to gravity $g = 9{\cdot}81$ m s^{-2}
Take the flux density of the horizontal component of the earth's magnetic field as $2{\cdot}0 \times 10^{-5}$ *teslas* $(Wb\,m^{-2})$

A

1. A conductor 2 cm long, carrying a current of 8 amp, lies at right-angles to the lines of induction in a field in which the flux density is 1·0 Vs m^{-2}.

Calculate the force exerted on the conductor and give a diagram to show its line of action and direction. (*Cam. Forces*)

2. A straight wire 100 cm long carries a current of 100 amps at right-angles to a uniform magnetic field of 1 weber metre^{-2}. Find the mechanical force on the wire and the power required to move it at 15 metres s^{-1} in a plane at right-angles to the field. (*S, part qn*)

3. A rectangular coil consisting of 100 turns, each 2·50 cm by 1·50 cm, is suspended so that it may turn about an axis through its centre parallel to one of its sides and perpendicular to a uniform magnetic field. When no current flows through the coil its plane is parallel to the field. When a current of 5·00 milliamperes is passed through the coil it comes to rest with its plane making an angle of 30° with the direction of the field. If the torsion constant of the suspension is $1{\cdot}44 \times 10^{-4}$ newton metre radian^{-1}, calculate the flux density of the field. (*Cam. Overseas*)

4. What is the flux density at a distance of 0·10 metre in air from a long straight conductor carrying a current of 6·5 amps? Hence calculate the force per metre on a similar parallel conductor distance 0·10 metre from the first and carrying a current of 3·0 amps. Explain how the expression for the force between two such conductors is used to define the ampere. (*C*)

5. Calculate the force per unit length between two parallel long straight wires 2 cm apart in air, each carrying a current of 5 amps. Show on a diagram the direction of the force acting on each wire. (*O*)

Other questions: Ch. 6: 3, 5 Ch. 9: 4 Ch. 11: 14(*a*)

B

6. A closed tube of square cross-section measures 1 cm × 1 cm × 10 cm. It is filled with mercury and placed in a magnetic field of 1 weber m^{-2} transverse to the axis of the tube. A current of 1000 amps is passed between opposite sides of the tube in the direction at right angles to the field. What is the pressure difference between the ends? (*Cam. Schol.*)

7. Explain in terms of the motion of free electrons what happens when an electric current flows through a metallic conductor.

A metal wire contains 5×10^{22} electrons per cm^3, and has a cross-sectional area of 1 mm^2. If the electrons move along the wire with a mean drift velocity of 1 mm s^{-1} calculate the current in amperes flowing in the wire.

If an electron of this velocity moves perpendicularly to a magnetic flux of density 0·1 weber m^{-2} what force does it experience? Hence calculate the total force on the electrons in a 1 cm length of wire when moving with a velocity of 1 mm per second, and compare this with the force on the wire calculated from the current flowing. (*O & C*)

8. A flat circular coil consisting of 150 turns of mean diameter 50 cm is fixed with its plane vertical and parallel to the magnetic meridian and a current of 2 amps is passed through it. A flat rectangular coil of 100 turns measuring 2 cm by 3 cm is situated at the centre of the larger coil with its longer sides vertical and its plane perpendicular to the meridian. Calculate the couple acting about a vertical axis on the smaller coil when a current of 4 milliamps is passed through it. (*L*)

9. A circular coil of radius 15 cm was mounted with its plane horizontal and another circular coil of 100 turns each of radius 2 cm was pivoted at its centre so that it could turn about a diameter of the large coil. The small coil carried a light counterbalanced pointer fixed at right angles to its plane. The coils were connected in series and a current passed through them. A rider of mass 0·04 g was adjusted on the pointer to keep the plane of the small coil vertical. The following results were obtained for d, the distance from rider to pivot and N, the number of turns in the large coil.

d in cm	14·7	13·4	12·0	10·6	9·3	8·0
N	100	90	80	70	60	50

Represent these results graphically and use the graph to calculate the current in the coils and the strength of the earth's vertical component.

State clearly any formulae used in the calculation, and list the probable sources of error in this experiment. (*S*)

10. A narrow solenoid 50 cm long, uniformly wound with 1000 turns, carries a current of 4·0 amps. A square coil of side 2·0 cm having 40 turns is suspended inside the solenoid. Calculate the maximum couple exerted on the coil if the current passing through it is 0·25 amp, and give a diagram showing the position of the coil when the couple has this value.

Discuss whether the couple on the coil would have the same mean value in the same position if the solenoid current were 4·0 amps a.c. and the coil current 0·25 amp a.c. of the same frequency. (*Cam. Forces*)

11. Two horizontal parallel straight conductors each 20 cm long are arranged in air one vertically above the other, and connected in series so that the current flowing through one is in the opposite direction to that flowing through the other. The lower conductor is fixed while the upper one is free to move vertically in guides under its own weight, the conductors always remaining parallel. If the upper conductor weighs 1·20 g, what is the approximate value of the current that will maintain the conductors at a distance of 0·75 cm apart? Neglect the effect of the earth's magnetic field. (*Cam. Overseas*)

12. A circular coil of 35 turns and mean radius

7·5 cm is set with its plane in the magnetic meridian. At the centre of the coil is pivoted a small magnetic needle free to move in a horizontal plane. When a current of 0·015 amp is passed through the coil, it is found that the coil must be rotated through 30° before the magnet is once again in the plane of the coil. Find the horizontal component of the earth's field. (*S, part qn*)

Other questions: Ch. 6: 6, 7 Ch. 8: 7, 8 Ch. 13: 14 Ch. 16: 22

C

13. A small freely supported magnet is placed at the centre of a circular coil of 500 turns of 15 cm mean diameter; what current is required to give a deflection of 45° when the axis of the coil is magnetic east and west? If the coil were turned through an angle of 10° about a vertical axis, would the current required to obtain the same deflection be more or less, and by how much? (A graphical solution is acceptable.) (*O & C*)

14. Two long straight wires lie parallel to each other a distance 5 cm apart. If one carries a current of 2 amps and the other a current in the opposite direction of 3 amps, find the force each wire exerts on the other, and the magnetic field at a point distance 5 cm from each. (*Ox. Schol.*)

Other questions: Ch. 6: 8 Ch. 14: 23 Ch. 16: 24

Chapter 6 Magnetic materials

Permeability of a vacuum $\mu_0 = 4\pi \times 10^{-7}\ NA^{-2}\ (H\ m^{-1})$

A

1. A narrow solenoid 50 cm long having an iron core is wound with 1600 turns of wire. When a current of 2 amps is passed the flux density in the iron is found to be 0·8 Vs m^{-2}. Calculate the permeability of the iron in these conditions. (*Cam. Forces*)

2. An iron ring of mean diameter 30 cm and of cross-sectional area 4 cm^2 is to be magnetized by a coil uniformly wound round its circumference. If the current in the coil is to be 1·5 amps, how many turns of wire are required to produce a total flux of 3·2 × 10^{-4} webers in the iron? Take the relative permeability of the iron to be 1000 under these conditions.

3. A small compass needle is freely pivoted so that it can swing in a horizontal plane. When a bar magnet is placed with its axis on the line passing through the centre of the compass perpendicular to the magnetic meridian, the needle is deflected through an angle of 35°. A second magnet is then placed with its axis on the same line and moved until the deflection is reduced to zero. If one of the magnets is now turned over end for end, what deflection will be produced?

4. Define magnetic flux Φ, flux density B magnetic field strength (magnetizing force) H, and permeability μ.

Calculate the values of Φ, B and H in a long air-cored solenoid of diameter 2·50 cm and of 2000 turns per metre when it is carrying a current of 0·10 amp. How would these values change if the solenoid were filled with a material of relative permeability 5000? (*Cam. Overseas*)

5. A coil of 50 turns of mean radius 5 cm is placed with its plane vertical in the earth's magnetic meridian and a short bar magnet is suspended at its centre so that it is free to rotate in a horizontal plane. Calculate the current in milliamperes required in the coil to cause a steady deflection of the magnet through 45° from the meridian. If the couple acting on the deflected magnet due to the current is 10^{-6} newton m, calculate the moment of the magnet.

(Assume the horizontal component of the earth's magnetizing force is 16 ampere m^{-1}.) (*O & C*)

B

6. A small magnet, suspended with its axis horizontal so as to be able to rotate freely about a vertical axis, is situated at the centre of a long horizontal solenoid, the axis of which lies at right-angles to the magnetic meridian. If the solenoid has 20 turns per cm, determine the value of the current passing through it which would cause the magnet to rotate through 50°.

(The horizontal component of the earth's magnetic field = 14 amp metre^{-1}.) (*N*)

7. Two long parallel straight wires are set up vertically in a plane at right-angles to the magnetic meridian, with their axes 10 cm apart. The wires carry, in the same direction, equal currents of 12·0 amps. Find the positions of neutral points in any horizontal plane.

(Horizontal component of earth's magnetic field = $50/\pi$ amp metre^{-1}.) (*L*)

C

8. A flat circular coil of 5 turns of diameter 14·0 cm is placed with its axis lying horizontally in the magnetic meridian. A small horizontal magnetic needle, suspended at the centre of the coil by a torsionless thread, performs 20 small rotational oscillations in 21·0 s when a current of 0·88 amp is passing through the coil. When the current is reversed the magnet also reverses its direction and now performs 20 oscillations in 36·8 s. Find the value of the horizontal component of the earth's magnetic field in the locality. (*L*)

9. A bar magnet suspended with its axis horizontal by a torsionless fibre has a time period of 2·50 s when making small oscillations in a horizontal plane. The magnet is then suspended by a fine wire in such a way that it rests with its axis horizontal, and in the meridian, when the wire is free from torsion. The time period is now T. It is then observed that when the top of the wire is turned through 360° the magnet sets with its axis at right-angles to the meridian. Calculate the value of T. (*N*)

10. In what respects do the magnetic properties of iron and steel differ? Derive an expression connecting magnetic induction, the magnetizing force, and the intensity of magnetization.

An iron ring has a cross-section of 3 cm^2 and a mean diameter of 25 cm. An air gap of 0·4 mm has been made by a cut across the section of the ring. The ring is wound with a coil of 200 turns through which a current of 2·0 amps passes. If the total magnetic flux is $2 \cdot 1 \times 10^{-4}$ webers find the relative permeability of the iron. (*S*)

11. A long thin iron rod is placed inside a long solenoid which has 20 turns of wire per cm and through which a current of 2 amps is passing. Calculate (*a*) the magnetizing field inside the solenoid; (*b*) the intensity of magnetization of the iron if its susceptibility is 20. (*O & C*)

12. Compare and contrast the magnetic properties of ferromagnetic, paramagnetic and diamagnetic substances, and give a qualitative account of the processes which occur in each of these three classes of magnetic material when subjected to an external magnetic field. (*L*)

Chapter 7 Electromagnetic induction

Permeability of a vacuum μ_0
$= 4\pi \times 10^{-7}\ H\ m^{-1}\ (NA^{-2})$

A

1. State the laws which govern the production of induced currents.

Explain clearly each of the following: (*a*) In measuring the resistance of the coil of an electromagnet by means of a Post Office box it is important to press the battery key before the galvanometer key; (*b*) when the key controlling the current in an electromagnet is opened a spark occurs at the breaking contact; (*c*) the core of a transformer is laminated. (*W*)

2. A 12-volt lamp and a coil of negligible resistance but appreciable inductance are connected in series across (*a*) 12-volt d.c. mains, (*b*) 12-volt a.c. mains. In case (*b*) the lamp is appreciably less bright than in case (*a*), and when a laminated soft iron core is pushed into the coil the lamp goes out in case (*b*) but is unaffected in case (*a*). Explain these observations. What would you expect to happen in case (*b*) if the iron core was absent and the frequency was gradually increased without altering the voltage? (*O & C*)

3. A magnetized needle is allowed to oscillate in turn over a glass sheet and then over a sheet of copper. Describe and explain what is observed. (*Cam. Overseas*)

4. A coil of self-inductance 10 henrys, in series with a wire of negligible resistance which fuses at a current of 10 amps, is connected to a 100-volt d.c. mains supply. How long after making the connection will the wire fuse?

Describe the changes of current and voltage before and after fusing. (*Ox. Schol.*)

5. What is the energy required to establish a current of 1·5 amps in a coil of inductance 1·6 henrys?

6. The mutual inductance between two adjacent air-cored coils is 0·20 henry. A current of 2·5 amps in one coil is reduced to zero at a uniform rate in one millisecond. What is the e.m.f. induced in the second coil?

7. A current of 5·0 amps flowing in a flat circular coil of 25 turns is found to produce a magnetic flux through the core of the coil of $3 \cdot 0 \times 10^{-5}$ webers. Calculate the inductance of the coil in microhenry.

8. Define a unit of inductance, and find an approximate value, in henrys, for the inductance of a uniformly wound solenoid 50 cm long and of diameter 2 cm if there are 2000 turns and no magnetic material is present.

The coils of a Post Office box are wound non-inductively. Describe this winding and explain the reason for using it. (*L*)

Other questions: Ch. 9: 4 Ch. 12: 2

B

9. An aeroplane with a wing-span of 25 metres is flying horizontally at a speed of 900 km per hour. Calculate the p.d. between the wing-tips if the vertical component of the flux density of the earth's field is $4 \cdot 0 \times 10^{-5}$ teslas. If a wire is connected between the wingtips, discuss whether a current will flow in it.

10. A fan having four blades each 20 cm long rotates at 3000 r.p.m. about a horizontal axis in the magnetic meridian at a place where the horizontal component of the earth's flux density is $2 \cdot 0 \times 10^{-5}$ $Vs\ m^{-2}$. Calculate the induced e.m.f. (*a*) between the tip of a blade and the axle, (*b*) between the tips of diametrically opposite blades, (*c*) between the tips of two blades at right-angles to one another. (*Cam. Forces*)

11. State Lenz's law and describe fully a method by which you could verify this law experimentally.

A horizontal metal disc of radius 10·0 cm is rotated about a central vertical axis at a region where the value of the earth's magnetic flux density is $5 \cdot 3 \times 10^{-5}\ Vs\ m^{-2}$ and the angle of dip is 70°. A sensitive galvanometer of resistance 150 ohms is connected between the centre of the disc and a brush pressing on the rim. Assuming the resistance of the disc to be negligible, what will be the current through the galvanometer when the disc is rotated at 1500 rev min^{-1}? If the system is frictionless, calculate the power required to maintain the motion. (*C*)

12. A closed circular loop of wire of radius 5 cm is placed with its plane at right angles to a uniform magnetic field in which the flux density is changing at the rate of 10^{-2} Wb m^{-2} per second. Calculate (*a*) the e.m.f. induced in the loop, (*b*) the energy dissipated in the loop in 10 seconds if its resistance is 2 ohms. (The self inductance of the loop may be ignored.)

Explain why there is a tension in the loop while the magnetic field is changing. How does it vary with time? (*O & C*)

13. Calculate the r.m.s. values of the e.m.f. produced by a coil having 50 turns of wire each of area 30 cm^2, when the coil is rotated at a uniform speed of 2100 r.p.m. about an axis perpendicular to a uniform magnetic field of flux density 0·80 Vs m^{-2}. (*Cam. Overseas*)

14. Describe simple experiments illustrating the phenomenon of self-induction, and explain what is meant by saying that the self-inductance of a coil is 1 henry.

A coil of self-inductance 10 henrys is joined to a battery of e.m.f. 20 volts, the total circuit resistance being 5 ohms. Calculate:

(*a*) The rate of growth of current at the instant at which the circuit is closed.

(*b*) The final value of the current.

(*c*) The mean e.m.f. induced if the circuit is broken in such a way that the current falls to zero in 0·1 s. (*Cam. Forces*)

15. A coil whose self-inductance is 0·05 henry and whose resistance is 100 ohms is connected to a battery. If a switch is thrown which disconnects the battery and replaces it by a short circuit, how long will it take for the current to fall to one quarter of its initial value?

16. Describe the construction of a simple form of alternating current transformer.

If the secondary coil is on open circuit explain, without calculation, the effect on the current flowing in the primary of (*a*) a fall in the supply frequency, (*b*) a reduction in the number of primary turns.

Calculate the current which flows in a resistance of 3 ohms connected to a secondary coil of 60 turns if the primary has 1200 turns and is connected to 240-volt a.c. supply, assuming that all the magnetic flux in the primary passes through the secondary and that there are no other losses. (*O & C*)

17. An air-cored solenoid 100 cm long, diameter 2·00 cm, is uniformly wound with 1000 turns of wire in which a direct current of 2·00 amps flows. Find the magnetic field strength at the middle of the solenoid. A short secondary coil of 200 turns is wound over the middle section of the solenoid. Find in volts the e.m.f. induced in the secondary coil if the current in the solenoid is uniformly reduced to zero and then raised to its former value but in the opposite direction, in 0·500 s. (*L*)

18. A current of maximum value 1 amp, alternating at 50 c/s with pure sine-wave form, flows in the primary of a mutual inductance of 0·20 henry. Find the maximum flux through the secondary of the mutual inductance, and the maximum e.m.f. induced in the secondary. (*N*)

19. A 50 c/s alternating current of 1 amp passes through a plane circular coil of radius 10 cm and 100 turns. Find the e.m.f. induced in a coplanar circular coil of 10 turns, each of radius 0·5 cm, with its centre at the centre of the larger coil. (*Ox. Schol.*)

20. A flat, circular coil of 2·0 cm diameter forms part of a circuit of 60 ohms resistance. There are 80 turns in the coil and when it is thrust suddenly into the space between the pole pieces of a large electromagnet so that its plane is perpendicular to the field, 123 microcoulombs of electricity flow round the circuit. Calculate the field strength of the magnet. (*L*)

21. A ballistic galvanometer gives a throw of 10 cm when a charge of 10^{-6} coulomb is rapidly passed through it. It is used to determine the field between the poles of an electromagnet by connecting it in series with a search coil and resistance to make the total resistance of the galvanometer circuit 1000 ohms. The search coil has 10 turns of wire wound on a cylindrical former 1 cm diameter. This is placed in the unknown field with its plane perpendicular to the field and, after the galvanometer has come to rest, the coil is quickly moved to a position of zero field. The throw of the galvanometer is 9 cm. What is the field in the gap of the electromagnet? (Galvanometer damping may be neglected.) (*C*)

22. Write down an expression for the magnetic field strength near the middle of a long solenoid, and indicate by a diagram how the direction of the field is related to the direction of current flow. By considering the solenoid to be divided into two halves, deduce what you can about the field at the end of a long solenoid.

A long solenoid consists of a single layer of turns of wire in the form of a helical spring. A small circular coil is placed coaxially inside the solenoid near its mid point, and connected to a centre-zero galvanometer. Explain carefully why the galvanometer deflects when the current in the solenoid is switched on.

If the galvanometer deflects through 10 divisions to the right when the current is switched on in this way, what would happen in each case if (*a*) the current were suddenly reversed, or (*b*) the circular coil were rapidly moved to one end of the solenoid, keeping it coaxial, or (*c*) the solenoid were suddenly compressed to two-thirds of its original length, the current remaining constant? (*O & C*)

23. Explain the differences in structure and action between a ballistic and an aperiodic galvanometer.

A ballistic galvanometer of resistance 15·0 ohms and sensitivity 5·0 divisions per microcoulomb is connected in series with a resistance of 100 ohms and a secondary coil of 500 turns and of resistance 50 ohms. This coil is wound round the middle of a long solenoid of radius 3·0 cm having 10 turns cm^{-1} and carrying a current of 0·60 amp. Assuming no damping, calculate the deflection produced in the galvanometer when the current in the solenoid is switched off. (*L*)

24. A small solenoid consisting of 100 turns of wire and having a cross-sectional area of 4 cm^2 is placed inside a larger solenoid which has 20 turns per cm and carries a current of 1 amp. The small solenoid is connected to a ballistic galvanometer and the total resistance in the circuit is 500 ohms. What charge will pass through the galvanometer when the current in the larger solenoid is switched off? (*S, part qn*)

25. An earth inductor with a resistance of 20 ohms and 200 turns, each of radius 15 cm, is connected in series with a ballistic galvanometer of 40 ohms resistance, which gives a throw of 10 divisions per microcoulomb. When the coil is set up in a vertical plane at right-angles to the magnetic meridian and turned sharply about a vertical axis through 180°, the galvanometer gives a throw of 70 divisions. Calculate the value of the horizontal component of the earth's magnetic field. (*O*)

26. A primary coil of 600 turns is wound uniformly on an iron ring of mean radius 8·0 cm and cross-sectional area 4 cm^2. A secondary coil of 5 turns is wound on the top of the primary, but insulated from it, and is connected with a ballistic galvanometer, the total resistance of the secondary circuit being 400 ohms. What quantity of electricity will be discharged through the galvanometer when a current of 3 amps is reversed in the primary coil, assuming that the relative permeability of the iron is 500? (*O & C*)

27. Explain the special features that are necessary in a moving coil galvanometer intended for ballistic use.

A ballistic galvanometer is connected in series with a search coil and the secondary winding of a mutual inductance. When a current is reversed in the primary winding of the inductance a charge of 90 microcoulombs flows through the galvanometer. After switching off the primary current the search coil (which has 200 turns of mean diameter 1·00 cm) is placed in and perpendicular to a magnetic field. The deflection of the galvanometer caused by the rapid removal of the search coil from the magnetic field is the same as was observed when the primary current was reversed. The total resistance of the galvanometer circuit is 250 ohms. Calculate the flux density of the magnetic field. Draw a complete circuit diagram of the arrangement. (*N*)

Other questions: Ch. 11: 37 Ch. 12: 9, 10, 16, 17

C

28. A solenoid with its axis perpendicular to the magnetic meridian is 1 metre long, has 1500 turns, and carries a current of 2 amps. A copper disc 4 cm in diameter is rotated at a uniform speed of 300 revolutions per minute about a thin axle which lies along the axis of the solenoid, so that the disc is well inside the uniform solenoid field. Calculate the p.d. between the axle of the disc and its rim.

A lead from a rubbing contact with the axle of the disc is taken through a sensitive galvanometer to a contact A on a length of low resistance wire W connected in series with the solenoid. A second lead from a rubbing contact on the rim of the disc is taken to a contact B which slides along the wire W. The distance between A and B is adjusted until no current flows in the galvanometer. Calculate the resistance in ohms of the wire between A and B. (*O*)

29. A straight piece of wire 10 cm long is spun about an axis normal to the wire and passing through one of its ends. The wire completes 5 revolutions per second and a uniform magnetic field of 10^{-2} Wb m^{-2} is applied parallel to the axis of rotation. Find:

(i) the e.m.f. between the ends of the wire;

(ii) what this e.m.f. would be if the wire were twice as long;

(iii) the e.m.f. between the ends of the wire if it is bent into a right angle at its mid-point, both arms remaining normal to the mgnetic field; and

(iv) the e.m.f. between the ends of the wire if it is bent into a circular loop with the ends just failing to make contact. (*S, part qn*)

30. A superconducting solenoid (which has no resistance), immersed in liquid helium, is connected to a 2-volt battery of negligible internal resistance. After 50 s the current carried is 5 amps, and the field in the solenoid has reached a very high value. If the solenoid now 'goes normal', so that a large resistance appears and its stored energy is suddenly converted to heat, what volume of helium gas at n.t.p. will boil off?

(Latent heat of boiling of helium: 22 joule g^{-1}.
Molecular weight of helium = 4;
The volume of 1 kg-mole of a gas at n.t.p. = 22·4 m^3.) (*Cam. Schol.*)

31. What is meant by (i) the *self-inductance* of a coil, (ii) the *mutual inductance* of two coils? Describe how you would construct a standard inductance for use in the laboratory.

The primary and secondary coils of a transformer have negligible resistance and are wound so that all the magnetic flux due to one coil is linked by the turns of the other coil. The current through the primary coil is increased at a uniform rate from zero to 10 amps in 2·5 s. With the secondary coil open-circuited, a steady potential difference of 0·5 volts appears across the secondary, and a steady potential difference of 0·04 volts across the primary. Find (i) the self-inductance of the primary coil, (ii) the ratio of the number of turns on the primary and secondary coils, (iii) the self-inductance of the secondary coil. (*O & C*)

32. A coil of wire of 100 turns forms the perimeter of a rectangle with sides 10 cm by 200 cm. Estimate the magnetic flux density at the centre of the rectangle when a current of 1 amp flows around the coil, explaining any approximations you make. A second small coil with five turns of area 1 cm^2 is placed at the centre of and coplanar with the rectangular coil. Calculate approximately (*a*) the mutual inductance between the two coils, (*b*) the r.m.s. value in volts of the electromotive force induced in the small coil when an alternating current of r.m.s. value 1 ampere and frequency 50 kilocycles per second is passed through the rectangular coil. (*O & C*)

33. Define and explain the terms *self inductance* and *mutual inductance.*

Calculate the self inductance of a 100 turn solenoid A wound on a former 40 cm long and 2 cm radius. A second solenoid B of 200 turns wound on a former 20 cm long and 1 cm radius, is mounted inside A in a symmetrical manner. What is the mutual inductance of the two solenoids? The solenoid A is connected to a 10,000 cycles s^{-1} generator which maintains 1 volt r.m.s. across the coil. What is the generator current and the open circuit voltage across the coil B? Neglect the resistances of the coils and end effects. (*O & C*)

34. Describe and explain how a B/H curve for a specimen of steel may be plotted experimentally using a ballistic galvanometer. What is meant by the term 'magnetic hysteresis'?

A specimen of steel in the form of a ring of radius 10 cm and cross-sectional area 2 cm^2 has a magnetizing coil wound around it. A second coil of 10 turns is also wound around the specimen and connected to a ballistic galvanometer. The specimen is originally unmagnetized, and when a current i is passed through the magnetizing coil, a charge of 30 microcoulombs flows through the galvanometer. The current is then reversed, and a charge of 58 microcoulombs flows in the opposite direction. When the current is finally switched off, a charge of 17 microcoulombs flows in the same direction as when the current was first switched on.

Explain these observations, and estimate the final flux density in the steel if the total resistance of the ballistic galvanometer circuit is 100 ohms. (*O & C*)

Other questions: Ch. 9: 12 Ch. 12: 35, 38, 42

Chapter 8 Measuring instruments

A

1. Explain, giving a circuit diagram, how you would use a potentiometer method to test the calibration of an ammeter of range up to 2 amps.

When checked by this method, a certain moving-coil ammeter appeared to read approximately 2% low at all parts of the scale. What possible causes can you suggest for this? (*Cam. Forces*)

2. Two moving-coil galvanometers, P and Q, are alike in all respects except that P's coil has 10 turns of resistance 2 ohms and Q's coil has 100 turns of resistance 180 ohms. Compare (*a*) their current sensitivities, (*b*) their voltage sensitivities. (*O & C*)

3. A milliammeter with a full-scale deflection of 10 milliamps has a resistance of 7·3 ohms. What resistance would be necessary, and how would it be connected, so that the instrument could be used (*a*) as an ammeter up to 1·5 amps, (*b*) as a voltmeter up to 150 volts? (*O & C*)

4. Give a labelled diagram which shows the construction of a moving-coil ammeter.

If such an instrument gives a full-scale deflection for a current of 15 milliamps and has a resistance of 5 ohms, how would you adapt it so that it could be used (i) as a voltmeter reading to 1·5 volts, (ii) as an ammeter reading to 1·5 amps? In case (i) above what voltage would be indicated if the instrument were connected to the terminals of a cell of e.m.f. 1·40 volts and internal resistance 5 ohms? (*W*)

5. An ammeter of 0·02 ohm resistance has a linear scale and requires a current of 0·5 amp to produce a full-scale deflection. How can it be adapted to operate as an ammeter reading to 10 amps? Show that the scale will still be linear. (*S, part qn*)

6. The coil of a milliammeter has a resistance of 100 ohms and a potential of 100 millivolts between the terminals produces a full-scale deflection. Describe and explain quantitatively how the meter could be made to read (*a*) currents up to 1 amp, (*b*) voltages up to 10 volts. (*O & C*)

Other question: Ch. 5: 3

B

7. Describe the construction and mode of action of a sensitive moving-coil galvanometer.

Derive an equation for the deflection produced by a given current in terms of the constants of the instrument.

A rectangular coil of 50 turns 4·0 cm long and 2·0 cm wide is suspended in a radial field of $7{\cdot}5 \times 10^{-2}$ weber per square metre. Using a lamp and scale positioned 1 m from the mirror on the coil, a current of 1 μA produces a deflection of 5 mm. What couple would be required to turn the coil through 1 degree? (*A.E.B.*)

8. A galvanometer with a linear scale has a coil consisting of 50 turns of wire wound on a former 2 cm square, mounted in a magnetic field of flux density 0·3 weber m^{-2}. A current of 10 microamps produces a deflection of 5°. Calculate the couple required to produce a twist of 1 radian in the suspension. (*O & C*)

9. Two galvanometers, identical in all other respects, are fitted by the makers with coils in the one case of 50 turns and 5 ohms resistance, and in the other of 600 turns and 720 ohms resistance. Which will give the greater deflection when connected to a battery of 1·5 volts e.m.f. and 80 ohms internal resistance? *(O & C)*

10. Describe the structure and explain the action of a moving-coil galvanometer. Discuss the factors which contribute to high sensitivity in this type of galvanometer.

A galvanometer of resistance 95 ohms, shunted by a resistance of 5 ohms, is joined in series with a resistance of 2×10^4 ohms and an accumulator of e.m.f. 2·0 volts. What is the sensitivity of the galvanometer if it shows a deflection of 50 divisions? *(L)*

11. An accumulator of e.m.f. 2 volts and negligible internal resistance is connected across two resistors AB and BC in series. The resistance of AB is 1000 ohms and that of BC is 1 ohm. A sensitive moving-coil galvanometer of resistance 200 ohms is connected across BC, and registers a deflection of 35 divisions. What is the current in microamps corresponding to a deflection of one division? *(O)*

12. A galvanometer coil wound with copper wire has resistance 5·0 ohms at 0°C. The instrument is converted into an ammeter by means of a constantan shunt of resistance 0·10 ohm, and the ammeter reads correctly at 0°C. What will it indicate for a current of 1·00 amp at 20°C?

(Temperature coefficient of resistance of copper $= 4{\cdot}0 \times 10^{-3}$ °C^{-1}; the temperature coefficient of resistance of constantan is negligible.) *(Cam. Forces)*

Other questions: Ch. 2: 15, 18, 19 Ch. 7: 20, 24, 25 Ch. 11: 37

C

13. Two galvanometers are identical in design except for the number of turns on their moving coils: the same mass of copper is used in winding each coil, but galvanometer A has twice as many turns as B. If A has a resistance of 10 ohms, which galvanometer will give the greater deflection when used with a thermocouple whose resistance is 4 ohms? (Ignore the resistance of the coil suspensions.) *(Cam. Schol.)*

14. An ammeter reads high by 5%. The resistance of its coil is 10 ohms. What resistances are required to make it read correctly and still present a resistance of 10 ohms? *(Ox. Schol.)*

15. A cell of known e.m.f. E and negligible internal resistance is connected in series with two known resistances P and Q. A moving-coil galvanometer is connected in series with a known variable resistance X and the two are connected in parallel with P. Values of the deflection θ of the galvanometer are observed for various values of X while P and Q are kept constant. Assuming that θ is proportional to the current passing through the galvanometer show how these observations could be used to plot a linear graph from which the resistance and current sensitivity of the galvanometer could be obtained. *(L)*

Other questions: Ch. 4: 26 Ch. 14: 21

Chapter 9 Generators and motors

A

1. A rectangular coil is pivoted about an axis perpendicular to a uniform magnetic field. Draw a graph showing how the induced e.m.f. in the coil varies with its angular position when the coil is rotated at a uniform speed. If the peak value of the induced e.m.f. is 200 volts, calculate (*a*) its r.m.s. value, (*b*) its instantaneous value when the plane of the coil is at 60° to the direction of the magnetic field. *(C)*

2. A resistance is sometimes connected so that it is in series with a d.c. motor when the motor is started, but completely cut out of the circuit when the motor has achieved its full running speed. Explain the purpose of this arragement. *(Cam. Overseas)*

3. Describe one simple type of d.c. motor and explain its action. Why is a series-wound d.c. motor specially suitable for electric traction? *(Cam. Forces)*

4. State an equation giving the force F exerted on a wire of length l carrying a current I at right-angles to a magnetic field of flux density B. Give the units of all the quantities in your equation and draw a diagram showing the direction of the force in relation to field and current.

The armature of a motor contains a rectangular coil of area 20 cm^2 having 50 turns and carrying 2·5 amps.

(*a*) Calculate the flux density of the field in which the coil is placed if the maximum torque on it is 0·30 newton metre and draw a diagram showing the position of the coil when this torque is developed.

(*b*) Calculate the back e.m.f. in the coil in this position if it is rotating at 2400 r.p.m. *(Cam. Forces)*

Other question: Ch. 5: 2.

B

5. A d.c. electric motor is used to raise ore from a mine at the rate of 1 ton per min, the distance through which the material is raised being 200 ft. If the efficiency of the motor is 80%, what current will it take from 506-volt mains?

(1 h.p. = 550 ft lbf s^{-1} or 746 watts.) *(C)*

6. What is meant by saying that the difference of potential between two points is 1 volt?

The following are connected in series and joined to a direct-current supply:

(*a*) a 10-ohm resistor;

(*b*) an accumulator having an internal resistance 0·1 ohm, the positive terminal of the accumulator being connected to the positive of the supply;

(*c*) a series-wound d.c. electric motor, the total resistance of which is 2 ohms;

(*d*) an ammeter.

The ammeter reads 5 amps, while the p.d. across the accumulator, measured with a high-resistance voltmeter, is 13·5 volts, and that across the motor is 35 volts.

Calculate the rate at which electrical energy is converted in (*a*), (*b*) and (*c*), and find the rate of production of heat in each case. (*Cam. Forces*)

7. Describe in outline the mode of working of (*a*) an a.c. generator, (*b*) a d.c. generator. How does the current taken by a shunt-wound motor automatically adjust itself to the external load?

A shunt-wound motor takes 15 amps from 60-volt mains. If the resistance of the armature is 0·2 ohm and that of the field coil is 30 ohms, calculate the armature current and back e.m.f. (*Cam. Schol.*)

8. A shunt-wound motor has field windings of resistance 50 ohms, and an armature of resistance 2 ohms. When connected across a 100-volt supply, it takes a current of 10 amps. Calculate (*a*) the back e.m.f. generated in the armature, (*b*) the power available for mechanical use. (*O*)

9. A shunt-wound electric motor, driven by a 20-volt d.c. supply, uses an armature current of 5·0 amps while developing $\frac{1}{8}$ horse-power. Calculate the resistance of the armature.

(1 horse-power may be taken to be equal to 750 watts.) (*L*)

10. Explain the importance of back e.m.f.'s in the theory of the d.c. electric motor. Describe the main features and applications of series- and shunt-wound d.c. motors.

A shunt-wound d.c. motor takes 10 amps from the 100-volt d.c. mains when the rotation speed of the armature is 3000 rev min^{-1}. The armature and field windings have resistances of 1 ohm and 100 ohms respectively. Calculate (*a*) the back e.m.f. in the armature, (*b*) the rate, in watts, at which electrical energy is dissipated as heat in the windings, and (*c*) the torque, in newton metre, exerted by the armature on the load. Give a qualitative account of the effect of inserting 100 ohms into the field circuit while the motor is running. Neglect friction. (*O & C*)

11. With the aid of a labelled diagram describe the essential features of a shunt-wound electric motor.

When a shunt-wound motor runs 'light' (i.e. off-load) at 12 revolutions per second on a 240-volt d.c. supply, the armature current is 0·3 amp. When driving a mechanism (i.e. on-load) the armature current increases to 16·0 amps, the potential difference across the motor remaining the same. If the resistance of the armature is 3 ohms, calculate the number of revolutions per second that the motor makes on-load. (*A.E.B.*)

Other questions: Ch. 3: 14 Ch. 7: 13 Ch. 12: 8, 11

C

12. A circular coil of negligible resistance, with 100 turns each of mean area 2 cm^2, rotates at 50 revolutions per second about a horizontal diameter in the vertical flux between the poles of a magnet. The coil is suitably connected through a split-ring commutator and brushes to a moving-coil meter of resistance 80 ohms. The brushes are set so that the current flows in one direction only in the meter. If the meter gives a steady reading of 5 milliamps, calculate the charge which passes around the circuit for one half revolution of the coil. Hence calculate the flux density of the field.

How would you use the device to find the direction of the flux at a point in a magnetic field? (*O & C*)

13. Show that the mechanical power supplied by a series-wound electric motor to which a constant e.m.f. E is applied is a maximum when the motor is running at such a speed that the back e.m.f. is $\frac{1}{2}E$. (*N*)

14. A 250-volt shunt-wound motor has an armature resistance of 0·2 ohm, and runs at 500 r.p.m. on no load, taking an armature current of 5 amps. How is the energy dissipated? Calculate the speed and estimate the power output when running on load and taking an armature current of 50 amps. How will the energy losses differ from the case of no load? (*L*)

15. Describe a simple type of d.c. motor and explain the principle of its operation.

A shunt-wound motor is joined to 210-volt d.c. mains. The armature current is 1·0 amp when the speed is 2500 r.p.m., but this rises to 5·0 amps when the speed falls to 2000 r.p.m. as a result of driving a greater load. Find, in the second case, the rate at which electrical energy is converted (*a*) into heat in the armature, (*b*) into mechanical energy. (*Cam. Forces*)

16. A shunt-wound motor takes 22$\frac{1}{2}$ amps from the 100-volt mains and runs at 1250 rev per min. Calculate the back e.m.f. if the resistances of the armature and field circuits are 0·4 ohm and 40 ohms respectively. The field circuit resistance is then increased to 60 ohms. Find (*a*) the new back e.m.f. and (*b*) the new steady speed of the motor, assuming that the torque is the same as before and that the flux is proportional to the field current. (*C*)

17. Show under what conditions (*a*) maximum external power, and (*b*) maximum external energy, can be obtained from an electric motor driven by a battery of finite storage capacity, of fixed internal resistance R and of e.m.f. E.

A lead-acid accumulator consists of six cells in series each of e.m.f. 2 volts and of internal resistance 0·01 Ω. Calculate the maximum external power available.

One cell in this accumulator starts to develop a higher resistance than 0·01 Ω. What is the maximum value of the internal resistance of this cell such that for maximum external power it should be short-circuited? *(Cam. Schol.)*

Chapter 10 Electrostatics

Additional data required:
Charge of an electron $= -1{\cdot}60 \times 10^{-19}$ *coulomb*
Mass of an electron $= 9{\cdot}1 \times 10^{-31}$ *kg*
Permittivity of a vacuum $\varepsilon_0 = 8{\cdot}85 \times 10^{-12}$ *F* m^{-1}
Acceleration due to gravity $g = 9{\cdot}81$ *m* s^{-2}

A

1. How would you demonstrate experimentally that—

(*a*) The electric intensity inside a charged hollow conductor containing no other charged bodies is zero?

(*b*) Positive and negative charges produced by friction are always equal in amount?

(*c*) The surface density of electricity is greatest on a conductor where it is most sharply curved?

'The gold-leaf electroscope is an instrument for measuring potential rather than charge.' Comment on this statement. *(Cam. Overseas)*

2. An electroscope is placed on a block of paraffin wax and the cap and case joined by a wire. The electroscope is charge and the wire is then disconnected from the cap without being earthed. Finally the cap is earthed. Describe and explain what happens at each stage of the experiment *(L)*

3. What is meant in electrostatics by an induced charge? How would you verify experimentally the equality in magnitude of inducing and total induced charge?

Describe *one* type of electrostatic generator pointing out clearly how its operation depends on induced charges. *(S)*

4. What is meant by the electrostatic potential at a point?

Illustrate, by a suitably labelled sketch in each case, how it is possible to set up electrostatic systems in which (*a*) a conductor at earth potential carries a net positive charge (*b*) a conductor at earth potential has regions of both negative and positive charge, (*c*) a conductor with no net charge is at a positive potential with respect to earth.

How would you verify experimentally that both negative and positive charges are present in case (*b*)? *(O & C)*

5. A potential difference of 15 volts is maintained between two electrodes in a vacuum tube. An electron is emitted with negligible velocity from the negative electrode; calculate its velocity when it reaches the positive electrode.

6. Calculate the radius of a water drop which would just remain suspended in the earth's electric field of 3 volts per cm when charged with 1 electron. *(S, part qn)*

7. A vacuum tube contains two plane parallel electrodes 7·5 mm apart. If a p.d. of 150 volts is maintained between them, what is the electric field intensity in the gap? What is the force acting on an electron in the gap?

An electron is emitted at negligible velocity from the negative electrode. How long does it take to cross the gap?

8. Give an account of a method by which the charge associated with an electron has been measured.

Calculate the potential difference in volts necessary to be maintained between two horizontal conducting plates, one 0·50 cm above the other, so that a small oil drop, of mass $1{\cdot}31 \times 10^{-11}$ g with two electrons attached to it, remains in equilibrium between them. Which plate would be at the positive potential? *(L)*

9. A charged oil drop of mass 10^{-11} g is observed to remain stationary in the space between two horizontal charged plates, the electrical forces being sufficient to counteract its weight. Find the charge on the drop if this occurs when the p.d. between the plates is 3000 volts and their distance apart 1 cm. *(Cam. Forces)*

10. Two small conducting spheres, each of mass 10 mg, are suspended from the same point by non-conducting strings of length 10 cm. They are given equal and similar charges, until the strings are equally inclined at 30° to the vertical. Calculate the charge on each sphere. *(N)*

11. If there is an electric intensity of 300 volts per metre at the surface of the earth, what is the charge in coulombs per square kilometre on the earth's surface? *(O & C)*

Other questions: Ch. 13: 4, 5 Ch. 16: 7

B

12. Explain the meaning of the terms electric potential and electric field strength.

A positively charged sphere is placed on the axis of an insulated metal rod some distance from one of its ends. Account for the resulting charge distribution over the metal rod.

How do you reconcile the charge distribution on the metal rod with the fact that the only free charges in a metal are electrons?

Sketch freehand graphs showing the form of the variation of (*a*) the electric potential and (*b*) the electric field strength along the axis of the metal rod from the surface of the glass sphere, through the rod to a distant point.

How and why is the electric field between the glass sphere and metal rod changed if the metal rod is earthed? *(O & C)*

13. The casing of an ordinary gold leaf electroscope is permanently earthed. A negatively charged rod is held near the cap of the electroscope and the cap is then earthed. The earth connection to the cap is then broken and the charged rod removed. Discuss this process from the point of view of charge distribution and potential distribution, and show the variation in the electrostatic field of force in suitable diagrams. *(W)*

14. A hollow can is placed on the cap of an uncharged electroscope whose case is earthed and a small positively charged metal sphere suspended by a nylon thread is lowered into the can without touching it and then withdrawn. Secondly, the sphere is lowered into the can as before and while it is inside, the can is earthed momentarily. The sphere is then withdrawn. Thirdly, the can and electroscope are discharged and the sphere is again lowered into the can, allowed to touch the bottom, and withdrawn. The sphere is then tested for charge on another uncharged electroscope.

State what happens to the electroscope deflection in these experiments and explain how the sign and magnitude of the potential of the leaves vary. How would the results differ if the sphere were made of insulating material with the same charge as that on the metal sphere? *(S, part qn)*

15. In Millikan's oil drop experiment the drops are observed to move (under the action of constant forces due to gravity and any uniform electric field that may be present) with constant velocity. Explain why their velocity, rather than their acceleration, is constant.

In this experiment an oil drop was prevented from falling or rising by applying a potential difference of 5750 volts between parallel horizontal plates 1·5 cm apart. Assuming that the drops each carried one electron charge ($1{\cdot}60 \times 10^{-19}$ coulomb) what was the radius of the drop?

(Density of oil = 0·92 g cm^{-3}.) *(O & C)*

16. Explain what is meant by the electrical potential at a point and obtain the expression for the potential at a point distance x from an isolated positive point charge of magnitude Q. Deduce the relationship between the strength of the electric field and the potential gradient at a given point in the field.

(*a*) The electric field near the surface of the earth 100 V m^{-1}. Calculate the surface density of charge on the earth's surface.

(*b*) If the potential gradient exceeds 30,000 V cm^{-1} a breakdown discharge takes place in air at s.t.p. Calculate the maximum radius of a sphere which will discharge to the air if its potential is 600 V. *(A.E.B.)*

17. A charged oil drop is prevented from falling under gravity by a vertical electric field between two horizontal metal plates charged to a difference of potential of 6920 volts, the distance between the plates being 1·30 cm. When the field is cut off the drop falls in air with a uniform velocity of $1{\cdot}90 \times 10^{-2}$ cm s^{-1}. Assuming Stokes's formula $R = 6\pi a\eta v$ for the viscous resistance R to the motion of a sphere of radius a travelling with uniform velocity v through a fluid of coefficient of viscosity η, calculate (*a*) the radius of the drop, (*b*) the charge on the drop.

(Density of oil = 0·90 g cm^{-3}. Coefficient of viscosity of air = $1{\cdot}80 \times 10^{-4}$ g cm^{-1} s^{-1}. The density of the air may be neglected in comparison with that of the oil.) *(O & C)*

18. In a certain experiment a very small drop of oil was introduced between two horizontal and parallel plates separated by a vertical distance of 20 mm, the apparatus being open to the air. The drop was observed to fall at a constant velocity over a distance of 15 mm in 15·1 s, but when a potential difference of 7500 volts was applied across the plates, it was found to rise, covering the same distance in times of 65·4, 65·4, 52·2, 52·2 s in four successive measurements. Given that under the conditions of the experiment the viscosity of air was 1830×10^{-7} g cm^{-1} s^{-1}, and the density of the oil 0·923 g cm^{-3}, deduce all you can from the observations.

(A sphere of radius a moving with steady velocity V through a fluid of viscosity η experiences a retarding force $F = 6\pi\eta aV$.) *(Cam. Schol.)*

Other questions: Ch. 13: 18 Ch. 15: 9

C

19. A parallel-plate air condenser is made from two discs, each of radius 8 cm, placed 1 cm apart. The upper plate is earthed and hung from a balance. The lower plate is initially at earth potential and is then raised to 15,000 volts. What is the change in the apparent weight of the upper plate? Ignore edge effects. *(Cam. Schol.)*

20. Assuming that a metal contains n 'free' conduction electrons per unit volume, explain what occurs when an electric field is set up in the wire by connecting a battery across its ends. Indicate the physical meaning of the concept of 'resistance'.

If the value of n for copper is 10^{23} per cm^3, and its resistivity is $1{\cdot}6 \times 10^{-6}$ ohm cm, calculate the mean drift velocity of the electrons under an electric field of 1 volt per cm. *(O & C)*

21. Use the data in (i) and (ii) below to obtain two independent values for the electromotive force in volts of a Daniell cell. Derive any relation used in (i) and state any assumptions that you make in (ii).

(i) The potential difference between the terminals of 284 Daniell cells in series, when applied between two parallel coaxial insulated circular metal discs, each of radius 7·50 cm and separated in air by a distance of 0·200 cm, produced a force of attraction between them of 0·180 gf.

(ii) 3·00 g of zinc dust was stirred into copper sulphate solution in a vacuum flask and the temperature of the contents rose from 10·00°C to 14·20°C. After the apparatus had cooled, a potential difference of 10·0 volts was maintained for 6 min 40 s across a platinum heating coil contained in the flask with a current of 2·50 amps flowing, and the temperature rose from 11·00°C to 15·53°C.

(Take the chemical equivalent of zinc as 32·5.) (*L*)

Other question: Ch. 11: 48

Chapter 11 Capacitance

Additional data required:
Permittivity of a vacuum $\varepsilon_0 = 8{\cdot}85 \times 10^{-12}\ F\ m^{-1}$
Faraday's constant $F = 9{\cdot}65 \times 10^7\ C\ kg\text{-}equiv^{-1}$
The volume of a kg-mole of any gas at n.t.p. $= 22{\cdot}4\ m^3$
1 *calorie* = 4·2 *joules*

A

1. A parallel-plate condenser is made up of eleven metal plates of area 5 cm by 3 cm, separated by sheets of mica of dielectric constant 6 and thickness 0·2 mm. Calculate its capacity in microfarads. (*O*)

2. Obtain an expression for the energy of a charged condenser and find the energy of a parallel-plate condenser whose circular plates, each 50 cm in diameter, are separated by 2·0 mm of sulphur, whose dielectric constant is 4·0, when the potential difference between the plates is 240 volts. (*L*)

3. A condenser of capacity of 100 microfarads is charged to a potential of 1000 volts. It is discharged through a coil placed in a calorimeter of total mass 5 g and average specific heat 0·5 cal gm^{-1} °C^{-1}. Calculate the rise of temperature in the calorimeter. (*Cam. Schol.*)

4. A metal sphere 4 metres in diameter, is charged to a potential of 3 million volts by a Van der Graaff induction machine. Calculate the heat generated when the sphere is earthed through a long resistance wire. (*O & C*)

5. A 20-microfarad capacitor is charged by joining to a 3000-volt supply, and is then discharged. Calculate the mean power of the discharge, if the charge remaining after 10 microseconds is negligible. (*Cam. Forces*)

6. If a steady potential difference of 250 volts is maintained across the combination of a condenser of capacitance 3 microfarads in series with one of 0·5 microfarad, calculate (*a*) the potential difference across each condenser, (*b*) the charge on each condenser, (*c*) the energy stored in each condenser. (*A.E.B.*)

7. How can three condensers of capacities 3, 6 and 9 μF respectively be arranged to give a system of capacity 11 μF? (*Cam. Schol.*)

8. A 2-μF capacitor is required to work at 1000 volts d.c. If the capacitor has to be replaced, but the only units available were of 2 μF with permissible d.c. voltage of 400, how would you proceed? (*Cam. Forces*)

9. Two condensers of capacitances 0·5 μF and 0·3 μF are joined in series. What is their capacitance? What value of capacitance joined in parallel with this combination would give a capacitance of 0·5 μF? What will be the energy of the system when connected to a 240-volt d.c. supply? (*W*)

10. A sphere of diameter 1 metre is charged up to 200 volts. How many times per second will it have to be discharged and recharged in order to give a mean discharge current of 1 microamp? (*Cam. Schol.*)

11. An electroscope has a capacity of 20 × 10^{-6} microfarads and its leaves diverge 25 divisions when charged to a potential of 600 volts. If in this position the divergence of the leaves decreases at the rate of one division in a minute owing to imperfect insulation, find the leakage current and the resistance of the insulation. You may assume that the divergence of the leaves is proportional to their potential. (*N*)

12. Derive from first principles an expression for the capacitance of a parallel-plate condenser for which the extent of the plates is much greater than their separation.

An air condenser consists of two parallel plates with an overlapping area of 1600 cm^2 and separated by 1·5 mm. A potential difference of 1200 volts is maintained across these plates. Find the change of electrical energy when the space between the plates is filled by an insulator of dielectric constant (i.e. relative permittivity) 5.

How is the difference between these initial and final amounts of electric energy explained? (*A.E.B.*)

13. Define *capacitance* and deduce an expression for the energy stored in a charged capacitor.

A capacitor of 10·0 microfarads is momentarily connected across a 100-volt d.c. supply. Calculate the charge on the capacitor and the energy stored. A second capacitor of 5·0 microfarads is now switched in parallel with the first. Calculate the final potential difference across the capacitors and the energy now stored in the system. Comment on the change of energy. (*Cam. Overseas*)

14. Three identical components P, Q, R are joined to a battery as shown.

Answer the following, giving your reasons:

(*a*) If P, Q, and R are solenoids, how are the intensities of the magnetic fields inside them related to one another?

(*b*) If P, Q, and R are resistors, how are the rates at which heat is produced in them related to one another?

(*c*) If P, Q, and R are electrolytic cells, how are the

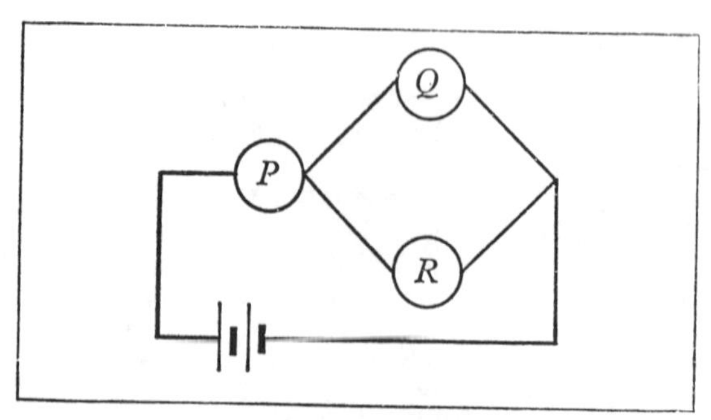

Fig. Q 11.14

masses of the same element deposited in them in the same time related to one another?

(*d*) If P, Q, and R are capacitors, how are the potential differences across them related to one another?

Describe how you would check your statement by experiment in *one* of the cases. (*Cam. Forces*)

B

15. To what potential difference must a 1 microfarad capacitor be charged in order to be capable of supplying on a single discharge as much energy as a 2-volt 10-ampere-hour accumulator? (*Cam. Forces*)

16. Define *capacitance*. What factors affect the capacitance of a conducting object insulated from earth?

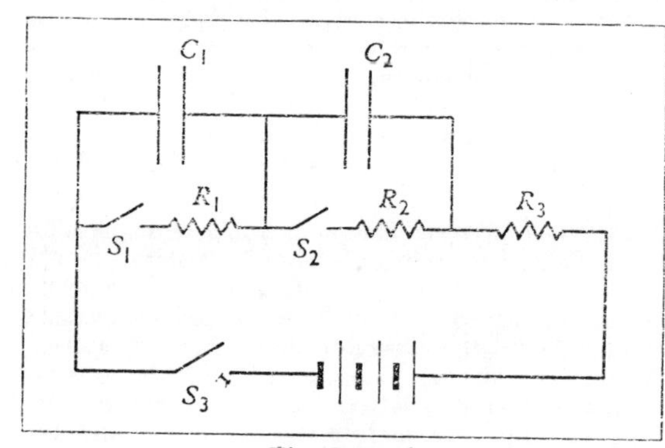

Fig. Q 11.16

Two capacitors and three resistors are connected as shown in the diagram to an accumulator of e.m.f. 6·0 volts and of negligible internal resistance. The values of the components are as follows: $C_1 = 4{\cdot}00$ microfarads; $C_2 = 10{\cdot}0$ microfarads; $R_1 = R_3 =$ 100 ohms; $R_2 =$ 50 ohms. Initially all the switches are open and the capacitors are uncharged. Calculate the charges on the capacitors when (*a*) switch S_3 alone is closed, (*b*) switches S_1 and S_2 are then also closed. (*C*)

17. Explain what is meant by *capacitance*, and define the *farad*.

Fifty similar 1-microfarad capacitors are charged by joining them separately to a battery of e.m.f. 600 volts. By a suitable mechanism they are then joined in series, with the positive plate of one to the negative of the next and so on. A 1-megohm resistor is then joined across the ends of the series chain.

(*a*) Calculate the initial current through the resistor.

(*b*) Calculate the total energy dissipated in the resistor from the moment of connection until the current ceases.

(*c*) Is a capacitor in the middle of the chain discharged in this process? If so, what happens to the charges on its plates? (*Cam. Forces*)

18. In the figure, C is a parallel-plate condenser having two plates, each of area 10 sq. cm, separated by a ceramic material of thickness 0·1 mm, and dielectric constant 1200; R is a 10-megohm resistance, and the battery has a constant e.m.f. of 60 volts. At a certain instant after the switch S has been closed, the microammeter M registers $3{\cdot}2 \times 10^{-6}$ amp. Find the potential difference between the plates of C, and the charge on each plate, at this instant. (*O*)

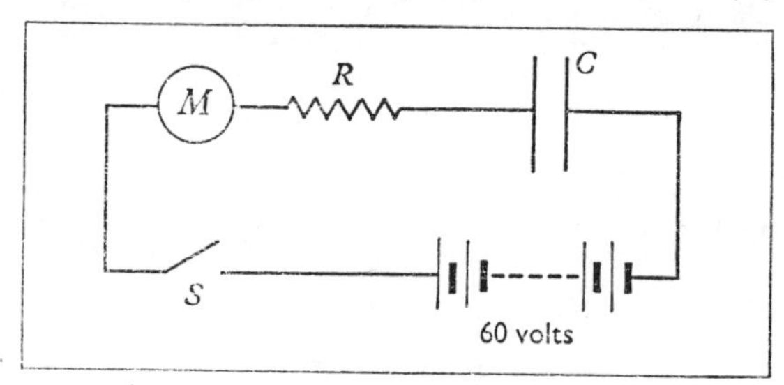

Fig. Q 11.18

19. Describe *one* form of capacitor in practical use, and give an example of an application for which it is suitable, stating your reasons.

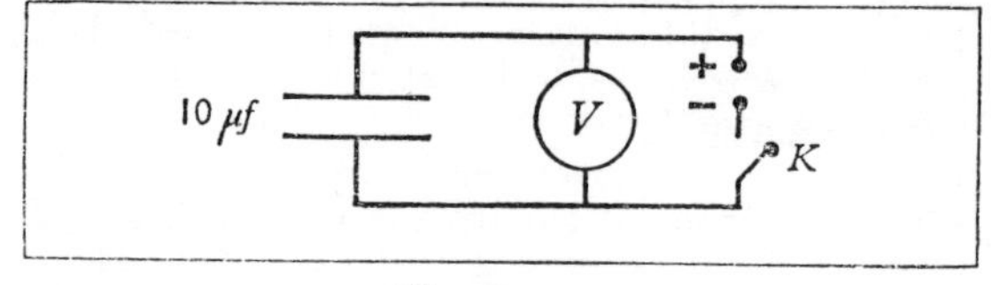

Fig. Q 11.19

The figure shows a 10-microfarad capacitor joined to an electrostatic voltmeter V. A high-tension source is joined to the capacitor when the key K is depressed, and the voltmeter reads 2250 volts. The key is released and the voltmeter reading falls slowly. What is the most likely reason for this? After 1·5 minutes the voltmeter reads 1950 volts; find (*a*) the charge lost, (*b*) the mean current during

this period, (*c*) an approximate value for the resistance which would produce the same results in a perfect apparatus. Where should we suppose such a fictitious resistance to be connected?
(*Cam. Overseas*)

20. Define *electric intensity* and *potential* at a point in an electric field and state the connection between them.

Derive an expression for the capacitance of an isolated conducting sphere. It may be assumed, that when such a sphere is charged, the electric intensity at points outside the sphere is equal to that which would be produced by the entire charge placed at the centre with no sphere present.

An isolated conducting sphere is surrounded by air which becomes conducting when the potential gradient at any point exceeds a certain value. Explain how the maximum potential to which the sphere may be raised depends on the radius. If, for a sphere of radius 50 cm, the maximum potential is 9×10^5 volts, calculate the potential gradient for breakdown. (*N*)

21. A 1-microfarad parallel-plate capacitor that can just withstand a p.d. of 6000 volts uses a dielectric having relative permittivity 5, which breaks down if the electric field intensity in it exceeds 3×10^7 volts metre^{-1}. Find (*a*) the thickness of dielectric required, (*b*) the effective area of each plate, (*c*) the energy stored per unit volume of dielectric. (*Cam. Forces*)

22. Spherical drops of water, diameter 2·0 mm, from a vessel maintained at a potential of 100 volts above earth potential, fall through a small hole into a thin-walled insulated metal sphere, diameter 5·0 cm. Find the energy of the sphere when it is just full. (*L*)

23. A 3·0-microfarad capacitor is charged by joining it to an 800-volt battery through a non-inductive resistor. How much energy is taken from the battery and how much is stored in the capacitor? What happens to any difference between these amounts? The charged capacitor is joined to a tube containing gas at low pressure, which conducts only so long as the p.d. across it is 200 volts or more. What quantity of electricity will pass through the tube, and how much energy will be lost by the capacitor? (*Cam. Forces*)

24. A condenser consisting of two parallel circular plates each of diameter 20 cm placed 0·1 cm apart in air is permanently connected to an electrostatic voltmeter of capacitance *C*. The system is charged to a potential difference of 80 volts and is then isolated. When the plate separation is increased to 0·5 cm the voltmeter reading rises to 300 volts. Explain the change in potential difference and calculate the value of *C*. (*O & C*)

25. Two parallel-plate condensers are made with the same dimensions. One has air as the dielectric, and the other glass. The air condenser is charged to a p.d. of 200 volts. When the two condensers are connected in parallel, the common p.d. is 36 volts. Calculate the dielectric constant of glass. (*O*)

26. A 0·05-microfarad condenser is charged by a battery and then connected across the terminals of an electrometer. The resulting scale deflection is 10 cm. When the battery itself is connected to the electrometer the deflection is 10·5 cm. Assuming these deflections to be proportional to the p.d. across the electrometer terminals, what is the capacity of the electrometer and how much energy is required to charge it to a potential of 20 volts?
(*S, part qn*)

27. One thousand identical spherical drops of water carrying equal charges coalesce to form one large spherical drop. Is the electrostatic energy the same before and after coalescence?

If not, in what ratio has it changed? (The capacitance of a spherical conductor is proportional to its radius.) (*Cam. Forces*)

28. A condenser of capacitance 4 microfarads, charged to a potential difference of 100 volts, shares its charge with another condenser of capacitance 2 microfarads charged to a potential difference of 400 volts.

Calculate (*a*) the charge on each condenser after sharing, (*b*) the loss of energy, in joules, as a result of the sharing. (*A.E.B.*)

29. Define the *farad*, and describe briefly a practical form of capacitor (condenser).

Deduce expressions for the effective capacitance of two capacitors (*a*) in parallel, (*b*) in series.

Two capacitors of capacitance 0·10 and 0·20 microfarad connected in series are joined to a d.c. supply of 100 volts. Calculate the charge on each capacitor. Without discharging the capacitors, they are disconnected from the supply and connected in parallel so that the plates with charges of like sign are connected together. What is now the potential difference across the capacitors? (*Cam. Overseas*

30. Describe the structure and explain the mode of action of a moving-coil galvanometer suitable for ballistic use.

A charged condenser A of capacitance 0·5 μF gave a throw of 27·0 scale divisions when discharged through a ballistic galvanometer. A was then recharged to the same potential and a second condenser B was alternately connected across A, removed and discharged, three times. The charge remaining on A caused a throw of 8·0 scale divisions. Find the capacitance of B. (*L*)

31. Explain what is meant by the terms *potential* and *capacity* applied to a conductor. How could you give to a flat copper plate, 10 cm square, a charge of 10^{-8} coulombs, if you had available a 120-volt battery and another similar plate?

A condenser is connected to a battery and to a galvanometer through a vibrating reed in such a way that the condenser is fully charged by the battery and then fully discharged through the galvanometer, 20 times per second. If the e.m.f. of this battery is

120 volts, and the galvanometer registers an apparently steady current of 0·6 milliamp, what is the capacity of the condenser? *(O & C)*

32. Derive an expression for the energy of a condenser of capacitance C when the potential difference between its terminals is V.

A multi-plate air condenser of capacitance $3{\cdot}0 \times 10^{-4}$ microfarad is connected in a circuit with a battery of e.m.f. 120 volts, a sensitive galvanometer and vibrating switch, in such a way that the condenser is fully charged by the battery then fully discharged through the galvanometer 100 times a second. Draw a diagram of the circuit and calculate (*a*) the value of the apparently steady current through the galvanometer, (*b*) the mean rate at which the battery supplies energy, (*c*) the mean rate at which energy is obtained from the condenser.

Describe briefly how you would use an arrangement of this kind to measure the dielectric constant of oil. *(N)*

33. A large condenser was connected in series with a high-tension battery, a resistance of 1 megohm, a microammeter and a plug key. Readings of the microammeter taken at the instant the key was put in, and at ten-second intervals afterwards, were as follows:

Time (s)	0	10	20	30	40	50
Current (μA)	100	61	37	22	13·5	8·2

Time (s)	60	70	80	90	100	120
Current (μA)	4·9	3·0	1·8	0·7	0·4	very small

Find the voltage of the battery and, from a suitable graph, a value for the capacitance of the condenser. ($1\ \mu A = 10^{-6}$ A) *(L)*

34. A condenser of capacity 1 microfarad is charged to 200 volts. If its leakage resistance is 10^8 ohms, in what time will the p.d. fall to 100 volts when the condenser is isolated? *(Cam. Schol.)*

35. A condenser of capacity C is charged to a potential V, and a resistance R is then connected across its plates. Derive from first principle an expression to show how the energy expended in the resistance varies with time.

A condenser of capacity 100 μF is charged to a voltage of 600 volts and is then discharged through a resistance of 2×10^6 ohms.

(*a*) What is the initial value of the current flowing through the resistance at the commencement of discharge?

(*b*) How much energy would have been dissipated in the resistance after a time equal to the time constant of the circuit? *Note:* Time constant of a circuit in seconds is the product of the capacitance in farads and the resistance in ohms.

(*c*) What is the value of the current through the resistance after 50 s? *(S)*

36. A parallel-plate capacitor is filled with a 'leaky' dielectric of resistivity $1{\cdot}50 \times 10^{12}$ ohm m and relative permittivity 4·0. It is joined across a battery of e.m.f. 100 volts and then isolated. What is the p.d. across the capacitor 60 seconds later?

37. Describe the differences in structure and action between a non-ballistic and a ballistic moving-coil galvanometer.

A corrected deflection of 24 scale divisions of a ballistic galvanometer is obtained *either* by charging a condenser of 3·0 μF capacity to a potential difference of 2·0 volts, and discharging it through the galvanometer, *or* by connecting the ballistic galvanometer in series with a flat circular coil of 80 turns each of diameter 1 cm, the combined resistance of coil and galvanometer being 2000 ohm, and quickly thrusting the coil into a strong magnetic field so that the plane of the coil is perpendicular to the direction of the field. State the sensitivity of the galvanometer and calculate the strength of the magnetic field. (The strength of the earth's magnetic field may be neglected.) *(L)*

Other question: Ch. 16: 8, 9

C

38. What do you understand by the term electrical energy?

A condenser of capacitance C is charged to a potential V_0 in the following three ways:

(*a*) it is connected to a source of potential difference the value of which is slowly increased from zero to V_0;

(*b*) it is connected to a source of constant potential difference V_0 by means of a fixed resistance R;

(*c*) it is connected directly to a source of constant potential difference V_0.

Discuss quantitatively what happens to the energy given out by the source on each case. The internal resistance of the source may be neglected. *(Cam. Schol.)*

39. Two isolated parallel plates separated by a distance d carry equal and opposite charges $\pm Q$ per unit area. The gas between the plates is irradiated with a short burst of X-rays in such a way that a layer of the gas of negligible thickness, midway between the plates, is partially ionized, there being n electrons and n positive ions per unit area of the layer. The electrons drift towards the positive plate with velocity v_1 and positive ions drift in the opposite direction with velocity v_2 ($<v_1$). If the charge on the electron has magnitude e and edge effects can be neglected, how does the potential difference between the plates vary with time from the instant of irradiation until all the charges have reached the plates? The relative permittivity of the gas may be taken as 1. *(O & C)*

40. It has been suggested that the charge on the electron differs from that on the proton by one part in 10^{18}. Calculate the intrinsic charge of 224 cm^3 of argon at 10 atmospheres pressure and 0°C, and describe how you would try to check your result experimentally by allowing the gas to leave an insulated spherical container. What do you think would be the main difficulties you would have to overcome? (Argon has atomic number 18.) (*Ox. Schol.*)

41. A parallel plate air-spaced condenser of capacitance 5×10^{-4} microfarads is charged to a potential difference of 100 volts. It is then isolated and immersed in oil of dielectric constant 2·5. Calculate the new value of the potential difference, and explain as fully as you can why the presence of the oil causes the potential difference to change.

What change in energy stored in the condenser occurs on immersion?

In an experiment to measure the dielectric constant of the oil, the above procedure was followed but with an electrometer connected across the plates of the condenser the whole time, to measure the initial and final values of the potential difference. The electrometer itself had a capacitance of 10^{-4} microfarads. What value would have been obtained for the dielectric constant if the electrometer capacitance had been neglected? (*O & C*)

42. A parallel-plate condenser is placed with its plates vertical in a non-conducting outer vessel, and one terminal is permanently earthed. A battery of 50 volts potential difference is connected across it and then removed. Oil of dielectric constant 4·6 is then poured into the vessel so that the air space between the plates is gradually replaced by oil. Find the potential difference across the condenser when the space is two-thirds filled with oil. (*N*)

43. A moving coil galvanometer has a current sensitivity of 200 millimetres per microamp and a resistance of 60 ohms and a time of swing on open circuit of 2 s. There is also a resistance of 1000 ohms in series. The galvanometer and a 1 μF condenser and a voltage source of 0·01 volt are connected with a rotary switch in such a way that the condenser is charged to 0·01 volt and then connected to the galvanometer in every switching cycle. Calculate how many times per second the switching must be done to obtain a galvanometer deflection of 10 cm.

What would be the effect of raising the series resistance from 1000 ohms to 1 megohm? (*Cam. Schol.*)

44. By means of a switch a condenser A, of capacitance 2·0 microfarads, is charged from a cell and then discharged through a galvanometer causing an initial deflection of 50 divisions. The condenser is then recharged from the same cell and, by means of another two-way switch, its charge is shared with another condenser B. The second switch is then also used to discharge B, not through the galvanometer. The process of alternately charging B from A and discharging B is repeated and carried out, in all, 10 times. The residual charge on A then produces a deflection of 25 divisions in the galvanometer. Draw a circuit diagram and calculate the capacitance of B. (*N*)

45. Describe methods for measuring the capacities of condensers of the order of (*a*) 10 μF, (*b*) 10^{-5} μF.

In the circuit shown in the diagram the neon lamp

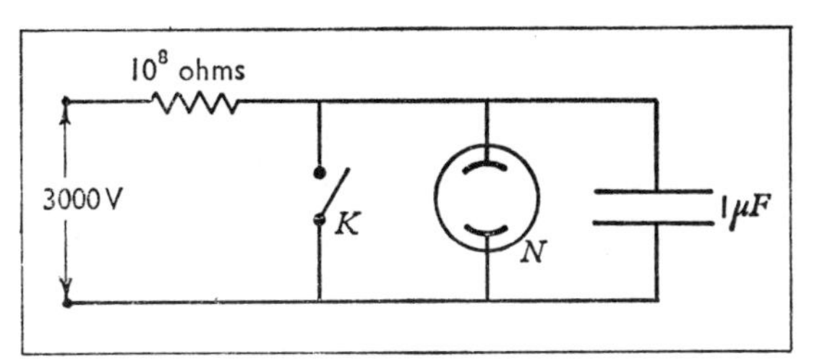

Fig. Q 11.45

N strikes when the voltage across it rises to 110 volts and extinguishes when it falls to 85 volts. When it is extinguished its resistance may be considered infinite, and when lit very small. If the key K is opened at time $t = 0$, calculate approximately the time when the lamp first strikes and the frequency of subsequent strikings. (*Cam. Schol.*)

46. A device to develop short pulses across a 100-Ω resistor consists of a 0·01-μF condenser which is connected alternately to a 100-volt supply for 0·01 s and to the 100-Ω resistor for 0·01 s. The time during which the condenser is connected neither to the supply nor to the load resistor is negligible.

What is the shape of the pulse across the load resistor? What is its peak amplitude? What is the average current through the load resistor? (*Ox. Schol.*)

47. It is found that a poor 50 μF condenser (parallel plates) failed to retain its charge due to the dielectric layer having become imperfect and possessing finite resistance. During a test in which the condenser was charged and then disconnected, the potential difference across its terminals was observed to fall to half its initial value in 10 minutes. If the dielectric constant of the layer is known to be 6, what is the resistivity of the layer? (*Cam. Schol*)

48. A 1-μF condenser consisting of two parallel plates separated by 0·01 cm is charged to a potential of 100 volts. What is the force between the plates? (*Cam. Schol.*)

49. A capacitor, consisting of two parallel plates in contact with an insulating slab of dielectric of thickness 0·1 cm and relative permittivity 4, has one plate earthed and the other at a potential of +100 volts. The positive plate is then moved away from the dielectric to produce an air gap of 0·07 cm giving a plate separation of 0·17 cm. Calculate the potential of the positive plate in this new position assuming that it remains insulated. (*O & C*)

50. The plates of a parallel-plate air condenser are 2 cm apart and are connected to an electrometer. The condenser is then charged. When a slab of sulphur 1·0 cm thick is inserted between the plates and the distance between them is altered by 0·75 cm the electrometer reading remains unchanged. Assuming that the area of the face of the slab is greater than that of the plates, what is the dielectric constant of the sulphur? (*L*)

51. Define *capacitance* of a condenser and obtain an expression for the energy required to establish a potential difference V between the plates of a condenser of capacitance C. Quote a consistent set of units for the physical quantities in your answer.

The plates of a parallel plate condenser are 1·0 cm apart. They are charged to a potential difference of 600 volts and then disconnected from the supply. If the distance between the plates is then increased to 1·5 cm find (*a*) the new potential difference between them, (*b*) the fractional change in the energy of the condenser. Explain the changes in (*a*) and (*b*).

Find the thickness of a sheet of glass of dielectric constant (relative permittivity) 6 which, placed between the plates and parallel to them when they are 1·5 cm apart, restores the original potential difference of 600 volts. (*L*)

52. A parallel plate air condenser of separation 1·0 cm has its plates maintained at a constant potential difference. Find the thickness of a slab of glass of dielectric constant $\kappa = 6{\cdot}0$ which must be inserted between the plates so that the electric field intensity in the remaining air space is increased to five times its original value. (*L*)

53. How would you measure three of the following:

(*a*) a resistance of 10^{11} ohms;

(*b*) a resistance of 0·001 ohm;

(*c*) the specific resistance of N/10 sulphuric acid solution;

(*d*) a capacitance of 100 μF;

(*e*) the capacitance of two sixpences separated by a piece of mica 1 mm thick? (Dielectric constant or relative permittivity = 6.)

State clearly the reason why you have selected the particular method you have used in each case. (*Ox. Schol.*)

Other question: Ch. 12: 40(*b*)

Chapter 12 Alternating currents

Additional data required:
Velocity of light = $3{\cdot}00 \times 10^8\ m\ s^{-1}$

A

1. An a.c. supply, which may be assumed to be of a sinusoidal type, is applied to (*a*) a resistance, (*b*) a condenser, (*c*) a large self-inductance of negligible resistance. Draw graphs to illustrate the variation of the applied potential difference and the resulting current with time in each case.

What is the effect on the current in each case of (i) increasing the frequency of the supply, (ii) increasing the resistance, the capacitance, and the inductance respectively. Give reasons for your answers. (*W*)

2. Describe and explain an instrument for measuring alternating current. What do you understand by the r.m.s. value of an alternating current? How is it related to the peak value in the case of a sinusoidal current?

When the coils of an electromagnet are connected to a 240-volt d.c. supply, the current taken is 10 amps. When connected to a 240-volt a.c. supply the current taken is only 1 amp. Explain why there is a smaller current on a.c., and calculate the resistance of the coils. Using the same time axis draw curves showing how the current through the coils and applied voltage vary with time when the supply is a.c. (*C*)

3. An iron-cored inductance of negligible resistance is connected across the 230-volt, 50-cycle mains supply. An alternating-current ammeter in series reads 0·5 amp. Find the value of the inductance in henrys.

How do you account for the fact that the power dissipation in the inductance is zero? (*O*)

4. An 8-microfarad condenser is placed across the 200 volts r.m.s. a.c. 50-cycle electric supply. Calculate the r.m.s. current which flows in the circuit. What is the peak value of the voltage across the condenser? (*Cam. Schol.*)

5. The e.m.f. of an a.c. generator is 100 volts r.m.s. Its internal impedance is 50 ohms and is independent of frequency. It is connected in turn to (*a*) a pure inductance, (*b*) a pure capacitance, and (*c*) a resistance of 50 ohms in series with a pure capacitance. Calculate the r.m.s. values of the current flowing and the voltage across the generator in each case when the frequency is very low i.e. $f \ll R/L$ and $1/RC$ and also when it is very high, i.e. $f \gg R/L$ and $1/RC$. (*O & C*)

6. A tuning coil of inductance 180 μH is to be connected in series with a capacitor to form a circuit which resonates at 720 kc/s. What value of capacitance is required? (*A.E.B.*)

7. Derive an expression for the resonant frequency of a series tuned circuit.

A 150-μH inductor is tuned by a parallel variable capacitor having maximum and minimum capacitances 500 and 20 pF, respectively. If stray capacitances amount to 40 pF, calculate the maximum and minimum frequencies to which the circuit can be tuned. (*Cam. Forces*)

Other questions: Ch. 2: 5, 11 Ch. 7: 2 Ch. 9: 1

B

8. Derive an expression for the e.m.f. induced in a coil rotating about an axis in its own plane and at right-angles to a magnetic field. State the meanings of the symbols you use. Describe *in outline* how this effect is used in a simple a.c. generator.

Such a generator, in which the field is kept constant, is joined in turn to (*a*) a pure resistance, (*b*) a pure inductance, (*c*) a pure capacitance. Discuss and illustrate by sketch graphs how the current flowing will vary with the speed of the generator in each case. Neglect the resistance and inductance of the coil of the generator itself. (*Cam. Forces*)

9. Describe fully how you would calibrate a

moving-iron ammeter to read the r.m.s. value of sinusoidal currents alternating at 50 cycles per second.

Such an instrument is used to measure the current flowing in an air-cored solenoid connected to a 50 c/s alternating voltage supply. State, with reasons, what changes in current you would expect to observe if (*a*) a laminated soft iron cylinder, (*b*) a copper cylinder, were inserted in the solenoid.

Why should the ammeter not be placed near the solenoid when measuring the current? (*O & C*)

10. The loss due to hysteresis in the iron core of a transformer is 20 watts under normal load conditions, the frequency of the a.c. supply being 50 c/s. What change, if any, do you expect in this loss (*a*) if the current taken from the secondary winding is doubled by reducing the resistance joined to the secondary, (*b*) if the transformer is used with a supply of frequency 100 c/s, and the primary voltage is doubled? Neglect any effect due to resistance of the windings. (*Cam. Forces*)

11. An alternating e.m.f. is represented by the equation $E = E_0 \sin \omega t$. Explain the physical significance of E_0 and ω. Express the variations in E in terms of the periodic time and of the frequency.

An osglim lamp lights when the voltage applied to it exceeds 170 volts and goes out when the voltage drops to 140 volts. If such a lamp is connected to a 240-volt alternating-current supply, for what fraction of the time is the lamp lit? (*W*)

12. Explain what is meant by the *power factor* of a circuit. Why is it desirable that the power factor of an installation should not differ much from unity?

A lamp taking 0·40 amp at 200 volts is supplied from 250-volt 50-cycle a.c. mains using a series capacitor. Calculate (*a*) the power factor of the circuit, (*b*) the capacitance of the capacitor, (*c*) the peak voltage across the capacitor.

In actual practice the power factor is found to be a little higher than the calculated value. Suggest an explanation. (*Cam. Forces*)

13. An a.c. motor developing $\frac{1}{4}$ horse-power is found to take 1·5 amps from 220-volt mains. The makers give the efficiency at this load as 75%. How do you reconcile these figures?

(1 horse-power = 746 watts.) (*Cam. Forces*)

14. A coil of inductance 0·50 henry and resistance 220 ohms is joined to an alternating supply of frequency 50 Hz. What is its power factor?

15. A non-inductive circuit of resistance 5·0 ohms carries sinusoidal alternating current of r.m.s. value 4·0 amps and frequency 50 cycle s^{-1}.

(i) Explain the significance of *r.m.s.* in this statement.

(ii) What is the peak value of the current?

(iii) What is the instantaneous current flowing 0·0006 s after it changes direction?

(iv) How much heat is developed in the circuit in 1 minute?

(v) How much heat would have been developed in 1 minute if part of the circuit had been made into a coil of inductance 0·010 henry, and the same potential difference applied? (*L*)

16. A transformer takes a current of 5·0 amps from 210-volt mains when the secondary supplies 18·0 amps through a resistive load, the p.d. across the load being 50 volts. In these circumstances the power factor of the primary circuit is 0·99. The resistance of the primary winding is 2·0 ohms, and that of the secondary winding 0·10 ohm. Calculate (*a*) the efficiency of the transformer at this load, (*b*) the power loss in the iron core. (*Cam. Forces*)

17. A power station has an output of 5000 kilowatts at 400 volts. The power is transformed up to 120,000 volts for transmission along a line of total resistance 25 ohms, and is then transformed down to 200 volts. If each transformer has an efficiency of 96%, what power output and current will be available in the 200-volt circuit? (*O & C*)

18. A coil with an inductance of 10 millihenrys and resistance 10 ohms is connected across an a.c. supply of 10 volts r.m.s. at a frequency of 200 cycles per second. Calculate the current in the circuit. (*A.E.B.*)

19. A condenser of capacitance 2 microfarads in series with a resistance of 1000 ohms is connected across a 50 cycles per second alternating supply of e.m.f. 240 volts r.m.s. Calculate (*a*) the current through the circuit, (*b*) the potential difference across the condenser. (*A.E.B.*)

20. The instantaneous e.m.f. E of a source is represented by $E = 300 \sin 100\pi t$ volts, where t represents time in seconds. If this e.m.f. is applied to a choke of resistance 10 ohms and self-inductance 100 millihenrys, what is (*a*) the maximum value of the current, (*b*) the root mean square value of the current through the choke? (*A.E.B.*)

21. An alternating e.m.f. given by $E = E_0 \sin \omega t$ is applied to a condenser of capacitance C. Draw rough graphs to show the variation of the potential across the condenser, the charge on the condenser, and the current flowing.

A 60-watt 120-volt lamp may be run off a 240-volt a.c. supply by placing in series with the lamp either a suitable resistance or a suitable condenser. Find the value of the suitable resistance and the capacitance of the suitable condenser. The frequency of the mains is 50 c/s. Compare these two methods of reducing the current through the lamp. (*W*)

22. Define *coefficient of self-induction* and hence derive a formula for the energy expended in establishing a direct current through an inductance. How is this energy dissipated when the current is switched off?

The current registered by a hot-wire ammeter, when a 6-volt battery and a solenoid are connected in series with it, is 2 amps. When the battery is replaced by a 40-volt a.c. supply whose frequency is 50 cycles per second, the current registered by the ammeter is 4 amps. Calculate the inductance of the solenoid and the rate of dissipation of energy in each case. (*S*)

23. A resistor of 10 ohms and a capacitor of 400 μF are connected to a 60-volt a.c. supply. If the current measured is 5 amps find the frequency of the supply and the phase angle between the current and the applied p.d. Draw the vector diagram. (*S, part qn*)

24. A non-inductive resistor of resistance 20 ohms and a coil are connected in series across a source of potential difference of r.m.s. value 8 volts alternating at 50 c/s. The r.m.s. potential differences across the resistor and coil are 4 volts and 6 volts respectively. Find the current in the circuit and, by calculation or by drawing a vector diagram on graph paper, determine the resistance and inductance of the coil. (*N*)

25. A coil is joined in series with a non-inductive resistor of resistance 800 ohms across a potential difference of 100 volts alternating at 50 c/s. The potential difference across the ends of the coil is found to be 45 volts and across the resistor 80 volts. Find the inductance and resistance of the coil by drawing a vector diagram on graph paper. (*N*)

26. A 10-ohm resistor and a coil having resistance and self-inductance are joined in series with a 50-cycle second^{-1} alternating current supply, the r.m.s. voltage between the terminals of which is 91. The r.m.s. voltages across the resistor and the coil are found to be 70 and 35 respectively. Calculate values for the following quantities associated with the coil; its impedance, its resistance, its reactance, its self-inductance, its power factor and the power dissipated in it.

Explain the action of a choking coil. (*L*)

27. Explain why it is not possible to measure the power in an alternating current circuit using only a voltmeter and an ammeter. What is meant by the power factor of a circuit?

A furnace with a non-reactive winding rated at 1 kW and designed for 100-volt mains, is to be operated from a 240-volt a.c. supply of frequency 50 cycle second^{-1}. Find the values of the components needed if this is to be done using (*a*) a resistor, (*b*) a condenser, (*c*) a choke coil which has a resistance of 5·0 ohms. (*L*)

28. A condenser *C*, an inductance *L*, and a resistance *R* are connected in series across a 50-cycle alternating supply, and the current flowing through the circuit is recorded as 1·5 amps, while the p.d. across the whole circuit is 130 volts. What do these readings represent?

A voltmeter connected in turn across the terminals of *C*, *L* and *R* separately registers 30 volts, 80 volts and 120 volts respectively (Fig. Q 12.28). Explain, with the help of a diagram, the phase relationship between these voltages, and show how they can be reconciled with the reading of 130 volts across the whole circuit. Calculate the capacity of *C* and the inductance of *L*. (*O*)

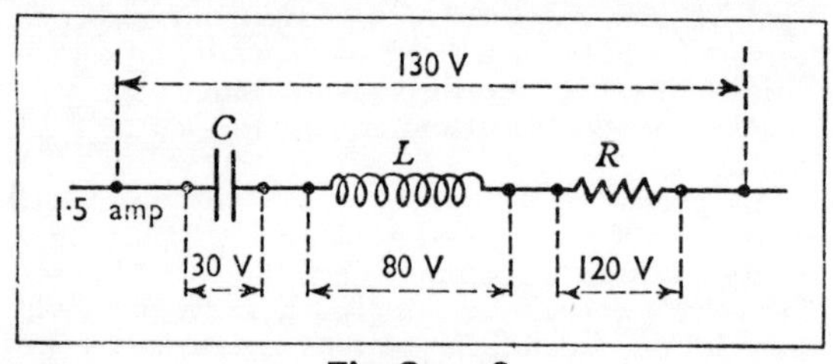

Fig. Q 12.28

29. Show that the heating effect of a sinusoidal alternating current of peak value I_0 passing through a resistance is the same as that of a steady current $I_0/\sqrt{2}$.

Describe an ammeter which gives r.m.s. values when used in an a.c. circuit.

A 12 μF condenser is connected in series with an inductance of 2 henrys and a resistance of 180 ohms and a 250-volt, 50 cycles second^{-1} a.c. supply. What is the value of the current that flows? (*S*)

30. Explain the action and use of a *choke*.

A coil of self-inductance *L* and ohmic resistance *R* is found to have an impedance of 500Ω at 800 c/s and 800Ω at 1600 c/s. Find the values of *R* and *L*. (Assume $2\pi = 6{\cdot}25$.) (*A.E.B.*)

31. A circuit contains an inductance coil of inductance *L* and resistance *R* in series with a condenser of capacitance *C*. Deduce from first principles an expression for the magnitude *Z* of the impedance of this circuit in terms of *L*, *R*, *C* and the frequency *f* of the a.c. supply.

If, in this circuit, *L* is 0·5 millihenry, *R* is 10 ohms and *C* is 1·0 microfarad, find the magnitude of the impedance if the frequency of the supply is 5000 cycle/second.

Find also the resonant frequency of a series circuit with these component values. (*A.E.B.*)

32. A circuit consists of a condenser of capacitance 0·0004 microfarad in series with a coil of inductance 200 microhenrys and resistance 5 ohms. If an alternating e.m.f. of 10 volts r.m.s. at a frequency of 500 kc/s is connected across this circuit, calculate the r.m.s. current that will flow through it and the r.m.s. alternating potential difference across the inductance. Find also the resonant frequency of this circuit. (*A.E.B.*)

33. A capacitor of 0·0003 μF and an inductive coil of self-inductance 150 microhenrys and resistance 6 ohms are available. The power factor of the capacitor may be assumed to be zero. Calculate:

(*a*) the impedance of these two components when connected in series across an a.c. supply of frequency 2 megacycles per second;

(*b*) the resonant frequency of these two components in a series tuned circuit;

(*c*) the value of the additional capacitance it would be necessary to connect in series with these two components to double the resonant frequency.

(Assume $\pi^2 = 10$) (*A.E.B.*)

34. A variable tuning capacitor has a maximum capacitance of 0·0005 μF and a minimum capacitance of one-tenth of this value. If it is used in series with a fixed inductor coil to form a tuned circuit, what must be the inductance of this coil in microhenrys if the circuit is to be resonant, at the maximum setting of the capacitor, to a radio-frequency signal of wavelength 600 metres? Calculate also the resonant wavelength for this circuit at the minimum setting of the capacitor.

If the inductor coil has a resistance of 5 ohms, what will be the voltage magnification provided by the circuit when tuned to a frequency of one megacycle per second? *(A.E.B.)*

Other questions: Ch. 5: 10 Ch. 7: 16, 18, 19 Ch. 14: 14

C

35. Describe in detail the action of a transformer. Show by vector diagrams, or otherwise, that when on load the primary current increases and the greater the load the more nearly the current and e.m.f. in the primary are in phase. What losses occur in a transformer, and how by suitable design are they reduced? *(S)*

36. A sinusoidal alternating current has a peak value of 25·0 amps and increases from zero to 20·0 amps in 0·00295 s. What is the frequency, and what impedance would be offered to this supply by a coil of resistance 10·0 ohms and inductance 0·05 henry? *(L)*

37. Explain the significance of the symbols in the equation for an alternating current $i = i_0 \sin 2\pi ft$. Draw sketch graphs of i against t and i^2 against t for one complete cycle. From your graphs or otherwise derive a relation between the root mean square current I and i_0.

A resistor of resistance 1000 ohm and a capacitor of capacitance 2×10^{-6} farad are connected in series across the 200 volt 50 c/s supply. Calculate the impedance of this combination and find the potential difference across (*a*) the resistor, (*b*) the capacitor, (*c*) the supply, at the instant when the current has its maximum value. *(O & C)*

38. The primary of an a.c. transformer has a large self-inductance and a low resistance; the mutual inductance between primary and secondary is 0·3 henry. If a 50 c/s current of 2 amps r.m.s. value flows in the primary, what will be the r.m.s. voltage developed in the secondary on open circuit? How will the phase of this voltage be related to the phase of the voltage across the primary? *(O & C)*

39. A box X and coil Y are connected in series with a variable frequency a.c. supply of constant e.m.f. 10 volt. X contains a capacitance 1·0 μF in series with a resistance 32 ohms, and Y has a self-inductance 5·1 millihenry and resistance 68 ohm. The frequency is adjusted until maximum current flows in X and Y. Determine the impedances of X and Y at this frequency and hence find the potential differences across X and Y separately Show these potential differences on a vector diagram indicating their phase relations with the applied e.m.f. *(L)*

40. Calculate the following:

(*a*) the resonant frequency of a series circuit consisting of an inductance of 2 millihenrys in series with a capacitance of 0·001 microfarad;

(*b*) the time constant of a capacitor of 0·15 microfarad across which there is a resistor of 250 kilohms;

(*c*) the *Q*-factor at a frequency of 100 kilocycles per second of a coil which has an inductance of 100 microhenrys and a resistance of 3 ohms;

(*d*) the impedance at 50 cycles per second of a capacitance of 2 microfarads in series with a resistance of 1 kilohm. *(A.E.B.)*

41. A coil of self-inductance 200 μH and resistance 5Ω is connected in series with a capacitor of capacitance 0·00030 μF.

Calculate

(*a*) the resonant frequency of this circuit,

(*b*) the impedance of this circuit at a frequency half the resonant frequency,

(*c*) the voltage amplification provided by this circuit at resonance. *(A.E.B.)*

42. A current of 2 amps passing through a circuit of self-inductance 200 millihenrys and negligible resistance is suddenly interrupted by opening a switch. Sparking is liable to occur if the potential across the switch exceeds 1000 volts. Show how a condenser may be used to prevent sparking and calculate the capacity that will suffice. Justify carefully the method of calculation you adopt. *(Cam. Schol.)*

43. Write an account of tuned circuits, including the series acceptor circuit and the parallel rejector circuit.

Explain briefly the importance of tuned circuits in radio reception.

Calculate the dynamic resistance at resonance of a rejector circuit consisting of a capacitor of capacitance 0·0003 microfarad and negligible power factor in parallel with a coil of inductance 150 microhenrys and resistance 5 ohms.

Evaluate approximately the resonant frequency of this circuit. *(A.E.B.)*

44. A coil of resistance R and inductance L is connected in series with a pure capacitance C to the output terminals of a variable frequency oscillator of negligible internal impedance. The current I in the circuit is varied by varying the frequency while the voltage between the output terminals is maintained constant. Assuming an expression for the impedance of the series circuit deduce an expression for the frequency f_0 when the current has its maximum value I_{max}. If f_1 and f_2 are the two frequencies for which $I/I_{max} = 1/\sqrt{2}$, show that the reactance of the circuit at f_1 and f_2 is numerically equal to the resistance. Hence show that

$$f_2 - f_1 = \frac{R}{2\pi L}$$

In an experiment in which C = 1·00 μF the following observations are made of I/I_{max} at different values of f:

f (c/s)	650	1000	1200	1250	1300
I/I_{max}	0·08	0·18	0·47	0·65	0·92
f	1350	1400	1450	1500	1800
I/I_{max}	0·99	0·75	0·56	0·44	0·12

Plot these results and use the curve to obtain values of L and R. What would be the power factor of (*a*) the coil, (*b*) the circuit at f_0? (*L*)

Other question: Ch. 14: 18

Chapter 13 Electrons and ions

Additional data required:
Planck's constant $h = 6{\cdot}62 \times 10^{-34}$ *J s*
Velocity of light $c = 3{\cdot}00 \times 10^{8}$ *m s*$^{-1}$
Charge of the electron $e = -1{\cdot}60 \times 10^{-19}$ *coulomb*
Mass of the electron $m = 9{\cdot}1 \times 10^{-31}$ *kg*
∴ *Specific charge of the electron* e/m
$= -1{\cdot}76 \times 10^{11}$ *C kg*$^{-1}$
Permittivity of a vacuum $\varepsilon_0 = 8{\cdot}85 \times 10^{-12}$ *F m*$^{-1}$
Acceleration due to gravity $g = 9{\cdot}81$ *m s*$^{-2}$
1 *Angstrom unit* (Å) $= 10^{-10}$ *m*

A

1. Draw a labelled diagram of a vacuum photocell.

Write notes on the following points in connection with this type of cell:

(*a*) The composition of a typical photocathode.

(*b*) The threshold wavelength of the incident light in relation to the photoelectric work function.

(*c*) The effect of altering the intensity of illumination of the photocathode.

(*d*) The characteristics of photoelectric current against anode potential for two different constant levels of illumination of the photocathode. (*A.E.B.*)

2. What is the energy of 1 quantum of sodium D light (wavelength $5{\cdot}89 \times 10^{-7}$ m)? A 200-watt sodium vapour street light has an efficiency of 30% (i.e. 30% of the supplied energy is emitted as the D light). How many quanta of light does it emit per second?

3. The ionization potential of hydrogen is 13·6 volts.

(*a*) Calculate the energy (in joules) and speed of the slowest electron that can ionize a hydrogen atom when it collides with it.

(*b*) Calculate the longest wavelength of electromagnetic radiation that could produce ionization in hydrogen.

(*c*) The lowest two excited states of a hydrogen atom are 10·2 and 12·0 eV above the ground state. Calculate *three* wavelengths of radiation that could be produced by transitions between these states and the ground state.

4. Describe a method by which the charge per unit mass of an electron has been determined.

An electron is accelerated from rest through a potential difference of 200 volts. Find the velocity that it acquires. (*L*)

5. Describe and give the theory of an experiment to determine the value of e/m for an electron.

If this value is actually $1{\cdot}76 \times 10^{11}$ amp s kg^{-1} calculate the velocity of an electron of energy 10,000 electron volts moving in a region of zero potential. (*L*)

6. A milliammeter connected in series with a hydrogen discharge tube indicates a current of 10^{-3} ampere. The number of electrons passing the cross-section of the tube at a particular point is $4{\cdot}0 \times 10^{15}$ per second. Find the number of protons that pass the same cross-section per second.

(Assume the electronic charge to be $-1{\cdot}60 \times 10^{-19}$ coulomb.) (*L*)

7. What do you understand by an electron?

Describe how a beam of electrons is produced and accelerated in a cathode-ray tube. Why must the envelope be highly evacuated? What factors determine the current of the electron beam?

Electrons in a certain cathode-ray tube are accelerated through a potential difference of 2kV between the cathode and the screen. Calculate the velocity with which they strike the screen. Assuming they lose all their energy on impact and given that 10^{12} electrons pass per second, calculate the power dissipation. (*O & C*)

B

8. Discuss the facts of the photoelectric effect.

The maximum wavelength that can cause emission of electrons from a metal plate is 5000 Å. Find the maximum velocity with which electrons are emitted if light of 4000 Å is incident on its surface. (*S*)

9. If a sodium surface in vacuum is illuminated with a monochromatic beam of ultra-violet light with a wavelength of 2×10^{-5} cm, what is the maximum velocity of emission of the electrons if the work function of the sodium is 2·46 volts? (*A.E.B.*)

10. Describe two types of photocell and indicate their application with reference to their characteristics.

The maximum wavelength that can cause electron emission from a metal is 5200 Å. Calculate the work function of the metal in electron volts and the maximum speed of electrons emitted by light of wavelength 4000 Å. (*A.E.B.*)

11. Explain Einstein's application of the quantum theory of radiation to the understanding of photoelectric emission.

Light of wavelength 4300 Å is incident on (*a*) a

nickel surface of work function 5 eV, (*b*) a potassium surface of work function 2·3 eV. Determine by calculation if electrons will be emitted and, if so, the maximum velocity of the emitted electrons in each case. (*A.E.B.*)

12. When ultraviolet light of wavelength 1500 Å falls in a vacuum on the clean surface of a well-insulated sheet of a certain metal the metal attains a potential of +6·6 volts relative to its surroundings. For wavelengths of 3000 Å and 4500 Å the corresponding potentials are +2·4 and +1·0 volt respectively. What theory can be advanced to explain these data?

Explain what is meant by *threshold wavelength* and *work function* and use the above data to determine numerical values for these quantities and for Planck's constant. (Take the charge on the electron to be $1{\cdot}6 \times 10^{-19}$ coulomb, and the speed of light to be 3×10^{8} m s^{-1}.)

Show how the theory put forward explains (i) the effect of increasing the intensity of the light falling on the metal surface, (ii) the effect of illuminating the surface simultaneously with light of all three wavelengths, the surface being initially at earth potential in each case. (*O & C*)

13. A beam of electrons accelerated by a potential difference of 1000 V in a cathode-ray tube passes between a pair of electrostatic deflecting plates which produce a field of 500 volt cm^{-1} perpendicular to the initial direction of the beam. What intensity of magnetic field, superimposed on the electric field and acting over the same region, will allow the beam to pass undeviated? How should the magnetic field be orientated? Describe how this principle has been used to measure the velocity of a beam of electrons. (*O & C*)

14. In one type of cathode-ray tube, electrons are accelerated in a narrow beam through a potential difference of 3000 volts. Calculate the velocity with which the electrons are then moving.

If the electrons then pass into a uniform magnetic field orientated at right angles to the beam, calculate the flux density B needed to deflect the beam so that the electrons begin to move along a circular arc of radius 10 cm, given that the force acting on a charge e moving with velocity v normally to the field is Bev. (*O & C*)

15. A heated filament provides the electrons in a high-vacuum cathode-ray tube. A potential difference of 2000 volts is maintained between the anode and the cathode. The stream of electrons, after falling through this potential difference, passes for 2 cm between two parallel metallic plates, separated by a distance of 1·2 cm and maintained at a potential difference of 120 volts. The path of the electrons on entry into the space between the two deflecting plates is parallel with the plates. Find the change in direction of the electron stream in its passage between the plates. Edge effects are to be neglected. (*W*)

16. A parallel beam of electrons, accelerated through a potential difference of 1000 volts, is injected horizontally halfway between two horizontal conducting plates 4 cm apart charged to a difference of potential of 100 volts. Find the horizontal distance from the point of injection to the point of impact of the electrons on one of the plates.

Describe in general terms the effect on the beam of a horizontal magnetic field at right angles to the beam.

Find the value of such a field if the path of the beam remains horizontal. (*Ox. Schol.*)

17. Outline a method of measuring *either* the charge e on an electron, *or* the ratio of e to the mass m of an electron. Why is it thought that all electrons have the same charge and mass?

In a cathode-ray tube electrons are accelerated through a potential difference of 1000 volts and focused into a narrow beam. Calculate the velocity of the electrons in the beam, and the number of electrons in a one-centimetre length of the beam if the current carried by the beam is one microamp.

Describe, giving quantitative details, *one* method of deflecting such a beam of electrons through an angle of 10°. (*O & C*)

18. An electron is travelling horizontally after having been accelerated from rest by a p.d. of 100 volts. How far would it fall under gravity in travelling 100 metres in a vacuum? (*C*)

Other question: Ch. 16: 22

C

19. According to Bohr an hydrogen atom may be regarded as a system consisting of an electron whose charge is $-1{\cdot}6 \times 10^{-19}$ coulomb and whose mass is $9{\cdot}1 \times 10^{-31}$ kg rotating in a circle of radius $5{\cdot}3 \times 10^{-11}$ m with a proton as centre. If the charge on the proton is $+1{\cdot}6 \times 10^{-19}$ coulomb calculate the magnetic field it experiences due to the rotating electron. (*O & C*)

20. An electron is injected with a velocity of $1{\cdot}5 \times 10^{8}$ cm s^{-1} into a uniform magnetic field of flux-density 0·10 V s $metre^{-2}$ *in vacuo*. The angle between the field and the initial direction of the electron is 10°. By considering the components of the velocity of the electron parallel to and perpendicular to the field, calculate the axial distance between successive turns of the helical path which the electron will follow. (*C*)

21. Show that a charged particle moving with constant speed at right-angles to a uniform magnetic field in a vacuum travels in a circular path and find an expression for the time taken to complete one revolution.

What is the path of a charged particle in a uniform magnetic field if its initial velocity is not at right angles to the field?

An electron of speed 2×10^{7} m s^{-1} is emitted from a point on the axis of a very long solenoid wound with 10 turns per cm and carrying a current of 2·5 amps. If the initial velocity makes an angle α with the solenoid axis, find at what distance from its starting point the particle next crosses the axis, and show that this distance is almost independent of α if α is small. (*O & C*)

Other questions: Ch. 11: 40, 45 Ch. 14: 19 Ch. 16: 24

Chapter 14 Electronics

A

1. A small step-down mains transformer and a 6-volt accumulator are joined in turn to a 6-volt bulb and are found to light it with equal brilliance. The bulb is disconnected and each source in turn is joined to the *y*-plates of a cathode-ray tube. Discuss the effects produced on the screen. (*Cam. Overseas*)

2. Describe the principles of a cathode-ray oscillograph.

When a potential of 100 volts d.c. is applied to the *Y*-plates of a C.R.O. the spot is deflected 1 cm. A sinusoidal alternating voltage is then applied to the *Y*-plates and a suitable time base to the *X*-plates. The wave thus presented is 5 cm in height from trough to crest. What is the r.m.s. value of the current taken when this same voltage is applied to a resistance of 1000 ohms? (*S*)

3. Draw a labelled diagram of a *diode valve* and describe how it can be used as a rectifier of alternating current, giving the appropriate circuit diagrams and explanatory graphs. (*A.E.B.*)

4. In obtaining the characteristics of a triode valve the following readings are recorded:

Grid voltage V_g *(volts)*	*Anode current* I_a *with anode potential* V_a *at 130 volts (mA)*	I_a *with* V_a *at 100 volts (mA)*
0	15	10
−2	13	8
−4	11	6
−6	9	4
−8	7	2
−12	3	0·4
−14	1·5	0·1
−16	0·7	0
−18	0·2	0

Plot the anode current against grid voltage characteristics.

Define the terms *mutual conductance* and *amplification factor*, and find the values of these constants at $V_a = 130$ volts and $V_g = -2$ volts. (*A.E.B.*)

5. Describe the action of a triode valve, stating the function of each electrode. Give a circuit diagram showing such a valve used as a voltage amplifier at audio-frequency.

In a single-stage amplifier the anode current flowing when there is no signal is 2·0 milliamps, the external resistance in the anode circuit of the valve being 100,000 ohms. On application of the signal the anode current fluctuates sinusoidally between 1·2 and 2·8 milliamps. What is the r.m.s. value of the alternating voltage developed across the resistance? (*Cam. Forces*)

6. A triode valve of anode slope resistance 20 kilohm is used with an anode load resistance of 50 kilohm. If an alternating signal is applied to the grid of 0·5 volt r.m.s., find the output voltage if the amplification factor of the valve is 15. Deduce from first principles the gain formula you use. (*A.E.B.*)

7. Explain, with a suitable circuit diagram and a chart, how a triode valve is used to amplify a small alternating voltage. Deduce a formula for the voltage gain in terms of the anode slope resistance, the amplification factor and the resistance of the anode load used.

A triode valve has an amplification factor of 70 and an anode slope resistance of 10,000 ohms. Find the value of the anode load resistance required in a simple amplifier circuit to produce an output r.m.s. voltage of 25 for an input r.m.s. voltage of 0·5. (*A.E.B.*)

Other questions: Ch. 10: 5, 7 Ch. 12: 6, 7 Ch. 13: 7

B

8. A voltage alternating at 50 c/s is connected across the *Y*-plates of a cathode-ray oscillograph. Sketch and explain the forms of the traces on the oscillograph screen when a linear time-base of frequency (i) 10 c/s, (ii) 100 c/s is connected across the *X*-plates. What would be the effect of disconnecting the time-base and connecting the *X* and *Y* plates in parallel? (*N*)

9. What is meant by the statement that the sensitivity of a certain cathode-ray tube for vertical deflection is 20 volts per cm?

With the help of a diagram explain the factors in the design of the tube which govern this sensitivity.

How would you use a cathode ray oscilloscope to do the following experiments?

(i) observe the waveform of the a.c. supply mains voltage (approximately 240 volt r.m.s., at 50 cycles per second);

(ii) check the calibration of an audio oscillator of variable frequency at a scale reading of about 1000 cycles per second, the frequency of the supply mains being assumed to be exactly 50 cycles per second. (*O & C*)

10. Describe the construction and working of a diode valve, and give a simple circuit in which it is used for rectification of an alternating current.

A diode valve with a high resistance connected in

series is joined to an alternating supply. The p.d. across the resistance is joined to the Y deflection plates of a cathode-ray oscillograph, while the alternating supply voltage is joined to the X deflection plates. Describe and explain the appearance on the screen. (*C*)

11. The current i through a given diode for various values of voltage V between anode and cathode is given in the following table.

V (volts)	0	50	100	150	200	250	300
i (mA)	0	8	30	55	70	76	78

Sketch a characteristic curve for this valve and explain its general shape.

When a diode with this characteristic is connected in series with an anode resistor to a 100-volt d.c. supply, a current of 8 mA is observed to flow. To what value must the supply voltage be raised in order to increase the current to 30 mA? (*O & C*)

12. Describe the main features of the construction and operation of a triode valve. Explain what is meant by (*a*) the anode resistance, (*b*) the mutual conductance, and (*c*) the amplification factor, and show how they are related.

A change of 0·8 mA in the anode current of a triode occurs when the anode potential is changed by 10 volts. What is the anode resistance? If the amplification factor is 8·0, what change in grid voltage would be required to cause a change of 4·0 mA in the anode current? (*Cam. Schol.*)

13. It is recommended that a triode valve be operated at an anode potential of 100 volts and a steady grid potential of -3 volts. At these potentials the steady anode current is 5 milliamps. The amplification factor of the valve is 40 and its anode slope resistance is 25,000 ohms. If an anode load resistance of 15,000 ohms is used, what e.m.f. is necessary for the H.T. supply? Draw the circuit diagram involved in using this valve as a voltage amplifier for an alternating input e.m.f. What is the maximum r.m.s. value of the input e.m.f. that can be accommodated without positive grid current flow? Calculate the r.m.s. value of the output e.m.f. when this maximum input is applied. (*A.E.B.*)

14. If the resistance of a moving-coil loudspeaker is 15 ohms and it is to be matched to an output power valve for which the optimum load is 2500 ohms, calculate the turns ratio of the coupling transformer required. (*A.E.B.*)

15. How may a triode valve be used (*a*) as an oscillator, (*b*) as a detector, (*c*) as an amplifier? Draw the necessary circuits and give approximate values for the components which should be used.

16. Explain why a triode valve is unsuitable for the amplification of high-frequency input signals. The high-frequency pentode valve was developed to overcome the difficulties encountered with triodes. Explain the action of the screen-grid in such a pentode valve and discuss also the reason for the introduction of the suppressor grid. (*A.E.B.*)

17. Explain what is meant by *interference of electromagnetic waves*. Why is it necessary that two trains of *light* waves must come from the same source if interference is to be observed? Discuss whether this is also necessary in the case of radio waves.

Radio waves pass from a transmitter to a receiver by two paths which differ in length, the length of the shorter path being 120 km. When the frequency used is changed from 2·5 to 2·4 Mc/s the received signal passes through 8 minima of intensity, being at maximum intensity at the initial and final frequencies. Calculate the length of the longer path.

(Velocity of electromagnetic waves $= 3{\cdot}0 \times 10^8$ metre s^{-1}) (*Cam. Forces*)

Other questions: Ch. 1: 6 Ch. 12: 33, 34, 40, 41, 43 Ch. 13: 15 Ch. 15: 7

C

18. Describe the construction and explain the principle of operation of a cathode-ray oscillograph.

A resistor PQ is joined in series with another component QR, and an a.c. supply is joined to P and R. The points P and Q are joined to the Y-plates of a cathode-ray oscillograph, while Q and R are joined to the X-plates. Describe and account for what will be seen on the screen if QR is (*a*) a resistor, (*b*) a capacitor, (*c*) an inductor with resistance, (*d*) a diode valve. (*Cam. Forces*)

19. Make a clear labelled diagram of a cathode-ray tube using electrostatic deflection and explain the function and mode of action of each part. The deflector plates in a given tube are 5 cm long and ½ cm apart. Calculate the sensitivity of the instrument (volts/cm) if the potential of the gun is -2000 volts and the distance between the centre of the deflector plates and the screen is 25 cm. What factors do you consider will affect this sensitivity when the potential difference between the deflector plates oscillates at high frequency? (*Ox. Schol.*)

20. The figure shows part of a circuit, in which d.c. currents and voltages are maintained as shown. The value of R is adjusted so that the point G is 5 V negative with respect to the point K. Find the value of R required to produce this voltage between G and K. What are then the voltages at A_1 and A_2? (*Cam. Schol.*)

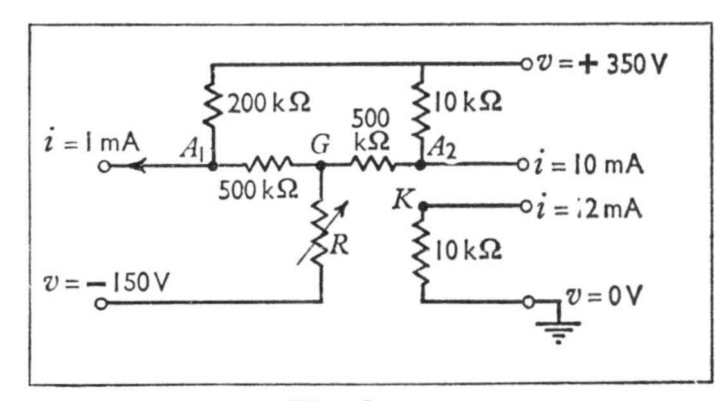

Fig. Q 14.20

21. Discuss the constructional details and the principle of operation of a thermionic diode. Measurements of V_a, the voltage across the diode, and I_a, the anode current, gave the following results:

V_a (volts)	0	10	20	30	40	50
I_a (mA)	0	2	6	10	17	25

V_a (volts)	60	70	80	90	100
I_a (mA)	35	47	63	80	100

Give a qualitative explanation of the general shape of the curve and calculate approximately, by means of a graphical method, the current indicated by a moving-coil meter of negligible resistance connected in series with the diode and an alternating potential of 50 volts r.m.s. at 50 c/s. *(O & C)*

22. A moving-coil loudspeaker has a cylindrical coil of diameter 2·0 cm with 40 turns, mounted in a radial magnetic field of flux-density 1·2 weber metre^{-2}. If the effective mass of the moving system is 4·0 gm, what is the amplitude of movement when a sinusoidal current of 0·10 amp r.m.s. at 1000 c/s flows through the coil? Neglect elastic forces due to the suspension. *(C)*

23. Draw the diagram and explain the action of a diode detector circuit suitable for the demodulation of a 400 kc/s signal carrying speech modulation.

Calculate a suitable value for the circuit time constant.

Why is automatic gain control desirable in a communications receiver? Explain briefly how a.g.c. is achieved. *(Cam. Forces)*

24. An observer at the equator wishes to study the 5 m radio waves emitted by a star overhead. He sets up two identical aerials 1 km apart on an east-west line and arranges for the signals they pick up from the star to be combined in a single tuned receiver. As the earth rotates the combined signal is found to oscillate in strength. Explain this effect and calculate the period of oscillation. *(Cam. Schol.)*

25. A radio receiver situated at the top of a cliff of height 50 metres overlooking the sea, is observing electromagnetic radiation of wavelength 10 cm coming from a transmitter carried by a balloon. Interference occurs between the direct beam and that reflected in the surface of the sea (taken to be plane). The balloon is released at a point 10 kilometres from the foot of the cliff and rises at a constant rate of 2 metres per second.

(*a*) What time elapses between successive minima in the signal at the receiver when the balloon starts to rise?

(*b*) How would you use this effect to estimate the height of the transmitter at any time? *(Ox. Schol.)*

26. Two vertical line sources of radiation, ABC and DEF, are placed with their mid-points B and E in the same horizontal plane and 0·5 metre apart. Each source radiates uniformly in this plane. The intensity of the radiation received along a line GH, parallel to BE, at the same horizontal level and distant 5 metre from BE, is measured and is found to be zero at O where the perpendicular bisector of BE meets GH, with a series of maxima and sharp minima on either side of O. The first minima on either side of O are 30 cm from O.

What deductions can be made concerning the radiators ABC and DEF?

When ABC is turned through 90°, to lie along BE, the intensity at points along GH shows negligible variation, but when DEF is rotated also in the same way the maxima and minima reappear. What further information does this give? *(O & C)*

27. Two vertical wireless aerials are spaced one-quarter of a wavelength apart and are fed with signals from the same transmitter but with a phase difference of 90°. How does the intensity of the transmitted radiation vary in a horizontal plane? What would be the effect of increasing the number of aerials, maintaining a constant distance and phase difference between one and the next? *(Ox. Schol.)*

Chapter 15 X-rays

Additional data required:
Charge of the electron $e = -1{\cdot}60 \times 10^{-19}$ *coulomb*
Mass of the electron $m = 9{\cdot}1 \times 10^{-31}$ *kg*
Planck's constant $h = 6{\cdot}62 \times 10^{-34}$ *J s*
Velocity of light $c = 3{\cdot}00 \times 10^{8}$ *m s*$^{-1}$
Faraday's constant $F = 9{\cdot}65 \times 10^{7}$ *C kg-equiv*$^{-1}$
Specific heat of water $= 4{\cdot}2$ *J °C*$^{-1}$ *g*$^{-1}$
(*i.e.* 1 *calorie* = 4·2 *joules*)

A

1. In what circumstances are X-rays produced? State *briefly* what you know about the nature and properties of these rays.

Draw a labelled diagram of a modern form of X-ray tube. Why is tungsten commonly used as the target material in such a tube? How would you expect the intensity and penetrating power of the X-rays from a tube to be altered by an increase in (*a*) the filament heating current, (*b*) the high tension across the tube? *(Cam. Overseas)*

2. In a commercial X-ray tube, a current of 15 milliamps of electrons strikes a target which is at a potential of 100 kilovolts positive with respect to the filament. What flow of water is required to keep the target cool assuming that all the heat is conducted away by the water, and that its rise in temperature should not be greater than 30°C? (You may assume that the entire kinetic energy of the electrons is converted into heat in the target.) *(O & C)*

3. Draw a labelled diagram of an X-ray tube. Explain its mode of operation, and indicate the electrical arrangements that are needed in order to work it from the normal mains supply.

If the potential difference applied across the tube is 5000 volts and the current through it is 2 milliamps, calculate the number of electrons striking the target per second, and the speed at which they strike it. (*O*)

4. Describe, with a labelled diagram, an X-ray tube of the high-vacuum (Coolidge) type which employs a thermionic filament.

An X-ray tube is operated with a potential difference of 10 kV between the target (anti-cathode) and the filament. What will be the minimum wavelength of the X-rays generated by this tube? (*A.E.B.*)

5. Calculate the minimum wavelength of the X-rays produced at the anode of a diode valve which is operated at a potential difference of 100 volts with respect to the cathode. (*A.E.B.*)

6. The spacing between the layers of atoms parallel to the cleavage face of calcite is $3{\cdot}03 \times 10^{-10}$ m. A monochromatic beam of X-rays is found to be strongly reflected from a crystal of calcite when the glancing angle it makes with the cleavage face is $14° 42'$; strong reflection does not occur for any smaller angle than this. Calculate the wavelength of the X-rays. At what other values of the glancing angle might strong reflection occur?

B

7. State briefly the methods that are available for the production and detection of electromagnetic waves of the following wavelengths: (i) 10^{-8} cm, (ii) 5×10^{-5} cm, (iii) 1 cm. (*Cam. Forces*)

8. Draw a labelled diagram of a high-vacuum X-ray tube. Sketch curves showing the manner in which the intensity of the X-rays varies with the wavelength for a given target with the target potential as the parameter.

What is meant by the characteristic X-ray for an element? Distinguish between those factors which give rise to the continuous X-ray spectrum and the characteristic line spectrum for a target metal like molybdenum. (*A.E.B.*)

9. Describe how X-rays are produced and give an account of their significance in modern physics.

The diffraction of X-rays by a crystal of rock-salt, NaCl, shows that in the crystal lattice the sodium and chlorine ions occupy alternate corners of cubes of side $2{\cdot}8 \times 10^{-8}$ cm. Given that the density of rock-salt is $2{\cdot}16$ g cm^{-3}, calculate the charge of the electron.

(Molecular weight of NaCl = 58·5.) (*Cam. Schol.*)

Chapter 16 Nuclear physics

Additional data required:
Charge of the electron $e = -1{\cdot}60 \times 10^{-19}$ *coulomb*
Avogadro's constant $N = 6{\cdot}02 \times 10^{26}$ *kg-atom*$^{-1}$
Velocity of light $c = 3{\cdot}00 \times 10^{8}$ *m s*$^{-1}$
Permittivity of a vacuum $\varepsilon_0 = 8{\cdot}85 \times 10^{-12}$ *F m*$^{-1}$
(*The electric space constant*)
Assume: 1 year $= 3{\cdot}15 \times 10^{7}$ *s*

A

1. Describe and explain simple experiments which demonstrate the distinctive properties of the radiations emitted by radioactive substances.

Write brief accounts of TWO *industrial* (*not medical*) uses to which radioactive substances may be put. (*N*)

2. Describe the nature and properties of the charged particles emitted from radioactive substances. Outline experiments to demonstrate *three* of these properties for one type of particle.

Discuss briefly the effect of the emission of such particles on (*a*) the atomic mass, (*b*) the atomic number of the element concerned.

What is the source of the energy liberated during the disintegration? (*N*)

3. '$_{11}Na^{24}$ is a *radioactive isotope* of sodium, which has a *half-life period* of 15 hours and disintegrates with the emission of *β-particles* and *γ-rays*. It emits β-particles that have energies of 4·2 *MeV*.'

Explain the meanings of the five terms that are italicized in the statement above. (*L*)

4. (*a*) What is the number of electrons in 1·0 kg of hydrogen.

(*b*) Show that most other light elements contain about half as many electrons per kg as hydrogen.

(*c*) How many electrons are there in 1·0 kg of uranium-238?

(The atomic number of uranium = 92)

5. Write brief notes on *three* of the following: (*a*) β-particles, (*b*) neutrons, (*c*) protons, (*d*) gamma radiation.

Discuss the reaction represented by

$$_7N^{14} + {}_2He^4 = {}_8O^{17} + {}_1H^1$$

explaining the meaning of the subscript and superscript numbers. (*L*)

6. In an experiment with a high-energy beam, hydrogen atoms each weighing $1{\cdot}67 \times 10^{-24}$ g strike a target with a velocity of 2×10^{9} cm s^{-1}. If 10^{15} atoms arrive each second, and the target is a lump of brass of 500 g thermally insulated, find how long it will take for the temperature of the brass to rise by 100°C.

(Specific heat of brass = 0·38 J °C^{-1} g^{-1}.)

7. Compare the velocities attained by a proton and an α-particle each of which has been accelerated from rest through the same potential difference. (*L*)

Other question: Ch. 1: 14

B

8. The *atomic number* and the *atomic mass* of aluminium are 13 and 27 respectively. What is the significance of these statements in relation to the structure of the aluminium atom?

A solid aluminium sphere has a radius of 10 cm. Find (*a*) the total number of electrons in the sphere (*b*) the fraction of these which are removed when the sphere is raised to a positive potential of 100 volts. (Density of aluminium = 2·7 g cm^{-3}) (*N*)

9. A radioactive source emits beta particles (electrons) at a substantially constant rate of $3{\cdot}7 \times 10^4$ per second. If the source, which is a metal sphere 1 mm in diameter, is electrically insulated, how long will it take for its potential to rise by 1 volt, assuming 90% of the beta particles emitted escape from the source? The capacitance in m.k.s. units of an isolated spherical conductor of radius r is $4\pi\varepsilon_0 r$. (*O & C*)

10. A Geiger-Müller tube is placed close to a source of beta particles of constant activity. Sketch a graph showing how the count-rate, measured using a suitable scaler or ratemeter, varies with the potential difference applied to the G.M. tube. Discuss how the form of the graph determines the choice of operating conditions for the tube.

Describe how you would investigate the absorption of beta particles by aluminium using a G.M. tube. Sketch a graph showing the results you would expect to obtain. How would the form of the graph change if (*a*) the same source was used with lead substituted for aluminium, (*b*) a different source emitting beta particles of higher energy was used, aluminium being the absorbing material? (*N*)

11. Describe how the nature of α-particles has been established experimentally.

The half-value period of the body polonium-210 is about 140 days. During this period the average number of α-emissions per day from a mass of polonium initially equal to 1 microgram is about 12×10^{12}. Assuming that one emission takes place per atom and that the approximate density of polonium is 10 g cm^{-3}, estimate the number of atoms in 1 cm^3 of polonium. (*N*)

12. Iodine-131 has a half-life of 8 days. A source containing this isotope has an initial activity of 2·0 curies.

(*a*) What is the activity of the source after 24 days?

(*b*) What time elapses before the activity of the source falls to 1·0 microcurie?

13. Radon is a monatomic gas of atomic mass 222 and with a radioactive constant equal to $2{\cdot}1 \times 10^{-6}$ s^{-1}. Calculate the number of α-particles emitted per second by 1 g of radon at n.t.p. when free from disintegration products. (*S, part qn*)

14. Calculate the mass of caesium-137 that has an activity of 5·0 microcuries.

(Half-life of caesium-137 = 30 years
1 curie is a disintegration rate of $3{\cdot}7 \times 10^{10}$ per second.)

15. Technetium-99 (atomic number 43, chemical symbol Tc) decays by emission of a negative β-particle into Ruthenium (chemical symbol Ru). Write down the nuclear equation representing this process.

A sample containing 0·100 μg of technetium is found to emit 135 β-particles per second. Calculate the half-life of technetium.

16. State the law governing the rate of decay of a radioactive substance, and explain the terms decay constant (λ) and half life (T). Show that these two quantities are related by the equation

$$\lambda T = \log_e 2$$

Describe briefly how the decay law may be verified experimentally for a source of half-life of about one hour.

Two radioactive sources A and B initially contain equal numbers of radioactive atoms. Source A has a half-life of 1 hour, and source B a half-life of 2 hours. What is the ratio of the rate of disintegration of source A to that of source B (*a*) initially, (*b*) after 2 hours, (*c*) after 10 hours? (*O & C*)

17. A compartment on a Geiger–Müller tube is filled with a solution containing 1·00 g of carbon extracted from one of the Dead Sea scrolls. The count rate recorded is 1000 per hour. When a similar solution containing 1·00 g of carbon extracted from a living plant is used instead, the count rate is 1200 per hour. Without any solution in the compartment the background count rate is found to be 300 per hour. Estimate the age of the scroll, if the half-life of carbon-14 is 5600 years.

18. Uranium-234 is formed as one of the decay products of uranium-238. The half-lives of the two isotopes are $2{\cdot}5 \times 10^5$ years and $4{\cdot}5 \times 10^9$ years, respectively. What proportion (by weight) of a sample of natural uranium would you expect to find in the form of uranium-234?

19. The ratio of the mass of lead-206 to the mass of uranium-238 in a certain rock is measured to be 0·45. Assuming that the rock originally contained no lead-206, estimate its age.

(Half-life of uranium = $4{\cdot}5 \times 10^9$ years.)

20. One of the decay products of thorium-232 is the monatomic radioactive gas radon-220. Calculate the mass of radon-220 in radioactive equilibrium with 25 g of thorium-232. What volume would this quantity of radon occupy at n.t.p.?

(Half-life of radon-220 = 54 s
Half-life of thorium-232 = $1{\cdot}4 \times 10^{10}$ years
The kg-molar volume of a gas at n.t.p. = 22·4 m^3)

21. The gamma-rays from cobalt-60 are reduced to half their initial intensity on passing through 1·2 cm of lead. What thickness of lead is required to reduce the intensity of the radiation to 10% of its initial value?

Calculate the mass absorption coefficient, if the density of lead is 11·3 g cm^{-3}.

22. A charged particle moves in a circle under the influence of a magnetic field of flux density 0·6 Wb m^{-2}. Show that the frequency of revolution is independent of the velocity of the particle, and find the value of this frequency for a proton.

Describe briefly any way in which this constant

frequency property has found practical application.

(e/M for the proton $= 9{\cdot}65 \times 10^7$ coulomb kg^{-1}.)

(*Ox. Schol.*)

23. (*a*) What quantity of energy (in joules) would have a mass of 1·00 kg?

(*b*) Express this quantity in electron-volts.

(*c*) At what speed is the mass of a particle twice its rest mass?

(*d*) What is the p.d. required to accelerate (i) an electron, (ii) an α-particle, to this speed?

(*e*) What is the smallest quantity of energy (in eV) that a γ-ray must have in order to give rise to an electron-positron pair?

(*f*) What is the wavelength of such a γ-ray?

(*g*) If a γ-ray of half this wavelength produces an electron-positron pair and its energy is equally shared between the particles, what is the energy (in eV) and speed of the particles?

(Mass of an electron $= 9{\cdot}1 \times 10^{-31}$ kg
Mass of an α-particle $= 6{\cdot}64 \times 10^{-27}$ kg
Planck's constant, $h = 6{\cdot}62 \times 10^{-34}$ J s.)

C

24. Give a brief account of the nature and properties of the radiations from naturally occurring radioactive substances.

A deposit of radium C is placed on a horizontal thin straight wire A above which a fine slit B is placed parallel to the wire and 6·0 cm from it. Alpha-particles from A pass through B and are detected at P on a photographic plate placed in a horizontal plane 6·0 cm vertically above B. The whole apparatus is enclosed in a highly evacuated chamber. A uniform magnetic field, of flux density 1·0 Wb m^{-2}, is now applied to the region between A and B in a direction parallel to the wire, while the space between B and P remains field free. Under these conditions the α-particles are detected on the plate 4·50 mm from P. Estimate the velocity with which the α-particles are emitted from the radium C, given that e/m for α-particles is $4{\cdot}80 \times 10^7$ C kg^{-1} (Relativistic effects may be neglected.) (*L*).

25. Give an account of the types of radioactive emissions found in nature and explain how you would distinguish between them.

A piece of timber has been recovered from an archaeological excavation and it is required to find its approximate age by measuring the radioactivity of the carbon-14 contained therein. For this purpose it may be assumed that the proportion of carbon-14 in the natural carbon of living wood is everywhere and at all times the same and that it begins to decay at death. If the number of disintegrations observed from 5 g of carbon prepared from the specimen is 21 per minute, how old is the specimen?

(The proportion of carbon-14 to natural carbon in living wood is 1·25 in 10^{12} and the half-value period of carbon-14 may be taken to be 5600 years. The mass number of natural carbon is 12.) (*N*)

26. How many α-particles are emitted altogether by an atom of uranium-238 as it decays progressively into lead-206?

A quantity of uranium ore contains 1·00 kg of uranium, which can be assumed to be in equilibrium with its decay products. Estimate the mass of helium generated per year in the ore.

(Half-life of uranium-238 $= 4{\cdot}5 \times 10^9$ years.)

27. Phosphorus-32 decays by β-emission into sulphur-32, the energy liberated in each disintegration as kinetic energy of the particles being 1·7 MeV. The atomic mass of sulphur-32 is 31·9721 atomic mass units. Calculate the atomic mass of phosphorus-32.

(1 atomic mass unit $= 1{\cdot}66 \times 10^{-27}$ kg.)

28. Using the information on atomic masses given below, show that a nucleus of uranium-238 can disintegrate with the emission of an alpha particle according to the reaction:

$$_{92}U^{238} \rightarrow {}_{90}Th^{234} + {}_2He^4.$$

Calculate (*a*) the total energy released in the disintegration, (*b*) the kinetic energy of the alpha particle, the nucleus being at rest before disintegration.

Mass of U^{238} = 238·12492 a.m.u.
Mass of Th^{234} = 234·11650 a.m.u.
Mass of He^4 = 4·00387 a.m.u.
1 a.m.u. (atomic mass unit) is equivalent to 930 MeV. (*N*)

Answers

Chapter 1 Atoms and electric currents (Questions p. 350)

1. 225 C; 1·4 × 10^{21} **2.** 240 J; (*a*) 720 J, (*b*) 240 J **3.** 18,000 J
4. 75 C; 1·25 A **5.** 85 cm hr^{-1} **6.** 7·8 × 10^{7}
7. $6{\cdot}2_5$ × 10^{-4} m s^{-1} **8.** 7 × 10^{8} V; 10^{4} **9.** 3·5 × 10^{5} J
10. 7·0 V **11.** 7 min; 38 min **12.** 5·6 × 10^{-13} kg m s^{-1}
13. 2(·5) × 10^{-10} m **14.** 2·9 : 1 (by weight)

Chapter 2 Resistance (Questions pp. 351–354)

1. (*a*) 4 V; 1·6 W, (*b*) 0·25 mA; 1·25 W; (*c*) 960 ohms; 484 ohms, (*d*) (i) 3 V, 0·8 ohm, 1·25 A, (ii) 1·5 V. 0·2 ohm, 0·83 A
2. (*a*) 8·3 ohms, (*b*) 11·7 ohms, (*c*) 12·5 ohms, (*d*) 1·7 ohms
3. 12·0 m **5.** (*a*) 1%, (*b*) 1000 V; 141,000 V **6.** 0·61 ohm
7. 500, 1000 and 2000 W **8.** 6·3 m **9.** 80 ohms; 1·18 cm
10. 1·52 × 10^{-4} °C^{-1} **11.** 2·2 A (2·24) **12.** 7
13. 4s. 1½d. profit **14.** 1·5 V; 0·5 ohm **15.** 17 ohms; 1·4 V (1·365)
16. 2·0 A; 1·9 A (1·856) **17.** 12·6 A **18.** 80 V; 64·5 V
19. 31 V; 10 V **20.** 21 mA **21.** (*a*) 6·7 V, (*b*) 8·0 V, (*c*) 0
22. 0·67 mile **23.** 27·79 ohms **24.** 2·6 × 10^{-3} cm
25. $3{\cdot}6_3$ × 10^{-4} m s^{-1} **26.** 6·11 × 10^{-3} °C^{-1} **27.** 0·24 W cm^{-2}
28. 4·1°C min^{-1} **29.** 5·4 **30.** $18{\cdot}7_5$ kW
31. 0·76 : 1 **32.** 2·5 W (2·53), $2{\cdot}2_5$ V **33.** (*a*) 1·082 kW, (*b*) 840°C
34. 2·2 J °C^{-1} g^{-1} (2·16) **35.** 350 J g^{-1}; 280 J min^{-1}
36. 0·54 cal g^{-1} °C^{-1}; 0·57 °C min^{-1} **37.** 2 : 3
38. 11·9 ohms; 0·16 ohm **39.** $E_1/E_2 = (R_1 + r_1)/(R_2 + r_2)$
40. 14 A **41.** $l_1/l_2 = (r_1/r_2)^{\frac{1}{2}}$; $W_1/W_2 = (r_1/r_2)^{\frac{3}{2}}$
42. $T \propto i^{\frac{2}{3}}/r$; 1·24; 0·48 **43.** 1400°C; 320 V (316)
44. 120 A (117) **45.** 6·4 A; 2·0 A, charging
47. 0·068 A; $0{\cdot}13_5$ A; 1·01 V

Chapter 3 Electrolysis and thermoelectricity (Questions pp. 354–356)

1. 5·6 A **2.** 18 min **3.** 0·954 g
4. 3(·1) × 10^{-7} kg C^{-1} **5.** 3·03 × 10^{6} C; 2·43 × 10^{7} J; 1s. 1½d. **6.** 3·41 × 10^{-21} J
7. 124 A; 115 times **8.** 6 ohms **9.** 1·03 × 10^{-8} kg C^{-1}
10. 470 J **11.** 200; 60·6 g
12. (ii) 100 A, (iii) 9 ohms, (iv) 22 V, (v) 1·4°C min^{-1} **13.** 2 ohms; 40%
14. 25 V battery + 100-ohm resistance; 50 ohms **15.** (i) 0·67 A, (ii) 8 A hr, 5 ohms
16. From A to D **17.** 0·058 cm^{3}

Chapter 4 Circuit measurements (Questions pp. 356–358)

1. 3·5 mV **2.** 4 ohms **3.** 9·1 ohms; 1·32 V
4. (*a*) 0·540 V, (*b*) 0·800 V **5.** 0·15 A low **6.** 883°C
7. increase to 22·5 ohms **8.** 2·49 mV; 1000 ohms in parallel with AB
9. 57·1 cm from left **10.** 3·21 ohms; 1·09 × 10^{-7} ohm m
11. 75 cm; (*a*) 90 cm, (*b*) 39(·5) cm **12.** (*a*) 55 cm; 55 cm, (*b*) 41(·25) cm; 27(·5) cm
13. 1·48 × 10^{-2} ohm m^{-1}; 1·73 × 10^{-8} ohm m **16.** 996 ohms; 0·2 cm
17. $2{\cdot}54_5$ mV **19.** 2235°C
20. 50·4°C **21.** (*a*) 3·94 × 10^{-3} °C, (*b*) 18·3°C, (*c*) $0{\cdot}2_4$°C
22. 0; 1 mA; 2 mA **23.** 1·00075 ohms **24.** 0·05°C
25. 12·5 μA **26.** 55°C **27.** 33 mA (32·7)

Chapter 5 Magnetic effects of a current (Questions pp. 358–360)

1. 0·16 N
2. 10^2 N; 1·5 kW
3. 0·402 T
4. $1·3 \times 10^{-5}$ T; $3·9 \times 10^{-5}$ N
5. $2·5 \times 10^{-4}$ N m^{-1}
6. 10^5 N m^{-2}
7. 8 A; $1·6 \times 10^{-23}$ N; 8×10^{-3} N
8. $1·8 \times 10^{-7}$ N m
9. $1·00_3$ A; $3·7 \times 10^{-5}$ T
10. $4·0 \times 10^{-5}$ N m
11. 47 A
12. $8·8 \times 10^{-6}$ T
13. 4·8 mA; 1·1 mA more or $0·6_5$ mA less
14. $2·4 \times 10^{-5}$ N m^{-1} (repulsion); $1·06 \times 10^{-5}$ T

Chapter 6 Magnetic materials (Questions pp. 360–361)

1. $\mu = 1·2_5 \times 10^{-4}$ N A^{-2} ($\mu_r = 100$)
2. 400
3. $54·_5°$
4. $1·23 \times 10^{-7}$ Wb; $2·5 \times 10^{-4}$ T, 200 A m^{-1}; Φ and B only multiplied by 5000
5. 32 mA; $8·8 \times 10^{-8}$ Wb m
6. 8·3 mA
7. In plane of wires, 4 cm and 30 cm from one wire
8. $B_0 = 2·01 \times 10^{-5}$ T ($H_0 = 16·0$ A m^{-1})
9. 2·27 s
10. 2500 (2470)
11. 4000 A m^{-1}; 0·10 Wb m^{-2}

Chapter 7 Electromagnetic induction (Questions pp. 361–364)

4. 1·0 s
5. 1·8 J
6. 500 V
7. 150 μH
8. 3(·2) mH
9. 0·25 V
10. (*a*) $0·12_6$ mV, (*b*) zero, (*c*) zero
11. $0·26_1$ μA; $1·0_2 \times 10^{-11}$ W
12. (*a*) 7·9 μV, (*b*) $3·1 \times 10^{-8}$ J
13. 18·7 V
14. (*a*) 2 A s^{-1}, (*b*) 4 A, (*c*) 400 V
15. 0·7 ms
16. 4 A
17. $2·51 \times 10^{-3}$ T; $6·3 \times 10^{-4}$ V
18. 0·2 Wb; 63 V
19. $0·15_5$ mV
20. $0·29_4$ T
21. 1·15 T
22. (*a*) 20 divs left, (*b*) 5 divs left, (*c*) 5 divs right
23. 32(·3) divs
24. 0·20 μC
25. $1·49 \times 10^{-5}$ T
26. 22(·5) μC
27. 1·43 T
28. 24 μV (23·7); $1·2 \times 10^{-5}$ ohm (1·18)
29. (i) 1·57 mV, (ii) 6·28 mV, (iii) $0·78_5$ mV, (iv) zero
30. $62·5 \times 10^3$ cm^3
31. (i) 10 mH, (ii) 12·5, (iii) 1·6 H (1·56)
32. $8·0 \times 10^{-4}$ T; (*a*) 0·4 μH, (*b*) 0·13 V
33. 40 μH; 20 μH; 0·4 A; 0·5 V
34. 0·55 T

Chapter 8 Measuring instruments (Questions pp. 364–365)

2. 1 : 10; 9 : 1
3. (*a*) 0·049 ohm in parallel, (*b*) $1·5 \times 10^4$ ohms in series
4. (i) 95 ohms in series, (ii) $5·0_5 \times 10^{-2}$ ohm in parallel; 1·33 V
5. $1·0_5 \times 10^{-3}$ ohm in parallel
6. (*a*) 0·100 ohm in parallel, (*b*) 9900 ohms in series
7. $2·1 \times 10^{-8}$ N m
8. $6·9 \times 10^{-7}$ N m
9. the 2nd (40 : 51)
10. 10 divs per μA
11. $0·28_4$ μA
12. 0·93 A
13. B (13 : 14)
14. *either* 200 ohms in parallel with meter, and 0·48 ohm in series with both; *or* 0·5 ohm in series with meter, and 210 ohms in parallel with both.
15. Plot $1/\theta$ against X

Chapter 9 Generators and motors (Questions pp. 365–367)

1. (*a*) 141 V, (*b*) 100 V
4. (*a*) 1·2 T, (*b*) 30 V
5. 25(·3) A
6. total energy conversion: (*a*) 250 W, (*b*) 67·5 W, (*c*) 175 W; heat production: (*a*) 250 W, (*b*) 2·5 W, (*c*) 50 W
7. 13 A; 57(·4) V
8. 84 V; 670 W (672)
9. 0·27 ohm
10. (*a*) 91 V, (*b*) 181 W, (*c*) 2·6 N m
11. $9·6_4$ s^{-1}
12. 50 μC; 0·1 T
14. 5 W in windings, 1245 W by friction; 480 r.p.m. (482); 10·8 kW; 500 W windings, 1200 W by friction
15. (*a*) 250 W, (*b*) 800 W
16. 92 V; (*a*) 88 V, (*b*) 1800 rev min^{-1} (1794)
17. 600 W; 0·022 ohm

Chapter 10 Electrostatics (Questions pp. 367–369)

5. $2·3 \times 10^6$ m s^{-1}
6. $1·0_5 \times 10^{-7}$ m
7. $2·0 \times 10^4$ V m^{-1}; $3·2 \times 10^{-15}$ N; $2·1 \times 10^{-9}$ s
8. $2·01 \times 10^3$ V; top
9. $3·3 \times 10^{-19}$ C
10. $8·0 \times 10^{-9}$ C (7·94)
11. $2·65 \times 10^{-3}$ C km^{-2}
15. $1·2 \times 10^{-4}$ cm (1·18)
16. (*a*) $8·85 \times 10^{-10}$ C m^{-2}, (*b*) 0·2 mm
17. (*a*) $1·32 \times 10^{-4}$ cm, (*b*) $1·6_0 \times 10^{-19}$ C
18. the charge changes half-way through from 21 electrons to 22 electrons
19. 20(·4) gf
20. 39 cm s^{-1}
21. (i) 1·06 V, (ii) 1 04 V

Chapter 11 Capacitance (Questions pp. 369–374)

1. 4×10^{-3} μF (3·98) **2.** $1{\cdot}0 \times 10^{-4}$ J
3. 4·8 °C **4.** 1000 J
5. 9×10^{6} W
6. (*a*) 36 V, 214 V, (*b*) $1{\cdot}0_7 \times 10^{-4}$ C, (*c*) $1{\cdot}9_2 \times 10^{-3}$ J, $11{.}_5 \times 10^{-3}$ J
7. 3 μF and 6 μF in series, with 9 μF in parallel with both
8. 3 groups in parallel, each of 3 in series
9. 0·19 μF (0·1875); 0·31 μF; $1{\cdot}4_4 \times 10^{-2}$ J **10.** 90 times per second
11. 8×10^{-12} A; $7{\cdot}5 \times 10^{13}$ ohms **12.** $2{\cdot}7 \times 10^{-3}$ J
13. $1{\cdot}00 \times 10^{-3}$ C; $5{\cdot}00 \times 10^{-2}$ J; 66·7 V, $3{\cdot}33 \times 10^{-2}$ J
14. The effects in P are twice those in Q and R, except in (*b*) where it is 4 times
15. $3{\cdot}8 \times 10^{5}$ V **16.** (*a*) both 17·1 μC, (*b*) 9·6 μC and 12·0 μC
17. (*a*) 3×10^{-2} A, (*b*) 9 J, (*c*) yes **18.** 28 V; 3·0 μC (2·97)
19. (*a*) $3{\cdot}0 \times 10^{-3}$ C, (*b*) $3{\cdot}3 \times 10^{-5}$ A, (*c*) $6{\cdot}3 \times 10^{7}$ ohms
20. potential $\propto$ radius; $1{\cdot}8 \times 10^{6}$ V m^{-1} **21.** (*a*) 0·2 mm, (*b*) 4·5 m^2, (*c*) $2{\cdot}0 \times 10^{4}$ J m^{-3}
22. $5{\cdot}4 \times 10^{-3}$ J **23.** 1·92 J; 0·96 J; $1{\cdot}8 \times 10^{-3}$ C; 0·90 J
24. 25(·2) pF **25.** 4·6 (4·56) **26.** $2{\cdot}5 \times 10^{-3}$ μF; 5×10^{-7} J
27. 100 : 1 **28.** (*a*) 8×10^{-4} C and 4×10^{-4} C, (*b*) 6×10^{-2} J
29. 6·7 μC; 44 V **30.** 0·25 μF **31.** 1·06 mm apart; 0·25 μF
32. (*a*) 3·6 μA, (*b*) $4{\cdot}3 \times 10^{-4}$ W, (*c*) $2{\cdot}2 \times 10^{-4}$ W
33. 100 V; 20 μF **34.** 69 s
35. $\frac{1}{2}CV^2[1 - e^{-2t/RC}]$; (*a*) 3×10^{-4} A, (*b*) 15·6 J, (*c*) $2{\cdot}34 \times 10^{-4}$ A
36. 32·3 V **37.** 4 divs per μC; $1{\cdot}9_1$ T **40.** $1{\cdot}74 \times 10^{-13}$ C
41. 40 V; $1{\cdot}5 \times 10^{-6}$ J; 2·25 **42.** 14·7 V **43.** 50 times per second
44. $0{\cdot}14_4$ μF **45.** 3·7 s; 70 min^{-1} **46.** 100 V; 50 μA
47. $1{\cdot}6 \times 10^{13}$ ohm m **48.** 50 N **49.** 380 V
50. $4{\cdot}_0$ **51.** 900 V; increases by 50%; 0·6 cm **52.** 0·96 cm

Chapter 12 Alternating currents (Questions pp. 374–378)

2. 24 ohms **3.** $1{\cdot}4_6$ H **4.** $0{\cdot}50_3$ A; 283 V
5. low *f*: (*a*) 2 A, 0, (*b*) 0, 100 V, (*c*) 0, 100 V
high *f*: (*a*) 0, 100 V, (*b*) 2 A, 0, (*c*) 1 A, 50 V
6. 270 pF **7.** 1·68 MHz; 0·559 MHz **10.** (*a*) no change, (*b*) doubled
11. 0·70 **12.** (*a*) 0·80, (*b*) 8·5 μF, (*c*) 212 V
13. power factor = 0·75 **14.** 0·81
15. (ii) 5·7 A, (iii) 1·06 A, (iv) $4{\cdot}8 \times 10^{3}$ J, (v) $3{\cdot}4 \times 10^{3}$ J
16. (*a*) $86{.}_5$%, (*b*) 58 W **17.** 4570 kW; 22,850 A **18.** $0{\cdot}62_3$ A
19. 0·128 A; 203 V **20.** (*a*) $9{\cdot}1_0$ A, (*b*) $6{\cdot}4_3$ A **21.** 240 ohms; 7·7 μF
22. $30{.}_4$ mH; 12 W; 48 W **23.** 60 Hz
24. 0·2 A; 7·5 ohms; 92·5 mH **25.** $1{\cdot}4_0$ H; $98{.}_5$ ohms
26. 5·0 ohms; 2·2 ohms; 5·0 ohms (4·98); $1{\cdot}6 \times 10^{-2}$ H; 0·44; 108 W
27. (*a*) 14 ohms, (*b*) 150 μF (146), (*c*) 60 mH (59·7) **28.** $1{\cdot}6 \times 10^{2}$ μF; 0·17 H
29. $0{\cdot}61_7$ A **30.** 346 ohms; 72(·1) mH **31.** 19 ohms; 7·1 kHz
32. 0·060 A (0·0598); 38 V (37·6); 563 kHz
33. (*a*) $1{\cdot}6_3 \times 10^{3}$ ohms, (*b*) $0{\cdot}74_5$ MHz, (*c*) 100 pF
34. 203 μH; 190 m; 255 **36.** 50·0 Hz; 18·6 ohms
37. 1880 ohms; (*a*) 283 V, (*b*) 0, (*c*) 283 V **38.** 190 V (188·5); in phase
39. 78 ohms, 99 ohms; 7·8 V; 9·9 V
40. (*a*) $1{\cdot}13 \times 10^{5}$ Hz, (*b*) $3{\cdot}7_5 \times 10^{-2}$ s, (*c*) 21, (*d*) 1880 ohms
41. (*a*) 0·65 MHz, (*b*) 1220 ohms, (*c*) 245 **42.** 0·8 μF across switch
43. $1{\cdot}0 \times 10^{5}$ ohms; 0·75 MHz **44.** 140 μH; 0·12 ohm

Chapter 13 Electrons and ions (Questions pp. 378–380)

2. $3{\cdot}37 \times 10^{-19}$ J; $1{\cdot}9 \times 10^{20}$ s^{-1}
3. (*a*) $2{\cdot}18 \times 10^{-18}$ J; $2{\cdot}19 \times 10^{6}$ m s^{-1}, (*b*) $9{\cdot}13 \times 10^{-8}$ m, (*c*) $1{\cdot}03_5 \times 10^{-7}$ m; $1{\cdot}22 \times 10^{-7}$ m; $6{\cdot}9 \times 10^{-7}$ m
4. $8{\cdot}4 \times 10^{6}$ m s^{-1} **5.** $5{\cdot}93 \times 10^{7}$ m s^{-1} **6.** $2{\cdot}2_5 \times 10^{15}$ s^{-1}
7. $2{\cdot}6_5 \times 10^{7}$ m s^{-1}; $3{\cdot}2 \times 10^{-4}$ W **8.** $4{\cdot}77 \times 10^{5}$ m s^{-1}
9. $1{\cdot}15 \times 10^{6}$ m s^{-1} **10.** 2·4 eV; $3{\cdot}9_7 \times 10^{5}$ m s^{-1} **11.** (*a*) no, (*b*) $4{\cdot}6 \times 10^{5}$ m s^{-1}
12. 7000 Å; 1·8 eV; $6{\cdot}7 \times 10^{-34}$ J s **13.** $2{\cdot}66 \times 10^{-3}$ T
14. $2{\cdot}45 \times 10^{7}$ m s^{-1}; $1{\cdot}96 \times 10^{-3}$ T **15.** 2·9°

16. 17·9 cm; $1{\cdot}33 \times 10^{-4}$ T **17.** $1{\cdot}87 \times 10^{7}$ m s^{-1}; $3{\cdot}33 \times 10^{3}$ **18.** $1{\cdot}4 \times 10^{-9}$ m
19. 12·4 T **20.** 0·52 mm **21.** 22·7 cm

Chapter 14 Electronics (Questions pp. 380–382)

2. $0{\cdot}17_7$ A **4.** 1·0 mA V^{-1}; 6·0 **5.** 57 V
6. 5·4 V (5·36) **7.** $2{\cdot}5 \times 10^{4}$ ohms **11.** 288 V
12. $1{\cdot}2_5 \times 10^{4}$ ohms; $6{\cdot}2_5$ V **13.** 175 V; < 2·1 V; < $31{\cdot}_5$ V
14. 13 : 1 **17.** 144 km **19.** 16 V cm^{-1}
20. 840 kΩ; 140 V; $247{\cdot}_4$ V **21.** $24{\cdot}_5$ mA **22.** $2{\cdot}7 \times 10^{-6}$ m
24. $68{\cdot}_8$ s **25.** 5 s
26. wavelength = 3·0 cm; sources in antiphase; equal amplitudes; etc.
27. intensity $\propto \cos^2\left[\frac{\pi}{4}(1 - \cos\theta)\right]$ at an angle θ with the plane of the aerials; the maximum would be narrower

Chapter 15 X-rays (Questions pp. 382–383)

2. $7{\cdot}1 \times 10^{2}$ g min^{-1} **3.** $1{\cdot}2_5 \times 10^{16}$ s^{-1}; $4{\cdot}2 \times 10^{7}$ m s^{-1}
4. $1{\cdot}24 \times 10^{-10}$ m **5.** $1{\cdot}24 \times 10^{-8}$ m
6. $1{\cdot}53_7 \times 10^{-10}$ m; 30° 30′; 49° 36′ **9.** $1{\cdot}6 \times 10^{-19}$ C (1·56)

Chapter 16 Nuclear physics (Questions pp. 383–385)

4. $6{\cdot}0 \times 10^{26}$; $2{\cdot}3 \times 10^{26}$ **6.** 57 s **7.** 1·41 : 1
8. (*a*) $3{\cdot}3 \times 10^{27}$, (*b*) $2{\cdot}1 \times 10^{-18}$ **9.** 10·4 s **11.** $3{\cdot}4 \times 10^{22}$ (3·36)
12. (*a*) 0·25 Ci, (*b*) 170 days (167) **13.** $5{\cdot}7 \times 10^{15}$ s^{-1} **14.** $5{\cdot}8 \times 10^{-11}$ kg (5·74)
15. $9{\cdot}9 \times 10^{4}$ yr **16.** (*a*) 2, (*b*) 1, (*c*) 1/16 **17.** 2030 yr old
18. $5{\cdot}5 \times 10^{-5}$ (5·47) **19.** $2{\cdot}7_2 \times 10^{9}$ yr
20. $2{\cdot}9 \times 10^{-18}$ kg; $2{\cdot}9 \times 10^{-19}$ m^3 **21.** 4·0 cm; $5{\cdot}8 \times 10^{-2}$ cm^2 g^{-1} **22.** $9{\cdot}22 \times 10^{6}$ s^{-1}
23. (*a*) $9{\cdot}0 \times 10^{16}$ J, (*b*) $5{\cdot}6_2 \times 10^{35}$ eV, (*c*) $2{\cdot}60 \times 10^{8}$ m s^{-1}, (*d*) (i) $5{\cdot}1_2 \times 10^{5}$ V, (ii) $1{\cdot}87 \times 10^{9}$ V, (*e*) $1{\cdot}02 \times 10^{6}$ eV, (*f*) $1{\cdot}21 \times 10^{-12}$ m, (*g*) $5{\cdot}1_2 \times 10^{5}$ eV; $2{\cdot}60 \times 10^{8}$ m s^{-1}
24. $1{\cdot}9_2 \times 10^{7}$ m s^{-1} **25.** $8{\cdot}9 \times 10^{3}$ yr **26.** 8; $2{\cdot}0_7 \times 10^{-11}$ kg
27. 31·9739 a.m.u. **28.** (*a*) 4·23 MeV, (*b*) 4·16 MeV

Appendix

Hints on using a slide-rule

It is best not to use too elaborate an instrument. The essential scales are the A, B, C and D scales and the reciprocal scale down the middle of the slide. Sine and tan scales are useful if they are on the *front* of the instrument. It is also useful to have the special scale for sines and tangents of small angles; and cube and log scales are also of some value. But log–log scales find little application in school work. Too elaborate an instrument is both confusing and unnecessarily expensive. In the Author's opinion all-plastic slide-rules are now more satisfactory (and usually cheaper) than the wooden ones with plastic facings.

The slide-rule is designed for the rapid calculation of expressions involving multiplication and division. Most calculations may be done with either the A and B pair of scales or the C and D pair. However, the divisions in the latter pair are more widely spaced, so that greater accuracy can be obtained with these. Also the reciprocal scale in the middle of the slide is designed to be used in conjunction with the C and D scales, and enables many short-cuts in calculations to be made. The A and B scales should therefore ordinarily be used only in calculations involving squares and square-roots. The accuracy obtainable with a good 10″ slide-rule is about 0·2%.

The basic expression that may be worked out with a single movement of the slide is one of the form

$$\frac{a \times c}{b}$$

Other simple expressions should be brought to this form by mentally adding 'ones' to the top or bottom lines. For example, think of

$$a \times c \quad \text{as} \quad \frac{a \times c}{1}$$

and

$$\frac{a}{b} \quad \text{as} \quad \frac{a \times 1}{b}$$

It is then not necessary to memorize separate rules for multiplication and division. The procedure is best described by considering examples:

(1) $$\frac{1{\cdot}92 \times 5{\cdot}45}{2{\cdot}71}$$

Set the cursor to 1·92 on the D scale.
Move the slide to bring 2·71 on the C scale under the cursor.
Move the cursor to 5·45 on the C scale.
Read off the answer (3·86) on the D scale under the cursor.

(2) $$6{\cdot}3 \times 1{\cdot}4$$

Set the cursor to 6·3 on the D scale.
Move the slide to bring 1 on the C scale under the cursor.
Move the cursor to 1·4 on the C scale.
Read off the answer (8·82) on the D scale under the cursor.

(3) $$\frac{9{\cdot}5}{7{\cdot}6}$$

Set the cursor to 9·5 on the D scale.
Move the slide to bring 7·6 on the C scale under the cursor.
Move the cursor to 1 on the C scale.
Read off the answer (1·25) on the D scale under the cursor.

The significant figures of a calculation are not affected by the positions of the decimal points in the numbers involved. The slide-rule only deals with the *significant* figures in a calculation. For instance, the point on the scale marked '2·5' stands equally for 0·25, 2500, 0·000025, $2{\cdot}5 \times 10^{13}$, etc. The position of the decimal point in the answer must be decided by a separate approximate calculation, which can usually be done in a moment in the head.

It follows also that the '10' can be used as a '1' whenever it is convenient to do so. Thus in the last example above, if the numerator were 6·5 instead of 9·5, the '1' on the C scale would come off the end of the D scale. The answer must then be read below the '10' on the C scale instead. Consider likewise the example:

(4) $4{\cdot}8 \times 6{\cdot}3$

First set the cursor to 4·8 on D.
If now we move the slide to bring 1 on C under the cursor, 6·3 on C is well beyond the end of the D scale. Therefore the 10 on C must be brought under the cursor instead. The answer (30·2) is then found below 6·3, as usual.

To some extent calculations can be saved from 'going off the end of the scale' by choosing carefully the order in which the figures are taken. But sometimes it is unavoidable. The procedure is then to interchange the positions of the '1' and '10' of the C scale, and carry on as before. The next example shows the technique.

(5) $$\frac{425 \times 134}{6{\cdot}75}$$

Set the cursor to 4·25 on D.
Bring 6·75 on C under the cursor. Now 1·34 on the C scale is beyond the left-hand end of the D scale. We therefore interchange the positions of the '10' and '1' as follows. Set cursor to 10 on C. Bring 1 on C under the cursor. Then read off the answer (8440) on D under 1·34 on C. (An approximate calculation shows that the result is somewhere near 8000.)

The basic slide-rule technique is readily extended to handle more elaborate calculations. Thus expressions of the form

$$\frac{a \times c \times e}{b \times d}$$

may be worked out with *two* movements of the slide (provided we can avoid 'going off the end of the scale'); and

$$\frac{a \times c \times e \times g}{b \times d \times f}$$

requires only *three* movements of the slide (with the same proviso). Note that there must always be *one more* number on the top line than on the bottom. As before, 'ones' (or 'tens') are mentally inserted in the expression to make this so. The procedure is shown in the next example.

(6) $$\frac{8\pi}{2{\cdot}65 \times 3{\cdot}86 \times 4{\cdot}23}$$

Set cursor to 8 on D. Bring 2·65 on C under cursor.
Set cursor to π on C. Bring 3·86 on C under cursor.
Set cursor to 1 on C. Bring 4·23 on C under cursor.
Set cursor to 1 on C. Read off the answer (0·581) on D.

The technique may be analysed as follows:

1. Take the numbers from the top line and bottom line alternately.
2. The D scale is used only for the *first number* and the *answer*. All other numbers are found on the C scale.
3. Move the *cursor* for *top line* figures, and the *slide* for *bottom line* figures. That is, we start the calculation by moving the cursor to a top line number on the D scale, and then move the slide and cursor alternately.

The use of the reciprocal scale

Apart from the obvious use of such a scale in giving reciprocals, its main purpose is to enable us to cut down the number of movements of the slide in many calculations. In this way accuracy is increased, since each movement of the slide must introduce a small error, and these errors accumulate in a long calculation. The ideal expression for slide-rule work is one in which there is one more number on the top line than on the bottom. When this is not so, we must usually insert extra 'ones' as needed; but an alternative technique is to *transfer* a number from one side of the fraction bar to the other by using the reciprocal scale. Thus any number on the top line can be treated as its reciprocal on the bottom line—or vice versa. For instance, in the calculation of example 6 given above two extra 'ones' had to be inserted in the top line. Instead we could transfer the 4·23 to the top line and think of the expression as

$$\frac{8\pi \times \frac{1}{4{\cdot}23}}{2{\cdot}65 \times 3{\cdot}86}$$

Set cursor to 8 on D. Bring 2·65 on C under cursor.
Set cursor to π on C. Bring 3·86 on C under cursor.
Set cursor to 4·23 on the reciprocal scale.
Read off the answer (0·581) on D.

The method is thus quicker and saves a movement of the slide.

By this means almost any expression involving

three numbers can be worked out with only one movement of the slide. Here is another example:

(7) $7{\cdot}86 \times 14{\cdot}7 \times 273$

This is re-arranged mentally as

$$\frac{7{\cdot}86 \times 273}{1/14{\cdot}7}$$

Set cursor to 7·86 on D.
Bring 14·7 on reciprocal scale under cursor.
Set cursor to 273 on C.
Read off the answer ($3{\cdot}15 \times 10^4$) on D.

The reciprocal scale can also be used to avoid the annoyance of 'going off the end of the scale'. It will be noticed that there is no possibility of this happening with a simple division of one number by another. Any denominator (on the C scale) can always be brought opposite any numerator (on the D scale). Thus in a calculation like that in example 4 above, all danger of this particular nuisance obtruding itself may be avoided by treating the expression as

$$\frac{4{\cdot}8}{1/6{\cdot}3}$$

Set cursor to 4·8 on D. Bring 6·3 on reciprocal scale under cursor. Then read off the answer (30·2) under the 'one' of the C scale.

Calculations involving squares and square-roots should be planned likewise to involve the minimum movement of the slide. Here are two examples:

(8) $8\sqrt{2}$

Set cursor to 2 on the A scale ($\sqrt{2}$ is then under the cursor on the D scale). Bring 8 on the reciprocal scale under cursor. Read off answer (11·31) on D under the 'one' of the C scale.

(9) What is the area of a circle of diameter 5·83 cm?

$$\text{The area} = \frac{\pi(5{\cdot}83)^2}{4}$$

Set cursor to π on the A scale. Bring 4 on the B scale under the cursor. Set cursor to 5·83 on C, and read off the answer (26·7 cm²) on A.

(Many slide-rules have special marks on the cursor to enable a few commonly recurring types of calculation like this one to be done by a single setting of the cursor. It is worth consulting the maker's instruction booklet to find out what is possible in this way.)

As the student gains familiarity with the slide-rule many other short-cuts will be discovered. We conclude with an example to illustrate another variation of technique.

(10) $\dfrac{1}{4\pi}$

Set cursor to 4 on D. Bring π on reciprocal scale under cursor. The number under the 'one' of the C scale is now 4π. The reciprocal of this is then to be found on the C scale above the 'ten' of the D scale (0·0796).

Although accuracies of 0·2% can be obtained if a slide-rule is used with great care, it is by no means always necessary to do this. If the figures in the calculation involve errors of 1% or more, it is a foolish waste of time to struggle to adjust the cursor or slide to a hair's breadth. The result of a calculation should in any case be adjusted to the number of places appropriate to the figures concerned. If the least accurate figure is known only to 2 significant figures, the answer should be corrected to 2 significant figures also. A fairly slapdash slide-rule technique is then quite adequate.

Useful expansions

When the angle θ is given in radians,

$$\sin\theta = \theta - \frac{\theta^3}{6} + \ldots$$

$$\tan\theta = \theta + \frac{\theta^3}{3} + \ldots$$

For sufficiently small angles both these expansions reduce to

$$\tan\theta \simeq \sin\theta \simeq \theta \text{ (in radians)}$$

The error in ignoring the other terms of the expansions is less than 1% for angles up to about 10°.

$$1 - \cos\theta = \frac{\theta^2}{2} - \frac{\theta^4}{24} + \ldots$$

$$e^x = 1 + x + \frac{x^2}{2} + \ldots$$

$$\log_e(1 + x) = x - \frac{x^2}{2} + \frac{x^3}{3} - \ldots$$

The last two expansions are of value only when x is small. However, it is never necessary to resort to tables of $\log_e y$ or e^x, if a set of common log tables or a slide-rule with a log scale is available. Thus

$$\log_e y = \log_e 10 . \log_{10} y = 2{\cdot}303 . \log_{10} y$$

The multiplier 2·303 is worth memorizing.

Also, if $y = e^x$

then $x = \log_e y = 2{\cdot}303 \log_{10} y$

$$\therefore\ y = e^x = \text{antilog}\left(\frac{x}{2{\cdot}303}\right)$$

The *binomial expansion* is of great value in many calculations in physics.

$$(1 + x)^n = 1 + nx + \frac{n(n-1)}{2}x^2 + \ldots$$

When x is small, the terms in x^2 and higher powers of x may be ignored. Here are some examples showing how this may be applied:

(i) $(1 + x)^2 = 1 + 2x + \ldots$

e.g. $(1{\cdot}0003)^2 = 1{\cdot}0006$

(ii) $\sqrt{(1 + x)} = (1 + x)^{\frac{1}{2}} = 1 + \frac{1}{2}x + \ldots$

e.g. $\sqrt{0{\cdot}996} = (1 - 0{\cdot}004)^{\frac{1}{2}} = 0{\cdot}998$

(iii) $\dfrac{1}{1 + x} = (1 + x)^{-1} = 1 - x + \ldots$

e.g. $\dfrac{1}{1{\cdot}0003} = 0{\cdot}9997$

Units and Conversion Factors

Derivations of the m.k.s. rationalized (SI) units* and their relationship with the units of the c.g.s. systems.

Quantity	*Page*	*m.k.s. (SI) unit*	*Derivation*	*Equivalent in c.g.s. units (unrationalized)*	
length		metre (m)		100 cm	
mass		kilogram (kg)		1000 g	
time		second (s)		1 s	
velocity		$m\ s^{-1}$		$100\ cm\ s^{-1}$	
acceleration		$m\ s^{-2}$		$100\ cm\ s^{-2}$	
force		newton (N)	$kg\ m\ s^{-2}$	10^5 dynes (dyn)	
couple (also moment of a force)		N m		10^7 dyn cm	
moment of inertia		$kg\ m^2$		$10^7\ g\ cm^2$	
energy (also work, heat)	11	joule (J)	N m	10^7 ergs	
power	12	watt (W)	$J\ s^{-1}$	$10^7\ erg\ s^{-1}$	
				Electromagnetic units (e.m.u.)	*Electrostatic units* (e.s.u.)
electric current	8	ampere (A)		10^{-1} e.m.u.	3×10^9 e.s.u.
electric charge (or quantity of electricity)	9	coulomb (C)	A s	10^{-1} e.m.u.	3×10^9 e.s.u.
potential difference (p.d.) (also e.m.f.)	11	volt (V)	$W\ A^{-1}$ (or $J\ C^{-1}$)	10^8 e.m.u.	$1/300$ e.s.u.
capacitance	176	farad (F)	$C\ V^{-1}$	10^{-9} e.m.u.	9×10^{11} e.s.u.
electric flux	170	C		$4\pi \times 10^{-1}$ e.m.u.	$12\pi \times 10^9$ e.s.u.
magnetic flux	107	weber (Wb)	V s	10^8 maxwells	$1/300$ e.s.u.
electric flux density D (or displacement)	171	$C\ m^{-2}$		$4\pi \times 10^{-5}$ e.m.u.	$12\pi \times 10^5$ e.s.u.
magnetic flux density B (or magnetic induction)	67	tesla (T) (or $Wb\ m^{-2}$)	$N\ A^{-1}\ m^{-1}$ ($Wb\ m^{-2}$)	10^4 gauss	$1/(3 \times 10^6)$ e.s.u.
frequency	27	hertz (Hz)	s^{-1}	1 c/s	1 c/s
inductance	112	henry (H)	$V\ s\ A^{-1}$	10^9 e.m.u.	$1/(9 \times 10^{11})$ e.s.u.
electric intensity E (or potential gradient)	157	$V\ m^{-1}$		10^6 e.m.u.	$1/(3 \times 10^4)$ e.s.u.
magnetizing force H (or magnetic intensity)	85	$A\ m^{-1}$		$4\pi \times 10^{-3}$ oersted	$12\pi \times 10^7$ e.s.u.
†magnetic moment	83	Wb m		$(1/4\pi) \times 10^{10}$ e.m.u.	$1/12\pi$ e.s.u.
†magnetization	93	$Wb\ m^{-2}$		$(1/4\pi) \times 10^4$ e.m.u.	$1/(12\pi \times 10^6)$ e.s.u.
†pole strength	83	Wb		$(1/4\pi) \times 10^8$ e.m.u.	$1/(12\pi \times 10^2)$ e.s.u.
magnetomotive force (m.m.f.)	88	A		$4\pi \times 10^{-1}$ gilbert	$12\pi \times 10^9$ e.s.u.
reluctance	89	$A\ Wb^{-1}$		$4\pi \times 10^{-9}$ e.m.u.	$36\pi \times 10^{11}$ e.s.u.
resistance (also reactance, impedance)	14	ohm (Ω)	$V\ A^{-1}$	10^9 e.m.u.	$1/(9 \times 10^{11})$ e.s.u.
conductance	14	ohm^{-1} (Ω^{-1}) (or mho) (℧)	$A\ V^{-1}$	10^{-9} e.m.u.	9×10^{11} e.s.u.
resistivity	19	ohm m (Ω m)		10^{11} e.m.u.	$1/(9 \times 10^9)$ e.s.u.
conductivity	20	$\Omega^{-1}\ m^{-1}$		10^{-11} e.m.u.	9×10^9 e.s.u.
magnetic space constant, μ_0	70	$4\pi \times 10^{-7}\ H\ m^{-1}$		1 e.m.u.	$1/(9 \times 10^{20})$ e.s.u.
electric space constant, ε_0	163 261	$\dfrac{1}{36\pi \times 10^9}\ F\ m^{-1}$		$1/(9 \times 10^{20})$ e.m.u.	1 e.s.u.

* The system of units and formulae employed in this book has gone through several modifications in the course of its development, and has accordingly been known by a variety of names—the Georgi system, the M.K.S.A. system (metre-kilogram-second-amp), etc. The present internationally agreed name for the accepted form of the system is 'SI' (standing for 'système internationale'). For further details the reader may consult B.S. 3763:1964.

† Kennelly system.

Index